Commentary artwork, including both drawings and photographs, is referred to as exhibits.

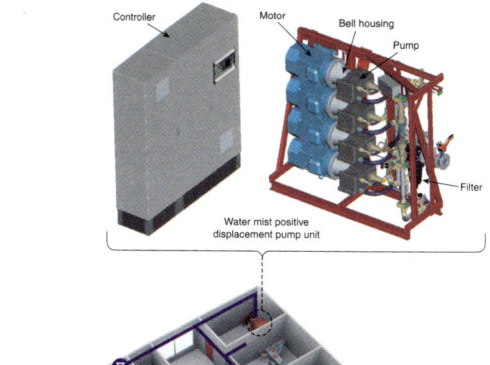

178 Part II • NFPA 20 • Chapte

EXHIBIT II.8.4 *Water Mist Positive Displacement Pumping Unit. (Courtesy of Marioff North America)*

2013   Stationary Fire Pumps Handbook

Section 4.10 • Calculation Results   **433**

TABLE 4.8.2 *Pumper Outlet Coefficients*

Pitot Pressure (Velocity Head)		
psi	bar	Coefficient
2	0.14	0.97
3	0.21	0.92
4	0.28	0.89
5	0.35	0.86
6	0.41	0.84
7 and over	0.48 and over	0.83

4.9.3 The readings obtained from a gauge so located, and the readings obtained from a gauge on a pitot tube held in the stream, are approximately the same.

4.10 Calculation Results

4.10.1 The discharge in gpm (L/min) for each outlet flowed is obtained from Table 4.10.1(a) and Table 4.10.1(b) or by the use of Equation 4.7.3.

> To determine the discharge flow, the pitot pressure, along with the outlet diameter and coefficient, must be known. The pitot pressure measured during the flow test in Exhibit III.4 is 50 psi (3.5 bar). From Table 4.10.1(a) and Table 4.10.1(b), a pitot pressure of 50 psi (3.5 bar) is 1319 gpm (4992 L/min) theoretical flow. After applying the outlet coefficient of 0.9, the actual flow measured during this flow test is 1187 gpm (4493 L/min).

◀ **DESIGN ALERT**

4.10.1.1 If more than one outlet is used, the discharges from all are added to obtain the total discharge.

4.10.1.2 The formula that is generally used to compute the discharge at the specified residual pressure or for any desired pressure drop is Equation 4.10.1.2:

$$Q_R = Q_F \times \frac{h_r^{0.54}}{h_f^{0.54}} \qquad (4.10.1.2)$$

where:
Q_R = flow predicted at desired residual pressure
Q_F = total flow measured during test
h_r = pressure drop to desired residual pressure
h_f = pressure drop measured during test

> Applying the formula in 4.10.1.2, the total flow measured during the test is as previously calculated: 1187 gpm (4493 L/min). The pressure drop to the desired pressure in this case is 55 psi (3.8 bar) [75 psi − 20 psi = 55 psi (5.2 bar − 1.4 bar = 3.8 bar)]. The pressure drop measured during the test was 23 psi (1.6 bar) [75 psi − 52 psi = 23 psi (5.2 bar − 3.6 bar = 1.6 bar)]. The result of this calculation predicts a flow of 1901 gpm (7195.3 L/min) when the residual pressure is drawn down to the minimum recommended 20 psi (1.4 bar). As stated in 4.1.1, the ratings and uniform marking of hydrants should be based on a residual pressure of 20 psi (1.4 bar). Therefore, the rating for the hydrant in this example should be based on a flow of 1901 gpm (7195.3 L/min), not the 1187 gpm (4493 L/min) measured during the test.

◀ **DESIGN ALERT**

Stationary Fire Pumps Handbook 2013

◀ Standard figures and tables are printed in black.

◀ Throughout, icons in the margins alert the reader to points of special interest.

Stationary Fire Pumps Handbook

Stationary Fire Pumps Handbook

FOURTH EDITION

Edited by

Chad R.W. Duffy

Senior Fire Protection Specialist, NFPA

With the complete text of the 2013 edition of NFPA® 20, *Standard for the Installation of Stationary Pumps for Fire Protection*

National Fire Protection Association®
Quincy, Massachusetts

Product Manager: Debra Rose
Development and Production Editor: Jennifer Harvey
Copy Editor: Pamela Nolan
Permissions Editor: Josiane Domenici
Art Coordinator: Cheryl Langway

Cover Designer: Cameron, Inc.
Composition: Shepherd, Inc.
Manufacturing Manager: Ellen Glisker
Printing: R. R. Donnelley/Willard

Copyright © 2012
National Fire Protection Association®
One Batterymarch Park
Quincy, Massachusetts 02169-7471

All rights reserved.

Important Notices and Disclaimers: Publication of this handbook is for the purpose of circulating information and opinion among those concerned for fire and electrical safety and related subjects. While every effort has been made to achieve a work of high quality, neither the NFPA® nor the contributors to this handbook guarantee the accuracy or completeness of or assume any liability in connection with the information and opinions contained in this handbook. The NFPA and the contributors shall in no event be liable for any personal injury, property, or other damages of any nature whatsoever, whether special, indirect, consequential, or compensatory, directly or indirectly resulting from the publication, use of, or reliance upon this handbook.

This handbook is published with the understanding that the NFPA and the contributors to this handbook are supplying information and opinion but are not attempting to render engineering or other professional services. If such services are required, the assistance of an appropriate professional should be sought.

NFPA codes, standards, recommended practices, and guides ("NFPA Documents"), including the NFPA Documents that are the subject of this handbook, are made available for use subject to Important Notices and Legal Disclaimers, which appear at the end of this handbook and can also be viewed at *www.nfpa.org/disclaimers*.

Notice Concerning Code Interpretations: This fourth edition of *Stationary Fire Pumps Handbook* is based on the 2013 edition of NFPA® 20, *Standard for the Installation of Stationary Pumps for Fire Protection*. All NFPA codes, standards, recommended practices, and guides ("NFPA Documents") are developed in accordance with the published procedures of the NFPA by technical committees comprised of volunteers drawn from a broad array of relevant interests. The handbook contains the complete text of NFPA 20, NFPA 24, and NFPA 291; relevant extracts from the text of NFPA 13, NFPA 14, NFPA 22, *NFPA® 70*, and NFPA 1; and any applicable Formal Interpretations issued by the NFPA. These NFPA Documents are accompanied by explanatory commentary and other supplementary materials.

The commentary and supplementary materials in this handbook are not a part of the NFPA Documents and do not constitute Formal Interpretations of the NFPA (which can be obtained only through requests processed by the responsible technical committees in accordance with the published procedures of the NFPA). The commentary and supplementary materials, therefore, solely reflect the personal opinions of the editor or other contributors and do not necessarily represent the official position of the NFPA or its technical committees.

The following are registered trademarks of the National Fire Protection Association®:

National Fire Protection Association®
NFPA®
Building Construction and Safety Code® and NFPA 5000®
NFPA 72®
Life Safety Code® and NFPA 101®
NEC®, National Electrical Code®, and NFPA 70®
Fire Protection Handbook®

NFPA No.: 20HB13

ISBN: 978-1-455-90408-2
Library of Congress Card Control No.: 2012935568

Printed in the United States of America
12 13   5 4 3 2 1

Contents

Preface ix

Acknowledgments xi

Dedication xiii

About the Contributors xv

About the Editor xix

PART ONE
An Overview of Fire Pump Configurations 1

PART TWO
NFPA 20, *Standard for the Installation of Stationary Pumps for Fire Protection*, 2013 Edition, with Commentary 15

1 Administration 17
1.1 Scope 17
1.2 Purpose 18
1.3 Application 19
1.4 Retroactivity 19
1.5 Equivalency 20
1.6 Units 21

2 Referenced Publications 23
2.1 General 23
2.2 NFPA Publications 23
2.3 Other Publications 23
2.4 References for Extracts in Mandatory Sections 25

3 Definitions 27
3.1 General 27
3.2 NFPA Official Definitions 27
3.3 General Definitions 29

4 General Requirements 49
4.1 Pumps 49
4.2 Approval Required 49
4.3 Pump Operation 52
4.4 Fire Pump Unit Performance 53
4.5 Certified Shop Test 54
4.6 Liquid Supplies 55
4.7 Pumps, Drivers, and Controllers 58
4.8 Centrifugal Fire Pump Capacities 63
4.9 Nameplate 64
4.10 Pressure Gauges 64
4.11 Circulation Relief Valve 65
4.12 Equipment Protection 66
4.13 Pipe and Fittings 70
4.14 Suction Pipe and Fittings 72
4.15 Discharge Pipe and Fittings 84
4.16 Valve Supervision 89
4.17 Protection of Piping Against Damage Due to Movement 89
4.18 Relief Valves for Centrifugal Pumps 90
4.19 Pumps Arranged in Series 96
4.20 Water Flow Test Devices 97
4.21 Steam Power Supply Dependability 104
4.22 Shop Tests 105
4.23 Pump Shaft Rotation 105
4.24 Other Signals 106
4.25 Pressure Maintenance (Jockey or Make-Up) Pumps 107
4.26 Summary of Centrifugal Fire Pump Data 111
4.27 Backflow Preventers and Check Valves 112
4.28 Earthquake Protection 114
4.29 Packaged Fire Pump Assemblies 115
4.30 Pressure Actuated Controller Pressure Sensing Lines 117
4.31 Break Tanks 120
4.32 Field Acceptance Test of Pump Units 123

Contents

5 Fire Pumps for High-Rise Buildings 125

5.1 General 126
5.2 Equipment Access 126
5.3 Water Supply Tanks 126
5.4 Fire Pump Test Arrangement 127
5.5 Auxiliary Power 127
5.6 Very Tall Buildings 127

6 Centrifugal Pumps 131

6.1 General 131
6.2 Factory and Field Performance 137
6.3 Fittings 140
6.4 Foundation and Setting 145
6.5 Connection to Driver and Alignment 145

7 Vertical Shaft Turbine–Type Pumps 149

7.1 General 149
7.2 Water Supply 151
7.3 Pump 159
7.4 Installation 165
7.5 Driver 167
7.6 Operation and Maintenance 168

8 Positive Displacement Pumps 171

8.1 General 171
8.2 Foam Concentrate and Additive Pumps 175
8.3 Water Mist System Pumps 176
8.4 Water Mist Positive Displacement Pumping Units 177
8.5 Fittings 180
8.6 Pump Drivers 183
8.7 Controllers 185
8.8 Foundation and Setting 186
8.9 Driver Connection and Alignment 186
8.10 Flow Test Devices 186

9 Electric Drive for Pumps 189

9.1 General 189
9.2 Normal Power 191
9.3 Alternate Power 197
9.4 Voltage Drop 199
9.5 Motors 201
9.6 On-Site Standby Generator Systems 207
9.7 Junction Boxes 210
9.8 Listed Electrical Circuit Protective System to Controller Wiring 211
9.9 Raceway Terminations 212

10 Electric-Drive Controllers and Accessories 215

10.1 General 215
10.2 Location 222
10.3 Construction 228
10.4 Components 235
10.5 Starting and Control 251
10.6 Controllers Rated in Excess of 600 V 263
10.7 Limited Service Controllers 268
10.8 Power Transfer for Alternate Power Supply 269
10.9 Controllers for Additive Pump Motors 279
10.10 Controllers with Variable Speed Pressure Limiting Control or Variable Speed Suction Limiting Control 280

11 Diesel Engine Drive 291

11.1 General 291
11.2 Engines 292
11.3 Pump Room 310
11.4 Fuel Supply and Arrangement 313
11.5 Engine Exhaust 318
11.6 Diesel Engine Driver System Operation 319

12 Engine Drive Controllers 323

12.1 Application 323
12.2 Location 325
12.3 Construction 326
12.4 Components 332
12.5 Battery Recharging 344
12.6 Battery Chargers 345
12.7 Starting and Control 348
12.8 Air-Starting Engine Controllers 362

13 Steam Turbine Drive 365

13.1 General 365
13.2 Turbine 368
13.3 Installation 370

14 Acceptance Testing, Performance, and Maintenance 373

14.1 Hydrostatic Tests and Flushing 373
14.2 Field Acceptance Tests 377
14.3 Record Drawings, Test Reports, Manuals, Special Tools, and Spare Parts 402
14.4 Periodic Inspection, Testing, and Maintenance 402
14.5 Component Replacement 403

Annex A Explanatory Material 407

Annex B Possible Causes of Pump Troubles 409

Annex C Informational References 415

Annex D Material Extracted by NFPA 70, Article 695 417

PART THREE

Water Supply Requirements, Water Demand, Hydrants, Tanks, and Piping 421

Section 1 Complete Text of NFPA 291, *Recommended Practice for Fire Flow Testing and Marking of Hydrants*, 2013 Edition, with Commentary 423
Section 2 Complete Text of NFPA 24, *Standard for the Installation of Private Fire Service Mains and Their Appurtenances*, 2013 Edition, with Commentary 447
Section 3 Section 18.4 and Annex E, NFPA 1, *Fire Code*, 2012 Edition, with Commentary 525
Section 4 Extracts with Commentary from NFPA 22, *Standard for Water Tanks for Private Fire Protection*, 2008 Edition 533
Section 5 Extracts with Commentary from NFPA 14, *Standard for the Installation of Standpipe and Hose Systems*, 2010 Edition 565
Section 6 Extracts with Commentary from NFPA 13, *Standard for the Installation of Sprinkler Systems*, 2013 Edition 583

PART FOUR

Supplements 591

1 Interrelationship of NFPA Standards Pertaining to a Fire Pump Installation 593
2 Commissioning and Inspection, Testing, and Maintenance Forms for Fire Pumps 603
3 *NEC®* Article 695, Fire Pumps 629
4 Technical/Substantive Changes from the 2010 to the 2013 Edition of NFPA 20 645

NFPA 20 Index 659

Important Notices and Legal Disclaimers 667

Preface

Fire pumps have been used to supply flow and pressure to fire protection systems for over 100 years. The first NFPA standard on automatic sprinkler systems was published in 1896 and included information on steam and rotary fire pumps that is still valid today. Among its requirements were 2½ in. outlets for testing purposes, equipment protection (the pump had to be located in a brick or stone enclosure and cut off from the main building by fire doors), and a weekly running test. The standard established the minimum size for fire pumps to be not less than 500 gpm rated capacity and required a 60 minute water supply. A spring-type pressure relief valve and pressure gauge were also required. These requirements amounted to less than one page of text for the installation of a fire pump.

These early pumps were not the primary water supply for sprinkler or standpipe systems and were started manually. Pumps were permitted to take suction by lift either from a connected water main or by means of connecting a primer pipe to a water tank of not less than 200 gallons capacity. The first pumps were usually powered by steam; gasoline engine–driven pumps were first mentioned in the standard in 1913. At first unreliable, these spark-ignited engines evolved into the reliable diesel engine–driven pumps of modern times.

Today, fire pumps are considered to be a primary component of the fire protection water supply and are started automatically. Modern fire pumps are connected to a reliable driver of either an electric motor or a diesel engine (some steam-driven units are still in service) and are designed to start and operate under the most demanding conditions. NFPA 20, *Standard for the Installation of Stationary Pumps for Fire Protection,* has undergone 30 revisions and has evolved into a comprehensive installation standard consisting of 14 chapters and 113 pages — far more comprehensive than the first standard on fire pumps.

In 1998 NFPA and the National Fire Sprinkler Association collaborated on the first edition of what was then called the *Fire Pump Handbook.* Kenneth E. Isman, P.E., of NFSA and Milosh Puchovsky, P.E., formerly of the NFPA staff, provided a depth of knowledge and expertise to create a handbook that became invaluable to those in the field. NFPA acknowledges the work of Mr. Isman and Mr. Puchovsky in their authorship of the first edition of the *Fire Pump Handbook,* portions of which have been used in the preparation of this book.

The purpose of this handbook, in addition to providing commentary on the requirements of NFPA 20, is to include in one document a complete handbook of all NFPA documents that establish water supply requirements for fixed suppression systems, regardless of the type of water supply. Part I provides examples of possible fire pump configuration based on the requirements of NFPA 20, and discusses the purpose of its components. Part II of this handbook contains the requirements for the installation of fire pumps from NFPA 20, with additional explanation in the form of commentary. Part III covers hydrant systems and how hydrant demand necessitates a fire pump installation and water tanks and private water supplies as they relate to the installation of fire pumps and suppression systems. It also discusses both automatic standpipe systems, which, due to their high flow and pressure demands, usually require the assistance of a fire pump, and sprinklers, which are the most common type of system installed and in many configurations create a pressure demand that necessitates a fire pump. Part III includes requirements and guidance from NFPA 291, *Recommended Practice for Fire Flow Testing and Marking of Hydrants,* NFPA 24, *Standard for the Installation of Private Fire Service Mains and Their Appurtenances*; NFPA 1, *Fire Code*; NFPA 22, *Standard for Water Tanks for Private Fire Protection*; NFPA 14, *Standard for the Installation of Standpipe and Hose Systems* and NFPA 13, *Standard for the Installation of Sprinkler Systems.*

The handbook also includes four supplements: a poster accompanied by a descriptor discussing why the configuration was chosen and to help the user understand that multiple standards need to be referenced when designing and installing a fire pump assembly; commissioning and inspection, testing, and maintenance forms which are intended to assist the user in managing a fire pump installation and to use during routine ITM procedures; extracted Article 695 from the *National Electrical Code® Handbook*; and a table of significant revisions from the 2010 to 2013 edition of NFPA 20.

I would like to thank all of the contributors to this project for their input and guidance on the preparation of this material.

Chad R.W. Duffy

Acknowledgments

Based on the laws of physics, fire pumps have always been — and have become more so as the technology has advanced — a necessary component in meeting the demands of liquid-based fire suppression systems. The supply is the backbone of any suppression system, a system which provides enhanced protection against injury, loss of life, and property damage. It is an honor and a privilege to be working on the design standards handbook that helps mitigate the effects of fire.

Producing a technical handbook is a significant undertaking, which requires an astounding amount of coordination and attention to detail. For their efforts on this handbook, I would like to recognize these individuals: Kerry Bell, Brad Cronin, David Fuller, Bill Harvey, John Kovacik, Stephan Laforest, Chris Lula, James Nasby, Gayle Pennel, and Milosh Puchovsky. I would also like to thank, from the NFPA staff, Debra Rose, senior product manager, and thank Jennifer Harvey, associate project manager, whose upbeat personality and sense of humor provides much needed comedic relief throughout the process. In addition, a special thanks to Mark Earley, chief electrical engineer, for his assistance on the electrical chapters, and Josiane Domenici, permissions editor, for her attention to detail and guidance throughout the process. Thanks are also due to M.E.P.CAD for use of their AutoSPRINK software, which was used to draw the 3-D images in this handbook.

Last, but certainly not least, I would like to thank my family. Thank you to my wife, Heather, for her patience, understanding, and support while much of my time beyond the standard workday was consumed with the deadlines of this project. To my daughter, Ellie, whose ecstatic excitement every time she sees me, brings me such joy and is a constant reminder why I do what I do for a living.

Chad Duffy

Dedication

I would like to dedicate this book to two influential individuals in my life, my father, Robert Duffy, and my college professor, Stuart Evans, "Stu." My father has been a positive influence throughout my life; his lessons of hard work and dedication instilled in me the desire to take on all challenges set before me. Stu is an inspiration to all his students, providing them education and guidance far beyond his responsibilities as a college professor, taking time out of his personal life to make sure that his students succeed in their education and quest to enter the fire protection community. Stu is also a great advocate for the fire protection community, making the Seneca College Fire Protection Program to what it is today. The teaching and guidance of these two men has allowed me to acquire my position with NFPA and be a part of this book's development and teachings for years to come.

About the Contributors

Kerry Bell

Kerry Bell is a Principal Engineer at UL specializing in fire sprinkler and pump equipment. Mr. Bell obtained his BS degree in fire protection and safety engineering from Illinois Institute of Technology and an MBA degree from Northern Illinois University. He has been a registered professional engineer in the state of Illinois since 1981. Mr. Bell is a member of NFPA and ASTM International, and a Fellow member of the Society of Fire Protection Engineers. He is currently a member of more than ten NFPA committees including the Technical Committees on Residential Sprinkler Systems, Sprinkler System Discharge Criteria and Fire Pumps, Correlating Committee on Automatic Sprinkler Systems, Chairman of the Technical Committee on Water Spray Fixed Systems and a member of the Standards Council since 2005.

Bradford T. Cronin, CFPS

Brad Cronin is a fire fighter with the Newport, Rhode Island Fire Department. He has experience performing fire code inspection and plans review as a licensed Fire Marshal for the State of Rhode Island. Mr. Cronin serves as an NFPA instructor and author, serves on and chairs NFPA technical committees and a UL standards technical panel. He is a Certified Fire Protection Specialist has BS Degrees from both Saint Michael's College and University of Cincinnati in Business Administration and Fire Safety and Engineering Technology.

David B. Fuller

David Fuller is a Senior Engineering Technical Specialist for FM Global. Based in the company's Norwood, MA, offices, Mr. Fuller works in the Engineering Standards group and is responsible for providing technical property loss prevention guidance in areas of his expertise to FM Global's 1500 engineers. He is the company subject matter expert for fire protection system installation, maintenance, and testing; fire protection in cold storage facilities; fire pumps; and fire protection equipment corrosion. Mr. Fuller is a member of National Fire Protection Association and currently serves on the NFPA committees for sprinkler installation, fire pumps, inspection test and maintenance, and foam-water sprinklers. He received a BS in Electrical Engineering from Northeastern University, Boston, MA.

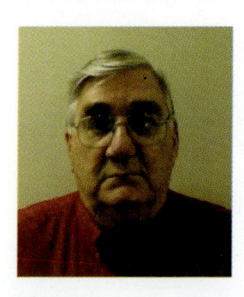

Bill M. Harvey, SET

Bill Harvey is CEO of Harvey & Associates, Inc., a fire sprinkler, fire alarm, and electrical contracting firm. He is a 53-year veteran of the fire protection industry. Mr. Harvey serves on the NFPA 20 technical committee, representing the American Fire Sprinkler Association as a principal member for 10 years. He also serves on UL STP Committees, UL 448 and UL 1247 for Fire Pump and Diesel Engines Drivers. His work history includes 48 years in design, installation, and performance testing of fire pumps for fire protection systems.

John Kovacik

John Kovacik is Underwriters Laboratories' Primary Designated Engineer (PDE) for Circuit Breakers located at the Northbrook office of UL. He represents UL on a number of outside committees, including NFPA, NEMA, and IEC. Mr. Kovacik is UL's principal member on the National Electrical Code Technical Correlating Committee and is responsible for coordinating UL's participation in *NEC* activities. He is also a member of the NFPA 20 Technical Committee on Fire Pumps. Mr. Kovacik holds a BSEE from Bradley University.

Stephan L. Laforest, CET

Stephan Laforest is the President of Summit Sprinkler Design Services, Inc. of Milton, Vermont. He has been involved in fire sprinkler system contracting since 1997, after graduating from the Fire Protection Engineering Technology Program at Seneca College in Toronto, Ontario, Canada. Mr. Laforest has experience in design, estimating, and project management. His past projects have included residential high rise towers, casinos, large storage and warehouse facilities, custom homes, apartment complexes and retail occupancies.

Christopher J. Lula, PE

Chris Lula (Chemical Engineering) is the Manager of Application Engineering for Layne / Verti-Line and Aurora vertical turbine pumps at Pentair's Kansas Plant. Mr. Lula has been an application engineer, developing pumps for not only the vertical turbine fire-pump market, but for industrial, commercial, municipal, and agricultural markets as well. Mr. Lula manages both the domestic and international applications and is considered one of Pentair's vertical turbine pump subject matter experts. In addition, he was a process and design engineer at the National Nuclear Security Administration's Kansas City Plant, the principal production site within the Department of Energy weapon complex. He holds a BS in Chemical Engineering from the University of Kansas.

James S. Nasby

James Nasby has been in responsible charge of fire pump controller design since 1972. He is a member of the NFPA 20 technical committee and former member of NFPA 110 and NFPA 70, the *NEC*®, committees for the past two cycles. He holds a BSEE from IIT and is a member of the NFPA, the SFPE, and the IEEE. Mr. Nasby has contributed to the last two editions of the *Fire Protection Handbook* as well as the NFPA *Pumps for Fire Protection Systems* textbook. He also serves on UL Standards Technical Panels for Fire Pump Controllers, Fire Pump Motors, and Engine & Turbine GenSets. His previous position was Director of Engineering for Master Control Systems, Inc., where he designed both diesel and electric fire pump controllers and battery chargers.

Gayle Pennel

Gayle Pennel is a registered professional engineer and has worked at Schirmer Engineering Corporation since 1990. He specializes in water supplies and fire protection systems evaluation and design. Projects have ranged from multiple-zone high-rise buildings to exhibition centers, refineries, and failure analysis. Mr. Pennel has over 40 years' fire protection experience and has served on the NFPA 20 Technical Committee on Fire Pumps for over ten years and also serves on the NFPA 25 technical committee.

Milosh Puchovsky, PE, FSFPE

Milosh Puchovsky is Professor of Practice in the Department of Fire Protection Engineering at Worcester Polytechnic Institute, where he focuses on fire protection systems, occupant safety, and building and fire regulations. He currently serves on several NFPA technical committees, including those on sprinkler systems, fire pumps, and commissioning, and serves on the Fire Council of Underwriters Laboratories. Mr. Puchovsky has experience as a consulting engineer, a forensics expert, a code consultant, and an insurance loss control specialist. He is a Fellow of the Society of Fire Protection Engineers (SFPE), serves as Vice President of Education for SFPE, and serves on SFPE's Professional Engineering Licensing Committee.

Joseph W. Spokis Jr.

Joe Spokis is an intern at the National Fire Protection Association, where he is responsible for the Technical/Substantive Changes and Artwork for the NFPA 13, 13D/R, and 20 Handbooks. Mr. Spokis presently holds a BS in Civil Engineering and an MS in Fire Protection Engineering, both from Worcester Polytechnic Institute. As an intern, Mr. Spokis held other responsibilities, such as assisting staff liaisons with advisory service calls, researching questions regarding the codes and standards, creating presentations for seminars, and updating the Glossary of Terms database.

About the Editor

Chad R.W. Duffy

Chad Duffy is a senior fire protection specialist at the National Fire Protection Association, where he is responsible for NFPA documents addressing stationary fire pumps, standpipes, water spray fixed systems, and other water based fire protection standards. He holds a fire protection engineering technologist degree from Seneca College, Toronto, Canada and a NICET certification.

Since 1997, Mr. Duffy has worked in the fire protection industry as a systems designer/project manager, and has owned his own design, installation, and maintenance firm for six years. He has extensive experience with system design software, and the design, installation, and maintenance of fire protection systems across the United States. His project experience ranges from small wood structures, office buildings, and condominium towers to some of the largest casinos built to date.

PART ONE

An Overview of Fire Pump Configurations

Part I of this handbook provides examples of various fire pump configurations and explains the purpose of the components of those pumps. Each component and its location in a fire pump assembly affects the fire pump's performance and its ability to meet the demands of the fire suppression system. When designing a fire pump assembly, the fire pump room must be evaluated to determine the best orientation of the pump and its components, and all the installation requirements of this standard must be met. If the pump and components are not arranged with adequate space and clearances allowing for proper operation, maintenance, and mobility — if the controllers were located at the back of the pump room, for instance — the pump operator could be at risk of injury. Locating the controllers close to the pump room exit and providing the required clearances allows the operator to quickly engage the emergency shutoff and evacuate the room. Exhibit I.1 depicts a fire pump controller installed in the back corner of a fire pump room, a location that could put an operator at risk of injury. Note that the access and egress are obstructed by piping and components, making it difficult to escape during an emergency.

EXHIBIT I.1 *Pump Room That Could Put the Pump Operator at Risk of Injury.*

On the following pages is a list of components that corresponds with the pump illustrations in this section. For each component, the list provides the component's designation, where the requirement for that component is located in Part II, and a brief description of its purpose, all to provide the user an understanding of why each component is required.

The pump configurations illustrated in this section are for educational purposes only and by no means constitute a preferred layout for the fire pump types shown. It is the responsibility of the system designer to ensure that all pump designs comply with the requirements of this standard.

Explanation of the Components of the Pumps Shown in Part I

1. OS&Y Gate Valve (Pump Suction) – 4.14.5.1
   The OS&Y gate valve in the suction piping of a fire pump serves two purposes. As liquid flows into a fire pump, it needs to be as free of turbulence as possible, to avoid introducing air pockets into the impeller and to avoid imbalanced loads on the impeller. When a gate valve is in the fully open position, the clapper is retracted into the body of the valve, leaving the liquid passageway clear of any obstruction and effectively enabling laminar flow. The OS&Y Valve also provides a way to isolate the fire pump from the liquid supply so a repair(s) can be made to the fire pump.

2. Eccentric Reducer (Pump Suction) – 4.14.6.4
   An eccentric reducer is used on the suction side of a fire pump assembly to reduce the likelihood of air pockets entering the pump impeller. In most pump installations, the suction pipe is larger than the pump suction opening; an eccentric reducer installed with the flat side on the top is used to reduce the suction size pipe to match the pump suction opening. If the suction pipe is the same as the pump suction opening, a reducer is not required.

3. Suction Pressure Gauge – 4.10.1
   When there is a possibility of a suction pressure below 20 psi, the suction pressure gauge is required to be a compound gauge capable of registering negative pressures. This gauge provides the pump operator the ability to monitor the suction pressure to assure that operating pressures comply with Section 4.14.3.1, which — except when taking suction from a tank — does not permit the suction pressure to drop below 0 psi while the pump is operating at 150 percent of its rated capacity. If a fire pump starts to draw a negative suction pressure, there is a possibility that both the fire pump and the suction piping could cavitate. Negative suction pressures in underground pipes can also cause infiltration of groundwater.

4. Discharge Pressure Gauge – 4.10.1
   The discharge pressure gauge provides the pump operator the ability to observe the discharge pressure exerted from the fire pump. It is beneficial to use liquid-filled gauges on both the suction and discharge side of the fire pump, because they dampen pressure fluctuations, making the gauges easier to read.

5. Automatic Air Release – 6.3.3
   The air release, when required, is typically part of the fire pump material distributed from the pump supplier. Air in the impeller can cause damage, so it is prudent to have a method of releasing that air, should it develop.

6. Relief Valve – 4.18
   The relief valve is not intended to control pressure when a fire pump has been overdesigned; it is intended to relieve pressure when a diesel engine is turning faster than normal, or when failure of the variable speed controls causes a pump to operate at rated speed. Section 4.18.1.1 provides guidance on determining if a relief valve is required.

7. Relief Cone – 4.18.5

   A method of detecting flow is required when a pressure relief valve is installed in closed loop systems. A sight glass in a cone downstream of the fire pump provides a way to observe water discharge. In practice, the sight glass becomes dirty or foggy, and it becomes difficult to observe flow through the cone. Should the sight glass become distorted to the point in which the flow can no longer be observed it can be replaced without having to remove the waste cone.

8. Check Valve (Discharge) – 4.15.6

   The check valve in the discharge piping restricts the pressure downstream of the fire pump, and keeps the pressurized liquid from going back through the fire pump. Backflow through a fire pump can spin the fire pump in reverse, causing damage. Pressure surges can develop and oscillate when the pump starts or stops, or when a significant change in the flow rate occurs. The discharge check valve is necessary for a jockey pump to maintain pressure on the system.

9. Indicating Gate or Butterfly Valve (Test Header) – 4.16.2

   This valve is typically supervised in the closed position and is only opened to deliver water to the test outlets or flowmeter during pump testing procedures. When this valve supplies a test header located on the exterior of the pump room, the valve becomes critically important in order to keep the test connections and supply piping from freezing. When this valve supplies a flowmeter, it should be installed per the manufacturer's recommendations.

10. Test Header – 4.20.1.1

    Though a test header is not specifically required by NFPA 20, a means of testing the pump is. One of the most commonly used means for testing is a test header. The test header provides the pump operator the ability to measure actual performance by flowing water through the fire pump, to take readings at appropriate flow rates, and to develop a test curve to verify that the pump is still producing and performing in accordance with the original manufacturer's pump curve.

11. Indicating Gate or Butterfly Valve (Fire Pump Discharge Control Valve) – 4.15.7

    The fire pump discharge control valve is for isolation purposes. This valve, combined with the pump suction control valve, provides the ability to isolate the fire pump, fire pump discharge check valve, and pump test header piping and components for repair, replacement, and testing, while keeping the pump suction and bypass piping in service.

12. Flowmeter – 4.20.2

    A flowmeter provides one method of measuring the flow rate from the fire pump. If the flowmeter is arranged in a loop piped back to the pump suction, the pump performance (but not the water supply performance) can be tested without discharging water. NFPA 25 allows "non-discharge" performance testing of fire pumps two out of every three years. Every third year, a "discharge" performance is required. A "discharge" performance test also tests the water supply to the fire pump.

13. Indicating Gate or Butterfly Valve (Flowmeter) – A.20.1.2(1)

    This valve should be installed per the manufacturer's recommendations. This valve is utilized as an isolation valve if the flowmeter needs to be repaired or replaced. It is also used as a throttling valve to control the flow rate during the performance test procedure, to achieve the 100 percent rated flow and 150 percent rated flow conditions.

14. Fire Pump Controller – 4.7.5

    Fire pump controllers are used to monitor and to start and stop fire pumps. Electric drive controllers monitor power availability and pump status, and control the power to the electric motor. Engine drive controllers monitor power availability and engine status, and send electronic signals to the engine driver. When a controller is set for automatic operation, a mercoid switch or pressure transducer is used to tell the controller to start the fire pump when the system pressure drops to a preset level. Criteria for electric drive controllers can be found in Chapter 10 and for engine drive controllers in Chapter 12.

15. Pressure Maintenance Pump (Jockey Pump) Controller – Used with a pressure maintenance pump installation

    Though a pressure maintenance pump is not specifically required by NFPA 20, a means of maintaining system pressure other than the fire pump is. If a pressure maintenance pump is used, it must have its own controller (the exception to this is water mist positive displacement pumping units). The jockey pump controller is used to start the jockey pump when the pressure in the fire protection system decreases to a preset level, and to stop the jockey pump when the pressure in the fire protection system increases to a preset level. In order for the jockey pump to work correctly, the start pressure of the jockey pump must be higher than the start pressure of the fire pump.

16. Sensing Line (Fire Pump Controller) – 4.30.1

    One end of the fire pump sensing line is connected to a pressure sensor (mercoid switch or pressure transducer) within the controller. The other end of the sensing line is connected to the fire pump piping between the discharge check valve and discharge control valve of the fire pump assembly. This provides the controller the ability to sense pressure change in the fire protection system when the fire pump is fully in service, but also allows testing of the pressure settings even if the pump discharge control valve is closed.

17. Sensing Line (Pressure Maintenance Pump) – 4.30.1

    This sensing line performs the same function as the fire pump sensing line, but for the pressure maintenance pump.

18. Pressure Maintenance Pump – 4.25

    Though NFPA 20 does not require the use of a pressure maintenance pump, it does require a means of maintaining system pressure, and a pressure maintenance pump is most commonly used to accomplish this requirement. The pressure maintenance pump maintains the pressure in the fire suppression system and prevents the fire pump from operating unless there is a significant flow of water. The pressure settings for the pressure maintenance pump and fire pump should be such that the starting of the fire pump will not create water hammer.

19. Isolation Valve (Pressure Maintenance Pump – Suction) – 4.25.5.3

    This valve is not required to be listed. Its primary function is to provide isolation from upstream water and pressures, so that the pressure maintenance pump or pressure maintenance discharge check valve can be repaired or replaced.

20. Check Valve (Pressure Maintenance Pump – Discharge) – 4.25.5.4

    This valve is required to hold pressure on the fire protection system so that when the fire pump is running, the pressure that it exerts does not flow back through the pressure maintenance pump, causing it to spin in reverse and damaging the pump. It may be necessary to use two check valves when dealing with high-pressure fire pumps. This is not a requirement, but these valves are typically brass and could fail under the repetitive starting and running pressures from high-pressure fire pump(s), so a backup is suggested.

21. Isolation Valve (Pressure Maintenance Pump – Discharge) – 4.25.5.4

    This valve is not required to be listed. It is used in conjunction with the pressure maintenance pump suction control valve to isolate the pressure maintenance pump or discharge check valve when necessary for servicing.

22. Check Valve (Fire Department Connection) – See NFPA 13 and NFPA 14
    In a fire pump configuration, the fire department connection (FDC) should be installed downstream of the fire pump discharge check valve and control valve. This check valve is utilized to prevent the FDC from experiencing pressures exerted from the fire pump and to reduce the possibility of freezing pipe.
23. Fire Department Connection – See NFPA 13 and NFPA 14
    The FDC provides a point of connection where pressures and flows can be supplemented by the fire department engine pumper during fire fighting operations.
24. Diesel Fuel Tank – 11.4
    The diesel fuel tank is required to provide the appropriate amount of fuel to assure the proper running duration time of the diesel engine driven fire pump. A designer should consult the manufacturer's data sheets to determine the outlet designation, to ensure that proper piping is installed from the diesel fuel tank.
25. Engine Exhaust – 11.5
    The engine exhaust is used to disperse the hot combustion gases produced from the diesel engine to a safe location outside the pump room. It is important to verify with the pump supplier or manufacturer that the exhaust pipe is the appropriate size for the distance and number of changes in direction, in order to avoid engine back pressure that exceeds manufacturer's recommendations.
26. Batteries – 11.2.7.2.1
    In a diesel engine drive configuration the batteries provide the cranking power to the starter upon receiving notification from the fire pump controller. In addition, the batteries provide backup power for the controller if the primary (or secondary) power supply fails.
27. Indicating Gate or Butterfly Valve (Bypass) – 4.16.1
    Indicating valves located before and after the bypass check valve on a bypass are normally open valves; they provide isolation of the bypass check valve should it require repair or replacement. An indicating valve in the bypass can be used in conjunction with the pump suction control valve to isolate the fire suppression system from the water supply.
28. Check Valve (Bypass) – 4.15.6
    The bypass check valve allows the water supply to provide some pressure to the fire suppression system should the fire pump fail to function. It also prevents the discharge from a pressure maintenance pump or a fire pump from entering the water supply.

Description

The diagram opposite illustrates a possible configuration for a diesel driven horizontal split case fire pump assembly, based on the requirements set forth in this standard. The water supply is an above ground suction tank; see NFPA 22, *Standard for Water Tanks for Private Fire Protection*, for tank requirements. For this example, a diesel drive has been used due to the lack of reliable power in the project area. Guidance on reliable power is found in Chapter 9 of this standard.

The discharge arrangement consists of a pressure relief valve, check valve, test header, discharge control valve, and a fire department connection (FDC). Isolation valves on the discharge side of the fire pump assembly can be of the indicating gate valve type or the butterfly valve type. Project specifications should be reviewed prior to designing the fire pump assembly. In this fire pump arrangement, a flow meter was used to conserve water from the tank during testing procedures. There is also a test header attached downstream of the meter. This test header serves two purposes: It will be used every three years for the annual test per the requirements of NFPA 25, *Standard for the Inspection, Testing, and Maintenance of Water-Based Fire Protection Systems*; and, per revised requirements in this edition of NFPA 20, the header will be used to verify that the readings from the flow meter are accurate (which will help determine whether the unit needs to be calibrated). Requirements specific to centrifugal fire pumps can be found in Chapter 6.

Legend

#	Description	#	Description
1	OS&Y Gate Valve (Suction Control Valve)	14	Fire Pump Controller
2	Eccentric Reducer	15	Pressure Maintenance Pump Controller (Jockey Pump)
3	Suction Pressure Gauge	16	Pressure Sensing Line (Fire Pump)
4	Discharge Pressure Gauge	17	Pressure Sensing Line (Jockey Pump)
5	Automatic Air Release	18	Pressure Maintenance Pump (Jockey Pump)
6	Relief Valve	19	Isolation Valve (Jockey Pump Suction)
7	Relief Cone	20	Check Valve (Jockey Pump Discharge)
8	Check Valve (Pump Discharge)	21	Isolation Valve (Jockey Pump Discharge)
9	Indicating Gate Valve or Butterfly Valve (Test Header)	22	Check Valve (FDC)
10	Test Header	23	Fire Department Connection
11	Indicating Gate Valve or Butterfly Valve (Discharge Control Valve)	24	Diesel Fuel Tank
12	Flow Meter	25	Diesel Drive Exhaust
13	Indicating Gate Valve or Butterfly Valve (Flow Meter)	26	Batteries

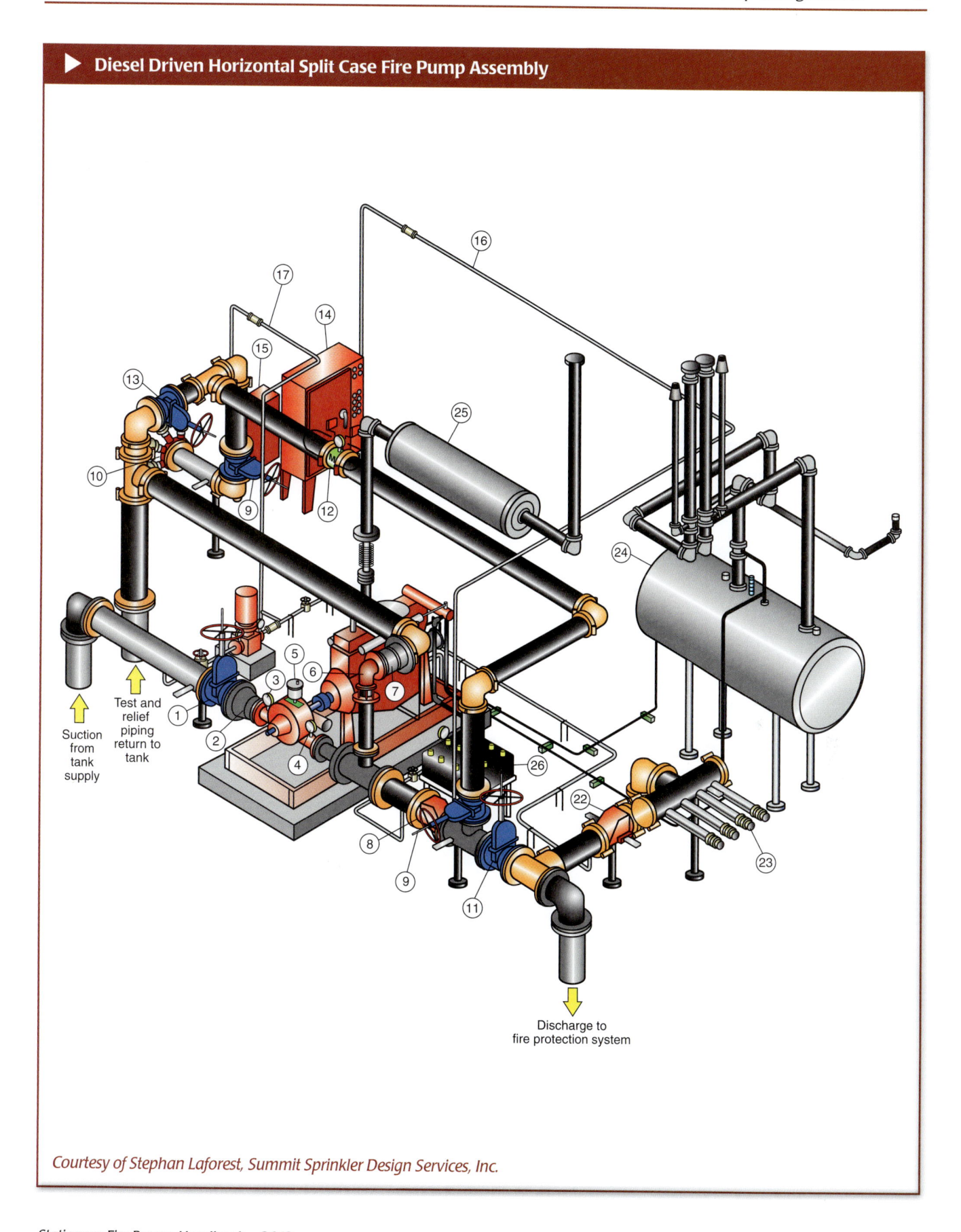

Part I • An Overview of Fire Pump Configurations 7

▶ **Diesel Driven Horizontal Split Case Fire Pump Assembly**

Suction from tank supply

Test and relief piping return to tank

Discharge to fire protection system

Courtesy of Stephan Laforest, Summit Sprinkler Design Services, Inc.

Stationary Fire Pumps Handbook 2013

Description

The diagram opposite illustrates a possible configuration for a diesel driven vertical turbine fire pump assembly, based on the requirements set forth in this standard. The water supply is an underground poured in place concrete tank; see NFPA 22, *Standard for Water Tanks for Private Fire Protection*, for tank requirements. Vertical turbine fire pumps differ from other fire pumps in that they have a shaft that is submersed in a static water supply, rather than a connection point for suction piping. This shaft contains impellers which create lift, so positive head from a water supply is not required. Vertical turbine fire pumps are used to take suction from wells and ground level storage reservoirs such as tanks, lakes, ponds, and rivers. Raw water sources require special considerations to prevent sediment and corrosion agents from entering the fire protection system.

The discharge arrangement consists of a check valve, test header, and a discharge control valve. Although this is a diesel driven fire pump and as such often requires a relief valve, one was not needed in this example because 121 percent of the fire pump net-rated shutoff (churn) pressure did not exceed the rating of the system components (this complies with the requirements of 4.18.1.2). Due to the lack of an available water supply, a submersible jockey pump was required; this pump can be seen in the diagram next to the vertical turbine fire pump shaft. Requirements specific to vertical turbine fire pumps can be found in Chapter 7.

Legend

8	Check Valve (Pump Discharge)	17	Pressure Sensing Line (Jockey Pump)
9	Indicating Gate Valve or Butterfly Valve (Test Header)	18	Submersible Pressure Maintenance Pump (Jockey Pump)
10	Test Header	20	Check Valve (Jockey Pump Discharge)
11	Indicating Gate Valve or Butterfly Valve (Discharge Control Valve)	21	Isolation Valve (Jockey Pump Discharge)
14	Fire Pump Controller	24	Diesel Fuel Tank
15	Pressure Maintenance Pump Controller (Jockey Pump)	25	Diesel Drive Exhaust
16	Pressure Sensing Line (Fire Pump)	26	Batteries

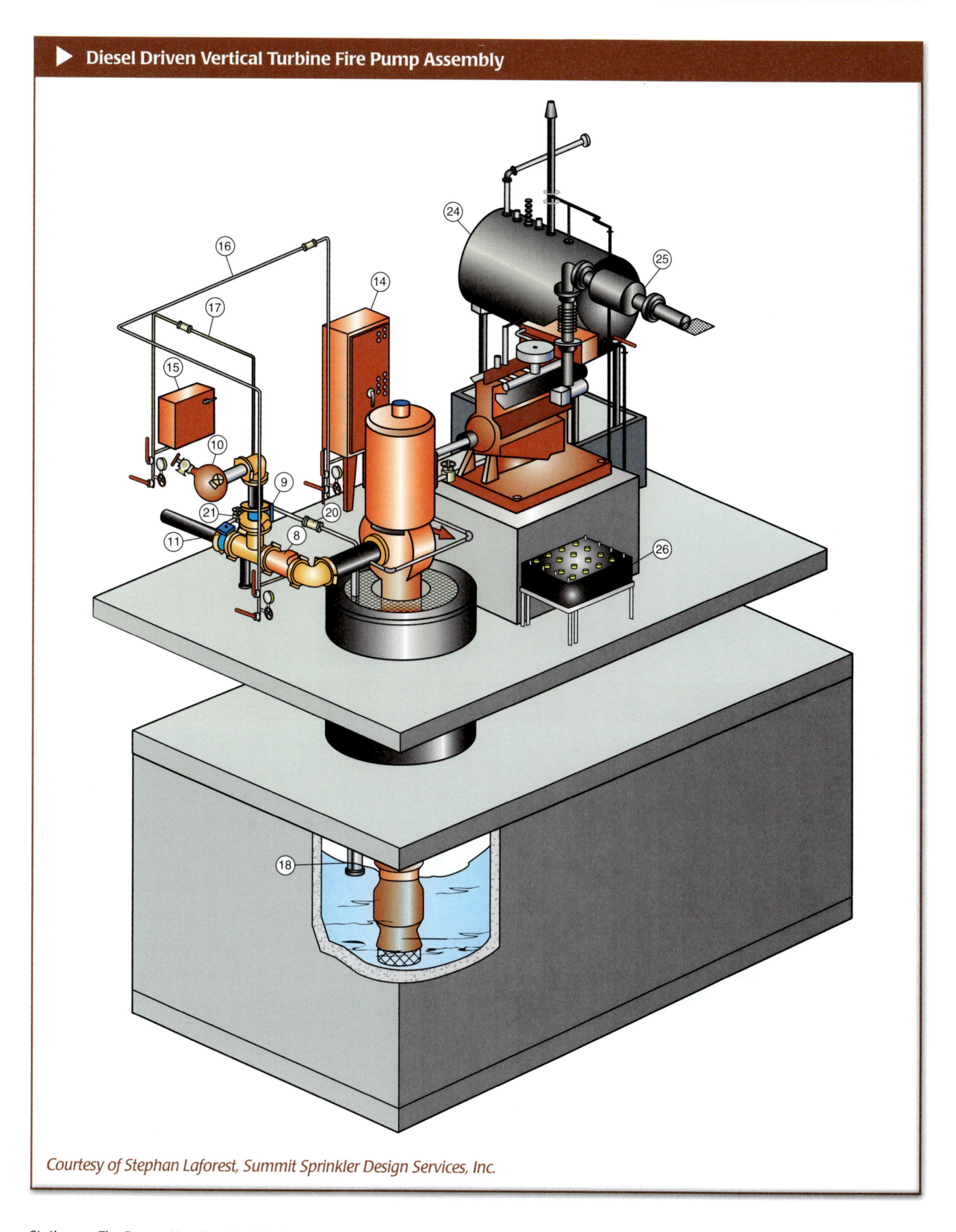

Part I • An Overview of Fire Pump Configurations

▶ Diesel Driven Vertical Turbine Fire Pump Assembly

Courtesy of Stephan Laforest, Summit Sprinkler Design Services, Inc.

Stationary Fire Pumps Handbook 2013

Part I • An Overview of Fire Pump Configurations

▶ Description

The diagram opposite illustrates a possible configuration for an electric motor driven horizontal split case fire pump assembly, based on the requirements set forth in this standard. The fire pump is supplied from a municipal water supply having insufficient pressure but adequate flow; see NFPA 24, *Standard for the Installation of Private Fire Service Mains and Their Appurtenances,* for on-site underground supply piping requirements.

The discharge arrangement consists of a check valve, test header, discharge control valve, and a bypass. A bypass is provided here, unlike in the other examples in this section, because the supply pressure is of material value to the fire suppression system even without the pump (see 4.14.4.1). In this example the suction pipe size is the same as the pump suction opening, so an eccentric reducer is not needed. Requirements specific to centrifugal fire pumps can be found in Chapter 6.

▶ Legend

①	OS&Y Gate Valve (Suction Control)	⑮	Pressure Maintenance Pump Controller (Jockey Pump)
③	Suction Pressure Gauge	⑯	Pressure Sensing Line (Fire Pump)
④	Discharge Pressure Gauge	⑰	Pressure Sensing Line (Jockey Pump)
⑤	Automatic Air Release	⑱	Pressure Maintenance Pump (Jockey Pump)
⑧	Check Valve (Pump Discharge)	⑲	Isolation Valve (Jockey Pump Suction)
⑨	Indicating Gate Valve or Butterfly Valve (Test Header)	⑳	Check Valve (Jockey Pump Discharge)
⑩	Test Header	㉑	Isolation Valve (Jockey Pump Discharge)
⑪	Indicating Gate Valve or Butterfly Valve (Discharge Control Valve)	㉗	Indicating Gate Valve or Butterfly Valve (Bypass)
⑭	Fire Pump Controller	㉘	Check Valve (Bypass)

2013 Stationary Fire Pumps Handbook

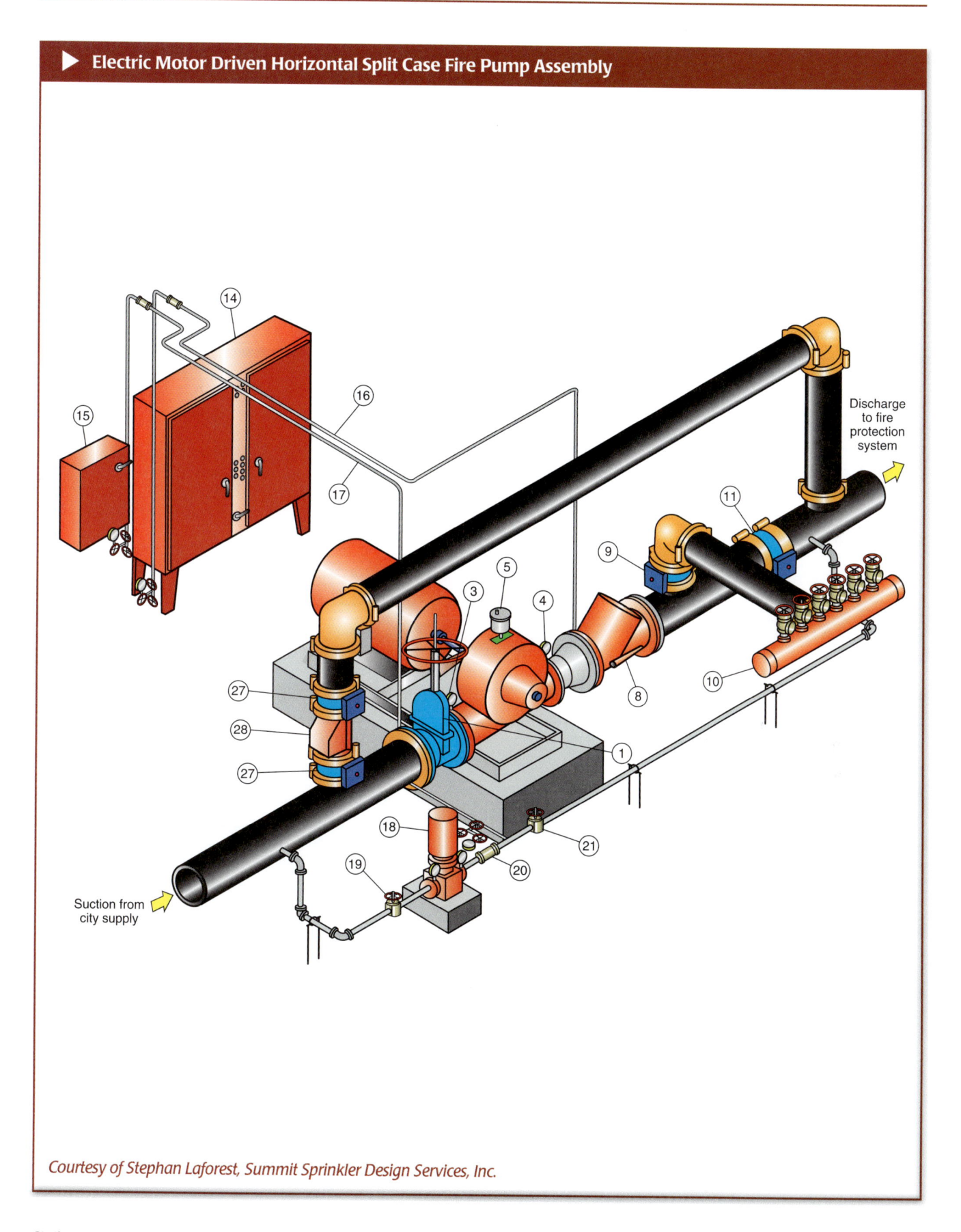

Part I • An Overview of Fire Pump Configurations   11

▶ **Electric Motor Driven Horizontal Split Case Fire Pump Assembly**

Discharge to fire protection system

Suction from city supply

Courtesy of Stephan Laforest, Summit Sprinkler Design Services, Inc.

Stationary Fire Pumps Handbook   2013

Description

The diagram opposite illustrates a possible configuration for an electric driven vertical inline fire pump assembly, based on the requirements set forth in this standard. The water supply is an above ground suction tank; see NFPA 22, *Standard for Water Tanks for Private Fire Protection*, for tank requirements. Vertical inline fire pumps can be used in a variety of building types. However, they are limited in capacity (approximately 1500 gpm) and not as commonly used as horizontal split case fire pumps. When space is limited, a vertical inline fire pump assembly might be preferred, because it is relatively compact and does not require a lot of space.

The discharge arrangement consists of a check valve, test header, and a discharge control valve. In this example, the pump suction pipe is the same size as the fire pump suction opening, so an eccentric reducer is not required. Requirements specific to centrifugal fire pumps can be found in Chapter 6.

Legend

1	OS&Y Gate Valve (Suction Control Valve)	15	Pressure Maintenance Pump Controller (Jockey Pump)
3	Suction Pressure Gauge	16	Pressure Sensing Line (Fire Pump)
4	Discharge Pressure Gauge	17	Pressure Sensing Line (Jockey Pump)
8	Check Valve (Pump Discharge)	18	Pressure Maintenance Pump (Jockey Pump)
9	Indicating Gate Valve or Butterfly Valve (Test Header)	19	Isolation Valve (Jockey Pump Suction)
10	Test Header	20	Check Valve (Jockey Pump Discharge)
11	Indicating Gate Valve or Butterfly Valve (Discharge Control Valve)	21	Isolation Valve (Jockey Pump Discharge)
14	Fire Pump Controller		

Part I • An Overview of Fire Pump Configurations

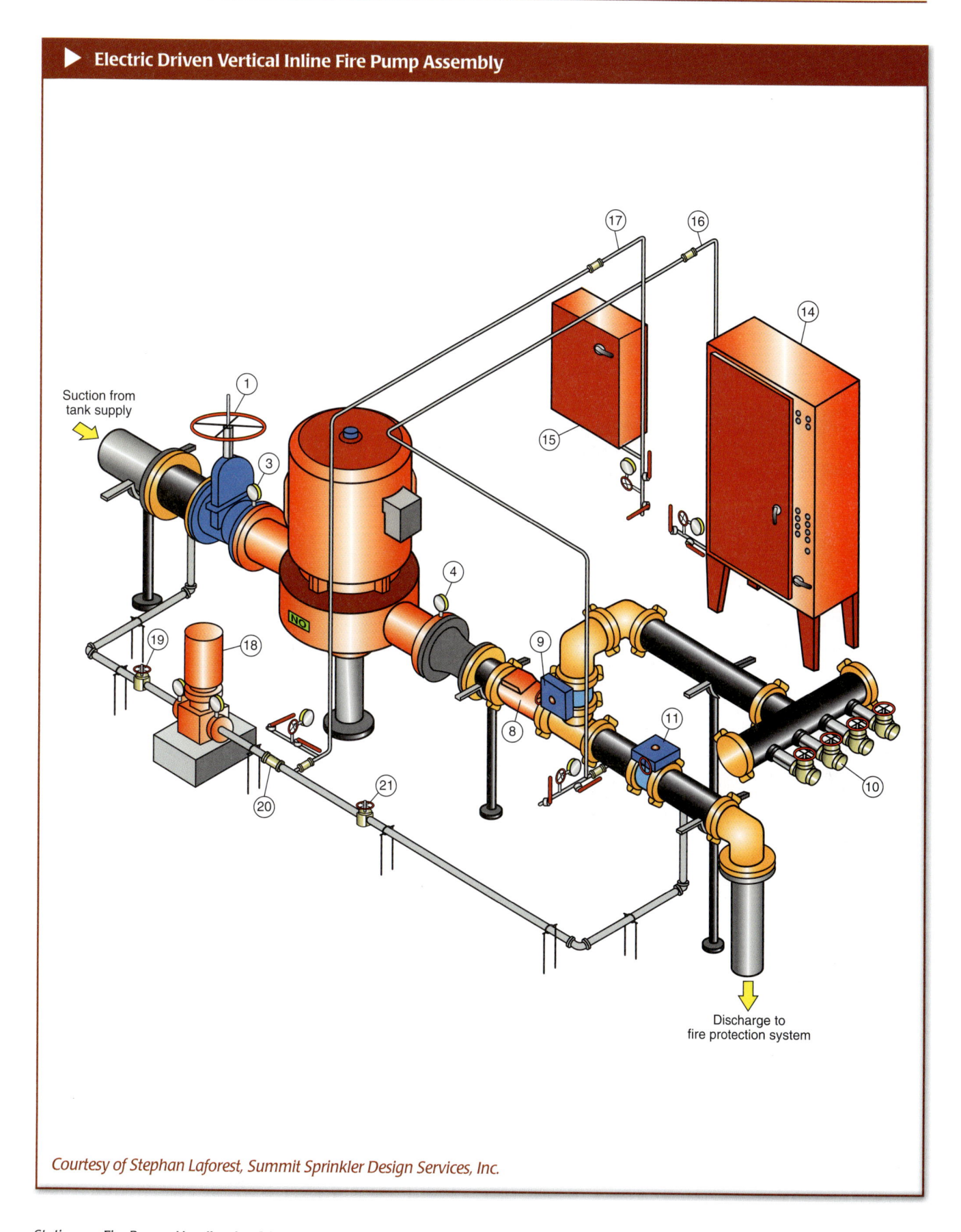

▶ Electric Driven Vertical Inline Fire Pump Assembly

Suction from tank supply

Discharge to fire protection system

Courtesy of Stephan Laforest, Summit Sprinkler Design Services, Inc.

Stationary Fire Pumps Handbook 2013

PART TWO

NFPA® 20, *Standard for the Installation of Stationary Pumps for Fire Protection*, with Commentary

Part II of this handbook includes the complete text of the 2013 edition of NFPA 20, *Standard for the Installation of Stationary Pumps for Fire Protection*. The standard consists of mandatory core chapters (Chapters 1 through 14) and nonmandatory annex material (Annex A). The mandatory provisions found in Chapters 1 through 14 were prepared by the Technical Committee on Fire Pumps within the framework of the NFPA consensus code- and standard-development process. Because these provisions are designed to be suitable for adoption into law, or for reference by other codes and standards, the text is concise, without extended explanation.

An asterisk (*) following a paragraph number within a chapter indicates that nonmandatory material pertaining to that paragraph appears in Annex A. The material found in Annex A of the standard was also generated by the fire pump committee within the NFPA code- and standard-development process. Annex A material is designed to assist users in interpreting the mandatory provisions. Annex A is not considered part of the requirements of the standard; it is advisory or informational. For the convenience of users of this handbook, Annex A text has been located to appear immediately following the mandatory paragraph to which it pertains.

The explanatory commentary accompanying the standard was prepared by the handbook editors. The commentary immediately follows the text of NFPA 20 to which it is related and is easily identified by its blue shaded background. Designed to help users understand and apply the requirements of the standard, the commentary provides detailed explanations of the reasoning behind requirements, examples of calculations, applications of requirements, and illustrations of key fire pump components.

This edition of the handbook includes two features: frequently asked questions (FAQ) and authority having jurisdiction (AHJ) frequently asked questions (AHJ FAQ). The FAQs are based on the questions most commonly asked of the NFPA 20 staff. The AHJ FAQs are questions most commonly asked of the AHJ. In addition, "design alert" and "calculation" icons indicate material of special interest.

CHAPTER 1

Administration

Chapter 1 of NFPA 20, *Standard for the Installation of Stationary Pumps for Fire Protection,* covers the administrative requirements for the selection and installation of fire pumps. In addition to the scope and purpose of the standard, Chapter 1 provides guidance on the ongoing use of existing fire pumps and the use of new equipment or technologies that might not be covered in the standard. Chapter 1 also provides guidance on the appropriate units and trade sizes used in the standard.

1.1* Scope

▶ **FAQ** Does NFPA 20 contain requirements mandating when a fire pump must be installed?

NFPA 20 is an installation standard for stationary fire pumps. As an installation standard, NFPA 20 does not specify when or if fire pumps must be installed to supplement an existing water supply. The purpose of the standard is to specify how to install a fire pump properly when one is needed and which components, equipment, and power supplies are acceptable for use in a fire pump installation. In other words, NFPA 20 indicates *how* to properly arrange and install a fire pump and its supporting equipment. The standard does not identify *when* a fire pump is required or needed.

Stationary fire pumps, such as the centrifugal or positive displacement types, are pumps that are permanently installed in a building. The term *stationary* is used to distinguish between the types of pumps described in NFPA 20 and the motorized types used by public and private fire departments.

The need for a fire pump is usually determined through an analysis of the fire protection system being considered, the water supply required for the system, and the water supply available, as indicated in Part III of this handbook. Fire pumps are principally used where pressure from the attached water supply is insufficient. Where capacity of the water supply is also in question, means in addition to a fire pump, such as water tanks, also need to be considered.

When considering whether a fire pump is necessary, the user should refer to the other NFPA standards that address various types of water-based fire protection systems. These standards include the following documents:

NFPA 11, *Standard for Low-, Medium-, and High-Expansion Foam*

NFPA 13, *Standard for the Installation of Sprinkler Systems*

NFPA 13D, *Standard for the Installation of Sprinkler Systems in One- and Two-Family Dwellings and Manufactured Homes*

NFPA 13R, *Standard for the Installation of Sprinkler Systems in Low-Rise Residential Occupancies*

NFPA 14, *Standard for the Installation of Standpipe and Hose Systems*

> NFPA 15, *Standard for Water Spray Fixed Systems for Fire Protection*
> NFPA 16, *Standard for the Installation of Foam-Water Sprinkler and Foam-Water Spray Systems*
> NFPA 750, *Standard on Water Mist Fire Protection Systems*
>
> Other NFPA documents address water supplies and their components and piping arrangements. These documents include the following:
>
> NFPA 22, *Standard for Water Tanks for Private Fire Protection*
> NFPA 24, *Standard for the Installation of Private Fire Service Mains and Their Appurtenances*

A.1.1 For more information, see NFPA 25, *Standard for the Inspection, Testing, and Maintenance of Water-Based Fire Protection Systems*, and *NFPA 70, National Electrical Code*, Article 695.

1.1.1 This standard deals with the selection and installation of pumps supplying liquid for private fire protection.

1.1.2 The scope of this document shall include liquid supplies; suction, discharge, and auxiliary equipment; power supplies, including power supply arrangements; electric drive and control; diesel engine drive and control; steam turbine drive and control; and acceptance tests and operation.

> In addition to the selection of the fire pump, NFPA 20 includes minimum requirements for the attached liquid supply, power supply arrangement, type of driver, and acceptance test procedures.

1.1.3 This standard does not cover system liquid supply capacity and pressure requirements, nor does it cover requirements for periodic inspection, testing, and maintenance of fire pump systems.

> For liquid supply capacity and pressure requirements, see the appropriate system installation standard (such as NFPA 13). For inspection, testing, and maintenance requirements for existing fire pump installations, see NFPA 25, *Standard for the Inspection, Testing, and Maintenance of Water-Based Fire Protection Systems*.

1.1.4 This standard does not cover the requirements for installation wiring of fire pump units.

> See *NFPA 70®, National Electrical Code®*, for requirements related to the installation of wiring for fire pumps.

1.2 Purpose

The purpose of this standard is to provide a reasonable degree of protection for life and property from fire through installation requirements for stationary pumps for fire protection based upon sound engineering principles, test data, and field experience.

▶ **FAQ** Is the purpose of a fire pump to create water, or is its purpose only to boost pressure?

Fire pump installations are a critical and essential component of the liquid supply for a fire protection system, because they provide the necessary system flow and pressure. The performance of any liquid-based fire protection system is dependent on the availability, adequacy, and reliability of the water supply to which it is connected. Fire pumps installed in accordance with NFPA 20 and inspected, tested, and maintained in accordance with NFPA 25 ensure that the available liquid supply will have the necessary operating pressure and flow when needed during an emergency. An important point to recognize, however, is that fire pumps cannot create water or increase the capacity of a liquid supply; they can only boost or supply the pressure and flow from an available liquid source.

NFPA 20 provides the minimum requirements needed for the satisfactory operation of all types of listed fire pumps. Even though the overall level of performance intended might be enhanced by exceeding the minimum requirements stated, the provisions of NFPA 20 allow for proper and effective fire pump performance.

1.3 Application

1.3.1 This standard shall apply to centrifugal single-stage and multistage pumps of the horizontal or vertical shaft design and positive displacement pumps of the horizontal or vertical shaft design.

▶ **FAQ** Which types of fire pumps does NFPA 20 cover?

NFPA 20 covers every type of listed fire pump. The most commonly used fire pump in fire protection is the centrifugal type, a type that includes horizontal split case, end suction, and in-line pumps. Other types of fire pumps include the piston plunger, positive displacement, rotary lobe, rotary vane, and vertical lineshaft turbine types.

The title of the 1999 edition of NFPA 20 was revised to indicate that the standard deals with all stationary pumps and not just centrifugal fire pumps. At that time, the standard began to include positive displacement pumps that pump water additives such as foam concentrate, and Section 1.3 was added to indicate the intended application of both centrifugal and positive displacement fire pumps.

1.3.2 Requirements are established for the design and installation of single-stage and multistage pumps, pump drivers, and associated equipment.

1.4 Retroactivity

The provisions of this standard reflect a consensus of what is necessary to provide an acceptable degree of protection from the hazards addressed in this standard at the time the standard was issued.

▶ **FAQ** Are fire pumps prescribed by older editions of NFPA 20 considered unsafe?

The retroactivity statement in Section 1.4 appears in most NFPA documents. The section's purpose is to reinforce the premise that any fire pump installed in accordance with the applicable edition of NFPA 20 is considered to be compliant with the standard for its lifetime, as long as the associated hazards remain unchanged and the pump is properly inspected, tested, and maintained. Therefore, an existing installation is not required to be reviewed for compliance with every new edition of the standard.

> Omission of the retroactivity statement would require the never-ending task of updating and revising a fire pump installation every time a new edition of NFPA 20 was published. Although newer editions contain information that may provide a greater level of safety or a more effective means of accomplishing a certain objective than that prescribed by older editions, the provisions of older editions should not necessarily be interpreted as unsafe. In those instances where a severe deficiency is discovered, Section 1.4 allows latitude for any authority having jurisdiction (AHJ) to require an upgrade.

1.4.1 Unless otherwise specified, the provisions of this standard shall not apply to facilities, equipment, structures, or installations that existed or were approved for construction or installation prior to the effective date of the standard. Where specified, the provisions of this standard shall be retroactive.

1.4.2 In those cases where the authority having jurisdiction determines that the existing situation presents an unacceptable degree of risk, the authority having jurisdiction shall be permitted to apply retroactively any portions of this standard deemed appropriate.

1.4.3 The retroactive requirements of this standard shall be permitted to be modified if their application clearly would be impractical in the judgment of the authority having jurisdiction, and only where it is clearly evident that a reasonable degree of safety is provided.

1.5 Equivalency

Nothing in this standard is intended to prevent the use of systems, methods, or devices of equivalent or superior quality, strength, fire resistance, effectiveness, durability, and safety over those prescribed by this standard.

> The increasing availability of specially listed materials and products, along with the continuing development of new pump-related technologies, necessitates some of the language in Section 1.5. This section allows for products or arrangements not specifically addressed by the standard to be used, provided that it can be demonstrated that the use of these products or arrangements does not lower the level of safety provided by the standard or alter the intent of the standard. This statement also alerts the user of NFPA 20 that specific requirements or limitations are often associated with specialized products and that these limitations are not addressed by the standard. Therefore, the listing information and the relevant manufacturers' literature will need to be reviewed.

1.5.1 Technical documentation shall be submitted to the authority having jurisdiction to demonstrate equivalency.

1.5.2 The system, method, or device shall be approved for the intended purpose by the authority having jurisdiction.

> Subsection 1.5.2 clarifies that the AHJ determines whether equivalency has been demonstrated based on technical documentation that has been submitted.

1.6 Units

1.6.1 Metric units of measurement in this standard are in accordance with the modernized metric system known as the International System of Units (SI).

1.6.2 *Liter* and *bar* in this standard are outside of but recognized by SI.

1.6.3 Units are listed in Table 1.6.3 with conversion factors.

1.6.4 Conversion. The conversion procedure is to multiply the quantity by the conversion factor and then round the result to an appropriate number of significant digits.

1.6.5 Trade Sizes. Where industry utilizes nominal dimensions to represent materials, products, or performance, direct conversions have not been utilized and appropriate trade sizes have been included.

TABLE 1.6.3 System of Units

Name of Unit	Unit Abbreviation	Conversion Factor
Meter(s)	m	1 ft = 0.3048 m
Foot (feet)	ft	1 m = 3.281 ft
Millimeter(s)	mm	1 in. = 25.4 mm
Inch(es)	in.	1 mm = 0.03937 in.
Liter(s)	L	1 gal = 3.785 L
Gallon(s) (U.S.)	gal	1 L = 0.2642 gal
Cubic decimeter(s)	dm^3	1 gal = 3.785 dm^3
Cubic meter(s)	m^3	1 ft^3 = 0.0283 m^3
Cubic foot (feet)	ft^3	1 m^3 = 35.31 ft^3
Pascal(s)	Pa	1 psi = 6894.757 Pa; 1 bar = 10^5 Pa
Pound(s) per square inch	psi	1 Pa = 0.000145 psi; 1 bar = 14.5 psi
Bar	bar	1 Pa = 10^{-5} bar; 1 psi = 0.0689 bar

Note: For additional conversions and information, see IEEE/ASTM SI10, *Standard for Use of the International System of Units (SI): The Modern Metric System.*

References Cited in Commentary

National Fire Protection Association, 1 Batterymarch Park, Quincy, MA 02169-7471.

NFPA 11, *Standard for Low-, Medium-, and High-Expansion Foam,* 2010 edition.
NFPA 13, *Standard for the Installation of Sprinkler Systems,* 2013 edition.
NFPA 13D, *Standard for the Installation of Sprinkler Systems in One- and Two-Family Dwellings and Manufactured Homes,* 2013 edition.
NFPA 13R, *Standard for the Installation of Sprinkler Systems in Low-Rise Residential Occupancies,* 2013 edition.
NFPA 14, *Standard for the Installation of Standpipe and Hose Systems,* 2010 edition.
NFPA 15, *Standard for Water Spray Fixed Systems for Fire Protection,* 2012 edition.
NFPA 16, *Standard for the Installation of Foam-Water Sprinkler and Foam-Water Spray Systems,* 2011 edition.
NFPA 22, *Standard for Water Tanks for Private Fire Protection,* 2008 edition.
NFPA 24, *Standard for the Installation of Private Fire Service Mains and Their Appurtenances,* 2013 edition.
NFPA 25, *Standard for the Inspection, Testing, and Maintenance of Water-Based Fire Protection Systems,* 2011 edition.
NFPA 70®, National Electrical Code®, 2011 edition.
NFPA 750, *Standard on Water Mist Fire Protection Systems,* 2010 edition.

CHAPTER 2

Referenced Publications

> This chapter lists mandatory referenced publications. Annex C lists nonmandatory referenced publications. Because the information is located immediately after Chapter 1, Administration, the user is presented with a complete list of publications needed for effective use of the standard before reading the specific requirements. The provisions of the mandatory publications referenced by NFPA 20, *Standard for the Installation of Stationary Pumps for Fire Protection,* are also requirements. Regardless of whether a requirement actually resides within NFPA 20 or is mandatorily referenced within NFPA 20 and appears only in the referenced publication, the requirement must be met to achieve compliance with NFPA 20.

2.1 General

The documents or portions thereof listed in this chapter are referenced within this standard and shall be considered part of the requirements of this document.

2.2 NFPA Publications

National Fire Protection Association, 1 Batterymarch Park, Quincy, MA 02169-7471.

NFPA 13, *Standard for the Installation of Sprinkler Systems,* 2013 edition.

NFPA 22, *Standard for Water Tanks for Private Fire Protection,* 2008 edition.

NFPA 24, *Standard for the Installation of Private Fire Service Mains and Their Appurtenances,* 2013 edition.

NFPA 25, *Standard for the Inspection, Testing, and Maintenance of Water-Based Fire Protection Systems,* 2011 edition.

NFPA 37, *Standard for the Installation and Use of Stationary Combustion Engines and Gas Turbines,* 2010 edition.

NFPA 51B, *Standard for Fire Prevention During Welding, Cutting, and Other Hot Work,* 2009 edition.

NFPA 70®, National Electrical Code®, 2011 edition.

NFPA 70E®, Standard for Electrical Safety in the Workplace®, 2012 edition.

NFPA *101®, Life Safety Code®,* 2012 edition.

NFPA 110, *Standard for Emergency and Standby Power Systems,* 2013 edition.

NFPA 1963, *Standard for Fire Hose Connections,* 2009 edition.

2.3 Other Publications

2.3.1 AGMA Publications. American Gear Manufacturers Association, 1001 N. Fairfax Street, 5th Floor, Alexandria, VA 22314-1560.

AGMA 390.03, *Handbook for Helical and Master Gears,* 1995.

2.3.2 ANSI Publications. American National Standards Institute, Inc., 25 West 43rd Street, 4th Floor, New York, NY 10036.

ANSI B15.1, *Safety Standard for Mechanical Power Transmission Apparatus*, 2000.

ANSI/IEEE C62.1, *IEEE Standard for Gapped Silicon-Carbide Surge Arresters for AC Power Circuits*, 1989.

ANSI/IEEE C62.11, *IEEE Standard for Metal-Oxide Surge Arresters for Alternating Current Power Circuits (>1 kV)*, 2005.

ANSI/IEEE C62.41, *IEEE Recommended Practice for Surge Voltages in Low-Voltage AC Power Circuits*, 1991.

2.3.3 ASCE Publications. American Society of Civil Engineers, 1801 Alexander Bell Drive, Reston, VA 20191-4400.

SEI/ASCE 7, *Minimum Design Loads for Buildings and Other Structures*, 2010.

2.3.4 ASME Publications. American Society of Mechanical Engineers, Three Park Avenue, New York, NY 10016-5990.

ASME *Boiler and Pressure Vessel Code*, 2010.

2.3.5 AWS Publications. American Welding Society, 550 NW Le Jeune Road, Miami, FL 33126.

AWS D1.1, *Structural Welding Code — Steel*, 2010.

2.3.6 HI Publications. Hydraulic Institute, 6 Campus Drive, First Floor North, Parsippany, NJ 07054-4406.

Hydraulic Institute Standards for Centrifugal, Rotary and Reciprocating Pumps, 14th edition, 1983.

HI 3.6, *Rotary Pump Tests*, 2010.

2.3.7 IEEE Publications. Institute of Electrical and Electronics Engineers, Three Park Avenue, 17th Floor, New York, NY 10016-5997.

IEEE/ASTM SI10, *Standard for Use of the International System of Units (SI): The Modern Metric System*, 2002.

2.3.8 ISO Publications. International Organization for Standardization, 1, ch. de la Voie-Creuse, Case postale 56, CH-1211 Geneva 20, Switzerland.

ISO 15540, *Fire Resistance of Hose Assemblies*, 1999.

2.3.9 NEMA Publications. National Electrical Manufacturers Association, 1300 North 17th Street, Suite 1752, Rosslyn, VA 22209.

NEMA MG-1, *Motors and Generators*, 2010.

2.3.10 UL Publications. Underwriters Laboratories Inc., 333 Pfingsten Road, Northbrook, IL 60062-2096.

ANSI/UL 142, *Standard for Steel Aboveground Tanks for Flammable and Combustible Liquids*, 2006, revised 2010.

ANSI/UL 508, *Standard for Industrial Control Equipment*, 1999, revised 2010.

ANSI/UL 1449, *Standard for Surge Protective Devices*, 2006, revised 2011.

2.3.11 Other Publications.

Merriam-Webster's Collegiate Dictionary, 11th edition, Merriam-Webster, Inc., Springfield, MA, 2003.

2.4 References for Extracts in Mandatory Sections

NFPA 14, *Standard for the Installation of Standpipe and Hose Systems*, 2010 edition.

NFPA 37, *Standard for the Installation and Use of Stationary Combustion Engines and Gas Turbines,* 2010 edition.

NFPA 70®, National Electrical Code®, 2011 edition.

NFPA 110, *Standard for Emergency and Standby Power Systems,* 2013 edition.

NFPA 1451, *Standard for a Fire Service Vehicle Operations Training Program,* 2007 edition.

NFPA 5000®, Building Construction and Safety Code®, 2012 edition.

CHAPTER 3

Definitions

Chapter 3 contains definitions of terms that are used in NFPA 20, *Standard for the Installation of Stationary Pumps for Fire Protection*. For the purpose of applying this standard, these definitions take precedence over definitions provided in other documents and should be used in conjunction with subjects covered throughout the standard. Where definitions are not provided, definitions in other NFPA codes and standards apply. For definitions not found in any NFPA document, the ordinary meanings of terms as defined in *Merriam-Webster's Collegiate® Dictionary* apply. For definitions of other fire protection–related terms that do not appear in either NFPA 20 or *Merriam-Webster's Collegiate Dictionary*, see the *NFPA Glossary of Terms*, available at www.nfpa.org.

3.1 General

The definitions contained in this chapter shall apply to the terms used in this standard. Where terms are not defined in this chapter or within another chapter, they shall be defined using their ordinarily accepted meanings within the context in which they are used. *Merriam-Webster's Collegiate Dictionary*, 11th edition, shall be the source for the ordinarily accepted meaning.

3.2 NFPA Official Definitions

3.2.1* Approved. Acceptable to the authority having jurisdiction.

A.3.2.1 Approved. The National Fire Protection Association does not approve, inspect, or certify any installations, procedures, equipment, or materials; nor does it approve or evaluate testing laboratories. In determining the acceptability of installations, procedures, equipment, or materials, the authority having jurisdiction may base acceptance on compliance with NFPA or other appropriate standards. In the absence of such standards, said authority may require evidence of proper installation, procedure, or use. The authority having jurisdiction may also refer to the listings or labeling practices of an organization that is concerned with product evaluations and is thus in a position to determine compliance with appropriate standards for the current production of listed items.

▶ **FAQ** What is the difference between the terms *approved* and *listed*?

As used in this standard, the term *approved* means only that the product is acceptable to the authority having jurisdiction (AHJ); use of the term *approved* does not imply that the product has been evaluated by a testing laboratory. In fact, a component or equipment that is required to be approved is not necessarily also required to be listed. The term *approved* is not the same as the term *listed*, which is defined in 3.2.3. The term *listed* indicates that a specific component or piece of equipment has been evaluated by an approved testing laboratory, while the term *approved*, which is defined in 3.2.1, indicates acceptance by the AHJ. In general, all components in a fire pump installation are required by NFPA 20 to be approved, but only specific components, such as the fire pump itself, are required to be listed.

3.2.2* Authority Having Jurisdiction (AHJ). An organization, office, or individual responsible for enforcing the requirements of a code or standard, or for approving equipment, materials, an installation, or a procedure.

A.3.2.2 Authority Having Jurisdiction (AHJ). The phrase "authority having jurisdiction," or its acronym AHJ, is used in NFPA documents in a broad manner, since jurisdictions and approval agencies vary, as do their responsibilities. Where public safety is primary, the authority having jurisdiction may be a federal, state, local, or other regional department or individual such as a fire chief; fire marshal; chief of a fire prevention bureau, labor department, or health department; building official; electrical inspector; or others having statutory authority. For insurance purposes, an insurance inspection department, rating bureau, or other insurance company representative may be the authority having jurisdiction. In many circumstances, the property owner or his or her designated agent assumes the role of the authority having jurisdiction; at government installations, the commanding officer or departmental official may be the authority having jurisdiction.

> The term *authority having jurisdiction (AHJ)* is intended to include anyone responsible for enforcing the requirements of NFPA 20. On many construction projects, multiple AHJs — such as the building owner's representative, a code official (e.g., a fire marshal), and an insurance representative — are present. All of the aforementioned AHJs are involved in the plan review and acceptance process and can issue an approval or comment on plan reviews and acceptance tests.

3.2.3* Listed. Equipment, materials, or services included in a list published by an organization that is acceptable to the authority having jurisdiction and concerned with evaluation of products or services, that maintains periodic inspection of production of listed equipment or materials or periodic evaluation of services, and whose listing states that either the equipment, material, or service meets appropriate designated standards or has been tested and found suitable for a specified purpose.

> Components critical to proper system operation are required to be listed. One testing laboratory uses the designation *classified* as an indication that a specific product meets its evaluation criteria. Materials with this designation meet the intent of the term *listed*.

A.3.2.3 Listed. The means for identifying listed equipment may vary for each organization concerned with product evaluation; some organizations do not recognize equipment as listed unless it is also labeled. The authority having jurisdiction should utilize the system employed by the listing organization to identify a listed product.

3.2.4 Shall. Indicates a mandatory requirement.

> The term *shall* indicates a requirement of this standard and mandates that a specific provision of NFPA 20 be followed. When the term *shall* is used with a specific provision of the standard, compliance with that provision is mandatory. Any exceptions to a requirement of this standard are specifically stated. The use of the term *shall* in an exception indicates that the exception is applicable only under the conditions specified in the exception. Mandatory requirements of NFPA 20 are found in the main body of the standard (see Chapters 1 through 14).

3.2.5 Should. Indicates a recommendation or that which is advised but not required.

> The term *should* indicates a recommendation of the standard. When the term *should* is used with a provision of the standard, the provision is not intended to be mandatory. Recommended provisions are limited to Annexes A and B of NFPA 20. The term *should* identifies a

good idea or a better practice. If the recommendation is not followed, the fire pump is still expected to perform satisfactorily. Any paragraph number preceded by a letter (e.g., A.1.1) is an annex item. Terms such as *should* and *recommended* are prevalent in annex paragraphs.

3.2.6 Standard. A document, the main text of which contains only mandatory provisions using the word "shall" to indicate requirements and which is in a form generally suitable for mandatory reference by another standard or code or for adoption into law. Nonmandatory provisions are not to be considered a part of the requirements of a standard and shall be located in an appendix, annex, footnote, informational note, or other means as permitted in the *Manual of Style for NFPA Technical Committee Documents*.

3.3 General Definitions

3.3.1 Additive. A liquid such as foam concentrates, emulsifiers, and hazardous vapor suppression liquids and foaming agents intended to be injected into the water stream at or above the water pressure.

The term *additive* was added to the standard in the 1999 edition, because the document scope was expanded to include positive displacement pumps, which are used for pumping foam concentrate and emulsifiers.

3.3.2 Aquifer. An underground formation that contains sufficient saturated permeable material to yield significant quantities of water.

3.3.3 Aquifer Performance Analysis. A test designed to determine the amount of underground water available in a given field and proper well spacing to avoid interference in that field. Basically, test results provide information concerning transmissibility and storage coefficient (available volume of water) of the aquifer.

A thorough analysis of the water supply for a fire protection system is critical to determine the supply's adequacy and reliability. See the extracts from and commentary for NFPA 291, *Recommended Practice for Fire Flow Testing and Marking of Hydrants*, in Part III of this handbook for water supply testing requirements.

3.3.4 Automatic Transfer Switch. See 3.3.50.2.1.

3.3.5 Branch Circuit. See 3.3.7.1.

3.3.6 Break Tank. A tank providing suction to a fire pump whose capacity is less than the fire protection demand (flow rate times flow duration).

▶ **FAQ** | What is the purpose of a break tank?

A break tank, as its name suggests, provides a break or separation between a fire pump and a city water supply. Traditionally, a break tank is a small tank intended to isolate the fire pump from the water supply. However, as defined and used in this standard, a break tank may be quite large.

If the discharge from an automatic tank fill valve is located a sufficient distance above the highest tank water level, the air gap offers backflow protection to avoid contamination of the city water supply. This arrangement provides the highest level of protection against backflow contamination. A break tank may also be used to eliminate pressure fluctuations at the pump

◀ **DESIGN ALERT**

Stationary Fire Pumps Handbook 2013

suction. Severe fluctuations in suction pressure make it difficult to properly size a fire pump without overpressurization of the fire protection system.

A break tank may also be used when the water source is reliable but lacks sufficient volume to supply the full fire protection demand. The combination of the water source and stored water can meet the full fire protection demand.

The definition of the term *break tank* and break tank requirements were added in the 2007 edition to provide guidance for sizing and using break tanks. A break tank (or full-size tank) is required in some municipalities where direct connection of a fire pump to a water main is not permitted.

3.3.7 Circuit.

3.3.7.1 Branch Circuit. The circuit conductors between the final overcurrent device protecting the circuit and the outlet(s). [**70:** Art. 100]

3.3.7.2 Fault Tolerant External Control Circuit. Those control circuits either entering or leaving the fire pump controller enclosure, which if broken, disconnected, or shorted will not prevent the controller from starting the fire pump from all other internal or external means and can cause the controller to start the pump under these conditions.

3.3.8 Circulation Relief Valve. See 3.3.57.5.1.

3.3.9 Corrosion-Resistant Material.
Materials such as brass, copper, Monel®, stainless steel, or other equivalent corrosion-resistant materials.

The list of corrosion-resistant materials in this definition is not intended to be comprehensive. The causes of corrosion vary, and the examples given in this definition are effective against certain types of water conditions that can cause corrosion. However, in the context of this standard, a corrosion-resistant material must resist the type of corrosion anticipated. [Monel® is a type of nickel alloy used in the manufacturing of pipe and certain metal components used within a pump (i.e. pump shaft).]

3.3.10 Diesel Engine. See 3.3.13.1.

3.3.11 Disconnecting Means.
A device, or group of devices, or other means by which the conductors of a circuit can be disconnected from their source of supply. [**70:** Art. 100]

Disconnecting means are usually devices used within fire pump controllers for electrically driven fire pumps. Means of disconnecting are usually accomplished through a circuit breaker or a locked-rotor overcurrent protection device. See 10.4.3, 10.6.3, and 10.6.8.

3.3.12 Drawdown.
The vertical difference between the pumping water level and the static water level.

The term *drawdown* is usually associated with vertical shaft turbine–type pumps that take water from a supply below the discharge flange of the pump, such as a well, river, or pond. When these pumps operate at peak capacity, the water level of the supply will usually drop a certain distance. The level of the water before the fire pump starts is referred to as the *static liquid level*. The level to which the water drops when the pump is operating at 150 percent of its rated capacity is referred to as the *pumping liquid level*. The difference between these two levels is referred to as the *drawdown*. See Figure A.7.2.2.1.

3.3.13 Engine.

3.3.13.1* Diesel Engine. An internal combustion engine in which the fuel is ignited entirely by the heat resulting from the compression of the air supplied for combustion.

> ▶ **FAQ** Are diesel engines the only type of internal combustion engine permitted by NFPA 20?

Diesel engines have proven to be very reliable and effective, and they are currently the only type of internal combustion engine permitted by NFPA 20 for driving fire pumps (See Chapter 11). Diesel engines used to power fire pumps must be listed for such use and, therefore, only specific engines are permitted. (See Chapter 9.) Spark-ignited internal combustion engines have not been permitted as a means of driving fire pumps since 1974. Exhibit II.3.1 illustrates a diesel engine–driven fire pump.

A.3.3.13.1 Diesel Engine. The oil-diesel engine operates on fuel oil injected near top dead center of the compression stroke. The combustion is effected within the working cylinder and not in external chambers.

3.3.13.1.1 Internal Combustion Engine. Any engine in which the working medium consists of the products of combustion of the air and fuel supplied.

The diesel engine is the only type of internal combustion engine permitted by NFPA 20 to drive fire pumps. See Chapter 11.

EXHIBIT II.3.1 *Diesel Engine–Driven Fire Pump. (Courtesy of Liberty Mutual Property)*

3.3.14 Fault Tolerant External Control Circuit. See 3.3.7.2.

3.3.15 Feeder. All circuit conductors between the service equipment, the source of a separately derived system, or other power supply source and the final branch-circuit overcurrent device. [**70:** Art. 100]

3.3.16 Fire Pump Alarm. A supervisory signal indicating an abnormal condition requiring immediate attention.

3.3.17 Fire Pump Controller. A group of devices that serve to govern, in some predetermined manner, the starting and stopping of the fire pump driver and to monitor and signal the status and condition of the fire pump unit.

> Different types of controllers are available for electric motor–driven and diesel engine–driven fire pumps. Controllers must be listed for fire pump service and incorporate both automatic and manual features for pump operation. Controllers for electrically driven fire pumps are available for full voltage and reduced voltage starting. Limited-service controllers for electrically driven fire pumps are also qualified for certain applications but are not required by NFPA 20 to meet the same requirements as full-service controllers. Chapter 10 addresses controllers for electric motor–driven pumps, and Chapter 12 addresses controllers for diesel engine–driven pumps. Exhibit II.3.2 illustrates an electric fire pump controller, and Exhibit II.3.3 illustrates the internal components of a diesel engine fire pump controller.

3.3.18 Fire Pump Unit. An assembled unit consisting of a fire pump, driver, controller, and accessories.

> Exhibit II.3.4 illustrates an example of a fire pump unit.

EXHIBIT II.3.2 Electric Fire Pump Controller.

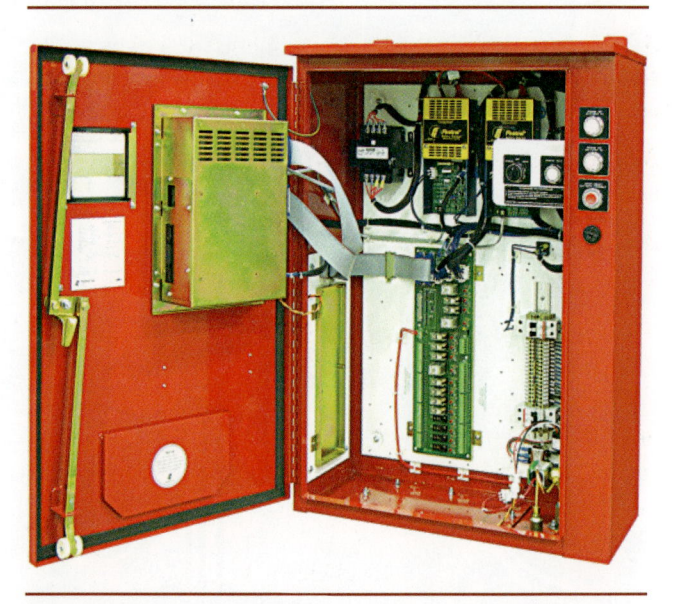

EXHIBIT II.3.3 Diesel Engine Fire Pump Controller. (Photograph courtesy of Firetrol Products, ASCO Power Technologies)

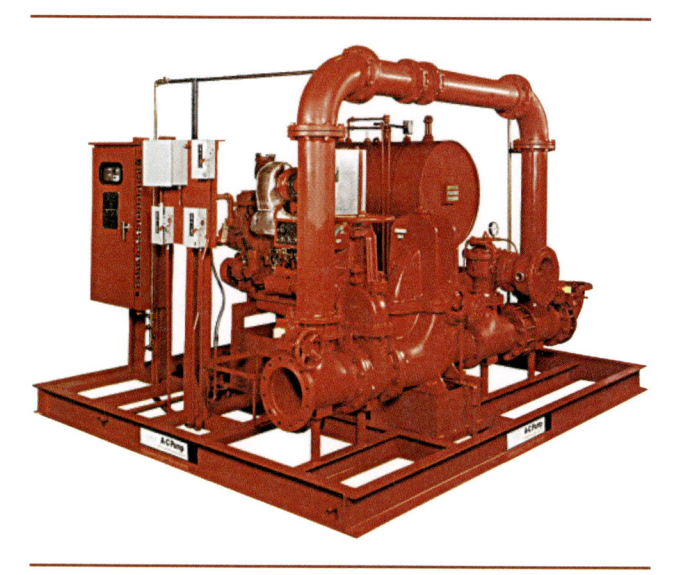

EXHIBIT II.3.4 *Fire Pump Unit. (Photograph courtesy of Xylem, Inc.)*

3.3.19 Flexible Connecting Shaft. A device that incorporates two flexible joints and a telescoping element.

Subsection 11.2.3 indicates where a flexible connecting shaft is required.

3.3.20 Flexible Coupling. A device used to connect the shafts or other torque-transmitting components from a driver to the pump, and that permits minor angular and parallel misalignment as restricted by both the pump and coupling manufacturers.

Subsections 6.5.1 and 11.2.3 indicate where a flexible coupling is needed.

3.3.21 Flooded Suction. The condition where water flows from an atmospheric vented source to the pump without the average pressure at the pump inlet flange dropping below atmospheric pressure with the pump operating at 150 percent of its rated capacity.

3.3.22 Groundwater. That water that is available from a well, driven into water-bearing subsurface strata (aquifer).

3.3.23* Head. A quantity used to express a form (or combination of forms) of the energy content of water per unit weight of the water referred to any arbitrary datum.

In the context of fire pumps, the pressure component is usually expressed in terms of feet (ft) or meters (m) of head rather than in pounds per square inch (psi) or bar.

A.3.3.23 Head. The unit for measuring head is the foot (meter). The relation between pressure expressed in pounds per square inch (bar) and pressure expressed in feet (meters) of head is expressed by the following formulas:

$$\text{Head in feet} = \frac{\text{Pressure in psi}}{0.433 \text{ specific gravity}}$$

$$\text{Head in meters} = \frac{\text{Pressure in bar}}{0.098 \text{ specific gravity}}$$

In terms of foot-pounds (meter-kilograms) of energy per pound (kilogram) of water, all head quantities have the dimensions of feet (meters) of water. All pressure readings are converted into feet (meters) of the water being pumped. *(See Figure A.3.3.23.)*

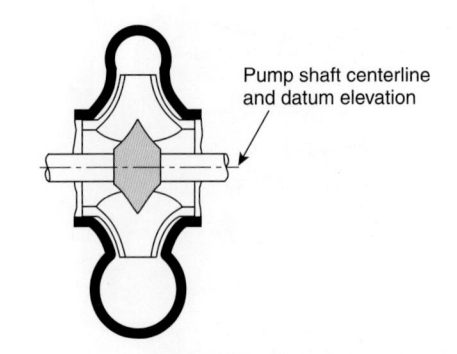

Horizontal double-suction pump

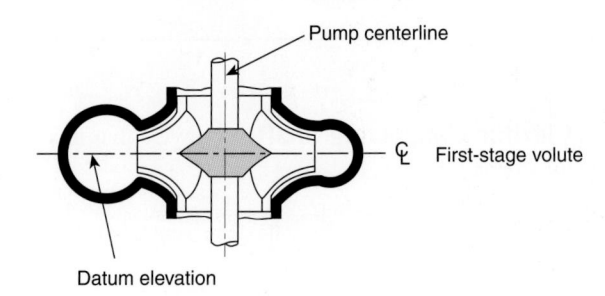

Vertical double-suction pump

Notes:
(1) For all types of horizontal shaft pumps (single-stage double-suction pump shown). Datum is same for multistage, single- (end) suction ANSI-type or any pump with a horizontal shaft.
(2) For all types of vertical shaft pumps (single-stage vertical double-suction pump shown). Datum is same for single- (end) suction, in-line, or any pump with a vertical shaft.

FIGURE A.3.3.23 Datum Elevation of Two Stationary Pump Designs.

3.3.23.1 Net Positive Suction Head (NPSH) (h_{sv}). The total suction head in feet (meters) of liquid absolute, determined at the suction nozzle, and referred to datum, less the vapor pressure of the liquid in feet (meters) absolute.

> In NFPA 20, the two types of net positive suction head (NPSH) that need to be considered are termed the *NPSH required* and the *NPSH supplied*. The NPSH required by the pump is determined by the pump manufacturer and is a function of both the speed and capacity of the pump. Curves indicating the NPSH versus flow can be obtained from the manufacturer. The NPSH supplied is the pressure head at the pump inlet that causes water to flow through the pump into the eye of the impeller. The NPSH supplied is a function of the water supply.
>
> For any pump installation, the NPSH supplied must be at least equal to the NPSH required for the operating conditions specified. If the NPSH supplied is less than the NPSH required, some of the incoming water vaporizes and forms bubbles. These bubbles can collapse within the pump, producing noise and vibration and causing significant damage to the pump. This phenomenon is referred to as *cavitation*.

3.3.23.2 Total Discharge Head (h_d). The reading of a pressure gauge at the discharge of the pump, converted to feet (meters) of liquid, and referred to datum, plus the velocity head at the point of gauge attachment.

> Figure A.3.3.23.3.1 and Figure A.3.3.23.3.2 illustrate the components of total discharge head.

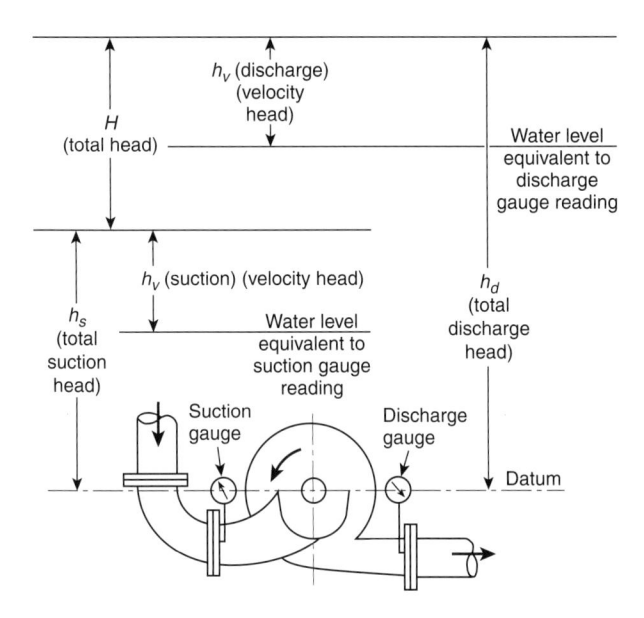

Note: Installation with suction head above atmospheric pressure shown.

FIGURE A.3.3.23.3.1 *Total Head of All Types of Stationary (Not Vertical Turbine–Type) Fire Pumps.*

3.3.23.3 Total Head.

3.3.23.3.1* Total Head (H), Horizontal Pumps. The measure of the work increase, per pound (kilogram) of liquid, imparted to the liquid by the pump, and therefore the algebraic difference between the total discharge head and the total suction head. Total head, as determined on test where suction lift exists, is the sum of the total discharge head and total suction lift. Where positive suction head exists, the total head is the total discharge head minus the total suction head.

A.3.3.23.3.1 Total Head (H), Horizontal Pumps. See Figure A.3.3.23.3.1. (Figure A.3.3.23.3.1 does not show the various types of pumps applicable.)

> Total head is the energy (pressure) added to the water supply by the pump. It is expressed in feet (ft) or meters (m). For horizontal pumps, where a positive suction pressure exists, the total head is usually calculated by subtracting the pressure at the suction gauge from the pressure at the discharge gauge. Where the suction and discharge flanges are not of the same diameter, the velocity head for the volume of water passing through the flanges must be calculated using the formula identified in the definition of velocity head (see 3.3.23.6). Where the suction and discharge flanges have the same diameter, the incoming and outgoing velocities are not different.

3.3.23.3.2* Total Head (H), Vertical Turbine Pumps. The distance from the pumping liquid level to the center of the discharge gauge plus the total discharge head.

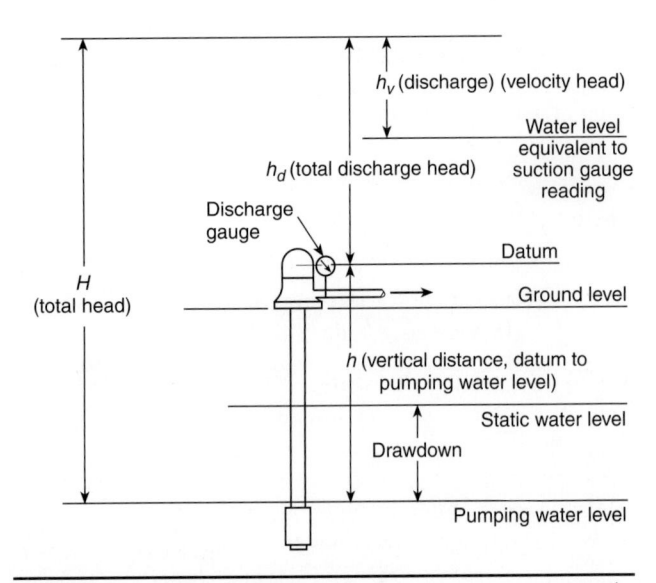

FIGURE A.3.3.23.3.2 *Total Head of Vertical Turbine–Type Fire Pumps.*

A.3.3.23.3.2 Total Head (H), Vertical Turbine Pumps. See Figure A.3.3.23.3.2.

> As noted in the commentary following A.3.3.23.3.1, total head is the energy (pressure) added to the water supply by the pump and is expressed in feet (ft) or meters (m). For vertical turbine pumps, a discharge gauge is located at the top of the pump column pipe. While the discharge pressure gauge indicates the pressure available to the fire protection system, the total pressure provided by the pump is determined by adding the reading of the discharge gauge to the suction lift (from the pumping water level up to the discharge flange) and the friction loss though the pipe column.

3.3.23.4 Total Rated Head. The total head developed at rated capacity and rated speed for a centrifugal pump.

> The total rated head is the amount of energy given to water as it passes through a specific pump when that pump is operating at its rated speed and at its rated flow.

3.3.23.5 Total Suction Head. Suction head exists where the total suction head is above atmospheric pressure. Total suction head, as determined on test, is the reading of a gauge at the suction of the pump, converted to feet (meters) of liquid, and referred to datum, plus the velocity head at the point of gauge attachment.

> Figure A.3.3.23.3.1 provides a graphical description of total suction head.

3.3.23.6* Velocity Head (h_v). The kinetic energy of a unit weight of fluid moving with velocity (v) determined at the point of the gauge connection.

A.3.3.23.6 Velocity Head (h_v). For the purpose of this standard, the velocity head is calculated using the average velocity, which can be obtained by dividing the flow in cubic feet per second (or cubic meters per second) by the actual area of pipe cross section in square feet (or square meters). The velocity head is the vertical distance a body would have to fall to acquire the velocity *(v)*. The velocity head *(h_v)* is expressed by the following formula:

$$h_v = \frac{v^2}{2g}$$

where:

v = velocity in the pipe [ft/sec (m/sec)]

g = acceleration due to gravity: 32.17 ft/sec² (9.807 m/sec²) at sea level and 45 degrees latitude

3.3.24 High-Rise Building. A building where the floor of an occupiable story is greater than 75 ft (23 m) above the lowest level of fire department vehicle access. [*5000*, 2012]

> The definition of the term *high-rise building* was added to the 2007 edition to assist the user in determining the appropriate protection criteria for fire pump rooms in high-rise buildings. Fire pump rooms must be protected by 2-hour fire-rated enclosures in high-rise buildings.

3.3.25 In Sight From (Within Sight From, Within Sight). Where this Code specifies that one equipment shall be "in sight from," "within sight from," or "within sight of," and so forth, another equipment, the specified equipment is to be visible and not more than 15 m (50 ft) distant from the other. [**70:** Art. 100]

> The definition of the phrase *in sight from* was added to the 2013 edition to provide enforceable language when this phrase is used.

3.3.26 Internal Combustion Engine. See 3.3.13.1.1.

3.3.27 Isolating Switch. See 3.3.50.1.

3.3.28 Liquid. For the purposes of this standard, liquid refers to water, foam-water solution, foam concentrates, water additives, or other liquids for fire protection purposes.

3.3.29 Liquid Level.

3.3.29.1 Pumping Liquid Level. The level, with respect to the pump, of the body of liquid from which it takes suction when the pump is in operation. Measurements are made the same as with the static liquid level.

> The term *pumping liquid level* is usually associated with vertical turbine–type pumps that take water or additives from an unpressurized source below the discharge flange of the pump, such as a well, river, or pond. See the commentary following the definition of *static liquid level* (3.3.29.2) for more information.

3.3.29.2 Static Liquid Level. The level, with respect to the pump, of the body of liquid from which it takes suction when the pump is not in operation. For vertical shaft turbine–type pumps, the distance to the liquid level is measured vertically from the horizontal centerline of the discharge head or tee.

> The term *static liquid level* is usually associated with vertical turbine–type pumps that take water or additives from an unpressurized source below the discharge flange of the pump, such as a well, river, or pond. When these pumps operate at peak capacity, the liquid level of the supply usually drops a certain distance. The level of the liquid before the fire pump

> starts is referred to as the *static liquid level*. The level to which the liquid drops when the pump is operating at 150 percent of its rated capacity is referred to as the *pumping liquid level* (see 3.3.29.1). The difference between these two levels is referred to as the *drawdown* (see 3.3.12). See Figure A.7.2.2.1.

3.3.30 Loss of Phase. The loss of one or more, but not all, phases of the polyphase power source.

3.3.31 Manual Transfer Switch. See 3.3.50.2.2.

3.3.32 Maximum Pump Brake Horsepower. The maximum brake horsepower required to drive the pump at rated speed. The pump manufacturer determines this by shop test under expected suction and discharge conditions. Actual field conditions can vary from shop conditions.

> When selecting a driver for a pump, the maximum brake horsepower demand of the pump when operating at its rated speed must be determined. The pump driver must be capable of supplying this horsepower. A pump's maximum brake horsepower can be obtained from the horsepower curve supplied by the pump manufacturer. The maximum demand of most pumps usually occurs between 140 percent and 170 percent of the pump's rated capacity.
>
> Exhibit II.3.5 illustrates the range in brake horsepower required to operate a hypothetical 750 gpm (2839 L/min) rated horizontal split-case fire pump. Notice that, as the pump produces larger flows, the brake horsepower required increases and the pressure produced by the pump decreases. Pump manufacturers provide similar curves for each of their pumps.

3.3.33 Motor.

3.3.33.1 Dripproof Guarded Motor. A dripproof machine whose ventilating openings are guarded in accordance with the definition for dripproof motor.

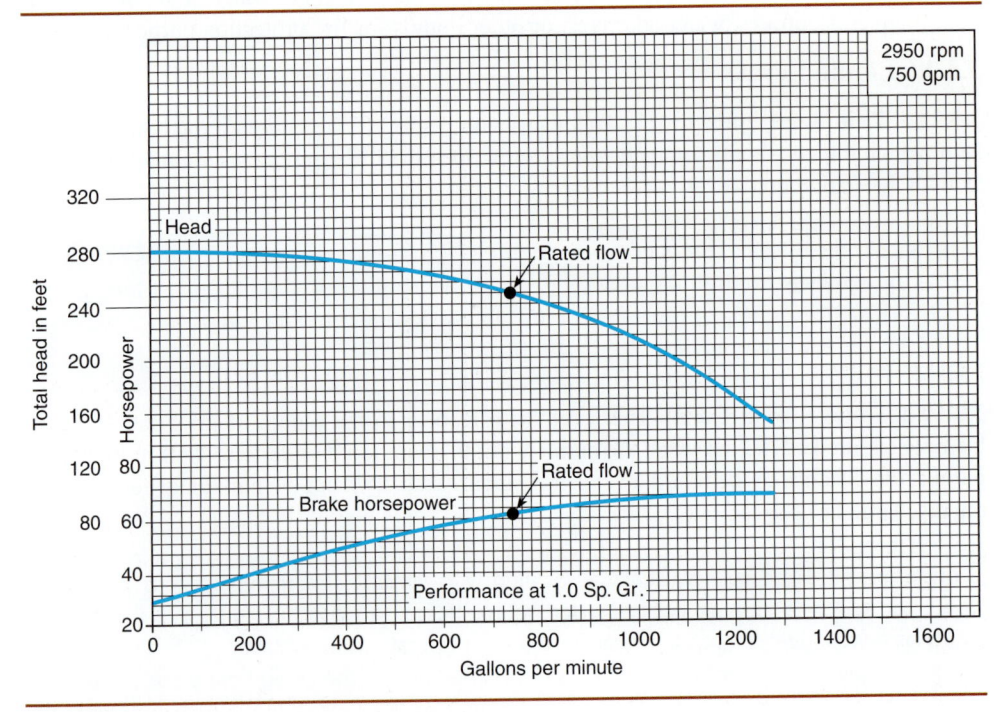

EXHIBIT II.3.5 *Brake Horsepower Curve.*

3.3.33.2 Dripproof Motor. An open motor in which the ventilating openings are so constructed that successful operation is not interfered with when drops of liquid or solid particles strike or enter the enclosure at any angle from 0 to 15 degrees downward from the vertical.

3.3.33.3 Dust-Ignition-Proof Motor. A totally enclosed motor whose enclosure is designed and constructed in a manner that will exclude ignitible amounts of dust or amounts that might affect performance or rating and that will not permit arcs, sparks, or heat otherwise generated or liberated inside of the enclosure to cause ignition of exterior accumulations or atmospheric suspensions of a specific dust on or in the vicinity of the enclosure.

3.3.33.4 Electric Motor. A motor that is classified according to mechanical protection and methods of cooling.

> Electric motors are a reliable, effective means of supplying power to fire pumps; however, at the time of publication of this handbook, only listed motors were permitted to be used. Electric motors must be selected so that they are of sufficient size to turn the pump at its rated speed under the required range of operating conditions. When an electric motor is considered, the power supply to the motor must be from a reliable source. Power supplies that can be compromised due to storms, fires, or other accidents can leave electric motors without power and render the fire pump useless. Where reliable sources cannot be secured, backup supplies, such as an emergency generator or diesel engine, are necessary. See Chapter 9 for more information.
>
> Electric motors used to drive fire pumps are required to withstand severe changes in operating conditions and should be arranged to accommodate large electric loads that might cause the motor to be destroyed. This strategy is contrary to the design and arrangement of most other electric motor applications. However, the reliability of the fire pump is the primary concern. The fire pump must continue to operate during a fire emergency for as long as is necessary. Exhibit II.3.6 shows a horizontal split-case fire pump driven by an electric motor.

EXHIBIT II.3.6 Electric Motor–Driven Fire Pump. (Photograph courtesy of Peerless Pump Company)

3.3.33.5 Explosionproof Motor. A totally enclosed motor whose enclosure is designed and constructed to withstand an explosion of a specified gas or vapor that could occur within it and to prevent the ignition of the specified gas or vapor surrounding the motor by sparks, flashes, or explosions of the specified gas or vapor that could occur within the motor casing.

3.3.33.6 Guarded Motor. An open motor in which all openings giving direct access to live metal or rotating parts (except smooth rotating surfaces) are limited in size by the structural parts or by screens, baffles, grilles, expanded metal, or other means to prevent accidental contact with hazardous parts. Openings giving direct access to such live or rotating parts shall not permit the passage of a cylindrical rod 0.75 in. (19 mm) in diameter.

3.3.33.7 Open Motor. A motor having ventilating openings that permit passage of external cooling air over and around the windings of the motor. Where applied to large apparatus

Stationary Fire Pumps Handbook 2013

without qualification, the term designates a motor having no restriction to ventilation other than that necessitated by mechanical construction.

3.3.33.8 Totally Enclosed Fan-Cooled Motor. A totally enclosed motor equipped for exterior cooling by means of a fan or fans integral with the motor but external to the enclosing parts.

3.3.33.9 Totally Enclosed Motor. A motor enclosed so as to prevent the free exchange of air between the inside and the outside of the case but not sufficiently enclosed to be termed airtight.

3.3.33.10 Totally Enclosed Nonventilated Motor. A totally enclosed motor that is not equipped for cooling by means external to the enclosing parts.

3.3.34 Net Positive Suction Head (NPSH) (h_{sv}). See 3.3.23.1.

3.3.35 On-Site Power Production Facility. The normal supply of electric power for the site that is expected to be constantly producing power.

3.3.36* On-Site Standby Generator. A facility producing electric power on site as the alternate supply of electrical power. It differs from an on-site power production facility, in that it is not constantly producing power. [**70:**695.2]

A.3.3.36 On-Site Standby Generator. It differs from an on-site power production facility in that it is not constantly producing power.

3.3.37 Pressure-Regulating Device. A device designed for the purpose of reducing, regulating, controlling, or restricting water pressure. [**14**, 2010]

3.3.38 Pump.

3.3.38.1 Additive Pump. A pump that is used to inject additives into the water stream.

3.3.38.2 Can Pump. A vertical shaft turbine–type pump in a can (suction vessel) for installation in a pipeline to raise water pressure.

> Vertical shaft turbine–type pumps are used primarily to pump water from nonpressurized water sources such as wells, ponds, and rivers. Can pumps are an application of vertical turbine–type pumps that boosts pressure from a pressurized water source such as a public water main. A cylinder or canister into which the pressurized water flows is constructed. The can pump, when placed into this cylinder, boosts the pressure of the incoming water supply. A can pump is essentially a vertical turbine–type pump that is used as a booster pump. Exhibit II.3.7 illustrates a typical vertical can pump installation.

3.3.38.3 Centrifugal Pump. A pump in which the pressure is developed principally by the action of centrifugal force.

> Water in centrifugal pumps enters the suction inlet and passes to the center of the impeller. Rotation of the impeller drives the water by centrifugal force to the rim, where it discharges.

3.3.38.4 End Suction Pump. A single suction pump having its suction nozzle on the opposite side of the casing from the stuffing box and having the face of the suction nozzle perpendicular to the longitudinal axis of the shaft.

> Exhibit II.3.8 illustrates an end suction pump.

3.3.38.5 Fire Pump. A pump that is a provider of liquid flow and pressure dedicated to fire protection.

EXHIBIT II.3.7 Vertical Can Pump.

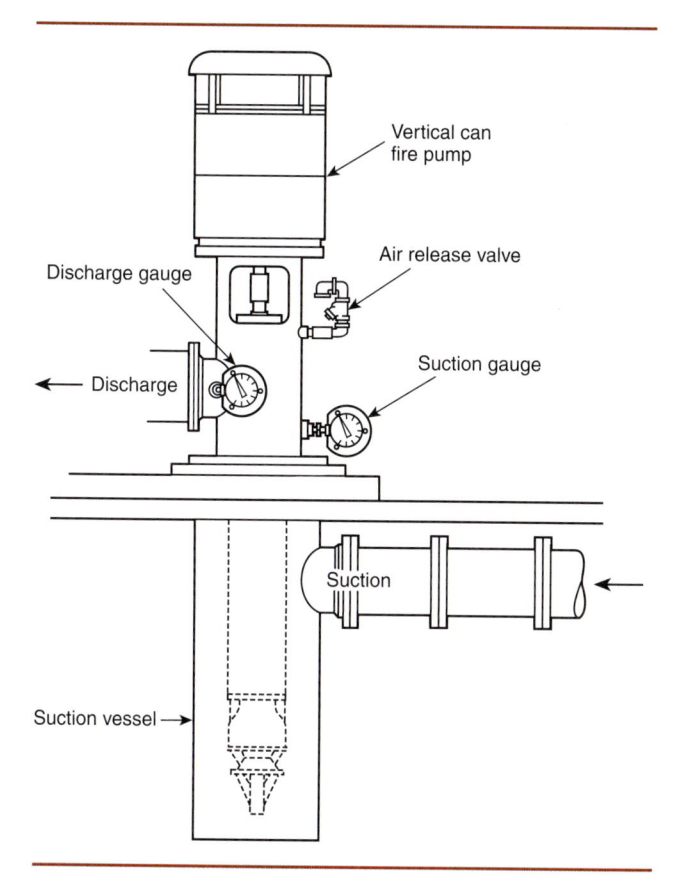

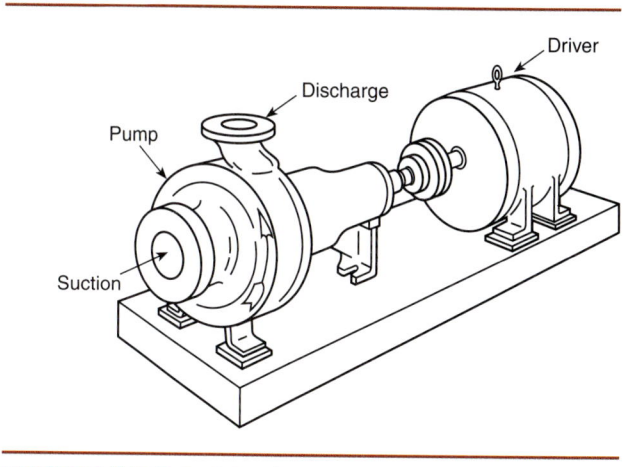

EXHIBIT II.3.8 End Suction Pump Driven by an Electric Motor.

3.3.38.6 Foam Concentrate Pump. See 3.3.38.1, Additive Pump.

3.3.38.7 Gear Pump. A positive displacement pump characterized by the use of gear teeth and casing to displace liquid.

3.3.38.8 Horizontal Pump. A pump with the shaft normally in a horizontal position.

Chapter 6 addresses the installation of horizontal pumps.

3.3.38.9 Horizontal Split-Case Pump. A centrifugal pump characterized by a housing that is split parallel to the shaft.

With this type of pump, the case in which the shaft and impeller are housed is split in the middle and can be separated. This type of housing allows for easy access for repair or replacement of internal components. These pumps are not self-priming and must be supplied with water at the suction flange at some positive pressure. Exhibit II.3.9 illustrates a horizontal split-case pump. (See Chapter 6 for information on the installation of this type of pump.)

3.3.38.10 In-Line Pump. A centrifugal pump whose drive unit is supported by the pump having its suction and discharge flanges on approximately the same centerline.

Exhibit II.3.10 shows a diagram of an in-line pump. See Chapter 6 for more information.

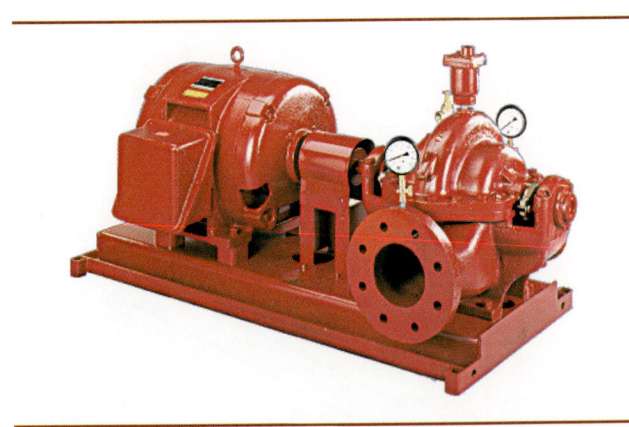

EXHIBIT II.3.9 Horizontal Split-Case Pump. (Courtesy of Hydraulics Institute)

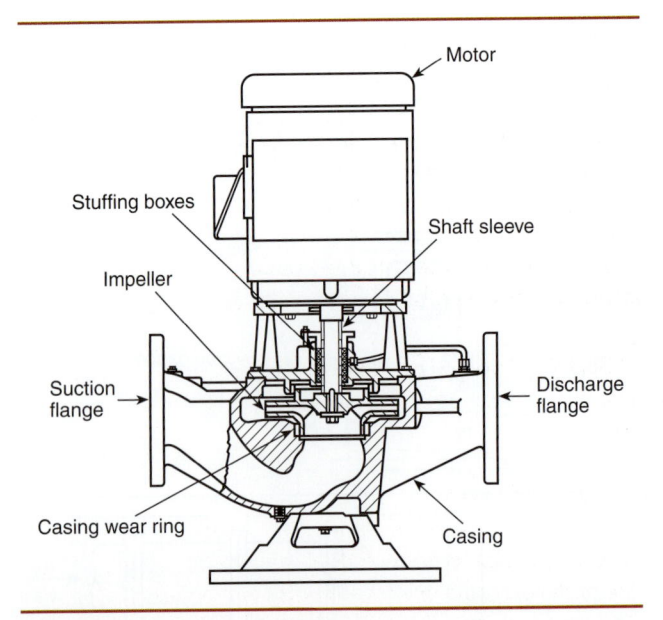

EXHIBIT II.3.10 In-Line Pump.

3.3.38.11 Packaged Fire Pump Assembly. Fire pump unit components assembled at a packaging facility and shipped as a unit to the installation site. The scope of listed components (where required to be listed by this standard) in a pre-assembled package includes the pump, driver, controller, and other accessories identified by the packager assembled onto a base with or without an enclosure.

> See Section 4.29 for packaged fire pump assembly requirements and associated commentary.

3.3.38.12 Piston Plunger Pump. A positive displacement pump characterized by the use of a piston or plunger and a cylinder to displace liquid.

3.3.38.13 Positive Displacement Pump. A pump that is characterized by a method of producing flow by capturing a specific volume of fluid per pump revolution and reducing the fluid void by a mechanical means to displace the pumping fluid.

3.3.38.14 Pressure Maintenance (Jockey or Make-Up) Pump. A pump designed to maintain the pressure on the fire protection system(s) between preset limits when the system is not flowing water.

> See Section 4.25 for pressure maintenance (jockey or make-up) pump requirements and associated commentary.

3.3.38.15 Rotary Lobe Pump. A positive displacement pump characterized by the use of a rotor lobe to carry fluid between the lobe void and the pump casing from the inlet to the outlet.

3.3.38.16 Rotary Vane Pump. A positive displacement pump characterized by the use of a single rotor with vanes that move with pump rotation to create a void and displace liquid.

3.3.38.17 Vertical Lineshaft Turbine Pump. A vertical shaft centrifugal pump with rotating impeller or impellers and with discharge from the pumping element coaxial with the shaft. The pumping element is suspended by the conductor system, which encloses a system of vertical shafting used to transmit power to the impellers, the prime mover being external to the flow stream.

Vertical lineshaft turbine pumps were originally designed to pump water from wells; however, vertical lineshaft turbine pumps can be used wherever a nonpressurized water source serves as the water supply. These pumps are most often used where horizontal fire pumps are prohibited from use because horizontal pumps would operate under a suction lift condition.

Although vertical lineshaft turbine pumps move water up to the discharge flange of the pump, they do not operate under a suction lift condition. The impellers for these pumps are located within the water supply and force water up to the discharge flange. Because the impellers sit in the water source, these pumps do not require priming (see Chapter 7). Exhibit II.3.11 shows a typical vertical lineshaft turbine pump.

3.3.38.18* Water Mist Positive Displacement Pumping Unit. Multiple positive displacement pumps designed to operate in parallel that discharges into a single common water mist distribution system.

The definition of, and requirements associated with, the term *water mist positive displacement pumping unit* was added to the 2013 edition to address newer technology that allows multiple small-volume positive displacement pumps operating in parallel to supply water mist fire protection systems. These pumps are mounted in a common rack and are listed as a unit. The concepts involved are significantly different from those used for traditional stand-alone pumps, or even for traditional pumps operating in series or parallel that still must be treated as individual pumps. Because of the low volume of an individual pump, the unit is allowed to serve as its own pressure maintenance pump. One (or more) of the individual pumps may operate with a variable speed driver to control the discharge pressure. The flow rate from the pump unit is not a smooth curve, but rather a "step" function that increases whenever another pump is activated or when a variable speed drive changes speed.

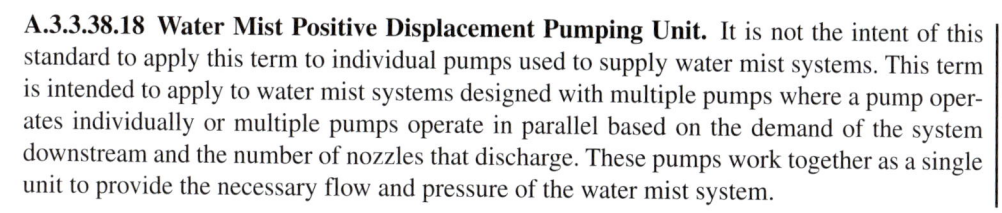

EXHIBIT II.3.11 *Vertical Lineshaft Turbine Pump. (Photograph courtesy of Patterson Pump)*

A.3.3.38.18 Water Mist Positive Displacement Pumping Unit. It is not the intent of this standard to apply this term to individual pumps used to supply water mist systems. This term is intended to apply to water mist systems designed with multiple pumps where a pump operates individually or multiple pumps operate in parallel based on the demand of the system downstream and the number of nozzles that discharge. These pumps work together as a single unit to provide the necessary flow and pressure of the water mist system.

3.3.39 Pumping Liquid Level. See 3.3.29.1.

3.3.40 Qualified Person. A person who, by possession of a recognized degree, certificate, professional standing, or skill, and who, by knowledge, training, and experience, has demonstrated the ability to deal with problems related to the subject matter, the work, or the project. [**1451**, 2007]

3.3.41 Record Drawing. A design, working drawing, or as-built drawing that is submitted as the final record of documentation for the project.

3.3.42* Series Fire Pump Unit. All fire pump units located within the same building that operate in a series arrangement where the first fire pump takes suction directly from a water supply and each sequential pump takes suction under pump pressure from the preceding pump. Two pumps that operate in series through a tank(s) or break tank(s) are not considered part of a series fire pump unit.

This definition does not include pumps that may operate in series but are separated by a distribution system that supplies other uses between the pumps. An example of a series pumping arrangement that is not included within this definition is a fire pump that takes suction from a municipal supply that is supplied by a pump. The primary application for fire pumps operating in series is high-rise buildings, where the pressure required to supply fire protection systems on upper floors exceeds the pressure of the components in the fire protection systems on the lower floors. One pump supplies the lower floors, and a second pump, in series with the first pump, supplies upper floors.

A.3.3.42 Series Fire Pump Unit. Pumps that fill tanks are not considered to be in series with the pumps supplied by those tanks. Water utilities and "campus type" water distribution systems that supply a fire pump within a building can have pumps that operate independently of, but are necessary to, the operation of a fire pump within the building. These pumps are not included within the definition of series fire pump unit, but the arrangement of these pumps should be reviewed as part of evaluating the water supply.

3.3.43* Service. The conductors and equipment for delivering electric energy from the serving utility to the wiring system of the premises served. [70: Art. 100]

A.3.3.43 Service. For more information, see *NFPA 70, National Electrical Code*, Article 100.

3.3.44* Service Equipment. The necessary equipment, usually consisting of a circuit breaker(s) or switch(es) and fuse(s) and their accessories, connected to the load end of service conductors to a building or other structure, or an otherwise designated area, and intended to constitute the main control and cutoff of the supply. [70: Art. 100]

A.3.3.44 Service Equipment. For more information, see *NFPA 70, National Electrical Code*, Article 100.

See also Chapters 9 and 10.

3.3.45 Service Factor. A multiplier of an ac motor that, when applied to the rated horsepower, indicates a permissible horsepower loading that can be carried at the rated voltage, frequency, and temperature. For example, the multiplier 1.15 indicates that the motor is permitted to be overloaded to 1.15 times the rated horsepower.

The motor for a fire pump installation must be of sufficient capacity so that it is not overloaded beyond the limit of the service factor at the pump's maximum brake horsepower when operating at its rated speed. For example, a 75 hp (56 kW) motor with a service factor of 1.12 could safely meet the demand of 84 hp (63 kW) [75 hp (56 kW) × 1.12]. The service factor is a function of the type of motor and is determined by the manufacturer (see 9.5.2.2).

3.3.46 Set Pressure. As applied to variable speed pressure limiting control systems, the pressure that the variable speed pressure limiting control system is set to maintain.

3.3.47* Signal. An indicator of status.

A.3.3.47 Signal. A response to signals is expected within 2 hours.

3.3.48 Speed.

3.3.48.1 Engine Speed. The speed indicated on the engine nameplate.

3.3.48.2 Motor Speed. The speed indicated on the motor nameplate.

3.3.48.3 Rated Speed. The speed for which the fire pump is listed and that appears on the fire pump nameplate.

3.3.49 Static Liquid Level. See 3.3.29.2.

3.3.50 Switch.

3.3.50.1 Isolating Switch. A switch intended for isolating an electric circuit from its source of power. It has no interrupting rating, and it is intended to be operated only after the circuit has been opened by some other means.

> Isolating switches typically are used within controllers for electric motor–driven pumps. They are used to isolate the power source from the controller and must be operable from the outside of the controller (see 10.4.2).

3.3.50.2 Transfer Switch.

3.3.50.2.1 Automatic Transfer Switch (ATS). Self-acting equipment for transferring the connected load from one power source to another power source. [**110**, 2013]

> Where an emergency or backup power source is supplied for electric motors driving fire pumps, an automatic transfer switch is used to transfer power spontaneously from the primary supply to the backup supply when the primary power supply fails. The transfer switch may be provided as an integral part of the fire pump controller, or it may be installed as a separate piece of equipment. Where the transfer switch is used independently from the controller, the transfer switch is required to be listed. Where the transfer switch is integrated into the controller, the entire assembly is required to be listed (see Section 10.8).

3.3.50.2.2 Manual Transfer Switch. A switch operated by direct manpower for transferring one or more load conductor connections from one power source to another.

> ▶ FAQ Does NFPA 20 allow the use of manual transfer switches?
>
> Where an emergency or backup power source is supplied for electric motors driving fire pumps, a manual power transfer switch allows for the manual (nonautomatic) transfer of power from the normal power supply to the secondary supply. However, NFPA 20 prohibits the use of manual transfer switches (see 10.8.2). In some facilities, a common practice is to have multiple sources of power to improve reliability. As a result, the intention of the Technical Committee on Fire Pumps is to prohibit the use of manual transfer switches.

3.3.51 Total Discharge Head (h_d). See 3.3.23.2.

3.3.52 Total Head (H), Horizontal Pumps. See 3.3.23.3.1.

3.3.53 Total Head (H), Vertical Turbine Pumps. See 3.3.23.3.2.

3.3.54 Total Rated Head. See 3.3.23.4.

3.3.55 Total Suction Head (h_s). See 3.3.23.5.

3.3.56 Total Suction Lift (h_l). Suction lift that exists where the total suction head is below atmospheric pressure. Total suction lift, as determined on test, is the reading of a liquid manometer at the suction nozzle of the pump, converted to feet (meters) of liquid, and referred to datum, minus the velocity head at the point of gauge attachment.

> ◀ DESIGN ALERT
>
> NFPA 20 no longer permits fire pumps to operate under a suction lift condition — that is, a negative pressure is not permitted at the suction flange. Prior to the 1974 edition of the standard, however, horizontal shaft pumps were permitted to take suction lift. The difficulty in taking a suction lift with a horizontal shaft fire pump is that the pump must remain primed with water at all times. The priming mechanism was required to have sufficient capacity to remove

air from the pump and suction pipe within 3 minutes. At one point, NFPA 20 contained 7 methods for priming pumps and suction pipes. This provision was removed from the standard beginning with the 1974 edition, because failure of the priming mechanism results in failure of the pump to move water. Exhibit II.3.12 illustrates one priming method that was permitted in previous editions of the standard. This exhibit is included for information only, as some of these installations may still be in service.

EXHIBIT II.3.12 Centrifugal Fire Pump Installation Where Pump Has Suction Lift.

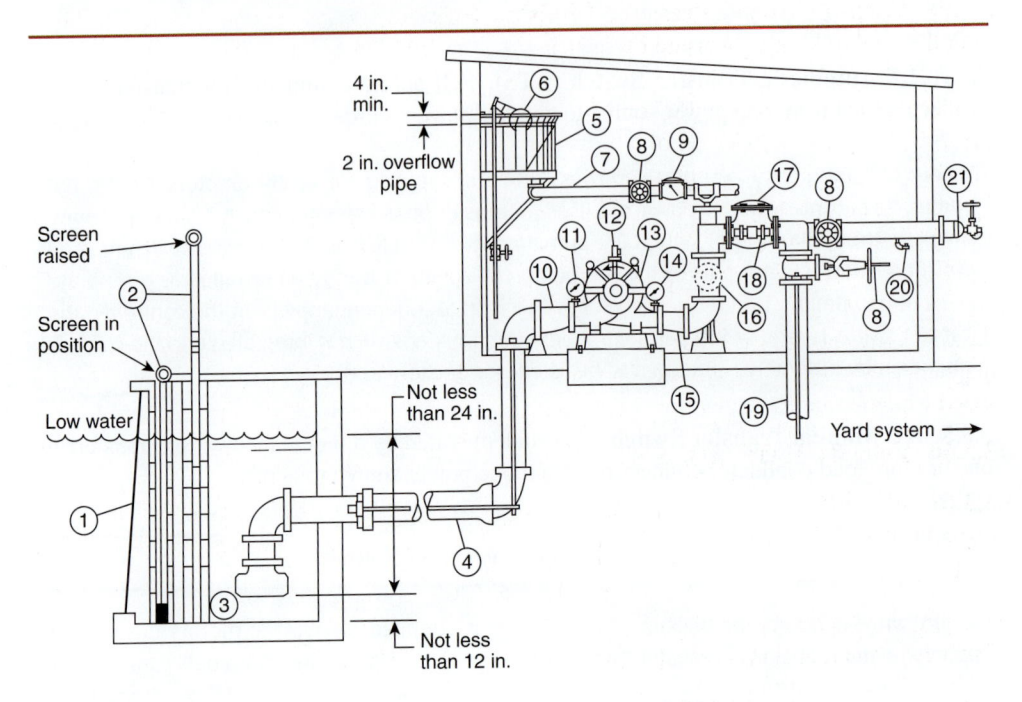

1. Trash rack (½ in. flat or ¾ in. round steel bars spaced 2 to 3 in. apart)
2. Double screens
3. Approved foot valve
4. Suction pipe from water supply to pump
5. Priming tank
6. Automatic float valve
7. Priming connection
8. OS&Y gate valves
9. Priming check valve
10. Eccentric reducer
11. Suction gauge
12. Umbrella cock
13. Horizontal fire pump
14. Discharge gauge
15. Concentric reducer
16. Relief valve (if required)
17. Discharge check valve
18. Priming bypass
19. Discharge pipe
20. Drain valve or ball drip
21. Hose valve manifold with hose valves

3.3.57 Valve.

3.3.57.1 Dump Valve. An automatic valve installed on the discharge side of a positive displacement pump to relieve pressure prior to the pump driver reaching operating speed.

3.3.57.2 Low Suction Throttling Valve. A pilot-operated valve installed in discharge piping that maintains positive pressure in the suction piping, while monitoring pressure in the suction piping through a sensing line.

3.3.57.3 Pressure Control Valve. A pilot-operated pressure-reducing valve designed for the purpose of reducing the downstream water pressure to a specific value under both flowing (residual) and nonflowing (static) conditions. [14, 2010]

3.3.57.4 Pressure-Reducing Valve. A valve designed for the purpose of reducing the downstream water pressure under both flowing (residual) and nonflowing (static) conditions. [14, 2010]

3.3.57.5 Relief Valve. A device that allows the diversion of liquid to limit excess pressure in a system.

3.3.57.5.1 Circulation Relief Valve. A valve used to cool a pump by discharging a small quantity of water. This valve is separate from and independent of the main relief valve.

3.3.57.6 Unloader Valve. A valve that is designed to relieve excess flow below pump capacity at set pump pressure.

3.3.58 Variable Speed Pressure Limiting Control. A speed control system used to limit the total discharge pressure by reducing the pump driver speed from rated speed.

> Variable speed pressure limiting control (VSPLC) systems are similar to conventional fire pump systems but with the addition of a pressure-sensing line that prevents the pump from over-pressurizing the fire protection system. Prevention of overpressurization is accomplished by altering the speed of the driver. A VSPLC can be used in place of a break tank or a pressure-reducing or pressure relief valve when the pressure varies in the attached water supply. The VSPLC can be driven by either an electric or a diesel engine.

3.3.59 Variable Speed Suction Limiting Control. A speed control system used to maintain a minimum positive suction pressure at the pump inlet by reducing the pump driver speed while monitoring pressure in the suction piping through a sensing line.

3.3.60 Velocity Head (h_v). See 3.3.23.6.

3.3.61 Wet Pit. A timber, concrete, or masonry enclosure having a screened inlet kept partially filled with water by an open body of water such as a pond, lake, or stream.

References Cited in Commentary

National Fire Protection Association, 1 Batterymarch Park, Quincy, MA 02169-7471.

NFPA 291, *Recommended Practice for Fire Flow Testing and Marking of Hydrants*, 2013 edition.
NFPA Glossary of Terms, 2008 (available online at www.nfpa.org).

Merriam-Webster, Inc., 47 Federal Street, P.O. Box 281, Springfield, MA 01102.

Merriam-Webster's Collegiate® Dictionary, 11th edition, 2003.

CHAPTER 4

General Requirements

Chapter 4 contains general requirements for fire pumps. It is important to note that NFPA 20, *Standard for the Installation of Stationary Pumps for Fire Protection,* does not establish the requirement to install a fire pump. Such requirements are based on the hydraulic demand of suppression systems. In some cases, where supplemental water supplies are needed, other types of water supplies can be used, such as elevated water tanks or pressure tanks. Where a fire pump(s) is part of the fire protection system, the design, installation, and acceptance testing requirements should be in accordance with this standard.

The chapter begins with design and operational requirements for fire pumps, including information related to required approvals, pump operation and performance, and water supplies. Standardized fire pump rated capacities are also established. The chapter concludes with specifications and installation requirements for all pump components and equipment, except for pump controllers and drivers.

4.1 Pumps

This standard shall apply to centrifugal single-stage and multistage pumps of the horizontal or vertical shaft design and positive displacement pumps of the horizontal or vertical shaft design.

Positive displacement pumps were added to the 1999 edition to provide guidance on the design and installation of pumps that are typically used for foam systems. Previously, no guidance for the design and installation of positive displacement pumps existed. Requirements for multiple positive displacement pumps designed and listed to operate as a single unit were added to the 2013 edition and referred to by the term *water mist positive displacement pumping unit*.

4.2* Approval Required

A.4.2 Because of the unique nature of fire pump units, the approval should be obtained prior to the assembly of any specific component.

4.2.1 Stationary pumps shall be selected based on the conditions under which they are to be installed and used.

The fire protection system designer should verify the fire pump suction and discharge pressures over the operating range of the fire pump. The fire pump should provide sufficient discharge pressure to supply the fire protection system demands, but the discharge pressure should not subject the fire protection system components to pressures in excess of their listed pressure rating.

The fire pump should be located in a suitable environment. See Section 4.12, Equipment Protection, for further guidance.

4.2.2 The pump manufacturer or its authorized representative shall be given complete information concerning the liquid and power supply characteristics.

> Pump capacities are based on the calculated system demand. Pressure boost or output of the pump is determined by the difference between the pressures available from the attached water supply and the pressure required by the fire protection system. The characteristics of that water supply — whether positive pressure or static — must be determined in order to select the correct type of pump and its performance characteristics.
>
> The available power supply for electric pumps must be suitable for the fire pump controller. The power supply must be analyzed for reliability, capacity, and suitability. This information must be made available to the pump manufacturer or manufacturer's representative for analysis.
>
> While fire pumps are traditionally associated with water, they are also used for pumping foam concentrate, foam water, water with other water additives, seawater, and possibly other liquids. The characteristics of the liquid are part of the critical information needed by the pump manufacturer.

4.2.3 A complete plan and detailed data describing pump, driver, controller, power supply, fittings, suction and discharge connections, and liquid supply conditions shall be prepared for approval.

> Ordinarily, upon purchase of a fire pump, the manufacturer or supplier provides the buyer with product data for the pump, the driver, the controller, and all accessories. This data package should include the pump's rated capacity and speed in addition to other performance characteristics. The data package should be submitted to the registered design professional (RDP) and authority having jurisdiction (AHJ) for approval before shipment of the pump and accessories. The reviewing authority should be contacted prior to submittal to determine the number of plans, calculations, or product data needed for review. The submitter should include an additional number of copies of the package to be returned documenting the approval or comments from the AHJ.
>
> In addition to this data package, a complete plan of the proposed installation, drawn to scale, should be submitted, indicating, as a minimum, the information outlined in Exhibit II.4.1.

4.2.3.1 Plans shall be drawn to an indicated scale, on sheets of uniform size, and shall indicate, as a minimum, the items from the following list that pertain to the design of the system:

(1) Name of owner and occupant
(2) Location, including street address
(3) Point of compass
(4) Name and address of installing contractor
(5) Pump make and model number
(6) Pump rating _____ gpm @ _____ psi _____ rpm
(7) Suction main size, length, location, weight, type of material, and point of connection to water supply, as well as size and type of valves, valve indicators, regulators, meters, and valve pits, and depth to top of pipe below grade
(8) Water supply capacity information including the following:
  (a) Location and elevation of static and residual test gauge with relation to the riser reference point
  (b) Flow location
  (c) Static pressure, psi (bar)
  (d) Residual pressure, psi (bar)
  (e) Flow, gpm (L/min)

PUMP INSTALLATION PLAN CHECKLIST

General
- ❏ Name of owner or occupant
- ❏ Location including street address
- ❏ Point of compass
- ❏ Name and address of designer and installing contractor
- ❏ Listed pump, make, model number, driver type, and rated capacity
- ❏ Type of system supplied by pump
- ❏ Design standard used including edition

Water Supply Characteristics
- ❏ Flow test data not more than 5 years old
- ❏ Underground main of adequate size
- ❏ Water storage tank of adequate capacity with automatic refill connection

Suction Piping
- ❏ Proper size
- ❏ Galvanized or painted on the inside for corrosion protection
- ❏ Isolation valve (OS&Y) in the proper location
- ❏ Backflow prevention or other device in proper location
- ❏ Elbows in the proper orientation or more than 10 pipe diameters away from suction flange of pump
- ❏ Eccentric reducer (if needed) installed correctly
- ❏ Pump bypass
- ❏ Suction and discharge pressure gauges
- ❏ Circulation relief valve

Discharge Piping
- ❏ Proper size
- ❏ Check valve
- ❏ Discharge isolation valve

Fire Pump Controller
- ❏ Listed for type of pump served
- ❏ If electric, type and arrangement of power supply

Water Flow Test Devices
- ❏ Test header or flowmeter
- ❏ Proper number of 2½ in. hose valves on test header
- ❏ Test header piping of proper size
- ❏ Proper size of flowmeter (if provided)

Jockey Pump
- ❏ Jockey pump bypasses fire pump
- ❏ Separate and dedicated sensing line for jockey pump
- ❏ Sensing lines ½ in. diameter and of brass, copper, or stainless steel piping
- ❏ No shutoff valves in sensing lines

Isolation Valves
- ❏ All isolation valves supervised in the open position
- ❏ Test header and flowmeter valves supervised in the closed position

Diesel Fire Pump
- ❏ Relief valve (if provided) with no isolation valves
- ❏ Two storage batteries provided with charger
- ❏ Cooling system from heat exchanger or cooling water supply from pump discharge
- ❏ Diesel tank located above ground
- ❏ Diesel tank of sufficient capacity 1 gallon per horsepower (gal/hp) plus 10 percent

Notes:
1. For all types of pumping equipment, a complete bill of material should be provided. This list should include the make, model, and part numbers of all components.
2. For all electrically operated pumps, provide an electrical schematic drawing that depicts which components are proposed for the power supply from the utility connection to the pump motor controller.
3. Ratings of all equipment and components and settings of breakers, fuses, switches, and transformers should be indicated. Size and length of all circuit conductors should also be noted.

EXHIBIT II.4.1 *Checklist for Plan of Proposed Installation.*

     (f) Date
     (g) Time
     (h) Name of person who conducted the test or supplied the information
     (i) Other sources of water supply, with pressure or elevation

> See 4.6.1.1 for a discussion of water supply evaluation.

  (9) Pump driver details including manufacturer, horsepower, voltage, or fuel system details
(10) Controller manufacturer, type, and rating

(11) Suction and discharge pipe, fitting, and valve types
(12) Test connection piping and valves
(13) Flow meter details (if used)
(14) Jockey pump and controller arrangement, including sensing line details

4.2.4 Each pump, driver, controlling equipment, power supply and arrangement, and liquid supply shall be approved by the authority having jurisdiction for the specific field conditions encountered.

4.3 Pump Operation

4.3.1 In the event of fire pump operation, qualified personnel shall respond to the fire pump location to determine that the fire pump is operating in a satisfactory manner.

> A qualified person is one who is familiar with the purpose and function of the fire pump and related equipment. This person should be trained in the operation of the pump. The regularly scheduled nonflow test required by NFPA 25, *Standard for the Inspection, Testing, and Maintenance of Water-Based Fire Protection Systems,* is an excellent training opportunity that can be used to demonstrate the proper operation of pumping equipment. One of the reasons for placing a fire pump in a protected enclosure, as required in Section 4.12, is to provide protection for both the pump operator and the pumping equipment. Consideration should be given to locating the fire pump room at grade level, with direct access to the exterior of the building, to afford quick and easy egress for the pump operator and access to the pumping equipment for fire department personnel.

4.3.2 System Designer.

4.3.2.1 The system designer shall be identified on the system design documents.

4.3.2.2 Acceptable minimum evidence of qualifications or certification shall be provided when requested by the authority having jurisdiction.

4.3.2.3 Qualified personnel shall include, but not be limited to, one or more of the following:

(1) Personnel who are factory trained and certified for fire pump system design of the specific type and brand of system being designed
(2)* Personnel who are certified by a nationally recognized fire protection certification organization acceptable to the authority having jurisdiction
(3) Personnel who are registered, licensed, or certified by a state or local authority

A.4.3.2.3(2) Nationally recognized fire protection certification programs include, but are not limited to, those programs offered by the International Municipal Signal Association (IMSA) and the National Institute for Certification in Engineering Technologies (NICET). Note: These organizations and the products or services offered by them have not been independently verified by the NFPA, nor have the products or services been endorsed or certified by the NFPA or any of its technical committees.

4.3.2.4 Additional evidence of qualification or certification shall be permitted to be required by the AHJ.

4.3.3 System Installer.

4.3.3.1 Installation personnel shall be qualified or shall be supervised by persons who are qualified in the installation, inspection, and testing of fire protection systems.

4.3.3.2 Minimum evidence of qualifications or certification shall be provided when requested by the authority having jurisdiction.

4.3.3.3 Qualified personnel shall include, but not be limited to, one or more of the following:

(1) Personnel who are factory trained and certified for fire pump system design of the specific type and brand of system being designed
(2)* Personnel who are certified by a nationally recognized fire protection certification organization acceptable to the authority having jurisdiction
(2) Personnel who are registered, licensed, or certified by a state or local authority

A.4.3.3.3(2) See A.4.3.2.3(2).

4.3.3.4 Additional evidence of qualification or certification shall be permitted to be required by the AHJ.

4.3.4* Service Personnel Qualifications and Experience.

A.4.3.4 Service personnel should be able to do the following:

(1) Understand the requirements contained in this standard and in NFPA 25, *Standard for the Inspection, Testing, and Maintenance of Water-Based Fire Protection Systems*, and the fire pump requirements contained in *NFPA 70, National Electrical Code*
(2) Understand basic job site safety laws and requirements
(3) Apply troubleshooting techniques and determine the cause of fire protection system trouble conditions
(4) Understand equipment-specific requirements, such as programming, application, and compatibility
(5) Read and interpret fire protection system design documentation and manufacturers' inspection, testing, and maintenance guidelines
(6) Properly use tools and equipment required for testing and maintenance of fire protection systems and their components
(7) Properly apply the test methods required by this standard and NFPA 25, *Standard for the Inspection, Testing, and Maintenance of Water-Based Fire Protection Systems*

4.3.4.1 Service personnel shall be qualified and experienced in the inspection, testing, and maintenance of fire protection systems.

4.3.4.2 Qualified personnel shall include, but not be limited to, one or more of the following:

(1) Personnel who are factory trained and certified for fire pump system design of the specific type and brand of system being designed
(2)* Personnel who are certified by a nationally recognized fire protection certification organization acceptable to the authority having jurisdiction
(3) Personnel who are registered, licensed, or certified by a state or local authority
(4) Personnel who are employed and qualified by an organization listed by a nationally recognized testing laboratory for the servicing of fire protection systems

A.4.3.4.2(2) See A.4.3.2.3(2).

4.3.4.3 Additional evidence of qualification or certification shall be permitted to be required by the AHJ.

4.4 Fire Pump Unit Performance

4.4.1* The fire pump unit, consisting of a pump, driver, and controller, shall perform in compliance with this standard as an entire unit when installed or when components have been replaced.

A.4.4.1 A single entity should be designated as having unit responsibility for the pump, driver, controller, transfer switch equipment, and accessories. *Unit responsibility* means the accountability to answer and resolve any and all problems regarding the installation, compatibility, performance, and acceptance of the equipment. Unit responsibility should not be construed to mean purchase of all components from a single supplier.

> Unit responsibility should be the responsibility of the installer until the equipment is accepted and officially turned over to the building owner. A representative of the installing contractor (usually a designer, a project manager, or an engineer) is normally involved with the fire pump installation from purchase of the equipment to acceptance testing. On projects where a formal commissioning program is used, the responsibility for the fire pump may be given to a commissioning agent.
>
> When individual components have been replaced, the contractor who installed the replacement components must verify that the entire unit functions as intended — see NFPA 25 for component replacement testing requirements after the initial acceptance test.
>
> A water mist positive displacement pumping unit and a series fire pump unit should also meet the requirements of Section 4.4.1.

▶ **FAQ** Must a fire pump, driver, and all related accessories/components be purchased from a single source?

> NFPA 20, *Standard for the Installation of Stationary Pumps for Fire Protection*, does not require the purchase of all components of a fire pump from a single source. However, consideration should be given to the listing and compatibility of all system components.

4.4.2 The complete fire pump unit shall be field acceptance tested for proper performance in accordance with the provisions of this standard. *(See Section 14.2.)*

> Upon completion of the installation and prior to final acceptance, the installation contractor should coordinate the field acceptance test, which includes participation of all interested parties, such as the pump manufacturer's representative and the fire pump controller manufacturer's representative, the building owner or his or her representative, the RDP, and the AHJ. The acceptance test should demonstrate to the building owner's representative and the AHJ that the pump performs as intended and complies with the requirements of this standard and any project specification. Ordinarily, the services of both the pump manufacturer and controller manufacturer are part of the fire pump purchase order and include the necessary labor hours to complete the acceptance test.
>
> A water mist positive displacement pumping unit and a series fire pump unit should also meet the requirements of 4.4.2.

4.5 Certified Shop Test

4.5.1 Certified shop test curves showing head capacity and brake horsepower of the pump shall be furnished by the manufacturer to the purchaser.

> The certified shop test provides verification that the fire pump was tested by the manufacturer prior to shipment to the job site and performs properly under ideal conditions. This test report should be submitted to the RDP and AHJ for review and approval. The data collected during the acceptance test are compared to the certified shop test report to reveal differences in pump performance from the shop test to the acceptance test. Variations in the pump speed can cause discrepancies between the certified shop test curve and the field results. Minor differences between these test results may be corrected by applying affinity laws as referenced in Chapter 14.

4.5.1.1 For water mist positive displacement pumping units, certified shop test data, including flow, pressure, and horsepower, shall be provided for each independent pump.

> While a water mist positive displacement pumping unit is acceptance tested as a unit, individual fire pump performance curves are useful for evaluating performance degradation and replacement of individual pumps within the unit.

4.5.1.2 For water mist positive displacement pumping units, certified shop test data, including flow, pressure, and horsepower, shall also be provided for the fire pump unit with variable speed features deactivated.

4.5.1.2.1 The certified fire pump unit shop test data shall be developed by activating the individual fire pumps in the same operating sequence that the controller will utilize.

> The variable speed feature is deactivated for part of the acceptance testing for water mist positive displacement pumping units. This curve is also useful for evaluating performance degradation and replacement of individual pumps within the unit.

4.5.1.3 For water mist positive displacement pumping units with variable speed features, certified shop test data, including flow, pressure, and horsepower, shall also be provided for the fire pump unit with variable speed features activated.

4.5.1.3.1 The certified fire pump unit shop test data shall be developed by activating the individual fire pumps in the same operating sequence that the controller will utilize.

> A water mist positive displacement pumping unit normally operates with the variable speed features activated. Acceptance testing of the unit includes operating the unit with the variable speed features activated.

4.5.2 The purchaser shall furnish the data required in 4.5.1 to the authority having jurisdiction.

> The certified shop test is performed when the fire pump and driver are assembled at the manufacturing facility. In some cases, the commissioning agent or project engineer may wish to witness the shop test. In such cases, the shop test should be coordinated with the manufacturer, purchaser, and commissioning agent. The certified shop test results should be submitted for review and approval as part of the commissioning process.

4.6 Liquid Supplies

4.6.1* Reliability.

> Determining and verifying reliability requires a complex evaluation that should take into account life safety, economic, and environmental risks. For a more thorough discussion regarding water supply adequacy and reliability, see Part III of this handbook.

A.4.6.1 For water supply capacity and pressure requirements, see the following documents:

(1) NFPA 1, *Fire Code*
(2) NFPA 13, *Standard for the Installation of Sprinkler Systems*
(3) NFPA 14, *Standard for the Installation of Standpipe and Hose Systems*
(4) NFPA 15, *Standard for Water Spray Fixed Systems for Fire Protection*

(5) NFPA 16, *Standard for the Installation of Foam-Water Sprinkler and Foam-Water Spray Systems*
(6) NFPA 750, *Standard on Water Mist Fire Protection Systems*

4.6.1.1 The adequacy and dependability of the water source are of primary importance and shall be fully determined, with due allowance for its reliability in the future.

> Water supplies should be thoroughly investigated to identify how system operation and variations in water usage may affect the waterflow test. Many water utilities use a combination of elevated tanks and pumps in the distribution system, which affects the available flows and pressures. Tank filling, which may occur at night, can cause changes in system pressures. The need for increased usage in the future, especially in rapidly developing areas, should be evaluated. These variations should be taken into account to ensure that the fire pump will provide sufficient discharge pressure to supply the fire protection system demands under reasonable, anticipated, worst-case conditions while not subjecting the fire protection system components to pressures in excess of their listed pressure rating.

4.6.1.2 Where a water flow test is used to determine the adequacy of the attached water supply, the test shall have been completed not more than 12 months prior to the submission of working plans, unless otherwise permitted by the authority having jurisdiction.

4.6.2* Sources.

A.4.6.2 Where the suction supply is from a factory-use water system, pump operation at 150 percent of rated capacity should not create hazardous process upsets due to low water pressure.

4.6.2.1 Any source of water that is adequate in quality, quantity, and pressure shall be permitted to provide the supply for a fire pump.

> Appropriate measures should be taken to protect the fire protection system from excessive corrosion, tabulation, and debris when using raw water sources such as ponds, lakes, and rivers. Debris, biological life, mineral content, and other pollutants can cause rapid deterioration of the fire protection system.

4.6.2.2 Where the water supply from a public service main is not adequate in quality, quantity, or pressure, an alternative water source shall be provided.

4.6.2.3 The adequacy of the water supply shall be determined and evaluated prior to the specification and installation of the fire pump.

4.6.2.3.1 Where the maximum flow available from the water supply cannot provide a flow of 150 percent of the rated flow of the pump, but the water supply can provide the greater of 100 percent of rated flow or the maximum flow demand of the fire protection system(s), the water supply shall be deemed to be adequate. In this case, the maximum flow shall be considered the highest flow that the water supply can achieve.

> Paragraph 4.6.2.3 does not apply to wells (see 7.2.7). It is the intent of this standard to size fire pumps based on the flow requirements of the attached fire protection system. The water supply, when operating under reasonable, anticipated, worst-case conditions, must be capable of maintaining a positive pump suction pressure while providing the rated pump flow rate at a residual pressure that is acceptable to the AHJ. The water supply, when operating under reasonable, anticipated, worst-case conditions, must also be capable of maintaining a positive pump suction pressure while providing the maximum fire protection system demand at a residual pressure that is acceptable to the AHJ. If the water supply meets these conditions, but

cannot provide sufficient flow to meet 150 percent of the rated flow of the pump, the water supply may be used to supply the fire pump and still meet the requirements of this standard. This commentary is consistent with the requirements of 14.2.6.2.8.

4.6.2.3.2 Where the water supply cannot provide 150 percent of the rated flow of the pump, a placard shall be placed in the pump room indicating the minimum suction pressure that the fire pump is allowed to be tested at and also indicating the required flow rate.

This information is intended to help evaluate the fire pump performance when flow testing required for the acceptance test or by NFPA 25 is conducted.

4.6.2.4 For liquids other than water, the liquid source for the pump shall be adequate to supply the maximum required flow rate for any simultaneous demands for the required duration and the required number of discharges.

4.6.3 Level. The minimum water level of a well or wet pit shall be determined by pumping at not less than 150 percent of the fire pump rated capacity.

4.6.4* Stored Supply.

A.4.6.4 Water sources containing salt or other materials deleterious to the fire protection systems should be avoided.

Where the authority having jurisdiction approves the start of an engine-driven fire pump on loss of ac power supply, the liquid supply should be sufficient to meet the additional cooling water demand.

Automatically starting a diesel fire pump on the loss of electrical power is not a recommended practice. Paragraph 11.2.7.2.1.5 requires that batteries be sized to start the diesel engine after 72 hours without recharging.

4.6.4.1 A stored supply plus reliable automatic refill shall be sufficient to meet the demand placed upon it for the design duration.

A stored supply of water for any fire protection system should be designed and installed in accordance with NFPA 22, *Standard for Water Tanks for Private Fire Protection.* The required quantity of water is determined by multiplying the system demand in gallons per minute (liters per minute) [gpm (L/min)] by the discharge duration specified by the appropriate system installation standard. The stored water available for fire protection is the water between the discharge and the overflow connection(s) on the storage tank. The total water available for fire protection is the stored fire protection water plus the water automatically refilled during the discharge duration. Even if the total water demand is stored, an automatic refill connection is preferable and should be designed and installed with sufficient flow capacity to refill the stored supply within 8 hours in accordance with NFPA 22.

◀ **CALCULATION**

4.6.4.2 A reliable method of replenishing the supply shall be provided.

4.6.5 Head.

As the term is used in 4.6.5, *head* refers to the pressure available at the fire pump suction at the maximum pump test flow. A positive head is required for a horizontal centrifugal fire pump (except where a tank is the suction supply) when the tank is at the lowest level available to maintain discharge and the pump is operating at 150 percent of rated flow.

4.6.5.1 Except as provided in 4.6.5.2, the head available from a water supply shall be figured on the basis of a flow of 150 percent of rated capacity of the fire pump.

> The capacity of a fire pump should be selected based upon the anticipated system demand (see Section 4.8). However, consideration should be given to performance of the pump when operating at overload (150 percent of rated capacity). Oversizing a fire pump can place an undue demand on the attached water supply, while undersizing the fire pump can cause difficulty in meeting system demand. Further, some water authorities prohibit flowing a quantity of water that causes residual pressures to drop below 20 psi (1.4 bar) at any point in the water system. Note that 4.14.3.1 limits suction pressure to not less than 0 psi (0 bar). Because of the aforementioned limitations, careful consideration of pump sizing must be exercised. It is also important to note that failure to operate a pump at 150 percent of rated capacity during the acceptance test does not constitute an unacceptable installation, provided that the pump meets 100 percent of rated capacity and also meets the system demand (see 14.2.6.2.6). The entire system must be considered when the water utility limits minimum residual pressure [typically 20 psi (1.4 bar)]. Residual pressures at sites located at higher elevations may be less than the residual pressure at sites at lower elevations.

4.6.5.2 Where the water supply cannot provide a flow of 150 percent of the rated flow of the pump but the water supply can provide the greater of 100 percent of the rated flow or the flow demand of the fire protection system(s), the head available from the water supply shall be permitted to be calculated on the basis of the maximum flow available as allowed by 4.6.2.3.1.

> This paragraph is consistent with 4.6.2.3.1 and reaffirms that the attached water supply need not be capable of flowing 150 percent of the fire pump rated capacity. The attached water supply must, however, be capable of providing sufficient flow to exceed the fire protection system demand.

4.6.5.3 The head described in 4.6.5.2 shall be as indicated by a flow test.

> When evaluating the ability of the fire pump to supply the system demand, the head available from a storage tank should be based on a water level measured at the lowest discharge level (tank nearly empty).
>
> Flow test data should never be more than 12 months old (see 4.6.1.2), since many factors affecting the flow test results may change during that time. See Part III of this handbook for information regarding fire flow testing.

4.7 Pumps, Drivers, and Controllers

4.7.1* Fire pumps shall be dedicated to and listed for fire protection service.

A.4.7.1 This subsection does not preclude the use of pumps in public and private water supplies that provide water for domestic, process, and fire protection purposes. Such pumps are not fire pumps and are not expected to meet all the requirements of this standard. Such pumps are permitted for fire protection if they are considered reliable by the analysis mandated in Section 4.6. Evaluating the reliability should include at least the levels of supervision and rapid response to problems as are typical in municipal water systems.

If a private development (campus) needs a fire protection pump, this is typically accomplished by installing a dedicated fire pump (in accordance with this standard) in parallel with a domestic pump or as part of a dedicated fire branch/loop off a water supply.

> Subsection 4.7.1 is intended to prohibit the use of fire pumps for other purposes, as A.4.7.1 points out. However, other pumps, such as those used for domestic or industrial purposes, can be used to supply fire protection water, provided that they are evaluated for reliability and are properly supervised. Fire pumps that are installed in parallel with domestic pumps on a combined fire/domestic water distribution system require special consideration. The domestic pumps should operate at or close to the fire pump pressures. The operation of one or two fire protection devices may not lower the system pressure sufficiently to activate the fire pump. In such cases, the domestic pumps may not provide adequate pressure for the initial demand in the fire protection system.

4.7.2 Acceptable drivers for pumps at a single installation shall be electric motors, diesel engines, steam turbines, or a combination thereof.

4.7.3* A pump shall not be equipped with more than one driver.

A.4.7.3 It is not the intent of this subsection to require replacement of dual driver installations made prior to the adoption of the 1974 edition of this standard.

4.7.4 Each fire pump shall have its own dedicated driver unless otherwise permitted in 8.6.3.1.

> Pumps in a water mist positive displacement pumping unit are allowed to share a common driver, because they are listed as a unit and they are operating as a unit — as multiple low volume pumps operating in parallel to supply a fire protection system.

4.7.5 Each driver or water mist positive displacement pumping unit shall have its own dedicated controller.

> The concept of a water mist positive displacement pumping unit with multiple positive displacement pumps operating in parallel to serve the same fire protection system was introduced in the 2013 edition. Because these pumps work as a unit to serve the same system, a single controller is allowed for the unit.

4.7.6* The driver shall be selected in accordance with 9.5.2 (electric motors), 11.2.2 (diesel engines), or 13.1.2 (steam turbines) to provide the required power to operate the pump at rated speed and maximum pump load under any flow condition.

A.4.7.6 For centrifugal and turbine pumps, the maximum brake horsepower required to drive the pump typically occurs at a flow beyond 150 percent of the rated capacity. For positive displacement pumps, the maximum brake horsepower required to drive the pump typically occurs when the relief valve is flowing 100 percent of the rated pump capacity. Pumps connected to variable speed drivers can operate at lower speeds, but the driver needs to be selected based upon the power required to drive the pump at rated speed and maximum pump load under any flow condition.

4.7.7* Maximum Pressure for Centrifugal Pumps.

A.4.7.7 A pressure relief valve is not an acceptable method of reducing system pressure under normal operating conditions. Prior to the 2003 edition, NFPA 20 did not strictly prohibit the use of pressure relief valves for management of excessive pressure at churn or low water flow conditions and only recognized it as a poor design practice as part of the 1999 edition. Existing installations designed under previous editions of NFPA 20 and containing pressure relief valves designed for such purposes might still be in service.

4.7.7.1 The net pump shutoff (churn) pressure plus the maximum static suction pressure, adjusted for elevation, shall not exceed the pressure for which the system components are rated.

> When selecting the fire pump, the maximum pump discharge pressure should not subject any system component to a pressure that exceeds the component pressure rating. The maximum discharge pressure normally occurs at zero flow (churn) and is calculated by adding the suction pressure to the net pump pressure. The maximum reasonably anticipated suction pressure should be used for this calculation. All listed fire protection components have a pressure rating of 175 psi (12.1 bar) or higher. If the system components have been properly selected, some fire protection systems may be rated for 300 psi (20.7 bar). Pressures in excess of the component rating can damage system components. The maximum pressure permitted in a standpipe system per NFPA 14, *Standard for the Installation of Standpipe and Hose Systems*, currently is 350 psi (24.1 bar), although NFPA 14 does not restrict the pressure in express risers in standpipe systems. See Exhibit II.4.2 for a check valve rated for 300 psi (20.7 bar).

EXHIBIT II.4.2. Check Valve Listed for 300 psi (20.7 bar). (Courtesy of Aon Fire Protection Engineering)

4.7.7.2* Pressure relief valves and pressure regulating devices in the fire pump installation shall not be used as a means to meet the requirements of 4.7.7.1.

> The use of pressure relief valves is limited in this standard to the following situations:
> - Diesel engine drivers where pressures developed may exceed the pressure rating of system components if an overspeed condition occurs
> - Variable speed drivers where pressures developed may exceed the pressure rating of system components if a failure causes the driver to revert to rated speed
>
> A pump should never be deliberately oversized so that a pressure reducing valve is needed. Pressure reducing valves and pressure relief valves are maintenance intensive and should be incorporated into the design of a fire pump system only when necessary and allowed by this standard. Pressure reducing valves are not permitted to be installed in the fire pump system piping. Where pump discharge pressures exceed 175 psi (12.1 bar), flanged cast-iron valves and fittings between the fire pump discharge flange and the discharge isolation valve must be extra heavy pattern. It is not the intent of Section 4.7.7.2 to limit the use of pressure reducing valves or pressure regulating valves downstream of the discharge isolation valve. The use of pressure regulating valves downstream of the discharge isolation valve is

covered in other standards. In many cases, high-pressure output is needed to meet the design requirements of systems installed in high-rise buildings and in standpipe systems in particular.

A.4.7.7.2 It is not the intent of this subsection to restrict the use of pressure reducing valves downstream of the discharge isolation valve for the purpose of meeting the requirements of 4.7.7.

4.7.7.3 Variable Speed Pressure Limiting Control.

4.7.7.3.1 Variable speed pressure limiting control drivers, as defined in this standard, shall be acceptable to limit system pressure.

◀ **DESIGN ALERT**

Variable speed pressure limiting control (VSPLC) or variable speed drive (VSD) was first introduced into NFPA 20 in the 2003 edition. As its name suggests, the pump speed is changed to prevent overpressurization of the fire protection system. A VSD fire pump and controller can be used when high flow and high pressure are needed but, due to the nature of fire pump performance, pressures at lower flows may be at levels in excess of the rating of system components. Examples of such systems are early suppression fast response (ESFR) sprinkler systems and standpipe systems. For example, a standpipe system with a system demand of 1000 gpm (3785 L/min) at 150 psi (10.3 bar) would require a fire pump rated for 1000 gpm (3785 L/min) at 160 psi (11.0 bar) [assuming 0 psi (0 bar) on the suction flange of the fire pump]. A fire pump of this rated capacity can produce as much as 224 psi (15.5 bar) — well in excess of the 175 psi (12.1 bar) rating of the system component. In this case, a VSD with a set pressure of 165 psi (11.4 bar) would prevent overpressurization of the system while meeting the fire protection system demand. Exhibit II.4.3 shows an electric motor variable speed pump drive controller with a transfer switch. Exhibit II.4.4 illustrates constant speed pump operation, and Exhibit II.4.5 illustrates variable speed pump operation.

4.7.7.3.2* The set pressure plus the maximum pressure variance of the variable speed pressure limiting controlled systems during variable speed operation and adjusted for elevation shall not exceed the pressure rating of any system component.

Paragraph 4.7.7.3.2 is intended to be a performance requirement to address the pressure tolerance rather than establishing a prescriptive requirement for a specific tolerance. This paragraph and A.4.7.7.3.2 also provide guidance for the fire protection system designer on how to deal with the effects of the variable speed pressure tolerance.

A.4.7.7.3.2 This requirement is intended to take into consideration the set pressure tolerance performance of the variable speed pressure limiting control as stated by the manufacturer.

EXHIBIT II.4.3 *Electric Motor Variable Speed Pump Drive Controller with Transfer Switch. (Courtesy of Master Control Systems, Inc.)*

EXHIBIT II.4.4 Constant Speed Pump Operation (pressure on left side of graph; pump speed on right side of graph). (Courtesy of Aon Fire Protection Engineering)

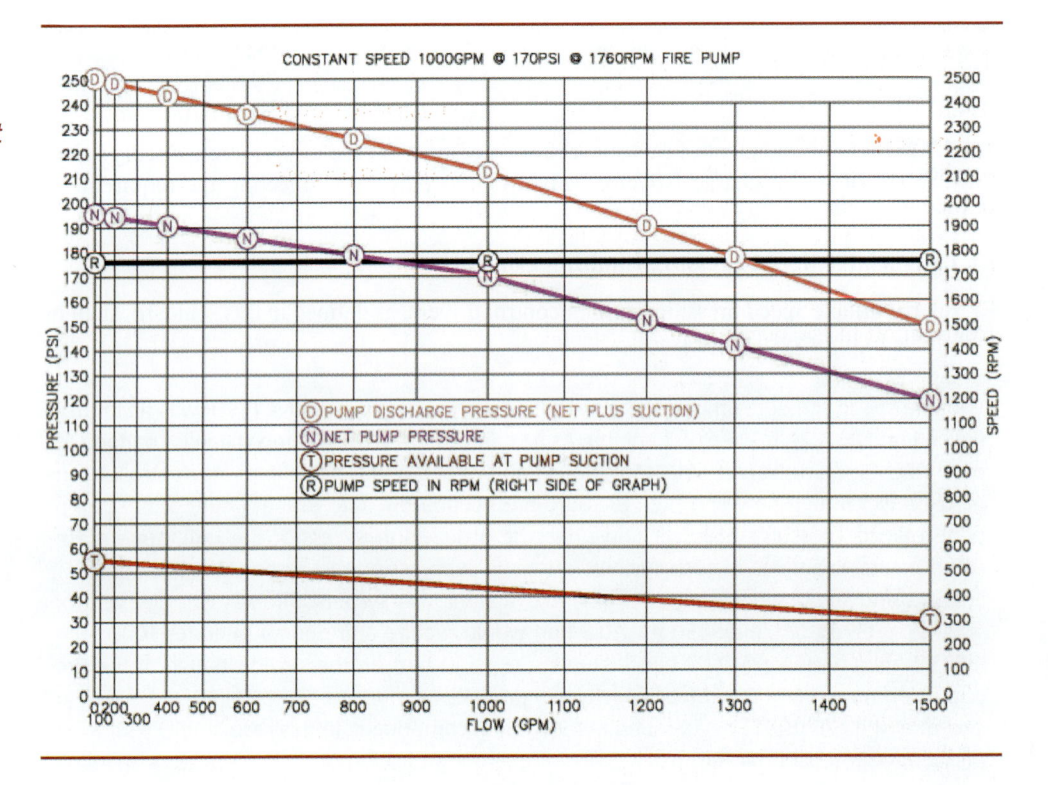

EXHIBIT II.4.5 Variable Speed Pump Operation (pressure on left side of graph; pump speed on right side of graph). (Courtesy of Aon Fire Protection Engineering)

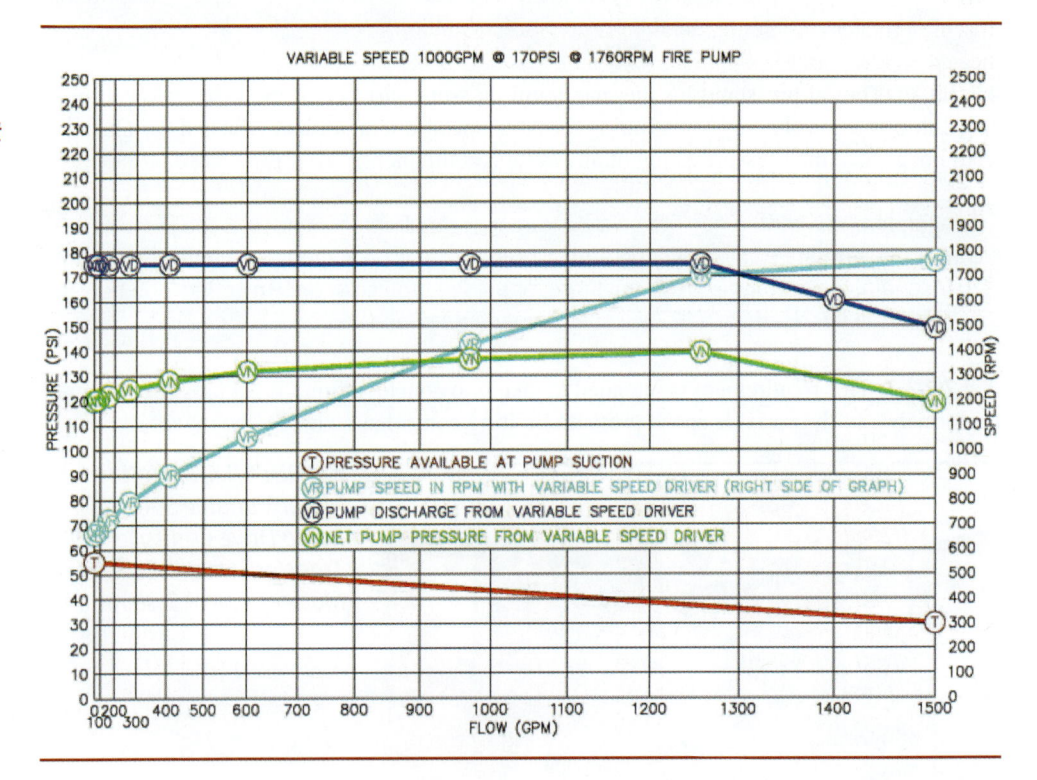

4.8* Centrifugal Fire Pump Capacities

A.4.8 The performance of the pump when applied at capacities over 140 percent of rated capacity can be adversely affected by the suction conditions. Application of the pump at capacities less than 90 percent of the rated capacity is not recommended.

The selection and application of the fire pump should not be confused with pump operating conditions. With proper suction conditions, the pump can operate at any point on its characteristic curve from shutoff to 150 percent of its rated capacity.

> **AHJ FAQ**
>
> The required flow for a fire protection system is higher than the rated flow for the fire pump selected by the designer. Is this allowed?
>
> *ANSWER:* Yes, a fire pump can satisfy any point along its performance curve. This includes points up to its overload point (150 percent of rated flow).

Selecting a pump so that the system demand falls between 90 percent and 140 percent of rated capacity ensures that the pump is not oversized and conversely is not operating at its maximum output. While NFPA 20 recommends selecting a pump of a size such that the system demand falls between 90 percent and 140 percent, some insurance companies limit pump sizing to 120 percent of rated capacity. The intent of this limitation is to avoid operating the pump at or near the overload point, which, over time, may compromise pump performance. After years of service, a fire pump may begin to show signs of wear and may not be capable of reaching 150 percent of rated capacity. If such a pump were selected based on performance at 150 percent capacity, a deficiency would eventually result. Selecting a fire pump of such a size that the system demand falls below 90 percent results in a fire pump that is too large for the system. The results of such a design include a potential overpressurization of the system and excessive cost.

4.8.1 A centrifugal fire pump for fire protection shall be selected so that the greatest single demand for any fire protection system connected to the pump is less than or equal to 150 percent of the rated capacity (flow) of the pump.

4.8.2* Centrifugal fire pumps shall have one of the rated capacities in gpm (L/min) identified in Table 4.8.2 and shall be rated at net pressure of 40 psi (2.7 bar) or more.

> **AHJ FAQ**
>
> Plans for a campus-style arrangement of buildings show a fire pump house with a single fire pump serving several buildings. Can this arrangement be approved?
>
> *ANSWER:* As long as the greatest single demand of any of the fire protection systems that the fire pump serves is met by the fire pump, the plan is acceptable.

TABLE 4.8.2 Centrifugal Fire Pump Capacities

gpm	L/min	gpm	L/min
25	95	1,000	3,785
50	189	1,250	4,731
100	379	1,500	5,677
150	568	2,000	7,570
200	757	2,500	9,462
250	946	3,000	11,355
300	1,136	3,500	13,247
400	1,514	4,000	15,140
450	1,703	4,500	17,032
500	1,892	5,000	18,925
750	2,839		

A.4.8.2 In countries that use the metric system, there do not appear to be standardized flow ratings for pump capacities; therefore, the metric conversions listed in Table 4.8.2 are soft conversions.

4.8.3 Centrifugal fire pumps with ratings over 5000 gpm (18,925 L/min) shall be subject to individual review by either the authority having jurisdiction or a listing laboratory.

4.9 Nameplate

> The nameplate provides specific performance data on the fire pump. This information must be maintained for the life of the system. Without the nameplate, it may be difficult to determine the performance characteristics of the pump. See Exhibit II.4.6 for an illustration of a typical fire pump nameplate.

4.9.1 Pumps shall be provided with a nameplate.

4.9.2 The nameplate shall be made of and attached using corrosion resistant material.

EXHIBIT II.4.6 Fire Pump Nameplate. (Courtesy of Aon Fire Protection Engineering)

4.10 Pressure Gauges

> Suction and discharge gauges are used to measure pressure on both sides of the fire pump when in operation and during the regularly scheduled nonflow testing required by NFPA 25. The gauges should be of sufficient quality to provide a reasonably accurate pressure reading, since they are used for testing purposes on a weekly and annual basis. These gauges should be calibrated each year to maintain accuracy.

4.10.1 Discharge.

4.10.1.1 A pressure gauge having a dial not less than 3.5 in. (89 mm) in diameter shall be connected near the discharge casting with a nominal 0.25 in. (6 mm) gauge valve.

4.10.1.2 The dial shall indicate pressure to at least twice the rated working pressure of the pump but not less than 200 psi (13.8 bar).

4.10.1.3 The face of the dial shall read in bar, pounds per square inch, or both with the manufacturer's standard graduations.

4.10.2* Suction.

A.4.10.2 For protection against damage from overpressure, where desired, a gauge protector should be installed.

4.10.2.1 Unless the requirements of 4.10.2.4 are met, a gauge having a dial not less than 3.5 in. (89 mm) in diameter shall be connected to the suction pipe near the pump with a nominal 0.25 in. (6 mm) gauge valve.

4.10.2.1.1 Where the minimum pump suction pressure is below 20 psi (1.3 bar) under any flow condition, the suction gauge shall be a compound pressure and vacuum gauge.

4.10.2.2 The face of the dial shall read in inches of mercury (millimeters of mercury) or psi (bar) for the suction range.

4.10.2.3 The gauge shall have a pressure range two times the rated maximum suction pressure of the pump.

4.10.2.4 The requirements of 4.10.2 shall not apply to vertical shaft turbine–type pumps taking suction from a well or open wet pit.

4.11 Circulation Relief Valve

4.11.1 Automatic Relief Valve.

4.11.1.1 Unless the requirements of 4.11.1.7 are met, each pump(s) shall have an automatic relief valve listed for the fire pump service installed and set below the shutoff pressure at minimum expected suction pressure.

> When a centrifugal fire pump is operating at churn, energy is continuously imparted to the water in the impeller, causing the water to heat. For electrical drive fire pumps and radiator-cooled engine–driven fire pumps, a listed circulation relief valve is needed to provide cooling water when the pump is operating at churn. The pipe connection for this valve must be located on the discharge side of the pump to cause flow through the pump casing and should discharge outdoors or to a floor drain where discharge can be observed by the pump operator. This valve should be installed in the vertical position, because installation in the horizontal position may cause the valve to fail at an accelerated rate due to obstructing material collecting in the valve seat. Failure or the omission of this valve can result in overheating and subsequent damage to the fire pump. Exhibit II.4.7 illustrates a ¾ in. (19 mm) circulation (casing) relief valve. Exhibit II.4.8 illustrates a cooling line to a diesel engine installed downstream of the pump, which eliminates the need for a circulation relief valve.

4.11.1.2 The valve shall be installed on the discharge side of the pump before the discharge check valve.

4.11.1.3 The valve shall provide flow of sufficient water to prevent the pump from overheating when operating with no discharge.

4.11.1.4 Provisions shall be made for discharge to a drain.

4.11.1.5 Circulation relief valves shall not be tied in with the packing box or drip rim drains.

EXHIBIT II.4.7 Circulation (Casing) Relief Valve.

EXHIBIT II.4.8 Cooling Line for a Diesel Engine Installed Downstream of the Pump, Eliminating the Need for a Circulation Relief Valve. (Courtesy of Aon Fire Protection Engineering)

4.11.1.6 The automatic relief valve shall have a nominal size of 0.75 in. (19 mm) for pumps with a rated capacity not exceeding 2500 gpm (9462 L/min) and have a nominal size of 1 in. (25 mm) for pumps with a rated capacity of 3000 gpm to 5000 gpm (11,355 L/min to 18,925 L/min).

4.11.1.7 The requirements of 4.11.1 shall not apply to engine-driven pumps for which engine cooling water is taken from the pump discharge.

> **AHJ FAQ**
>
> **The plans for a centrifugal fire pump do not show an automatic circulation relief valve. Are all centrifugal pumps required to have an automatic circulation relief valve?**
>
> ANSWER: An automatic circulation relief valve on a centrifugal fire pump is provided to keep the pump from overheating at churn. If an engine-driven centrifugal fire pump takes its cooling water from the pump discharge, water will be flowing at churn and, therefore, an automatic circulation relief valve is not required.

Water taken from the discharge side of a fire pump for engine cooling provides the same cooling effect as a circulation relief valve. See Chapter 11 for more details.

4.12* Equipment Protection

Section 4.12 provides requirements and guidance on the proper placement of a fire pump and related equipment. In this case, equipment protection is also intended to provide a relatively safe environment for the equipment operator. In accordance with Section 4.3, a qualified fire pump operator will report to the fire pump room to confirm that the equipment is operating properly in the event of operation of the plant. By locating the fire pump room near an exterior wall and enclosing the equipment in a fire-rated room, the operator is provided with a means of egress, and both the operator and the equipment are protected from fire exposure. Although not specifically addressed in this section, damage to fire pump systems can also occur from careless installation and maintenance procedures. Exhibit II.4.9 shows the effect of one such incident.

▶ **FAQ** Does the standard require a separate fire pump room for each fire pump installation?

EXHIBIT II.4.9 Fire Pump Damage from Careless Installation and Maintenance Procedures. (Courtesy of Aon Fire Protection Engineering)

The intent of this section, by requiring fire separation, is to locate a fire pump and its related equipment in a dedicated fire pump room. For installations with multiple pumps, a single fire pump room is needed. Fire pumps must not be located in mechanical equipment spaces, which would expose the equipment and operator to damage and potential injury.

AHJ FAQ

The plans submitted for a fire pump installation show the pump located in the middle of a storage room. Can the designer be required to alter the location?

ANSWER: Yes, there are several requirements that deal with the protection of the fire pump. The pump needs to be separated with fire resistance–rated construction in accordance with Table 4.12.1.1.2 and cannot be used for any other purposes other than those that are essential to the fire pump operation. In addition, since the fire department usually responds to the pump room during an emergency, the AHJ is allowed input regarding the location of, and access to, the room. A new section (4.12.2.1.1) in the 2013 edition mandates a rated enclosure to the pump room when the room is not directly accessed from the exterior of the building.

A.4.12 Special consideration needs to be given to fire pump installations installed belowgrade. Light, heat, drainage, and ventilation are several of the variables that need to be addressed. Some locations or installations might not require a pump house. Where a pump room or pump house is required, it should be of ample size and located to permit short and properly arranged piping. The suction piping should receive first consideration. The pump house should preferably be a detached building of noncombustible construction. A one-story pump room with a combustible roof, either detached or well cut off from an adjoining one-story building, is acceptable if sprinklered. Where a detached building is not feasible, the pump room should be located and constructed so as to protect the pump unit and controls from falling floors or machinery and from fire that could drive away the pump operator or damage the pump unit or controls. Access to the pump room should be provided from outside the building. Where the use of brick or reinforced concrete is not feasible, metal lath and plaster is recommended for the construction of the pump room. The pump room or pump house should not be used for storage purposes. Vertical shaft turbine–type pumps might necessitate a removable panel in the pump house roof to permit the pump to be removed for inspection or repair. Proper clearances to equipment should be provided as recommended by the manufacturer's drawings.

4.12.1* General Requirements. The fire pump, driver, controller, water supply, and power supply shall be protected against possible interruption of service through damage caused by explosion, fire, flood, earthquake, rodents, insects, windstorm, freezing, vandalism, and other adverse conditions.

A.4.12.1 A fire pump that is inoperative for any reason at any time constitutes an impairment to the fire protection system. It should be returned to service without delay.

Rain and intense heat from the sun are adverse conditions to equipment not installed in a completely protective enclosure. At a minimum, equipment installed outdoors should be shielded by a roof or deck.

4.12.1.1* Indoor Fire Pump Units.

A.4.12.1.1 Most fire departments have procedures requiring operation of a fire pump unit during an incident. Building designers should locate the fire pump room to be easily accessible during an incident.

4.12.1.1.1 Fire pump units serving high-rise buildings shall be protected from surrounding occupancies by a minimum of 2-hour fire-rated construction or physically separated from the protected building by a minimum of 50 ft (15.3 m).

> A minimum of 50 ft (15.3 m) of separation or a 2-hour fire-rated enclosure will protect the building's fire pump and personnel from an exposure fire.

4.12.1.1.2* Indoor fire pump rooms in non-high-rise buildings or in separate fire pump buildings shall be physically separated or protected by fire-rated construction in accordance with Table 4.12.1.1.2.

TABLE 4.12.1.1.2 Equipment Protection

Pump Room/House	Building(s) Exposing Pump Room/House	Required Separation
Not sprinklered	Not sprinklered	2 hour fire-rated
Not sprinklered	Fully sprinklered	or
Fully sprinklered	Not sprinklered	50 ft (15.3 m)
Fully sprinklered	Fully sprinklered	1 hour fire-rated or 50 ft (15.3 m)

> For the purposes of Table 4.12.1.1.2, a building with a fire protection system(s) supplied from the fire pump is considered to be a building exposing the pump room/house. There may also be other buildings or hazards that expose the pump room/house.

A.4.12.1.1.2 The purpose for the "Not Sprinklered" column in Table 4.12.1.1.2 is to provide guidance for unsprinklered buildings. This does not permit sprinklers to be omitted from pump rooms in fully sprinklered buildings.

4.12.1.1.3 The location of and access to the fire pump room shall be preplanned with the fire department.

4.12.1.1.4* Except as permitted in 4.12.1.1.5, rooms containing fire pumps shall be free from storage, equipment, and penetrations not essential to the operation of the pump and related components.

> It is the intent of the standard to prohibit the use of a fire pump room for the purposes of storage. It is not the intent of the standard to prohibit the installation of other types of equipment such as domestic water distribution equipment as indicated in 4.12.1.1.5. No materials or equipment that add to the combustibility of the space should be placed in the fire pump room. This provision is intended to provide protection to equipment as well as operating personnel.

A.4.12.1.1.4 Equipment that increases the fire hazard (such as boilers) and is not related to fire protection systems should not be in a fire pump room.

4.12.1.1.5 Equipment related to domestic water distribution shall be permitted to be located within the same room as the fire pump equipment.

4.12.1.1.6 The pump room or pump house shall be sized to fit all of the components necessary for the operation of the fire pump and to accommodate the following:

(1) Clearance between components for installation and maintenance
(2) Clearance between a component and the wall for installation and maintenance
(3) Clearance between energized electrical equipment and other equipment in accordance with *NFPA 70, National Electrical Code*
(4) Orientation of the pump to the suction piping to allow compliance with 4.14.6.3

4.12.1.2 Outdoor Fire Pump Units.

4.12.1.2.1 Fire pump units that are outdoors shall be located at least 50 ft (15.3 m) away from any buildings and other fire exposures exposing the building.

4.12.1.2.2 Outdoor installations shall be required to be provided with protection against possible interruption, in accordance with 4.12.1.

4.12.1.3 Fire Pump Buildings or Rooms with Diesel Engines. Fire pump buildings or rooms enclosing diesel engine pump drivers and day tanks shall be protected with an automatic sprinkler system installed in accordance with NFPA 13, *Standard for the Installation of Sprinkler Systems*.

4.12.2 Equipment Access.

> Fire pump rooms should be accessible under fire conditions. Where a fire pump room cannot be located on grade with direct access from outside of the building, a protected space with a fire rating in accordance with Table 4.12.1.1.2 must be provided. The space must be accessible through an enclosed passageway from an exterior exit or from an enclosed stairway.

4.12.2.1 The location of and access to the fire pump room(s) shall be pre-planned with the fire department.

4.12.2.1.1 Fire pump rooms not directly accessible from the outside shall be accessible through an enclosed passageway from an enclosed stairway or exterior exit.

4.12.2.1.2 The enclosed passageway shall have a fire-resistance rating not less than the fire-resistance rating of the fire pump room.

4.12.3 Heat.

4.12.3.1 An approved or listed source of heat shall be provided for maintaining the temperature of a pump room or pump house, where required, above 40°F (4°C).

4.12.3.2 The requirements of 11.6.5 shall be followed for higher temperature requirements for internal combustion engines.

4.12.4 Normal Lighting. Artificial light shall be provided in a pump room or pump house.

4.12.5 Emergency Lighting.

4.12.5.1 Emergency lighting shall be provided in accordance with NFPA *101, Life Safety Code*.

4.12.5.2 Emergency lights shall not be connected to an engine starting battery.

Stationary Fire Pumps Handbook 2013

4.12.6 Ventilation. Provision shall be made for ventilation of a pump room or pump house.

For an engine drive pump, the ventilation must be adequate to provide combustion air. See 11.3.2.

4.12.7* Drainage.

A.4.12.7 Pump rooms and pump houses should be dry and free of condensate. To accomplish a dry environment, heat might be necessary.

4.12.7.1 Floors shall be pitched for adequate drainage of escaping water away from critical equipment such as the pump, driver, controller, and so forth.

4.12.7.2 The pump room or pump house shall be provided with a floor drain that will discharge to a frost-free location.

Where a reduced pressure zone (RPZ) backflow prevention device is installed in a fire pump room, the discharge from the middle chamber should be piped outside, or the floor drain should be sized to accommodate the full discharge from the RPZ.

4.12.8 Guards. Couplings and flexible connecting shafts shall be installed with a coupling guard in accordance with Section 8 of ANSI B15.1, *Safety Standard for Mechanical Power Transmission Apparatus.*

4.13 Pipe and Fittings

4.13.1* Steel Pipe.

The type of steel piping used for fire pump installation should be identical to that specified for the suppression system. See Commentary Table II.4.1 for acceptable pipe types.

COMMENTARY TABLE II.4.1 *Pipe or Tube Materials and Dimensions*

Materials and Dimensions	Standard
Ferrous Piping (Welded and Seamless)	
Specification for black and hot-dipped zinc-coated (galvanized) welded and seamless steel pipe for fire protection use	ASTM A 795
Specification for pipe, steel, black and hot-dipped, zinc-coated, welded and seamless	ANSI/ASTM A 53
Welded and seamless wrought steel pipe	ANSI/ASME B36.10M
Specification for electric-resistance-welded steel pipe	ASTM A 135
Copper Tube (Drawn, Seamless)	
Specification for seamless copper tube	ASTM B 75
Specification for seamless copper water tube	ASTM B 88
Specification for general requirements for wrought seamless copper and copper-alloy tube	ASTM B 251
Specification for liquid and paste fluxes for soldering of copper and copper alloy tube	ASTM B 813
Specification for filler metals for brazing and braze welding	AWS A5.8
Specification for solder metal, Section 1: Solder alloys containing less than 0.2% lead and having solidus temperatures greater than 400°F	ASTM B 32
Alloy materials	ASTM B 446

Source: NFPA 13 Standard for the Installation of Sprinkler Systems 2013, Table 6.3.1.1.

4.12.1.1.5 Equipment related to domestic water distribution shall be permitted to be located within the same room as the fire pump equipment.

4.12.1.1.6 The pump room or pump house shall be sized to fit all of the components necessary for the operation of the fire pump and to accommodate the following:

(1) Clearance between components for installation and maintenance
(2) Clearance between a component and the wall for installation and maintenance
(3) Clearance between energized electrical equipment and other equipment in accordance with *NFPA 70, National Electrical Code*
(4) Orientation of the pump to the suction piping to allow compliance with 4.14.6.3

4.12.1.2 Outdoor Fire Pump Units.

4.12.1.2.1 Fire pump units that are outdoors shall be located at least 50 ft (15.3 m) away from any buildings and other fire exposures exposing the building.

4.12.1.2.2 Outdoor installations shall be required to be provided with protection against possible interruption, in accordance with 4.12.1.

4.12.1.3 Fire Pump Buildings or Rooms with Diesel Engines. Fire pump buildings or rooms enclosing diesel engine pump drivers and day tanks shall be protected with an automatic sprinkler system installed in accordance with NFPA 13, *Standard for the Installation of Sprinkler Systems*.

4.12.2 Equipment Access.

> Fire pump rooms should be accessible under fire conditions. Where a fire pump room cannot be located on grade with direct access from outside of the building, a protected space with a fire rating in accordance with Table 4.12.1.1.2 must be provided. The space must be accessible through an enclosed passageway from an exterior exit or from an enclosed stairway.

4.12.2.1 The location of and access to the fire pump room(s) shall be pre-planned with the fire department.

4.12.2.1.1 Fire pump rooms not directly accessible from the outside shall be accessible through an enclosed passageway from an enclosed stairway or exterior exit.

4.12.2.1.2 The enclosed passageway shall have a fire-resistance rating not less than the fire-resistance rating of the fire pump room.

4.12.3 Heat.

4.12.3.1 An approved or listed source of heat shall be provided for maintaining the temperature of a pump room or pump house, where required, above 40°F (4°C).

4.12.3.2 The requirements of 11.6.5 shall be followed for higher temperature requirements for internal combustion engines.

4.12.4 Normal Lighting. Artificial light shall be provided in a pump room or pump house.

4.12.5 Emergency Lighting.

4.12.5.1 Emergency lighting shall be provided in accordance with NFPA *101, Life Safety Code*.

4.12.5.2 Emergency lights shall not be connected to an engine starting battery.

4.12.6 Ventilation. Provision shall be made for ventilation of a pump room or pump house.

> For an engine drive pump, the ventilation must be adequate to provide combustion air. See 11.3.2.

4.12.7* Drainage.

A.4.12.7 Pump rooms and pump houses should be dry and free of condensate. To accomplish a dry environment, heat might be necessary.

4.12.7.1 Floors shall be pitched for adequate drainage of escaping water away from critical equipment such as the pump, driver, controller, and so forth.

4.12.7.2 The pump room or pump house shall be provided with a floor drain that will discharge to a frost-free location.

> Where a reduced pressure zone (RPZ) backflow prevention device is installed in a fire pump room, the discharge from the middle chamber should be piped outside, or the floor drain should be sized to accommodate the full discharge from the RPZ.

4.12.8 Guards. Couplings and flexible connecting shafts shall be installed with a coupling guard in accordance with Section 8 of ANSI B15.1, *Safety Standard for Mechanical Power Transmission Apparatus*.

4.13 Pipe and Fittings

4.13.1* Steel Pipe.

> The type of steel piping used for fire pump installation should be identical to that specified for the suppression system. See Commentary Table II.4.1 for acceptable pipe types.
>
> **COMMENTARY TABLE II.4.1** *Pipe or Tube Materials and Dimensions*
>
> | Materials and Dimensions | Standard |
> |---|---|
> | **Ferrous Piping (Welded and Seamless)** | |
> | Specification for black and hot-dipped zinc-coated (galvanized) welded and seamless steel pipe for fire protection use | ASTM A 795 |
> | Specification for pipe, steel, black and hot-dipped, zinc-coated, welded and seamless | ANSI/ASTM A 53 |
> | Welded and seamless wrought steel pipe | ANSI/ASME B36.10M |
> | Specification for electric-resistance-welded steel pipe | ASTM A 135 |
> | **Copper Tube (Drawn, Seamless)** | |
> | Specification for seamless copper tube | ASTM B 75 |
> | Specification for seamless copper water tube | ASTM B 88 |
> | Specification for general requirements for wrought seamless copper and copper-alloy tube | ASTM B 251 |
> | Specification for liquid and paste fluxes for soldering of copper and copper alloy tube | ASTM B 813 |
> | Specification for filler metals for brazing and braze welding | AWS A5.8 |
> | Specification for solder metal, Section 1: Solder alloys containing less than 0.2% lead and having solidus temperatures greater than 400°F | ASTM B 32 |
> | Alloy materials | ASTM B 446 |
>
> *Source: NFPA 13 Standard for the Installation of Sprinkler Systems 2013, Table 6.3.1.1.*

A.4.13.1 The exterior of aboveground steel piping should be kept painted.

4.13.1.1 Steel pipe shall be used aboveground except for connection to underground suction and underground discharge piping.

4.13.1.2 Where corrosive water conditions exist, steel suction pipe shall be galvanized or painted on the inside prior to installation with a paint recommended for submerged surfaces.

> In most cases, black steel, when used with any water, eventually corrodes to some extent. All suction pipes are recommended to be protected from corrosion by galvanizing or painting to protect the pump. If the suction pipe corrodes, rust or scale could dislodge from the inside wall of the pipe and enter the pump, potentially damaging the pump impeller. Corrosion is intensified where air pockets form. The pump installation should avoid trapped sections of pipe where air pockets may form.

4.13.1.3 Thick bituminous linings shall not be used.

4.13.2* Joining Method.

> The type of fittings used for the fire pump installation should be identical to those specified for the suppression system. The fittings used in sprinkler systems have been evaluated for that use and have been found to be acceptable. Using the same type of fittings for a system's fire pump installation eliminates performance and compatibility issues. See Commentary Table II.4.2 for acceptable fitting types.

COMMENTARY TABLE II.4.2 *Fittings Materials and Dimensions*

Materials and Dimensions	Standard
Gray Iron	
Gray iron threaded fittings, Classes 125 and 250	ANSI/ASME B16.4
Gray iron pipe flanges and flanged fittings, Classes 25, 125, and 250	ANSI/ASME B16.1
Malleable Iron	
Malleable iron threaded fittings, Classes 150 and 300	ANSI/ASME B16.3
Factory-made wrought buttwelding fittings	ANSI/ASME B16.9
Buttwelding ends	ANSI/ASME B16.25
Specification for piping fittings of wrought carbon steel and alloy steel for moderate and high temperature service	ASTM A 234
Pipe flanges and flanged fittings, NPS ½ through NPS 24	ANSI/ASME B16.5
Forged fittings, socket-welding and threaded	ANSI/ASME B16.11
Wrought copper and copper alloy solder joint pressure fittings	ANSI/ASME B16.22
Cast copper alloy solder joint pressure fittings	ANSI/ASME B16.18

Source: NFPA 13 Standard for the Installation of Sprinkler Systems, 2013, Table 6.4.1.

> Chlorinated polyvinyl chloride (CPVC) and copper pipe and fittings should not be used on the suction side of a fire pump due to potential settlement of either the foundation on which the fire pump rests or the water supply pipe.

A.4.13.2 Flanges welded to pipe are preferred.

4.13.2.1 Sections of steel piping shall be joined by means of screwed, flanged mechanical grooved joints or other approved fittings.

4.13.2.2 Slip-type fittings shall be permitted to be used where installed as required by 4.14.6 and where the piping is mechanically secured to prevent slippage.

4.13.3 Concentrate and Additive Piping.

4.13.3.1 Foam concentrate or additive piping shall be a material that will not corrode in this service.

> Pure foam concentrate is frequently corrosive; therefore, foam concentrate piping is normally installed with either stainless steel or brass pipe. Galvanized pipe should never be used for foam concentrate piping, because the concentrate may cause the galvanizing to "flake" and detach from the pipe wall.

4.13.3.2 Galvanized pipe shall not be used for foam concentrate service.

4.13.4 Drain Piping. Drain pipe and its fittings that discharge to atmosphere shall be permitted to be constructed of metallic or polymeric materials.

4.13.5 Piping, Hangers, and Seismic Bracing. Pipe, fittings, hangers, and seismic bracing for the fire pump unit, including the suction and discharge piping, shall comply with the applicable requirements of NFPA 13, *Standard for the Installation of Sprinkler Systems*.

4.13.6* Cutting and Welding. Torch cutting or welding in the pump house shall be permitted as a means of modifying or repairing pump house piping when it is performed in accordance with NFPA 51B, *Standard for Fire Prevention During Welding, Cutting, and Other Hot Work*.

A.4.13.6 When welding is performed on the pump suction or discharge piping with the pump in place, the welding ground should be on the same side of the pump as the welding.

4.14 Suction Pipe and Fittings

4.14.1* Components.

A.4.14.1 The exterior of steel suction piping should be kept painted.

Buried iron or steel pipe should be lined and coated or protected against corrosion in conformance with AWWA C104, *Cement-Mortar Lining for Cast-Iron and Ductile-Iron Pipe and Fittings for Water*, or equivalent standards.

> Underground pipe frequently is coated with a material that prevents the formation of debris or tuberculation. Steel pipe should never be used for underground piping, particularly on the suction side of the pump.

4.14.1.1 The suction components shall consist of all pipe, valves, and fittings from the pump suction flange to the connection to the public or private water service main, storage tank, or reservoir, and so forth, that feeds water to the pump.

4.14.1.2 Where pumps are installed in series, the suction pipe for the subsequent pump(s) shall begin at the system side of the discharge valve of the previous pump.

4.14.2 Installation. Suction pipe shall be installed and tested in accordance with NFPA 24, *Standard for the Installation of Private Fire Service Mains and Their Appurtenances*.

> Trapped pipe sections that allow the formation of air pockets should be avoided. Air pockets intensify corrosion and may interfere with the flow of water into the pump impeller. Automatic air release valves should in installed at the high points on trapped pipe sections.

4.14.3 Suction Size.

4.14.3.1 Unless the requirements of 4.14.3.2 are met, the size of the suction pipe for a single pump or of the suction header pipe for multiple pumps (designed to operate together) shall be such that, with all pumps operating at maximum flow (150 percent of rated capacity or the maximum flow available from the water supply as discussed in 4.6.2.3.1), the gauge pressure at the pump suction flanges shall be 0 psi (0 bar) or higher.

> The suction pipe must be sized for all fire pumps that are designed to operate simultaneously to meet the fire protection system demand. Redundant pumps installed to increase system reliability are not considered when selecting the pipe size for simultaneous fire pump operation.
>
> It is important to note that horizontal shaft fire pumps are not permitted to draft water. The suction pipe must also be large enough so that a substantial pressure loss does not occur when the pump is supplied by a water storage tank. For such applications, the suction pressure is permitted to drop to not less than −3 psi (−0.2 bar). This negative pressure is permitted only after the tank has provided the maximum system demand (flow rate and duration); the remaining water in the tank is level with, or higher than, the pump; and the pump is operating at 150 percent of rated flow. This negative pressure allowance is intended to compensate for the friction loss through the suction pipe, fittings, and gate valve. The allowable negative pressure of −3 psi (−0.2 bar) is based on flowing 150 percent of the rated pump capacity after the tank has provided the full system demand (flow rate and duration); the initial head, or any head caused by a water level above the full system demand, is not included in the suction pressure calculation and should be subtracted from the acceptance test results when verifying the acceptability of the suction supply.

4.14.3.2* The requirements of 4.14.3.1 shall not apply where the supply is a suction tank with its base at or above the same elevation as the pump, and the gauge pressure at the pump suction flange shall be permitted to drop to −3 psi (−0.2 bar) with the lowest water level after the maximum system demand and duration have been supplied.

A.4.14.3.2 It is permitted that the suction pressure drop to −3 psi (−0.2 bar) for a centrifugal pump that is taking suction from a grade level storage tank where the pump suction elevation is at or below the water level in the water storage tank at the end of the required water flow duration. This negative suction pressure is to allow for the friction loss in the suction piping when the pump is operating at 150 percent capacity.

4.14.3.3 The size of that portion of the suction pipe located within 10 pipe diameters upstream of the pump suction flange shall be not less than that specified in Section 4.26.

> The size of the suction pipe is based on limiting water velocity to not more than 15 ft/sec (4.57 m/sec) to limit turbulent flow in the pipe. Turbulent flow generates air bubbles in the water, which adversely affect pump efficiency. As the pump impeller is imparting energy to the water (which is relatively incompressible), the introduction of air bubbles (which are easily compressible) can create small regions of high pressure due to collapsed air bubbles and causes inefficient pump operation.

4.14.4* Pumps with Bypass.

A.4.14.4 The following notes apply to Figure A.4.14.4:

(1) A jockey pump is usually required with automatically controlled pumps.
(2) If testing facilities are to be provided, also see Figure A.4.20.1.2(a) and Figure A.4.20.1.2(b).
(3) Pressure-sensing lines also need to be installed in accordance with 10.5.2.1 or 12.7.2.1. See Figure A.4.30(a) and Figure A.4.30(b).

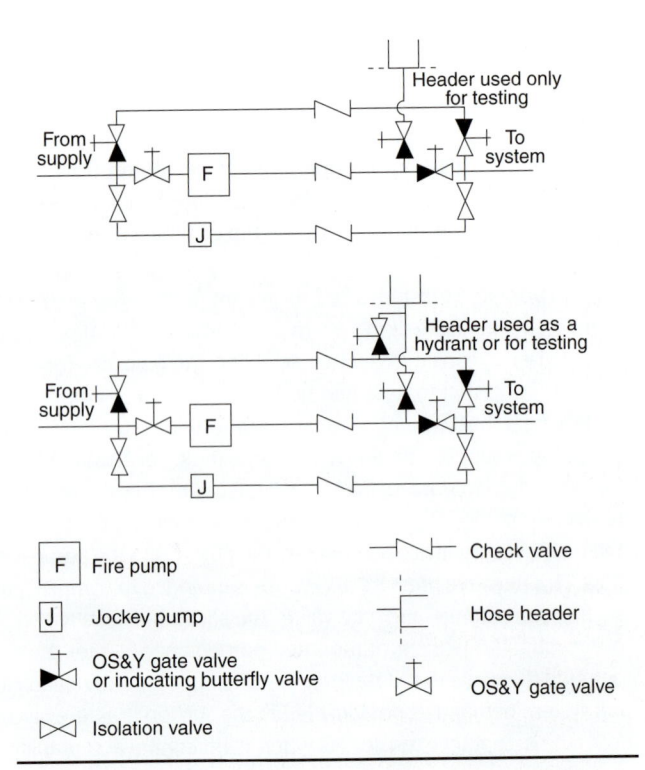

FIGURE A.4.14.4 Schematic Diagram of Suggested Arrangements for a Fire Pump with a Bypass, Taking Suction from Public Mains.

4.14.4.1 Where the suction supply is of sufficient pressure to be of material value without the pump, the pump shall be installed with a bypass. *(See Figure A.4.14.4.)*

> In most cases, a pump that is connected to a public or private water supply should include a bypass. Only in rare cases, where the pressure available is so low that the water supply is of no value without the pump, should a fire pump be installed without a bypass. When a pump is supplied by a suction tank, a bypass is not needed [see Figure A.6.3.1(a)], because a suction tank will not provide sufficient pressure to be of value without the fire pump operating. The valves on the bypass are required to be normally open so that the attached water supply is available automatically. In this case, "normally open" refers to the valve being in the open position at all times. The bypass valves should be closed only for system maintenance. The function of the pump bypass is sometimes mistakenly thought to be for use only when the fire pump is out of service. This is not the use intended by the standard, because the water from the bypass must be available automatically if the pump fails to start.

AHJ FAQ

A fire pump installation connected to an on-site storage tank does not have a fire pump bypass installed. Is this omission permitted?

ANSWER: The purpose of a fire pump bypass is to allow water to flow into the system without passing through the impeller if the pump fails. The water available through this bypass will not meet the demand of the fire protection system. If it could, a pump would not have to be installed. The bypass provides some degree of safety if a pump experiences a locked rotor condition or other catastrophic failure. In the case of an on-site storage tank, the pressure available would not be of any significant value without the pump operating, and, therefore, a bypass can be omitted.

4.14.4.2 The size of the bypass shall be at least as large as the pipe size required for discharge pipe as specified in Section 4.26.

4.14.5* Valves.

A.4.14.5 Where the suction supply is from public water mains, the gate valve should be located as far as is practical from the suction flange on the pump. Where it comes from a stored water container, the gate valve should be located at the outlet of the container. A butterfly valve on the suction side of the pump can create turbulence that adversely affects the pump performance and can increase the possibility of blockage of the pipe.

> The required placement of the valve in the suction line is intended to limit turbulence. While the "gate" in an outside screw and yoke (OS&Y) valve is outside of the waterway when fully open and does not interfere with flow, the "butterfly" in a butterfly valve is always suspended in the pipe and causes turbulent flow. While previous editions of NFPA 20 prohibited the use of butterfly valves in the suction pipe, the standard now permits the use of this type of valve if the valve is located at a sufficient distance from the suction flange of the pump. By locating the butterfly valve 50 ft (15.3 m) or more away from the suction flange of the pump, the turbulence can be eliminated before the water reaches the suction flange.

4.14.5.1 A listed outside screw and yoke (OS&Y) gate valve shall be installed in the suction pipe.

4.14.5.2 No control valve other than a listed OS&Y valve and the devices as permitted in 4.27.3 shall be installed in the suction pipe within 50 ft (15.3 m) of the pump suction flange.

4.14.6* Installation.

A.4.14.6 See Figure A.4.14.6. *(See Hydraulic Institute Standards for Centrifugal, Rotary and Reciprocating Pumps for additional information.)*

> The layout and design of the suction pipe must be carefully considered in order to avoid the generation of air bubbles in the suction pipe and to avoid unbalanced flow into the pump suction flange. Where a reduction in pipe diameter is necessary, an eccentric reducer must be used to prevent the development of an air pocket, as illustrated in Figure A.4.14.6. The flat portion of the reducer must be on top, with the actual reduction in pipe diameter taking place on the bottom of the fitting. Installing the reducer in this position will prevent an air pocket from forming. Placing an elbow with the centerline in the horizontal plane within 10 pipe diameters of the suction flange creates an unbalanced flow of water into the impeller. An unbalanced flow causes an axial load to be placed on the pump shaft and bearings, causing excessive wear to the bearings and subsequent damage to the pump over time and/or severe cavitation if not corrected. The elbow shown in Exhibit II.4.10 will eventually cause damage to the pump and may result in pump shaft-bearing failure. A butterfly valve installed within 10 pipe diameters will also cause turbulent flow and cavitation in the pump and should be avoided.
>
> In the lower portion of Figure A.4.14.6, the correct method for designing a turn in the suction piping is illustrated. Although not required by NFPA 20, the suction elbow should be of the long turn radius type. An elbow in the vertical position (either directed upward or downward) should be used, and the elbow causing a turn in the pipe parallel with the pump shaft should be installed only when located greater than 10 pipe diameters away from the suction flange. For example, an elbow installed in the horizontal plane in a suction pipe with a diameter of 8 in. (200 mm) should be located 80 in. (2000 mm) away from the suction flange of the pump.

◀ **DESIGN ALERT**

4.14.6.1 General. Suction pipe shall be laid carefully to avoid air leaks and air pockets, either of which can seriously affect the operation of the pump.

> Trapped pipe sections that allow the formation of air pockets should be avoided. Air pockets intensify corrosion and may interfere with the flow of water into the pump impeller. Automatic air release valves should be installed on trapped pipe sections.

4.14.6.2 Freeze Protection.

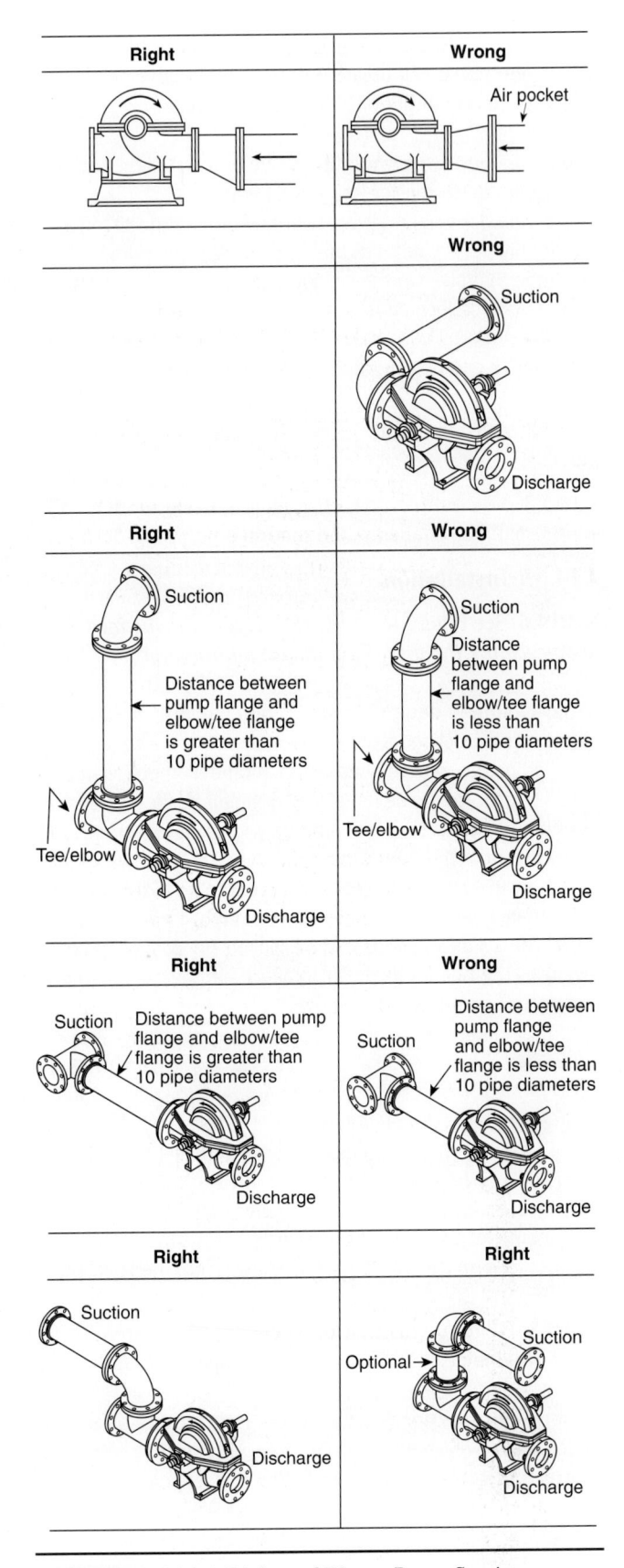

FIGURE A.4.14.6 Right and Wrong Pump Suctions.

EXHIBIT II.4.10 *Pump Suction Elbow Installed in the Incorrect Plane.*

4.14.6.2.1 Suction pipe shall be installed below the frost line or in frostproof casings.

4.14.6.2.2 Where pipe enters streams, ponds, or reservoirs, special attention shall be given to prevent freezing either underground or underwater.

4.14.6.3 Elbows and Tees.

4.14.6.3.1 Unless the requirements of 4.14.6.3.2 are met, elbows and tees with a centerline plane parallel to a horizontal split-case pump shaft shall not be permitted. *(See Figure A.4.14.6.)*

> Any fitting that causes a change in the direction of flow, including elbows, tees, and crosses, when installed in the horizontal plane, should be located a minimum of 10 pipe diameters away from the suction flange of the pump. While fittings in the vertical position usually do not cause an unbalanced flow into the pump suction flange, such an installation should be avoided if possible to prevent turbulence.

4.14.6.3.2 The requirements of 4.14.6.3.1 shall not apply to elbows and tees with a centerline plane parallel to a horizontal split-case pump shaft where the distance between the flanges of the pump suction intake and the elbow and tee is greater than 10 times the suction pipe diameter.

4.14.6.3.3 Elbows and tees with a centerline plane perpendicular to the horizontal split-case pump shaft shall be permitted at any location in the pump suction intake.

An elbow perpendicular to the pump shaft will not produce an unbalanced flow condition. Regardless of the position of the elbow in a fire pump suction pipe, when designing a change in direction in the suction pipe of a pump, the elbows used should be of the long radius type to permit as smooth a transition in the change in direction as possible. A long radius elbow provides a smoother turn in direction, further limiting turbulence.

▶ **FAQ** Does NFPA 20 require the installation of a long radius elbow in the suction pipe?

Although a long radius elbow is *not* required by NFPA 20, the use of such a fitting reduces the likelihood of turbulence in the suction pipe. In all cases, even a long radius elbow should never be installed in the parallel plane with the pump shaft.

4.14.6.4 Eccentric Tapered Reducer or Increaser. Where the suction pipe and pump suction flange are not of the same size, they shall be connected with an eccentric tapered reducer or increaser installed in such a way as to avoid air pockets.

4.14.6.5 Strain Relief. Where the pump and its suction supply are on separate foundations with rigid interconnecting pipe, the pipe shall be provided with strain relief. *(See Figure A.6.3.1(a).)*

DESIGN ALERT ▶ Strain relief can be accomplished by means of a flexible joint such as that depicted in Exhibit II.4.11 or by a series of grooved couplings as shown in Exhibit II.4.12. Flanged fittings are not flexible. When using grooved couplings for strain relief, rigid grooved couplings should not be used. Deflection for most nonrigid grooved couplings is on the order of 3½ in. to 4½ in./20 ft (89 mm to 114 mm/6m) length of pipe in sizes 6 in. to 8 in. (150 mm to 200 mm). A series of these grooved couplings in the suction pipe provides flexibility if the pump or tank foundations begin to settle and should also provide for limited movement due to expansion and contraction. The purpose of providing such strain relief is to prevent damage to the pump suction flange, since misalignment or stress placed on the pump suction flange can cause the flange to crack. The fire pump design should avoid the use of separate foundations for the pump and the driver, particularly in earthquake-prone areas.

Seismic design should follow the design requirements found in ASCE/SEI 7, *Minimum Design Loads for Buildings and Other Structures,* and NFPA 13, *Standard for the Installation of Sprinkler Systems.* Prior to the design of the installation, the locally adopted building code must be checked for the seismic requirements for the fire pump piping installation.

A means of joint restrain other than thrust blocks should be used where a flanged spigot passes through the floor. A thrust block may hold the pipe in place while the building settles and cause strain on the pipe and fittings inside the building that are directly connected to the flanged spigot.

4.14.7 Multiple Pumps. Where a single suction pipe supplies more than one pump, the suction pipe layout at the pumps shall be arranged so that each pump will receive its proportional supply.

Multiple pump installations are used to meet one of the following design scenarios:

1. Two or more pumps are installed in parallel to meet the minimum water demand of the fire protection system.
2. Two or more pumps are installed in parallel to meet the minimum water demand of the fire protection system, and one or more redundant pumps are installed to increase reliability.

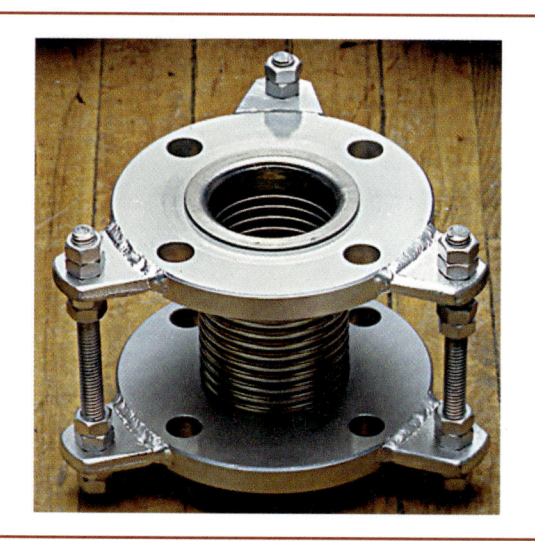

EXHIBIT II.4.11 *Flexible Joint for Use on Suction Pipe. (Courtesy of Metraflex Company)*

EXHIBIT II.4.12 *Grooved Fittings and Couplings and Flanged Fittings. (Courtesy of Aon Fire Protection Engineering)*

3. Two or more pumps are installed in series to meet the pressure demand of the fire protection system.
4. Two or more pumps are installed, and each of the pumps is individually required to meet the minimum water supply and pressure demand required by the fire protection system. One pump will serve as the primary source; the other(s) will serve as a backup or reserve supply.

In the first situation, the suction piping is required to be sized assuming that all of the installed pumps are operating simultaneously. In the second situation, the suction piping is required to be sized assuming that all of the installed pumps, except the redundant pumps, are operating simultaneously. In the third situation, the piping is sized for the flow of both pumps operating, but the flow will equal that of a single pump. The fourth arrangement will have piping sized for either (but not for both or multiple pumps) of the pumps operating.

4.14.8* Suction Screening.

Suction screens are located in a suction pit at the pump end of a channel, reservoir, pond, or lake that allows water to flow from the source to the pump, as indicated in Figure A.7.2.2.2. The figure shows a vertical shaft turbine–type pump. A horizontal pump may also be used, if the suction pressure is positive. Any debris entering the channel can accumulate in the area in front of the suction crib. If a fire occurs, it is possible that fire pump operation will draw some of this settled material onto the first screen, reducing the flow area and restricting flow to the fire protection system. The standard requires two screens to be installed to allow removal of the first screen if it becomes obstructed. The opening size in the screens is intended to be small to provide protection to the distribution portion of the fire protection system. For example, the openings would be restricted to less than ½ in. (12.7 mm) for a sprinkler having an orifice of ½ in. (12.7 mm). Caution should be exercised when removing a debris-loaded screen, because the removal may cause the obstructing material to fall off and thereby obstruct the second screen.

A.4.14.8 In the selection of screen material, consideration should be given to prevention of fouling from aquatic growth. Antifouling is best accomplished with brass or copper wire.

4.14.8.1 Where the water supply is obtained from an open source such as a pond or wet pit, the passage of materials that might clog the pump shall be obstructed.

Where a pipe is used between the suction pipe and the water source, the end of the pipe should turn vertically downward to prevent debris from settling into the suction pipe. A vortex plate and a screen should be attached to the water source end of the pipe.

4.14.8.2 Double intake screens shall be provided at the suction intake.

4.14.8.3 Screens shall be removable, or an in situ cleaning shall be provided.

4.14.8.4 Below minimum water level, these screens shall have an effective net area of opening of 1 in.^{2} for each 1 gpm (170 mm^2 for each 1 L/min) at 150 percent of rated pump capacity.

The following example illustrates how to correctly size a suction screen for a fire pump.

▶ **EXAMPLE**

CALCULATION ▶

A 750 gpm (2839 L/min) vertical shaft turbine pump requires a screen to protect the wet pit that takes water from an open lake or river. What size is needed for the screen?

Solution: Since the pump is rated for 750 gpm (2839 L/min), the pump delivers 1125 gpm (4258 L/min) at overload (150 percent of 750 and 2839). Therefore, the screen must have an area of 1125 in.^{2} (0.72 m^2). Since 4.14.8.6 requires a 0.50 in. (12.7 mm) mesh and a No. 10 B&S gauge wire, the total screen area must be 1800 in.^{2} (1.1 m^2) (1.6 × 1125 in.^{2}). This area is equivalent to a screen area of 12.5 ft^2 (1.16 m^2).

▶ **FAQ**   Does NFPA 20 prohibit the installation of any device in the suction pipe of a fire pump?

> The installation of devices in the suction pipe of a fire pump is restricted to the devices listed in 4.14.9.2. The installation of any device or fitting should be made at least 10 pipe diameters upstream of the pump suction flange to avoid unnecessary turbulence.

4.14.8.5 Screens shall be so arranged that they can be cleaned or repaired without disturbing the suction pipe.

4.14.8.6 Mesh screens shall be brass, copper, Monel, stainless steel, or other equivalent corrosion-resistant metallic material wire screen of 0.50 in. (12.7 mm) maximum mesh and No. 10 B&S gauge.

4.14.8.7 Where flat panel mesh screens are used, the wire shall be secured to a metal frame sliding vertically at the entrance to the intake.

4.14.8.8 Where the screens are located in a sump or depression, they shall be equipped with a debris-lifting rake.

4.14.8.9 Periodically, the system shall be test pumped, the screens shall be removed for inspection, and accumulated debris shall be removed.

4.14.8.10 Continuous slot screens shall be brass, copper, Monel, stainless steel, or other equivalent corrosion-resistant metallic material of 0.125 in. (3.2 mm) maximum slot and profile wire construction.

4.14.8.11 Screens shall have at least 62.5 percent open area.

4.14.8.12 Where zebra mussel infestation is present or reasonably anticipated at the site, the screens shall be constructed of a material with demonstrated resistance to zebra mussel attachment or coated with a material with demonstrated resistance to zebra mussel attachment at low velocities.

> With the introduction of zebra mussels into many United States waterways, extra care should be exercised with the use of suction screens. While the introduction of zebra mussels was first observed in the western end of Lake Erie, the organisms have since spread to all of the Great Lakes and the Mississippi River, Ohio River, and Illinois River, including approximately 22 major lakes and tributaries. The mussels can spread rapidly once established in a water source, with each female producing approximately 300,000 to 1,000,000 larvae annually. The larvae, in turn, can grow from the size of fine sand to an adult size of 6 in. (150 mm) in length in 2 to 3 years. They attach to many materials as well as to each other, forming obstructions to piping and discharge device orifices.
>
> NFPA 20 requires screens to be constructed of brass, copper, Monel™, stainless steel, or other corrosion-resistant material. Copper-based materials have been tested and proven resistant to the attachment of zebra mussels. Pure copper is resistant to attachments until corrosion products form on its surface. Some copper alloys, Monel, and stainless steel have been tested and found to offer no resistance to the attachment of zebra mussels. In areas where zebra mussel infestation is present, screens should be constructed with materials or coated with a corrosion-resistant material that has demonstrated long-term resistance to their attachment in low-velocity conditions. Materials that require regular maintenance to remove zebra mussels should not be used.

4.14.8.13 The overall area of the screen shall be 1.6 times the net screen opening area. *(See screen details in Figure A.7.2.2.2.)*

4.14.9* Devices in Suction Piping.

A.4.14.9 The term *device* as used in this subsection is intended to include, but not be limited to, devices that sense suction pressure and then restrict or stop the fire pump discharge.

Due to the pressure losses and the potential for interruption of the flow to the fire protection systems, the use of backflow prevention devices is discouraged in fire pump piping. Where required, however, the placement of such a device on the discharge side of the pump is to ensure acceptable flow characteristics to the pump suction. It is more efficient to lose the pressure after the pump has boosted it, rather than before the pump boosts it. Where the backflow preventer is on the discharge side of the pump and a jockey pump is installed, the jockey pump discharge and sensing lines need to be located so that a cross-connection is not created through the jockey pump.

4.14.9.1 No device or assembly, unless identified in 4.14.9.2, that will stop, restrict the starting of, or restrict the discharge of a fire pump or pump driver shall be installed in the suction piping.

> The purpose of the requirement in 4.14.9.1 is to prohibit the use of certain devices in the suction pipe of a fire pump that would cause turbulence or excess friction loss or cut off waterflow to the pump when flowing at 150 percent of rated capacity. If the water supply to a fire pump is shut off while the pump is running, a catastrophic failure will occur. Only an OS&Y control valve is permitted to be installed in the suction pipe of a fire pump to allow isolation of the fire pump for maintenance and repair. Additional valves are permitted only where a piece of equipment such as a backflow prevention device is installed at least 10 pipe diameters upstream of the pump suction flange. All control valves, which should be of the approved indicating type, are required to be supervised, preferably by a constantly attended monitoring station. The problem is that they could be shut off before or during operation of the fire pump.
>
> It is important to note that valves that have been shut off continue to be the leading cause of fire protection system failure. A vigorous inspection program and constant electronic surveillance of all water supply control valves are critical to the continued operation of any fire protection system, particularly water supplies such as fire pumps.

4.14.9.2 The following devices shall be permitted in the suction piping where the following requirements are met:

(1) Check valves and backflow prevention devices and assemblies shall be permitted where required by other NFPA standards or the authority having jurisdiction and installed in accordance with Section 4.27.

> Check valves and backflow prevention devices are generally prohibited in the suction pipe due to concerns over turbulence and friction loss; however, some jurisdictions require the installation of one of these devices in the suction pipe when backflow contamination is a possibility.
>
> Backflow preventers and their isolation valves contribute to turbulence and friction loss, usually more so than check valves. It is common for reduced pressure backflow preventers to cause more than 10 psi (0.7 bar) of friction loss at the overload point (150 percent of pump rated capacity).

CALCULATION ▶

> For example, consider a fire protection system connected to a public main where the pump is at an elevation 20 ft (6.1 m) higher than the street main and, at maximum pump flow, the residual pressure in the public main is 20 psi (138 kPa). Taking into account a 9 psi (62.1 kPa) loss in pressure due to elevation [0.433 psi × 20 ft = 8.7 psi (0.098 bar × 6.1 m = 0.6 bar = 60 kPA)] and 5 psi (34.5 kPa) friction loss in the suction pipe, the pressure at the suction flange will be 6 psi (41.4 kPa) [20 psi − 9 psi − 5 psi = 6 psi (138 kPa − 62.1 kPa − 34.5 kPa = 41.4 kPa)]. This situation is acceptable in accordance with 4.14.3.1. However, if a backflow preventer were installed in this suction pipe with a friction loss of 10 psi (69 kPa), the system would become inoperative [20 psi − 9 psi − 5 psi − 10 psi = −4 psi (138 kPa − 62.1 kPa − 34.5 kPa − 69 kPa = −27.6 kPa)]. The pump would not be able to obtain water at a positive pressure and would be severely damaged if operated at maximum flow.

For these reasons, the Technical Committee on Fire Pumps prefers that backflow preventers be installed on the discharge side of the pump. In the preceding example, the pump would receive water with adequate pressure at 6 psi (41.4 kPa), boost the pressure, and force it through the backflow preventer, where it would then lose 10 psi (69 kPa). Although this arrangement solves one concern, installation of backflow prevention on the discharge side of pumps can also cause other concerns. Three of the more common concerns are as follows:

1. *Maximum pressure.* Backflow preventers are listed for maximum pressures of 175 psi (1207 kPa). The pressure on the discharge side of the pump frequently exceeds this value.
2. *Water authorities.* Water authorities sometimes assume that pumps are a possible cause of contamination of the public water supply, and, therefore, they require the backflow preventer on the suction side. However, the water authorities are incorrect in this assumption. As indicated by the American Water Works Association, fire pumps do not cause contamination problems for potable water supplies. Their manual AWWA M14, *Recommended Practice for Backflow Prevention and Cross-Connection Control,* states that "booster pumps do not affect the potability of the system." (See the definition of *Class 2 Fire Protection Systems* in AWWA M14.)
3. *Jurisdiction.* In some communities, the backflow preventer is the traditional point where the jurisdiction of one labor trade group ends and another begins. Typically, work from the water supply to the backflow preventer is done by mechanical contractors, and work from the backflow preventer to the fire protection devices is done by fire protection contractors. If the backflow preventer is installed on the discharge side of the fire pump, the fire protection contractor in many jurisdictions is prohibited from working on the fire pump because of its location. This situation can result in the installation of the fire pump by inadequately trained personnel inexperienced with fire pump functions and operations. Installation of fire pumps by unqualified personnel should be a cause of concern.

NFPA 20 does not mandate the location of backflow preventers. Instead, a technical preference for the installation of the backflow preventer on the discharge side of the pump is indicated. However, this arrangement is not always possible, so Section 4.27 is included, which contains all of the regulations to follow if the backflow preventer is to be installed on the suction side.

(2) Where the authority having jurisdiction requires positive pressure to be maintained on the suction piping, a pressure sensing line for a low suction pressure control, specifically listed for fire pump service, shall be permitted to be connected to the suction piping.

Prior to the 2003 edition, NFPA 20 appeared to permit the installation of low suction throttling valves to be installed in the suction pipe of a fire pump. The 2003 edition of the standard clarified that the industry term for a flow control valve is actually a *low suction throttling valve* and that such a device is required to be installed on the discharge side of a fire pump, with the sensing line connected to the suction piping. Since the 2003 edition, only a pressure-sensing line serving a low suction throttling valve is referenced. The sensing line is connected to the suction pipe but does not obstruct the waterway. A significant friction loss may be associated with the low suction throttling valve, and this loss must be accounted for in determining the pressure available for the fire protection system.

(3) Devices shall be permitted to be installed in the suction supply piping or stored water supply and arranged to activate a signal if the pump suction pressure or water level falls below a predetermined minimum.

This list item permits the installation of a device such as a pressure switch to monitor the pressure in the suction pipe (where the suction pipe is connected to the municipal water supply). The intent of this requirement is to permit monitoring of the suction pressure to activate a signal indicating that the pressure has dropped below a predetermined level. This list item does not permit shutting off the pump due to low pressure. In the 2007 edition, the technical committee added a definition for the term *signal* in Chapter 3, which is defined as "an indicator of status." The purpose of the signal is to provide an indication that the suction pressure is low; however, the signal does not allow the pump to be shut off due to low suction pressure conditions.

(4) Suction strainers shall be permitted to be installed in the suction piping where required by other sections of this standard.

(5) Other devices specifically permitted or required by this standard shall be permitted.

4.14.10* Anti-Vortex Plate. Where a tank is used as the suction source for a fire pump, the discharge outlet of the tank shall be equipped with an assembly that controls vortex flow in accordance with NFPA 22, *Standard for Water Tanks for Private Fire Protection*.

When water discharges at or near the bottom of an abovegrade storage tank or holding basin, a vortex forms as the water level lowers, and without an anti-vortex plate, air will be introduced into the suction pipe. A similar phenomenon occurs when a sink full of water empties through its drain. If this vortex were allowed to develop in a suction tank, it would result in turbulence (air bubbles) in the suction pipe and cause damage to the pump. To prevent this from happening, vortex plates (called *anti-vortex plates* in NFPA 22) are placed in the bottom of the tank and connected to the discharge flange of the tank.

A vortex plate is a flat, square or round sheet of steel. The length of the sides or the diameter of the plate must be at least twice the diameter of the suction pipe, although it is recommended that the plate be at least 48 in. (1219 mm) along each side. A hole that is the same size as the suction pipe is located in the center of the plate.

The plate is mounted parallel to the floor of the tank at a height at least half the diameter of the suction pipe but not less than 6 in. (152 mm) above the floor and is connected to a long turn elbow that connects to the discharge flange installed in the wall of the tank. Water that flows out of the tank drops under the plate, up through the hole, up into the elbow, and out the discharge flange into the suction pipe. By forcing the water to follow this path, the creation of a vortex is prevented, which is why the NFPA 22 technical committee changed the name of this device to anti-vortex plate. See Exhibit II.4.13 for vortex plate details.

A.4.14.10 For more information, see the *Hydraulic Institute Standards for Centrifugal, Rotary and Reciprocating Pumps*. *(See Figure A.4.14.10.)*

4.15 Discharge Pipe and Fittings

4.15.1 The discharge components shall consist of pipe, valves, and fittings extending from the pump discharge flange to the system side of the discharge valve.

4.15.2 The pressure rating of the discharge components shall be adequate for the maximum total discharge head with the pump operating at shutoff and rated speed but shall not be less than the rating of the fire protection system.

The standard makes reference to "total discharge head" rather than working pressure to require that discharge components be rated for the actual pressure as measured at the discharge gauge of the fire pump. The intent of this subsection is to require the installation of

Section 4.15 • Discharge Pipe and Fittings   85

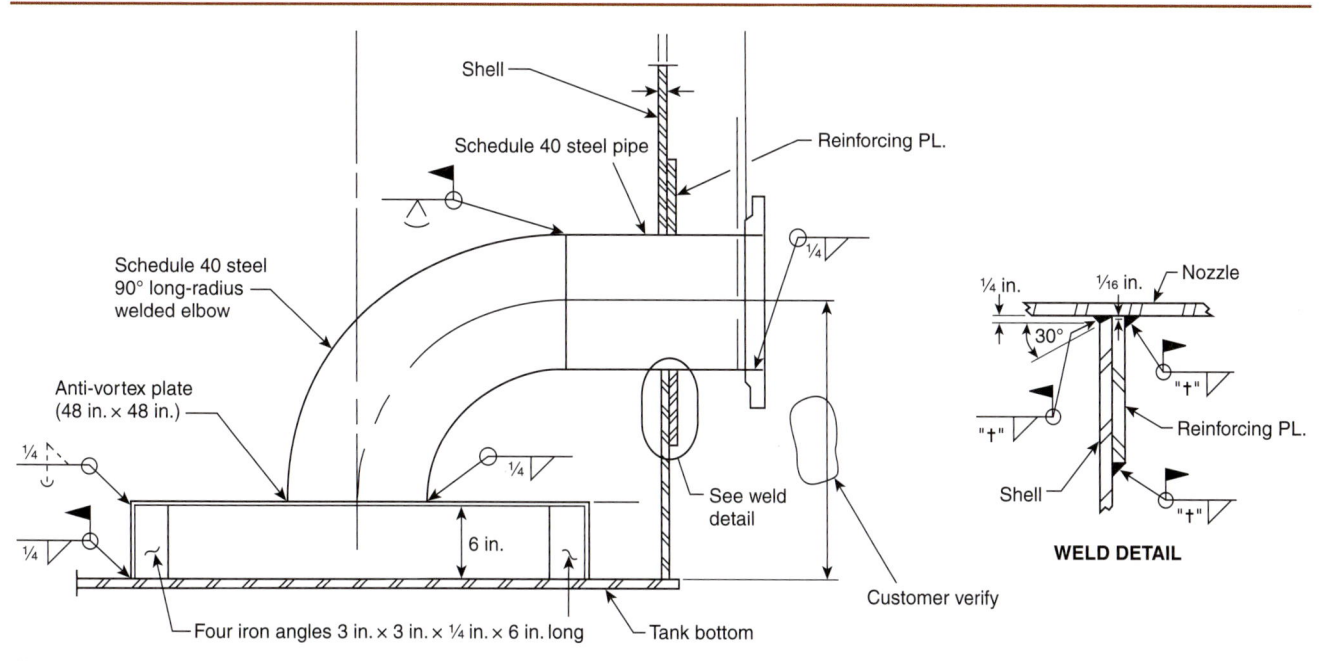

EXHIBIT II.4.13 *Suction Nozzle with Anti-Vortex Plate for Suction Tanks.*

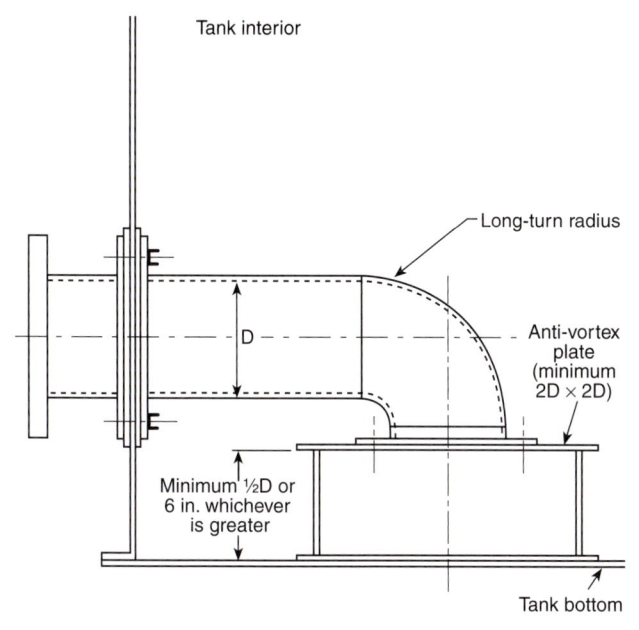

FIGURE A.4.14.10 *Anti-Vortex Plate Assembly.*

discharge components that are rated for the maximum pressure that occurs with a VSPLC turned off and the pump running at rated speed. The maximum pressure is normally the shut-off or churn pressure developed by the pump, plus any suction head at the inlet to the pump. For the purposes of applying 4.15.2, the maximum reasonably foreseeable suction pressure should be used in determining the total discharge pressure.

Stationary Fire Pumps Handbook   2013

4.15.3* Steel pipe with flanges, screwed joints, or mechanical grooved joints shall be used above ground.

> Subsection 4.15.3 prohibits the use of plain or beveled end locking valves or fittings in the discharge piping.

A.4.15.3 Flanges welded to the pipe are preferred.

> Threaded flanges are more subject to leakage and resulting corrosion over time than are welded flanges.

4.15.4 All pump discharge pipe shall be hydrostatically tested in accordance with NFPA 13, *Standard for the Installation of Sprinkler Systems*.

4.15.5* The size of pump discharge pipe and fittings shall not be less than that given in Section 4.26.

> The discharge pipe in this case is considered to be the pipe between the fire pump discharge flange and the discharge isolation valve specified in 4.15.7. Piping downstream of the discharge isolation valve is beyond the scope of NFPA 20 and should be designed and installed in accordance with the standard governing the system it supplies, such as NFPA 13 or NFPA 14.

A.4.15.5 The discharge pipe size should be such that, with the pump(s) operating at 150 percent of rated capacity, the velocity in the discharge pipe does not exceed 20 ft/sec (6.1 m/sec).

4.15.6* A listed check valve or backflow preventer shall be installed in the pump discharge assembly.

> **AHJ FAQ**
>
> **The plans for a fire pump installation show a backflow preventer on the discharge side of the pump but do not show a check valve. Is this an acceptable installation?**
>
> *ANSWER:* A means to keep water in the fire protection system when the pump is maintained is the reason behind having a check valve installed. When a backflow preventer is included on the discharge side of the pump, it can serve as the check valve.

> The check valve or backflow preventer is necessary to prevent water from flowing back into the fire pump. Backflow on pump shutdown can be a source of pressure surges. Spring loaded, anti-slam check valves reduce pressure surges by minimizing backflow.

A.4.15.6 Large fire protection systems sometimes experience severe water hammer caused by backflow when the automatic control shuts down the fire pump. Where conditions can be expected to cause objectionable water hammer, a listed anti-water-hammer check valve should be installed in the discharge line of the fire pump. Automatically controlled pumps in tall buildings could give trouble from water hammer as the pump is shutting down.

Where a backflow preventer is substituted for the discharge check valve, an additional backflow preventer might be necessary in the bypass piping to prevent backflow through the bypass.

Where a backflow preventer is substituted for the discharge check valve, the connection for the sensing line is permitted to be between the last check valve and the last control valve if the pressure sensing line connection can be made without altering the backflow valve or violating its listing. This method can sometimes be done by adding a connection through the test port on the backflow valve. In this situation, the discharge control valve is not necessary, because the last control valve on the backflow preventer serves this function.

Where a backflow preventer is substituted for the discharge check valve and the connection of the sensing line cannot be made within the backflow preventer, the sensing line should be connected between the backflow preventer and the pump's discharge control valve. In this situation, the backflow preventer cannot substitute for the discharge control valve because the sensing line must be able to be isolated.

4.15.7* A listed indicating gate or butterfly valve shall be installed on the fire protection system side of the pump discharge check valve.

> Isolation valves on the discharge side of the fire pump do not affect the operation of the pump, so approved indicating valves that are not OS&Y type are permitted. The exception to this statement is when two or more pumps are installed in series (see 4.15.8) and the discharge from one pump is part of the suction of the next pump. Therefore, OS&Y gate valves must be used as both the suction valve and the discharge valve for the lead pump(s) installed in series. See Exhibit II.4.14 for an example of OS&Y gate valves.

A.4.15.7 See A.4.15.6 for circumstances where a backflow preventer can be substituted for the discharge control valve.

4.15.8 Where pumps are installed in series, a butterfly valve shall not be installed between pumps.

4.15.9 Low Suction Pressure Controls.

4.15.9.1 Low suction throttling valves or variable speed suction limiting controls for pump drivers that are listed for fire pump service and that are suction pressure sensitive shall be permitted where the authority having jurisdiction requires positive pressure to be maintained on the suction piping.

> The use of low suction pressure throttling valves is discouraged but permitted if required by an AHJ. A low pressure throttling valve reduces the pressure available to the fire protection system and provides another potential point of failure. When used, a low pressure throttling valve is installed in the discharge piping, with the sensing line for the device connected to the suction piping. Exhibit II.4.15 shows a low suction throttling valve.

EXHIBIT II.4.14 OS&Y Gate Valves.

EXHIBIT II.4.15 Low Suction Throttling Valve. (Courtesy of Aon Fire Protection Engineering)

4.15.9.2 When a low suction throttling valve is used, it shall be installed according to manufacturers' recommendations in the piping between the pump and the discharge check valve.

4.15.9.3 The size of the low suction throttling valve shall not be less than that given for discharge piping in Section 4.26.

> The intent of this paragraph is to require that the size of the low suction throttling valve be not less than that specified in Table 4.26(a) and Table 4.26(b) for discharge piping. This valve must be sized properly so that the system design is not compromised.

4.15.9.4 The friction loss through a low suction throttling valve in the fully open position shall be taken into account in the design of the fire protection system.

CALCULATION ▶

> Low suction throttling valves should operate in the fully open position as long as the suction pressure is above the valve pressure setting. Friction loss though a fully open low suction throttling valve is usually determined using a friction loss graph or can be calculated using the valve's *Cv* value in the following formula:
>
> $$F = (Q/Cv)^2$$
>
> Where:
> F = friction loss (psi)
> Q = flow through valve (gpm)
> Cv = valve friction loss factor [(gpm/psi)$^{0.5}$]

4.15.9.5 System design shall be such that the low suction throttling valve is in the fully open position at system design point.

> The residual suction pressure at the system demand must be above the low suction control valve pressure setting.

4.15.10* Pressure Regulating Devices. No pressure regulating devices shall be installed in the discharge pipe except as permitted in this standard.

> Subsection 4.15.10 was added to the 2003 edition to prohibit the use of pressure regulating valves in the discharge piping, except where allowed by NFPA 20 to address special issues. The intent of the technical committee is to encourage matching the pump to the system demands without overpressurizing the system and then using pressure regulating valves to compensate. Although data has not been collected to evaluate the reliability of pressure devices that operate in a primarily static condition, the committee believes pressure regulating valves operating in this environment are not sufficiently reliable for use and should be avoided when possible. The valves require regular maintenance, especially in a static environment, to prevent strainers in pilot-operated valves from clogging and causing the valve to fail open. These valves can also fail in the closed position, especially if the seats are not exercised regularly. It is important to note that the scope of NFPA 20 stops at the discharge control valve, and nothing in NFPA 20 prohibits the use of pressure regulating devices downstream of the discharge isolation valve. Any pipe, fitting, valve, or other device downstream of the discharge isolation valve is beyond the scope of NFPA 20, and such pipe, fittings, valves, or other devices should be installed in accordance with the appropriate listing and installation standard, such as NFPA 13 or NFPA 14.
>
> Fire pumps taking suction from a long dead-end main, and fire pumps upstream of large dry-pipe valves, may be subject to pressure surges (water hammer). A surge analysis can be conducted to check the pressure surge potential with different design options for minimizing pressure surges.

A.4.15.10 See 4.7.7.2.

4.16* Valve Supervision

A.4.16 Isolation valves and control valves are considered to be identical when used in conjunction with a backflow prevention assembly.

4.16.1 Supervised Open. Where provided, the suction valve, discharge valve, bypass valves, and isolation valves on the backflow prevention device or assembly shall be supervised open by one of the following methods:

(1) Central station, proprietary, or remote station signaling service
(2) Local signaling service that will cause the sounding of an audible signal at a constantly attended point
(3) Locking valves open
(4) Sealing of valves and approved weekly recorded inspection where valves are located within fenced enclosures under the control of the owner

> The four methods of supervision represented in 4.16.1 help to ensure that a common mode of sprinkler system failure is reduced to a minimum. Approximately one-third of system failures are attributable to closed or partially closed sprinkler system control valves. The methods listed are in descending order of preference.
>
> Central station, proprietary, or remote station supervision methods are preferred over the other methods listed. Supervisory signals received at such alarm facilities are usually forwarded to maintenance staff who are responsible for the protected property. This method also allows for much more efficient notification of the fire department that will respond (the signal is not considered a fire alarm signal and, therefore, does not require an immediate response). Early notification that a control valve has been closed for any reason may alter the fire department response and fire-fighting tactics. However, central station, proprietary, or remote station supervision methods have not been mandated by NFPA 20.
>
> All bypass valves are required to be supervised open. Where a bypass is installed, water is expected to flow through the bypass into the fire protection system automatically, that is, without human intervention, in the event that the fire pump does not start. When the control valves are open, water can flow through the bypass automatically. A check valve, therefore, needs to be installed in the bypass line so that, when the pump is running, water does not circulate through the bypass back to the pump.

4.16.2 Supervised Closed. Control valves located in the pipeline to the hose valve header shall be supervised closed by one of the methods allowed in 4.16.1.

> For test headers that are installed outside, or where freezing is a danger, an isolation valve as required by 4.20.3.3.1 must be installed. This valve is normally closed and is only opened for testing purposes. The valve is required to be supervised in the closed position to prevent accidental freezing of the test header. The hose header is required to have an automatic drain valve (typically a ball drip valve) where it is subject to freezing conditions.

4.17* Protection of Piping Against Damage Due to Movement

A clearance shall be provided around pipes that pass through walls, ceilings, or floors of the fire pump room enclosure.

A.4.17 Pipe breakage caused by movement can be greatly lessened and, in many cases, prevented by increasing flexibility between major parts of the piping. One part of the piping should never be held rigidly and another free to move without provisions for relieving the strain. Flexibility can be provided by the use of flexible couplings at critical points and by allowing clearances at walls and floors. Fire pump suction and discharge pipes should be treated the same as sprinkler risers for whatever portion is within a building. *(See NFPA 13, Standard for the Installation of Sprinkler Systems.)*

Holes through pump room fire walls should be packed with mineral wool or other suitable material held in place by pipe collars on each side of the wall. Pipes passing through foundation walls or pit walls into ground should have clearance from these walls, but holes should be watertight. Space around pipes passing through pump room walls or pump house floors can be filled with asphalt mastic.

4.17.1 Unless the requirements of 4.17.2 through 4.17.4 are met, where pipe passes through walls, ceilings, or floors of the fire pump room enclosure, the holes shall be sized such that the diameter of the hole is nominally 2 in. (50 mm) larger than the pipe.

4.17.2 Where clearance is provided by a pipe sleeve, a nominal diameter 2 in. (50 mm) larger than the nominal diameter of the pipe shall be acceptable.

4.17.3 No clearance is required if flexible couplings are located within 1 ft (305 mm) of each side of the wall, ceiling, or floor.

4.17.4 Where protection of piping against damage caused by earthquakes is required, the provisions of Section 4.28 shall apply.

4.17.5 Where required, the clearance shall be filled with flexible material that is compatible with the piping materials and maintains any required fire resistance rating of the enclosure.

> In the 2010 edition, a 1 in. (25 mm) clearance was required. In the 2013 edition, clearance requirements were modified, establishing consistency with the seismic requirements of NFPA 13.

4.18 Relief Valves for Centrifugal Pumps

4.18.1* General.

A.4.18.1 The pressure is required to be evaluated at 121 percent of the net rated shutoff pressure because the pressure is proportional to the square of the speed that the pump is turned. A diesel engine governor is required to be capable of limiting the maximum engine speed to 110 percent, creating a pressure of 121 percent. Since the only time that a pressure relief valve is required by the standard to be installed is when the diesel engine is turning faster than normal, and since this is a relatively rare event, it is permitted for the discharge from the pressure relief valve to be piped back to the suction side of the pump.

> A circulation relief valve located downstream of the pressure reducing valve is required by 4.18.7.1 whenever the pressure relief valve is piped back to suction. This is in addition to the circulation relief valve required by Section 4.11, which is designed to provide cooling water under churn conditions. The purpose of the circulation relief valve required in 4.18.7.1 is to provide cooling of the fire pump when water is flowing through the pressure relief valve back to the pump suction. The pressure setting on this circulation relief valve should be well below the pressure setting of the pressure relief valve, but above the maximum suction pressure.

4.18.1.1* Pressure relief valves shall be used only where specifically permitted by this standard.

Section 4.18 • Relief Valves for Centrifugal Pumps

> ▶ **FAQ** — Does NFPA 20 permit the installation of main pressure relief valves in electric fire pump systems?
>
> Paragraph 4.18.1.1 states that main relief valves should only be used where specifically permitted by this standard. No section in this standard specifically permits the use of a main pressure relief valve on an electric fire pump, except where a variable speed driver is used. Variable speed drivers are required to default to constant rated speed operation in the event the variable speed driver fails. If operating at constant rated speed can result in system overpressurization, a pressure relief valve is required. The pressure relief valve setting must be above the set pressure of the variable speed driver. The use of a main pressure relief valve to trim excess pressure is considered to be poor design and should be avoided. Several methods are available to cope with excessive pressures, such as the following:
>
> 1. A break tank (see Section 4.31)
> 2. A variable speed pressure limiting control device (see 11.2.4.3)
> 3. Other pressure regulating devices downstream of the fire pump discharge control valve

A.4.18.1.1 In situations where the required system pressure is close to the pressure rating of the system components and the water supply pressure varies significantly over time, to eliminate system overpressurization, it might be necessary to use one of the following:

(1) A tank between the water supply and the pump suction, in lieu of directly connecting to the water supply piping
(2) A variable speed pressure limiting control device

4.18.1.2 Where a diesel engine fire pump is installed and where a total of 121 percent of the net rated shutoff (churn) pressure plus the maximum static suction pressure, adjusted for elevation, exceeds the pressure for which the system components are rated, a pressure relief valve shall be installed.

Prior to the 1996 edition, NFPA 20 required the installation of pressure relief valves for all diesel engine fire pumps. This requirement was based on the assumption that, if engines ran too fast (a condition known as *overspeed*), the fire protection system would be exposed to pressures in excess of the pressure ratings of the system components. Because an overspeed shutdown device is required, the technical committee believes that a pressure relief valve is not needed on all diesel fire pump installations. Pumps that create pressures less than the pressure rating of the fire protection system components [typically 175 psi (912.1 bar)] at 110 percent of rated speed do not need a pressure relief valve. The example that follows illustrates the procedure used to determine if a pressure relief valve is needed.

> 🛡 **AHJ FAQ**
>
> **The plans for a diesel engine–driven centrifugal pump do not show the installation of a pressure relief valve. Is this omission permitted?**
>
> ANSWER: Yes, this is acceptable in some installations. The pressure relief valve requirement on diesel engine–driven pumps is intended to prevent the piping from overpressurization if the fire pump malfunctions and runs at a higher speed than anticipated. If 121 percent of the pump's churn pressure is added to the maximum static pressure of the water supply, and the total does not exceed the maximum working pressure of the system components, a pressure relief valve is not required.

▶ EXAMPLE

A fire pump rated for 1500 gpm (5677 L/min) and 100 psi (6.9 bar) at 1750 rpm has a shutoff pressure of 120 psi (8.3 bar). The shutoff pressure produces 145 psi (10 bar) at 110 percent of rated speed. If a maximum of 45 psi (3.1 bar) static pressure is available from the city water supply, the total pressure when the pump is running at churn at 110 percent of rated speed is 190 psi (13.1 bar).

$$\text{Pressure increase at 110 percent of rated speed} = 110\%^2 = 121\%$$

$$120 \text{ psi (8.3 bar)} \times 1.21 = 145 \text{ psi (10 bar)}$$

$$145 \text{ psi} + 45 \text{ psi (10 bar} + 3.1 \text{ bar)} = 190 \text{ psi (13.1 bar)}$$

EXHIBIT II.4.16 *Pressure Relief Valve. (Courtesy of Aon Fire Protection Engineering)*

> In this case, a pressure relief valve (see Exhibit II.4.16) is needed if the fire protection system components are rated at 175 psi (912.1 bar). A pressure relief valve is not required if the fire protection system components are rated for 200 psi (13.8 bar) or higher.

4.18.1.3 Where an electric variable speed pressure limiting controller or a diesel pressure limiting driver is installed, and the maximum total discharge head adjusted for elevation with the pump operating at shutoff and rated speed exceeds the pressure rating of the system components, a pressure relief valve shall be installed.

> The maximum discharge pressure for a variable speed pump is the churn pressure at rated speed plus the maximum suction pressure. It is not expected that a diesel engine governor will fail and cause overspeed conditions at the same time that the variable speed driver fails; therefore, the maximum pressure for the diesel engine drive fire pump should be evaluated at rated speed.

4.18.2 Size. The relief valve size shall be determined by one of the methods specified in 4.18.2.1 or 4.18.2.2.

4.18.2.1* The relief valve shall be permitted to be sized hydraulically to discharge sufficient water to prevent the pump discharge pressure, adjusted for elevation, from exceeding the pressure rating of the system components.

> Sizing the relief valve piping hydraulically can reduce the loss of water to the fire protection system if the pressure relief valve fails to close and conserves water during the weekly operating test when the pump is running at churn. The intent of the technical committee is to allow a performance-based option for the sizing of the relief valve piping for this reason. The purpose of a performance-based design option is to establish a set of stated goals. In this case, the goal is conservation of water discharge from the pressure relief valve. Performance-based design options are used in place of prescriptive design options that, in the case of pressure relief valves, may not consider water conservation or short runs of piping where smaller pipe diameters can be used to advantage through hydraulic calculations.

> When the net pressure from the fire pump plus the suction pressure will not subject any system component to a pressure that exceeds the pressure rating of the component, the pressure relief valve piping should be sized for a minimum flow equal to or greater than the flow through the fire pump.

A.4.18.2.1 See Figure A.4.18.2.1.

4.18.2.2 If the relief valve is not sized hydraulically, the relief valve size shall not be less than that given in Section 4.26. *(See also 4.18.7 and A.4.18.7 for conditions that affect size.)*

4.18.3 Location. The relief valve shall be located between the pump and the pump discharge check valve and shall be so attached that it can be readily removed for repairs without disturbing the piping.

4.18.4 Type.

4.18.4.1 Pressure relief valves shall be either a listed spring-loaded or a pilot-operated diaphragm type.

> Listed pressure relief valves are required by the listing standard to limit the pressure upstream of the valve to 125 percent of the pressure relief valve setting. Therefore, a pressure relief valve set at 175 psi (12.1 bar) could allow pressures as high as 218 psi (15 bar) and still pass the listing requirement. Spring-operated pressure relief valves do not fully open until the pressure exceeds the set pressure. Pilot-operated valves should be capable of limiting the maximum pressure closer to the set pressure. The operating characteristics of pressure relief valves should be reviewed before selecting a valve.

4.18.4.2 Pilot-operated pressure relief valves, where attached to vertical shaft turbine pumps, shall be arranged to prevent relieving of water at water pressures less than the pressure relief setting of the valve.

4.18.5* Discharge.

A.4.18.5 The relief valve cone should be piped to a point where water can be freely discharged, preferably outside the building. If the relief valve discharge pipe is connected to an underground drain, care should be taken that no steam drains enter near enough to work back through the cone and into the pump room.

> The discharge of the relief valve should preferably be piped to a water storage tank for safe discharge or to the outside to discharge at a safe point. If discharge to a tank or a safe point outside is not possible, discharge to a drain of adequate size and capacity to accept the maximum flow from any discharge should be arranged.

4.18.5.1 The relief valve shall discharge into an open pipe or into a cone or funnel secured to the outlet of the valve.

4.18.5.2 Water discharge from the relief valve shall be readily visible or easily detectable by the pump operator.

4.18.5.3 Splashing of water into the pump room shall be avoided.

4.18.5.4 If a closed-type cone is used, it shall be provided with means for detecting motion of water through the cone.

> Relief valve discharge is permitted to be piped back to the suction side of a fire pump (closed loop). Where this piping arrangement is used, a method to detect flow must be provided. This

SAMPLE PRESSURE RELIEF VALVE CALCULATION
DISCHARGE TO ATMOSPHERE
For a 1,500 gpm at 100 psi Fire Pump

1. Pressure rating of the system components		175
2. Maximum pump over speed		110%
3. Pump size		1,500
4. Rated pump pressure		100

	Normal Static Pressure (at rated speed)	Maximum Pressure Static (at Max Pump Over speed)
5. Pump net pressure (rated pressure at 100%)	100	121.0
6. Pump net churn pressure (0 flow)	120	145.2
7. Pump net pressure @ 150% of rated flow	65	78.7
8. Maximum static pressure at pump suction	50	50
9. Available flow at pump suction	1,320	1,320
10. Residual pressure at pump suction	45	45
11. Maximum pump discharge pressure at churn	170	195.2
12. Maximum allowable net discharge pressure		125
13. Pump flow rate at maximum net pressure adjusted to normal speed [#12/(#2*#2)] = 103.3 psi		1,360.4
14. Required flow rate through the pressure relief valve (pump flow rate at 125 psi and overspeed or [#13*#2])		1,496.5
15. Set pressure for pressure relief valve		175
16. Pressure relief valve size		4
17. Pressure relief valve pipe size		4.026
18. Nozzle (pipe) discharge coefficient		0.9
19. C factor		120
20. Pressure relief valve Cv (P=[Q/Cv]^2)		240

	Type Fitting	Number of Fittings	Equivalent Length	Total Equivalent Length
21. **Pressure Relief Valve Fittings**	45°	1	4	4
	Ells	2	10	20
	LRE	0	6	0

22. Pressure relief valve pipe length	30
23. Total equivalent length	54
24. Friction loss per foot in pipe at flow #14	0.594
25. Total loss in pressure relief valve piping (#23 × #24)	32.1
26. Friction loss in pressure relief valve at estimated flow (valve wide open) ([#14/#20]^2)	38.9
27. Pressure at pressure relief piping discharge (#1 – #26 – #26)	104.1
28. Elevation difference (0.433*Elev difference in feet)	0
29. Required pressure at relief piping discharge (pitot pressure at a flow of #14) ({#14/[29.83 × #18 × #17^2]}^{2})	11.8

Conclusion: The discharge pressure at the pressure relief piping (with the pressure relief valve wide open) exceeds the pitot pressure required for the flow; therefore the pressure relief components are adequately sized.

FIGURE A.4.18.2.1 Sample Pressure Relief Valve Calculation.

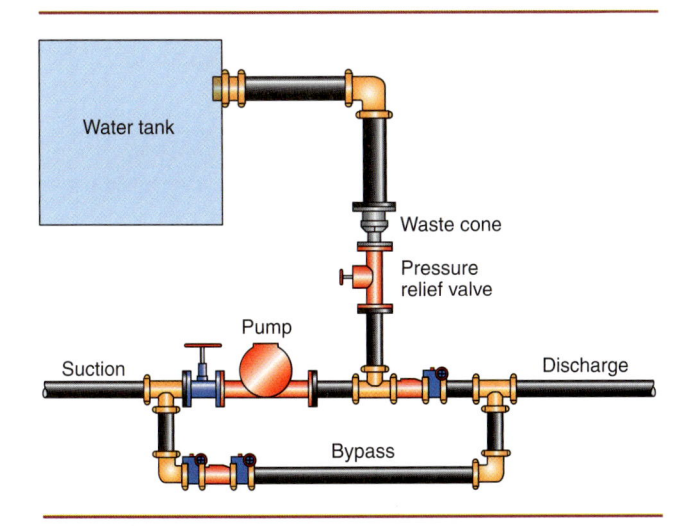

EXHIBIT II.4.17 Pressure Relief Valve Piped to Drain. (Courtesy of Stephan Laforest, Summit Sprinkler Design Services, Inc.)

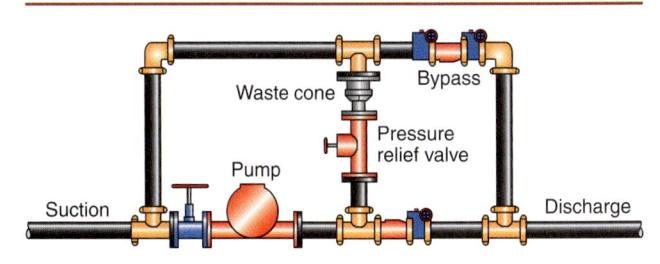

EXHIBIT II.4.18 Pressure Relief Valve Piped to Pump Suction (Closed). (Courtesy of Stephan Laforest, Summit Sprinkler Design Services, Inc.)

> method may be a flow connected to a local bell or a closed waste cone (which is pictured in Exhibit II.4.16). When this method is used, a circulation relief valve must be installed for cooling purposes (see A.4.18.1 for additional details). The pressure setting of the circulation relief valve should be well below the pressure relief valve setting but above the maximum suction pressure. For relief valve arrangements (open and closed discharge), see Exhibit II.4.17 and Exhibit II.4.18.

4.18.5.5 If the relief valve is provided with means for detecting motion (flow) of water through the valve, then cones or funnels at its outlet shall not be required.

4.18.6 Discharge Piping.

4.18.6.1 Except as permitted in 4.18.6.2, the relief valve discharge pipe shall be of a size not less than that given in Section 4.26.

4.18.6.2 The discharge pipe shall be permitted to be sized hydraulically to discharge sufficient water to prevent the pump discharge pressure, adjusted for elevation, from exceeding the pressure rating of the system components.

4.18.6.2.1 If the pipe employs more than one elbow, the next larger pipe size shall be used.

4.18.6.3 Relief valve discharge piping returning water back to the supply source, such as an aboveground storage tank, shall be run independently and not be combined with the discharge from other relief valves.

4.18.7* Discharge to Source of Supply. Where the relief valve is piped back to the source of supply, the relief valve and piping shall have sufficient capacity to prevent pressure from exceeding that for which system components are rated.

A.4.18.7 Where the relief valve discharges back to the source of supply, the back pressure capabilities and limitations of the valve to be used should be determined. It might be necessary to increase the size of the relief valve and piping above the minimum to obtain adequate relief capacity due to back pressure restriction.

4.18.7.1 Where a pressure relief valve has been piped back to suction, a circulation relief valve sized in accordance with 4.11.1.6 and discharged to atmosphere shall be provided downstream of the pressure relief valve.

> Where a pressure relief valve discharge is piped back to the suction side of a fire pump (closed loop), a circulation relief valve must be installed for cooling purposes (see A.4.18.1 for additional detail). The pressure setting of the circulation relief valve should be well below the pressure relief valve setting but above the maximum suction pressure.

4.18.7.2 Where pump discharge water is piped back to pump suction and the pump is driven by a diesel engine with heat exchanger cooling, a high cooling water temperature signal at 104°F (40°C) from the engine inlet of the heat exchanger water supply shall be sent to the fire pump controller, and the controller shall stop the engine provided that there are no active emergency requirements for the pump to run.

> This requirement is intended to prevent damage to the engine under nonemergency conditions; under emergency conditions, the engine should continue to run, even if it results in damage to the engine.

4.18.7.2.1 The requirements of 4.18.7.2 shall not apply when pump discharge water is being piped back to a water storage reservoir.

> The temperature change from energy added by the fire pump to water that is recirculated through a water storage tank should be minimal.

4.18.8* Discharge to Suction Reservoir. Where the supply of water to the pump is taken from a suction reservoir of limited capacity, the drain pipe shall discharge into the reservoir at a point as far from the pump suction as is necessary to prevent the pump from drafting air introduced by the drain pipe discharge.

A.4.18.8 When discharge enters the reservoir below minimum water level, there is not likely to be an air problem. If it enters over the top of the reservoir, the air problem is reduced by extending the discharge to below the normal water level.

4.18.9 Shutoff Valve. A shutoff valve shall not be installed in the relief valve supply or discharge piping.

4.19 Pumps Arranged in Series

4.19.1 Series Fire Pump Unit Performance.

4.19.1.1 A series fire pump unit (pumps, drivers, controllers, and accessories) shall perform in compliance with this standard as an entire unit.

4.19.1.2 Within 20 seconds after a demand to start, pumps in series shall supply and maintain a stable discharge pressure (±10 percent) throughout the entire range of operation.

4.19.1.2.1 The discharge pressure shall be permitted to be adjusted and restabilized whenever the flow condition changes.

> The requirement for stable discharge pressures applies while conditions affecting flow (primarily valve positions and open orifices) stay the same. The discharge pressures will change and restabilize when the conditions affecting flow change.

4.19.1.3 The complete series fire pump unit shall be field acceptance tested for proper performance in accordance with the provisions of this standard. *(See Section 14.2.)*

4.19.2 Fire Pump Arrangement.

4.19.2.1 No more than three pumps shall be allowed to operate in series.

> The addition of each pump in a series results in some decrease in overall reliability. High pressures, increasing complexity, and decreasing reliability make it undesirable to have more than three pumps operating in series.

◀ **DESIGN ALERT**

4.19.2.2 No pump in a series pump unit shall be shut down automatically for any condition of suction pressure.

4.19.2.3 No pressure reducing or pressure regulating valves shall be installed between fire pumps arranged in series.

4.19.2.4 The pressure at any point in any pump in a series fire pump unit, with all pumps running at shutoff and rated speed at the maximum static suction supply, shall not exceed any pump suction, discharge, or case working pressure rating.

4.20 Water Flow Test Devices

4.20.1 General.

> Acceptance testing includes testing at rated speed without flow through a pressure relief valve (pressure relief valves may be required on variable speed pumps and diesel engine drive pumps). The pipe, valves, and fittings in the pump discharge piping must be rated for the maximum discharge pressure. If the discharge pressure exceeds 175 psi (12.1 bar), flanged fittings must be of extra heavy pattern.

4.20.1.1* A fire pump installation shall be arranged to allow the test of the pump at its rated conditions as well as the suction supply at the maximum flow available from the fire pump.

> Every fire pump installation needs a method for performing the acceptance test required in Chapter 14. NFPA 25 permits the following three methods for testing fire pumps:
>
> 1. *Test header.* This device is connected to the discharge side of the pump and has a number of hose outlets as illustrated in Exhibit II.4.19. When testing the pump, the hose is connected to the outlets with water discharged in a safe location as illustrated in Exhibit II.4.20 and Exhibit II.4.21. Flow readings are usually taken from the end of the hose with a Pitot tube or other flow measuring device.
> 2. *Flowmeter.* A special pipe is run from the discharge side of the pump back to the water supply (or to some other acceptable discharge point) with a flowmeter and control valve in the line. When testing the pump, the control valve is partially opened (with the pump already running) to achieve the 100 percent flow condition. The valve is then opened more to achieve the 150 percent flow condition.
> 3. *Closed loop metering.* This method consists of a bypass line with a flowmeter and control valves (before and after the flowmeter), which leads directly from the pump

discharge to the pump suction. This test is run the same way as method 2, but the water does not come from the supply. The water is recirculated in a small loop through the pump, as illustrated in Exhibit II.4.22. It should be noted that A.4.20.2.1 recommends that metering devices discharge to drain, to the water supply tank, or to the outside. It should also be noted that acceptance testing must include operating the pump at 100 percent and 150 percent without recirculating the water back to the pump suction (that is, the water must discharge though a test header or other safe location).

Because closed loop metering (method 3) does not test the water supply's ability to get water to the pump, this method cannot be used for an acceptance test. NFPA 25 also limits the application of this method. Once every 3 years, testing must be performed in accordance with method 1 or method 2. Therefore, the designer must make sure that equipment used to perform the annual flow test in accordance with method 1 or method 2 is included in the original design. Otherwise, the acceptance test cannot be completed, and, 3 years after installation, the building owner will be unable to comply with NFPA 25.

▶ **FAQ** If the fire pump has no means to discharge water, must a test header or flowmeter be provided for a test that cannot be completed?

Under no circumstances can a fire pump be placed into service without an acceptance test. NFPA 25 requires an annual test of the fire pump by means of observing discharge of water at least once every 3 years. A flowmeter in a closed loop arrangement permits testing the pump without discharging water; however, water discharge is required for acceptance testing at least once every 3 years thereafter.

EXHIBIT II.4.19 Test Header. (Courtesy of Aon Fire Protection Engineering)

EXHIBIT II.4.20 Test Header with Hose Attached. (Courtesy of Aon Fire Protection Engineering)

EXHIBIT II.4.21 *Fire Hose Connected to Hose Monsters. (Courtesy of Aon Fire Protection Engineering)*

EXHIBIT II.4.22 *Closed Loop Metering. (Courtesy of Aon Fire Protection Engineering)*

A.4.20.1.1 The two objectives of running a pump test are to make sure that the pump itself is still functioning properly and to make sure that the water supply can still deliver the correct amount of water to the pump at the correct pressure. Some arrangements of test equipment do not permit the water supply to be tested. Every fire pump installation needs to have at least one arrangement of test equipment where the water supply can be tested. Inspection, testing, and maintenance standards (NFPA 25, *Standard for the Inspection, Testing, and Maintenance of Water-Based Fire Protection Systems*) require the pump test to be run at least once every 3 years using a method that tests the water supply's ability to provide water to the pump.

4.20.1.2* Where water usage or discharge is not permitted for the duration of the test specified in Chapter 14, the outlet shall be used to test the pump and suction supply and determine that the system is operating in accordance with the design.

A.4.20.1.2 Outlets can be provided through the use of standard test headers, yard hydrants, wall hydrants, or standpipe hose valves.

The following notes apply to Figure A.4.20.1.2(a) and Figure A.4.20.1.2(b):

(1) The distance from the flowmeter to either isolation valve should be as recommended by the meter manufacturer.
(2) There should be a distance of not less than 5 diameters of suction pipe for top or bottom suction connection to the fire pump suction flange. There should be a distance of not less than 10 diameters of suction pipe for side connection (not recommended) to the fire pump suction flange.
(3) Automatic air release should be provided if piping forms an inverted "U," trapping air.
(4) The fire protection system should have outlets available to test the fire pump and suction supply piping. *(See A.4.20.3.1.)*
(5) The closed loop meter arrangement will test only net pump performance. It does not test the condition of the suction supply, valves, piping, and so forth.
(6) Return piping should be arranged so that no air can be trapped that would eventually end up in the eye of the pump impeller.
(7) Turbulence in the water entering the pump should be avoided to eliminate cavitation, which would reduce pump discharge and damage the pump impeller. For this reason, side connection is not recommended.

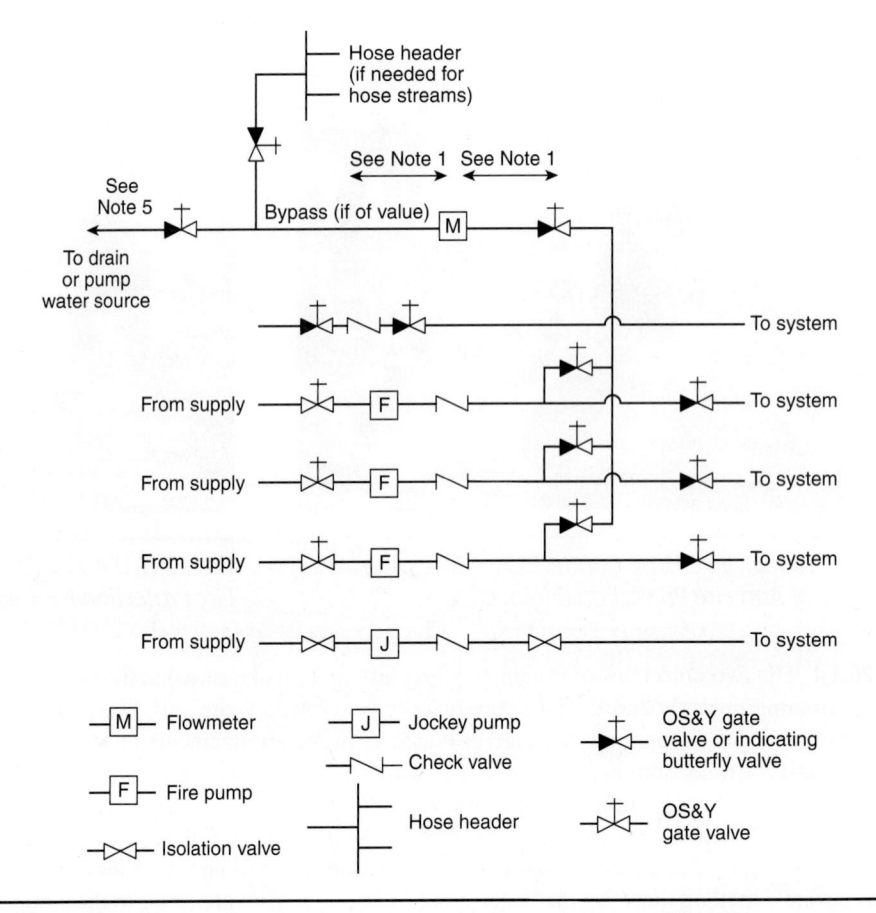

FIGURE A.4.20.1.2(a) *Preferred Arrangement for Measuring Fire Pump Water Flow with Meter for Multiple Pumps and Water Supplies. Water is permitted to discharge to a drain or to the fire pump water source. (See the text for information on the notes.)*

(8) Prolonged recirculation can cause damaging heat buildup, unless some water is wasted.
(9) The flowmeter should be installed according to manufacturer's instructions.
(10) Pressure sensing lines also need to be installed in accordance with 10.5.2.1. *[See Figure A.4.30(a) and Figure A.4.30(b).]*

4.20.1.3 The flow shall continue until the flow has stabilized. *(See 14.2.6.4.)*

4.20.1.4* Where a test header is installed, it shall be installed on an exterior wall or in another location outside the pump room that allows for water discharge during testing.

A.4.20.1.4 The hose valves of the fire pump test header should be located on the building exterior. This is because the test discharge needs to be directed to a safe outdoor location, and to protect the fire pumps, controllers, and so forth, from accidental water spray. In instances where damage from theft or vandalism is a concern, the test header hose valves can be located within the building but outside of the fire pump room if, in the judgment of the AHJ, the test flow can be safely directed outside the building without undue risk of water spray to the fire pump equipment.

4.20.2 Meters and Testing Devices.

4.20.2.1* Metering devices or fixed nozzles for pump testing shall be listed.

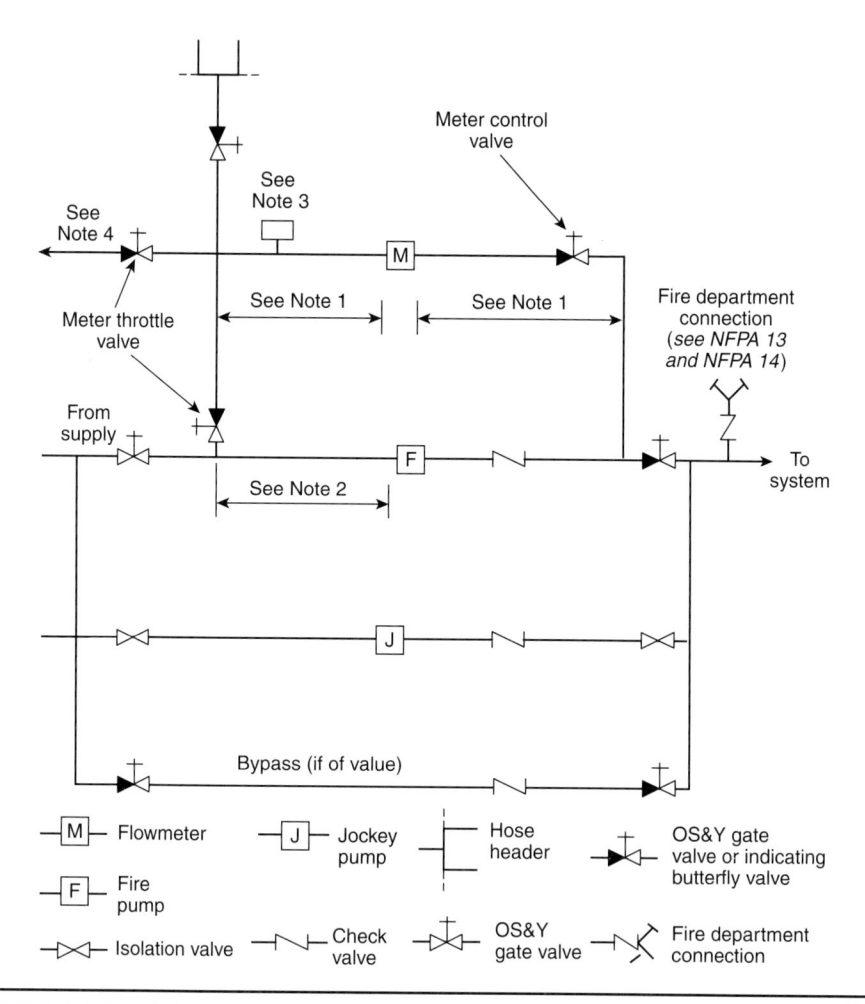

FIGURE A.4.20.1.2(b) *Typical Arrangement for Measuring Fire Pump Water Flow with Meter. Discharge from the flowmeter is recirculated to the fire pump suction line. (See the text for information on the notes.)*

A.4.20.2.1 Metering devices should discharge to drain.

In the case of a limited water supply, the discharge should be back to the water source (e.g., suction tank, small pond). If this discharge enters the source below minimum water level, it is not likely to create an air problem for the pump suction. If it enters over the top of the source, the air problem is reduced by extending the discharge to below the normal water level.

4.20.2.2 Metering devices or fixed nozzles shall be capable of water flow of not less than 175 percent of rated pump capacity.

4.20.2.3 All of the meter system piping shall be permitted to be sized hydraulically but shall not be smaller than as specified by the meter manufacturer.

> Sizing the meter piping hydraulically can reduce the cost of installation without sacrificing system performance. The intent of the technical committee is to allow a performance-based option for sizing the flowmeter piping. In a closed loop system, the meter piping should be sized so that, at 150 percent flow, the friction loss through the meter, meter piping valves (fully open), and fittings is less than the net pressure of the fire pump and so that sufficient pressure is available to perform an acceptance test by discharging and measuring flow through the pump test header or other means.

> If the meter piping discharges to atmosphere, the meter piping should be sized so that, at 150 percent flow, the friction loss through the meter, meter piping valves (fully open) and fittings is less than the net pressure of the fire pump plus the residual suction pressure and so that sufficient pressure is available to perform an acceptance test by discharging and measuring flow through the pump test header or other means.

4.20.2.4 If the meter system piping is not sized hydraulically, then all of the meter system piping shall be sized as specified by the meter manufacturer but not less than the meter device sizes shown in Section 4.26.

4.20.2.5 For nonhydraulically sized piping, the minimum size meter for a given pump capacity shall be permitted to be used where the meter system piping does not exceed 100 ft (30.5 m) equivalent length.

4.20.2.6 For nonhydraulically sized piping, where meter system piping exceeds 100 ft (30.5 m), including length of straight pipe plus equivalent length in fittings, elevation, and loss through meter, the next larger size of piping shall be used to minimize friction loss.

4.20.2.7 The primary element shall be suitable for that pipe size and pump rating.

4.20.2.8 The readout instrument shall be sized for the pump-rated capacity. *(See Section 4.26.)*

4.20.2.9 When discharging back into a tank, the discharge nozzle(s) or pipe shall be located at a point as far from the pump suction as is necessary to prevent the pump from drafting air introduced by the discharge of test water into the tank.

4.20.2.10* Where a metering device is installed in a loop arrangement for fire pump flow testing, an alternate means of measuring flow shall be provided.

A.4.20.2.10 The testing arrangement should be designed to minimize the length of fire hose needed to discharge water safely [approximately 100 ft (30 m)]. Where a flow test meter is installed, an alternate means of testing, such as hydrants, hose valves, test header(s), and so forth is needed as an alternate means of testing the performance of the fire pump, and to verify the accuracy of the metering device.

4.20.2.10.1 The alternate means of measuring flow shall be located downstream of and in series with the flow meter.

> The alternate means of measuring flow allows verification of the accuracy of a flow test meter, acceptance testing, and verification of the suction supply every 3 years as required by NFPA 25.

4.20.2.10.2 The alternate means of measuring flow shall function for the range of flows necessary to conduct a full flow test.

4.20.2.10.3 An appropriately sized test header shall be an acceptable alternate means of measuring flow.

> Locating a test header downstream of the flowmeter allows verification of the accuracy of a flow test meter, acceptance testing, and verification of the suction supply every 3 years as required by NFPA 25.

4.20.3 Hose Valves.

4.20.3.1* General.

A.4.20.3.1 The hose valves should be attached to a header or manifold and connected by suitable piping to the pump discharge piping. The connection point should be between the discharge check valve and the discharge gate valve. Hose valves should be located to avoid any possible water damage to the pump driver or controller. If there are other adequate pump testing facilities, the hose valve header can be omitted when its main function is to provide a method of pump and suction supply testing. Where the hose header also serves as the equivalent of a yard hydrant, this omission should not reduce the number of hose valves to less than two.

4.20.3.1.1 Hose valves shall be listed.

4.20.3.1.2 The number and size of hose valves used for pump testing shall be as specified in Section 4.26.

4.20.3.1.3 Where outlets are being utilized as a means to test the fire pump in accordance with 4.20.1.1, one of the following methods shall be used:

(1)* Hose valves mounted on a hose valve header with supply pipe sized in accordance with 4.20.3.4 and Section 4.26
(2) Wall hydrants, yard hydrants, or standpipe outlets of sufficient number and size to allow testing of the pump

A.4.20.3.1.3(1) Outlets are typically provided through a standard test header. The test header is usually connected to the pump system between the discharge check valve and the discharge control valve for the pump so that the fire protection system can be isolated from the pump during testing if desired. However, the objective of testing the pump can be achieved with other arrangements as well.

4.20.3.2 Thread Type. Thread types shall be in compliance with one of the following:

(1) Hose valve(s) shall have the NH standard external thread for the valve size specified, as stipulated in NFPA 1963, *Standard for Fire Hose Connections*.
(2) Where local fire department connections do not conform to NFPA 1963 and the connection is to be utilized as a wall hydrant, the authority having jurisdiction shall designate the threads to be used.

4.20.3.3 Location.

4.20.3.3.1 A listed indicating butterfly or gate valve shall be located in the pipeline to the hose valve header.

4.20.3.3.2 A drain valve or automatic ball drip shall be located in the pipeline at a low point between the valve and the header. *[See Figure A.6.3.1(a) and Figure A.7.2.2.1.]*

4.20.3.3.3 The valve required in 4.20.3.3.1 shall be at a point in the line close to the pump. *[See Figure A.6.3.1(a).]*

4.20.3.4 Pipe Size. The pipe size shall be in accordance with one of the following two methods:

(1) Where the pipe between the hose valve header and the connection to the pump discharge pipe is over 15 ft (4.5 m) in length, the next larger pipe size than that required by 4.20.3.1.3 shall be used.
(2)* This pipe is permitted to be sized by hydraulic calculations based on a total flow of 150 percent of rated pump capacity, including the following:
   (a) This calculation shall include friction loss for the total length of pipe plus equivalent lengths of fittings, control valve, and hose valves, plus elevation loss, from the pump discharge flange to the hose valve outlets.
   (b) The installation shall be proven by a test flowing the maximum water available.

A.4.20.3.4(2) See Figure A.4.20.3.4(2).

104 Part II • NFPA 20 • Chapter 4 General Requirements

SAMPLE PUMP TEST HEADER SIZE CALCULATION

Pump size		1500	
Number of test hose streams		6	
Size of hose		2½	
Feet of hose per test hose		50	
Nozzle size		1.75	
Nozzle coefficient		0.97	
Pump test header pipe size		8.071	
C factor		120	

	Type Fitting	Number	Equiv. Length	Total Equiv. Length
Pump Test Header Pipe Fittings	45°	1	9	9
	E	1	18	18
	LRE	0	13	0
	T	1	35	35
	BV	0	12	0
	GV	1	4	4
	SW	1	45	45

Pump test header pipe length	30	
Total equivalent length	141	
Maximum test flow	2250	
Friction loss per ft in pipe	0.0392	
Total loss in pump test header pipe		5.5
Flow in each hose	375	
Friction loss in 100 ft of hose	28.125	
Total friction loss in hose		14.1
Equivalent pipe length 2½ in. valve	7	
Friction loss in 2½ in. pipe	0.4561	
Friction loss through 2½ in. valve		3.2
Required pitot pressure		18
Elevation difference		0
Required pump discharge		**40.8**

FIGURE A.4.20.3.4(2) Sample Pump Test Header Calculation.

4.21 Steam Power Supply Dependability

Existing steam-driven fire pump installations are rare, and new installations of steam-driven fire pumps are very rare. However, NFPA 20 still recognizes steam turbine drivers with a reliable source of steam as acceptable fire pump drivers.

4.21.1 Steam Supply.

4.21.1.1 Careful consideration shall be given in each case to the dependability of the steam supply and the steam supply system.

2013 Stationary Fire Pumps Handbook

4.21.1.2 Consideration shall include the possible effect of interruption of transmission piping either on the property or in adjoining buildings that could threaten the property.

4.22 Shop Tests

> Shop test results are used for comparison to acceptance testing results during the field acceptance test (for annual testing requirements, see NFPA 25). See Chapter 14 for more information regarding shop and acceptance tests.

4.22.1 General. Each individual pump shall be tested at the factory to provide detailed performance data and to demonstrate its compliance with specifications.

4.22.2 Preshipment Tests.

4.22.2.1 Before shipment from the factory, each pump shall be hydrostatically tested by the manufacturer for a period of not less than 5 minutes.

4.22.2.2 The test pressure shall not be less than one and one-half times the sum of the pump's shutoff head plus its maximum allowable suction head, but in no case shall it be less than 250 psi (17.24 bar).

4.22.2.3 Pump casings shall be essentially tight at the test pressure.

4.22.2.4 During the test, no objectionable leakage shall occur at any joint.

4.22.2.5 In the case of vertical turbine–type pumps, both the discharge casting and pump bowl assembly shall be tested.

4.23* Pump Shaft Rotation

Pump shaft rotation shall be determined and correctly specified when fire pumps and equipment involving that rotation are ordered.

> The rotation of a fire pump dictates the entire layout of piping and equipment in the fire pump room. The correct rotation must be determined before attempting a piping layout and design, and the rotation should be specified on the fire pump purchase order. It is important to note that not all fire pump manufacturers view rotation from the same perspective. Therefore, the manufacturer's catalog should be consulted for the proper procedure for determining pump rotation. Some diesel engine drives are only available using a single rotation direction.

◀ **DESIGN ALERT**

A.4.23 Pumps are designated as having right-hand, or clockwise (CW), rotation or left-hand, or counterclockwise (CCW), rotation. Diesel engines are commonly stocked and supplied with clockwise rotation.

Pump shaft rotation can be determined as follows:

(1) *Horizontal Pump Shaft Rotation.* The rotation of a horizontal pump can be determined by standing at the driver end and facing the pump. *[See Figure A.4.23(a).]* If the top of the shaft revolves from the left to the right, the rotation is right-handed, or clockwise (CW). If the top of the shaft revolves from right to left, the rotation is left-handed, or counterclockwise (CCW).

(2) *Vertical Pump Shaft Rotation.* The rotation of a vertical pump can be determined by looking down at the top of the pump. If the point of the shaft directly opposite revolves from left to right, the rotation is right-handed, or clockwise (CW). *[See Figure A.4.23(b).]* If the point of the shaft directly opposite revolves from right to left, the rotation is left-handed, or counterclockwise (CCW).

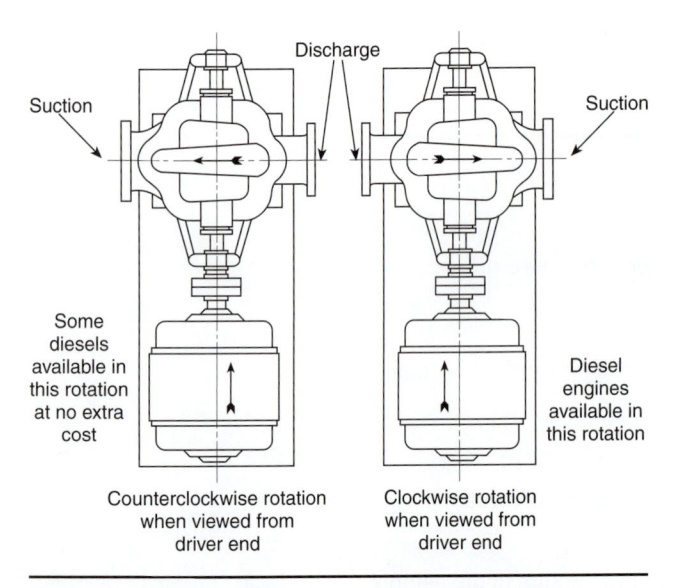

FIGURE A.4.23(a) Horizontal Pump Shaft Rotation.

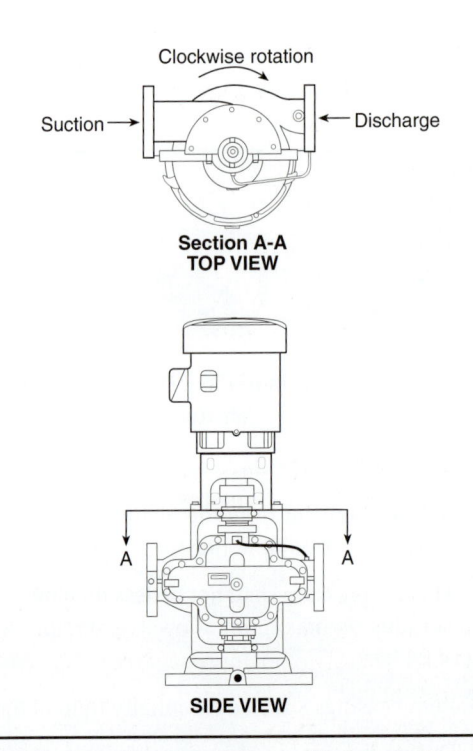

FIGURE A.4.23(b) Vertical Pump Shaft Rotation.

EXHIBIT II.4.23 Pump Shaft Rotation Marking. (Courtesy of Aon Fire Protection Engineering)

See Exhibit II.4.23 for an example of a pump marked with the shaft rotation direction.

4.24* Other Signals

Where required by other sections of this standard, signals shall call attention to improper conditions in the fire pump equipment.

A.4.24 In addition to those conditions that require signals for pump controllers and engines, there are other conditions for which such signals might be recommended, depending upon local conditions. Some of these conditions are as follows:

(1) Low pump room temperature
(2) Relief valve discharge
(3) Flowmeter left on, bypassing the pump
(4) Water level in suction supply below normal
(5) Water level in suction supply near depletion
(6) Steam pressure below normal

Such additional signals can be incorporated into the trouble signals already provided on the controller, or they can be independent.

4.25* Pressure Maintenance (Jockey or Make-Up) Pumps

Depending on factors such as the nature of the water supply or leakage through the system piping, pressure drops or fluctuations can occur within the fire protection system. These pressure fluctuations can cause the fire pump to operate even though the fire protection system has not been activated — that is, a sprinkler did not activate, or a standpipe valve was not opened. It is not the intent to operate the fire pump frequently for short durations to maintain system pressure. Fire pumps are not designed for, nor are they intended for, this purpose. Subsection 4.25.6 prohibits the use of fire pumps as pressure maintenance devices.

The benefits of a pressure maintenance pump go beyond maintaining system pressure. A pressure maintenance pump will provide a higher pressure for the first sprinklers that activate and may improve their effectiveness, thus limiting the total number of sprinklers that activate. The pump will also reduce the likelihood of a pressure surge and thereby improve the effectiveness of waterflow alarms by avoiding false alarms created by pressure surges.

It is important to note that pressure maintenance pumps (also called make-up pumps or jockey pumps) are not required by NFPA 20. NFPA 20 provides requirements for jockey pumps only when they are installed; however, some method of maintaining system pressure, other than through the fire pump(s), should be provided. The pump, as illustrated in Exhibit II.4.24, is the most frequently used method to maintain system pressure. For systems that use the underground piping for both domestic and fire protection purposes, the operating pressure must be coordinated, and the domestic pumps can operate as pressure maintenance pumps.

Since pressure maintenance pumps are not required, this section was revised for the 2010 edition to indicate that valves and other components are not required to be listed. The suction isolation valve no longer needs to be an OS&Y, and isolation valves are not required to be supervised. These changes were made because failure of the pressure maintenance pump will not compromise the fire protection features of the fire pump system. Therefore, the level of protection required of a fire pump is not necessary for a pressure maintenance pump.

▶ **FAQ** Is a pressure maintenance (jockey or make-up pump) required on all fire pump installations?

A fire pump cannot be used to maintain system pressure (see 4.25.6). A nonautomatic starting fire pump may not require a pressure maintenance pump. For a fire pump that starts automatically on a pressure drop, a pressure maintenance pump is required unless a means (other than a fire pump) of maintaining system pressure above the fire pump start pressure is provided. Some combination fire protection/water distribution systems have domestic pumps that operate in parallel with the fire pump and provide domestic water at a pressure above the fire pump start pressure setting.

EXHIBIT II.4.24 *Pressure Maintenance Pump.*

> Although not recommended, it is also possible that the water pressure in the piping to a deluge system supplied by a fire pump may be maintained at the pressure provided by the water department. The large flow that occurs when the deluge system operates drops the residual pressure below the fire pump start pressure setting; alternatively, the fire pump may be started electronically by a signal from the detection system. This scenario is not recommended for the following reasons:
>
> 1. The normal pressure is substantially below the pump discharge pressure, and activating the fire pump against the lower initial pressure will cause a higher pressure surge.
> 2. The residual pressure may not drop sufficiently to activate the fire pump.
> 3. Fluctuations in the water supply pressure could cause the fire pump to activate without the deluge system tripping.

A.4.25 Pressure maintenance (jockey or make-up) pumps should be used where it is desirable to maintain a uniform or relatively high pressure on the fire protection system.

A domestic water pump in a dual-purpose water supply system can function as a means of maintaining pressure.

4.25.1 Pressure maintenance pumps shall not be required to be listed. Pressure maintenance pumps shall be approved.

4.25.1.1* The pressure maintenance pump shall be sized to replenish the fire protection system pressure due to allowable leakage and normal drops in pressure.

A.4.25.1.1 The sizing of the pressure maintenance pump requires a thorough analysis of the type and size of system the pressure maintenance pump will serve. Pressure maintenance pumps on fire protection systems that serve large underground mains need to be larger than pressure maintenance pumps that serve small aboveground fire protection systems. Underground mains are permitted by NFPA 24, *Standard for the Installation of Private Fire Service Mains and Their Appurtenances*, to have some leakage *(see 10.10.2.2.6 of NFPA 24)*, while aboveground piping systems are required to be tight when new and should not have significant leakage.

For situations where the pressure maintenance pump serves only aboveground piping for fire sprinkler and standpipe systems, the pressure maintenance pump should be sized to provide a flow less than a single fire sprinkler. The main fire pump should start and run (providing a pump running signal) for any waterflow situation where a sprinkler has opened, which will not happen if the pressure maintenance pump is too large.

One guideline that has been successfully used to size pressure maintenance pumps is to select a pump that will make up the allowable leakage rate in 10 minutes or 1 gpm (3.8 L/min), whichever is larger.

4.25.2 Pressure maintenance pumps shall have rated capacities not less than any normal leakage rate.

> **AHJ FAQ**
>
> **The submittal for a fire pump installation does not show a listing for the jockey pump. Is a nonlisted pump permitted?**
>
> *ANSWER:* Yes. Pressure maintenance pumps are not a critical component in the functioning of a fire protection system, and, therefore, a failure of such a device does not cause the system to fail. The failure of the pump will eventually cause the fire pump to run and, hopefully, to be identified and corrected at that time. Based on the fact that the pump is not a critical component of the system, it is required only to be approved and is not required to be listed.

◀ **CALCULATION**

> Pressure maintenance pumps, also called jockey pumps, are generally low-flow, high-pressure pumps. For sprinkler systems, jockey pumps are usually sized to flow a quantity of water less than or equal to that required by a single sprinkler. When sized in this manner, if a sprinkler opens on the system, the jockey pump will not be able to keep up with system demand, the pressure will continue to fall, and the fire pump will start, usually after an additional 5 psi (0.3 bar) has been lost to friction following the start of the jockey pump. See Chapter 14, Annex A.14.2.5, for recommendations on starting pressures for jockey pumps and fire pumps. The pressure differential between the pressure maintenance pump and the fire pump should be a minimum of 10 psi (0.7 bar) to avoid false starting of the fire pump due to pressure fluctuations in the system piping.
>
> A general rule of thumb for sizing jockey pumps supplying underground piping has been to use 1 percent of the fire pump rated capacity and add 10 psi (0.7 bar) to the pressure rating of the fire pump. For example, a fire pump with a rated capacity of 1000 gpm at 100 psi (3785 L/min at 6.9 bar) should be provided with a jockey pump of 10 gpm at 110 psi (37.8 L/min at 7.6 bar) rated capacity. An exception to this general rule is when older underground systems leak excessively. In such a case, the jockey pump capacity should be increased further, based on the leakage rate of the underground system.

4.25.3 Pressure maintenance pumps shall have discharge pressure sufficient to maintain the desired fire protection system pressure.

4.25.4* Excess Pressure.

A.4.25.4 A centrifugal-type pressure maintenance pump is preferable.
The following notes apply to a centrifugal-type pressure maintenance pump:

(1) A jockey pump is usually required with automatically controlled pumps.
(2) Jockey pump suction can come from the tank filling supply line. This situation would allow high pressure to be maintained on the fire protection system even when the supply tank is empty for repairs.
(3) Pressure sensing lines also need to be installed in accordance with 10.5.2.1. *[See Figure A.4.30(a) and Figure A.4.30(b).]*

4.25.4.1 Where a centrifugal-type pressure maintenance pump has a total discharge pressure with the pump operating at shutoff exceeding the working pressure rating of the fire protection equipment, or where a turbine vane (peripheral) type of pump is used, a relief valve sized to prevent overpressuring of the system shall be installed on the pump discharge to prevent damage to the fire protection system.

> Unless a pressure relief valve is provided, a positive displacement pressure maintenance pump that fails to shut off for any reason will continue to build up pressure until the motor seizes or something breaks. In general, the pressure maintenance pump is not electrically supervised, and no one will be warned of its continued operation.

4.25.4.2 Running period timers shall not be used where jockey pumps are utilized that have the capability of exceeding the working pressure of the fire protection systems.

> A running period timer could cause the pump to continue to operate after the pressure demand is satisfied, which could cause a significant discharge of water though the pressure relief valve.

4.25.5 Piping and Components for Pressure Maintenance Pumps.

4.25.5.1 Steel pipe shall be used for suction and discharge piping on pressure maintenance pumps, which includes packaged prefabricated systems.

4.25.5.2 Valves and components for the pressure maintenance pump shall not be required to be listed.

4.25.5.3 An isolation valve shall be installed on the suction side of the pressure maintenance pump to isolate the pump for repair.

4.25.5.4 A check valve and isolation valve shall be installed in the discharge pipe.

4.25.5.5* Indicating valves shall be installed in such places as needed to make the pump, check valve, and miscellaneous fittings accessible for repair.

A.4.25.5.5 See Figure A.4.25.5.5.

4.25.5.6 The pressure sensing line for the pressure maintenance pump shall be in accordance with Section 4.30.

4.25.5.7 The isolation valves serving the pressure maintenance pump shall not be required to be supervised.

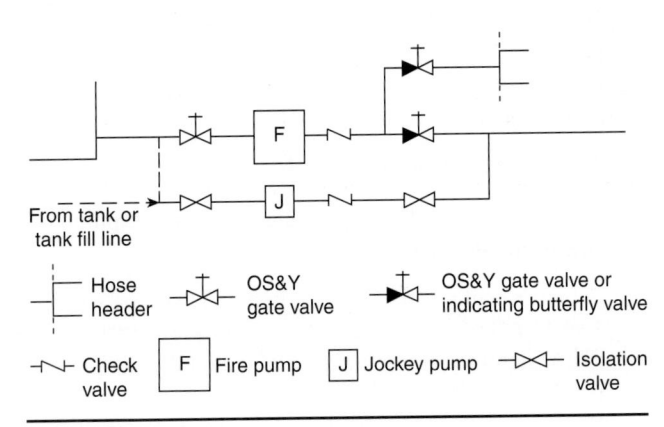

FIGURE A.4.25.5.5 *Jockey Pump Installation with Fire Pump.*

4.25.6 Except as permitted in Chapter 8, the primary or standby fire pump shall not be used as a pressure maintenance pump.

4.25.7 The controller for a pressure maintenance pump shall be listed but shall not be required to be listed for fire pump service.

4.25.8 The pressure maintenance pump shall not be required to have secondary or standby power.

4.26 Summary of Centrifugal Fire Pump Data

The size indicated in Table 4.26(a) and Table 4.26(b) shall be used as a minimum.

TABLE 4.26(a) Summary of Centrifugal Fire Pump Data (U.S. Customary)

Pump Rating (gpm)	Suction[a,b,c]	Discharge[a]	Relief Valve	Relief Valve Discharge	Meter Device	Number and Size of Hose Valves	Hose Header Supply
25	1	1	¾	1	1¼	1 — 1½	1
50	1½	1¼	1¼	1½	2	1 — 1½	1½
100	2	2	1½	2	2½	1 — 2½	2½
150	2½	2½	2	2½	3	1 — 2½	2½
200	3	3	2	2½	3	1 — 2½	2½
250	3½	3	2	2½	3½	1 — 2½	3
300	4	4	2½	3½	3½	1 — 2½	3
400	4	4	3	5	4	2 — 2½	4
450	5	5	3	5	4	2 — 2½	4
500	5	5	3	5	5	2 — 2½	4
750	6	6	4	6	5	3 — 2½	6
1000	8	6	4	8	6	4 — 2½	6
1250	8	8	6	8	6	6 — 2½	8
1500	8	8	6	8	8	6 — 2½	8
2000	10	10	6	10	8	6 — 2½	8
2500	10	10	6	10	8	8 — 2½	10
3000	12	12	8	12	8	12 — 2½	10
3500	12	12	8	12	10	12 — 2½	12
4000	14	12	8	14	10	16 — 2½	12
4500	16	14	8	14	10	16 — 2½	12
5000	16	14	8	14	10	20 — 2½	12

Notes:
(1) The pressure relief valve shall be permitted to be sized in accordance with 4.18.2.1.
(2) The pressure relief valve discharge shall be permitted to be sized in accordance with 4.18.6.2.
(3) The flowmeter device shall be permitted to be sized in accordance with 4.19.2.2.
(4) The hose header supply shall be permitted to be sized in accordance with 4.19.3.4.
[a]Actual diameter of pump flange is permitted to be different from pipe diameter.
[b]Applies only to that portion of suction pipe specified in 4.14.3.4.
[c]Suction pipe sizes in Table 4.26(a) and Table 4.26(b) are based on a maximum velocity at 150 percent rated capacity to 15 ft/sec (4.6 m/sec) in most cases.

TABLE 4.26(b) *Summary of Centrifugal Fire Pump Data (Metric)*

Pump Rating (L/min)	Suction[a,b,c]	Discharge[a]	Relief Valve	Relief Valve Discharge	Meter Device	Number and Size of Hose Valves	Hose Header Supply
95	25	25	19	25	32	1 — 38	25
189	38	32	32	38	50	1 — 38	38
379	50	50	38	50	65	1 — 65	65
568	65	65	50	65	75	1 — 65	65
757	75	75	50	65	75	1 — 65	65
946	85	75	50	65	85	1 — 65	75
1,136	100	100	65	85	85	1 — 65	75
1,514	100	100	75	125	100	2 — 65	100
1,703	125	125	75	125	100	2 — 65	100
1,892	125	125	75	125	125	2 — 65	100
2,839	150	150	100	150	125	3 — 65	150
3,785	200	150	100	200	150	4 — 65	150
4,731	200	200	150	200	150	6 — 65	200
5,677	200	200	150	200	200	6 — 65	200
7,570	250	250	150	250	200	6 — 65	200
9,462	250	250	150	250	200	8 — 65	250
11,355	300	300	200	300	200	12 — 65	250
13,247	300	300	200	300	250	12 — 65	300
15,140	350	300	200	350	250	16 — 65	300
17,032	400	350	200	350	250	16 — 65	300
18,925	400	350	200	350	250	20 — 65	300

Notes:
(1) The pressure relief valve shall be permitted to be sized in accordance with 4.18.2.1.
(2) The pressure relief valve discharge shall be permitted to be sized in accordance with 4.18.6.2.
(3) The flow meter device shall be permitted to be sized in accordance with 4.19.2.2.
(4) The hose header supply shall be permitted to be sized in accordance with 4.19.3.4.

[a]Actual diameter of pump flange is permitted to be different from pipe diameter.
[b]Applies only to that portion of suction pipe specified in 4.14.3.4.
[c]Suction pipe sizes in Table 4.26(a) and Table 4.26(b) are based on a maximum velocity at 150 percent rated capacity to 15 ft/sec (4.6 m/sec) in most cases.

4.27 Backflow Preventers and Check Valves

4.27.1 Check valves and backflow prevention devices and assemblies shall be listed for fire protection service.

4.27.2 Relief Valve Drainage.

4.27.2.1 Where the backflow prevention device or assembly incorporates a relief valve, the relief valve shall discharge to a drain appropriately sized for the maximum anticipated flow from the relief valve.

> Relief valves on reduced pressure backflow preventers (shown in Exhibit II.4.25) can discharge a large volume of water. The drain for these devices needs to be adequately sized to handle this flow. When reduced pressure backflow preventers are used, the pressure maintenance pump sensing line connection should be made on the supply side of the reduced pressure backflow preventers. If the pressure maintenance pump is permitted to pressurize the system side of a reduced pressure backflow preventer, the pressure relief valve remains open, continually discharging water into the drain.
>
> The air gap on a reduced pressure backflow preventer is necessary for two reasons. First, without an air gap, the backflow preventer becomes a potential cross connection that could contaminate the water supply from the drain. Second, the drain sometimes leads to an unheated area, in which case the cold temperatures could freeze the inside of the backflow preventer if it is not separated by an air gap.

EXHIBIT II.4.25 *Reduced Pressure Backflow Preventer. (Courtesy of Aon Fire Protection Engineering)*

4.27.2.2 An air gap shall be provided in accordance with the manufacturer's recommendations.

4.27.2.3 Water discharge from the relief valve shall be readily visible or easily detectable.

4.27.2.4 Performance of the requirements in 4.27.2.1 through 4.27.2.3 shall be documented by engineering calculations and tests.

4.27.3 Devices in Suction Piping. Where located in the suction pipe of the pump, check valves and backflow prevention devices or assemblies shall be located a minimum of 10 pipe diameters from the pump suction flange.

> Installation of a backflow preventer on the suction side of a fire pump should be avoided if possible, because the water discharging from a backflow prevention device is extremely turbulent. Before this water enters the fire pump, turbulent flow must be minimized. The distance of 10 pipe diameters is considered acceptable in the water control industries for performing this task. Exhibit II.4.26 illustrates a double check backflow preventer assembly that should have been located 10 pipe diameters from the fire pump suction flange.

EXHIBIT II.4.26 Double Check Valve Assembly Installed on Suction Side of Fire Pump. (Courtesy of Aon Fire Protection Engineering)

4.27.3.1 Where a backflow preventer with butterfly control valves is installed in the suction pipe, the backflow preventer is required to be at least 50 ft (15.2 m) from the pump suction flange (as measured along the route of pipe) in accordance with 4.14.5.2.

4.27.4 Evaluation.

4.27.4.1 Where the authority having jurisdiction requires the installation of a backflow prevention device or assembly in connection with the pump, special consideration shall be given to the increased pressure loss resulting from the installation.

> Water must enter the suction flange of the pump at a positive pressure. See the commentary following 4.14.3.1 for more information on this subject. The friction loss from the backflow prevention device and two accompanying isolation OS&Y gate valves is significant and must be included in the system calculations.

4.27.4.2 Where a backflow prevention device is installed, the final arrangement shall provide effective pump performance with a minimum pump suction pressure of 0 psi (0 bar) at the gauge at 150 percent of rated capacity.

4.27.4.3 If available suction supplies do not permit the flowing of 150 percent of rated pump capacity, the final arrangement of the backflow prevention device shall provide effective pump performance with a minimum suction pressure of 0 psi (0 bar) at the gauge at the maximum allowable discharge.

4.27.4.4 The discharge shall exceed the fire protection system design flow.

4.27.4.5 Determination of effective pump performance shall be documented by engineering calculations and tests.

4.28 Earthquake Protection

> The potential for seismic activity is higher than previously thought for most areas of the United States and for a significant portion of the world. Fire protection systems, including fire pump installations, need to be designed so that they can function during and following a seismic event, because post-earthquake fires are relatively common.

> In general, fire protection systems are designed to handle horizontal forces equal to one-half of their weight (0.5 g, where g is force of gravity). This design covers most of the areas where earthquakes are an infrequent, but still possible, occurrence. In some areas, the building code allows systems to be designed to withstand horizontal forces less than one-half the weight of the component. In these situations, the lower value specified by the building code is acceptable. For more active and intense earthquake areas, fire protection systems need to be designed to withstand larger horizontal and vertical forces. In such cases, the design factors will be provided in the building code. NFPA 13 should be used for guidance related to seismic support. The building codes specify very strict earthquake design parameters for vitally necessary buildings such as hospitals, fire stations, schools, and other similar occupancies.

4.28.1 General.

4.28.1.1 Where local codes require fire protection systems to be protected from damage subject to earthquakes, 4.28.2 and 4.28.3 shall apply.

4.28.1.2* Horizontal seismic loads shall be based on NFPA 13, *Standard for the Installation of Sprinkler Systems*, SEI/ASCE 7, *Minimum Design Loads for Buildings and Other Structures*, local, state, or international codes, or other sources acceptable to the authority having jurisdiction.

A.4.28.1.2 NFPA 13, *Standard for the Installation of Sprinkler Systems*, contains specific requirements for seismic design of fire protection systems. It is a simplified approach that was developed to coincide with SEI/ASCE 7, *Minimum Design Loads for Buildings and Other Structures*, and current building codes.

4.28.2 Pump Driver and Controller. The fire pump, driver, diesel fuel tank (where installed), and fire pump controller shall be attached to their foundations with materials capable of resisting lateral movement from horizontal loads.

4.28.2.1 Pumps with high centers of gravity, such as vertical in-line pumps, shall be mounted at their base and braced above their center of gravity.

4.28.3 Piping and Fittings.

4.28.3.1 Pipe and fittings shall be protected in accordance with NFPA 13, *Standard for the Installation of Sprinkler Systems*.

4.28.3.2 Where the system riser is also a part of the fire pump discharge piping, a flexible pipe coupling shall be installed at the base of the system riser.

4.29 Packaged Fire Pump Assemblies

> See Exhibit II.4.27 for an illustration of a packaged fire pump system.

4.29.1 A packaged fire pump assembly, with or without an enclosure, shall meet all of the following requirements:

(1) The components shall be assembled and affixed onto a steel framing structure.
(2) Welders shall be qualified in accordance with the Section 9 of ASME *Boiler and Pressure Vessel Code* or with the American Welding Society AWS D1.1, *Structural Welding Code — Steel*.
(3) The assembly shall be listed for fire pump service.

EXHIBIT II.4.27 *Packaged Fire Pump System. (Courtesy of Aon Fire Protection Engineering)*

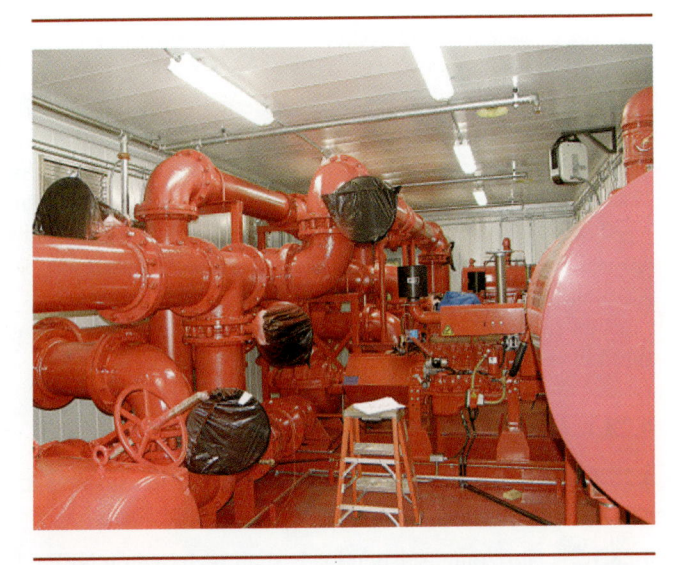

(4) The total assembly shall be engineered and designed by a system designer as referenced in 4.3.2.
(5) All plans and data sheets shall be submitted and reviewed by the authority having jurisdiction, with copies of the stamped approved submittals used in the assembly and for record keeping.

4.29.2 All electrical components, clearances, and wiring shall meet the minimum requirements of the applicable *NFPA 70, National Electrical Code*, articles.

4.29.3 Packaged and prefabricated skid unit(s) shall meet all the requirements in this standard, including those described in Sections 4.12 through 4.17.

4.29.4 Careful consideration shall be given to the possible effects of system component damage during shipment to the project site.

4.29.4.1 The structural integrity shall be maintained with minimal flexing and movement.

4.29.4.2 The necessary supports and restraints shall be installed to prevent damage and breakage during transit.

4.29.5 The packaged fire pump shall have the correct lifting points marked to ensure safe rigging of the unit.

4.29.6 All packaged pump house or pump skids, or both, shall meet the requirements of Section 4.28.

4.29.7 Suction and discharge piping shall be thoroughly inspected, including checking all flanged and mechanical connections per manufacturers' recommendations, after the pump house or skid unit is set in place on the permanent foundation.

4.29.8 The units shall be properly anchored and grouted in accordance with Section 6.4.

4.29.9* The interior floor shall be solid with grading to provide for proper drainage for the fire pump components.

A.4.29.9 Figure A.4.29.9 illustrates a typical foundation detail for a packaged fire pump assembly.

4.29.9.1 The structural frame for a packaged pump shall be mounted to an engineered footing designed to withstand the live loads of the packaged unit and the applicable wind loading requirements.

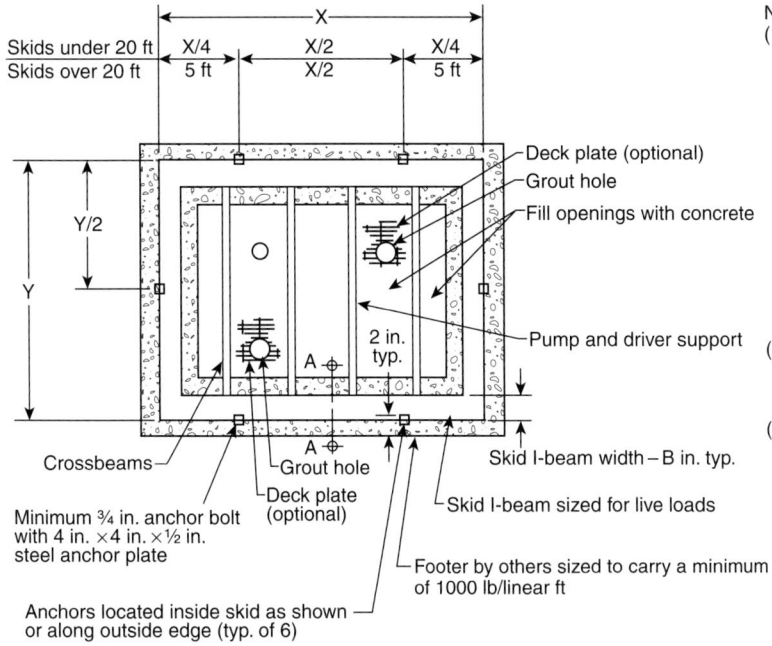

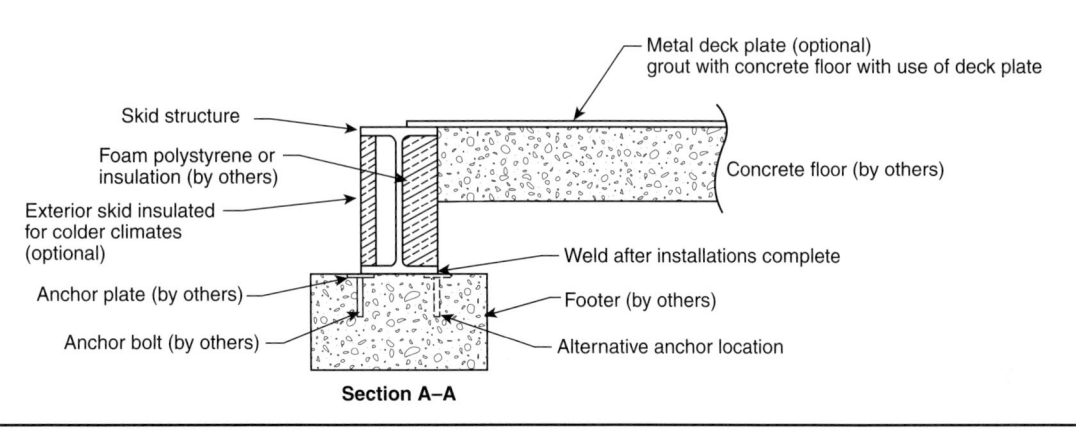

FIGURE A.4.29.9 Typical Foundation Detail for Packaged Fire Pump Assembly.

4.29.9.2 The foundation footings shall include the necessary anchor points required to secure the package to the foundation.

4.29.10 A high skid-resistant, solid structural plate floor with grout holes shall be permitted to be used where protected from corrosion and drainage is provided for all incidental pump room spillage or leakage.

4.30* Pressure Actuated Controller Pressure Sensing Lines

> The pressure sensing line specified in Section 4.30 is intended to control the starting of the pump motor and the running cycle of the fire pump. The sensing line specified in Section 4.30 is in addition to the pressure sensing lines required by 10.5.2.1.7.5 and 11.2.4.3.4. The pressure sensing lines required by 10.5.2.1.7.5 and 11.2.4.3.4 are intended to control the variable speed pressure limiting control function of the fire pump controller.

A.4.30 See Figure A.4.30(a) and Figure A.4.30(b).

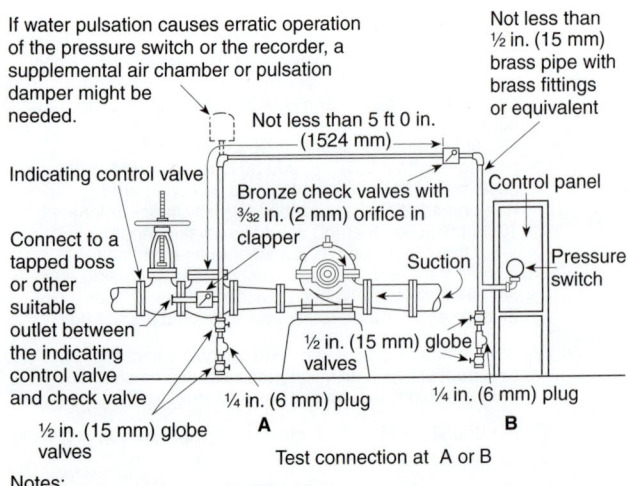

FIGURE A.4.30(a) *Piping Connection for Each Automatic Pressure Switch (for Electric-Driven and Diesel Fire Pump and Jockey Pumps).*

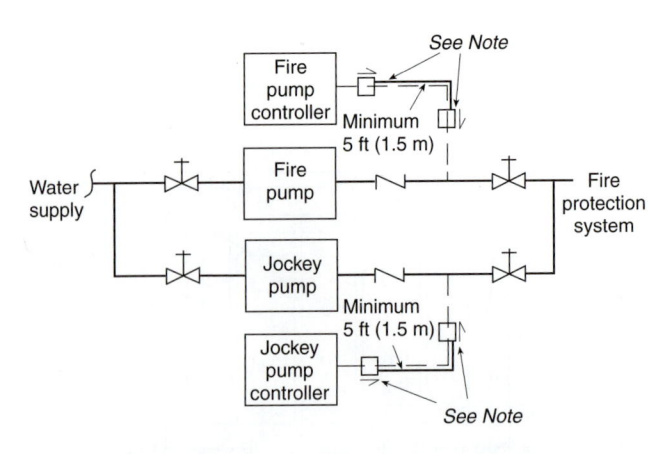

FIGURE A.4.30(b) *Piping Connection for Pressure Sensing Line (Diesel Fire Pump).*

Note: Check valves or ground-face unions complying with 10.5.2.1.

Notes:
(1) Solenoid drain valve used for engine-driven fire pumps can be at A, B, or inside controller enclosure.
(2) If water is clean, ground-face unions with noncorrosive diaphragms drilled for 3/32 in. orifices can be used in place of the check valves.

4.30.1 For all pump installations, including jockey pumps, each controller shall have its own individual pressure sensing line.

> ▶ **FAQ** Can the sensing line for a fire pump and pressure maintenance pump be combined and installed as a single pipe?

> Each pump controller, including the pressure maintenance pump controller, must have its own dedicated sensing line. Except for water mist positive displacement pumping units, each pump, including the pressure maintenance pump in multiple pump installations, must also be provided with a separate dedicated sensing line. Providing separate dedicated pressure sensing lines adds reliability to the starting and control of the pumps. For example, if one line becomes obstructed, the other line is not affected. See Exhibit II.4.28 for an illustration of pressure sensing lines for a diesel-driven fire pump that are combined and, therefore, not compliant with 4.30.1.

4.30.2 The pressure sensing line connection for each pump, including jockey pumps, shall be made between that pump's discharge check valve and discharge isolation valve.

> Subsection 4.30.5 does not allow a shutoff valve in the pressure sensing line. Subsection 4.30.2 requires the pressure sensing line to be connected between the fire pump discharge check valve and the pump side of the discharge isolation valve. The following are the two reasons for this requirement:
>
> 1. This arrangement allows the fire pump to be tested and the pressure settings to be verified with the fire protection system isolated.
> 2. This arrangement allows servicing of the pressure sensing line without draining the entire fire protection system.

EXHIBIT II.4.28 *Combined Sensing Lines.*

4.30.3* The pressure sensing line shall be brass, rigid copper pipe Types K, L, or M, or Series 300 stainless steel pipe or tube, and the fittings shall be of ½ in. (15 mm) nominal size.

A.4.30.3 The use of soft copper tubing is not permitted for a pressure sensing line because it is easily damaged.

4.30.4 Check Valves or Ground-Face Unions.

4.30.4.1 Where the requirements of 4.30.4.2 are not met, there shall be two check valves installed in the pressure sensing line at least 5 ft (1.52 m) apart with a nominal 0.09375 in. (2.4 mm) hole drilled in the clapper to serve as dampening. *[See Figure A.4.30(a) and Figure A.4.30(b).]*

> The check valves referred to in 4.30.4.1 are intended to absorb the pressure surge created when the fire pump starts. Without this feature, the pressure switch inside the fire pump controller could be damaged. The check valves should be installed as shown in Figure A.4.30(a) to restrict flow from the pump to the controller but should open upon flow from the controller to the pump.

4.30.4.2 Where the water is clean, ground-face unions with noncorrosive diaphragms drilled with a nominal 0.09375 in. (2.4 mm) orifice shall be permitted in place of the check valves.

4.30.4.3 There shall be two inspection test valves attached to the pressure sensing line that shall consist of a tee, a valve, a second tee with the branch plugged, and a second valve. *[See Figure A.4.30(a) and Figure A.4.30(b).]*

4.30.5 Shutoff Valve. There shall be no shutoff valve in the pressure sensing line.

4.30.6 Pressure Switch Actuation. Pressure switch actuation at the low adjustment setting shall initiate the pump starting sequence (if the pump is not already in operation).

4.31 Break Tanks

Where a break tank is used to provide the pump suction water supply, the installation shall comply with Section 4.31.

DESIGN ALERT ▶

> As defined in 3.3.6, a *break tank* is not intended to provide a storage capacity equal to the entire system demand (required flow rate times duration) as specified by a fire protection system installation standard such as NFPA 13 or NFPA 14. The break tank is also not intended to be a pressurized source, except for the initial pressure that is developed by the depth of water in the tank. A break tank must be equipped with an automatic fill connection to a source of water that is considered reliable in normal circumstances. If a tank is required because the water source is not sufficiently reliable, or a tank is required by the applicable codes and standards, the tank should be sized for the entire system demand. If the water source is reliable but cannot supply the required flow rate, a break tank can be used. For example, in an area subject to earthquakes, a city supply would not be considered reliable, and the tank would be sized for the required total facility water supply. Using a break tank to augment a reliable water source that is inadequate in volume allows a smaller tank. A break tank can also be used as a backflow prevention device by providing an air gap between the automatic fill valve and the highest water level.
>
> The fire pump technical committee added information on the use of break tanks for the 2007 edition of NFPA 20, because guidance was needed on the design and testing of this type of water tank, which does not have the capacity to meet the full fire protection system demand. The tank design, construction, and installation must comply with the requirements of NFPA 22. The following are the primary purposes for using break tanks:
>
> 1. To augment a reliable water source that cannot supply the required flow rate
> 2. To address situations where the municipality or water purveyor prohibits the direct connection of a fire pump to the water main
> 3. To serve in situations where the pressure fluctuation causes overpressurization of the fire protection system
>
> See Exhibit II.4.29 for an altitude-type (pressure actuated) automatic fill valve and Exhibit II.4.30 for a typical break tank with an automatic fill valve and bypass.

4.31.1 Application. Break tanks shall be used for one or more of the following reasons:

(1) As a backflow prevention device between the water supply and the fire pump suction pipe

> When acting only as a backflow device, the tank will typically be the minimum size allowed and will require two automatic fill valves and lines, each sized to match the maximum flow through the fire pump (150 percent of pump rated capacity). The second fill valve and line is a backup and does not have to operate simultaneously with the first fill line.

(2) To eliminate fluctuations in the water supply pressure and provide a steady suction pressure to the fire pump

> When used to level out pressure fluctuations, the break tank will typically be the minimum size allowed and will require two automatic fill valves and lines, each sized to match the maximum flow through the fire pump (150 percent of pump rated capacity). The second fill valve and line is a backup and does not have to operate simultaneously with the first fill line.

Section 4.31 • Break Tanks **121**

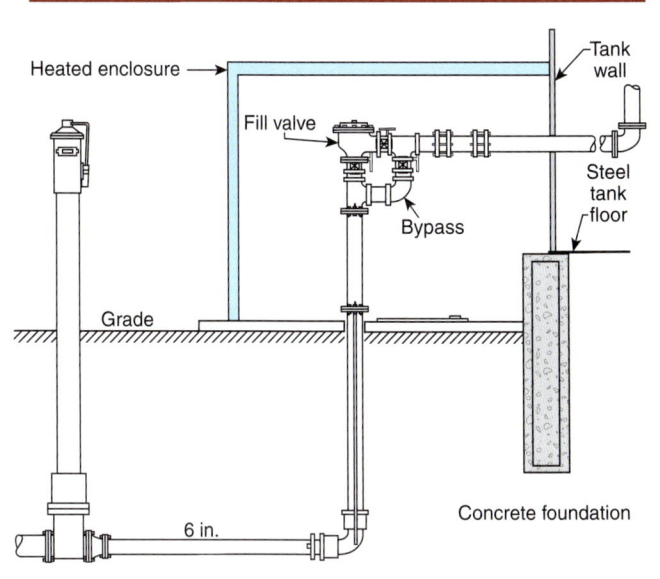

EXHIBIT II.4.30 Break Tank Installation with Automatic Fill Valve and Bypass. (Courtesy of Aon Fire Protection Engineering)

EXHIBIT II.4.29 Altitude-Type (Pressure Actuated) Automatic Fill Valve. (Courtesy of Aon Fire Protection Engineering)

(3) To provide a quantity of stored water on site where the normal water supply will not provide the required quantity of water required by the fire protection system

> When augmenting a reliable water supply, the tank will typically be sized to make up the difference between the available flow rate and the system demand. As an example, if the system demand is 1000 gpm (3785 L/min) for 90 minutes and the water supply provides 500 gpm (1892 L/min) with a 10 percent safety factor, the tank would be sized for [1000 gpm (3785 L/min) − 500 gpm (1892 L/min) × (90 minutes)], or 45,000 gal (170,344 L). The tank would then hold a 45-minute water supply and be required to have one automatic fill line, sized for a flow rate of 500 gpm (1892 L/min).

4.31.2 Break Tank Size. The tank shall be sized for a minimum duration of 15 minutes with the fire pump operating at 150 percent of rated capacity.

> A redundant automatic fill line is required to ensure that the tank does not run out of water before the fire department can respond and verify the operation of the tank fill valve. It is

CALCULATION ▶

anticipated that the fire department will respond and review the tank and tank fill within 30 minutes. An example of sizing a tank for 15 minutes under the requirements of 4.31.2 is the following:

A 1000 gpm (3785 L/min) fire pump flows 1500 gpm (5678 L/min) at 150 percent of rated demand. A 15-minute supply, therefore, equals 15 min × 1500 gpm (5678 L/min) or 22,500 gal (85,163 L).

This requirement is based on the fire pump rating, not the fire protection demand.

▶ FAQ Does NFPA 20 specify the flow rate of an automatic fill connection for all water storage tanks?

The tank size and refill rate is based on the NFPA 20 requirement that the stored supply plus automatic refill must meet the fire protection system demand. Note that the intent of NFPA 20 is to establish the requirements for the application, sizing, and refill of break tanks only. For capacity and filling requirements for water storage tanks for private fire protection systems, see NFPA 22.

4.31.3 Refill Mechanism. The refill mechanism shall be listed and arranged for automatic operation.

4.31.3.1 If the break tank capacity is less than the maximum system demand for 30 minutes, the refill mechanism shall meet the requirements in 4.31.3.1.1 through 4.31.3.1.5.

The requirement in 4.31.3.1 is intended to provide the fire department with time to respond and verify the conditions.

4.31.3.1.1 Dual automatic refill lines, each capable of refilling the tank at a minimum rate of 150 percent of the fire pump(s) capacity, shall be installed.

The requirement for a redundant automatic tank fill is intended to provide the fire department with time to respond and check the fill valve operation.

4.31.3.1.2 If available supplies do not permit refilling the tank at a minimum rate of 150 percent of the rated pump capacity, each refill line shall be capable of refilling the tank at a rate that meets or exceeds 110 percent of the maximum fire protection system design flow.

4.31.3.1.3 A manual tank fill bypass designed for and capable of refilling the tank at a minimum rate of 150 percent of the fire pump(s) capacity shall be provided.

4.31.3.1.4 If available supplies do not permit refilling the tank at a minimum rate of 150 percent of the rated pump capacity, the manual fill bypass shall be capable of refilling the tank at a rate that meets or exceeds 110 percent of the maximum fire protection system design flow.

4.31.3.1.5 A local visible and audible low liquid level signal shall be provided in the vicinity of the tank fill mechanism.

In addition to the local audible and visual signal, it is recommended that a supervisory signal be transmitted to a constantly attended location.

4.31.3.2 If the break tank is sized to provide a minimum duration of 30 minutes of the maximum system demand, the refill mechanism shall meet the requirements in 4.31.3.2.1 and 4.31.3.2.2.

4.31.3.2.1 The refill mechanism shall be designed for and capable of refilling the tank at 110 percent of the rate required to provide the total fire protection system demand [110% × (total demand − tank capacity) / duration].

> The 110 percent fill rate is intended to provide some allowance for future water supply degradation. As an example, a system demand of 1000 gpm (3785 L/min) for 90 minutes would require 90,000 gal (340,687 L) of water. If the tank is sized for 45,000 gal (170,344 L), the water supply must provide the following:
>
> 90,000 gal (340,687 L) − 45,000 gal (170,344 L) tank or 45,000 gal (170,344 L) in 90 minutes
>
> 45,000 gal (170,344 L) / 90 min = 500 gpm (1892 L/min)
>
> Applying the 10 percent safety factor to 500 gpm (1892 L/min) requires a tank refill rate of 550 gpm (2082 L/min).

4.31.3.2.2 A manual tank fill bypass shall be designed for and capable of refilling the tank at 110 percent of the rate required to provide the total fire protection system demand [110% × (total demand − tank capacity) / duration].

4.31.3.3 The pipe between the municipal connection and the automatic fill valve shall be installed in accordance with NFPA 24, *Standard for the Installation of Private Fire Service Mains and Their Appurtenances*.

4.31.3.4 The automatic filling mechanism shall be maintained at a minimum temperature of 40°F (4.4°C).

4.31.3.5 The automatic filling mechanism shall activate a maximum of 6 in. (152 mm) below the overflow level.

4.31.4 Installation Standard. The break tank shall be installed in accordance with NFPA 22, *Standard for Water Tanks for Private Fire Protection*.

4.32 Field Acceptance Test of Pump Units

Upon completion of the entire fire pump installation, an acceptance test shall be conducted in accordance with the provisions of this standard. *(See Chapter 14.)*

References Cited in Commentary

National Fire Protection Association, 1 Batterymarch Park, Quincy, MA 02169-7471.

NFPA 13, *Standard for the Installation of Sprinkler Systems*, 2013 edition.
NFPA 14, *Standard for the Installation of Standpipe and Hose Systems*, 2010 edition.
NFPA 22, *Standard for Water Tanks for Private Fire Protection*, 2008 edition.
NFPA 25, *Standard for the Inspection, Testing, and Maintenance of Water-Based Fire Protection Systems*, 2011 edition.

American Society of Civil Engineers, 1801 Alexander Bell Drive, Reston, VA 20191-4400.

ASCE/SEI 7, *Minimum Design Loads for Buildings and Other Structures*, 2010.

American Society of Mechanical Engineers, Three Park Avenue, New York, NY 10016-5990.

ANSI/ASME B16.1, *Gray Iron Pipe Flanges and Flanged Fittings, Classes 25, 125, and 250*, 2010.
ANSI/ASME B16.3, *Malleable Iron Threaded Fittings, Classes 150 and 300*, 2011.
ANSI/ASME B16.4, *Gray Iron Threaded Fittings, Classes 125 and 250*, 2011.

ANSI/ASME B16.5, *Pipe Flanges and Flanged Fittings,* NPS 1/2 through NPS 24, 2009.
ANSI/ASME B16.9, *Factory-Made Wrought Buttwelding Fittings,* 2007.
ANSI/ASME B16.11, *Forged Fittings, Socket-Welding and Threaded,* 2011.
ANSI/ASME B16.18, *Cast Copper Alloy Solder Joint Pressure Fittings,* 2012.
ANSI/ASME B16.22, *Wrought Copper and Copper Alloy Solder Joint Pressure Fittings,* 2001 (Revised 2005).
ANSI/ASME B16.25, *Buttwelding Ends,* 2007.
ANSI/ASME B36.10M, *Welded and Seamless Wrought Steel Pipe,* 2004 (Revised 2010).

ASTM International, 100 Barr Harbor Drive, P.O. Box C700, West Conshohocken, PA 19428-2959.

ANSI/ASTM A 53, *Specification for Pipe, Steel, Black and Hot-Dipped, Zinc-Coated, Welded and Seamless,* 2012.
ASTM A 135, *Specification for Electric-Resistance-Welded Steel Pipe,* 2009.
ASTM A 234, *Specification for Piping Fittings of Wrought Carbon Steel and Alloy Steel for Moderate and High Temperature Service,* 2011a.
ASTM A 795, *Specification for Black and Hot-Dipped Zinc-Coated (Galvanized) Welded and Seamless Steel Pipe for Fire Protection Use,* 2008.
ASTM B 32, *Specification for Solder Metal,* 2008.
ASTM B 75, *Specification for Seamless Copper Tube,* 2011.
ASTM B 88, *Specification for Seamless Copper Water Tube,* 2009.
ASTM B 251, Specification for General Requirements for Wrought Seamless Copper and Copper-Alloy Tube, 2010.
ASTM B 446, *Specification for Nickel-Chromium-Molybdenum-Columbium Alloy (UNS N06625), Nickel-Chromium-Molybdenum-Silicon Alloy (UNS N06219), and Nickel-Chromium-Molybdenum-Tungsten Alloy (UNS N06650)*, Red and Bar,* 2003 (2008, e1).
ASTM B 813, *Specification for Liquid and Paste Fluxes for Soldering of Copper and Copper Alloy Tube,* 2010.

American Welding Society, 550 NW LeJeune Road, Miami, FL 33126.

AWS A5.8, *Specification for Filler Metals for Brazing and Braze Welding,* 2004.

American Water Works Association, 6666 West Quincy Avenue, Denver, CO 80235.

AWWA M14, *Recommended Practice for Backflow Prevention and Cross-Connection Control,* 2004.

CHAPTER 5

Fire Pumps for High-Rise Buildings

Chapter 5 was added in the 2010 edition of NFPA 20, *Standard for the Installation of Stationary Pumps for Fire Protection,* to delineate special requirements for fire pumps in high-rise buildings. A special chapter on high-rise buildings was removed from NFPA 20 in 1996 after a review revealed that only one requirement in the chapter was unique to high-rise buildings — standby power was required whenever a fire pump served a portion of a building that was beyond the pumping capacity of the fire department. This requirement was moved to another chapter, and the high-rise chapter was eliminated.

Sections 5.1 through 5.4 apply to high-rise buildings as defined in 3.3.24. Sections 5.5 and 5.6 are additional requirements that apply when the building height exceeds the pumping capacity of the fire department apparatus.

▶ **FAQ** Why was a chapter on fire pumps in high-rise buildings, which was removed from earlier editions of NFPA 20, reinstated in the 2010 edition?

The chapter was reinstated in the 2010 edition because some of the most complex fire pump installations exist in high-rise buildings. The chapter does not, however, provide a complete guideline for fire protection design in high-rise buildings. Tanks, standpipes, and fire pumps are highly interrelated and, therefore, must be coordinated and require the use of multiple standards.

The technical committee for NFPA 20 understands that the high-rise chapter has significant implications for high-rise fire protection and may also impact cost and some current design practices. The requirements were developed after reviewing current code requirements, current design practices, fire-fighting operations, maintenance implications, and overall reliability and risk exposure.

NFPA 20 is a standard of minimum requirements. The requirements of NFPA 20 attempt to create a balance between installation cost and risk. Typically, high-value and high-risk facilities are built with higher performance requirements than are mandated in NFPA 20. It is also appropriate within minimum requirements standards such as NFPA 20 to provide for higher performance requirements, especially whenever the known life safety risk is high.

The following characteristics of high-rise buildings are the basis of the requirements in this chapter:

1. High-rise buildings require some of the most complex fire pump installations.
   a. Most series fire pump arrangements are found in high-rise buildings.
   b. High-rise buildings may contain multiple vertical zone standpipe systems, tanks, and automatic refill valves.
2. More than in any other occupancy type, high-rise occupants are dependent on the building fire pump(s) to function reliably during a fire. High-rise evacuation plans, which can incorporate "refuge floors," depend on the building's automatic and manual fire-fighting systems to be fully operational.
3. High-rise buildings are high-value buildings.
4. Direct access to the fire pump(s) is needed during fire-fighting operations.
5. The occupancy fire load can be significant in many high-rise buildings.

6. For buildings that exceed 300 ft (91 m) in height, fire-fighting and fire department operations in most jurisdictions are totally reliant on fire pumps within the building.
7. For buildings that exceed 500 ft (152.4 m) in height, it can be impractical to evacuate the occupants, in which case the occupants must be "protected in place."

▶ **FAQ** What is the intent of this chapter?

The requirements in this chapter are intended to provide a highly reliable fire pump installation as part of the fire protection system(s). For high-rise buildings with floors above the pumping capacity of the fire department, the fire pump and water supply arrangement requirements are intended to be such that, even with any single piece of equipment impaired, the full fire protection demand can be met.

5.1 General

5.1.1 Application.

5.1.1.1 This chapter applies to all fire pumps within a building wherever a building is defined as high-rise in accordance with 3.3.24.

5.1.1.2 The provisions of all other chapters of this standard shall apply unless specifically addressed by this chapter.

5.2* Equipment Access

Location and access to the fire pump room shall be preplanned with the fire department.

In general, preplanning and coordination with the fire service are critical to successfully fighting a fire in a high-rise building. The fire service should have specific knowledge of the location, type, pressures, and operation of the building fire pumps.

A.5.2 The location of a pump room in a high-rise building requires a great deal of consideration. Personnel are required to be sent to the pump room to monitor the operation of the pump during fire-fighting activities in the building. The best way to protect these people who are being sent to the pump room is to have the pump room directly accessible from the outside, but that is not always possible in high-rise buildings. In many cases, pump rooms in high-rise buildings will need to be located many floors above grade or at a location below grade, or both.

For the circumstance where the pump room is not at grade level, this standard requires protected passageways of a fire resistance rating that meets the minimum requirements for exit stairwells at the level of the pump room from the exit stairwell to the pump room. Many codes do not allow the pump room to open directly onto an exit stairwell, but the distance between an exit stairwell and the pump room on upper or lower floors needs to be as short as possible with as few openings to other building areas as possible to afford as much protection as possible to the people going to the pump room and staying in the pump room during a fire in the building.

In addition to the need for access, pump rooms need to be located such that discharge from pump equipment including packing discharge and relief valves is adequately handled.

5.3 Water Supply Tanks

5.3.1 Where provided, water tanks shall be installed in accordance with NFPA 22, *Standard for Water Tanks for Private Fire Protection*.

5.3.2 When a water tank serves domestic and fire protection systems, the domestic supply connection shall be connected above the level required for fire protection demand.

> Connecting the domestic water supply above the level required for fire protection demand prevents domestic water usage from depleting the fire protection water reserve.

5.4 Fire Pump Test Arrangement

Where the water supply to a fire pump is a tank, a listed flowmeter or a test header discharging back into the tank with a calibrated nozzle(s) arranged for the attachment of a pressure gauge to determine pitot pressure shall be required.

> Holding a pitot tube in a hose stream discharging into a water tank in tight spaces is difficult, can create safety issues, and, therefore, should be avoided. Permanently affixed flowmeters or calibrated nozzles with fixed attachment points avoid this problem.
>
> High-rise buildings tend to be located in congested areas where testing may present a hazard to the public and test times may be restricted. Annual flow testing in accordance with NFPA 25, *Standard for the Inspection, Testing, and Maintenance of Water-Based Fire Protection Systems*, verifies that the pump system is fully functional. Omission of this testing can result in an unreliable pump. There are many high-rise buildings where pump flow testing is difficult or impractical. Locating the pump test header on an outside wall facilitates testing and allows quicker setup and takedown time, while minimizing hose usage. Special consideration should be given to the pressure rating of the test header and the pressure developed by the pump, which, in most cases, will exceed 175 psi (12.1 bar).

5.5 Auxiliary Power

Where electric motors are used and the height of the structure is beyond the pumping capability of the fire department apparatus, a reliable emergency source of power in accordance with Section 9.6 shall be provided for the fire pump installation.

> While it is appropriate to provide standby power for all electric motor fire pumps serving a high-rise building, it is specifically required if the fire department cannot provide an alternative source of fire protection water.

5.6* Very Tall Buildings

> Where the building height exceeds the pumping capacity of the fire department apparatus, an adequate auxiliary water supply is needed. Subsection 5.6.1 provides guidance on arranging water supply tanks for this purpose, and 5.6.2 provides guidance on backup fire pumps.
>
> The auxiliary means of providing the full fire protection water demand allows for service and maintenance of the primary system without the loss of fire protection. Exhibit II.5.1 shows one possible means of meeting this requirement using a redundant fire pump fed from a water tank. Gravity feed from a tank located on a higher level might be another way to meet this requirement.

A.5.6 When trying to determine the pumping capability of the fire department, the concern is for the height of the building. There are some buildings that are so tall that it is impossible for the fire department apparatus to pump into the fire department connection at the street and overcome the elevation loss and friction loss in order to achieve 100 psi at the hose outlets

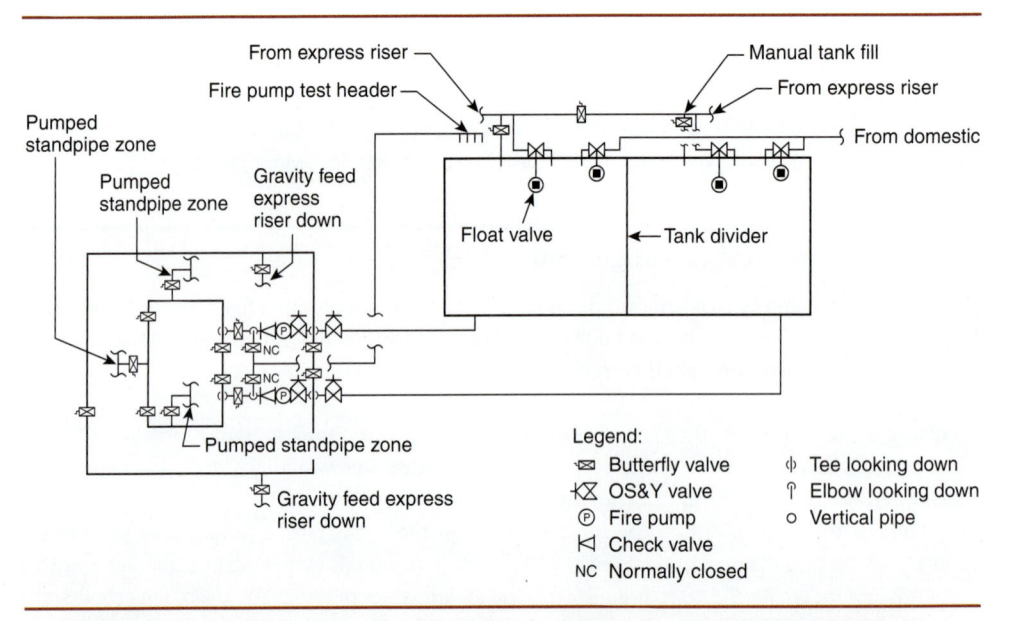

EXHIBIT II.5.1 *Redundant Fire Pump Fed from a Water Tank. (Courtesy of Schirmer Engineering Corp.)*

up in the building. In these cases, the fire protection system in the building needs to have additional protection, including sufficient water supplies within the building to fight a fire and sufficient redundancy to be safe. Since fire departments all purchase different apparatus with different pumping capabilities, this standard has chosen to address this concern with performance-based criteria rather than a specified building height. Most urban fire departments have the capability of getting sufficient water at sufficient pressure up to the top of 200 ft (61 m) tall buildings. Some have the ability to get water as high as 350 ft (107 m). The design professional will need to check with the local fire department to determine its capability.

5.6.1 Water Supply Tanks for Very Tall Buildings.

Section 5.6.1 is intended to ensure the following:

1. The stored water supply will meet the full fire protection water demand.
2. The tanks are automatically refilled at the minimum rate of the fire protection water demand.
3. With any single tank or tank compartment (and automatic refill valve for that tank) out of service, the full fire protection water demand can still be met through stored water and automatic tank refill.

5.6.1.1 Where the primary supply source is a tank, two or more water tanks shall be provided.

This paragraph allows maintenance and servicing of a tank without impairing the building fire protection.

5.6.1.1.1 A water tank shall be permitted to be divided into compartments such that the compartments function as individual tanks.

5.6.1.1.2 The total volume of all tanks or compartments shall be sufficient for the full fire protection demand.

5.6.1.1.3 Each individual tank or compartment shall be sized so that at least 50 percent of the fire protection demand is stored with any one compartment or tank out of service.

5.6.1.2 An automatic refill valve shall be provided for each tank or tank compartment.

5.6.1.3 A manual refill valve shall be provided for each tank or tank compartment.

5.6.1.4 Each refill valve shall be sized and arranged to independently supply the system fire protection demand.

5.6.1.5 The automatic and manual fill valve combination for each tank or tank compartment shall have its own connection to one of the following:

(1) A standpipe riser that is supplied with a backup fire pump
(2) A reliable domestic riser sized to meet the requirements of 5.6.1.4

5.6.1.5.1* Each connection shall be made to a different riser.

◀ **DESIGN ALERT**

> Exhibit II.5.1 shows one way of meeting the requirements of Section 5.6 with a divided water tank with refill valves supplying a two-zone standpipe system with backup fire pumps. Exhibit II.5.2 shows a divided water tank with refill valves supplying a two-zone standpipe system, with the lower zone gravity fed and the upper zone supplied by a fire pump with a backup fire pump.
> The refill valves do not have to be designed to operate simultaneously. When operating independently, the refill rate must be the minimum of the system's fire protection demand. Under conditions where the refill valves are operating simultaneously, the total refill must be the minimum of the system's fire protection demand.
> Exhibit II.5.3 shows a tank with duplicate automatic refill.

A.5.6.1.5.1 The different connections should be arranged so that the tank refill rate required in 5.6.1.4 can be maintained even with the failure of any single valve, pipe, or pump.

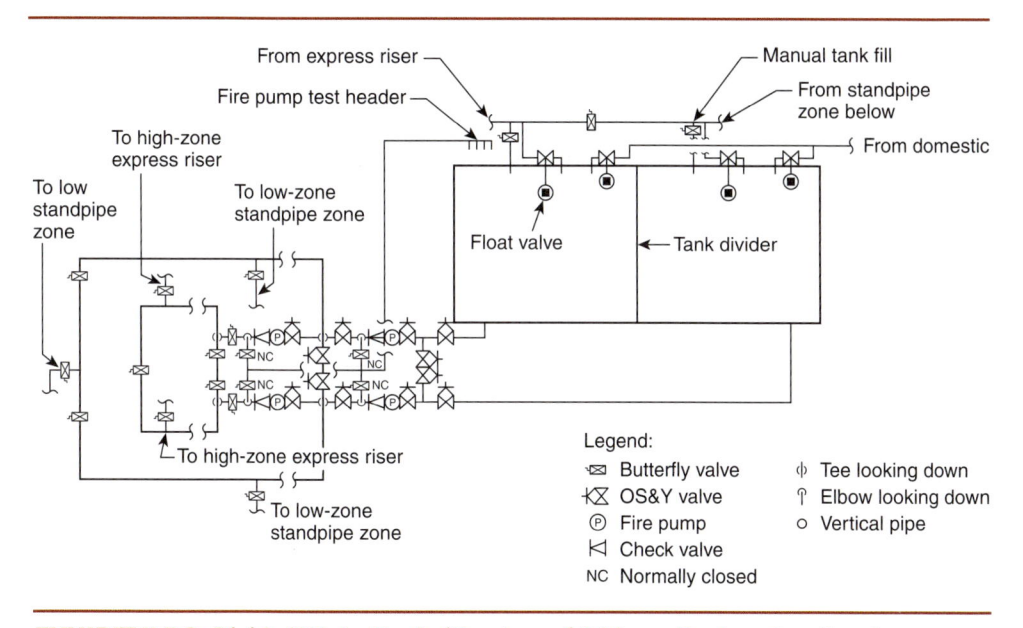

EXHIBIT II.5.2 *Divided Water Tank. (Courtesy of Schirmer Engineering Corp.)*

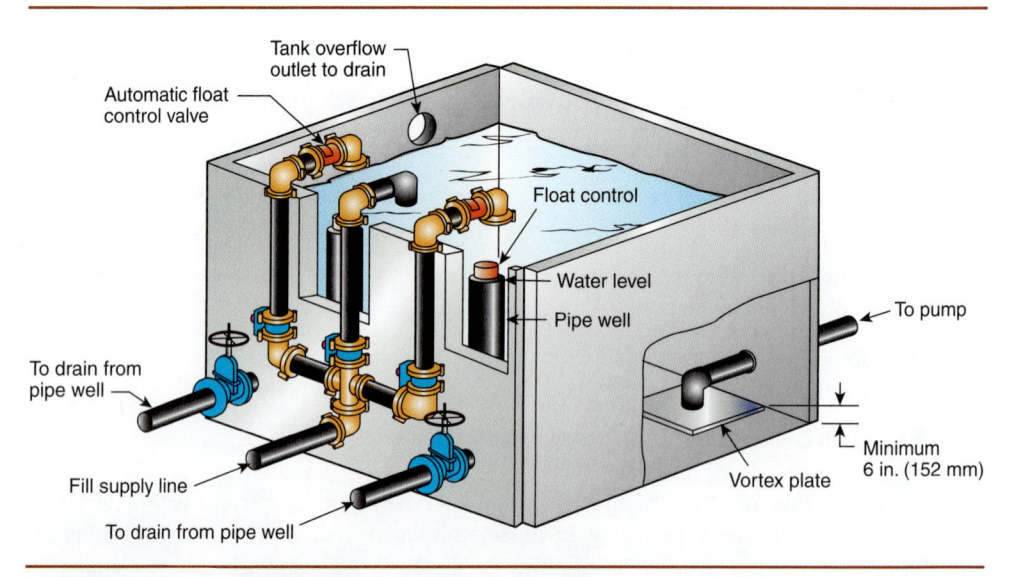

EXHIBIT II.5.3 *Tank with Duplicate Automatic Refill.*

5.6.2 Fire Pump Backup. Fire pumps serving zones that are partially or wholly beyond the pumping capability of the fire department apparatus shall be provided with one of the following:

(1) A fully independent and automatic backup fire pump unit(s) arranged so that all zones can be maintained in full service with any one pump out of service.
(2) An auxiliary means that is capable of providing the full fire protection demand and that is acceptable to the authority having jurisdiction.

References Cited in Commentary

National Fire Protection Association, 1 Batterymarch Park, Quincy MA 02169-7471.

NFPA 25, *Standard for the Inspection, Testing, and Maintenance of Water-Based Fire Protection Systems,* 2011 edition.

CHAPTER 6

Centrifugal Pumps

Chapter 6 covers the basic physical and performance characteristics of centrifugal fire pumps for fire protection service. Centrifugal fire pumps are generally compact, reliable, and easy to maintain. The hydraulic characteristics of the pump and variety of drivers available have made the centrifugal pump the most common of all pumps for fire protection.

A desirable feature of the centrifugal pump is its relationship between flow and pressure at a constant speed — that is, when flow is decreased, pressure increases. A centrifugal pump functions by converting kinetic energy to pressure and velocity energy through centrifugal force. As water enters the eye of the impeller, centrifugal force from the rotation of the impeller forces water through the impeller to the discharge outlet of the pump.

Centrifugal fire pumps are not appropriate for suction lift applications and are limited to applications where the water source can provide a positive pressure to the suction side of the pump.

6.1 General

6.1.1* Types.

▶ **FAQ** What is an impeller?

An impeller is the rotating element of the pump that imparts energy in the form of increased pressure to the water passing through the pump. Exhibit II.6.1 illustrates a typical centrifugal pump impeller.

▶ **FAQ** What is an overhung impeller design?

In an overhung impeller design, an overhung impeller is located at the end of the impeller shaft. The shaft is supported on only one side of the impeller.

▶ **FAQ** What is an impeller between bearings design?

In an impeller between bearings design, the pump shaft extends through the impeller, and shaft bearings are located on both sides the impeller to provide support for the loads imposed by the impeller, as well as by the shaft.

A.6.1.1 See Figure A.6.1.1(a) through Figure A.6.1.1(h).

6.1.1.1 Centrifugal pumps shall be of the overhung impeller design and the impeller between bearings design.

6.1.1.2 The overhung impeller design shall be close coupled or separately coupled single- or two-stage end-suction-type *[see Figure A.6.1.1(a) and Figure A.6.1.1(b)]* or in-line-type *[see Figure A.6.1.1(c) through Figure A.6.1.1(e)]* pumps.

EXHIBIT II.6.1 Centrifugal Pump Impeller. (Courtesy of Peerless Pump Fire Segment)

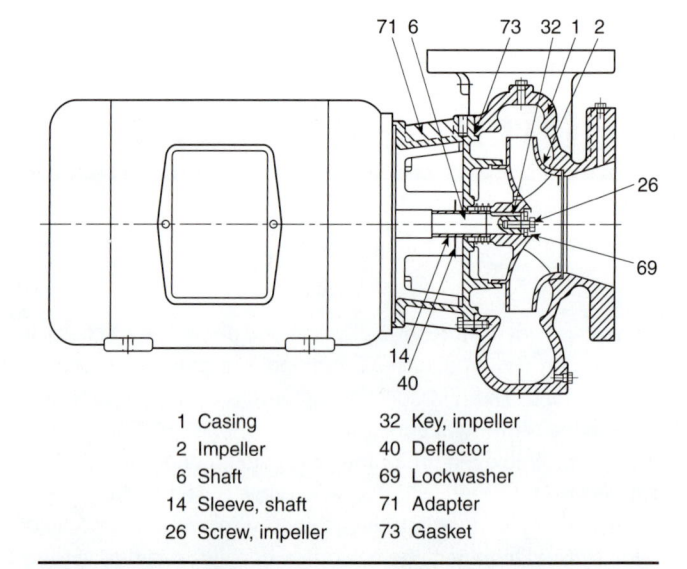

1 Casing	32 Key, impeller
2 Impeller	40 Deflector
6 Shaft	69 Lockwasher
14 Sleeve, shaft	71 Adapter
26 Screw, impeller	73 Gasket

FIGURE A.6.1.1(a) Overhung Impeller — Close Coupled Single Stage — End Suction.

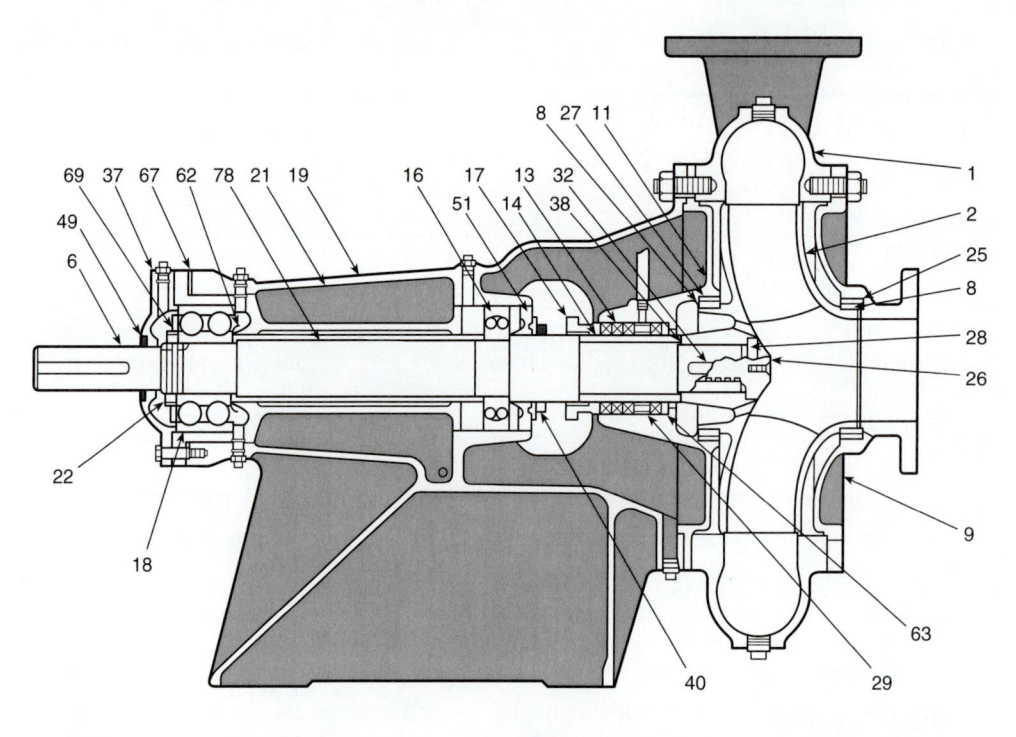

1 Casing	16 Bearing, inboard	27 Ring, stuffing-box cover	49 Seal, bearing cover, outboard
2 Impeller	17 Gland	28 Gasket	51 Retainer, grease
6 Shaft, pump	18 Bearing, outboard	29 Ring, lantern	62 Thrower (oil or grease)
8 Ring, impeller	19 Frame	32 Key, impeller	63 Busing, stuffing-box
9 Cover, suction	21 Liner, frame	37 Cover, bearing, outboard	67 Shim, frame liner
11 Cover, stuffing-box	22 Locknut, bearing	38 Gasket, shaft sleeve	69 Lockwasher
13 Packing	25 Ring, suction cover	40 Deflector	78 Spacer, bearing
14 Sleeve, shaft	26 Screw, impeller		

FIGURE A.6.1.1(b) Overhung Impeller — Separately Coupled Single Stage — Frame Mounted.

Section 6.1 • General 133

1 Casing	13 Packing	40 Deflector
2 Impeller	14 Sleeve, shaft	71 Adapter
11 Cover, seal chamber	17 Gland, packing	73 Gasket, casing

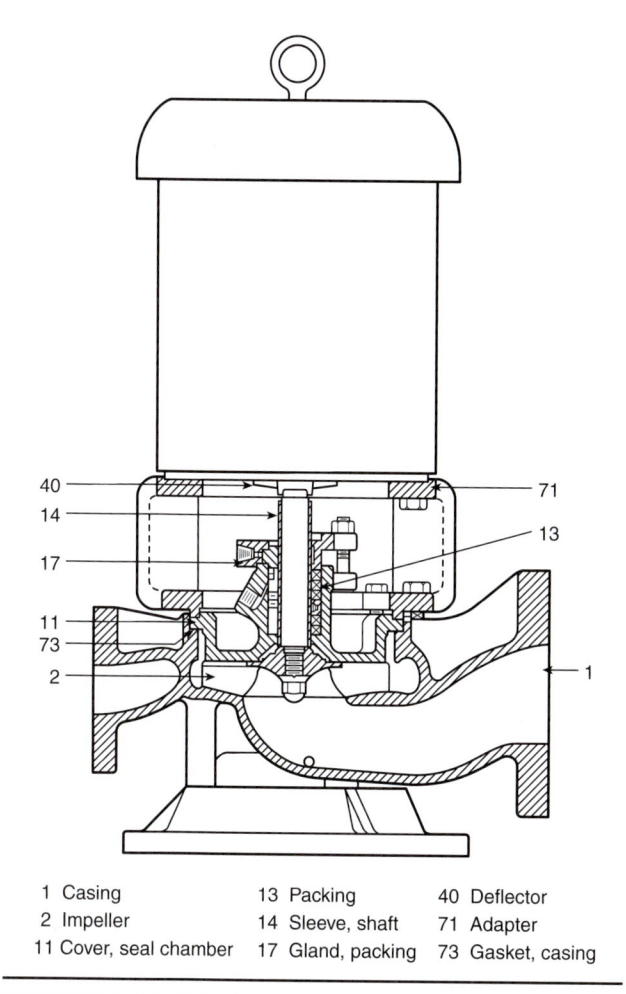

FIGURE A.6.1.1(c) *Overhung Impeller — Close Coupled Single Stage — In-Line (Showing Seal and Packaging).*

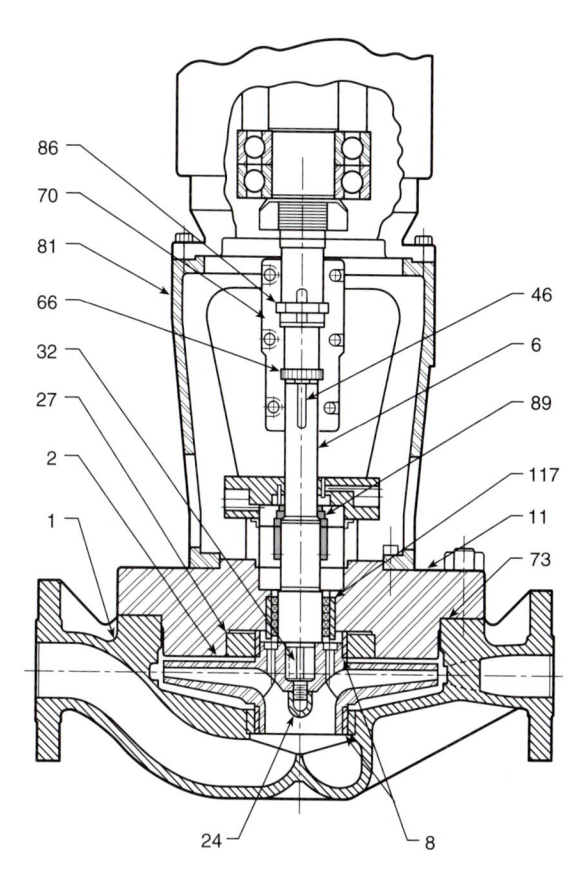

1 Casing	46 Key, coupling
2 Impeller	66 Nut, shaft adjusting
6 Shaft, pump	70 Coupling, shaft
8 Ring, impeller	73 Gasket
11 Cover, seal chamber	81 Pedestal, driver
24 Nut, impeller	86 Ring, thrust, split
27 Ring, stuffing-box cover	89 Seal
32 Key, impeller	

FIGURE A.6.1.1(d) *Overhung Impeller — Separately Coupled Single Stage — In-Line — Rigid Coupling.*

▶ **FAQ** What do the terms *close coupled, separately coupled,* and *multistage pump* mean?

The driver on a close coupled pump is attached to the pump housing, and the impeller of the pump is assembled onto the shaft of the driver. On a separately coupled pump, the pump and the driver have separate drive shafts that must be connected. A multistage pump has multiple impellers; as water passes from one impeller to the next, the pressure increases. Exhibit II.6.2 illustrates a typical two-stage centrifugal pump.

▶ **FAQ** What is an end suction–type pump?

An end suction–type pump is a device in which water enters the end of the pump perpendicular to the plane of the impeller.

Stationary Fire Pumps Handbook   2013

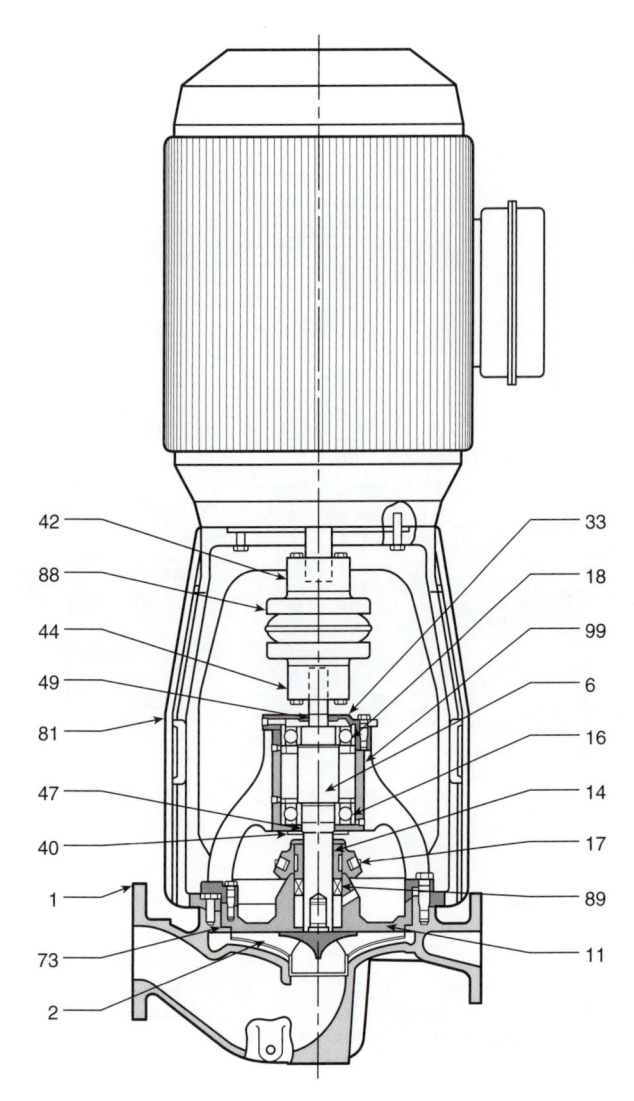

1A Casing, lower half	22 Locknut
1B Casing, upper half	31 Housing, bearing inboard
2 Impeller	32 Key, impeller
6 Shaft	33 Housing, bearing outboard
7 Ring, casing	35 Cover, bearing inboard
8 Ring, impeller	37 Cover, bearing outboard
14 Sleeve, shaft	40 Deflector
16 Bearing, inboard	65 Seal, mechanical stationary element
18 Bearing, outboard	
20 Nut, shaft sleeve	80 Seal, mechanical rotating element

FIGURE A.6.1.1(f) *Impeller Between Bearings — Separately Coupled — Single Stage — Axial (Horizontal) Split Case.*

1 Casing	42 Coupling half, driver
2 Impeller	44 Coupling half, pump
6 Shaft, pump	47 Seal, bearing cover, inboard
11 Cover, seal chamber	49 Seal, bearing cover, outboard
14 Sleeve, shaft	73 Gasket
16 Bearing, inboard	81 Pedestal, driver
17 Gland	88 Spacer, coupling
18 Bearing, outboard	89 Seal
33 Cap, bearing, outboard	99 Housing, bearing
40 Deflector	

FIGURE A.6.1.1(e) *Overhung Impeller — Separately Coupled Single Stage — In-Line — Flexible Coupling.*

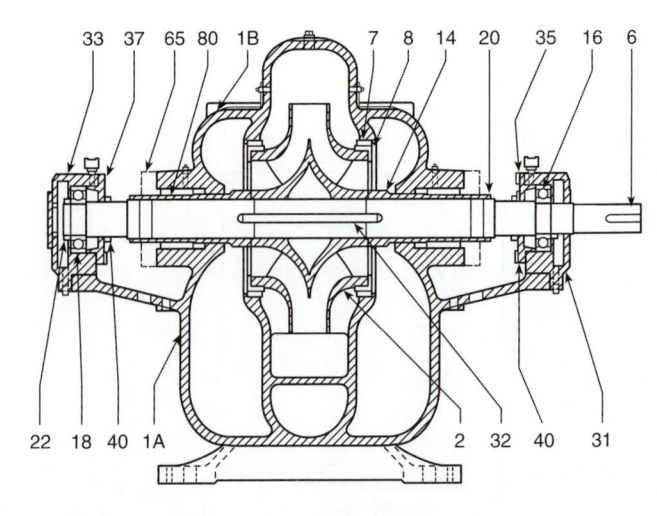

EXHIBIT II.6.2 Two-Stage Centrifugal Pump. *(Courtesy of Peerless Pump Fire Segment)*

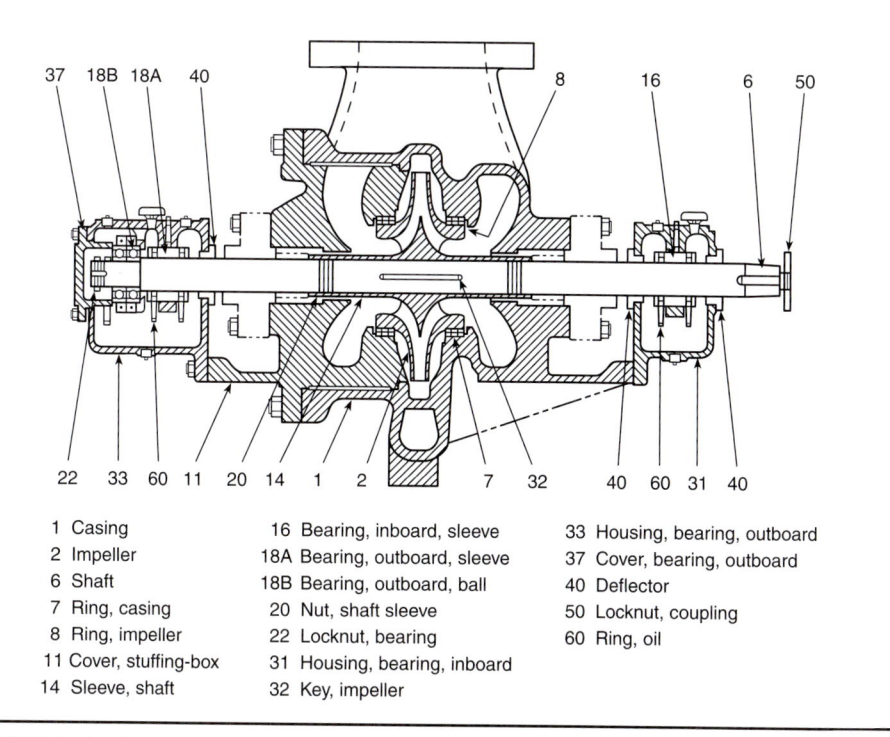

1 Casing	16 Bearing, inboard, sleeve	33 Housing, bearing, outboard
2 Impeller	18A Bearing, outboard, sleeve	37 Cover, bearing, outboard
6 Shaft	18B Bearing, outboard, ball	40 Deflector
7 Ring, casing	20 Nut, shaft sleeve	50 Locknut, coupling
8 Ring, impeller	22 Locknut, bearing	60 Ring, oil
11 Cover, stuffing-box	31 Housing, bearing, inboard	
14 Sleeve, shaft	32 Key, impeller	

FIGURE A.6.1.1(g) *Impeller Between Bearings — Separately Coupled — Single Stage — Radial (Vertical) Split Case.*

Kinetic
- Centrifugal*
  - Overhung impeller
    - Close coupled, single and two stage
      - End suction (including submersibles) — Figure A.6.1.1(a)
      - In-line — Figure A.6.1.1(c)
    - Separately coupled, single and two stage
      - In-line — Figures A.6.1.1(d) and (e)
      - Frame mounted — Figure A.6.1.1(b)
      - Centerline support — Not shown
      - Frame mounted — Not shown
      - Wet pit volute — Not shown
      - Axial flow impeller (propeller) volute type (horizontal or vertical) — Not shown
  - Impeller between bearings
    - Separately coupled, single stage
      - Axial (horizontal) split case — Figure A.6.1.1(f)
      - Radial (vertical) split case — Figure A.6.1.1(g)
    - Separately coupled, multistage
      - Axial (horizontal) split case — Not shown
      - Radial (vertical) split case — Not shown
  - Turbine type
    - Vertical type, single and multistage
      - Deep well turbine (including submersibles) — Not shown
      - Barrel or can pump — Not shown
      - Short setting or close coupled — Not shown
    - Axial flow impeller (propeller) or mixed flow type (horizontal or vertical) — Not shown
- Regenerative turbine — Impeller overhung or between bearings
  - Single stage — Not shown
  - Two stage — Not shown
- Special effect
  - Reversible centrifugal — Not shown
  - Rotating casing (Pitot) — Not shown

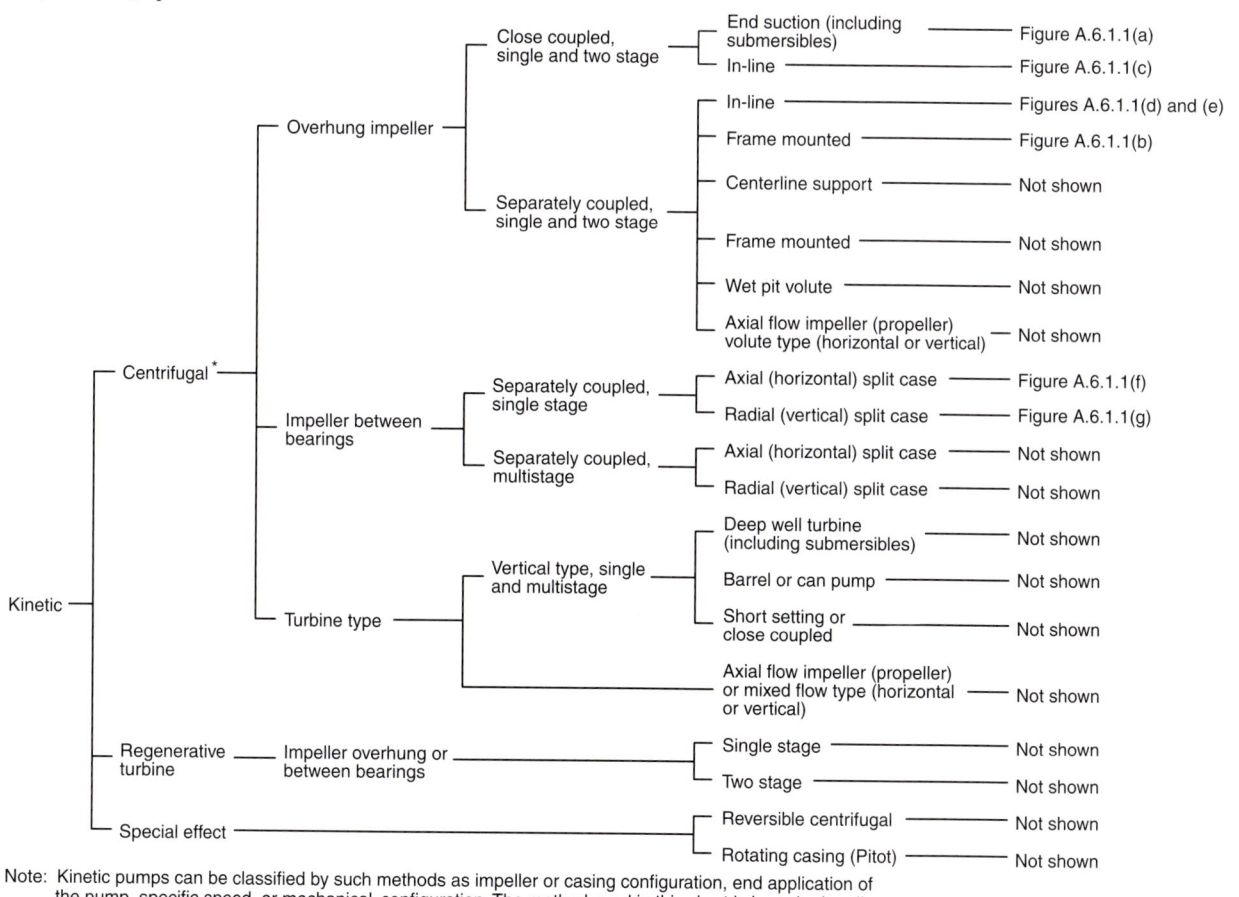

Note: Kinetic pumps can be classified by such methods as impeller or casing configuration, end application of the pump, specific speed, or mechanical configuration. The method used in this chart is based primarily on mechanical configuration.

*Includes radial, mixed flow, and axial flow designs.

FIGURE A.6.1.1(h) *Types of Stationary Pumps.*

6.1.1.3 The impeller between bearings design shall be separately coupled single-stage or multistage axial (horizontal) split-case-type *[see Figure A.6.1.1(f)]* or radial (vertical) split-case-type *[see Figure A.6.1.1(g)]* pumps.

▶ **FAQ** What is an axial (horizontal) split-case-type pump?

An axial (horizontal) split-case-type pump is a device in which the impeller shaft is located in a horizontal plane and the impeller rotates in a vertical plane. Exhibit II.6.3 illustrates a typical single-stage axial (horizontal) split-case-type centrifugal pump.

EXHIBIT II.6.3 Single-Stage Axial (Horizontal) Split-Case-Type Centrifugal Pump. (Courtesy of Schirmer Engineering Corp.)

EXHIBIT II.6.4 Radial (Vertical) Split-Case-Type Centrifugal Pump. (Courtesy of Schirmer Engineering Corp.)

> ▶ FAQ What is a radial (vertical) split-case-type pump?

A radial (vertical) split-case-type pump is a device in which the impeller shaft is located in a vertical plane and the impeller rotates in a horizontal plane. Exhibit II.6.4 illustrates a typical radial (vertical) split-case-type centrifugal pump.

6.1.2* Application. Centrifugal pumps shall not be used where a static suction lift is required.

> ▶ FAQ Why are centrifugal pumps not allowed where a suction lift is required?

A centrifugal pump is not self-priming — that is, it will not pump unless water covers the impeller. Once primed, a centrifugal pump can theoretically operate with a suction lift up to atmospheric pressure. Earlier editions of NFPA 20, *Standard for the Installation of Stationary Pumps for Fire Protection,* allowed centrifugal pumps to operate under suction lift when they were provided with a tank of priming water. This arrangement did not prove to be sufficiently reliable and was removed from the standard beginning with the 1974 edition. (Problems included leaky foot valves that caused the loss of priming water.)

Unlike positive displacement pumps, which continue to build pressure when operating against a closed system (until something breaks or the power source can no longer drive the pump), a centrifugal pump reaches a maximum pressure and can operate against a closed system indefinitely, provided that adequate water is discharged to cool the pump. When operating against a closed system, the energy the pump is imparting to the water is converted to heat, so water must be discharged to keep the pump from overheating.

A.6.1.2 The centrifugal pump is particularly suited to boost the pressure from a public or private supply or to pump from a storage tank where there is a positive static head.

6.2* Factory and Field Performance

A.6.2 Listed pumps can have different head capacity curve shapes for a given rating. Figure A.6.2 illustrates the extremes of probable curve shapes. Shutoff head will range from a minimum of 101 percent to a maximum of 140 percent of rated head. At 150 percent of rated capacity, head will range from a minimum of 65 percent to a maximum of just below rated head. Pump manufacturers can supply expected curves for their listed pumps.

> ▶ FAQ How does a fire pump characteristic curve impact fire protection system design?

The suction pressure from the water supply must be added to the fire pump characteristic curve to determine the fire pump discharge curve. See Exhibit II.6.5.

Fire protection systems can operate at any point along the pump discharge curve; however, all system components must be rated for the maximum pressure developed by the fire pump.

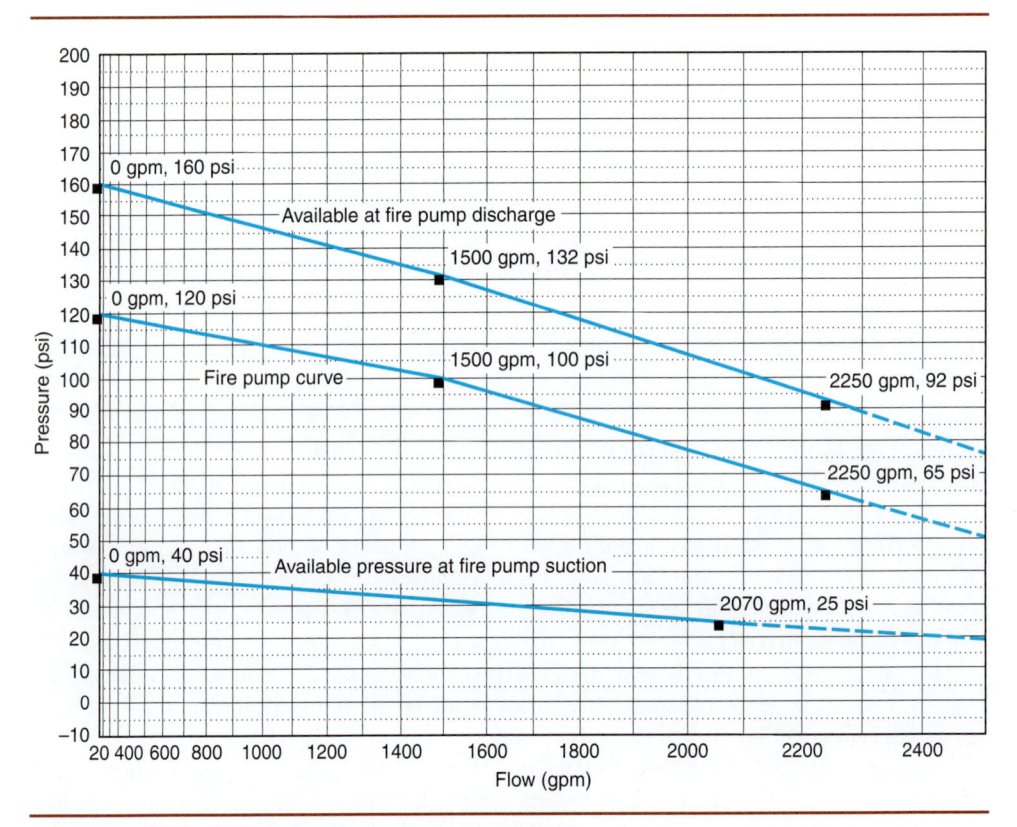

EXHIBIT II.6.5 Sample Fire Pump Discharge Curve.

6.2.1 Pumps shall furnish not less than 150 percent of rated capacity at not less than 65 percent of total rated head. *(See Figure A.6.2.)*

CALCULATION ▶ A fire pump is required to have a characteristic curve that meets the following three criteria:

1. The shutoff head (churn) is limited to 140 percent of the rated head.
2. The rated head at the rated flow, commonly referred to as the pump pressure and capacity ratings, is marked on the pump nameplate.
3. A minimum of 65 percent of the rated head is provided at 150 percent of the rated capacity.

It is recommended that the maximum system demand flow correlate to a point on the pump curve between 90 percent and 140 percent of the rated pump flow (capacity), but in no case is the maximum system demand flow permitted to exceed 150 percent of pump rated flow. NFPA 25, *Standard for the Inspection, Testing, and Maintenance of Water-Based Fire Protection Systems,* allows 5 percent degradation in performance before requiring an investigation.

The actual fire pump curve typically provides less pressure than the maximum limit established by NFPA 20 (140 percent of rated pressure at churn) when flowing less than 100 percent of rated flow and provides a higher pressure than the minimum limit (65 percent of rated pressure at 150 percent of rated flow) when flowing more than the rated flow. Although NPFA 20 requires the pump performance to be measured against the original pump curve when evaluating pump performance, the test results are sometimes compared against the previously

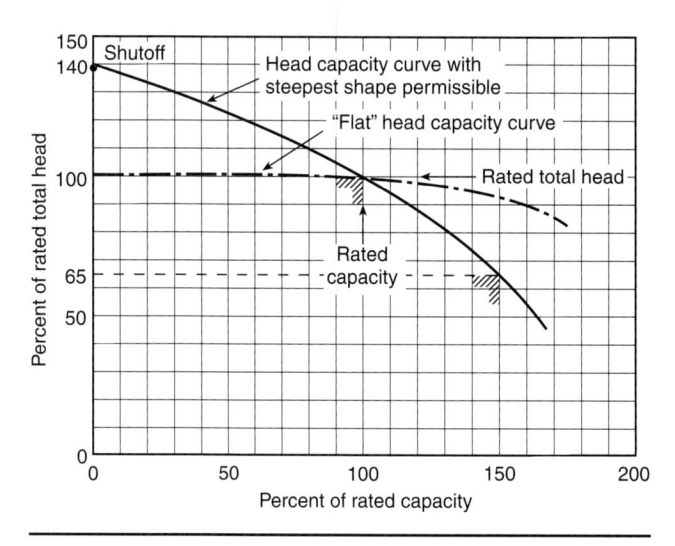

FIGURE A.6.2 *Pump Characteristics Curves.*

---

mentioned criteria when the original pump curve is not readily available. When designing to flow rates that exceed 100 percent of the rated flow, designers should consider the pump performance characteristic limitations referenced in NFPA 20 (65 percent of rated pressure at 150 percent of rated flow).

6.2.2 The shutoff head shall not exceed 140 percent of rated head for any type pump. *(See Figure A.6.2.)*

Historically, vertical turbine pumps have a higher churn pressure than horizontal centrifugal pumps. Earlier editions of NFPA 20 limited the horizontal centrifugal pump's churn pressure (shutoff head) to 120 percent of the rated head. Although the distinction was removed from NFPA 20 in the 1990 edition, several horizontal centrifugal fire pumps still churn at less than 120 percent of the rated pressure.

The fire pump characteristic curve is required to meet the minimum and maximum flow and pressure requirements of Section 6.2, as indicated in Figure A.6.2. The fire pump output can vary significantly. For an example of fire pump outputs, see Commentary Table II.6.1.

The pump should be carefully selected to meet the fire protection system demand. Based upon daily and seasonal fluctuations in the flow demand on a water supply, adjustments should be made to the water supply measurements taken during the flow test. With these adjustments, the pump discharge pressure should not be less than the required system demand pressure and should not exceed the pressure rating of the system components at the churn pressure. To plot a fire pump curve using the sample fire pump discharge curve in Figure A.6.2, the first rated capacity of the pump (which should be a known value) should be plotted. In this case, the value is 1500 gpm at 100 psi (5677 L/min at 6.8 bar). The maximum pump output, commonly referred to as *overload*, can then be plotted as 2250 gpm at 65 psi (8516 L/min at 4.5 bar) (150 percent of rated flow at 65 percent of rated pressure). The shutoff or churn point can be estimated at midrange, or 120 percent of rated pressure at 0 gpm (0 bar) flow, or may be obtained from the manufacturer's shop test curve, if available. In this case, 120 psi (8.3 bar) was used.

COMMENTARY TABLE II.6.1 Fire Pump Selection Table

Item	Value	Comments
1. Maximum system demand flow (gpm)	1700	Determined by hydraulic calculations based on design criteria.
2. Required system demand pressure at the pump discharge (psi)	107	Determined by hydraulic calculations based on design criteria.
3. Pump suction pressure at zero (0) flow (psi)	40	Based on the available water supply pressure at the pump inlet.
4. Pump suction pressure at maximum system demand flow (psi)	30	Based on a flow test of the available water supply; the residual pressure at the pump inlet at a flow of 1700 gpm is determined.
5. Rated pump flow [capacity(gpm)]	1500	Maximum system demand flow should be between 90% and 140% of pump's rated capacity, but, in no case, greater than 150% of the pump's rated capacity.
6. Net pump pressure rating (psi)	100	Selected so that the discharge pressure (net pump pressure plus suction pressure) provides not less than the pressure required for the fire protection system(s).
7. Net pump churn pressure (psi)	120	Review pump curves from manufacturer to approximate churn pressure (120% of rated pressure used for this example).
8. Net pump pressure at 150% of rated flow (psi)	65	Must be a minimum of 65% of rated pressure. Actual pump curve may be higher.
9. Net pump pressure at maximum system demand flow (psi)	92	From pump curve at maximum system demand flow (1700 gpm used for this example).
10. Pump discharge pressure at zero (0) flow (psi)	160	Items 3 + 7: Pump suction pressure plus net pump pressure at zero (0) flow (psi). The pump discharge pressure, after adjusting for elevation, cannot exceed the rating of the system components.
11. Pump discharge pressure at maximum system demand flow (psi)	122	Items 4 + 9: Must be higher than the maximum system demand plus any required safety factors.

For pumps attached to a city water supply, the city water flow test data must be adjusted to the fire pump curve (elevation difference and friction loss). These data should be plotted next and are shown on the sample fire pump discharge curve in Exhibit II.6.5 as 40 psi (2.8 bar) static pressure and 2070 gpm at 25 psi (7835 L/min at 1.7 bar) residual pressure. To determine the combined fire pump/city water supply curve, the pressures (churn, rated capacity, and overload) from the city water supply must be added to each point on the fire pump curve. Performing this calculation on the sample curve results in the following data: 160 psi (11 bar) at 0 flow (churn); 132 psi at 1500 gpm (9.1 bar at 5677 L/min) (rated capacity); and 92 psi at 2250 gpm (6.3 bar at 8516 L/min) (overload). This procedure can be used for any size fire pump that is connected to a city water supply.

6.3 Fittings

6.3.1* Where necessary, the following fittings for the pump shall be provided by the pump manufacturer or an authorized representative:

(1) Automatic air release valve

> See 6.3.3 for more information on automatic air release.

(2) Circulation relief valve

> A circulation relief valve (see Exhibit II.6.6) allows a small amount of water to discharge from the pump to keep the fire pump from overheating when operating at no flow or very low flows.

EXHIBIT II.6.6 *Circulation Relief Valve.*

(3) Pressure gauges

> The pressure gauge on the suction side of the fire pump should be a compound gauge to register negative pressures whenever the water supply is a ground level storage tank or has a low residual pressure. Liquid-filled gauges dampen pressure fluctuations, making the gauges easier to read during pump testing.

A.6.3.1 See Figure A.6.3.1(a) and Figure A.6.3.1(b).

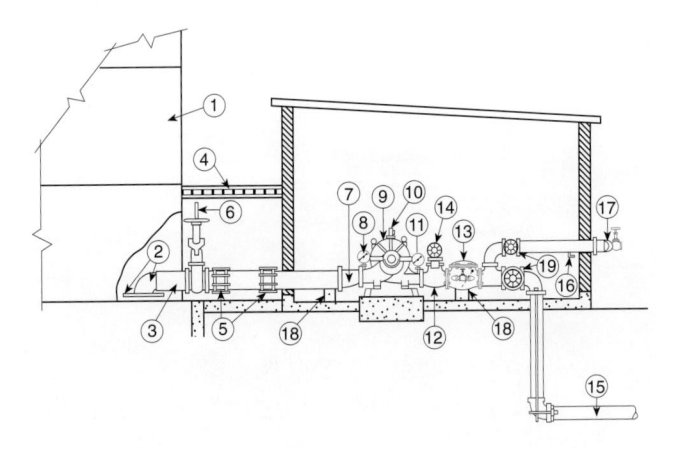

1. Aboveground suction tank
2. Entrance elbow and square steel vortex plate with dimensions at least twice the diameter of the suction pipe. Distance above the bottom of tank is one-half the diameter of the suction pipe with minimum of 6 in. (152 mm).
3. Suction pipe
4. Frostproof casing
5. Flexible couplings for strain relief
6. OS&Y gate valve (see 4.14.5 and A.4.14.5)
7. Eccentric reducer
8. Suction gauge
9. Horizontal split-case fire pump
10. Automatic air release
11. Discharge gauge
12. Reducing discharge tee
13. Discharge check valve
14. Relief valve (if required)
15. Supply pipe for fire protection system
16. Drain valve or ball drip
17. Hose valve manifold with hose valves
18. Pipe supports
19. Indicating gate or indicating butterfly valve

FIGURE A.6.3.1(a) *Horizontal Split-Case Fire Pump Installation with Water Supply Under a Positive Head.*

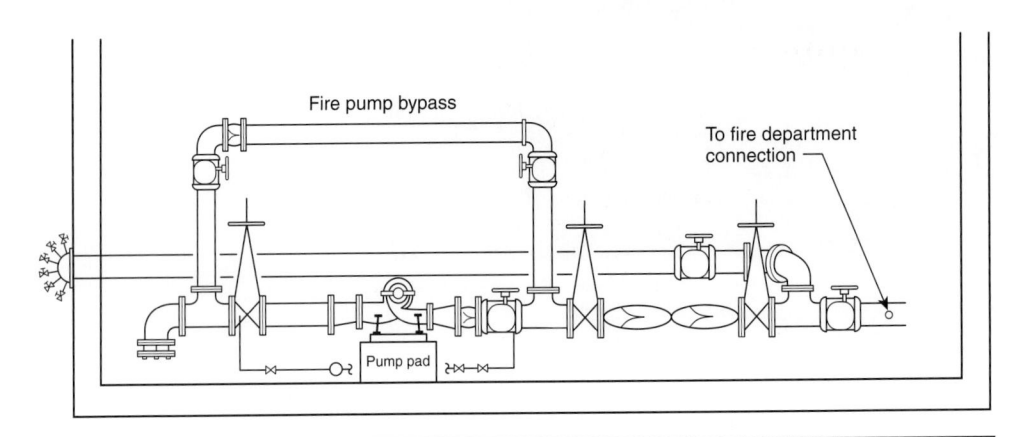

FIGURE A.6.3.1(b) *Backflow Preventer Installation.*

Section 6.3 • Fittings 143

6.3.2 Where necessary, the following fittings shall be provided:

(1) Eccentric tapered reducer at suction inlet

When the pump suction pipe is larger than the pump suction flange, an eccentric tapered reducer (see Exhibit II.6.7) is required to minimize the possibility of air pockets forming at the pump suction. See 4.14.6.4.

(2) Hose valve manifold with hose valves

See 4.20.3 for more information on hose valves. Exhibit II.6.8 illustrates a hose valve manifold.

(3) Flow measuring device

See Exhibit II.6.9 for an illustration of a venturi-type flowmeter. See 4.20.2 for more information related to the requirements for installing flow measuring devices.

(4) Relief valve and discharge cone

See Section 4.18 for the allowable use and the requirements associated with the installation of pressure relief valves and discharge cones (see Exhibit II.6.10).

(5) Pipeline strainer

Earlier editions of NFPA 20 required a strainer in the suction lines of pumps that required removal of the driver to remove rocks or debris from the pump impeller. This requirement was removed from the 2007 edition. See 7.2.2.2.4, 7.3.4, 8.5.5, 11.2.8.5.3.2, and 11.2.8.6.2 for strainer requirements.

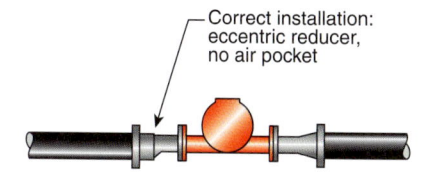

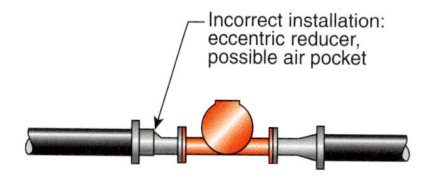

EXHIBIT II.6.7 *Correct and Incorrect Installation of an Eccentric Reducer. (Courtesy of Stephan Laforest, Summit Sprinkler Design Services, Inc.)*

EXHIBIT II.6.8 *Hose Valve Manifold.*

Stationary Fire Pumps Handbook 2013

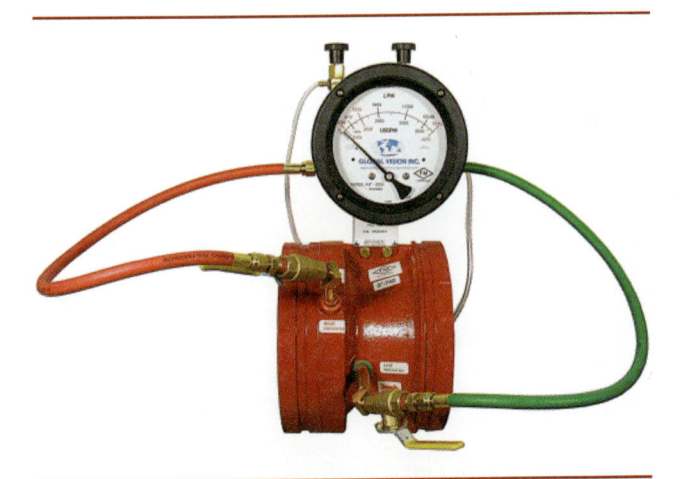

EXHIBIT II.6.9 Venturi-Type Flowmeter. *(Courtesy of Global Vision Inc.)*

EXHIBIT II.6.10 Pressure Relief Valve Piped Back to Suction. *(Courtesy of Peerless Pump Fire Segment)*

EXHIBIT II.6.11 Automatic Air Release Valve. *(Courtesy of Global Vision Inc.)*

6.3.3 Automatic Air Release.

6.3.3.1 Unless the requirements of 6.3.3.2 are met, pumps that are automatically controlled shall be provided with a listed float-operated air release valve having a nominal 0.50 in. (12.7 mm) minimum diameter discharged to atmosphere.

Air in the impeller can cause cavitation, damage the impeller, and negatively impact the pump performance. Exhibit II.6.11 illustrates an automatic air release valve.

6.3.3.2 The requirements of 6.3.3.1 shall not apply to overhung impeller–type pumps with top centerline discharge or that are vertically mounted to naturally vent the air.

6.4 Foundation and Setting

6.4.1* Overhung impeller and impeller between bearings design pumps and driver shall be mounted on a common grouted base plate.

A.6.4.1 Flexible couplings are used to compensate for temperature changes and to permit end movement of the connected shafts without interfering with each other.

6.4.2 Pumps of the overhung impeller close coupled in-line type *[see Figure A.6.1.1(c)]* shall be permitted to be mounted on a base attached to the pump mounting base plate.

6.4.3 The base plate shall be securely attached to a solid foundation in such a way that pump and driver shaft alignment is ensured.

6.4.4* The foundation shall be sufficiently substantial to form a permanent and rigid support for the base plate.

A.6.4.4 A substantial foundation is important in maintaining alignment. The foundation preferably should be made of reinforced concrete.

6.4.5 The base plate, with pump and driver mounted on it, shall be set level on the foundation.

6.5* Connection to Driver and Alignment

A.6.5 A pump and driver shipped from the factory with both machines mounted on a common base plate are accurately aligned before shipment. All base plates are flexible to some extent and, therefore, should not be relied upon to maintain the factory alignment. Realignment is necessary after the complete unit has been leveled on the foundation and again after the grout has set and foundation bolts have been tightened. The alignment should be checked after the unit is piped and rechecked periodically. To facilitate accurate field alignment, most manufacturers either do not dowel the pumps or drivers on the base plates before shipment or, at most, dowel the pump only.

After the pump and driver unit has been placed on the foundation, the coupling halves should be disconnected. The coupling should not be reconnected until the alignment operations have been completed.

The purpose of the flexible coupling is to compensate for temperature changes and to permit end movement of the shafts without interference with each other while transmitting power from the driver to the pump.

The two forms of misalignment between the pump shaft and the driver shaft are as follows:

(1) *Angular misalignment* — shafts with axes concentric but not parallel
(2) *Parallel misalignment* — shafts with axes parallel but not concentric

The faces of the coupling halves should be spaced within the manufacturer's recommendations and far enough apart so that they cannot strike each other when the driver rotor is moved hard over toward the pump. Due allowance should be made for wear of the thrust bearings. The necessary tools for an approximate check of the alignment of a flexible coupling are a straight edge and a taper gauge or a set of feeler gauges.

A check for angular alignment is made by inserting the taper gauge or feelers at four points between the coupling faces and comparing the distance between the faces at four points spaced at 90-degree intervals around the coupling. *[See Figure A.6.5(a).]* The unit will be in angular alignment when the measurements show that the coupling faces are the same distance apart at all points.

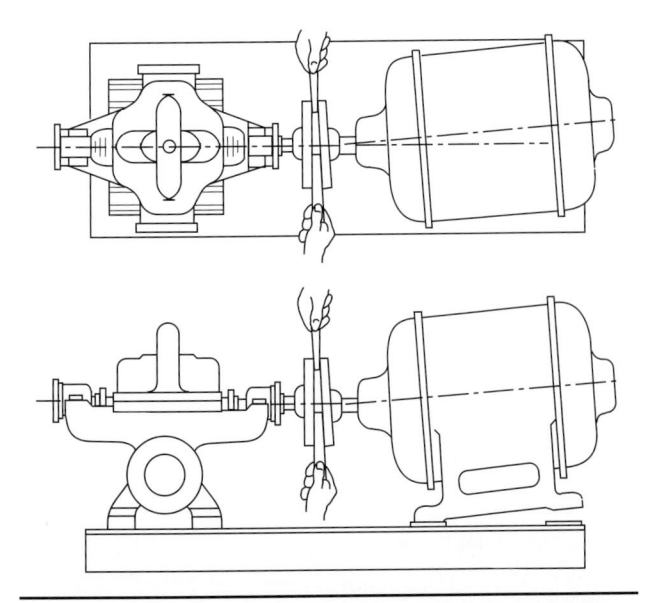

FIGURE A.6.5(a) *Checking Angular Alignment. (Courtesy of Hydraulic Institute, www.pumps.org.)*

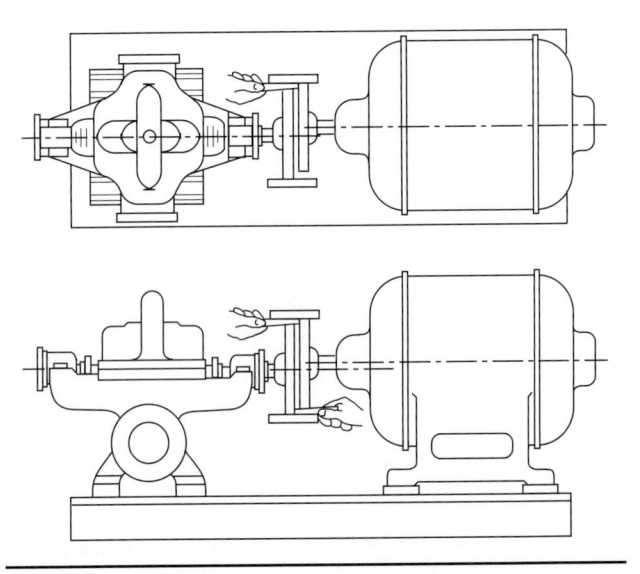

FIGURE A.6.5(b) *Checking Parallel Alignment. (Courtesy of Hydraulic Institute, www.pumps.org.)*

A check for parallel alignment is made by placing a straight edge across both coupling rims at the top, bottom, and both sides. *[See Figure A.6.5(b).]* The unit will be in parallel alignment when the straight edge rests evenly on the coupling rim at all positions.

Allowance might be necessary for temperature changes and for coupling halves that are not of the same outside diameter. Care should be taken to have the straight edge parallel to the axes of the shafts.

Angular and parallel misalignment are corrected by means of shims under the motor mounting feet. After each change, it is necessary to recheck the alignment of the coupling halves. Adjustment in one direction can disturb adjustments already made in another direction. It should not be necessary to adjust the shims under the pump.

The permissible amount of misalignment will vary with the type of pump, driver, and coupling manufacturer, model, and size.

The best method for putting the coupling halves in final accurate alignment is by the use of a dial indicator.

When the alignment is correct, the foundation bolts should be tightened evenly but not too firmly. The unit can then be grouted to the foundation. The base plate should be completely filled with grout, and it is desirable to grout the leveling pieces, shims, or wedges in place. Foundation bolts should not be fully tightened until the grout has hardened, usually about 48 hours after pouring.

After the grout has set and the foundation bolts have been properly tightened, the unit should be checked for parallel and angular alignment, and, if necessary, corrective measures taken. After the piping of the unit has been connected, the alignment should be checked again.

The direction of driver rotation should be checked to make certain that it matches that of the pump. The corresponding direction of rotation of the pump is indicated by a direction arrow on the pump casing.

The coupling halves can then be reconnected. With the pump properly primed, the unit should be operated under normal operating conditions until temperatures have stabilized. It then should be shut down and immediately checked again for alignment of the coupling. All alignment checks should be made with the coupling halves disconnected and again after they are reconnected.

After the unit has been in operation for about 10 hours, the coupling halves should be given a final check for misalignment caused by pipe or temperature strains. This check should be repeated after the unit has been in operation for about 3 months. If the alignment is correct, both pump and driver should be dowelled to the base plate. Dowel location is very important, and the manufacturer's instructions should be followed, especially if the unit is subject to temperature changes.

The unit should be checked periodically for alignment. If the unit does not stay in line after being properly installed, the following are possible causes:

(1) Settling, seasoning, or springing of the foundation and pipe strains distorting or shifting the machine
(2) Wearing of the bearings
(3) Springing of the base plate by heat from an adjacent steam pipe or from a steam turbine
(4) Shifting of the building structure due to variable loading or other causes
(5) If the unit and foundation are new, need for the alignment to be slightly readjusted from time to time

6.5.1 Coupling Type.

6.5.1.1 Separately coupled–type pumps with electric motor drivers shall be connected by a flexible coupling or flexible connecting shaft.

> The couplings referred to in 6.5.1.1 connect the pump to the driver—either an electric motor or a diesel engine. The two types of couplings used are a flexible coupling, as illustrated in Exhibit II.6.12, or a flexible shaft, as illustrated in Exhibit II.6.13.

6.5.1.2 All coupling types shall be listed for the service referenced in 6.5.1.1.

> Due to concerns related to the performance of flexible couplings in field installations, and due to the occurrence of some failures, flexible couplings were required to be listed in the 1993 edition of NFPA 20. While this listing requirement has existed for many years, no manufacturers obtained a listing until 2007. Currently, there are listed flexible couplings and connecting shafts available from several manufacturers. These products are required to be installed in accordance with the instructions provided by the manufacturer, which include information related to the maximum permitted angular and parallel misalignment. Requirements for listing flexible couplings are included in ANSI/UL 448A, *Standard for Flexible Couplings and Connecting Shafts for Stationary Fire Pumps*, and *Class Number 1336*.
>
> Flexible couplings and connecting shafts are required, and they must be selected in the proper size to accommodate the torque associated with the horsepower rating of the driver. A service factor, which is specified by the coupling or connecting shaft manufacturer, but is not to be less than those specified in Commentary Table II.6.2, should be included in the assessment to determine the appropriate rating of coupling or connecting shaft. This service factor is intended to account for conditions not considered in the flexible coupling or connecting shaft design, including vibratory torque loads. The manufacturer's instructions include detailed information as to how to apply the service factor.

6.5.2 Pumps and drivers on separately coupled–type pumps shall be aligned in accordance with the coupling and pump manufacturers' specifications and the *Hydraulic Institute Standards for Centrifugal, Rotary and Reciprocating Pumps. (See A.6.5.)*

Commentary Table II.6.2 *Service Factors for Determining Application of Torque*

Load Type	Driver Type		
	Electric Motor	Diesel engine — 5 or fewer cylinders	Diesel engine — 6 or more cylinders
Centrifugal Pump	1.00	2.00	1.50
Reciprocating Pump — double acting	2.00	3.00	2.50
Reciprocating Pump — 1 or 2 cylinders	2.25	3.25	2.75
Reciprocating Pump — 3 or more cylinders	1.75	2.75	2.25
Rotary — gear, lobe or vane	1.50	2.50	2.00

Note — The service factors for the load type specified in this Table are referenced in the Load Classification and Service Factors for Flexible Couplings Information Sheet, AGMA-A96, published by the American Gear Manufacturers Association.

EXHIBIT II.6.12 Flexible Coupling. (Courtesy of Xylem Inc.)

EXHIBIT II.6.13 Flexible Connecting Shaft. (Courtesy of Clarke Fire Protection)

References Cited in Commentary

National Fire Protection Association, 1 Batterymarch Park, Quincy, MA 02169-7471.

NFPA 25, *Standard for the Inspection, Testing, and Maintenance of Water-Based Fire Protection Systems*, 2011 edition.

American Gear Manufacturers Association, 100 North Fairfax Street, Suite 500, Alexandria, VA 22314-1587.

AGMA 922-A96, *Load Classification and Service Factors for Flexible Couplings Information Sheet*, 1996.

FM Approvals LLC, 1151 Boston-Providence Turnpike, P. O. Box 9102, Norwood, MA 02062.

Approval Standard for Flexible Fire Pump Couplings and Flexible Connecting Shafts for Fire Protection Service, Class Number 1336, September 2011.

Underwriters Laboratories Inc., 333 Pfingsten Road, Northbrook, IL 60062-2096.

ANSI/UL 448A, *Standard for Flexible Couplings and Connecting Shafts for Stationary Fire Pumps*, 2008.

CHAPTER 7

Vertical Shaft Turbine–Type Pumps

The vertical turbine pump is designed to provide suction lift from a static source of water; therefore, it differs in design and installation requirements from other pumps covered in this handbook. The installation type, either a well or wet pit, drives the requirements for design and installation. In either case, this chapter provides specific requirements for the design and construction of the pump and the design of the well or wet pit. This chapter also defines the necessary accessories for the installation and proper operation of the pump.

7.1* General

A.7.1 Satisfactory operation of vertical turbine–type pumps is dependent to a large extent upon careful and correct installation of the unit; therefore, it is recommended that this work be done under the direction of a representative of the pump manufacturer.

The actual installation and setting of a vertical turbine is critical to its operation and service life. Ideally, a vertical turbine should be level and plumb. This enables the pump shaft to hang vertically, theoretically making no contact at all with the pump sleeve bearings. A film of water then creates a protective barrier between the shaft and bearings during operation. This enables vertical turbines to provide very long service life.

If a vertical turbine pump is not level and plumb, the offset can cause the pump shaft to make contact with the sleeve bearings, resulting in potential vibration, premature wear, and ultimate failure of the pump. Therefore, it is critical that precise and correct installation is performed under the proper supervision of a representative of the pump manufacturer.

7.1.1* Suitability. Where the water supply is located below the discharge flange centerline and the water supply pressure is insufficient for getting the water to the fire pump, a vertical shaft turbine–type pump shall be used.

A.7.1.1 The vertical shaft turbine–type pump is particularly suitable for fire pump service where the water source is located below ground and where it would be difficult to install any other type of pump below the minimum water level. It was originally designed for installation in drilled wells but is permitted to be used to lift water from lakes, streams, open swamps, and other subsurface sources. Both oil-lubricated enclosed-line-shaft and water-lubricated open-line-shaft pumps are used. *(See Figure A.7.1.1.)* Some health departments object to the use of oil-lubricated pumps; such authorities should be consulted before proceeding with oil-lubricated design.

Horizontal pumps are limited by 4.14.3.2 to operate with a suction lift of −3 psi (−0.2 bar) at the lowest water level and at 150 percent of the rated flow. Therefore, the vertical shaft turbine–type pump is the best choice for locations that cannot provide positive pressure to the suction side of the fire pump. As shown in Figure A.7.2.2.1 and Figure A.7.2.2.2, vertical shaft turbine–type pumps have their impellers submerged in the water supply. The operating impellers force the water up the pipe column to the discharge flange and into the system piping. (See Exhibit II.3.11 for an example of a vertical turbine fire pump.)

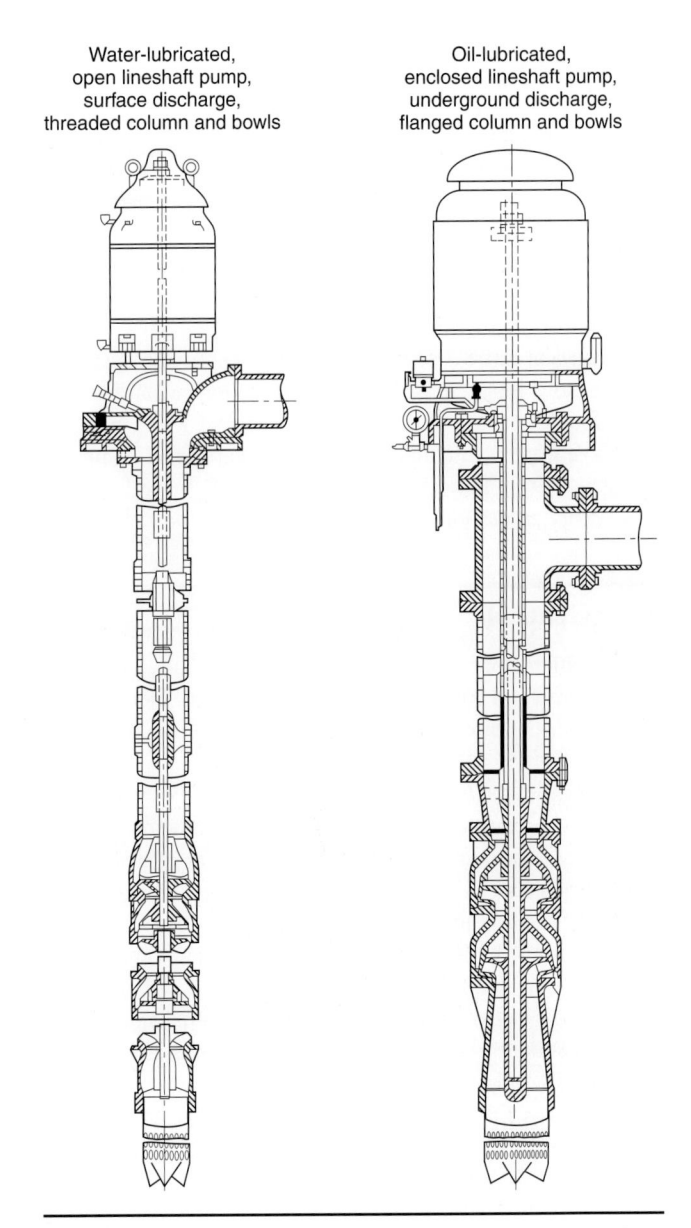

FIGURE A.7.1.1 *Water-Lubricated and Oil-Lubricated Shaft Pumps.*

7.1.2 Characteristics.

> Fire pumps are designed and built so that they can operate over a range of pressure and flow conditions in relation to their rating. Subsection 7.1.2 identifies the endpoints of this range, and all listed pumps must have characteristic curves somewhere within the range. The shutoff head corresponds to the maximum pressure the pump can produce when the pump is operating in a churn condition — that is, when no water is flowing. When the pump operates at churn pressure, the shutoff head or pressure cannot exceed 140 percent of the pump's pressure rating. For example, a pump with a rating of 100 psi at 1000 gpm (6.7 bar at 3785 L/min) cannot produce a shutoff head in excess of 140 psi (9.7 bar). A pump can be designed and manufactured so that the shutoff head falls below this value, such as at 100 psi (6.7 bar).

> The other endpoint identifies the minimum pressure the pump can produce. This minimum pressure occurs when the pump is operating at 150 percent of its capacity and does not fall below 65 percent of the pump's rated pressure. Consider a pump with a rating of 125 psi at 1000 gpm (8.6 bar at 3785 L/min). When the flow reaches 150 percent of the pump's capacity or 1500 gpm (5677 L/min), the minimum pressure cannot fall below 81.25 psi (5.6 bar). Any pressure between 81.25 psi and 125 psi (5.6 bar and 8.6 bar) would be acceptable. Each manufacturer's pump can be different, but all must meet the minimum criteria.

7.1.2.1 Pumps shall furnish not less than 150 percent of rated capacity at a total head of not less than 65 percent of the total rated head. *(See Figure A.6.2.)*

7.1.2.2 The total shutoff head shall not exceed 140 percent of the total rated head on vertical turbine pumps. *(See Figure A.6.2.)*

7.2 Water Supply

7.2.1 Source.

7.2.1.1* The water supply shall be adequate, dependable, and acceptable to the authority having jurisdiction.

> ◀ **CALCULATION**
>
> Water supplies need to be thoroughly evaluated. Without a sufficient and reliable water supply, the entire fire protection system is in jeopardy. A fire pump augments an existing water supply by increasing the pressure. The fire pump cannot create water; it only increases pressure. The amount of water necessary for a fire protection system in terms of flow, pressure, and capacity is obtained through other NFPA standards, such as NFPA 13, *Standard for the Installation of Sprinkler Systems,* or NFPA 14, *Standard for the Installation of Standpipe and Hose Systems;* an authority having jurisdiction (AHJ), such as an insurance company; and building codes. The amount of water available from a supply can be evaluated using NFPA 291, *Recommended Practice for Fire Flow Testing and Marking of Hydrants.* When the municipal water supply is inadequate, the supply must be augmented using tanks, reservoirs, lakes, or rivers.
>
> The components of a fire protection system are limited by their pressure rating. The use of the correct components in the design of a fire pump system is important to eliminate overpressure problems. The pump churn or shutoff pressure is often a restricting factor that is missed in the design process, which then requires compensation in the form of the addition of pressure relief or control devices that have less reliability than the fire pump or fire protection system components. Where vertical shaft turbine–type fire pumps draw water from naturally occurring water sources such as wells, lakes, rivers, or ponds, consideration should be given to factors such as the seasonal fluctuation of water level, the presence of snow and ice, the accumulation of trash, and chemical and/or organic factors that may cause corrosion and/or tuberculation.

▶ **FAQ** | How is the depth of a suction pit determined?

> The minimum water level required for vertical turbine pumps to draw suction must be considered for aboveground and belowground tanks. Generally, a suction pit needs to be constructed below the tank, so that the total capacity of the tank can be used by the pump. The depth of the suction pit depends on the minimum submergence required for the application.

A.7.2.1.1 Stored water supplies from reservoirs or tanks supplying wet pits are preferable. Lakes, streams, and groundwater supplies are acceptable where investigation shows that they can be expected to provide a suitable and reliable supply.

Lakes, streams, and groundwater supplies are subject to changes in operating water level, which, in turn, affect the output flow and pressure of the fire pump. In extreme circumstances (such as during a drought), these water supplies can run dry, which prevents the pump from providing any fire protection. Another aspect that must be considered is that these natural water supplies can carry abrasives (such as sand, silt, or gravel), as well as naturally occurring chemicals (such as calcium, salts, and sulfides), which can lead to failure from wear or corrosion to pump components.

7.2.1.2* The acceptance of a well as a water supply source shall be dependent upon satisfactory development of the well and establishment of satisfactory aquifer characteristics.

A.7.2.1.2 The authority having jurisdiction can require an aquifer performance analysis. The history of the water table should be carefully investigated. The number of wells already in use in the area and the probable number that can be in use should be considered in relation to the total amount of water available for fire protection purposes.

Lakes, streams, and groundwater supplies must be carefully analyzed for suitability and serviceability. Data not older than 10 years should be used to determine the minimum conditions regarding the water level expected. The following are some of the factors to be considered in the aquifer analysis:

- Water level in the source and how it changes from year to year and during the year
- Seasonal changes in overall water elevation
- Ground freezing conditions in cold climates
- Presence of trash and debris in the water source

7.2.2 Pump Submergence.

Pump submergence requirements must be reviewed for each application. Pump manufacturers publish minimum submergence requirements to prevent surface vortices. In addition, the Hydraulic Institute has published minimum submergence guidelines for vertical turbine pumps in the *Hydraulic Institute Standards for Centrifugal, Rotary & Reciprocating Pumps*. This minimum submergence is intended to prevent surface vortices only and is not intended to prevent cavitation. The designer must review the operating requirements of the pump and check the net positive suction head rate (NPSHR) at runout as well to ensure that the pump will not cavitate. The NPSHR may require additional water level over the minimum submergence required to prevent surface vortices.

7.2.2.1* Well Installations.

A.7.2.2.1 See Figure A.7.2.2.1.

7.2.2.1.1 Submergence of the pump bowls shall be provided for reliable operation of the fire pump unit.

7.2.2.1.2 Submergence of the second impeller from the bottom of the pump bowl assembly shall be not less than 10 ft (3.2 m) below the pumping water level at 150 percent of rated capacity. *(See Figure A.7.2.2.1.)*

7.2.2.1.3* The submergence shall be increased by 1 ft (0.3 m) for each 1000 ft (305 m) of elevation above sea level.

The increase specified in 7.2.2.1.3 is in addition to any other increases that are required elsewhere in the standard. Failure to include this increase can lead to cavitation in the pump when operating at higher pump flows.

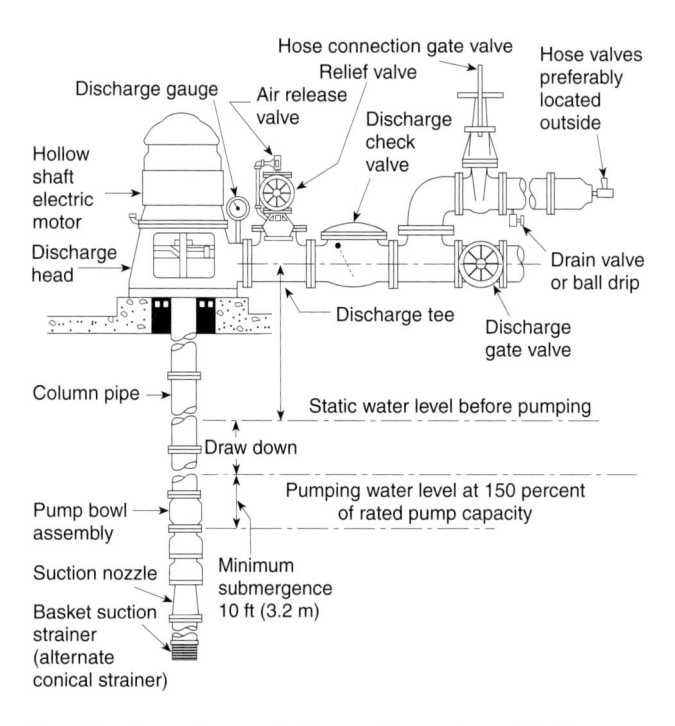

FIGURE A.7.2.2.1 *Vertical Shaft Turbine–Type Pump Installation in a Well.*

A.7.2.2.1.3 The acceptability of a well is determined by a 24-hour test that flows the well at 150 percent of the pump flow rating. This test should be reviewed by qualified personnel (usually a well drilling contractor or a person having experience in hydrology and geology). The adequacy and reliability of the water supply are critical to the successful operation of the fire pump and fire protection system.

A 10 ft (3.05 m) submergence is considered the minimum acceptable level to provide proper pump operation in well applications. The increase of 1 ft (0.30 m) for each 1000 ft (305 m) increase in elevation is due to loss of atmospheric pressure that accompanies elevation. Therefore, the net positive suction head (NPSH) available must be considered in selection of the pump. For example, to obtain the equivalent of 10 ft (3.05 m) of NPSH available at an elevation of 1000 ft (305 m), approximately 11 ft (3.35 m) of water is required.

Several other design parameters need to be considered in the selection of a vertical turbine pump, including the following:

(1) *Lineshaft lubrication when the pump is installed in a well.* Bearings are required to have lubrication and are installed along the lineshaft to maintain alignment. Lubrication fluid is usually provided by a fluid reservoir located aboveground, and the fluid is supplied to each bearing by a copper tube or small pipe. This lubrication fluid should use a vegetable-based material that is approved by the federal Clean Water Act to minimize water contamination.

(2) *Determination of the water level in the well.* When a vertical turbine pump is tested, the water level in the well needs to be known so that the suction pressure can be determined. Often the air line for determining the depth is omitted, so testing of the pump for performance is not possible. The arrangement of this device is shown in Figure A.7.3.5.3, and its installation should be included in the system design.

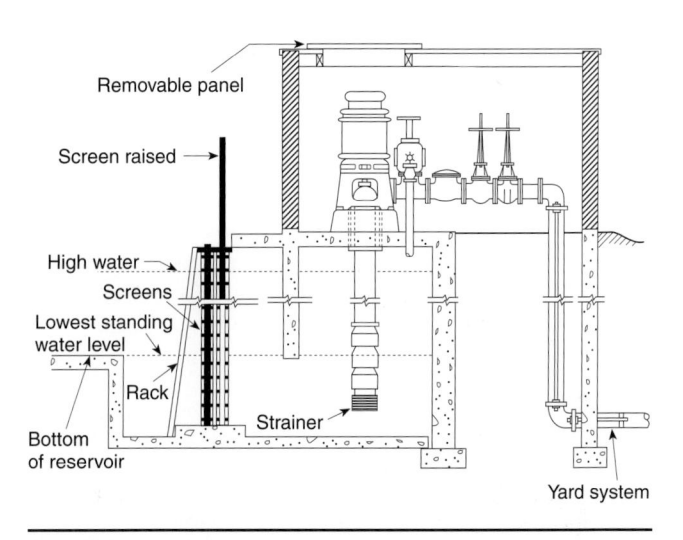

FIGURE A.7.2.2.2 *Vertical Shaft Turbine–Type Pump Installation in a Wet Pit.*

7.2.2.2* Wet Pit Installations.

A.7.2.2.2 The velocities in the approach channel or intake pipe should not exceed approximately 2 ft/sec (0.7 m/sec), and the velocity in the wet pit should not exceed approximately 1 ft/sec (0.3 m/sec). *(See Figure A.7.2.2.2.)*

The ideal approach is a straight channel coming directly to the pump. Turns and obstructions are detrimental because they can cause eddy currents and tend to initiate deep-cored vortices. The amount of submergence for successful operation will depend greatly on the approaches of the intake and the size of the pump.

The *Hydraulic Institute Standards for Centrifugal, Rotary and Reciprocating Pumps* recommends sump dimensions for flows 3000 gpm (11,355 L/min) and larger. The design of sumps for pumps with discharge capacities less than 3000 gpm (11,355 L/min) should be guided by the same general principles shown in the *Hydraulic Institute Standards for Centrifugal, Rotary and Reciprocating Pumps*.

> The quality of the water source is a concern if a pond, a lake, or other natural source is used as a water supply. Open ponds and lakes may contain fish, debris, leaves, or other floating materials that could cause problems with the pump's operation.
>
> The function of the intake structure — whether it is an open channel, a fully wetted tunnel, a sump, or a tank — is to supply an evenly distributed flow to the pump suction. An uneven distribution of flow, such as that caused by strong local currents, can result in the formation of surface or submerged vortices. With certain low values of submergence, an uneven distribution of flow can introduce air into the pump, causing reduced capacity, increased vibration, and additional noise. Uneven flow distribution can also increase or decrease the power consumption with a change in total developed head.
>
> Calculation of low average velocity, as recommended by A.7.2.2.2, is one means of judging the suitability of an intake structure. High velocities of isolated currents and swirls may be present; however, in general, very low average velocities are present.
>
> Water should not flow past one pump suction bell, suction pipe, or other intake to reach the next intake. If the intakes must be placed in line with the flow, it might be necessary to construct an open front cell around each pump intake or to place turning vanes under the intake to deflect the water upward. Streamlining should be used to reduce alternating vortices in the wake of an intake or other obstructions in the stream flow.

> The amount of submergence available is only one factor affecting vortex-free operation. The pump can have adequate submergence and still be subject to submerged vortices that may have an adverse effect on pump operation. Successful vortex-free operation depends greatly on the design approach used upstream of the sump. The design of any suction inlet structure should use the principles of the *Hydraulic Institute Standards for Centrifugal, Rotary & Reciprocating Pumps*.

7.2.2.2.1 To provide submergence for priming, the elevation of the second impeller from the bottom of the pump bowl assembly shall be such that it is below the lowest pumping water level in the open body of water supplying the pit.

7.2.2.2.2 For pumps with rated capacities of 2000 gpm (7570 L/min) or greater, additional submergence is required to prevent the formation of vortices and to provide required net positive suction head (NPSH) in order to prevent excessive cavitation.

7.2.2.2.3 The required submergence shall be obtained from the pump manufacturer.

7.2.2.2.4 The distance between the bottom of the strainer and the bottom of the wet pit shall be at least one-half of the pump bowl diameter but not less than 12 in. (305 mm).

7.2.3 Well Construction.

7.2.3.1 It shall be the responsibility of the groundwater supply contractor to perform the necessary groundwater investigation to establish the reliability of the supply, to develop a well to produce the required supply, and to perform all work and install all equipment in a thorough and workmanlike manner.

> Using a well is much more complex than using an underground tank as a water source. Reliability of the water supply is critical. The groundwater investigation required by 7.2.3.1 must be performed prior to well construction and purchase of the pump. A critically important consideration in the well analysis is the well's water level with the pump operating at churn (zero flow). The fire protection components may not be installed to withstand the developed pressures if a large drawdown of the water level occurs at zero flow and 150 percent flow. The well casing may have to be perforated to allow additional flow into the casing and reduce the drawdown.

7.2.3.2 The vertical turbine–type pump shall be designed to operate in a vertical position with all parts in correct alignment.

7.2.3.3 To support the requirements of 7.2.3.1, the well shall be of ample diameter and sufficiently plumb to receive the pump.

> The vertical shaft turbine–type pump should be installed vertically, plumbed for straightness, and checked for clearance between the pump bowls and well casing before the pump is started. The longer the pump column, the more critical the well casing and pump straightness become. If the well casing is not straight, it will push against the pump, causing the column and shaft to bend. The pump shaft could then make contact with the pump sleeve bearings, which can cause vibration and wear during operation, ultimately leading to premature pump failure. The depth of the well is no longer limited when the well is used as a fire protection water supply. Therefore, as the depth increases, the requirements for vertical alignment become more critical. See Exhibit II.7.1.
>
> The vertical shaft turbine–type pump should be mounted on a machined, level, grouted sole plate. The mating surfaces between the sole plate and discharge head need to be clean and free of dirt, rust, burrs, or scale.

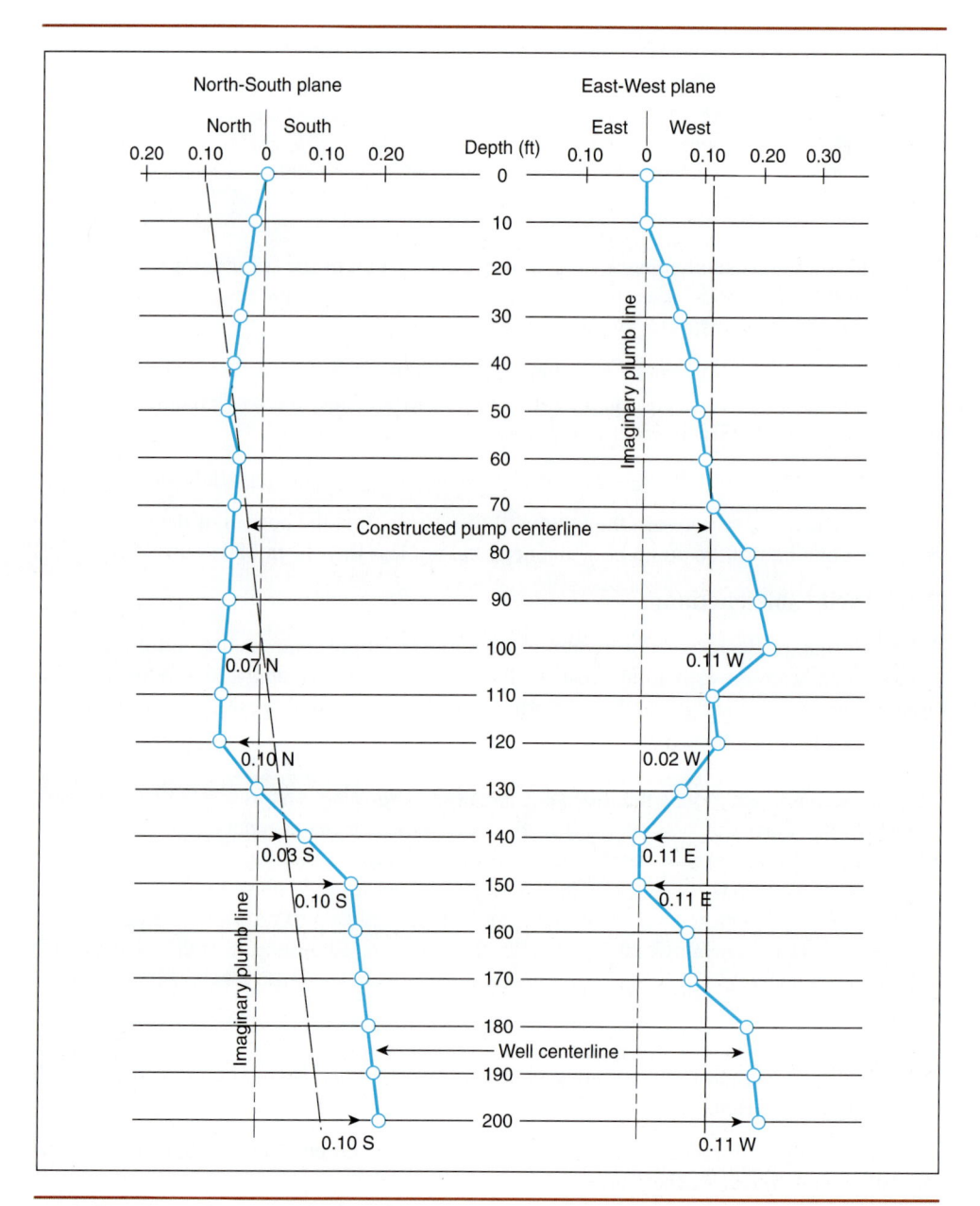

EXHIBIT II.7.1 Well and Pump Centerlines. (Reprinted from AWWA Standard A100-06, Water Wells, by permission. Copyright © 2006 by the American Water Works Association.)

7.2.4 Unconsolidated Formations (Sands and Gravels).

7.2.4.1 All casings shall be of steel of such diameter and installed to such depths as the formation could justify and as best meet the conditions.

7.2.4.2 Both inner and outer casings shall have a minimum wall thickness of 0.375 in. (9.5 mm).

7.2.4.3 Inner casing diameter shall be not less than 2 in. (51 mm) larger than the pump bowls.

7.2.4.4 The outer casing shall extend down to approximately the top of the water-bearing formation.

7.2.4.5 The inner casing of lesser diameter and the well screen shall extend as far into the formation as the water-bearing stratum could justify and as best meets the conditions.

> The outer casing must protect the pump bowls and pump column from damage under all conditions of service. The purpose of the outer casing is to keep the well walls from collapsing onto the line shaft and pump.

7.2.4.6 The well screen is a vital part of the construction, and careful attention shall be given to its selection.

7.2.4.7 The well screen shall be the same diameter as the inner casing and of the proper length and percent open area to provide an entrance velocity not exceeding 0.15 ft/sec (46 mm/sec).

7.2.4.8 The screen shall be made of a corrosion- and acid-resistant material, such as stainless steel or Monel.

7.2.4.9 Monel shall be used where it is anticipated that the chloride content of the well water will exceed 1000 parts per million.

7.2.4.10 The screen shall have adequate strength to resist the external forces that will be applied after it is installed and to minimize the likelihood of damage during the installation.

7.2.4.11 The bottom of the well screen shall be sealed properly with a plate of the same material as the screen.

> Proper selection of the screen material and size is as critical to the successful operation of the pump as is low entrance velocity of the water. If the screen fails or the screen size is incorrect, sand or larger rock particles could enter the well and cause damage to the pump impellers, bearings, and other pump components.

7.2.4.12 The sides of the outer casing shall be sealed by the introduction of neat cement placed under pressure from the bottom to the top.

> The well must be sealed by cement to prevent the ingress of sand or rock particles. This installation is usually a one-time event. Installation should be performed with extreme care by experienced personnel.

7.2.4.13 Cement shall be allowed to set for a minimum of 48 hours before drilling operations are continued.

7.2.4.14 The immediate area surrounding the well screen not less than 6 in. (152 mm) shall be filled with clean and well-rounded gravel.

7.2.4.15 This gravel shall be of such size and quality as will create a gravel filter to ensure sand-free production and a low velocity of water leaving the formation and entering the well.

7.2.4.16 Tubular Wells.

7.2.4.16.1 Wells for fire pumps not exceeding 450 gpm (1703 L/min) developed in unconsolidated formations without an artificial gravel pack, such as tubular wells, shall be acceptable sources of water supply for fire pumps not exceeding 450 gpm (1703 L/min).

7.2.4.16.2 Tubular wells shall comply with all the requirements of 7.2.3 and 7.2.4, except compliance with 7.2.4.11 through 7.2.4.15 shall not be required.

7.2.5* Consolidated Formations. Where the drilling penetrates unconsolidated formations above the rock, surface casing shall be installed, seated in solid rock, and cemented in place.

> The surface casing prevents sand and rock particles from unconsolidated areas above rock formations from entering the well and also prevents the well walls from collapsing.

A.7.2.5 Where wells take their supply from consolidated formations such as rock, the specifications for the well should be decided upon by the authority having jurisdiction after consultation with a recognized groundwater consultant in the area.

> It is critical that the water supply obtained from a well be of sufficient capacity and that it be reliable. Consideration should be given to seasonal and yearly fluctuations in the water supply.
>
> When installing vertical shaft turbine–type pumps in wells, consideration must be given to the formation of the well before installation. Installing a unit in a crooked well may bind and distort the pump column or pump–motor assembly, leading to potential malfunction. Well straightness should be within 1 in./100 ft (25.4 mm/30.5 m) and without double bend. The well should be "gauged" prior to installation by lowering a dummy assembly on a cable into the well. This assembly should be slightly longer and larger in diameter than the actual pump or pump–motor assembly. Gauging is also important when a stepped well casing is used — that is, when the lower part of the well casing has a smaller inside diameter than the upper part.
>
> Wells that have not been properly constructed or developed or wells that produce sand can be detrimental to the pump. If a well is suspected of producing an excessive amount of sand, a unit other than the fire pump must be used to clear the well. (See also 7.2.6.) If corrosives are present, pump material selection must be made to accommodate this condition. Special materials such as nickel-aluminum-bronze alloy, stainless steel, or other special alloys can be specified.

7.2.6 Developing a Well.

7.2.6.1 Developing a new well and cleaning it of sand or rock particles (not to exceed 5 ppm) shall be the responsibility of the groundwater supply contractor.

7.2.6.2 Such development shall be performed with a test pump and not a fire pump.

7.2.6.3 Freedom from sand shall be determined when the test pump is operated at 150 percent of rated capacity of the fire pump for which the well is being prepared.

7.2.7* Test and Inspection of Well.

A.7.2.7 Before the permanent pump is ordered, the water from the well should be analyzed for corrosiveness, including such items as pH, salts such as chlorides, and harmful gases such as carbon dioxide (CO_2) or hydrogen sulfide (H_2S). If the water is corrosive, the pumps should be constructed of a suitable corrosion-resistant material or covered with special protective coatings in accordance with the manufacturers' recommendations.

7.2.7.1 A test to determine the water production of the well shall be made.

7.2.7.2 An acceptable water measuring device such as an orifice, a venturi meter, or a calibrated pitot tube shall be used.

7.2.7.3 The test shall be witnessed by a representative of the customer, contractor, and authority having jurisdiction, as required.

7.2.7.4 The test shall be continuous for a period of at least 8 hours at 150 percent of the rated capacity of the fire pump with 15-minute-interval readings over the period of the test.

7.2.7.5 The test shall be evaluated with consideration given to the effect of other wells in the vicinity and any possible seasonal variation in the water table at the well site.

7.2.7.6 Test data shall describe the static water level and the pumping water level at 100 percent and 150 percent, respectively, of the rated capacity of the fire pump for which the well is being prepared.

7.2.7.7 All existing wells within a 1000 ft (305 m) radius of the fire well shall be monitored throughout the test period.

7.3 Pump

7.3.1* Vertical Turbine Pump Head Component.

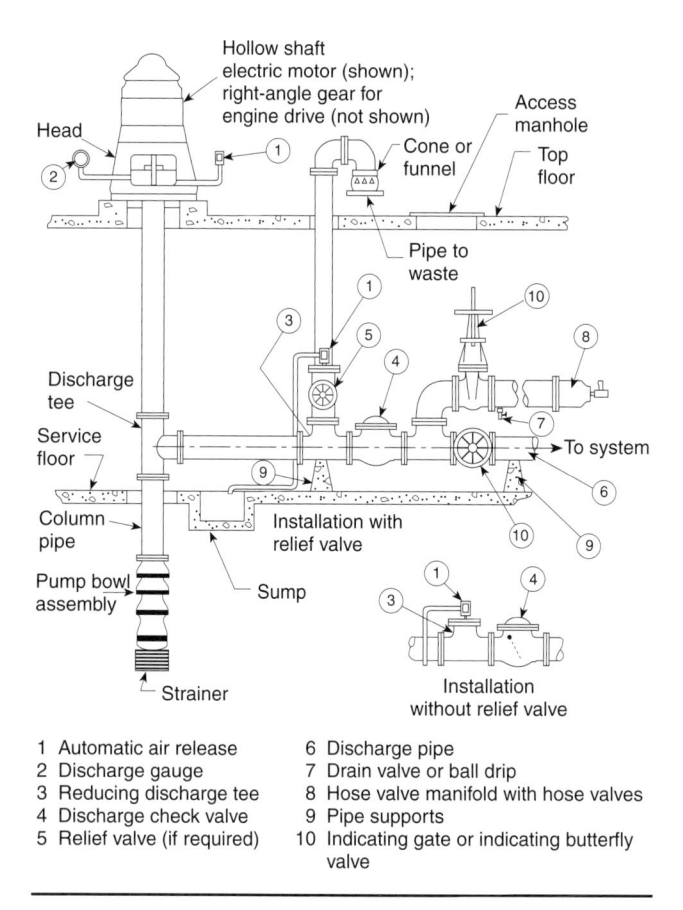

A.7.3.1 See Figure A.7.3.1.

FIGURE A.7.3.1 Belowground Discharge Arrangement.

7.3.1.1 The pump head shall be either the aboveground or belowground discharge type.

7.3.1.2 The pump head shall be designed to support the driver, pump, column assembly, bowl assembly, maximum down thrust, and the oil tube tension nut or packing container.

7.3.2 Column.

7.3.2.1* The pump column shall be furnished in sections not exceeding a nominal length of 10 ft (3 m), shall be not less than the weight specified in Table 7.3.2.1(a) and Table 7.3.2.1(b), and shall be connected by threaded-sleeve couplings or flanges.

TABLE 7.3.2.1(a) Pump Column Pipe Weights (U.S. Customary)

Nominal Size (in.)	Outside Diameter (O.D.) (in.)	Weight per Unit Length (Plain Ends) (lb/ft)
6	6.625	18.97
7	7.625	22.26
8	8.625	24.70
9	9.625	28.33
10	10.75	31.20
12	12.75	43.77
14	14.00	53.57

TABLE 7.3.2.1(b) Pump Column Pipe Weights (Metric)

Nominal Size (mm)	Outside Diameter (O.D.) (mm)	Weight per Unit Length (Plain Ends) (kg/m)
150	161	28.230
200	212	36.758
250	264	46.431
300	315	65.137
350	360	81.209

A.7.3.2.1 In countries that utilize the metric system, there do not appear to be standardized flow ratings for pump capacities; therefore, a soft metric conversion is utilized.

7.3.2.2 The ends of each section of threaded pipe shall be faced parallel and machined with threads to permit the ends to butt so as to form accurate alignment of the pump column.

7.3.2.3 All column flange faces shall be parallel and machined for rabbet fit to permit accurate alignment.

7.3.2.4 Where the static water level exceeds 50 ft (15.3 m) below ground, oil-lubricated-type pumps shall be used. *(See Figure A.7.1.1.)*

> When static water levels exceed 50 ft (15.3 m), the upper line shaft and stuffing box bearing on a water-lubricated open line shaft construction pump can run dry at start-up, causing severe damage and pump failure. Therefore, an oil-lubricated enclosed line shaft is employed to protect the upper pump bearings. Biodegradable food-grade oil is available for enclosed line shaft arrangements when authorities or health departments show concern about oil use. No other type of lubrication (such as grease or fresh water flush) is allowed.

7.3.2.5 Where the pump is of the enclosed line shaft oil-lubricated type, the shaft-enclosing tube shall be furnished in interchangeable sections not over 10 ft (3 m) in length of extra-strong pipe.

7.3.2.6 An automatic sight feed oiler shall be provided on a suitable mounting bracket with connection to the shaft tube for oil-lubricated pumps. *(See Figure A.7.1.1.)*

7.3.2.7 The pump line shafting shall be sized so critical speed shall be 25 percent above and below the operating speed of the pump.

> When a vertical shaft turbine–type pump ramps up to speed, it generally passes through one or more critical speeds. Critical speeds correspond to the speeds of the rotating pump shaft. At critical speeds, the pump shaft reaches its natural frequency of vibration and becomes unstable. A pump passing through a critical speed has no detrimental effect on the pump, because it passes through this point in 1 second or less. However, running the pump at a critical speed is not recommended for any pump; most pumps operate between two critical speeds. Experience has shown that a 25 percent margin is a safe range for pump operation. If a higher percentage were to be used, the results would be the gross oversizing of the pump shaft and much closer bearing spacing than is required in present pump designs.

7.3.2.8 Operating speed shall include all speeds from shutoff to the 150 percent point of the pump, which vary on engine drives.

7.3.2.9 Operating speed for variable speed pressure limiting control drive systems shall include all speeds from rated to minimum operating speed.

> In some cases, changing the size of the shafting alone will not move a critical speed outside the operating range. Column length, and, consequently, line shaft bearing spacing, greatly affects the location of the first and second lateral critical speeds. Often it is found that nominal 5 ft (1.5 m) column lengths are employed on vertical turbine pumps to move the first critical speed above the pump operating speed. This enables the variable frequency drive (VFD) controller to use the full speed range as needed without locking out a specific speed.

7.3.3 Bowl Assembly.

7.3.3.1 The pump bowl shall be of close-grained cast iron, bronze, or other suitable material in accordance with the chemical analysis of the water and experience in the area.

> The proper material for the pump bowl must be specified when the pump is ordered. The application, water conditions, and local conditions determine the material that is most suitable for the application.

7.3.3.2 Impellers shall be of the enclosed type and shall be of bronze or other suitable material in accordance with the chemical analysis of the water and experience in the area.

> Enclosed impeller design is specified by 7.3.3.2 because, with the hydraulic design of a vertical shaft turbine–type pump, shaft stretch can impact the operating efficiency of a semi-open impeller. The efficiency of semi-open impellers decreases as the depth of the pumping water level increases. Enclosed impellers also generate less downthrust than semi-open impellers.

7.3.4 Suction Strainer.

7.3.4.1 A cast or heavy fabricated, corrosion-resistant metal cone or basket-type strainer shall be attached to the suction manifold of the pump.

> Cone strainers are used on well installations because of their smaller diameter, enabling them to fit down the well casing. Basket strainers are larger in diameter but shorter in height, enabling vertical turbines to utilize the maximum available water in a wet pit.

7.3.4.2 The suction strainer shall have a free area of at least four times the area of the suction connections, and the openings shall be sized to restrict the passage of a 0.5 in. (12.7 mm) sphere.

> The openings in the strainer are sized to prevent clogging of the impeller, thus protecting the pump from potentially damaging debris in the wet pit or well.

7.3.4.3 For installations in a wet pit, this suction strainer shall be required in addition to the intake screen. *(See Figure A.7.2.2.2.)*

> The intake screen usually allows passage of objects with sphere sizes larger than 0.5 in. (12.7 mm). Screens serve to protect the pump from objects falling into the wet pit or biological growth that may occur in an open pond or stream. These objects may clog the impeller or suction strainer.

7.3.5 Fittings.

7.3.5.1 The following fittings shall be required for attachment to the pump:

(1) Automatic air release valve as specified in 7.3.5.2
(2) Water level detector as specified in 7.3.5.3
(3) Discharge pressure gauge as specified in 4.10.1
(4) Relief valve and discharge cone where required by 4.18.1
(5) Hose valve header and hose valves as specified in 4.20.3 or metering devices as specified in 4.20.2

> The pump installation must also include piping for interconnection of the fittings per the design.
> The installation of a flowmeter is preferred to measure the flow during the acceptance testing required by Chapter 14 and the periodic testing required by NFPA 25, *Standard for the Inspection, Testing, and Maintenance of Water-Based Fire Protection Systems*, although the installation of a test header can be beneficial for use as a fire hydrant.

7.3.5.2 Automatic Air Release.

> NFPA 20, *Standard for the Installation of Stationary Pumps for Fire Protection*, requires an air release valve because it is critical that air within the pump column and pump head be vented when the pump is started. The presence of trapped air can damage the vertical turbine pump and system components and impair operation of the fire protection system.
> It is equally important that, when the pump stops, a vacuum condition does not exist within the pump column. For this reason, the air release valve must be a type that also allows air to enter the pump column when the pump stops. See Exhibit II.7.2 for an illustration of an air release valve.

7.3.5.2.1 A nominal 1.5 in. (38 mm) pipe size or larger automatic air release valve shall be provided to vent air from the column and the discharge head upon the starting of the pump.

7.3.5.2.2 This valve shall also admit air to the column to dissipate the vacuum upon stopping of the pump.

EXHIBIT II.7.2 *Air Release Valve.*

7.3.5.2.3 This valve shall be located at the highest point in the discharge line between the fire pump and the discharge check valve.

7.3.5.3* **Water Level Detection.** Water level detection shall be required for all vertical turbine pumps installed in wells to monitor the suction pressure available at the shutoff, 100 percent flow, and 150 percent flow points, to determine if the pump is operating within its design conditions.

> NFPA 20 requires water level detectors because the water level must be determined in order to calculate the net pump discharge pressure. Water level detectors of other than the air line type can be used if they are approved for the intended use. The electronic type is preferred for pumps taking suction from below-grade tanks and suction cribs.

A.7.3.5.3 Water level detection using the air line method is as follows:

(1) A satisfactory method of determining the water level involves the use of an air line of small pipe or tubing of known vertical length, a pressure or depth gauge, and an ordinary bicycle or automobile pump installed as shown in Figure A.7.3.5.3. The air line pipe should be of known length and extend beyond the lowest anticipated water level in the well, to ensure more reliable gauge readings, and should be properly installed. An air pressure gauge is used to indicate the pressure in the air line. *(See Figure A.7.3.5.3.)*

(2) The air line pipe is lowered into the well, a tee is placed in the line above the ground, and a pressure gauge is screwed into one connection. The other connection is fitted with an ordinary bicycle valve to which a bicycle pump is attached. All joints should be made carefully and should be airtight to obtain correct information. When air is forced into the line by means of the bicycle pump, the gauge pressure increases until all of the water has been expelled. When this point is reached, the gauge reading becomes constant. The maximum maintained air pressure recorded by the gauge is equivalent to that necessary to support a column of water of the same height as that forced out of the air line. The length of this water column is equal to the amount of air line submerged.

Stationary Fire Pumps Handbook 2013

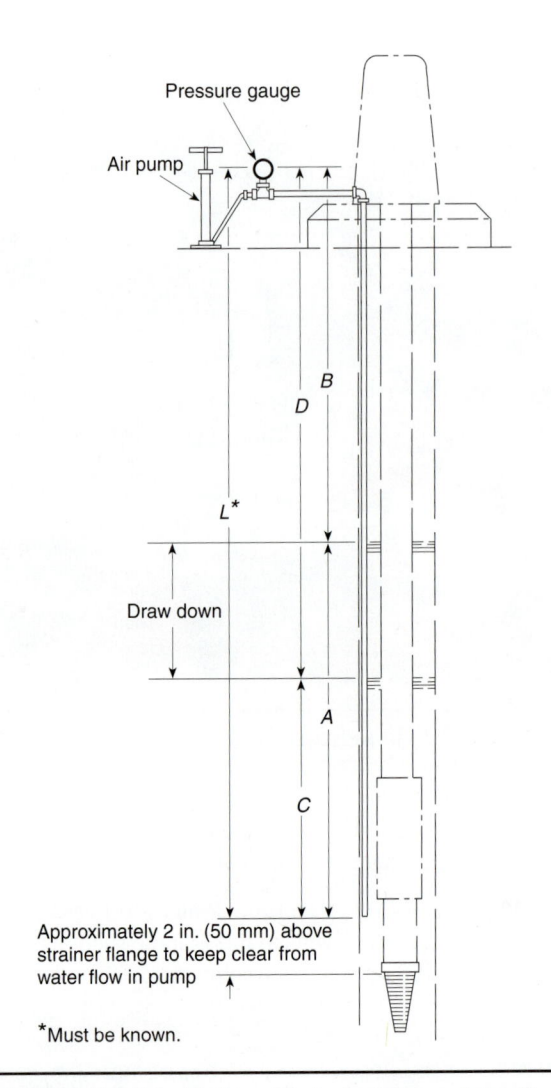

FIGURE A.7.3.5.3 *Air Line Method of Determining Depth of Water Level.*

(3) Deducting this pressure converted to feet (meters) (pressure in psi × 2.31 = pressure in feet, and pressure in bar × 10.3 = pressure in meters) from the known length of the air line will give the amount of submergence.

CALCULATION ▶

Example: The following calculation will serve to clarify Figure A.7.3.5.3.

Assume a length (*L*) of 50 ft (15.2 m).

The pressure gauge reading before starting the fire pump (p_1) = 10 psi (0.68 bar). Then *A* = 10 × 2.31 = 23.1 ft (0.68 × 10.3 = 7.0 m). Therefore, the water level in the well before starting the pump would be *B* = *L* − *A* = 50 ft − 23.1 ft = 26.9 ft (*B* = *L* − *A* = 15.2 m − 7 m = 8.2 m).

The pressure gauge reading when the pump is running (p_2) = 8 psi (0.55 bar). Then *C* = 8 × 2.31 = 18.5 ft (0.55 × 10.3 = 5.6 m). Therefore, the water level in the well when the pump is running would be *D* = *L* − *C* = 50 ft − 18.5 ft = 31.5 ft (*D* = *L* − *C* = 15.2 m − 5.6 m = 9.6 m).

The draw down can be determined by any of the following methods:

(1) *D* − *B* = 31.5 ft − 26.9 ft = 4.6 ft (9.6 m − 8.2 m = 1.4 m)
(2) *A* − *C* = 23.1 ft − 18.5 ft = 4.6 ft (7.0 m − 5.6 m = 1.4 m)
(3) $p_1 - p_2$ = 10 − 8 = 2 psi = 2 × 2.31 = 4.6 ft (0.68 − 0.55 = 0.13 bar = 0.13 × 10.3 = 1.4 m)

Section 7.4 • Installation   165

7.3.5.3.1 Each well installation shall be equipped with a suitable water level detector.

7.3.5.3.2 If an air line is used, it shall be brass, copper, or series 300 stainless steel.

7.3.5.3.3 Air lines shall be strapped to column pipe at 10 ft (3 m) intervals.

7.4* Installation

Exhibit II.7.3(a) through Exhibit II.7.3(d) show the process of installing a vertical turbine fire pump.

A.7.4 Several methods of installing a vertical pump can be followed, depending upon the location of the well and facilities available. Since most of the unit is underground, extreme

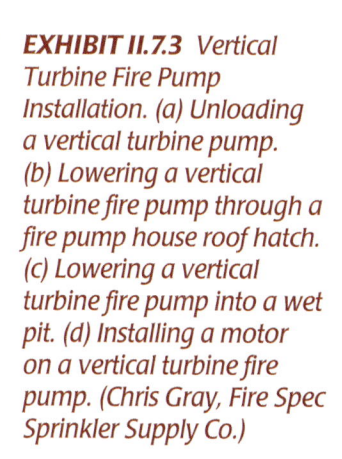

EXHIBIT II.7.3 *Vertical Turbine Fire Pump Installation. (a) Unloading a vertical turbine pump. (b) Lowering a vertical turbine fire pump through a fire pump house roof hatch. (c) Lowering a vertical turbine fire pump into a wet pit. (d) Installing a motor on a vertical turbine fire pump. (Chris Gray, Fire Spec Sprinkler Supply Co.)*

(a)
(b)
(c)
(d)

Stationary Fire Pumps Handbook   2013

care should be used in assembly and installation, thoroughly checking the work as it progresses. The following simple method is the most common:

(1) Construct a tripod or portable derrick and use two sets of installing clamps over the open well or pump house. After the derrick is in place, the alignment should be checked carefully with the well or wet pit to avoid any trouble when setting the pump.
(2) Attach the set of clamps to the suction pipe on which the strainer has already been placed and lower the pipe into the well until the clamps rest on a block beside the well casing or on the pump foundation.
(3) Attach the clamps to the pump stage assembly, bring the assembly over the well, and install pump stages to the suction pipe, until each piece has been installed in accordance with the manufacturer's instructions.

7.4.1 Pump House.

7.4.1.1 The pump house shall be of such design as will offer the least obstruction to the convenient handling and hoisting of vertical pump parts.

The pump house normally includes a roof hatch at least 4 ft × 4 ft (1219 mm × 1219 mm) in size to provide access to the pump room so that the line shaft can be removed for maintenance. When establishing clearances, *NFPA 70®*, *National Electrical Code®*, requirements, as well as normal working distances, need to be taken into consideration. External access needs to be large enough to remove a pump driver or other component. Heating, cooling, and combustion air units (for diesel engines) need to be included and sized for the location of the unit. Floor drainage and spill protection must also be included in any design to protect the environment.

7.4.1.2 The requirements of Sections 4.12 and 11.3 shall also apply.

7.4.2 Outdoor Setting.

7.4.2.1 If in special cases the authority having jurisdiction does not require a pump room and the unit is installed outdoors, the driver shall be screened or enclosed and protected against tampering.

The design of the pump and related equipment must still comply with the requirements of 4.12.1.

7.4.2.2 The screen or enclosure required in 7.4.2.1 shall be easily removable and shall have provision for ample ventilation.

7.4.3 Foundation.

In earthquake-prone areas, the foundation is required to be designed by a licensed professional engineer certified in the design of earthquake-resistant structures. The provisions for earthquake design are found in the applicable building code and in NFPA 13.

7.4.3.1 Certified dimension prints shall be obtained from the manufacturer.

7.4.3.2 The foundation for vertical pumps shall be built to carry the entire weight of the pump and driver plus the weight of the water contained in it.

7.4.3.3 Foundation bolts shall be provided to firmly anchor the pump to the foundation.

7.4.3.4 The foundation shall be of sufficient area and strength that the load per square inch (square millimeter) on concrete does not exceed design standards.

7.4.3.5 The top of the foundation shall be carefully leveled to permit the pump to hang freely over a well pit on a short-coupled pump.

7.4.3.6 On a well pump, the pump head shall be positioned plumb over the well, which is not necessarily level.

7.4.3.7 Sump or Pit.

7.4.3.7.1 Where the pump is mounted over a sump or pit, I-beams shall be permitted to be used.

7.4.3.7.2 Where a right-angle gear is used, the driver shall be installed parallel to the beams.

7.5 Driver

7.5.1 Method of Drive.

7.5.1.1 The driver provided shall be so constructed that the total thrust of the pump, which includes the weight of the shaft, impellers, and hydraulic thrust, can be carried on a thrust bearing of ample capacity so that it will have an average life rating of 5 years continuous operation.

7.5.1.2 All drivers shall be so constructed that axial adjustment of impellers can be made to permit proper installation and operation of the equipment.

7.5.1.3 Vertical shaft turbine pumps shall be driven by a vertical hollow shaft electric motor or vertical hollow shaft right-angle gear drive with diesel engine or steam turbine except as permitted in 7.5.1.4.

7.5.1.4 The requirements of 7.5.1.3 shall not apply to diesel engines and steam turbines designed and listed for vertical installation with vertical shaft turbine–type pumps, which shall be permitted to employ solid shafts and shall not require a right-angle gear drive but shall require a nonreverse ratchet.

7.5.1.5 Motors shall be of the vertical hollow-shaft type and comply with 9.5.1.9.

7.5.1.6 Mass Elastic System.

7.5.1.6.1 For drive systems that include a right angle gear drive, the pump manufacturer shall provide a complete mass elastic system torsional analysis to ensure there are no damaging stresses or critical speeds within 25 percent above and below the operating speed of the pump and drive.

7.5.1.6.2 The torsional analysis specified in 7.5.1.6.1 shall include the mass elastic characteristics for a wetted pump with the specific impeller trim, coupling, right-angle gear, flexible connecting shaft, and engine, plus the excitation characteristics of the engine.

7.5.1.6.3 For variable speed vertical hollow shaft electric motors, the pump manufacturer shall provide a complete mass elastic system torsional analysis to ensure there are no damaging stresses or critical speeds within 25 percent above and below the operating speed of the pump and drive.

7.5.1.6.4 For vertical turbine pumps using angle gear drives driven by a diesel engine, a torsional vibration damping type coupling shall be used and mounted on the engine side of the driver shaft.

7.5.1.6.4.1 The torsional vibration damping type coupling shall be permitted to be omitted when a mass elastic system torsional analysis is provided and accepted by the authority having jurisdiction.

7.5.1.7 Gear Drives.

7.5.1.7.1 Gear drives and flexible connecting shafts shall be acceptable to the authority having jurisdiction.

7.5.1.7.2 Gear drives shall be of the vertical hollow-shaft type, permitting adjustment of the impellers for proper installation and operation of the equipment.

7.5.1.7.3 The gear drive shall be equipped with a nonreverse ratchet.

7.5.1.7.4 All gear drives shall be listed and rated by the manufacturer at a load equal to the maximum horsepower and thrust of the pump for which the gear drive is intended.

7.5.1.7.5 Water-cooled gear drives shall be equipped with a visual means to determine whether water circulation is occurring.

7.5.1.8 Flexible Connecting Shafts.

7.5.1.8.1 The flexible connecting shaft shall be listed for this service.

7.5.1.8.2 The operating angle for the flexible connecting shaft shall not exceed the limits specified by the manufacturer for the speed and horsepower transmitted under any static or operating conditions.

> The operating angle is usually set for a minimum offset angle of 1 degree up to a maximum of 3 degrees. This angle is intended to prevent failure of the universal joints during operation.

7.5.2 Controls. The controllers for the motor, diesel engine, or steam turbine shall comply with specifications for either electric-drive controllers in Chapter 10 or engine drive controllers in Chapter 12.

> ▶ **FAQ** When should the use of a variable speed controller be considered?

> Coordination of the variable speed controller manufacturer, engineers, electric power supplier, owner, installing contractor, and the AHJ is essential in the proper selection of the controller and the required features. A full-service controller is preferred for electric motor drives. Use of a variable speed controller should be considered when high discharge pressures are required and electric costs are excessive.

7.5.3 Variable Speed Vertical Turbine Pumps.

7.5.3.1 The pump supplier shall inform the controller manufacturer of any and all critical resonant speeds within the operating speed range of the pump, which is from zero up to full speed.

7.5.3.2 When water-lubricated pumps with line shaft bearings are installed, the pump manufacturer shall inform the controller manufacturer of the maximum allowed time for water to reach the top bearing under the condition of the lowest anticipated water level of the well or reservoir.

7.6 Operation and Maintenance

7.6.1 Operation.

7.6.1.1* Before the unit is started for the first time after installation, all field-installed electrical connections and discharge piping from the pump shall be checked.

A.7.6.1.1 The setting of the impellers should be undertaken only by a representative of the pump manufacturer. Improper setting will cause excessive friction loss due to the rubbing of impellers on pump seals, which results in an increase in power demand. If the impellers are adjusted too high, there will be a loss in capacity, and full capacity is vital for fire pump service. The top shaft nut should be locked or pinned after proper setting.

7.6.1.2 With the top drive coupling removed, the drive shaft shall be centered in the top drive coupling for proper alignment and the motor shall be operated momentarily to ensure that it rotates in the proper direction.

> The following method may be used to mount and align vertical hollow shaft drivers:
>
> 1. Remove the coupling from the top of the hollow shaft and mount the driver on top of the discharge head/driver stand.
> 2. For designs requiring the pump head to be installed prior to mounting the driver (i.e., short cast heads with one-piece head shafts), lower the hollow shaft driver with care over the head shaft to ensure the latter is not damaged.
> 3. Check the driver for correct rotation, as given in the manufacturer's installation instructions.
> 4. Install the head shaft (if this has not already been done), and check that it is centered in the hollow shaft. If the head shaft is off-center, check for runout in the head shaft misalignment from the discharge head to the driver, or check whether the suspended pump is out of plumb. Shims can be placed under the discharge head to center the head shaft, but shims should not be placed between the motor and the discharge head.
> 5. Install the driver coupling and check the nonreverse ratchet for operability.
> 6. Install the coupling gib key and the adjusting nut, and raise the shaft assembly with the impeller(s) to the correct running position in accordance with the manufacturer's instructions.
> 7. Secure the adjusting nut to the clutch; double-check the driver hold-down bolts for tightness.
> 8. Most hollow shaft drivers have register fits. Therefore, further recentering of these drivers normally is not required, and dowels are not recommended.
> 9. Always consult the manufacturer's manual for proper installation and operation.

7.6.1.3 With the top drive coupling reinstalled, the impellers shall be set for proper clearance according to the manufacturer's instructions.

7.6.1.4* With the precautions of 7.6.1.1 through 7.6.1.3 taken, the pump shall be started and allowed to run.

A.7.6.1.4 Pumping units are checked at the factory for smoothness of performance and should operate satisfactorily on the job. If excessive vibration is present, the following conditions could be causing the trouble:

(1) Bent pump or column shaft
(2) Impellers not properly set within the pump bowls
(3) Pump not hanging freely in the well
(4) Strain transmitted through the discharge piping

Excessive motor temperature is generally caused either by a maintained low voltage of the electric service or by improper setting of impellers within the pump bowls.

7.6.1.5 The operation shall be observed for vibration while running, with vibration limits according to the *Hydraulic Institute Standards for Centrifugal, Rotary and Reciprocating Pumps*.

> The responsible installing contractor should generate a written report of the testing and running for inclusion in the acceptance test report.

7.6.1.6 The driver shall be observed for proper operation.

7.6.2 Maintenance.

7.6.2.1 The manufacturer's instructions shall be carefully followed in making repairs and dismantling and reassembling pumps.

7.6.2.2 When spare or replacement parts are ordered, the pump serial number stamped on the nameplate fastened to the pump head shall be included in order to make sure the proper parts are provided.

> The pump manufacturer must know the serial number in order to provide replacement parts that duplicate the original construction of the listed fire pump. This information ensures that the correct components are supplied, which maintains the listing of the pump if component replacement is required. Consult the operating manual for information regarding the ordering of parts.

7.6.2.3 Ample head room and access for removal of the pump shall be maintained.

> Accessibility for pump repairs and maintenance should be considered when the pump house or pump room is designed. Accessibility information is available from all major pump manufacturers to assist the designer with proper design.

References Cited in Commentary

National Fire Protection Association, 1 Batterymarch Park, Quincy, MA 02169-7471.

NFPA 13, *Standard for the Installation of Sprinkler Systems,* 2013 edition.
NFPA 14, *Standard for the Installation of Standpipe and Hose Systems,* 2010 edition.
NFPA 25, *Standard for the Inspection, Testing, and Maintenance of Water-Based Fire Protection Systems,* 2011 edition.
NFPA 70®, *National Electrical Code®,* 2011 edition.

NFPA 291, *Recommended Practice for Fire Flow Testing and Marking of Hydrants,* 2013 edition.

American Water Works Association, 6666 West Quincy Avenue, Denver, CO 80235.

ANSI/AWWA A100-06, *Standard for Water Wells,* 2006.

Hydraulic Institute, 6 Campus Drive, First Floor North, Parsippany, NJ 07054-4406.

Hydraulic Institute Standards for Centrifugal, Rotary & Reciprocating Pumps, 14th edition, 1983.

CHAPTER 8

Positive Displacement Pumps

The requirements in Chapter 8 for positive displacement fire pumps first appeared in the 1999 edition of NFPA 20, *Standard for the Installation of Stationary Pumps for Fire Protection*. Positive displacement pumps, sometimes referred to as PD pumps, are more apt to be found as part of water mist and foam-based fire protection system installations. Note that, in most instances, NFPA 20 uses the term *liquid* rather than the term *water*, due to the type of pumps addressed in Chapter 8. New to Chapter 8 are requirements for water mist positive displacement pumping units, which are assemblies of pumps, driver(s), and a controller as complete operating units. Requirements for pressure maintenance were also added to recognize the capability of water mist positive displacement pumping units to perform this function. The following relevant NFPA installation standards reference NFPA 20 for the installation of positive displacement pumps:

NFPA 11, *Standard for Low-, Medium-, and High-Expansion Foam*

NFPA 16, *Standard for the Installation of Foam-Water Sprinkler and Foam-Water Spray Systems*

NFPA 750, *Standard on Water Mist Fire Protection Systems*

8.1* General

A.8.1 All the requirements in Chapter 4 might not apply to positive displacement pumps.

▶ **FAQ** When the requirements in Chapter 8 conflict with those in Chapter 4, which chapter takes precedence?

For positive displacement fire pumps, in cases where requirements in Chapter 8 conflict with requirements in Chapter 4, the requirements in Chapter 8 take precedence.

8.1.1 Types. Positive displacement pumps shall be as defined in 3.3.38.13.

A *positive displacement pump* is defined in 3.3.38.13 as "a pump that is characterized by a method of producing flow by capturing a specific volume of fluid per pump revolution and reducing the fluid void by a mechanical means to displace the pumping fluid." Exhibit II.8.1 illustrates one style of positive displacement pump; Exhibit II.8.1 does not represent the only acceptable pump design or style.

▶ **FAQ** How do positive displacement pumps differ from (traditional) centrifugal pumps?

All fire pumps create pressure by imparting energy into the fluid being pumped. A positive displacement pump differs from (traditional) centrifugal pumps because a positive displacement pump imparts energy into the fluid by "pushing" the fluid, whereas a centrifugal pump imparts energy into the fluid through spinning action. In a common example, displacement forces keep a heavy boat from sinking through the pressure created by the displacement of water created by the bottom of the boat.

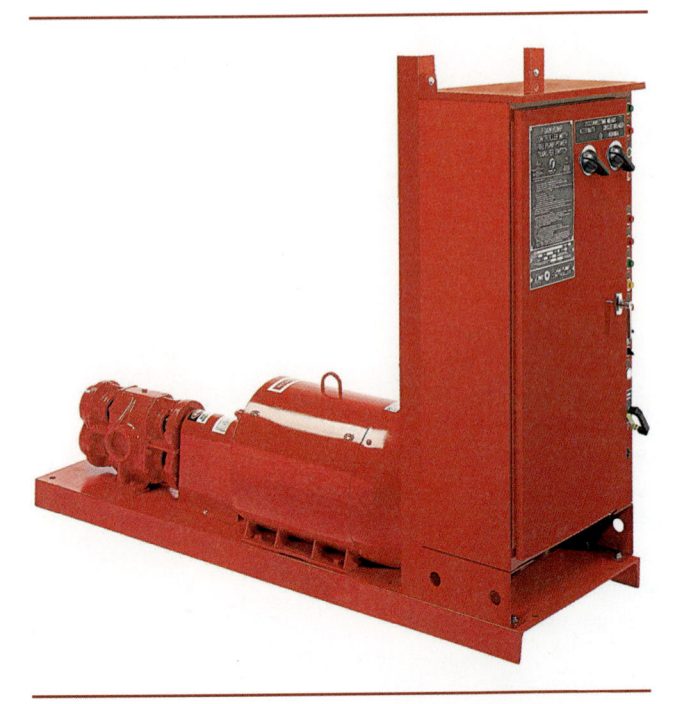

EXHIBIT II.8.1 Positive Displacement Pump with Electric Motor Driver. *(Courtesy of Pentair Water, North Aurora Operations)*

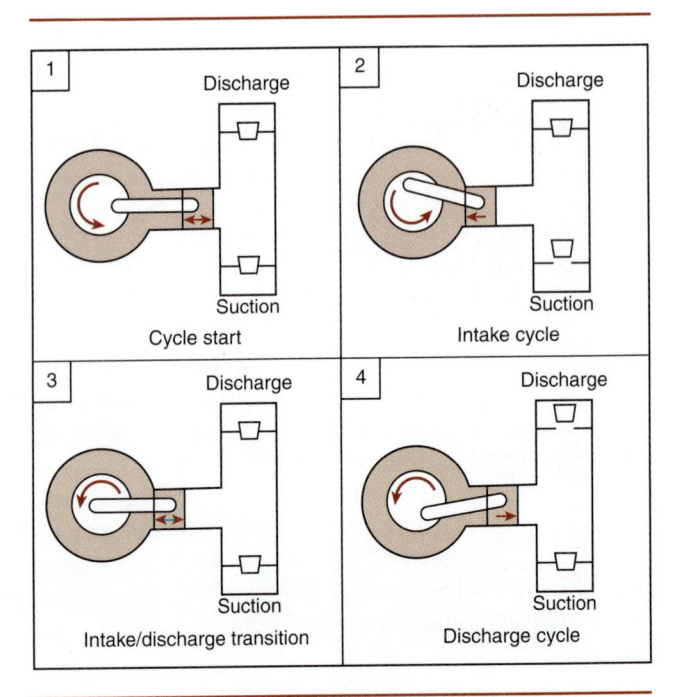

EXHIBIT II.8.2 Reciprocating Positive Displacement Pump – Pump Action.

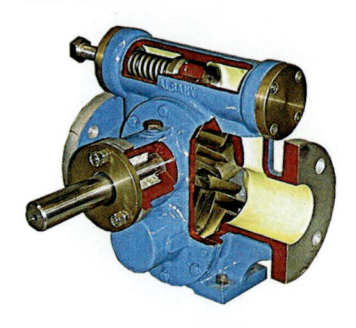

EXHIBIT II.8.3 One Type of Rotary Gear Positive Displacement Pump – Interior Cross Section. *(Courtesy of Fire Lion Global LLC, www.firelionglobal.com)*

▶ **FAQ** Certain positive displacement pumps involve a spinning action to create pressure. Why isn't this type of pump addressed in Chapter 6 on centrifugal pumps?

Two common displacement techniques are used in which a positive displacement pump imparts energy into the liquid. The first technique imparts energy through a two-way piston stroke in a chamber (see Exhibit II.8.2), which clearly differs from the mechanism used in the common centrifugal pump.

The second technique imparts energy through a series of gears or vanes rotating in a chamber (see Exhibit II.8.3). As the gears or vanes rotate inside the pump casing, a void is created by the gears or vanes passing by the suction port. Liquid is captured by the rotating elements, and, as the void volume is decreased on the discharge side of the pump, the liquid is "displaced" and exits the pump discharge port.

8.1.2* Suitability.

A.8.1.2 Special attention to the pump inlet piping size and length should be noted.

8.1.2.1 The positive displacement–type pump shall be listed for the intended application.

In order for the pump installation to comply with NFPA 20, the positive displacement pump and the system type must be listed for fire protection use as follows:

1. Water mist pumps:
   a. UL Listed Category – ZDPJ
   b. FM *Approval Standard for Water Mist Systems, Class Number 5560*

2013 Stationary Fire Pumps Handbook

> 2. Foam concentrate pumps:
>   a. UL Listed Category – GKWT
>   b. FM *Approval Standard for Positive Displacement Fire Pumps (Rotary Gear Type), Class Number 1313*

8.1.2.2* The listing shall verify the characteristic performance curves for a given pump model.

A.8.1.2.2 This material describes a sample pump characteristic curve and gives an example of pump selection methods. Characteristic performance curves should be in accordance with HI 3.6, *Rotary Pump Tests*.

Example: An engineer is designing a foam-water fire protection system. It has been determined, after application of appropriate safety factors, that the system needs a foam concentrate pump capable of 45 gpm at the maximum system pressure of 230 psi. Using the performance curve *(see Figure A.8.1.2.2)* for pump model "XYZ-987," this pump is selected for the application. First, find 230 psi on the horizontal axis labeled "differential pressure," then proceed vertically to the flow curve to 45 gpm. It is noted that this particular pump produces 46 gpm at a standard motor speed designated "rpm-2." This pump is an excellent fit for the application. Next, proceed to the power curve for the same speed of rpm-2 at 230 psi and find that it requires 13.1 hp to drive the pump. An electric motor will be used for this application, so a 15 hp motor at rpm-2 is the first available motor rating above this minimum requirement.

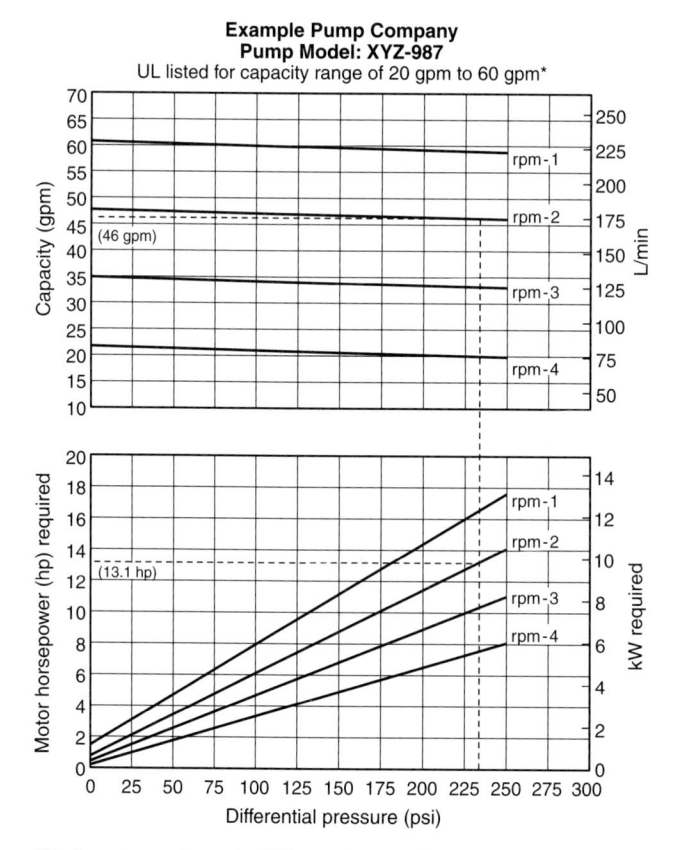

*Conforms to requirements of Chapter 8 on positive displacement foam concentrate and additive pumps.

FIGURE A.8.1.2.2 *Example of Positive Displacement Pump Selection.*

> Positive displacement pump performance curves determined through the pump's listing and approval tests must be provided by the manufacturer so that the fire protection system designer can properly select the correct size positive displacement pump.

8.1.3 Application.

8.1.3.1 Positive displacement pumps shall be permitted to pump liquids for fire protection applications.

> Paragraph 8.1.3.1 was changed for the 2007 edition to recognize that the requirements of Chapter 8 apply to any stationary positive displacement fire pump installation that must comply with NFPA 20, regardless of the liquid that the pump handles.

8.1.3.2 The selected pump shall be appropriate for the viscosity of the liquid.

> The selected pump is required to be verified for the maximum viscosity of the liquid to be pumped. Liquid viscosity maximums are a part of the listings and approvals process and are posted in UL Listed Category GKWT, UL Listed Category ZDPJ, and FM Approvals.
> All pumps have maximum viscosities listed as part of the listing process. Since foam concentrates and additives have widely ranging viscosities, it is important that users of the standard refer to the listing itself, which shows the maximum viscosity for a given pump.

8.1.4 Pump Seals.

8.1.4.1 The seal type acceptable for positive displacement pumps shall be either mechanical or lip seal.

8.1.4.2 Packing shall not be used.

> Note the difference in packing requirements between the positive displacement pump and the common horizontal split-case centrifugal pump covered in Chapters 4 and 6. Packing seals, which are found on the common horizontal split-case centrifugal pump, are prohibited for positive displacement pumps.

8.1.5* Pump Materials. Materials used in pump construction shall be selected based on the corrosion potential of the environment, fluids used, and operational conditions. *(See 3.3.9 for corrosion-resistant materials.)*

> Foam concentrates and solutions can be significantly more corrosive to common fire protection system materials than domestic water.

A.8.1.5 Positive displacement pumps are tolerance dependent. Corrosion can affect pump performance and function. *(See HI 3.5, Standard for Rotary Pumps for Nomenclature, Design, Application and Operation.)*

8.1.6 Dump Valve.

> A dump valve is required to be provided for diesel-driven foam concentrate or additive pumps to allow the positive displacement pump to bleed off excess pressure during the start cycle. It is not necessary for a dump valve to be used for electric-driven foam concentrate or additive pumps.
> Electric-driven foam pumps have never required a dump valve. The requirement originated because of the crank cycle of diesel engines trying to start the engine against a pressurized system. Dump valves are only needed for diesel-driven foam or additive pumps.

8.1.6.1 A dump valve shall be provided on all closed head systems to allow the positive displacement pump to bleed off excess pressure and achieve operating speed before subjecting the driver to full load.

> The driver for a positive displacement pump is not generally sized to overcome the backpressure created if none of the fire protection system discharge devices have activated while starting. Backpressure can result in a phenomenon similar to the "banana in the tailpipe" prank, whereby a car engine is prevented from being started by blocking the exhaust pipe. The requirement in 8.1.6.1 is meant to address the same problem.

8.1.6.2 The dump valve shall operate only for the duration necessary for the positive displacement pump to achieve operating speed.

8.1.6.3 Dump Valve Control.

8.1.6.3.1 Automatic Operation. When an electrically operated dump valve is used, it shall be controlled by the positive displacement pump controller.

> ▶ **FAQ** Why does NFPA 20 require the use of a dump valve control?
>
> Requiring that the dump valve control be part of the fire pump controller ensures that the fire pump operator can monitor all fire pump functions in one location.

8.1.6.3.2 Manual Operation. Means shall be provided at the controller to ensure dump valve operation during manual start.

8.1.6.4 Dump valves shall be listed.

8.1.6.5 Dump valve discharge shall be permitted to be piped to the liquid supply tank, pump suction, drain, or liquid supply.

> When the discharge is piped to the pump suction, the design must take into consideration that the maximum suction pressures are not to be exceeded.

8.2 Foam Concentrate and Additive Pumps

> The requirements in Section 8.2 are intended to apply to any positive displacement pump in a fire protection system that is pumping any liquid other than water-only solutions.

8.2.1 Additive Pumps. Additive pumps shall meet the requirements for foam concentrate pumps.

8.2.2* Net Positive Suction Head. Net positive suction head (NPSH) shall exceed the pump manufacturer's required NPSH plus 5 ft (1.52 m) of liquid.

> The *net positive suction head (NPSH)*, which is defined in 3.3.23.1, is the total liquid head-absolute (in feet) at the suction nozzle minus the head pressure of the liquid (in feet). To convert pressure in psi to head in feet, multiply the pressure in psi by 2.31 and divide by the specific gravity of the liquid.

A.8.2.2 Specific flow rates should be determined by the applicable NFPA standard. Viscose concentrates and additives have significant pipe friction loss from the supply tank to the pump suction.

8.2.3 Seal Materials. Seal materials shall be compatible with the foam concentrate or additive.

> Incompatibility of the seal material with the foam or additive being pumped could lead to degradation of the seal and inadequate performance of the pump.

8.2.4* Dry Run. Foam concentrate pumps shall be capable of dry running for 10 minutes without damage.

> The time of 10 minutes specified in 8.2.4 permits the fire pump operator time to shut down the positive displacement pump after the foam/additive tank has been depleted and before damage occurs to the pump. Positive displacement pumps typically rely on the fluid being pumped to provide pump lubrication.

A.8.2.4 This requirement does not apply to water mist pumps.

8.2.5* Minimum Flow Rates. Pumps shall have foam concentrate flow rates to meet the maximum foam flow demand for their intended service.

> The required flow rate is determined through other NFPA fire protection system installation standards, such as the following:
>
> NFPA 11, *Standard for Low-, Medium-, and High-Expansion Foam*
> NFPA 13, *Standard for the Installation of Sprinkler Systems*
> NFPA 14, *Standard for the Installation of Standpipe and Hose Systems*
> NFPA 15, *Standard for Water Spray Fixed Systems for Fire Protection*
> NFPA 16, *Standard for the Installation of Foam-Water Sprinkler and Foam-Water Spray Systems*

A.8.2.5 Generally, pump capacity is calculated by multiplying the maximum water flow by the percentage of concentration desired. To that product is added a 10 percent "over demand" to ensure that adequate pump capacity is available under all conditions.

> The "over demand" is not fixed at 10 percent, but the ten percent figure is intended to be a guide.

8.2.6* Discharge Pressure. The discharge pressure of the pump shall exceed the maximum water pressure under any operating condition at the point of foam concentrate injection.

A.8.2.6 Generally, concentrate pump discharge pressure is required to be added to the maximum water pressure at the injection point plus 25 psi (2 bar).

> The required positive displacement pump discharge pressure is based on the calculated discharge pressure of the foam system, which, in turn, is determined through other NFPA fire protection system installation standards, such as NFPA 11 and NFPA 16. The required discharge pressure of the foam concentrate pump has an added safety factor to ensure that foam concentrate enters the system.

8.3 Water Mist System Pumps

8.3.1* Positive displacement pumps for water shall have adequate capacities to meet the maximum system demand for their intended service.

A.8.3.1 It is not the intent of this standard to prohibit the use of stationary pumps for water mist systems.

> ▶ **FAQ** Can a (traditional) centrifugal fire pump be used in a water mist system?
>
> A listed (traditional) centrifugal fire pump meeting the system demand is permitted to be used in a water mist fire protection system. It is more common to see a positive displacement pump(s) supplying a water mist fire protection system, because positive displacement pumps can produce relatively higher pressures.

8.3.2 NPSH shall exceed the pump manufacturer's required NPSH plus 5 ft (1.52 m) of liquid.

> The NPSH is the total liquid head-absolute (in feet) at the suction nozzle minus the head pressure of the liquid (in feet). To convert pressure in psi to head in feet, multiply the pressure in psi by 2.31 and divide by the specific gravity of the liquid.

8.3.3 The inlet pressure to the pump shall not exceed the pump manufacturer's recommended maximum inlet pressure.

8.3.4 When the pump output has the potential to exceed the system flow requirements, a means to relieve the excess flow such as an unloader valve or orifice shall be provided.

> In order to meet the pressure requirements, a larger pump that produces excess flow might be needed. A common fire protection goal of reducing water discharge in the protected space often leads to the use of water mist fire protection systems.

8.3.5 Where the pump is equipped with an unloader valve, it shall be in addition to the safety relief valve as outlined in 8.5.2.

8.4 Water Mist Positive Displacement Pumping Units

8.4.1 Water mist positive displacement pumping units shall be dedicated to and listed as a unit for fire protection service.

> ▶ **FAQ** How do water mist positive displacement pumping units differ from other water mist system pumps or (traditional) centrifugal pumps?
>
> A *water mist positive displacement pumping unit* is defined in 3.3.38.18 as "multiple positive displacement pumps designed to operate in parallel that discharge into a single common water mist distribution system." For water mist positive displacement pumping units, it is not required that each driver have its own dedicated controller, since the unit includes redundant monitoring and protection schemes that ensure that a single point of failure will not prevent the system from performing as intended. Using current microprocessor-based control technologies and architecture, a single controller with redundancy schemes can provide equivalent or superior reliability to other fire suppression systems while providing the added benefit of a central control location. This central control location allows for advanced system diagnostics and status, and the option of remote monitoring advances fire-fighting efforts and system availability. In order for the pump installation to comply with NFPA 20, the positive

displacement pump and the system type must be listed for fire protection use. Exhibit II.8.4 illustrates a complete water mist fire suppression system, incorporating a water mist positive displacement pumping unit as part of the system.

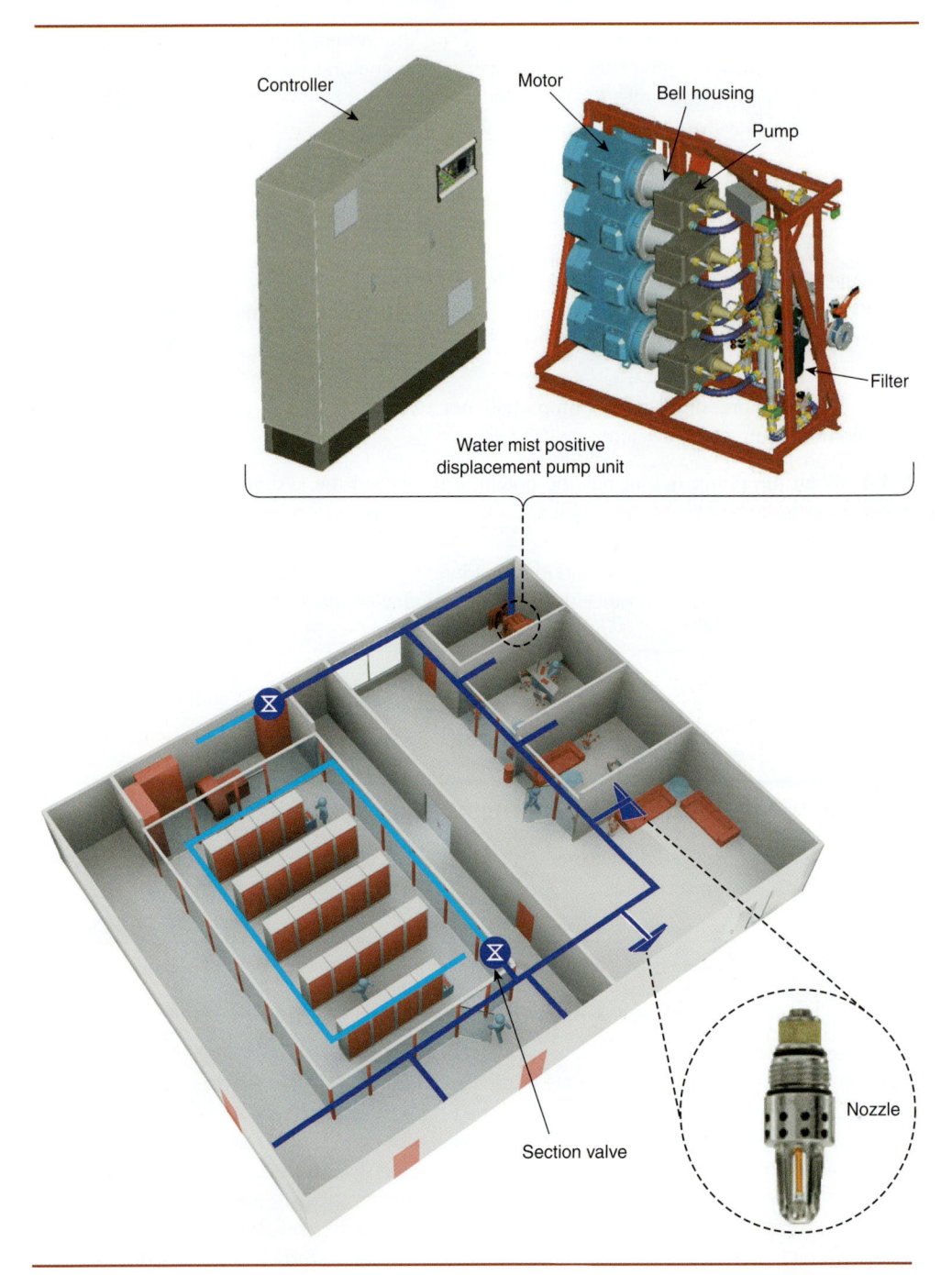

EXHIBIT II.8.4 *Water Mist Positive Displacement Pumping Unit. (Courtesy of Marioff North America)*

8.4.2 Except as provided in 8.4.3 through 8.4.8, all the requirements of this standard shall apply.

> Section 8.4 contains requirements for water mist positive displacement pumping units that supplement or modify the basic requirements of NFPA 20 and differentiate them from other types of pumps having separate drivers and controllers.

8.4.3 Water mist positive displacement pumping units shall include pumps, driver(s), and controller as a complete operating unit.

> ▶ **FAQ** What makes a water mist positive displacement pumping unit unique compared to other types of pumps addressed by NFPA 20?
>
> Water mist positive displacement pumping units are unique in that they have a single controller for multiple pumps and a driver(s).

8.4.4 The pump controller shall manage the performance of all pumps and drivers to provide continuous and smooth operation without intermittent pump cycling or discharge pressure varying by more than 10 percent during pump sequencing after rated pressure has been achieved.

8.4.5 Redundancy shall be built into the units such that failure of a line pressure sensor or primary control board will not prevent the system from functioning as intended.

> ▶ **FAQ** Why is it necessary to have redundancy?
>
> Redundant monitoring and protective functions are incorporated into water mist positive displacement pumping units. These redundancies are necessary, because there is only one controller in a system employing multiple pumps.

8.4.6 Where provided with a variable speed control, failure of the variable speed control feature shall cause the controller to bypass and isolate the variable speed control system.

> ▶ **FAQ** Why is it acceptable to bypass the variable speed control system?
>
> When a water mist positive displacement pumping unit is provided with a variable speed control, it is required to bypass and isolate the variable speed control in the event of failure of the variable speed control system. In this mode, the fire suppression system may be overpressurized, but it is more important to maintain waterflow than it is to prevent damage to the system. This mirrors the requirement of 10.10.3.1 for controllers with variable speed pressure limiting control.

8.4.7 The unit controller shall be arranged so that each pump can be manually operated individually without opening the enclosure door.

> Each pump must be able to be controlled manually from outside the controller enclosure; this operation is similar to that required for an electric motor or diesel engine fire pump controller and allows for bypassing of all automatic starting means, whether or not in failure mode. This is the same type of performance required for other types of fire pump equipment.

Stationary Fire Pumps Handbook 2013

8.4.8 The requirement in 10.3.4.3 shall apply to each individual motor and the entire unit.

> ▶ **FAQ** Why is it important that all line currents and all line voltages are able to be read from the exterior of the controller?

Paragraph 10.3.4.3 requires that a means be provided on the exterior of the controller to read all line currents and all line voltages within an accuracy of ±5 percent of motor nameplate voltage and current. It is important to measure these parameters to verify the correct operation of each pump motor and the entire pumping unit. Providing the means for reading these parameters on the exterior of the controller minimizes the exposure of electric shock hazard to test personnel.

8.5 Fittings

8.5.1 Gauges. A compound suction gauge and a discharge pressure gauge shall be furnished.

A compound suction gauge is a gauge that is capable of reading negative and positive pressures. The discharge pressure gauge is not required to be a compound gauge, because it is assumed that the discharge pressures will never be negative. Both gauges are needed to indicate the pressure boost across the pump.

8.5.2* General Information for Relief Valves.

A.8.5.2 Positive displacement pumps are capable of quickly exceeding their maximum design discharge pressure if operated against a closed discharge system. Other forms of protective devices (e.g., automatic shutdowns, rupture discs) are considered a part of the pumping system and are generally beyond the scope of the pump manufacturer's supply. These components should be safely designed into and supplied by the system designer or by the user, or both. *(See Figure A.8.5.2 for proposed schematic layout of pump requirements.)*

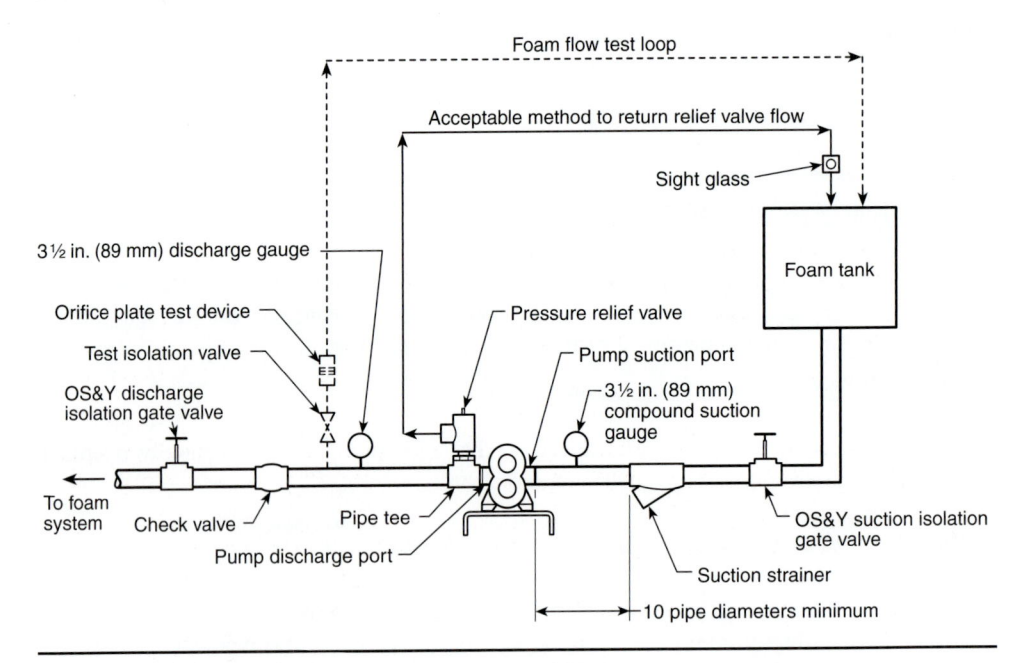

FIGURE A.8.5.2 *Typical Foam Pump Piping and Fittings.*

8.5.2.1 All pumps shall be equipped with a listed safety relief valve capable of relieving 100 percent of the rated pump capacity at a pressure not exceeding 125 percent of the relief valve set pressure.

> Without a properly sized pressure relief valve, positive displacement pumps continue to increase pressure when pumping against a fire protection system that has few or no openings until the driver stalls or the fire protection system fails because it has exceeded component failure pressures.

8.5.2.2 The pressure relief valve shall be set such that the pressure required to discharge the rated pump capacity is at or below the lowest rated pressure of any component.

8.5.2.3 The relief valve shall be installed on the pump discharge to prevent damage to the fire protection system.

8.5.3* Relief Valves for Foam Concentrate Pumps. For foam concentrate pumps, safety relief valves shall be piped to return the valve discharge to the concentrate supply tank.

> The relief valve discharge needs to be piped back to the supply so that the foam/additive supply is not depleted. The quantity of foam concentrate is based on a specific discharge duration and should not be depleted by pressure relief valve discharge. The routine testing of the pump installation with no flow is frequently performed where the relief valve discharges.

A.8.5.3 Only the return to source and external styles should be used when the outlet line can be closed for more than a few minutes. Operation of a pump with an integral relief valve and a closed outlet line will cause overheating of the pump and a foamy discharge of fluid after the outlet line is reopened.

8.5.4* Relief Valves for Water Mist Pumps.

A.8.5.4 Backpressure on the discharge side of the pressure relief valve should be considered. *(See Figure A.8.5.4 for proposed schematic layout of pump requirement.)*

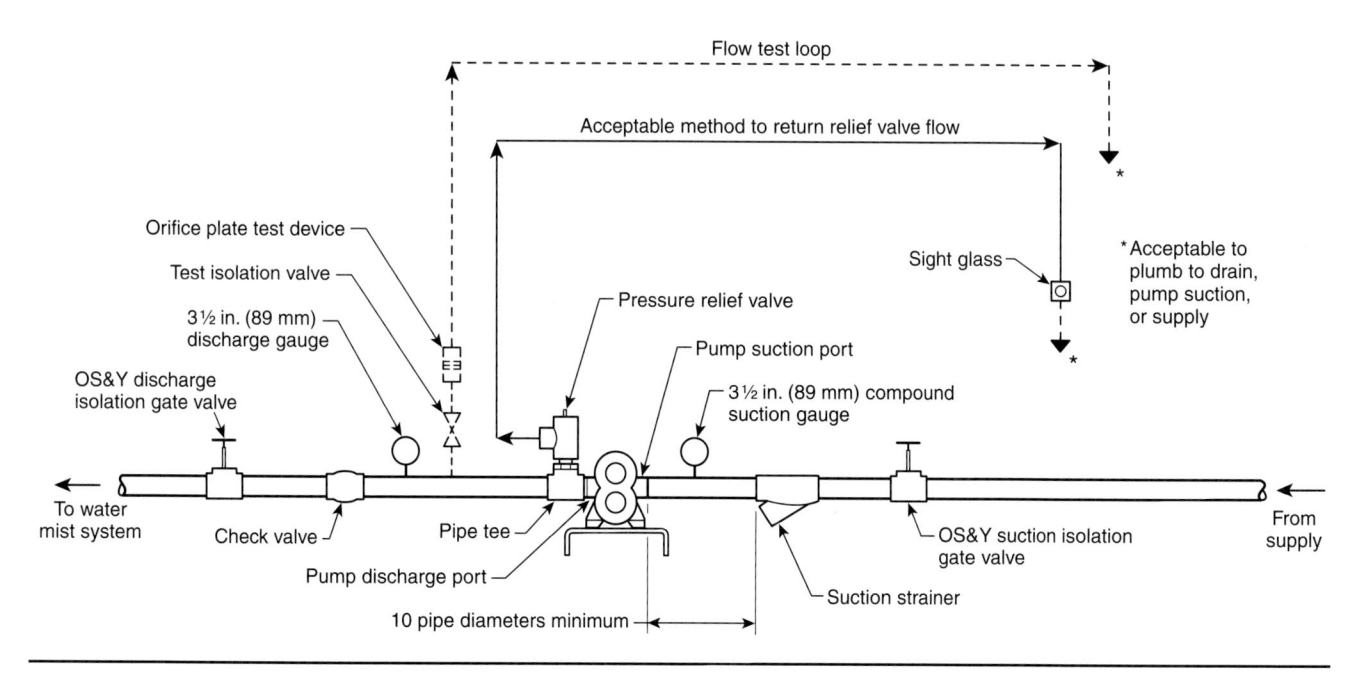

FIGURE A.8.5.4 Typical Water Mist System Pump Piping and Fittings.

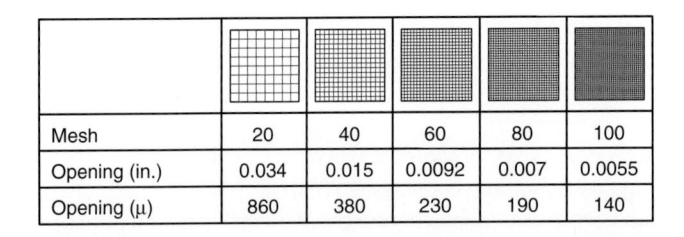

Mesh	20	40	60	80	100
Opening (in.)	0.034	0.015	0.0092	0.007	0.0055
Opening (μ)	860	380	230	190	140

FIGURE A.8.5.5 *Standard Mesh Sizes.*

8.5.4.1 For positive displacement water mist pumps, safety relief valves shall discharge to a drain or to a water supply at atmospheric pressure.

8.5.4.2 A means of preventing overheating shall be provided when the relief valve is plumbed to discharge to the pump suction.

> Two possible means of preventing overheating are the installation of a pressure relief valve as a circulation relief valve and discharge to a drain.

8.5.5* Suction Strainer.

> Positive displacement pumps are more susceptible than other types of pumps to damage from debris in the pumped fluid; therefore, a strainer in the suction piping is required for all positive displacement pumps.

A.8.5.5 Strainer recommended mesh size is based on the internal pump tolerances. *(See Figure A.8.5.5 for standard mesh sizes.)*

8.5.5.1 Pumps shall be equipped with a removable and cleanable suction strainer installed at least 10 pipe diameters from the pump suction inlet.

8.5.5.2 Suction strainer pressure drop shall be calculated to ensure that sufficient NPSH is available to the pump.

> See 3.3.23.1 for the definition of *net positive suction head (NPSH)$_{(hsv)}$* and the commentary following 8.2.2 for more detailed information on the term.

8.5.5.3 The net open area of the strainer shall be at least four times the area of the suction piping.

8.5.5.4 Strainer mesh size shall be in accordance with the pump manufacturer's recommendation.

8.5.6 Water Supply Protection. Design of the system shall include protection of potable water supplies and prevention of cross connection or contamination.

> The system designer must plan for backflow prevention devices in the design of the system. Many water purveyors require protection of water supplies when chemicals such as foam/additives are used in a fire protection system. NFPA 20 permits the use of a backflow prevention device or a break tank (see Chapter 4) when water supply protection is required by local authorities.

8.5.7 Pressure Maintenance.

8.5.7.1 Except as permitted in 8.5.7.2, the primary or standby fire pump shall not be used as a pressure maintenance pump.

8.5.7.2 Water mist positive displacement pumping units that are designed and listed to alternate pressure maintenance duty between two or more pumps with variable speed pressure limiting control, and that provide a supervisory signal wherever pressure maintenance is required more than two times in one hour, shall be permitted to maintain system pressure.

> ▶ **FAQ** Why are water mist positive displacement pumping units allowed to provide pressure maintenance in lieu of a jockey or make-up pump?

> For water mist positive displacement pumping units listed for pressure maintenance duty, the primary or standby fire pump can be used as a pressure maintenance pump, since there is a discrete signaling means to indicate system activation in response to a fire. Using current microprocessor-based control and variable-voltage variable-frequency drive technologies, a water mist system may be designed to use the primary motors/pumps to maintain system pressure. This function can be rotated from pump to pump to improve system reliability.

8.5.7.3 When in the pressure maintenance mode, water mist positive displacement pumping units used for pressure maintenance shall not provide more than half of the nozzle flow of the smallest system nozzle when the standby pressure is applied at the smallest nozzle.

> The requirement in 8.5.7.3 is intended to set a limit for the water mist positive displacement pumping unit relative to the amount of pressure it can create when in the pressure maintenance mode and to distinguish between the need for pressure maintenance and a fire event.

8.5.7.4 A single sensing line shall be permitted to be used for a water mist positive displacement pumping unit controller where the unit also serves for pressure maintenance on a water mist system.

> ▶ **FAQ** Why are multiple sensing lines not needed for a water mist positive displacement pumping unit?

> Multiple sensing lines are not needed for a water mist positive displacement pumping unit, because the unit is able to distinguish between the need for pressure maintenance and a fire event. This is based on the controller's ability to sense the pressure differential from the permissible limited flow demand of 8.5.7.3, while in the pressure maintenance mode, and the pressures required from a larger flow demand, such as those required in a fire scenario.

8.6 Pump Drivers

8.6.1* The driver shall be sized for and have enough power to operate the pump and drive train at all design points.

A.8.6.1 Positive displacement pumps are typically driven by electric motors, internal combustion engines, or water motors.

EXHIBIT II.8.5
Premanufactured Fire Pump Unit with Positive Displacement Pump and Diesel Engine Driver. (Courtesy of Pentair Water, North Aurora Operations)

▶ **FAQ** Does the motor driving a positive displacement pump have to be installed in accordance with Chapter 9?

A positive displacement pump is considered a type of fire pump. Therefore, the intent is that all NFPA 20 requirements apply to a fire pump installation involving a positive displacement pump (see Exhibit II.8.5), unless specific NFPA 20 provisions dictate otherwise. The driver must comply with Chapter 9, Chapter 11, or Chapter 13.

8.6.2 Reduction Gears.

8.6.2.1 If a reduction gear is provided between the driver and the pump, it shall be listed for the intended use.

8.6.2.1.1 Reduction gears shall meet the requirements of AGMA 390.03, *Handbook for Helical and Master Gears*.

If gears are designed improperly, or if gears are selected on speed change alone, the reduction gear can become a failure point, resulting in loss of the fire pump. The Technical Committee on Fire Pumps has received reports that gear failures have occurred in fire pump installations.

8.6.2.2 Gears shall be AGMA Class 7 or better, and pinions shall be AGMA Class 8 or better.

8.6.2.3 Bearings shall be in accordance with AGMA standards and applied for an L10 life of 15,000 hours.

8.6.2.4 For drive systems that include a gear case, the pump manufacturer shall provide a complete mass elastic system torsional analysis to ensure there are no damaging stresses or critical speeds within 25 percent above and below the operating speed of the pump(s) and driver.

8.6.2.4.1 For variable speed drives, the analysis of 8.6.2.4 shall include all speeds down to 25 percent below the lowest operating speed obtainable with the variable speed drive.

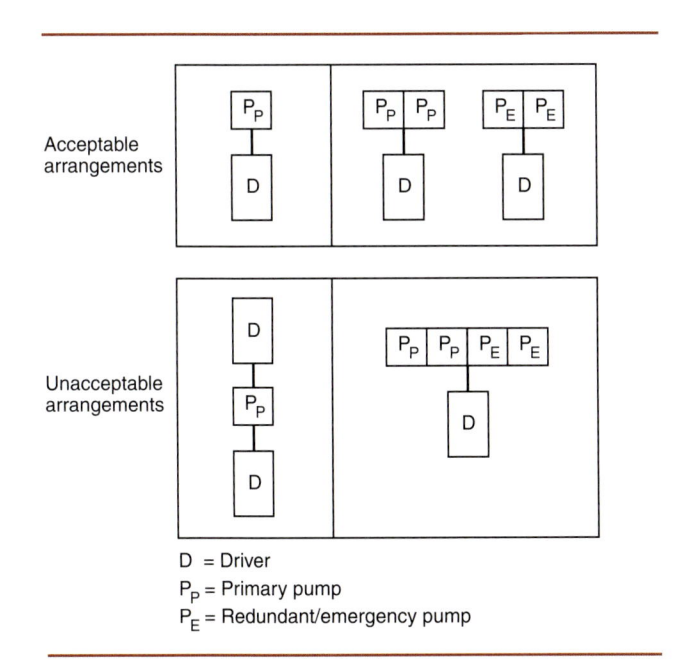

EXHIBIT II.8.6 *Acceptable and Unacceptable Common Driver Arrangements.*

D = Driver
P_P = Primary pump
P_E = Redundant/emergency pump

8.6.3 Common Drivers.

8.6.3.1 A single driver shall be permitted to drive more than one positive displacement pump.

Subsection 8.6.3.1 overrides the requirements in 4.7.3 and permits one motor/engine to drive multiple pumps. The flow demands of a water mist fire protection system, combined with low operating flow range of a positive displacement pump, commonly require that more than one pump be driven off the same motor/engine driver. Manufacturers typically supply this arrangement in a preassembled unit.

8.6.3.2 Redundant pump systems shall not be permitted to share a common driver.

▶ **FAQ** | Does the requirement specified in 8.6.3.2 conflict with the requirement in 8.6.3.1?

The intention of NFPA 20 is that the requirement in 8.6.3.2 does not conflict with the requirement in 8.6.3.1. NFPA 750 requires that a redundant pump installation be provided for a water mist fire protection system in case a failure occurs in the primary pump installation. For example, if two pumps are needed to meet the system demand, then four pumps are needed to comply with NFPA 750. If electric motors are installed as drivers, at least two motors are required — one motor for the primary set of pumps as, allowed by 8.6.3.1, and another motor for the redundant (emergency) set of pumps, as allowed by 8.6.3.1. The requirement in 8.6.3.2 prohibits a single motor to drive all four pumps. See Exhibit II.8.6 for an example of a comparison of acceptable and unacceptable common driver arrangements.

◀ **CALCULATION**

8.7* Controllers

See Section 8.4 and Chapters 10 and 12 for requirements for controllers.

> The requirements in NFPA 20 for controllers of positive displacement pumps are the same as those for (traditional) centrifugal pumps.

A.8.7 These controllers can incorporate means to permit automatic unloading or pressure relief when starting the pump driver.

8.8 Foundation and Setting

> The requirements in NFPA 20 for foundations and setting of positive displacement pumps are similar to those for (traditional) centrifugal pumps.

8.8.1 The pump and driver shall be mounted on a common grouted base plate.

8.8.2 The base plate shall be securely attached to a solid foundation in such a way that proper pump and driver shaft alignment will be maintained.

8.8.3 The foundation shall provide a solid support for the base plate.

8.9 Driver Connection and Alignment

8.9.1 The pump and driver shall be connected by a listed, closed coupled, flexible coupling or timing gear type of belt drive coupling.

> ▶ **FAQ** Does NFPA 20 permit the use of a belt drive to connect a positive displacement pump to a driver?

A positive displacement pump is permitted to be connected to the driver through a belt drive, which is not permitted for any other type of pump complying with NFPA 20.

8.9.2 The coupling shall be selected to ensure that it is capable of transmitting the horsepower of the driver and does not exceed the manufacturer's maximum recommended horsepower and operating speed.

8.9.3 Pumps and drivers shall be aligned once final base plate placement is complete.

8.9.4 Alignment shall be in accordance with the coupling manufacturer's specifications.

8.9.5 The operating angle for the flexible coupling shall not exceed the recommended tolerances.

8.10 Flow Test Devices

8.10.1 A positive displacement pump installation shall be arranged to allow the test of the pump at its rated conditions as well as the suction supply at the maximum flow available from the pump.

> A means of testing the suction supply by actually flowing liquid through the pump from the supply — not just through a test loop — must be available. Due to the viscosity of foam/additive solutions, significant friction loss can occur in the suction supply piping while under flow conditions during a fire.

8.10.2 Additive pumping systems shall be equipped with a flow meter or orifice plate installed in a test loop back to the additive supply tank.

> A test loop is required to discharge back to the foam/additive tank so that the foam/additive tank is not depleted during routine testing of the foam/additive supply.

8.10.3 Water pumping systems shall be equipped with a flowmeter or orifice plate installed in a test loop back to the water supply, tank, inlet side of the water pump, or drain.

References Cited in Commentary

National Fire Protection Association, 1 Batterymarch Park, Quincy, MA 02169-7471.

NFPA 11, *Standard for Low-, Medium-, and High-Expansion Foam,* 2010 edition.
NFPA 13, *Standard for the Installation of Sprinkler Systems,* 2013 edition.
NFPA 14, *Standard for the Installation of Standpipe and Hose Systems,* 2010 edition.
NFPA 15, *Standard for Water Spray Fixed Systems for Fire Protection,* 2012 edition.
NFPA 16, *Standard for the Installation of Foam-Water Sprinkler and Foam-Water Spray Systems,* 2011 edition.
NFPA 750, *Standard on Water Mist Fire Protection Systems,* 2010 edition.

FM Approvals LLC, 1151 Boston-Providence Turnpike, P.O. Box 9102, Norwood, MA 02062.

Approval Standard for Positive Displacement Fire Pumps (Rotary Gear Type), Class Number 1313, November 2007.
Approval Standard for Water Mist Systems, Class Number 5560, March 2009.

CHAPTER 9

Electric Drive for Pumps

Chapter 9 describes the sources and transmission of electric power for motors driving stationary fire pumps. Also covered are the performance and installation criteria for motors that provide the prime mover for stationary fire pumps. The transmission of power includes all equipment between the source(s) and the fire pump motor, with the exception of the controllers, transfer switches, and accessories, which are covered in Chapter 10. The focus of this chapter is the proper installation of electrical equipment associated with a stationary fire pump.

▶ **FAQ** What is the relationship between Chapter 9 of NFPA 20, *Standard for the Installation of Stationary Pumps for Fire Protection,* and Article 695 of NFPA 70®, *National Electrical Code*®?

Article 695 of *NFPA 70* is the authority on (is responsible for) the electrical installation of fire pumps. Article 695 also contains text on power supplies extracted from Chapter 9 of NFPA 20. Paragraph 1.1.2, which is part of the scope of NFPA 20, specifies that NFPA 20 has primary responsibility (authority) for power supplies for fire pumps. Adherence to Article 695 is required for the safe and reliable installation of both electric motor–driven and diesel engine–driven fire pumps, including the protection of power and control circuits. Article 700 and Article 701 of *NFPA 70* apply where emergency power or legally required standby power is utilized for fire pumps.

9.1 General

9.1.1 This chapter covers the minimum performance and testing requirements of the sources and transmission of electrical power to motors driving fire pumps.

Although performance and testing are critical, the emphasis of Chapter 9 is the proper selection and installation of the sources and equipment used to transmit power to the fire pump motor. The objective is the continuity of power to drive the fire pump motor in the event of a fire, as well as the reliability of the power sources.

9.1.2 This chapter also covers the minimum performance requirements of all intermediate equipment between the source(s) and the pump, including the motor(s) but excepting the electric fire pump controller, transfer switch, and accessories *(see Chapter 10).*

All equipment necessary to ensure the minimum performance of sources of electric power to the fire pump motor is subject to the requirements of this chapter. This equipment includes motors, generators, junction boxes, and any overcurrent protection (fuses or circuit breakers) and switches other than those within a fire pump controller.

9.1.3 All electrical equipment and installation methods shall comply with *NFPA 70, National Electrical Code*, Article 695, and other applicable articles.

> **▶ FAQ** Which articles of *NFPA 70*, other than Article 695, most affect the installation of fire pumps?

Some of the other significant articles in *NFPA 70* include Article 230, Services; Article 250, Grounding and Bonding; Article 700, Emergency Systems; and Article 708, Critical Operations Power Systems (COPS), among others.

9.1.4* All power supplies shall be located and arranged to protect against damage by fire from within the premises and exposing hazards.

Damage from fire exposure to conductors or equipment that supplies power to the fire pump could render the fire pump inoperable. Keeping this equipment safe from fire includes adequate installation and protection of overhead lines, if used.

A.9.1.4 Where the power supply involves an on-site power production facility, the protection is required for the facility in addition to the wiring and equipment.

The requirement for protection in 9.1.4 also applies to the building housing the on-site power production facility. This facility is an integral part of the fire protection for the premises.

9.1.5 All power supplies shall have the capacity to run the fire pump on a continuous basis.

> **▶ FAQ** Since fire pumps operate infrequently, is any of the fire pump equipment rated for less than continuous duty?

Although fire pumps are not expected to run continuously, they can run for long periods of time, such as over a holiday weekend or for many hours when fighting a high-rise building fire. Therefore, no reliable method exists for sizing the power supply based on a duty cycle other than that of continuous duty (100 percent duty cycle). For example, operation of the fire pump where there is a signal for a specific fire may not be a one-time event. In some instances, fire pumps may operate for days in order to prevent rekindling of a fire after the main extinguishing is completed. Also, in some cases, the fire pump is used for other purposes, sometimes by design, but not always. Whatever the specific circumstances, the integrity of the power supply is critical to both reliable and adequate fire pump operation; therefore, all power supplies should be rated for continuous duty.

9.1.6 All power supplies shall comply with the voltage drop requirements of Section 9.4.

In some cases, the capacity of a power supply may need to be increased in order to meet the voltage drop requirements of Section 9.4.

9.1.7* Phase converters shall not be used to supply power to a fire pump.

> **▶ FAQ** Are phase converters acceptable for use in a fire pump circuit?

Phase converters are used to provide three-phase power from a single-phase power source, such as is often found in rural locations. However, a phase converter is not considered a reliable source of power for a fire pump motor because of the imbalance in the voltage between the phases when there is no load on the equipment. Additionally, a phase converter would need to be run continuously in order for the fire pump controller to remain in the standby condition. When three-phase power is not available, a single-phase fire pump controller and motor would need to be used. Single-phase fire pump controllers are available to 10 hp max at 240 V.

A.9.1.7 Phase converters that take single-phase power and convert it to three-phase power for the use of fire pump motors are not recommended because of the imbalance in the voltage between the phases when there is no load on the equipment. If the power utility installs phase converters in its own power transmission lines, such phase converters are outside the scope of this standard and need to be evaluated by the authority having jurisdiction to determine the reliability of the electric supply.

9.1.8* Interruption.

A.9.1.8 Ground fault alarm provisions are not prohibited.

> Ground-fault alarms can alert users of the presence of a fault, but the power to the fire pump is not to be interrupted by the ground fault. It is imperative that the cause of the alarm be investigated immediately, so that the hazardous condition can be corrected. This requirement is intended to maintain fire protection before or during a fire. Likewise, arc-fault devices must sound an alarm but must not interrupt power to the pump.

9.1.8.1 No ground fault interruption means shall be installed in any fire pump control or power circuit.

9.1.8.2 No arc fault interruption means shall be installed in any fire pump control or power circuit.

9.2* Normal Power

A.9.2 See Figure A.9.2 for typical power supply arrangements from source to the fire pump motor.

> Figure A.9.2, which was formerly Figure A.9.3.2, showed Arrangement A and Arrangement B only. The figure has been revised to show the single upstream disconnect and possible overcurrent protection (OCP) when permitted by 9.2.3 in order to provide a more complete picture of the three common types of connection to the power supply. Arrangement A remains illustrative of a direct service connection, whereby the fire pump controller or the normal source side of a fire pump power transfer switch acts as the service entrance equipment. Arrangement B remains illustrative of a transformer connection, whereby the transformer primary overcurrent protection is attached to the service in compliance with 9.2.2(5). New Arrangement C is illustrative of the indirect connection of 9.2.2(3) and 9.2.3, whereby the fire pump controller or the normal source side of a fire pump power transfer switch connects to upstream service entrance equipment.

◀ **DESIGN ALERT**

9.2.1 An electric motor–driven fire pump shall be provided with a normal source of power as a continually available source.

> If a normal source of power is not continually available, an alternate source of power must be provided as described in Section 9.3 in order to main the liquid source for the fire protection system.

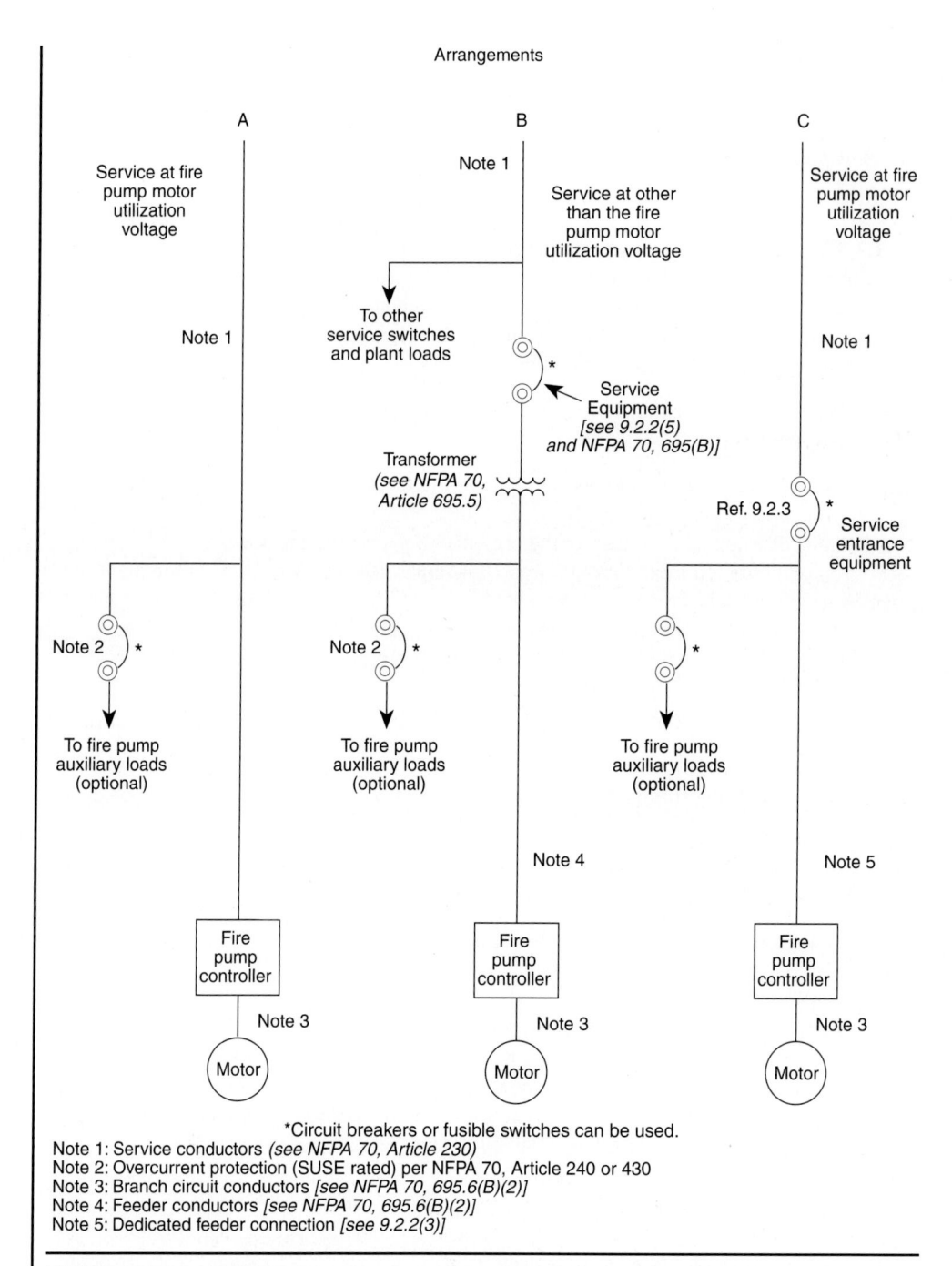

```
                                    Arrangements

               A                         B                        C
                                       Note 1
       Service at fire                                     Service at fire
       pump motor          Service at other                pump motor
       utilization         than the fire                   utilization
       voltage             pump motor                      voltage
                           utilization voltage

                           To other
                           service switches        *
         Note 1            and plant loads                   Note 1
                                           ← Service
                                             Equipment
                                             [see 9.2.2(5)
                                             and NFPA 70, 695(B)]
                           Transformer
                           (see NFPA 70,
                           Article 695.5)                              *
                                                          Ref. 9.2.3    Service
                                                                        entrance
                                                                        equipment

       Note 2  *           Note 2  *                        *

       To fire pump        To fire pump              To fire pump
       auxiliary loads     auxiliary loads           auxiliary loads
       (optional)          (optional)                (optional)

                                         Note 4              Note 5

       Fire                Fire                       Fire
       pump                pump                       pump
       controller          controller                 controller

              Note 3              Note 3                    Note 3
       Motor               Motor                      Motor
```

*Circuit breakers or fusible switches can be used.
Note 1: Service conductors *(see NFPA 70, Article 230)*
Note 2: Overcurrent protection (SUSE rated) per NFPA 70, Article 240 or 430
Note 3: Branch circuit conductors *[see NFPA 70, 695.6(B)(2)]*
Note 4: Feeder conductors *[see NFPA 70, 695.6(B)(2)]*
Note 5: Dedicated feeder connection *[see 9.2.2(3)]*

FIGURE A.9.2 *Typical Power Supply Arrangements from Source to Motor.*

9.2.2 The normal source of power required in 9.2.1 and its routing shall be arranged in accordance with one of the following:

(1) Service connection dedicated to the fire pump installation

> The intent of the requirement of 9.2.2(1) is that this service is to supply no loads other than those associated with the fire pump. This requirement is normally met by providing a separate service or a tap ahead of the main service equipment (disconnect) for the building that is

entered by the conductors supplying power to the fire pump. Frequently, the fire department shuts down power to the building during a fire to provide for fire fighter safety. In such situations, power must remain available for the fire pump.

(2) On-site power production facility connection dedicated to the fire pump installation

> **AHJ FAQ**
>
> **The plans for a fire pump installation show the building electrical service powering the fire pump. Is this permitted?**
>
> *ANSWER:* No, the fire pump must have a dedicated service connection. During a fire, the electric pump must remain energized; having the fire pump connected to the main service may result in shutdown during fire-fighting operations.

An on-site standby or emergency generator does not meet the requirement for an on-site power production facility in accordance 9.2.2(2). See the definitions of *on-site power production facility* and *on-site standby generator* in 3.3.35 and 3.3.36, respectively. An on-site power production facility produces power continuously (that is, 24 hours per day, 7 days per week, 365 days per year). In many cases, on-site power production sources are electric-generating stations dedicated to a particular facility or to a particular facility campus-style distribution system. Fire protection is lost when fire pump power is lost. Information on fire protection systems for on-site generating stations can be found in NFPA 850, *Recommended Practice for Fire Protection for Electric Generating Plants and High Voltage Direct Current Converter Stations.*

(3) Dedicated feeder connection derived directly from the dedicated service to the fire pump installation

> ▶ **FAQ**
>
> Can a fire pump feeder circuit supply loads other than those associated with the fire pump?

The dedicated service and feeder connection referred to in 9.2.2(3) are not permitted to supply loads other than those associated with the fire pump; these must be tapped ahead of the fire pump controller(s) and not from within any fire pump controller.

(4) As a feeder connection where all of the following conditions are met:

> ▶ **FAQ**
>
> Can a fire pump feeder circuit supply loads other than those associated with the fire pump if an alternate source of power is provided for the fire pump motor?

The feeder connection referred to in 9.2.2(4) may supply loads other than those associated with the fire pump. This practice is acceptable and not considered to reduce the reliability of the power supplied to the fire pump, since a backup source of power is required in 9.2.2(4)(b), and all feeder circuits supplying a fire pump(s) must be selectively coordinated. Most of these installations are fed from two separate utility connections via double-ended switch gear (main-tie-main) with automatic or manual switchover means.

  (a) The protected facility is part of a multibuilding campus-style arrangement.
  (b) A backup source of power is provided from a source independent of the normal source of power.
  (c) It is impractical to supply the normal source of power through the arrangement in 9.2.2(1), 9.2.2(2), or 9.2.2(3).
  (d) The arrangement is acceptable to the authority having jurisdiction.
  (e) The overcurrent protection device(s) in each disconnecting means is selectively coordinated with any other supply side overcurrent protective device(s).

> Selective coordination in the context of this chapter ensures that, when an overcurrent condition occurs in a fire pump branch circuit, power is not interrupted to any loads served by upstream protective devices. This avoids cascading faults, which kill other emergency loads, such as emergency lighting, annunciators, and alarm systems.

(5) Dedicated transformer connection directly from the service meeting the requirements of Article 695 of *NFPA 70, National Electrical Code*

▶ **FAQ** What is meant by a "dedicated transformer connection"?

A fire pump is permitted to be powered by a transformer if the transformer is dedicated to the fire pump and does not serve any other loads not associated with the fire pump. The exception is stated in 695.5(C) of *NFPA 70*: "Where a feeder source is provided in accordance with 695.3(C), transformers supplying the fire pump system shall be permitted to supply other loads." Section 695.3(C) applies to multi-building, campus-style complexes.

9.2.3 For fire pump installations using the arrangement of 9.2.2(1), 9.2.2(2), 9.2.2(3), or 9.2.2(5) for the normal source of power, no more than one disconnecting means and associated overcurrent protection device shall be installed in the power supply to the fire pump controller.

> The requirement of 9.2.3 allows the installation of a single disconnecting switch and fuse or a circuit breaker only if in compliance with one of the four arrangements specified. Some jurisdictions, such as large U.S. cities, prohibit this practice in their local codes, while other jurisdictions require it. Some jurisdictions revert to NFPA 20, while others permit the determination to be made by the local utility.

▶ **FAQ** Why does NFPA 20 place a limit on the number of disconnecting means allowed upstream of the fire pump controller?

The single upstream disconnect permitted in 9.2.3 affords disconnection of the fire pump feeder circuit for activities such as maintenance and is not intended to be operated under fire conditions. Additional disconnecting means would reduce the reliability of the power supplied to the fire pump, since these disconnects can be remote from the pump room and are not easily located. Supervising multiple disconnects can be difficult to manage, especially during a fire.

9.2.3.1 Where the disconnecting means permitted by 9.2.3 is installed, the disconnecting means shall meet all of the following:

(1) They shall be identified as being suitable for use as service equipment.

> It is vital for both personnel safety and fire protection reliability that the disconnecting means be listed and labeled as suitable for use as service entrance equipment, especially since very high available short-circuit currents exist in many installations.

(2) They shall be lockable in the closed position.

▶ **FAQ** Why does NFPA 20 require the disconnect external to the fire pump controller to be capable of being locked in the closed position?

> As generally provided by the manufacturer, most disconnects are suitable for locking in the "off" position, but not the "on" position. In order to provide the capability of locking in both positions, a locking accessory with such capability may need to be field installed on the disconnect device. Locking in the closed position prevents inadvertent loss of power to the fire pump motor.

(3)* They shall be located remote from other building disconnecting means.

> ▶ **FAQ** Why does NFPA 20 require the disconnects external to the fire pump controller to be located remote from other building disconnecting means?

> The disconnects are required to be located remotely so that they are not confused with disconnects for building loads and inadvertently opened, such as by the fire department staff in the event of a fire.

A.9.2.3.1(3) The disconnecting means should be located such that inadvertent simultaneous operation is not likely.

> It is important to avoid opening of multiple disconnects at the same time, which can result in disabling fire protection in more than one location or building.

(4)* They shall be located remote from other fire pump source disconnecting means.

A.9.2.3.1(4) The disconnecting means should be located such that inadvertent simultaneous operation is not likely.

(5) They shall be marked "Fire Pump Disconnecting Means" in letters that are no less than 1 in. (25 mm) in height and that can be seen without opening enclosure doors or covers.

> Disconnects need to be readily identifiable for maintenance and fire service personnel.

9.2.3.2 Where the disconnecting means permitted by 9.2.3 is installed, a placard shall be placed adjacent to the fire pump controller stating the location of this disconnecting means and the location of any key needed to unlock the disconnect.

> ▶ **FAQ** Why does NFPA 20 require a placard adjacent to the fire pump controller?

> A placard provides for easy identification of the single disconnect in the event the power provided to the controller needs to be disconnected.

9.2.3.3 Where the disconnecting means permitted by 9.2.3 is installed, the disconnect shall be supervised in the closed position by one of the following methods:

(1) Central station, proprietary, or remote station signal device
(2) Local signaling service that will cause the sounding of an audible signal at a constantly attended location
(3) Locking the disconnecting means in the closed position
(4) Sealing of disconnecting means and approved weekly recorded inspections where the disconnecting means are located within fenced enclosures or in buildings under the control of the owner

> **FAQ** What is the purpose of the supervision methods for the disconnecting means external to the fire pump controller?

All of the methods listed in 9.2.3.3(1) through (4) reduce the likelihood that the single disconnecting means will be open when a fire event occurs and power needs to be supplied to the fire pump motor.

9.2.3.4 Where the overcurrent protection permitted by 9.2.3 is installed, the overcurrent protection device shall be rated to carry indefinitely the sum of the locked rotor current of the largest pump motor and the full-load current of all of the other pump motors and accessory equipment.

The requirement in 9.2.3.4 ensures that, in the event that an overload occurs in the fire pump branch circuit, the circuit breaker in the fire pump controller is the only protective device to disconnect power to the fire pump motor. This arrangement is necessary, as the overload may only be temporary. In such a case, the circuit breaker in the fire pump controller may be reset and the fire pump motor restarted if needed. If protective devices upstream of the fire pump controller open during an overload condition, it would be virtually impossible to reset or replace them quickly enough to return the fire pump to service for the successful extinguishment of the fire.

When sizing the overcurrent protection based on the requirements of 9.2.3.4 or 9.3.2.4.1, compliance with the selective coordination requirement in 9.2.2(4)(e) also must be met. This compliance is critical, since the power source is often capable of producing very high fault currents, and since heavy short circuits do occur in fire pump circuits, particularly in the motor. Note that the overcurrent device must be sized to hold at least the motor locked-rotor current in amperes (LRA) continuously, along with any other connected loads. This value is typically 600 percent or more of the motor full-load current in amperes (FLA). However, the wiring in this circuit is required to be only 125 percent of the motor FLA, so the lugs of the device are often too large for the wire size (i.e., the marked minimum wire size rating of the lugs may exceed the desired wire size). This problem is easily resolved by the use of inexpensive, optional lug kits, which are available for smaller wire sizes.

The requirements in 9.2.3.4 have been revised to reduce the size of this device when multiple fire pumps are connected. Previous editions required carrying the LRA of all connected pumps — that is, the fire pumps and the jockey pump(s) — in addition to the full-load current of all connected loads.

9.2.3.4.1 Alternately, compliance with 9.2.3.4 shall be based on an assembly listed for fire pump service complying with the following:

(1) The overcurrent protection device shall not open within 2 minutes at 600 percent full load current.
(2) The overcurrent protection device shall not open with a restart transient of 24 times the full load current.
(3) The overcurrent protection device shall not open within 10 minutes at 300 percent full load current.
(4) Trip point for circuit breakers not field adjustable.

The alternate requirements of 9.2.3.4.1 change the duration of the motor rated locked-rotor current (LRA) from indefinitely to 2 minutes and 300 percent of the motor full-load current rating (FLA) for 10 minutes. However, the fire pump controller no-trip is not less than 300 percent of motor FLA. It is vital that this upstream overcurrent protection not trip before the circuit breaker in the fire pump controller. That breaker is prohibited from tripping at any current that is less than 300 percent of motor FLA, and it must also trip between 8 seconds and 20 seconds at motor LRA. Since the controller circuit breaker current can be higher, but not less than, 300 percent, the controller manufacturer should be consulted to determine how much above 300 percent the controller breaker is set to hold.

9.3 Alternate Power

9.3.1 Except for an arrangement described in 9.3.3, at least one alternate source of power shall be provided where the height of the structure is beyond the pumping capacity of the fire department apparatus.

> **AHJ FAQ**
>
> **The system design shows an electric fire pump without backup power. Can this design be approved?**
>
> *ANSWER:* Yes, if it is determined that the source of power is reliable and the height of the building is not outside the pumping capability of the fire department as specified in Section 5.5.

This requirement is vitally important for taller high-rise buildings, since supplying adequate sprinkler and hose pressure is beyond the capability of fire pump trucks. As a result, the fire protection of the upper portions of these buildings is solely dependant on the building's own fire pump(s). In most cases (such as those of pumps in series), fire protection is dependent on two, or usually three, fire pumps. This is known as "protection in place," since the fire protection for the upper reaches of the building must be completely self-contained. This self-containment is most important where evacuation is not possible or unadvisable, or where people are likely to be asleep.

The concept of the requirement in 9.3.1 can also apply to buildings with large square footage, such as factories or warehouses, where the central portion of the building is beyond the reach of fire department ladder or snorkel trucks, and/or where the required waterflow is beyond that which can be delivered into the fire department connection by fire engine pump trucks.

9.3.2* Other Sources. Except for an arrangement described in 9.3.3, at least one alternate source of power shall be provided where the normal source is not reliable.

> ▶ **FAQ**
>
> Are any guidelines available for determining the criteria that constitute a reliable power source?

A reliable power source is not defined by NFPA 20. However, fairly detailed guidance is provided in A.9.3.2 to help determine whether or not the normal power source can be considered reliable. For every project, the authority having jurisdiction (AHJ) should be consulted and verification received regarding whether the power source in question is reliable. Special attention should be paid to cases where power is brought in by overhead lines.

A.9.3.2 A reliable power source possesses the following characteristics:

(1) The source power plant has not experienced any shutdowns longer than 4 continuous hours in the year prior to plan submittal. NFPA 25, *Standard for the Inspection, Testing, and Maintenance of Water-Based Fire Protection Systems*, requires special undertakings (i.e., fire watches) when a water-based fire protection system is taken out of service for longer than 4 hours. If the normal source power plant has been intentionally shut down for longer than 4 hours in the past, it is reasonable to require a backup source of power.
(2) Power outages have not routinely been experienced in the area of the protected facility caused by failures in generation or transmission. The standard is not intended to require that the normal

source of power be infallible to deem the power reliable. NFPA 20 does not intend to require a back-up source of power for every installation using an electric motor–driven fire pump. Note that should the normal source of power fail in a rare event, the impairment procedures of NFPA 25 could be followed to mitigate the fire risk. If a fire does occur during the power loss, the fire protection system could be supplied through the fire department connection.

> The text of A.9.3.2 was revised in this edition to include a reference to NFPA 25, *Standard for the Inspection, Testing, and Maintenance of Water-Based Fire Protection Systems*, and a reference to the fire department connection. However, it should be noted that the fire department connection may or may not be able to provide fire protection to the entire facility, either because limited pressure is available (e.g., in taller high-rise buildings) or because flow is limited (e.g., in large facilities and airplane hangars). The text of A.9.3.2 also references situations where frequent interruption of a power source may warrant a backup standby power source.

(3) The normal source of power is not supplied by overhead conductors outside the protected facility. Fire departments responding to an incident at the protected facility will not operate aerial apparatus near live overhead power lines, without exception. A backup source of power is required in case this scenario occurs and the normal source of power must be shut off. Additionally, many utility providers will remove power to the protected facility by physically cutting the overhead conductors. If the normal source of power is provided by overhead conductors, which will not be identified, the utility provider could mistakenly cut the overhead conductor supplying the fire pump.

(4) Only the disconnect switches and overcurrent protection devices permitted by 9.2.3 are installed in the normal source of power. Power disconnection and activated overcurrent protection should only occur in the fire pump controller. The provisions of 9.2.2 for the disconnect switch and overcurrent protection essentially require disconnection and overcurrent protection to occur in the fire pump controller. If unanticipated disconnect switches or overcurrent protection devices are installed in the normal source of power that do not meet the requirements of 9.2.2, the normal source of power must be considered not reliable and a back-up source of power is necessary.

Typical methods of routing power from the source to the motor are shown in Figure A.9.2. Other configurations are also acceptable. The determination of the reliability of a service is left up to the discretion of the authority having jurisdiction.

9.3.3 An alternate source of power for the primary fire pump shall not be required where a backup engine-driven fire pump, backup steam turbine-driven fire pump, or backup electric motor–driven fire pump with independent power source meeting 9.2.2 is installed in accordance with this standard.

> The use of a backup engine-driven or backup steam turbine–driven fire pump increases the reliability of the fire suppression system for the premises being served and, therefore, an alternate source of power is not required.

9.3.4 Where provided, the alternate source of power shall be supplied from one of the following sources:

> Both the regular and the alternate source could be a service connection as specified in 9.2.2(1), since the two sources derive from separate substations and are fed by different power lines.

(1) A generator installed in accordance with Section 9.6
(2) One of the sources identified in 9.2.2(1), 9.2.2(2), 9.2.2(3), or 9.2.2(5) where the power is provided independent of the normal source of power

> **▶ FAQ** What is meant by "power is provided independent of the normal source of power"?

The phrase "power is provided independent of the normal source of power" in 9.3.4(2) means the alternate sources of power are not derived from the service providing the normal source of power.

9.3.5 Where provided, the alternate supply shall be arranged so that the power to the fire pump is not disrupted when overhead lines are de-energized for fire department operations.

> The alternate power supply must not be transmitted through overhead power lines in close proximity to the building being supplied so that fire department operations do not result in the de-energization of these lines.

9.3.6 Two or More Alternate Sources. Where the alternate source consists of two or more sources of power and one of the sources is a dedicated feeder derived from a utility service separate from that used by the normal source, the disconnecting means, overcurrent protective device, and conductors shall not be required to meet the requirements of Section 9.2 and shall be permitted to be installed in accordance with *NFPA 70, National Electrical Code*.

> The alternate source may consist of two or more sources of power. When one of those sources is a dedicated feeder derived from a utility service separate from that used by the normal source, the reliability of the power available to the fire pump controller is increased significantly, and the special requirements of Section 9.2 are considered unnecessary.

9.4* Voltage Drop

A.9.4 Normally, conductor sizing is based on appropriate sections of *NFPA 70, National Electrical Code*, Article 430, except larger sizes could be required to meet the requirements of *NFPA 70*, Section 695.7 (NFPA 20, Section 9.4). Transformer sizing is to be in accordance with *NFPA 70*, Section 695.5(A), except larger minimum sizes could be required to meet the requirements of *NFPA 70*, Section 695.7.

> Calculating starting voltage drops is more complex than calculating motor running voltage drops. The reason is that a motor running at, or near, full load has a power factor (PF) of approximately 0.85 (85 percent PF). As such, the calculation can usually be made with the help of Chapter 9, Table 9, of *NFPA 70* using the calculation method "effective Z at 0.85 PF." For starting voltage drop, the motor impedance has a PF of 30 percent to 40 percent (PF of 0.30 to 0.40). The PF of the source, the transformers, and all wiring — especially wiring to the controller and motor — must be calculated vectorially. This is especially important with larger size (higher horsepower) motors, since larger power sources and wiring tend to have lower PFs. The reactance and ac resistance of wiring is provided in the other columns of Table 9 of *NFPA 70*. The reactance and resistance are part of the variables needed for the complex impedance (vector) calculations. Another resource for calculating voltage drop is NEMA ICS 14, *Application Guide for Electric Fire Pump Controllers*, Section 2.1, Power Conductors Connected to the Fire Pump Controller and Motor.

9.4.1 Unless the requirements of 9.4.2 or 9.4.3 are met, the voltage at the controller line terminals shall not drop more than 15 percent below normal (controller-rated voltage) under motor-starting conditions.

▶ **FAQ** | Why is voltage drop critical in the operation of a fire pump?

Excessive voltage drop results in the motor magnetic contactor chattering, since the holding coils of magnetically operated devices are only required to hold the contacts closed at a minimum of 85 percent of rated voltage. While some controllers are designed and tested to allow a much greater voltage drop, the 85 percent minimum hold-in voltage is the minimum requirement for contactors in general.

9.4.2 The requirements of 9.4.1 shall not apply to emergency-run mechanical starting. *(See 10.5.3.2.)*

The most common field failure of electric fire pump controllers is a burnt-out contactor coil(s) or the burnt-out circuitry that is supposed to energize the coil(s). The emergency-run mechanical operator pushes in the contactor plunger mechanically (directly) to close the contacts in order to energize the motor. However, some contactor designs locate the coil over the open gap between the motor contactor plunger and the fixed magnetic core. When the coil melts into this gap, it prevents the mechanical operator from working.

▶ **FAQ** | Why do the voltage drop requirements of Section 9.4 not apply to the emergency-run mechanical starting operation of the fire pump controller?

Many controllers are of reduced voltage starting–type construction. These controllers are usually used where the capacity and/or voltage regulation of the normal or alternative source (gen-set) is insufficient to start the motor under full voltage across-the-line (direct-on-line) ATL (DOL) conditions. The emergency-run mechanical start operation of these controllers is accomplished by starting the motor at full available line voltage mechanically, so the consideration of controller contactor chattering or other controller malfunction is irrelevant. The motor will accelerate the pump to full speed, even at full pump load and even when the voltage drops more than 115 percent (within reasonable limits).

9.4.3 The requirements of 9.4.1 shall not apply to the bypass mode of a variable speed pressure limiting control *(see 10.10.3)*, provided a successful start can be demonstrated on the standby gen-set.

The voltage drop requirements of Section 9.4 do not apply to the bypass mode of a variable speed pressure limiting control (VSPLC) device for the same reasons they do not apply to the emergency-run mechanical starting operation of the fire pump controller. The requirement that a successful start can be demonstrated on the standby generator, if provided, is necessary, because the capacity of the standby generator may not be sufficient to allow for starting of the fire pump motor in this mode. If successful start cannot be demonstrated, the bypass mode is of no use in the event of failure of the normal power source. This risk of failure is also why nearly all such controllers make use of reduced voltage (reduced in-rush) starting in the bypass path, of which the two most common types are primary reactor start or soft start. However, it should be noted that the starting currents in the variable speed mode are lower than those of any other starting method, and the generator is often sized (selected) based on this fact (the generator capacity is often selected based on meeting the 115 percent maximum drop under "normal" motor starting conditions).

9.4.4 The voltage at the contactor(s) load terminals to which the motor is connected shall not drop more than 5 percent below the voltage rating of the motor when the motor is operating at 115 percent of the full-load current rating of the motor.

> The substantial revision to 9.4.4 is that the motor voltage drop measurement is now specified at the motor contactor(s) in the controller, rather than "at the motor terminals" as stated in previous editions. This revision means the voltage drop easier to measure and is also common practice. However, the revised location also makes the requirement somewhat more difficult to meet in marginal (close) situations, since the voltage drop in the wiring to the motor is now included in the voltage drop measurement. Fortunately, this is a small factor in most installations due the normally short wiring run between the controller and the motor.
> Note that the list of raceways (conduits) is also covered in 695.6(D) of *NFPA 70*.
>
> ▶ **FAQ** Is voltage drop critical at any time other than during motor starting?
>
> Voltage drop is critical during motor operation at the maximum allowable service factor (SF) of 1.15, because lower voltages result in higher motor running currents. Higher currents occur because the motor must draw more electrical current at lower available voltage in order to deliver the required mechanical power (torque times speed) to the pump shaft. A drop of only 5 percent results in 121 percent motor current, which results in 46 percent greater motor winding heating. The following equations illustrate an example:
>
> $$1.15 / 95\% = 1.21 \ (121\%)$$
>
> and
>
> $$121\% \text{ squared } (1.21^2) \text{ heating} = 1.46 \ (146\%)\%$$

9.4.4.1 Wiring from the controller(s) to the pump motor shall be in rigid metal conduit, intermediate metal conduit, electrical metallic tubing, liquidtight flexible metal conduit, or liquidtight flexible nonmetallic conduit Type LFNC-B, listed Type MC cable with an impervious covering, or Type MI cable.

9.4.4.2 Electrical connections at motor terminal boxes shall be made with a listed means of connection.

9.4.4.3 Twist-on insulation-piercing type and soldered wire connectors shall not be permitted to be used for this purpose.

9.5 Motors

9.5.1 General.

9.5.1.1 All motors shall comply with NEMA MG-1, *Motors and Generators*, shall be marked as complying with NEMA Design B standards, and shall be specifically listed for fire pump service. *(See Table 9.5.1.1.)*

> NEMA MG-1, *Motors and Generators*, covers ac and dc motors and generators. It establishes minimum requirements and also establishes definitions and accepted values of numerous parameters. Design B motors, as outlined by NEMA, are considered to be low starting current and high starting torque motors, such as squirrel cage motors, which are widely used. The applicable standard is UL 1004-5, *Fire Pump Motors*. These motors are tested for minimum performance and electrical characteristics.
> See Exhibit II.9.1 for a typical fire pump motor nameplate.

9.5.1.2 The requirements of 9.5.1.1 shall not apply to direct-current, high-voltage (over 600 V), large-horsepower [over 500 hp (373 kW)], single-phase, universal-type, or wound-rotor motors, which shall be permitted to be used where approved.

TABLE 9.5.1.1 Horsepower and Locked Rotor Current Motor Designation for NEMA Design B Motors

Rated Horsepower	Locked Rotor Current Three-Phase 460 V (A)	Motor Designation (NFPA 70, Locked Rotor Indicating Code Letter) "F" to and Including
5	46	J
7½	64	H
10	81	H
15	116	G
20	145	G
25	183	G
30	217	G
40	290	G
50	362	G
60	435	G
75	543	G
100	725	G
125	908	G
150	1085	G
200	1450	G
250	1825	G
300	2200	G
350	2550	G
400	2900	G
450	3250	G
500	3625	G

EXHIBIT II.9.1 Fire Pump Motor Nameplate. (Courtesy of Marathon Electric)

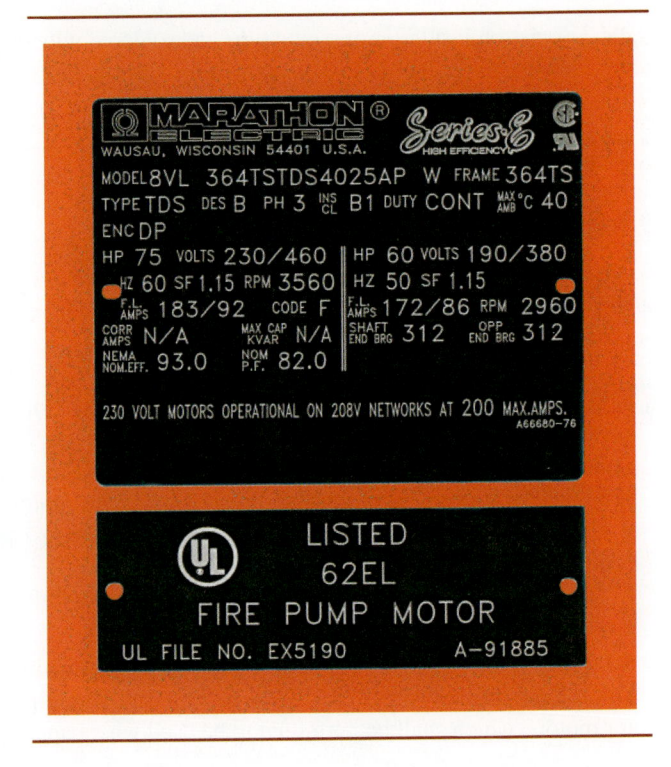

Table 9.5.1.1 establishes the acceptable range of NEMA starting codes. These codes indirectly limit the range of starting currents that the designated motors are allowed to draw. The table values limit starting currents (LRA) to a range of approximately 5.0 times to 6.6 times motor FLA (500 percent to 660 percent). This limit is absolutely essential, since fire pump controllers are designed, tested, listed, and approved for starting currents of 600 percent of FLA. Some motors draw 700 percent or 800 percent, but these motors result in unreliable and possibly unsafe installations. The starting current is essential, even for reduced voltage and variable speed applications, since the emergency-run mechanical operator always starts the motor in a full voltage ATL (DOL) mode.

▶ **FAQ** Do any types of fire pump motors exist that are not required to be listed?

Paragraph 9.5.1.1 does not apply to the motors specified in 9.5.1.2. Currently, no published standards cover these motors for fire pump applications. In addition, motors for use with variable speed controllers must meet the requirements of 9.5.1.4 (i.e., Part 30 and Part 31 of NEMA MG-1). As of the publication of this handbook, no such motors are listed as fire pump motors under ANSI/UL 1004-5, *Standard for Fire Pump Motors*. High voltage motors (now more typically known as medium voltage motors) are used worldwide. These motors are more likely to be specific to the application. Also, many variable speed high voltage motors are unsuitable for full voltage starting, which is mandatory for a fire pump motor due to the emergency-run mechanical operator. It was once common to find dc motors in downtown areas where buildings used to be supplied by either 200 VDC or 400 VDC service; however, most have been replaced with ac motors and controllers. Many wound rotor motors are still in use because their applications — reduced voltage starting and variable speed operation — are still needed in many high-rise buildings in New York City.

9.5.1.3 Part-winding motors shall have a 50-50 winding ratio in order to have equal currents in both windings while running at nominal speed.

The requirement in 9.5.1.3 specifies the criteria needed for proper sizing of contactors in a part winding fire pump controller.

9.5.1.4* Motors Used with Variable Speed Controllers.

A.9.5.1.4 Variable fire pump motors must be of the inverter duty type for the installation to be reliable. Inverter duty motors have higher insulation voltage rating, suitable temperature rise rating, and protection from bearing damage.

9.5.1.4.1 Motors shall meet the applicable requirements of NEMA MG-1, *Motors and Generators*, Part 30 or 31.

See NEMA MG-1, Part 30, "Application Considerations for Constant Speed Motors Used on a Sinusoidal Bus with Harmonic Content and General Purpose Motors Used with Adjustable-Voltage or Adjustable-Frequency Controls or Both"; and NEMA MG-1, Part 31, "Definite-Purpose Inverter-Fed Polyphase Motors."

9.5.1.4.2 Motors shall be listed, suitable, and marked for inverter duty.

Although 9.5.1.4.2 requires that the motor be listed, there are no listed inverter duty fire pump motors as of the publication of this handbook. It is more important for the motor to be labeled (marked) as suitable for inverter duty. Inverter duty motors differ from traditional motors in at least the following three ways:

1. The winding voltage rating is higher to accommodate the transient and harmonic voltages created by the variable frequency drive (VFD) [also known as adjustable speed drive (ASD) or adjustable frequency drive (AFD)] in the variable speed fire pump controller (VSD).

2. The motor can accommodate the higher winding temperatures that occur due to the high frequency modulated voltage supplied by the VFD.
3. The motor has a means to prevent rotor current flow through the motor bearings, which is created by the high frequency modulated voltage impressed on the motor by the VFD.

The two common methods of dealing with bearing currents are either to insulate both bearings or to supply grounding brushes, straps, fingers, or the like to bypass the generated harmonic currents around the bearings. See A.9.5.1.4, which is new to this edition, for more information.

▶ **FAQ** Why aren't all fire pump motors suitable for use with variable speed fire pump controllers?

Standard NEMA Design B motors may not be suitable for use with variable speed controllers, as their performance may be adversely affected by the output voltage waveforms produced by these controllers. Only those motors marked for inverter duty are suitable for use with variable speed controllers.

9.5.1.4.3 Listing shall not be required if 9.5.1.2 applies.

9.5.1.5* The corresponding values of locked rotor current for motors rated at other voltages shall be determined by multiplying the values shown by the ratio of 460 V to the rated voltage in Table 9.5.1.1.

CALCULATION ▶ For example, the locked-rotor current of a 100 hp motor for other voltages would be as follows:

Voltage (V)	208	230	380	400	415	575
LRA (LRC) (A):	1603	1540	878	834	804	580

A.9.5.1.5 The locked rotor currents for 460 V motors are approximately six times the full-load current.

9.5.1.6 Code letters of motors for all other voltages shall conform with those shown for 460 V in Table 9.5.1.1.

9.5.1.7 All motors shall be rated for continuous duty.

9.5.1.8 Electric motor–induced transients shall be coordinated with the provisions of 10.4.3.3 to prevent nuisance tripping of motor controller protective devices.

When motors are started at full voltage, such as by the use of the emergency-run mechanical operator, the first half line cycle peak current can be up to 12 times the motor rated FLA current. These peak starting currents tend to be higher on higher horsepower installations, since such installations tend to have higher reactance to resistance (X to R) ratios. In other words, they tend to be more inductive (reactive). Some controllers utilize circuit breakers that tolerate 13 times motor FLA to avoid false tripping.

When power to a running motor is momentarily interrupted, the resulting short-time transient current spike can be up to 18 times the motor FLA rating (3 times typical LRA values). This situation occurs when the controller, transfer switch, or utility switches connect a still-spinning motor to the power source, because a spinning motor generates a voltage [back electromagnetic force (EMF)] due to the motor's magnetic field. When the motor voltage phase angle differs from that of the power source, a transient related to the amount of the phase angle difference is created. Some controllers incorporate a restart delay to mitigate these potential high spikes in transient currents. A delay of 2 seconds to 3 seconds is sufficient for the motor back EMF to drop to safe values, since the typical magnetic field decays by 63 percent in 3/4 of a second to 1 1/2 seconds after being disconnected from line voltage.

▶ **FAQ** Why are electric motor–induced transient currents a concern in the operation of a fire pump motor?

Electric motor–induced transient currents are normal during motor starting. When power is briefly interrupted, these motor transient currents are even higher than the normal starting peak currents and may contain enough energy to cause the fire pump circuit breaker specified in 10.4.3 to trip. The instantaneous setting of this circuit breaker should be high enough to prevent the electric motor–induced transient currents from tripping the circuit breaker. The instantaneous setting of the circuit breaker is allowed to be as high as 20 times the full-load current rating of the motor.

9.5.1.9 Motors for Vertical Shaft Turbine–Type Pumps.

9.5.1.9.1 Motors for vertical shaft turbine–type pumps shall be dripproof, squirrel-cage induction type.

Exhibit II.9.2 illustrates an open dripproof induction motor.

EXHIBIT II.9.2 *Dripproof Induction Motor.*

9.5.1.9.2 The motor shall be equipped with a nonreverse ratchet.

Pumps must turn in only one direction. Backspin (spinning in reverse) of the pump is likely without the nonreverse ratchet. Backspin occurs when motor power is cut, due to the weight of the water column in the discharge pipe column. This situation is especially common in the case of vertical turbine pumps. Backspin of a turbine pump is likely to damage some of the pump bowls.

9.5.2 Current Limits.

9.5.2.1 The motor capacity in horsepower shall be such that the maximum motor current in any phase under any condition of pump load and voltage unbalance shall not exceed the motor-rated full-load current multiplied by the service factor.

9.5.2.2 The following shall apply to the service factor:

(1) The maximum service factor at which a motor shall be used is 1.15.

> The maximum motor load may occur at pump outputs at or above the 150 percent flow rate, but the flow rates are usually below that point. The motor must not be run into an overload condition, regardless of where on the pump curve this condition occurs. Note that fire water demand can vary due to both fused sprinklers and connected hoses.
>
> ▶ **FAQ** What is the required SF for fire pump motors?
>
> The requirement in 9.5.2.2(1) applies only to motors with an SF of 1.15. Fire pump motors are permitted, but not required, to have an SF of 1.15. Running the motor into its SF will increase the motor current by a percentage equal to that of the SF. Therefore, running the motor at 1.15 times its horsepower rating will increase the motor current to 115 percent of its rated FLA. Worse yet, if the motor voltage is at the minimum of 95 percent (with 5 percent voltage drop), the motor current will be equal to 1.15 / 0.95, or 121 percent of the motor FLA. Running a motor above its SF load (115 hp with a 100 hp motor having an SF of 1.15) will quickly overheat the windings. Exhibit II.9.3 shows damaged windings.

EXHIBIT II.9.3 Damaged Windings.

(2) Where the motor is used with a variable speed pressure limiting controller, the service factor shall not be used.

> ▶ **FAQ** Why is the SF not utilized for motors used with a VSPLC device?
>
> As discussed in the commentary to 9.5.2.2(1), motors driven by a VFD will exhibit extra heating due to the high frequency components of the modulated voltage and, subsequently, current, supplied by the VFD. Running the motor into its SF would overheat the windings.

9.5.2.3 These service factors shall be in accordance with NEMA MG-1, *Motors and Generators*.

9.5.2.4 General-purpose (open and dripproof) motors, totally enclosed fan-cooled (TEFC) motors, and totally enclosed nonventilated (TENV) motors shall not have a service factor larger than 1.15.

9.5.2.5 Motors used at altitudes above 3300 ft (1000 m) shall be operated or derated according to NEMA MG-1, *Motors and Generators*, Part 14.

> The requirement of 9.5.2.5 is especially important whenever the motor is run into its SF. The information provided in the commentary to 9.5.2.2(1) demonstrates that a motor running at 1.15 SF causes 32 percent additional heating of the windings (1.15^2). The thinner atmosphere reduces the cooling ability of the motor's fan.

9.5.3 Marking.

9.5.3.1 Marking of motor terminals shall be in accordance with NEMA MG-1, *Motors and Generators*, Part 2.

9.5.3.2 A motor terminal connecting diagram for multiple lead motors shall be furnished by the motor manufacturer.

> Achieving the proper wiring of multiple lead motors (more than three conductors) is difficult without the use of a wiring diagram that identifies the proper connections. This issue is generally confined to part winding and wye-delta motors. Table 7-6 in *Pumps for Fire Protection Systems* lists the 14 configurations of three, six, nine, and twelve lead motors. Figure 7-4 through Figure 7-17 of this publication are wiring diagrams for the 14 configurations.

9.6 On-Site Standby Generator Systems

9.6.1 Capacity.

9.6.1.1 Where on-site generator systems are used to supply power to fire pump motors to meet the requirements of 9.3.2, they shall be of sufficient capacity to allow normal starting and running of the motor(s) driving the fire pump(s) while supplying all other simultaneously operated load(s) while meeting the requirements of Section 9.4.

> The requirement in 9.6.1.1 prevents undersizing of the generator under all conditions of loading.
> Automatic load shedding used to be specifically permitted in NFPA 20; this allowance was eliminated starting with the 1999 edition, but it is not specifically prohibited. It remains permitted in 695.3(D)(1) of *NFPA 70*. The AHJ should be consulted in any case. Some load shedding is done during the entire time the fire pump(s) is in use. Other loads, such as elevators, are disconnected only during the starting cycle (sequential and acceleration times) of the fire pump(s). Some controllers have the option of being equipped to provide for one load or the other, or both.

9.6.1.2 A tap ahead of the on-site generator disconnecting means shall not be required.

▶ **FAQ** Why is a tap ahead of the on-site generator disconnecting means not required?

An on-site generator may supply loads other than those associated with the fire pump. The normal source power supply is required to be a service connection dedicated to the fire pump installation when the source supplies loads other than those associated with the fire pump;

however, this requirement does not apply to the alternate source power supply. Furthermore, the generator main circuit breaker is often an integral part of the generator set package. Wiring ahead of this breaker would likely void any warranties and any listings. As such, the fire pump load may be connected on the load side of the generator disconnecting means.

9.6.2* Power Sources.

A.9.6.2 Where a generator is installed to supply power to loads in addition to one or more fire pump drivers, the fuel supply should be sized to provide adequate fuel for all connected loads for the desired duration. The connected loads can include such loads as emergency lighting, exit signage, and elevators.

9.6.2.1 On-site standby generator systems shall comply with Section 9.4 and shall meet the requirements of Level 1, Type 10, Class X systems of NFPA 110, *Standard for Emergency and Standby Power Systems*.

▶ **FAQ** Why were the requirements of Level 1, Type 10, Class X systems in NFPA 110, *Standard for Emergency and Standby Power Systems*, chosen to define the performance of on-site generators?

Adherence to the requirements of NFPA 110 for Level 1, Type 10, Class X systems ensures that the alternate power source is available within 10 seconds. This time period is considered necessary to provide adequate water supply in the event of a fire. NFPA 110 also includes various requirements that have the goal of improving reliability, since the quality and reliability of gen-sets vary widely. This problem is evident whenever a power outage covering a wide area occurs and gen-sets fail to start. NFPA 110 was created to establish backup power requirements for electric fire pumps that more closely reflect the requirements and reliability of diesel fire pumps. The requirements apply to both the engine itself and its controls and alarms.

9.6.2.2 The generator shall run and continue to produce rated nameplate power without shutdown or derate for alarms and warnings or failed engine sensors, except for overspeed shutdown.

Level 1 emergency power supply systems (EPSS), as defined by NFPA 110 and Articles 700, 701, and 702 of *NFPA 70*, require the emergency backup unit (generator or other) to operate with limited safety or warning shutdowns. This requirement is applicable to the standby generator used to provide an alternate power source. Certain municipalities have required adherence to a level higher than Level 1 EPSS and do not allow the engine to shut down or derate power in the event of a failed sensor on an electronic control module (ECM) engine or in the case of high water temperature or low oil pressure. This requirement ensures that engines used as the electrical backup for electric drivers have a level of performance similar to that required in Chapter 11 of NFPA 20, which includes a requirement that is more stringent than that required for a Level 1 EPSS. Also, properly sized generators aren't likely to shut down due to either excess load power (frequency drop) or excess current flow. The change from the 2010 edition is the substitution of "generator" for "engine."

9.6.2.3 The generator fuel supply capacity shall be sufficient to provide 8 hours of fire pump operation at 100 percent of the rated pump capacity in addition to the supply required for other demands.

The need for 8 hours of operation has been demonstrated numerous times by long-lasting fires, especially in high-rise buildings.

9.6.3 Sequencing. Automatic sequencing of the fire pumps shall be permitted in accordance with 10.5.2.5.

> Paragraph 10.5.2.5 requires sequential starting of pumps in parallel or pumps in series.

9.6.4 Transfer of Power. Transfer of power to the fire pump controller between the normal supply and one alternate supply shall take place within the pump room.

> The requirement of 9.6.4 is important, because it enables the user to operate the transfer switch quickly in the event of power supply failure, particularly when the transfer switch is not part of the fire pump controller itself.
>
> ▶ **FAQ** Do all upstream transfer switches have to be located in the pump room?
>
> According to 9.6.4, the transfer switch that has the fire pump controller connected to its load terminals must be located in the pump room. There may be additional transfer switches ahead of such a switch, as in the case of hospital "hot- emergency- bus-" type power systems. However, the requirement applies only to the transfer switch immediately upstream of the fire pump motor. This limitation is specified by the reference in 9.6.4 to "one alternate supply."

9.6.5* Protective Devices. Protective devices installed in the on-site power source circuits at the generator shall allow instantaneous pickup of the full pump room load and shall comply with *NFPA 70, National Electrical Code*, Section 700.27.

> Article 445 of *NFPA 70* requires that the load side of the generator be supplied with overcurrent protection. The reference to 700.27 of *NFPA 70* was added to ensure that these protective devices are selectively coordinated with all overcurrent devices in the alternate power circuit. This coordination is important, since the emergency-run manual operator(s) of the fire pump controller(s) may be routinely engaged by the fire service during a fire. In this case, when power is lost and the load is transferred to the emergency gen-set, the motor(s) will be started at full voltage ATL (DOL) mode. Therefore, the overcurrent protective devices must not trip due to the resulting motor LRA current, including first half-cycle offset peak currents.

A.9.6.5 Generator protective devices are to be sized to permit the generator to allow instantaneous pickup of the full pump room load. This includes starting any and all connected fire pumps in the across-the-line (direct on line) full voltage starting mode. This is always the case when the fire pump(s) is started by use of the emergency-run mechanical control in 10.5.3.2.

> The requirement in 9.6.5 ensures that no nuisance tripping of the generator protective device(s) occurs when the full pump room load is connected to the generator by the transfer switch. The circuit breaker in the fire pump controller should be the only protective device to disconnect power to the fire pump motor under any conditions of loading other than short circuit. If protective devices upstream of the fire pump controller open when the full pump room load is transferred to the generator, or during an overload condition, it may be impossible to reset or replace them quickly enough to return the fire pump to operation for successful extinguishment of the fire.

9.7 Junction Boxes

Where fire pump wiring to or from a fire pump controller is routed through a junction box, the following requirements shall be met:

> Junction boxes are commonly mounted on fire pump controllers for the purpose of terminating cables. In some cases, these junction boxes are used to comply with Section 9.8 or to provide a way to tap power for other loads, such as a jockey pump, since tapping power from within the fire pump controller is strictly prohibited. See 10.3.4.5.1 and 12.3.5.3.1.

(1) The junction box shall be securely mounted.
(2)* Mounting and installation of a junction box shall not violate the enclosure type rating of the fire pump controller(s).

> ▶ **FAQ** If mounting the junction box requires modification of the fire pump controller enclosure, what are the concerns that need to be addressed?

> If the junction box is mounted to the fire pump controller enclosure, care should be used to ensure that the mounting process does not affect the environmental rating of the fire pump controller enclosure. Modification of the controller enclosure may be necessary in order to mount the junction box. For example, cutting or drilling an opening in the controller enclosure that would allow the entrance of water or expose live electrical parts could pose a shock hazard.

A.9.7(2) See also 10.3.3.

(3)* Mounting and installation of a junction box shall not violate the integrity of the fire pump controller(s) and shall not affect the short circuit rating of the controller(s).

> If the junction box is mounted to the fire pump controller enclosure, care should be used to ensure that the mounting process does not affect the strength and rigidity of the enclosure and its ability to contain an internal disturbance due to a short circuit within or downstream of the controller. Modification of the enclosure may be necessary in order to mount the junction box. An example is cutting or drilling an opening in the controller enclosure that would allow flammable material to escape during a short circuit.

A.9.7(3) See 10.1.2.1, controller short circuit (withstand) rating.

> ▶ **FAQ** Does a junction box require a specific environmental rating?

> A Type 2 enclosure provides a degree of protection for the enclosed equipment against incidental contact with falling dirt, falling liquids, and light splashing. Because the junction box and controller enclosure are in the same environment, the junction box requires an environmental rating at least equivalent to that of the controller enclosure. A junction box with an environmental rating more stringent than that of the controller enclosure is acceptable. (Refer to Commentary Table II.10.1, which follows 10.3.3, to determine the degree of protection against specific environmental conditions provided by an enclosure, based on its type rating.)

(4) As a minimum, a Type 2, dripproof enclosure (junction box) shall be used. The enclosure shall be listed to match the fire pump controller enclosure type rating.

(5) Terminals, junction blocks, and splices, where used, shall be listed.

> This requirement is based on section 695.6(I)(5) of *NFPA 70*. It also applies to diesel-driven fire pump controllers in accordance with 12.3.5.3.1 of NFPA 20.

(6) A fire pump controller or fire pump power transfer switch, where provided, shall not be used as a junction box to supply other equipment, including a pressure maintenance (jockey) pump(s). *(See 10.3.4.5.1 and 10.3.4.6.)*

9.8 Listed Electrical Circuit Protective System to Controller Wiring

9.8.1* Where single conductors (individual conductors) are used, they shall be terminated in a separate junction box.

> The single (individual) conductors in question are typically mineral insulated (Type MI) or similar fire-resistive cables. A junction box is required to terminate these conductors, because they are usually solid wire conductors, and the inlet terminals of fire pump controllers are rated for normal stranded building wire. (These terminal lugs usually are not rated for fine stranded wire conductors, and, if they are rated for such conductors, they must be so marked. Pull-out resistance is much lower in the case of fine stranded wire conductors in lugs for which they are not rated.) In addition, it is difficult to fit the (three) conductors into a raceway(s) (conduit) to prevent heating of the cabinet (enclosure) ferrous (steel) sheet metal, which would otherwise occur due to the eddy currents produced by the magnetic field that is generated by current-carrying conductors. (In other words, the cabinet sheet metal acts as a single turn, secondary to the wires passing through the metal.)
>
> ▶ **FAQ** What special requirements apply when single conductors (individual conductors) are used to provide power to the fire pump controller?
>
> NFPA 20 does not permit the single conductors to enter a fire pump controller enclosure separately (individually) because of the circulating (eddy) currents that occur due to the flow of current through the conductors. When the conductors are in a raceway (conduit), the magnetic fields of the conductors largely cancel one another. When the conductors are spread out and pass through a ferrous (steel) metal surface, such as a fire pump enclosure cabinet, they generate magnetic currents in that metal and, thereby, cause the metal to heat, sometimes to the point of becoming red-hot.

A.9.8.1 Cutting slots or rectangular cutouts in a fire pump controller will violate the enclosure type rating and the controller's short circuit (withstand) rating and will void the manufacturer's warranty. See also *NFPA 70, National Electrical Code*, Articles 300.20 and 322, for example, for further information.

> ▶ **FAQ** Is the modification of a fire pump controller enclosure permitted in order to comply with 9.8.1?
>
> No modification of a fire pump controller enclosure is permitted. This is why a junction box is required. Cutting slots or cutouts in the controller enclosure may affect the environmental rating of the enclosure, will affect the short-circuit rating of the controller, and would likely void both the manufacturer's warranty and the controller listing and approval. Furthermore, the manufacturer of the controller should be consulted for guidance on this issue before any modifications are made to the controller enclosure.

Stationary Fire Pumps Handbook 2013

9.8.1.1 The junction box shall be installed ahead of the fire pump controller, a minimum of 12 in. (305 mm) beyond the fire-rated wall or floor bounding the fire zone.

9.8.1.2 Single conductors (individual conductors) shall not enter the fire pump enclosure separately.

9.8.2* Where required by the manufacturer of a listed electrical circuit protective system, by *NFPA 70, National Electrical Code*, or by the listing, the raceway between a junction box and the fire pump controller shall be sealed at the junction box end as required and in accordance with the instructions of the manufacturer. *(See NFPA 70, National Electrical Code, Article 695.)*

> When fire-resistive cables pass through a fire zone, they will likely produce flammable gases and smoke. These cables are not allowed to enter the fire pump controller to prevent damage to the controller and to prevent an explosion when gases are ignited by the motor contactor(s) or other sparking or arcing components, such as control relays or circuit breakers.

A.9.8.2 When so required, this seal is to prevent flammable gases from entering the fire pump controller.

▶ **FAQ** Which type of sealing means is acceptable for compliance with 9.8.2?

> The provision in 9.8.2 has been revised in this edition to require sealing of the raceway between a junction box and the fire pump controller in all installations where a junction box is provided. Previously, sealing was only necessary where required by the manufacturer of a listed electrical circuit protective system, by *NFPA 70*, or by the listing.
>
> Types of acceptable sealing compounds are specified by the manufacturer of the listed electrical circuit protective system. Normally, acceptable sealing compounds are those that do not suffer degradation from exposure to gases and do not harden and/or become brittle with age.

9.8.3 Standard wiring between the junction box and the controller shall be considered acceptable.

> The provision in 9.8.3 is based on compliance with 9.8.1 and 9.8.2.

9.9 Raceway Terminations

9.9.1 Listed conduit hubs shall be used to terminate raceway (conduit) to the fire pump controller.

> Unfortunately, many fire pump controllers are installed using only star nuts to fasten conduits. This allows water to enter the controller, especially when the conduits enter the top of the enclosure. Many controllers have been ruined as a result.

▶ **FAQ** Which types of conduit hubs are acceptable for compliance with 9.9.1?

> Conduit hubs may be marked with a type rating, as in the case of a controller enclosure. However, most hubs are marked with designations such as "raintight" and "watertight." To obtain an equivalent type rating for conduit hubs, refer to Commentary Table II.10.1 to determine the degree of protection against specific environmental conditions provided by an enclosure,

> based on its type rating. Some product standards for enclosures allow an enclosure rated Type 3, Type 3S, Type 3SX, Type 3X, Type 4, or Type 4X to be marked "raintight." An enclosure rated Type 4 or Type 4X might be permitted to be marked "watertight."

9.9.2 The type rating of the conduit hub(s) shall be at least equal to that of the fire pump controller.

> Since the type rating of the controllers is required to be at least Type 2, the hub(s) is required to be at least a Type 2 listed hub(s). The hub(s) for a Type 4 enclosure must be a Type 4 hub(s), and so forth, to prevent voiding the manufacturer's warrantee.

9.9.3 The installation instructions of the manufacturer of the fire pump controller shall be followed.

> Following the manufacturer's instructions is important to avoid voiding the manufacturer's warrantee and/or the listing and/or approval of the equipment.

9.9.4 Alterations to the fire pump controller, other than conduit entry as allowed by *NFPA 70, National Electrical Code*, shall be approved by the authority having jurisdiction.

9.9.5 Where the raceway (conduit) between the controller and motor is not capable of conducting ground fault current sufficient to trip the circuit breaker when a ground fault occurs, a separate equipment grounding conductor shall be installed between the controller and motor.

> This requirement has been added to the 2013 edition to address the use of electrical metallic tubing (EMT) in Article 695 of *NFPA 70*. The overcurrent protection within the fire pump controller, and, particularly, any overcurrent protection ahead of the controller, is set to hold 300 percent or more of the motor FLA. However, the conductors are not required to be larger than 125 percent of the motor FLA. Therefore, faults need to be quite large before the overcurrent protection will trip Therefore, any ground fault will involve currents well above those that the conduit is normally expected to carry. If the conduit or its connectors, the couplings, or the hubs are too high in resistance, the overcurrent protection may not trip in time, resulting in far more extensive damage to the motor or conductors and possibly leaving the motor frame in an energized (electrically live) condition.
>
> ### References Cited in Commentary
>
> National Fire Protection Association, 1 Batterymarch Park, Quincy, MA 02169-7471.
>
> NFPA 25, *Standard for the Inspection, Testing, and Maintenance of Water-Based Fire Protection Systems*, 2011 edition.
> NFPA 70®, *National Electrical Code*®, 2011 edition.
> NFPA 110, *Standard for Emergency and Standby Power Systems*, 2013 edition.
> NFPA 850, *Recommended Practice for Fire Protection for Electric Generating Plants and High Voltage Direct Current Converter Stations*, 2010 edition.
> Isman, K. E. and M.T. Puchovsky, *Pumps for Fire Protection Systems*, 2002.
>
> National Electrical Manufacturers Association,1300 North 17th Street, Suite 1752, Rosslyn, VA 22209-38306.
>
> NEMA ICS 14, *Application Guide for Electric Fire Pump Controllers*, 2010.
> NEMA MG-1, *Motors and Generators*, 2011.
>
> Underwriters Laboratories Inc., 333 Pfingsten Road, Northbrook, IL 60062-2096.
>
> ANSI/UL 1004-5, *Standard for Fire Pump Motors*, 2008.

Electric-Drive Controllers and Accessories

CHAPTER 10

Chapter 10 describes the attributes of electric motor–driven fire pump controllers, including controllers with power transfer switches and separate power transfer switches, all for use in electric fire pump motor circuits. Also covered in this chapter are accessories used with the controllers to ensure the minimum performance required by NFPA 20. The focus of Chapter 10 is the design, construction, and performance of the controllers, transfer switches, and their accessories. This chapter also provides information to aid in specifying, installing, and operating this equipment.

10.1 General

10.1.1 Application.

10.1.1.1 This chapter covers the minimum performance and testing requirements for controllers and transfer switches for electric motors driving fire pumps.

Although jockey pumps (pressure maintenance pumps) are mentioned in the standard (in 10.3.4.6) they are not covered by NFPA 20. Best practice dictates using reliable listed control panels, since failure of the jockey pump or controller will eventually cause the fire pump(s) to start and take the place of the jockey pump.

10.1.1.2 Accessory devices, including fire pump alarm and signaling means, are included where necessary to ensure the minimum performance of the equipment mentioned in 10.1.1.1.

All equipment necessary to ensure the minimum performance of the fire pump controller and/or transfer switch, whether installed as part of such equipment or separately, is subject to the requirements of this chapter. In general, these are items in the so-called "critical path" for starting and running the fire pump(s) to provide fire water for sprinklers, hoses and standpipes.

10.1.2 Performance and Testing.

10.1.2.1 Listing. All controllers and transfer switches shall be specifically listed for electric motor–driven fire pump service.

▶ **FAQ** What is meant by the term *listed*?

Equipment that is listed meets the criteria specified in the definition for this term in 3.2.3. The listing identification for this equipment must include a reference to "Fire Pump(s)" or "Fire Pump Service."

10.1.2.2* Marking.

A.10.1.2.2 The phrase *suitable for use* means that the controller and the transfer switch have been prototype tested and have demonstrated by those tests their short-circuit withstandability and interrupting capacity at the stated magnitude of short-circuit current and voltage available at their line terminals. *(See ANSI/UL 508, Standard for Industrial Control Equipment, and ANSI/UL 1008, Standard for Transfer Switch Equipment.)*

A short-circuit study should be made to establish the available short-circuit current at the controller in accordance with IEEE 141, *Electric Power Distribution for Industrial Plants*, IEEE 241, *Electric Systems for Commercial Buildings*, or other acceptable methods.

> A short-circuit study is necessary to ensure that equipment will not present a risk of fire or shock hazard if a short circuit occurs in the fire pump motor circuit.
>
> Typical short circuit withstand ratings for equipment vary from 10 kA to 200 kA (10,000 to 200,000 amperes). The acronym WIC is often used in lieu of "short circuit rating." This abbreviation stands for withstand and interrupt current, and it refers to the two types of tests performed by the listing, approval, and/or testing laboratory. The first type of test is the "closing" test wherein the controller is connected to a high current power source with a short circuit wired to the controller contactor. The contactor is energized and the controller must pass seven test criteria. The equipment:
>
> 1. must not become a shock hazard
> 2. must not become a fire hazard
> 3. must pass a dielectric strength test
> 4. must not eject parts
> 5. must not allow the door(s) to blow open
> 6. must be able to be de-energized
> 7. must not blow a ground fuse
>
> The "withstand" test is a similar test, except the controller contactor is mechanically closed, and the power is applied to the controller. Multiple tests are often required. The power source must be capable of producing the desired rated short circuit current.
>
> The prototype testing is carried out in a "high fault" test laboratory and is witnessed by testing agencies. The equipment must meet the seven test criteria listed in the commentary earlier in this section, but the equipment need not be functional. Some controller designs do remain functional after testing.
>
> Note that the short-circuit study referred to is intended to establish that the available fault current does not exceed what the controller is labeled for. This is not necessarily the same as the study needed for arc-flash labeling. Exhibit II.10.1 shows a controller power switching device being tested in a laboratory.

After the controller and transfer switch have been subjected to a high fault current, they might not be suitable for further use without inspection or repair.

> The high fault current described here refers to an actual fault in the field, not prototype laboratory testing. Controllers are tested to meet the aforementioned seven criteria; but, controllers will need examination and probably repair after clearing a heavy fault condition. Some designs can and do clear faults without significant damage or degradation.
>
> ▶ **FAQ** What if no evidence exists that a fire pump controller and transfer switch subjected to a high fault current present a risk of fire or shock hazard?
>
> Equipment must be thoroughly inspected, and repaired if necessary, after a short circuit has occurred. A short circuit can cause damage to equipment that would render it inoperable or unsafe for further use. While the short-circuit event might not present a risk of fire or shock hazard, the equipment may sustain damage that would require repair before further use.

EXHIBIT II.10.1 *Evaluation Test of Prospective Fire Pump Controller Power Switching Devices. (Courtesy of Master Control Systems, Inc.)*

10.1.2.2.1 The controller and transfer switch shall be suitable for the available short-circuit current at the line terminals of the controller and transfer switch.

> ▶ **FAQ** Why is it necessary to know the available short-circuit current at the line terminals of the fire pump controller and transfer switch?

The compatibility of the short-circuit current rating marked on the equipment (required in 10.1.2.2.2) with the available short-circuit current at the line terminals of the equipment is covered in 10.1.2.2.1. The available short-circuit current determined by a short-circuit study, as described in 10.1.2.2, must not exceed the short-circuit current rating marked on the equipment. If the equipment (controller or transfer switch) is connected to a power supply that is capable of supplying short-circuit currents greater than what the equipment is capable of and marked for, the installation is both unsafe and unreliable. The facility and any exposed personnel are at significant risk.

10.1.2.2.2 The controller and transfer switch shall be marked "Suitable for use on a circuit capable of delivering not more than ____ amperes RMS symmetrical at ____ volts ac," or "____ amperes RMS symmetrical at ___ volts ac short-circuit current rating," or equivalent, where the blank spaces shown shall have appropriate values filled in for each installation.

The values to be inserted in the blank spaces in 10.1.2.2.2 are determined through prototype testing, which is described in A.10.1.2.2. Other marking that conveys the same information may be used. See Exhibit II.10.2 for typical combination fire pump controller/transfer switch nameplates. Exhibit II.10.2 (left) illustrates a marking where the combination fire pump controller/transfer switch has the same short-circuit rating for both the normal and alternate power circuits. Exhibit II.10.2 (right) illustrates a marking where the combination fire pump controller/transfer switch has different short-circuit ratings for the normal and alternate power circuits.

Transfer switches and combination fire pump controller transfer switch units must have two short circuit ratings, one for the normal source power supply side, and one for the emergency source power supply side. Special attention is needed for older transfer switches and transfer switch controllers, because they may lack an emergency source circuit breaker, which is now mandatory. It is important to know if the emergency source is a so-called "emergency bus," since these are normally fed by utility power, which may have a short-circuit current capability well above the generator(s) on the property.

Commentary Table II.10.1 *Common Short Circuit (WIC) Ratings for Full Service Controllers*

System Voltage:	208	240	400	480	600	2500	5000	7200	Vac
WIC Rating*:	42	42	30	30	22	–	–	–	kA
MVA Rating**:	–	–	–	–	–	200	400	600	MVA

*Full service low voltage controllers are available with WIC ratings up to 200kA. Limited service standard ratings are often less than those shown in the table.

**Medium voltage units are also typically rated in MVA instead of or in addition to kA. The values in the table correspond to approximately 50 kA short circuit current.

10.1.2.3 Preshipment. All controllers shall be completely assembled, wired, and tested by the manufacturer before shipment from the factory.

10.1.2.3.1 Controllers shipped in sections shall be completely assembled, wired, and tested by the manufacturer before shipment from the factory.

10.1.2.3.2 Such controllers shall be reassembled in the field, and the proper assembly shall be verified by the manufacturer or designated representative.

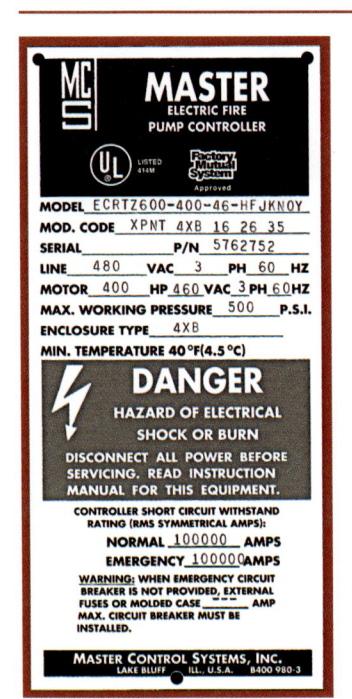

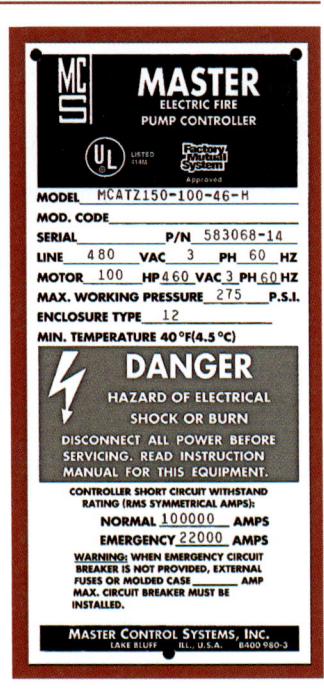

EXHIBIT II.10.2
Combination Fire Pump Controller/Transfer Switch Nameplates. (Courtesy of Master Control Systems, Inc.)

This paragraph was added to recognize that some fire pump controllers are shipped in sections due to size and construction. Proper reassembly at the installation site is ensured by requiring the reassembly to be verified by the manufacturer or designated representative.

▶ **FAQ** — Is a fire pump controller allowed to be assembled in the field from individual parts or subassemblies?

Fire pump controllers cannot be assembled in the field from individual parts or subassemblies. They must be completely assembled and tested at the manufacturer's factory in order to be marked with the listing and/or approval label or mark. Further, complete factory assembly ensures proper design, coordination, and performance of the controller. When it is impractical to ship as a complete assembly, the equipment is assembled, wired, and tested by the manufacturer prior to shipment. After testing, the equipment is then disassembled, shipped, and reassembled at the installation site. It must then be tested again to ensure proper coordination of the components and performance of the complete unit. The tests at the installation site must comply with the manufacturer's requirements.

10.1.2.4 Service Equipment Listing. All controllers and transfer switches shall be listed as "suitable for use as service equipment" where so used.

The terms *service* and *service equipment* are defined in *NFPA 70®*, *National Electrical Code®* (*NEC®*) in Article 100 and repeated in this standard in 3.3.41 and 3.3.42. Note that *service equipment* is also commonly called *service entrance equipment*. This equipment connects to the local utility service power directly. There is no disconnecting means nor any overcurrent protection (OCP) ahead of the equipment to protect or isolate it for servicing or repairing, which is why the equipment must be designed, built, tested, and labeled as suitable for service equipment use.

> Most currently listed and approved fire pump controllers are listed and approved as suitable for use as service equipment. These all ship with a separate "service disconnect," or similar label. When the equipment is actually used as service entrance equipment, this label must be affixed near the operator handle of the normal source isolating switch, and, if applicable, to the emergency source isolating switch if the power supply connected to the switch is also utility service. These labels are important to identify the switches, and as a warning that there is no upstream OCP nor any disconnecting or isolating means.

10.1.2.5 Additional Marking.

10.1.2.5.1 All controllers shall be marked "Electric Fire Pump Controller" and shall show plainly the name of the manufacturer, identifying designation, maximum operating pressure, enclosure type designation, and complete electrical rating.

> Labeling the maximum operating pressure is important for reliable and safe operation of the controller. While most systems tend to operate at or below 175 psi, most controllers are built and marked to handle 300 psi (275 psi for controllers built with a safety factor). However, some high-rise buildings have one or more zones which exceed these values. In these applications, a "high pressure" controller – typically rated at 500 psi or 600 psi (550 psi with safety factor) – must be used instead.

▶ **FAQ** What is the significance of marking the minimum operating temperature on a fire pump controller?

> Most controllers are marked with the minimum operating temperature of their environment. Controllers that are pressure actuated are connected to pipes containing water. Consequently, the controller must not be located in an area where temperatures at or below freezing may occur. Note that variable controllers must also be marked with a maximum rated operating temperature (see 10.10.2).

10.1.2.5.2 Where multiple pumps serve different areas or portions of the facility, an appropriate sign shall be conspicuously attached to each controller indicating the area, zone, or portion of the system served by that pump or pump controller.

> This is crucially important with multi-zone high-rise installations. Whether all of the pumps are in the same room or staged on different floors, knowing which pump or pumps service a given zone is important from both a service and testing standpoint, as well as during a fire. The three pumps shown in Exhibit II.10.3 all supply one of three zones in a high-rise building, and all feed from city water.

10.1.2.6 Service Arrangements. It shall be the responsibility of the pump manufacturer or its designated representative to make necessary arrangements for the services of a manufacturer's representative when needed for service and adjustment of the equipment during the installation, testing, and warranty periods.

> Although service and adjustment of the equipment is normally performed by the manufacturer of the controller and transfer switch, ultimately the pump manufacturer is responsible for ensuring that this work is performed. Note that it is the fire pump manufacturer who typically supplies the pump, the driver (motor or engine), the controller, valves, the jockey pump and its controller, and the like.

EXHIBIT II.10.3 *Three of Six Fire Pumps for a High-Rise Building. (Courtesy of Master Control Systems, Inc.)*

10.1.2.7 State of Readiness. The controller shall be in a fully functional state within 10 seconds upon application of ac power.

◀ **DESIGN ALERT**

> The requirement for state of readiness relates to controllers that are computer controlled. For such equipment, boot-up time may be needed before the controller is fully functional, and any delay in being fully functional must be minimized in order to ensure that the performance of the fire pump is not diminished.
>
> This state-of-readiness clause was added because some early design microprocessor based fire pump controllers were too slow to respond and take control of the fire pump. It is important for electric fire pump controllers to boot up promptly, and in a situation where pumps are in series in a three zone high-rise building, for example, this is important because the controllers will usually be reduced starting types since alternate (emergency) power from a generator will normally be mandated. In a hypothetical situation like this one, there is a water demand in the high zone from fused (open) sprinkler heads or hose(s), and the normal source power to the high zone pump has delay stack-up as follows:
>
> T = 0: The high-zone transfer switch senses the power failure and starts a time delay per 10.8.3.12(1).
>
> T = 3: At the end of this delay (a few seconds — 3 for example), the transfer switch signals the generator to start.
>
> T = 13: The generator starts and supplies power after 10 seconds, per NFPA-110, Type 10 per 9.6.2.1.
>
> T = 23: The three controllers boot up and become ready (10 seconds max.) per 10.1.2.7.
>
> T = 23: The high-zone controller signals the mid-zone controller to start the mid-zone pump.
>
> T = 23: The mid-zone controller signals the low-zone controller to start the low zone pump.

> T = 33: The low-zone controller accelerates the low zone pump motor (10 seconds max.) per 10.4.5.2.2.
>
> T = 48: The mid-zone pump delays starting for 5 to 10 seconds (5 seconds for example) per 10.5.2.5.4. (In some controllers, this delay applies to all start demands, including manual start pushbutton, and should also latch (lock-in) in the demand once the demand occurs.)
>
> T = 53: The high zone pump delays starting for 5 to 10 seconds (10 seconds for example) per 10.5.2.5.4.
>
> T = 58: The mid-zone controller accelerates the low zone pump motor (10 seconds max.) per 10.4.5.2.2. (Some controllers reach full speed in 3 seconds or less, giving full pressure before the next pump tries to start. If the actual accelerate time is 10 seconds, as is the case with some starting types, the next pump may not have full suction pressure for a few seconds as it accelerates.)
>
> T = 68: The high-zone controller accelerates the low zone pump motor (10 seconds max.) per 10.4.5.2.2.
>
> All of the delays in this scenario amount to a very long interruption of water to the fire sprinklers or fire hose(s). This is enough time for the fire plume to fuse a lot of additional heads beyond the original fire zone. Note that when pumps are not series connected, the maximum times involved are much less.

10.1.3* Design. All electrical control equipment design shall comply with *NFPA 70, National Electrical Code*, Article 695, and other applicable documents.

> The role of *NEC* Article 695 is emphasized in 10.1.3 (see also Supplement 3). NFPA 20 covers the minimum performance and testing requirements for controllers and transfer switches, while Article 695 covers the installation requirements of such equipment. Some of the requirements of Article 695 are extracted from NFPA 20 because those requirements are under the purview of the NFPA 20 Technical Committee on Fire Pumps. However, all of the installation requirements for controllers, transfer switches, and fire pump accessory equipment are located in Article 695.

A.10.1.3 All electrical control equipment design should also follow the guidelines within NEMA ICS 14, *Application Guide for Electric Pump Controllers*.

> NEMA ICS 14, *Application Guide for Electric Pump Controllers* and NEMA ICS 15, *Instructions for the Handling, Installation, Operation, and Maintenance of Electric Fire Pump Controllers Rated Not More Than 600V*, provides practical information concerning the general technical considerations in the installation of electric fire pump controllers. These guides are intended to be used by specifiers, purchasers, installers, and owners of fire pump controllers.

10.2 Location

10.2.1* Controllers shall be located as close as is practical to the motors they control and shall be within sight of the motors.

> Close proximity is best for maintenance and operation of fire pump installations, and also minimizes the voltage drop between the fire pump controller and fire pump motor.

A.10.2.1 If the controller must be located outside the pump room, a glazed opening should be provided in the pump room wall for observation of the motor and pump during starting. The pressure control pipe line should be protected against freezing and mechanical injury.

10.2.2 Controllers shall be located or protected so that they will not be damaged by water escaping from pumps or pump connections.

Controllers are installed in diverse locations in relation to the pump, varying according to where in the world the installation occurs. The requirement in 10.2.2 that controllers be suitably protected understands that the controller is essentially the core of the sprinkler system, and any threat or harm to the fire pump controller is a threat or harm to the fire protection of the facility. Controllers are sometimes installed outdoors. In outdoor installations enclosures must be suitably rated for the environment, such as a Type 4 enclosure. Controllers installed in salt air environments — as well as those installed in corrosive areas and areas subject to blowing sand or dust — must be corrosion resistant. Special means must be provided in areas where hazardous (explosive) atmospheres exist or may exist. See Section 10.3.3 for more information.

▶ **FAQ** Are any restrictions or guidance given on where a fire pump controller can be installed?

The location of the controller is critical with regard to exposure to water. Although the controller enclosure is rated Type 2, the enclosure protects equipment only from contact with water dripping from a source located above the controller. The spray pattern of valves should be determined, as well as potential leak points of pump connections.

10.2.3 Current-carrying parts of controllers shall be not less than 12 in. (305 mm) above the floor level.

This requirement for distance above floor level is to help keep water out of the electrical circuitry. A 4-inch "housekeeping pad," or equivalent means of keeping the controller bottom clear of puddles, is also good practice.

10.2.4 Working clearances around controllers shall comply with *NFPA 70, National Electrical Code*, Article 110.

Providing adequate working clearances is important everywhere; it is especially important in those pump rooms where space is limited. Subsection 10.2.4 references Article 110 of the *NEC*, 2011 edition. The relevant paragraphs from Section 110.26 of the *NEC*, which contain the working clearance requirements, are extracted following this paragraph. The accompanying commentary is modified from the *National Electrical Code® Handbook*, 2011 edition.

> **II. 600 Volts, Nominal, or Less**
>
> **110.26 Spaces About Electrical Equipment.** Access and working space shall be provided and maintained about all electrical equipment to permit ready and safe operation and maintenance of such equipment.
>
> **(A) Working Space.** Working space for equipment operating at 600 volts, nominal, or less to ground and likely to require examination, adjustment, servicing, or maintenance while energized shall comply with the dimensions of 110.26(A)(1), (A)(2), and (A)(3) or as required or permitted elsewhere in this *Code*.
>
> **(1) Depth of Working Space.** The depth of the working space in the direction of live parts shall not be less than that specified in Table 110.26(A)(1) unless the requirements of 110.26(A)(1)(a), (A)(1)(b), or (A)(1)(c) are met. Distances shall be measured from the exposed live parts or from the enclosure or opening if the live parts are enclosed.

Included in the clearance requirements in Table 110.26(A)(1) is the step-back distance from the face of the equipment. It provides requirements for clearances away from the equipment, based on the circuit voltage to ground and whether there are grounded or ungrounded objects in the step-back space or exposed live parts across from each other. The voltages to ground consist of two groups: 0 to 150, inclusive, and 151 to 600, inclusive.

Table 110.26(A)(1) Working Spaces

Nominal Voltage to Ground	Minimum Clear Distance		
	Condition 1	Condition 2	Condition 3
0–150	914 mm (3 ft)	914 mm (3 ft)	914 mm (3 ft)
151–600	914 mm (3 ft)	1.07 m (3 ft 6 in.)	1.22 m (4 ft)

Note: Where the conditions are as follows:

Condition 1 – Exposed live parts on one side of the working space and no live or grounded parts on the other side of the working space, or exposed live parts on both sides of the working space that are effectively guarded by insulating materials.

Condition 2 – Exposed live parts on one side of the working space and grounded parts on the other side of the working space. Concrete, brick, or tile walls shall be considered as grounded.

Condition 3 – Exposed live parts on both sides of the working space.

(a) *Dead-Front Assemblies.* Working space shall not be required in the back or sides of assemblies, such as dead-front switchboards or motor control centers, where all connections and all renewable or adjustable parts, such as fuses or switches, are accessible from locations other than the back or sides. Where rear access is required to work on nonelectrical parts on the back of enclosed equipment, a minimum horizontal working space of 762 mm (30 in.) shall be provided.

(b) *Low Voltage.* By special permission, smaller working spaces shall be permitted where all exposed live parts operate at not greater than 30 volts rms, 42 volts peak, or 60 volts dc.

(c) *Existing Buildings.* In existing buildings where electrical equipment is being replaced, Condition 2 working clearance shall be permitted between dead-front switchboards, panelboards, or motor control centers located across the aisle from each other where conditions of maintenance and supervision ensure that written procedures have been adopted to prohibit equipment on both sides of the aisle from being open at the same time and qualified persons who are authorized will service the installation.

(2) Width of Working Space. The width of the working space in front of the electrical equipment shall be the width of the equipment or 762 mm (30 in.), whichever is greater. In all cases, the work space shall permit at least a 90 degree opening of equipment doors or hinged panels.

(3) Height of Working Space. The work space shall be clear and extend from the grade, floor, or platform to a height of 2.0 m (6½ ft) or the height of the equipment, whichever is greater. Within the height requirements of this section, other equipment that is associated with the electrical installation and is located above or below the electrical equipment shall be permitted to extend not more than 150 mm (6 in.) beyond the front of the electrical equipment.

Exception No. 1: In existing dwelling units, service equipment or panelboards that do not exceed 200 amperes shall be permitted in spaces where the height of the working space is less than 2.0 m (6½ ft).

Exception No. 2: Meters that are installed in meter sockets shall be permitted to extend beyond the other equipment. The meter socket shall be required to follow the rules of this section.

(B) Clear Spaces. Working space required by this section shall not be used for storage. When normally enclosed live parts are exposed for inspection or servicing, the working space, if in a passageway or general open space, shall be suitably guarded.

The key to understanding Section 110.26 is the division of requirements for spaces about electrical equipment into two separate and distinct categories: working space and dedicated equipment space. The term *working space* generally applies to the protection of the worker, and *dedicated equipment space* applies to the space reserved for future access to electrical equipment and to protection of the equipment from intrusion by nonelectrical equipment. The performance requirements for all spaces about electrical equipment are set forth in the first sentence. Storage of materials that blocks access or prevents safe work practices must be avoided at all times. The intent of 110.26(A) is to provide enough space for personnel to perform any of the operations listed without jeopardizing worker safety. Examples of such equipment include panelboards, switches, circuit breakers, controllers, and controls on heating and air-conditioning equipment. It is important to understand that the word examination, as used in 110.26(A), includes such tasks as checking for the presence of voltage using a portable voltmeter.

Minimum working clearances are not required if the equipment is such that it is not likely to require examination, adjustment, servicing, or maintenance while energized. However, access and working space are still required by the opening paragraph of 110.26.

Included in the clearance requirements in Table 110.26(A)(1) is the step-back distance from the face of the equipment. It provides requirements for clearances away from the equipment, based on the circuit voltage to ground and whether there are grounded or ungrounded objects in the step-back space or exposed live parts across from each other. The voltages to ground consist of two groups: 0 to 150, inclusive, and 151 to 600, inclusive.

Examples of common electrical supply systems covered in the 0 to 150 volts to ground group are 120/240-volt, single-phase, 3-wire and 208Y/120-volt, 3-phase, 4-wire systems. Examples of common electrical supply systems covered in the 151 to 600 volts to ground group are 240-volt, 3-phase, 3-wire; 480Y/277-volt, 3-phase, 4-wire; and 480-volt, 3-phase, 3-wire (ungrounded and corner grounded) systems. Remember, where an ungrounded system is utilized, the voltage to ground (by definition) is the greatest voltage between the given conductor and any other conductor of the circuit. For example, the voltage to ground for a 480-volt ungrounded delta system is 480 volts. See Exhibit II.10.4 for the general working clearance requirements for each of the three conditions listed in Table 110.26(A)(1). For assemblies, such as switchboards or motor-control centers, that are accessible from the back and expose live parts, the working clearance dimensions are required at the rear of the equipment, as illustrated. Note that for Condition 3, where there is an enclosure on opposite sides of the working space, the clearance for only one working space is required.

The intent of this section is to point out that work space is required only from the side(s) of the enclosure requiring access. The general rule applies: Equipment that requires front, rear, or side access for the electrical activities described in 110.26(A) must meet the requirements of Table 110.26(A)(1). In many cases, equipment of "dead-front" assemblies requires only front access. For equipment that requires rear access for nonelectrical activity, however, a reduced working space of at least 30 in. must be provided. Exhibit II.10.5 shows a reduced working space of 30 in. at the rear of equipment to allow work on nonelectrical parts.

This section permits some relief for installations that are being upgraded. When assemblies such as dead-front switchboards, panelboards, or motor-control centers are replaced in an existing building, the working clearance allowed is that required by Table 110.26(A)(1), Condition 2. The reduction from a Condition 3 to a Condition 2 clearance is allowed only

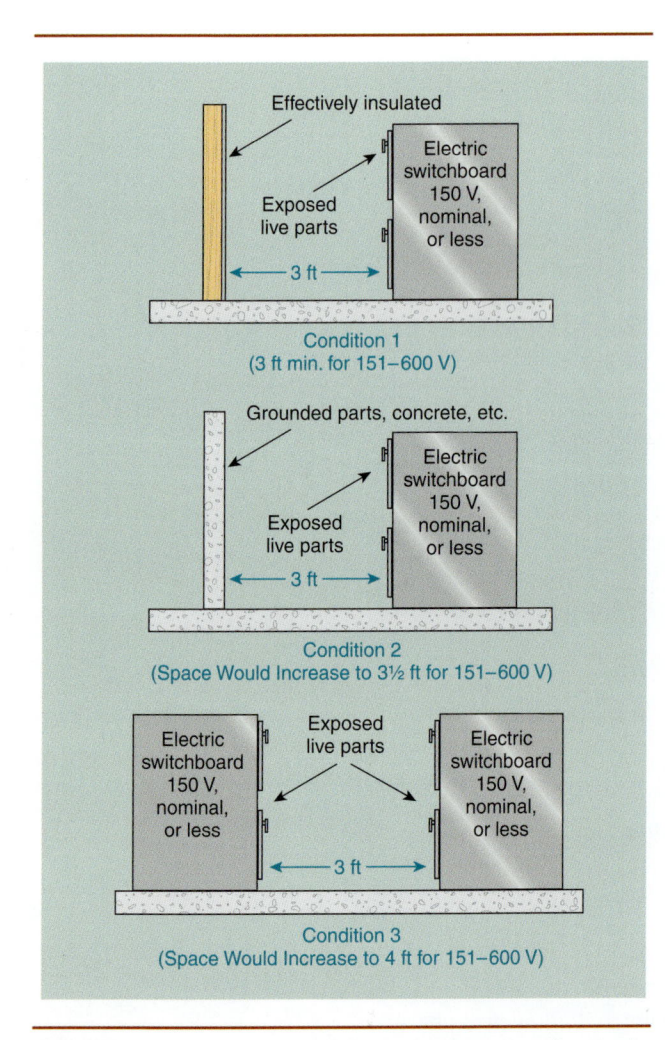

EXHIBIT II.10.4 Distances Measured from the Live Parts If the Live Parts Are Exposed or from the Enclosure Front If the Live Parts Are Enclosed.

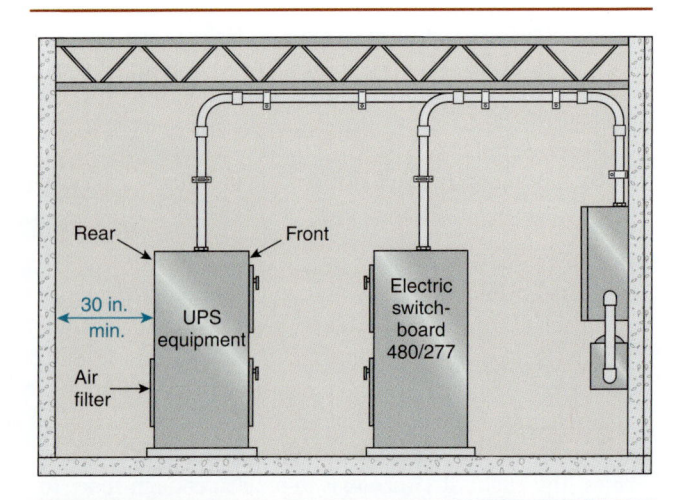

EXHIBIT II.10.5 Example of the 30-in. Minimum Working Space at the Rear of Equipment to Allow Work on Nonelectrical Parts.

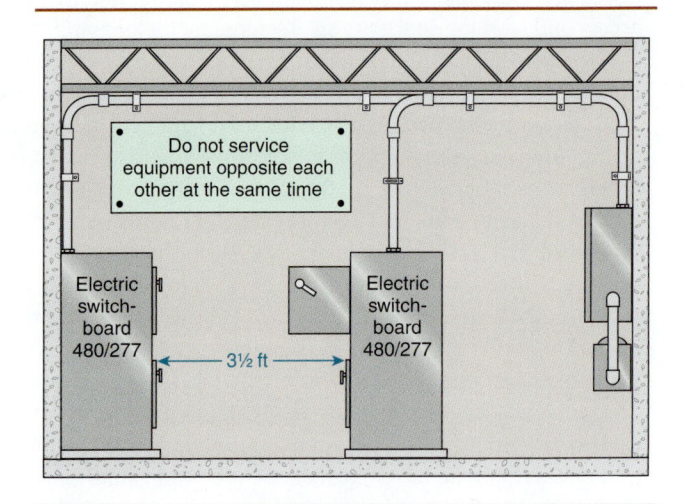

EXHIBIT II.10.6 Permitted Reduction from a Condition 3 to a Condition 2 Clearance According to NEC 110.26(A)(1)(c).

where a written procedure prohibits facing doors of equipment from being open at the same time and where only authorized and qualified persons service the installation. Exhibit II.10.6 illustrates this relief for existing buildings.

Regardless of the width of the electrical equipment, the working space cannot be less than 30 in. wide, as required by Table 110.26(A)(1). This space allows an individual to have at least shoulder-width space in front of the equipment. The 30-in. measurement can be made from either the left or the right edge of the equipment and can overlap other electrical equipment, provided the other equipment does not extend beyond the clearance required by Table 110.26(A)(1). If the equipment is wider than 30 in., the left-to-right space must be equal to the width of the equipment. Exhibit II.10.7 illustrates the 30-in. width requirement.

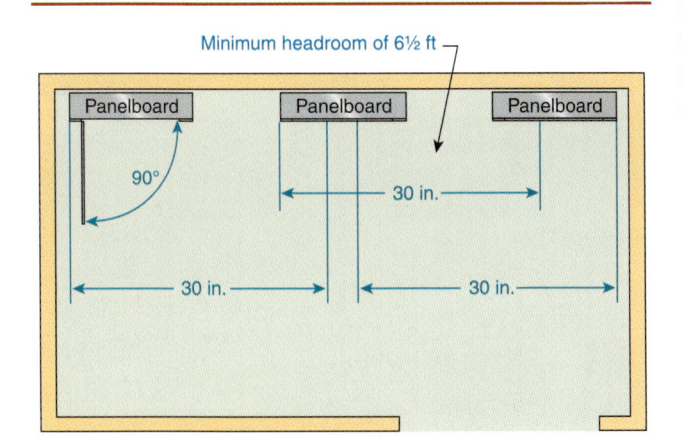

EXHIBIT II.10.7 The 30-in. Wide Front Working Space, Which Can Overlap Other Electrical Equipment.

EXHIBIT II.10.8 Illustration of the Depth of Working Space Which Must Be Sufficient to Allow a Full 90 Degree Opening of Equipment Doors for a Safe Working Approach.

Sufficient depth in the working space is also required to allow a panel or a door to open at least 90 degrees. If doors or hinged panels are wider than 3 ft, more than a 3-ft deep working space must be provided to allow a full 90 degree opening. (See Exhibit II.10.8.)

In addition to requiring a working space to be clear from the floor to a height of 6½ ft or to the height of the equipment, whichever is greater, 110.26(A)(3) permits electrical equipment located above or below other electrical equipment to extend into the working space not more than 6 in. This requirement allows the placement of a 12 in. × 12 in. wireway on the wall directly above or below a 6 in.-deep panelboard without impinging on the working space or compromising practical working clearances. The requirement prohibits large differences in depth of equipment below or above other equipment that specifically requires working space. To minimize the amount of space required for electrical equipment, large free-standing, dry-type transformers are commonly installed within the required work space for a wall-mounted panel-board, which compromises clear access to the panelboard and which is clearly not permitted by this section. Electrical equipment that produces heat or that otherwise requires ventilation also must comply with 110.3(B) and 110.13.

EXHIBIT II.10.9 *Equipment Location Free of Storage Allows the Equipment to Be Worked on Safely. (Courtesy of the International Association of Electrical Inspectors)*

> This exception is new in the 2011 Code. It recognizes that a meter socket installation could be in compliance with this section, but the installation of the meter into the socket could extend beyond the 6 in. permitted by 110.26(A)(3). The exception permits the meter to extend further, provided the meter socket does not exceed the 6 in. depth.
>
> Section 110.26(B), as well as the rest of 110.26, does not prohibit the placement of panelboards in corridors or passageways. For that reason, when the covers of corridor-mounted panelboards are removed for servicing or other work, access to the area around the panelboard should be guarded or limited to prevent injury to unqualified persons using the corridor.
>
> Equipment that requires servicing while energized must be located in an area that is not used for storage, as shown in Exhibit II.10.9.

10.3 Construction

10.3.1 Equipment. All equipment shall be suitable for use in locations subject to a moderate degree of moisture, such as a damp basement.

> In damp locations, the controllers should be equipped with cabinet heaters. Some designs incorporate a humidistat to control the heaters, and heaters operated at half voltage (240V ac heaters used at 120V ac, for example) which keeps the heater surface at a much lower and safer temperature since this is ¼ of the rated heater wattage. Additional heaters should be used for multi-bay enclosures. Condensation should never be allowed, particularly with modern controllers due to their wide use of electronic and microprocessor circuitry.
>
> ▶ **FAQ** *What ensures that the enclosure for fire pump equipment provides the minimum amount of protection required for the enclosed equipment?*
>
> Controllers listed for fire pump service fulfill the requirement in 10.3.1 and are suitable for use in environments that may be damp.

10.3.2 Mounting. All equipment shall be mounted in a substantial manner on a single noncombustible supporting structure.

Some controllers generate a considerable amount of heat when operating — in some cases at a level to ignite combustible materials — especially primary resistor starting controllers (see 10.4.5.3). Some controllers have open bottoms. The use of a noncombustible supporting structure to close the bottom maintains enclosure integrity.

10.3.3 Enclosures.

See Commentary Table II.10.2 to determine the degree of protection against specific environmental conditions provided by an enclosure based on its type rating.

COMMENTARY TABLE II.10.2 *Enclosure Selection*

For Outdoor Use

Provides a Degree of Protection Against the Following Environmental Conditions	3	3R	3S	3X	3RX	3SX	4	4X	6	6P
Incidental contact with the enclosed equipment	X	X	X	X	X	X	X	X	X	X
Rain, snow, and sleet	X	X	X	X	X	X	X	X	X	X
Sleet*	–	–	X	–	–	X	–	–	–	–
Windblown dust	X	–	X	X	–	X	X	X	X	X
Hosedown	–	–	–	–	–	–	X	X	X	X
Corrosive agents	–	–	–	X	X	X	–	X	–	X
Temporary submersion	–	–	–	–	–	–	–	–	X	X
Prolonged submersion	–	–	–	–	–	–	–	–	–	X

For Indoor Use

Provides a Degree of Protection Against the Following Environmental Conditions	1	2	4	4X	5	6	6P	12	12K	13
Incidental contact with the enclosed equipment	X	X	X	X	X	X	X	X	X	X
Falling dirt	X	X	X	X	X	X	X	X	X	X
Falling liquids and light splashing	–	X	X	X	X	X	X	X	X	X
Circulating dust, lint, fibers, and flyings	–	–	X	X	–	X	X	X	X	X
Settling airborne dust, lint, fibers, and flyings	–	–	X	X	X	X	X	X	X	X
Hosedown and splashing water	–	–	X	X	–	X	X	–	–	–
Oil and coolant seepage	–	–	–	–	–	–	–	X	X	X
Oil or coolant spraying and splashing	–	–	–	–	–	–	–	–	–	X
Corrosive agents	–	–	–	X	–	–	X	–	–	–
Temporary submersion	–	–	–	–	–	X	X	–	–	–
Prolonged submersion	–	–	–	–	–	–	X	–	–	–

*Mechanism shall be operable when ice covered.

Note: The term raintight is typically used in conjunction with Enclosure Types 3, 3S, 3SX, 3X, 4, 4X, 6, and 6P. The term rainproof is typically used in conjunction with Enclosure Types 3R and 3RX. The term watertight is typically used in conjunction with Enclosure Types 4, 4X, 6, 6P. The term driptight is typically used in conjunction with Enclosure Types 2, 5, 12, 12K, and 13. The term dusttight is typically used in conjunction with Enclosure Types 3, 3S, 3SX, 3X, 5, 12, 12K, and 13.

Source: NFPA 70®, National Electrical Code®, 2011, Table 110.20.

10.3.3.1* The structure or panel shall be securely mounted in, as a minimum, a National Electrical Manufacturers Association (NEMA) Type 2, dripproof enclosure(s) or an enclosure(s) with an ingress protection (IP) rating of IP31.

> A Type 2 enclosure provides a degree of protection against incidental contact with the enclosed equipment from falling dirt, falling liquids, and light splashing. This paragraph was revised to recognize the ingress protection (IP) rating system for enclosures. An enclosure with an IP rating of IP31 is considered to provide protection equivalent to the protection provided by a Type 2 enclosure.

A.10.3.3.1 For more information, see NEMA 250, *Enclosures for Electrical Equipment*.

> Controllers listed for fire pump service fulfill the requirement for NEMA Type 2 as a minimum. Some controllers may be marked with ratings other than Type 1. These enclosure types and the protection they provide are defined in NEMA 250, *Enclosures for Electrical Equipment*. The standard for the evaluation of fire pump enclosures is UL 50, *Enclosures for Electrical Equipment, Non-Environmental Considerations,* and UL 50E, *Enclosures for Electrical Equipment, Environmental Considerations*. Note that there are some minor differences in enclosure requirements between NEMA Standard NEMA 250 and the two cited UL standards.

10.3.3.2 Where the equipment is located outside, or where special environments exist, suitably rated enclosures shall be used.

> The requirement for a suitably rated enclosure in 10.3.3.2 includes enclosures better suited for splashing and spraying water, as well as corrosive and salt spray environs and the special requirements for areas where hazardous atmospheres exist or may exist. Where enclosures are well sealed, some jurisdictions permit using equipment purge techniques in accordance with NFPA 496, *Standard for Purged and Pressurized Enclosures for Electrical Equipment* to render fire pump controllers suitable for these installations.
>
> For corrosive environs, a Type 4X enclosure should be employed. There are four common types with the following unofficial designations (listed in the typical order of increasing price):
>
> **Type 4XA** Listed organic (paint) finish. Organic versions of Type 4XA require special UL listing and normally make use of UL approved (recognized or listed) corrosion resistant paint, usually baked on or two-part paint.
>
> **Type 4XB** 400 series (magnetic) stainless steel.
>
> **Type 4XC** 300 series (non-magnetic) stainless steel. Type 4XC is often type 316 stainless steel. Note that all welding of any stainless steel enclosure must employ compatible weld material (weld wire or rods) to avoid galvanic corrosion and alloy degradation.
>
> Type 4XCL Type 316L low carbon stainless steel.

▶ **FAQ** | How should equipment used in a special environment be protected?

> Refer to Section 110.28 of the *National Electrical Code®* or NEMA 250 to determine the correct enclosure-type rating required. Each enclosure type is associated with specific degrees of protection provided. The NEC® describes the protection provided by an enclosure according to type number. All controllers listed for fire pump service are marked with one or more type ratings.

10.3.3.3 The enclosure(s) shall be grounded in accordance with *NFPA 70, National Electrical Code*, Article 250.

> Grounding of this equipment is very important due to the high fault currents which are associated with this equipment. Depending on the pump or pressure sensing line as a ground is not good practice, because the piping is likely to be non-metallic and the motor coupling likely non-conductive.

10.3.4 Connections and Wiring.

10.3.4.1 All busbars and connections shall be readily accessible for maintenance work after installation of the controller.

> Modern controllers make use of a limited number of busbars. In some arrangements, the busbar is used to connect the isolating switch to the circuit breaker. Older controllers made extensive use busbars, and as a result, both front doors and rear access panels were available to allow work on the busbars, or they were of the unenclosed (open slate panel) type of construction. Many of these types of equipment are still in service.
>
> Some controllers have insulation or insulating parts which are removed for wiring, especially for the incoming power wiring, and which should be replaced prior to energizing the controller. Failure to do so creates a personnel hazard and may interfere with the ability to clear short circuits or other faults. This may also void the controller's short circuit rating(s) and the controller listing and warranty.

10.3.4.2 All busbars and connections shall be arranged so that disconnection of the external circuit conductors will not be required.

> This requirement, like the similar one in 10.3.4.1, was also mostly applicable to older types of construction.

10.3.4.3 Means shall be provided on the exterior of the controller to read all line currents and all line voltages with an accuracy within ±5 percent of motor nameplate voltage and current.

> This requirement was added to NFPA 20 in the 2007 edition. Previous editions of NFPA 20 required that provisions be made within the controller to permit the use of test instruments for measuring all line voltages and currents without disconnecting any conductors within the controller. However, this required the measurement to be performed inside the enclosure with the fire pump controller energized and the motor running. This was essentially "hot" work, exposing the person making the measurements to energized circuitry. This was dangerous and required great care. The new requirement in 10.3.4.3 for a means on the exterior of the controller to read all line currents and all line voltages mitigated this hazard and provided a safer environment for measuring all line currents and all line voltages. Considering all of the heightened concern about the hazards of arc flash, it made no sense to allow personnel to be exposed to shock and arc flash hazards when NFPA 20 mandated a safer method that mitigates these hazards. The requirement that provisions be made within the controller to permit the use of test instruments for measuring all line voltages and currents without disconnecting any conductors within the controller was deleted.
>
> ▶ **FAQ** How can line voltages and currents be measured without the risk of electric shock?

The requirement in 10.3.4.3 allows for reading all line currents and all line voltages with the equipment energized and the enclosure door(s) closed. All modern listed controllers have meters mounted on the outside of the controller; although an external connector(s) for portable meters is allowed as an alternative means. Note, however, that some controllers require opening the controller door to change recorder printer paper, or floppy disks or access to the data port (such as a USB connector). Caution is needed to avoid contact with line voltage on any nearby terminals.

10.3.4.4 Continuous-Duty Basis.

10.3.4.4.1 Unless the requirements of 10.3.4.4.2 are met, busbars and other wiring elements of the controller shall be designed on a continuous-duty basis.

10.3.4.4.2 The requirements of 10.3.4.4.1 shall not apply to conductors that are in a circuit only during the motor starting period, which shall be permitted to be designed accordingly.

Some factory-installed conductors, such as those connecting starting resistors, reactors, or an autotransformer, are generally smaller in size than the conductors carrying the line currents of the motor. This difference in size is permissible as these conductors only carry current during the motor starting period. The most critical of these are the starting resistors in primary resistor start controllers, since these resistors are needed only for 5 seconds out of every 80 seconds (see 10.4.5.3). Others (reactors and autotransformers) may be used for the full 10 seconds of 10.4.5.2.2. Separately, field wiring should be sized at 50 percent of 125 percent of FLA or part winding starting and 58 percent (57.7 percent) for Wye-Delta starting [see *NEC* 695.6(B)(1)].

10.3.4.5 Field Connections.

10.3.4.5.1 A fire pump controller shall not be used as a junction box to supply other equipment.

Allowing the fire pump controller to be used as a junction box to supply other equipment could result in the fire pump controller being rendered inoperable if a short circuit occurred in the equipment being supplied by the fire pump controller. However, it is permitted to tap the fire pump power supply, if this is done according to Chapter 9, including the use of a listed and suitable junction box for the purpose.

10.3.4.5.2 No undervoltage, phase loss, frequency sensitive, or other device(s) shall be field installed that automatically or manually prohibits electrical actuation of the motor contactor.

Although protection of equipment from undervoltage, phase loss, or other abnormal conditions is common in industrial control systems, fire pump controllers are considered expendable. This standard seeks to provide the least complex solution and minimize the number of items in the so-called "critical starting path" of the fire pump, to maintain the reliability of the motor contactor. The text of 10.3.4.5.2 was clarified in the 2010 edition so that the requirement applies to any field-installed device that automatically or manually prohibits electrical actuation of the motor contactor.

10.3.4.6 Electrical supply conductors for pressure maintenance (jockey or make-up) pump(s) shall not be connected to the fire pump controller.

Jockey (make-up) pump(s) are allowed to be supplied by the same source supplying the fire pump motor, but a suitable means for connection upstream of the fire pump controller must be provided. This arrangement allows for servicing the jockey or make-up pump(s) without disturbing the wiring connections in the fire pump controller. Also note that if the power source to the fire pump controller is utility service, any jockey pump power by the same source will likely need to be rated as service entrance equipment.

10.3.5 Protection of Control Circuits.

10.3.5.1 Circuits that are necessary for proper operation of the controller shall not have overcurrent protective devices connected in them.

This prohibition specifically applies to the control transformers of the controller. Any fuse could be replaced with the wrong size, and any other device might be tripped manually. This is another example of the expendable nature of fire pump controllers. The general concept when a fire pump is running to fight a fire is to "run it to destruction." Interruption of fire water is never advised.

10.3.5.2 The secondary of the transformer and control circuitry shall be permitted to be ungrounded except as required in 10.6.5.4.

This subsection has been applied for the many decades when it was common for the control voltage of fire pump controllers to be 120V ac. It was also common for the control transformer, or one of the control transformers, to provide auxiliary 120 V ac power for use with local alarms or similar systems. Most modern controllers utilize low voltage — 24V ac or less — for control power, but auxiliary 120 V ac power is an option with some manufacturers.

10.3.6* External Operation. All switching equipment for manual use in connecting or disconnecting or starting or stopping the motor shall be externally operable.

A.10.3.6 For more information, see *NFPA 70, National Electrical Code*.

This is accomplished by either one or two operator handles for the controller isolating switch and its circuit breaker.

All manual switches in the fire pump motor circuit must be readily accessible to allow for prompt energizing of the fire pump motor circuit.

10.3.7 Electrical Diagrams and Instructions.

10.3.7.1 An electrical schematic diagram shall be provided and permanently attached to the inside of the controller enclosure.

Proper maintenance of the controller requires knowledge of the electrical equipment and circuitry within the fire pump controller.

10.3.7.2 All the field wiring terminals shall be plainly marked to correspond with the field connection diagram furnished.

The marking requirement required by 10.3.7.2 is very important for part winding and wye-delta motors, especially since it is not unusual for fire pump motors to have 12 leads needing to be connected. A clear diagram is also important when making proper connection of alarm circuitry to the three or five (or more) alarm contacts, and for any external demand (calls to start signals) inputs or outputs.

> **▶ FAQ** What ensures that the electrical connections of field conductors are being terminated properly?

How to properly connect a controller in the field is not always obvious. The combination of properly identified field-wiring terminals and a field connection diagram ensures correct electrical connections.

10.3.7.3* Complete instructions covering the operation of the controller shall be provided and conspicuously mounted on the controller.

It is important to install fire pump controllers according to the manufacturer's instruction manual, and according to start-up or commissioning instructions, to ensure safe and reliable fire protection. This is also important for controllers that employ microprocessor control circuitry, since there are corresponding programming menus which need to be accessed to properly set up (commission) the controller.

A.10.3.7.3 Pump operators should be familiar with instructions provided for controllers and should observe in detail all their recommendations.

The identification of the steps needed for proper operation of the controller is important. These steps must be easily understood by the operator, whether the controller is being operated in a fire condition or for startup/periodic testing.

10.3.7.4 The installation instructions of the manufacturer of the fire pump controller shall be followed.

This requirement was added with this edition of the standard due to field problems caused by failure to properly carry out the steps and procedures of the manufacturers' instructions.

10.3.8 Marking.

10.3.8.1 Each motor control device and each switch and circuit breaker shall be marked to plainly indicate the name of the manufacturer, the designated identifying number, and the electrical rating in volts, horsepower, amperes, frequency, phases, and so forth, as appropriate.

> **▶ FAQ** Why is it important to mark control devices, switches, and circuit breakers with so much identification information?

In the event that replacement of critical components is necessary, the clear identification of what is being replaced is important so that the same or equivalent component can be installed.

10.3.8.2 The markings shall be so located as to be visible after installation.

This becomes more critical as hard wired control devices (such as relays) become smaller, and as markings on various sides of the components become more difficult to read after installation.

10.4 Components

10.4.1* Voltage Surge Arrester.

A.10.4.1 Operation of the surge arrester should not cause either the isolating switch or the circuit breaker to open. Arresters in ANSI/IEEE C62.11, *IEEE Standard for Metal-Oxide Surge Arresters for AC Power Circuits*, are normally zinc-oxide without gaps.

> The "gaps" referred to in A.10.4.1 are common when using the silicon-carbide (Si-C) types of surge arresters referred to in 10.4.1.1. A surge voltage of sufficient magnitude arcs across the gap and the Si-C semiconductor element absorbs the surge energy. Once the voltage decays sufficiently, or at the next zero crossing of the current, the arc extinguishes and prevents any further leakage or follow-on current. This allows the Si-C element to cool down and be ready for the next surge, spike, or transient event.

10.4.1.1 Unless the requirements of 10.4.1.3 or 10.4.1.4 are met, a voltage surge arrester complying with ANSI/IEEE C62.1, *IEEE Standard for Gapped Silicon-Carbide Surge Arresters for AC Power Circuits*, or ANSI/IEEE C62.11, *IEEE Standard for Metal-Oxide Surge Arresters for Alternating Current Power Circuits (>1 kV)*, shall be installed from each phase to ground. *(See 10.3.3.3.)*

> Surge arresters are provided in controllers to prevent power line surges from damaging components in the controller and/or rendering them inoperable. Typical failures due to a power line surge are burnout of indicating lamps and dielectric breakdown of the magnetic contactor holding coil.

10.4.1.2 The surge arrester shall be rated to suppress voltage surges above line voltage.

> Even when the surge arrester is rated above line voltage as required by this section, there will be some amount of increase of the voltage during a surge; no arrester can perfectly clamp excess voltage. However, properly rated devices should limit damage to the equipment. Some controller designs have a high "voltage withstand" capability which provides the best reliability against surges.

10.4.1.3 The requirements of 10.4.1.1 and 10.4.1.2 shall not apply to controllers rated in excess of 600 V. *(See Section 10.6.)*

> The typical voltages involved are 2300V ac or higher. These controllers are required to meet a minimum basic insulation level (BIL) so that they would be immune to normal surges, switching spikes, and transients.

10.4.1.4 The requirements of 10.4.1.1 and 10.4.1.2 shall not apply where the controller can withstand without damage a 10 kV impulse in accordance with ANSI/IEEE C62.41, *IEEE Recommended Practice for Surge Voltages in Low-Voltage AC Power Circuits*, or where the controller is listed to withstand surges and impulses in accordance with ANSI/UL 1449, *Standard for Surge Protective Devices*.

> The reference to UL 1449 was added as an alternate to the 10 kV (10,000 Volt) impulse test provision, since no manufacturer has been able to meet this requirement. This is partly due to the lack of the equipment to do this testing at the listing and approval agencies.

> **▶ FAQ** Are surge arresters required in all fire pump controllers?

> Either suitable surge arresters must be provided or the controller must be tested to determine that its construction is able to provide protection equivalent to that of a surge arrester.

10.4.2 Isolating Switch.

10.4.2.1 General.

10.4.2.1.1 The isolating switch shall be a manually operable motor circuit switch or a molded case switch having a horsepower rating equal to or greater than the motor horsepower.

> The isolating switch may have a horsepower rating but might not be suitable for either breaking a fault current or for closing in on a fault (short circuit). The controller operating instructions should always be followed to avoid injury or damage to equipment.

10.4.2.1.2* A molded case switch having an ampere rating not less than 115 percent of the motor rated full-load current and also suitable for interrupting the motor locked rotor current shall be permitted.

> An ampere rated isolating switch is not necessarily suitable for breaking or closing in on a fault, while the circuit breaker is. The 115 percent of FLA minimum ampere rating is the minimum, since the motor is allowed to draw 115 percent of FLA or more continuously. See 9.5.2.2(1) for more information.

A.10.4.2.1.2 For more information, see *NFPA 70, National Electrical Code.*

> A molded case switch used as an isolating switch need not be rated in horsepower if it has a suitable ampere rating and has been evaluated for switching the locked rotor current of the fire pump motor. The source of the 115 percent requirement is Section 430.110 of the *NEC*. This sizing rule accounts for motors running continuously in excess of their rated current.

10.4.2.1.3 A molded case isolating switch shall be permitted to have self-protecting instantaneous short-circuit overcurrent protection, provided that this switch does not trip unless the circuit breaker in the same controller trips.

> The self-protecting instantaneous short-circuit overcurrent protection referred to is a mechanical means for the molded case switch to open and trip at very high fault currents. Not all molded case switches incorporate this feature. Switches having this feature are said to have "blow-apart contacts." This self-protecting feature, when present, typically trips only on currents greater than 10 times the frame rating of the switch. These contacts are arranged in a "U" configuration, such that increasing current creates greater magnetic force which tends to separate the contacts. When the contacts begin to separate, the switch trips to finish opening the contacts. Commentary Table II.10.3 shows typical self protection trip currents for four common frame designations.

Commentary Table II.10.3 *Self Protection Trip Currents*

Frame Size*	Frame Rating	Min Trip		
"E"	100 to 150 amps	1000 amps		
"F"	250 amps	2500 amps "J"	400 amps	4000 amps
"L"	600 amps	6000 amps		

*These frame designations vary from manufacturer to manufacturer.

Some designs require that this trip value be no less than 13 times (13X) the motor FLA value on all three poles of both the isolating switch and the circuit breaker to avoid any false tripping during A-T-L (D-O-L) starting with first half-cycle offset currents. For more, see the commentary to 10.4.3.3.1(5) and 9.5.1.8.

▶ **FAQ** | Why is the isolating switch permitted to have any integral instantaneous short-circuit overcurrent protection?

Molded case switches are sometimes similar in construction to circuit breakers, except for the absence of the tripping elements. The instantaneous elements that remain provide short-circuit protection only. Most times, the instantaneous elements are the so-called self protecting "blow-apart" contacts in the device. These begin to open when enough current flows through the device. The magnetic interaction causes the contacts to begin to open. This causes the normal trip mechanism to activate and finish opening the contacts fully, if the magnetic forces weren't sufficient to do so. These instantaneous (blow-apart) elements also allow the fire pump controller to achieve a higher short-circuit rating than the interrupting capability of the circuit breaker in the fire pump controller. This higher short-circuit rating of the circuit breaker is accomplished by testing the combination of molded case switch and circuit breaker in a fire pump controller to demonstrate the controller can interrupt currents in excess of the interrupting capability of the circuit breaker alone. This occurs when both the isolating switch and the circuit breaker are similar in construction and have similar short circuit trip currents and opening (clearing) times. When properly designed, both devices open in unison (roughly simultaneously) to clear the fault faster than a single set of contacts in a single device can do.

10.4.2.2 Externally Operable. The isolating switch shall be externally operable.

Isolating switches must be readily accessible to allow for prompt energizing of the fire pump motor circuit. External operation is necessary to permit non-electrically qualified personnel to safely energize and de-energize the fire pump controller. With two handle designs, this is done after the circuit breaker has been de-energized (opened). With single handle designs, the pump motor and circuit breaker are de-energized nearly simultaneously.

10.4.2.3* Ampere Rating. The ampere rating of the isolating switch shall be at least 115 percent of the full-load current rating of the motor.

A.10.4.2.3 For more information, see *NFPA 70, National Electrical Code*.

▶ **FAQ** | How is the ampere rating of the isolating switch determined?

The source of the 115 percent requirement is Section 430.110 of the *NEC*. This sizing rule accounts for motors running continuously in excess of their rated current.

10.4.2.4 Warning.

10.4.2.4.1 Unless the requirements of 10.4.2.4.2 are met, the following warning shall appear on or immediately adjacent to the isolating switch:

WARNING
DO NOT OPEN OR CLOSE THIS SWITCH WHILE
THE CIRCUIT BREAKER (DISCONNECTING MEANS)
IS IN CLOSED POSITION.

> The isolating switch is not required to have a short-circuit interrupting rating. Only the fire pump controller circuit breaker is capable of manually closing on and interrupting a short-circuit current. Attempting to do so with an isolating switch could result in a fire and/or shock hazard. As a result, either the warning label or mechanical interlocking is required to prevent the isolating switch from being either opened or closed unless the circuit breaker is open. The intent is for the circuit breaker to make and break the power path, with the assumption that there is fault current involved, for both personnel safety and equipment protection.

10.4.2.4.2 Instruction Label. The requirements of 10.4.2.4.1 shall not apply where the requirements of 10.4.2.4.2.1 and 10.4.2.4.2.2 are met.

10.4.2.4.2.1 Where the isolating switch and the circuit breaker are so interlocked that the isolating switch can be neither opened nor closed while the circuit breaker is closed, the warning label shall be permitted to be replaced with an instruction label that directs the order of operation.

> ▶ **FAQ** Is the interlocking mechanism for the isolating switch and circuit breaker interlocked with the controller enclosure door and, if so, is it provided with a means to circumvent it to allow for testing?

> The interlock referenced in 10.4.2.4.2.1 ensures that the circuit breaker, and not the isolating switch, "makes" and "breaks" the current flowing in the fire pump power circuit. Most controllers are provided with interlocking mechanisms having either one or two operating handles. In most cases, the isolating switch operating handle mechanism has a two-way interlock with the enclosure door. The interlocking means is also provided with a means to circumvent the interlock to allow for testing and maintenance with the controller energized.

10.4.2.4.2.2 This label shall be permitted to be part of the label required by 10.3.7.3.

10.4.2.5 Operating Handle.

10.4.2.5.1 Unless the requirements of 10.4.2.5.2 are met, the isolating switch operating handle shall be provided with a spring latch that shall be so arranged that it requires the use of the other hand to hold the latch released in order to permit opening or closing of the switch.

> The use of a spring latch requires a two-hand operation to close or open the isolating switch. This safety feature minimizes inadvertent opening or closing of the switch. Instructions for operating the latch are normally included with the warning marking required in 10.4.2.4.1. This requirement is historic, since modern listed fire pump controllers do not make use of this option.

10.4.2.5.2 The requirements of 10.4.2.5.1 shall not apply where the isolating switch and the circuit breaker are so interlocked that the isolating switch can be neither opened nor closed while the circuit breaker is closed.

10.4.3 Circuit Breaker (Disconnecting Means).

10.4.3.1* General. The motor branch circuit shall be protected by a circuit breaker that shall be connected directly to the load side of the isolating switch and shall have one pole for each ungrounded circuit conductor.

> ▶ **FAQ** What types of circuit breakers are acceptable in a fire pump controller?
>
> At one time there were separately listed circuit breakers by General Electric, Westinghouse, and ITE (now part of Siemens). These have not been offered for several decades. All circuit breakers in modern controllers consist of an instantaneous only circuit breaker or "motor circuit protector," or a self-protected molded case switch, all of which contain a shunt trip coil (solenoid) controlled by current sensing circuitry fed by three current transformers (CTs). There were a number of controllers shipped with only two CTs for a period of time.

A.10.4.3.1 For more information, see *NFPA 70, National Electrical Code*, Article 100.

10.4.3.2 Mechanical Characteristics. The circuit breaker shall have the following mechanical characteristics:

(1) It shall be externally operable. *(See 10.3.6.)*
(2) It shall trip free of the handle.

> ▶ **FAQ** What does "trip free of the handle" mean?
>
> The term "trip free" means that upon sensing a high enough current for long enough, the circuit breaker acts to interrupt the circuit, even if the handle is jammed or stuck in the ON position. Tripping free can occur in fire pump controllers by way of the external handle and operator mechanism. Modern self-protecting isolating switches also have the ability to trip free.

(3) A nameplate with the legend "Circuit breaker — disconnecting means" in letters not less than ⅜ in. (10 mm) high shall be located on the outside of the controller enclosure adjacent to the means for operating the circuit breaker.

10.4.3.3* Electrical Characteristics.

A.10.4.3.3 Attention should be given to the type of service grounding to establish circuit breaker interrupting rating based on grounding type employed.

> The service grounding described in A.10.4.3.3 is usually the neutral of a 4-wire wye system – such as a 208/120V ac, or a 480/277V ac, system. However, it can also be one side of 120/240/208V ac 4-wire corner grounded delta system, which were once common and are still widely used. Corner ground, and ungrounded delta, at 240V ac or 480V ac, is also possible.
> The controller must be compatible with the system voltage and method of system grounding.

10.4.3.3.1 The circuit breaker shall have the following electrical characteristics:

(1) A continuous current rating not less than 115 percent of the rated full-load current of the motor

> **▶ FAQ** How is the ampere rating of the circuit breaker determined?

The ampere rating of the circuit breaker is on the listing label of the circuit breaker. It is usually also marked on the toggle handle of the breaker. The FLA of the motor is determined by Section 430.110 of the *NEC*. This sizing rule accounts for motors running continuously in excess of their rated horsepower by up to 115 percent (1.15 SF). This requirement is also applied to the isolating switch in 10.4.2.3.

(2) Overcurrent-sensing elements of the nonthermal type

This requirement was added as a result of older construction methods for a brand of fire pump controllers which used thermal sensing elements for measuring and timing the tripping of the circuit breaker. All modern controllers use a combination of current transformers, electronic analog and/or digital circuitry, and a shunt trip coil (solenoid) in the circuit breaker. Compliance with this requirement results in the breaker being capable of being reset immediately (without delay). If the circuit breaker trips while the pump motor is running under fire conditions, resetting the circuit breaker to try to put the motor back into service should be accomplished without waiting.

> **▶ FAQ** Why is a thermal magnetic circuit breaker not allowed to provide overcurrent protection for the fire pump motor?

Thermal magnetic breakers also contain thermal elements which require a cooling time before they can be reset. Hence, they would always violate the requirement in 10.4.3.3.1(2), and so cannot be used.

(3) Instantaneous short-circuit overcurrent protection

This characteristic is vital, fundamental, and the most important aspect of the short-circuit (WIC) rating of the controller. The opening time for the circuit breaker, and for many designs also for the isolating switch, determines how much short-circuit current the controller will be exposed to, and for how long.

(4)* An adequate interrupting rating to provide the suitability rating of the controller discussed in 10.1.2.2

> **▶ FAQ** How can the interrupting rating (AIC) of the circuit breaker be less than the short-circuit rating of the fire pump controller?

Generally, the higher the interrupt rating of the circuit breaker, the higher the short-circuit rating of the controller can be. However, the two ratings should never be confused or used interchangeably, since the short-circuit rating must be established by high current laboratory type (prototype) testing of a number of sample controllers of different horsepower and/or different voltage ratings. The controller actual short-circuit rating can either be a smaller or higher number than the circuit breaker AIC rating value. It can be higher when the isolating switch is also of the self-protecting type.

A.10.4.3.3.1(4) The interrupting rating can be less than the suitability rating where other devices within the controller assist in the current-interrupting process.

(5) Capability of allowing normal and emergency starting and running of the motor without tripping *(see 10.5.3.2)*

> Fire pump motors exhibit high inrush currents under starting conditions. These currents are maximized (as high as locked rotor). It is important that these inrush currents do not cause tripping of the circuit breaker. This is especially important on power supply circuits having a high X to R ratio (X/R), meaning circuits which are more inductive, as is typical of higher horsepower installations. The first half-cycle starting current inrush current can be up to 12 times motor FLA. Some controllers have no less than 13 times instant trip ratings on all three poles of the circuit breaker. The reference to 10.5.3.2 means that all controllers are started in the full voltage A-T-L (D-O-L) mode when the emergency-run mechanical operator is used to start the fire pump. Even higher transient currents can occur with open transition wye-delta starting, depending on the fire pump controller design. These higher transient currents can also occur when power is briefly interrupted with controllers that are not equipped with a restart delay function, and when there is momentary de-energization of the motor, such as when the automatic stop timer stops a pump supplying a small sprinkler demand (only a few sprinkler heads open). The result is often a rapid pressure drop and re-starting the motor after only a brief off period. The residual voltage will always be out of phase and frequency with the power supply resulting in transient (spike) currents of some magnitude.

(6) An instantaneous trip setting of not more than 20 times the full-load current

▶ **FAQ** Why is the instantaneous trip setting of the fire pump breaker allowed to be a maximum of 20 times the motor full-load current?

The limit of 20 times for the instantaneous trip setting of the fire pump breaker was increased from the value allowed in *NEC* Article 430 for ordinary motor branch circuits. This was done to allow higher settings in fire pump controllers to reduce the likelihood of the circuit breaker tripping under motor starting conditions. Either open transition wye delta starting or rapid re-starting of the pump motor can cause transients up to 17 times the motor FLA value, or higher when the transition is not properly controlled. Some controllers delay the transition for one or more seconds, which loses most of the motor RPM speed and results in across-the-line (D-O-L) currents after the transition to full voltage (delta) running. Some controllers control the transition by way of a lead-lag monitor to keep the transient (spike) currents within acceptable limits. Some designs also test each pole for each switching device (isolating switch and circuit breaker) to ensure that the instantaneous trip currents are no less than 13 times the motor's rated FLA current. This provides adequate tolerance of starting surge currents (first-half cycle offset currents), especially in systems with high X to R ratios, as is normally the case with large (high horsepower) motor circuits.

10.4.3.3.1.1* The circuit breaker shall not trip when starting a motor from rest in the across-the-line (direct-on-line) mode, whether or not the controller is of the reduced inrush starting type.

A.10.4.3.3.1.1 The isolating switch is not allowed to trip either. See also 10.4.2.1.3.

> This requirement was added to clarify that the circuit breaker needs to allow full voltage starting, since this is how the motor is started whenever the emergency-run mechanical operator is used and whenever power is interrupted and re-applied while the mechanical operator is engaged.
> Most controllers use a self-protecting molded case switch for both the isolating switch and the circuit breaker.

10.4.3.3.1.2* The circuit breaker shall not trip when power is interrupted from a running pump, or if the pump is restarted in less than 3 seconds after being shut down. If a control circuit preventing a re-start within 3 seconds is provided, this requirement shall not apply.

A.10.4.3.3.1.2 See also A.10.4.3.3.1.1

> This requirement was added to clarify that the circuit breaker should not trip when the motor stops, such as after the automatic stop (minimum run) timer times out while a pump demand still exists. This is usually the case when there is a small demand (such as when a few sprinkler heads are open and controlling a fire). The pump discharge pressure will usually be high enough to reset the pressure switch or sensor, which allows the automatic stop timer to stop the pump. When the pressure drops back below the pump start value, the controller will re-start the motor. For controllers lacking a restart delay provision, the motor will have some amount of back EMF, which will be both out of phase and not the same frequency as the line voltage. The result will be a transient of up to 17 times the motor FLA value.

10.4.3.3.2* Current limiters, where integral parts of the circuit breaker, shall be permitted to be used to obtain the required interrupting rating, provided all the following requirements are met:

(1) The breaker shall accept current limiters of only one rating.
(2) The current limiters shall hold 300 percent of full-load motor current for a minimum of 30 minutes.
(3) The current limiters, where installed in the breaker, shall not open at locked rotor current.
(4) A spare set of current limiters of correct rating shall be kept readily available in a compartment or rack within the controller enclosure.

A.10.4.3.3.2 Current limiters are melting link-type devices that, where used as an integral part of a circuit breaker, limit the current during a short circuit to within the interrupting capacity of the circuit breaker.

> ▶ **FAQ** Why are current limiters allowed when fuses may not be used to provide overcurrent protection in the fire pump controller?

Current limiters (current limiting fuses) were very common before circuit breakers with high current-interrupting ratings were available. Current limiters are intended to provide short-circuit protection only, and they must not open under any conditions of motor overload.

10.4.4 Locked Rotor Overcurrent Protection. The only other overcurrent protective device that shall be required and permitted between the isolating switch and the fire pump motor shall be located within the fire pump controller and shall possess the following characteristics:

> In fire pump controllers, the motor overload protection means are the same as the overcurrent elements; namely, the current transformers, the sensing circuitry and the shunt trip coil (solenoid) in the circuit breaker. Separate overload relays are prohibited under 10.4.4(5). This reduces the number of elements in the "critical starting" and running path of the fire pump motor, which serves to increase reliability.

(1) For a squirrel-cage or wound-rotor induction motor, the device shall be of the time-delay type having tripping times as follows:

This requirement sets the time and current values for the motor overload protection. Traditionally, this function has been an "inverse time" or reciprocal curve. With the values defined in 10.4.4(1)(a) and (b), the curve is defined and can be drawn, within limits. This reciprocal characteristic best suits protecting a motor, since it is similar in nature to the motor heating curve. This provides the best protection of the motor consistent with not over-protecting it at the expense of stopping a running fire pump providing fire water.

    (a) Between 8 seconds and 20 seconds at locked rotor current

The 8 to 20 seconds is longer than most combination motor starters, which will generally trip in 10 seconds or less. However, the intent is to keep the starting torque applied long enough to hopefully break away a stuck or seized pump. The motor may be warm at the end of this time, but not destroyed or inoperable.

    (b) Three minutes at a minimum of 300 percent of motor full-load current

Subsection 10.4.4(1)(b) used to read "(b) Calibrated and set at a minimum of 300 percent of motor full-load current." This meant that the circuit breaker timing circuit would not begin its timing means unless the current was above 300 percent of motor FLA. This allowed a running motor to continue running if one of the three phases was lost. Note that the actual "pick-up" (timing start) current is some value higher than 300 percent, since that is the minimum value.
    The requirement was changed for the 2013 edition to make testing and verification easier, since three minutes should normally be long enough to verify that the electronic sensing circuitry has not started timing and is not likely to trip the circuit breaker.

(2)   For a direct-current motor, the device shall be as follows:
    (a) Of the instantaneous type
    (b) Calibrated and set at a minimum of 400 percent of motor full-load current

The requirements in 10.4.4(2) are for a legacy from the era when dc motors were commonly used for fire pumps. See the commentary to 9.5.1.2 for more on this.

(3)* There shall be visual means or markings clearly indicated on the device that proper settings have been made.

The visual means referred to here are intended to clearly indicate the motor FLA value to which the breaker is set. It could be a dial or a display screen menu item.

A.10.4.4(3) It is recommended that the locked rotor overcurrent device not be reset more than two consecutive times if tripped due to a locked rotor condition without the motor first being inspected for excessive heating and to alleviate or eliminate the cause preventing the motor from attaining proper speed.

Motors that are subjected to more than two consecutive starts under locked rotor conditions may incur damage that would affect the motor's performance or render it inoperable.
    See Formal Interpretation 83–1.

(4)   It shall be possible to reset the device for operation immediately after tripping, with the tripping characteristics thereafter remaining unchanged.
(5)   Tripping shall be accomplished by opening the circuit breaker, which shall be of the external manual reset type.

Stationary Fire Pumps Handbook 2013

> **Formal Interpretation**
> **NFPA 20**
> **Stationary Pumps for Fire Protection**
> **2010 Edition**
>
> *Reference:* 10.4.3, 10.4.4
>
> F.I. 83–1
>
> *Question 1:* Is it the intent to allow continuous 300 percent of full load current electrical overloading of the fire pump feeder circuits, including transformers, disconnects or other devices on this circuit?
>
> *Answer:*
>
> a) Relative to protective devices in the fire pump feeder circuit, such devices shall not open under locked rotor currents (see 9.3.2.2).
>
> b) Relative to the isolating means and the circuit breaker of the fire pump controller, it is the intent of 10.4.3 to permit 300 percent of full load motor current to flow continuously through these devices until an electrical failure occurs. [This statement also applies to the motor starter of the fire pump controller, but this device is not in the feeder (see Section 3.3).]
>
> c) Relative to all devices other than those cited above, refer to NFPA 70 for sizing.
>
> *Question 2:* If the answer to Question 1 is no, what is meant by "setting the circuit breaker at 300 percent of full load current"?
>
> *Answer:* The phrase "setting the circuit breaker at 300 percent of full load current" means that the circuit breaker will not open (as a normal operation) at 300 percent of full load current. It does not mean that the circuit breaker can pass 300 percent of full load current without ultimately failing from overheating.
>
> *Question 3:* What is meant by "calibrated up to and set at 300 percent" of motor full load current?
>
> *Answer:* Question 2 answers the "set at 300 percent" of motor full load current. "Calibrated up to 300 percent" of motor full load current means that calibration at approximately 300 percent is provided by the manufacturer of the circuit breaker.
>
> *Issue Edition:* 1983
>
> *Reference:* 6–3.5, 7–4.3
>
> *Date:* January 1983
>
> Copyright © 2009 All Rights Reserved
> NATIONAL FIRE PROTECTION ASSOCIATION

10.4.5 Motor Starting Circuitry.

10.4.5.1 Motor Contactor. The motor contactor shall be horsepower rated and shall be of the magnetic type with a contact in each ungrounded conductor.

> **FAQ** Don't all contactors have a contact in each line conductor?

> The code requires three contacts for three phase controllers for safety reasons. There are some designs which utilize two pole contactors for three phase circuits. While this works on a functional basis, it leaves the motor windings electrically live all the time, since one phase is tied directly to the voltage source.

10.4.5.1.1 Running contactors shall be sized for both the locked rotor currents and the continuous running currents encountered.

> With A-T-L (D-O-L) starting, the running current is the motor SFA (FLA × the maximum 1.15 service factor), while the starting current is calculated to be at LRA (600 percent of FLA). With part winding starting, however, while the two contactors carry only half of the SFA current, the contactor which does the starting carries 65 percent of the motor FLA. So, two 50 hp contactors would not be large enough for a 100 hp part winding controller, due to the starting current. Some contactors are in the circuit only during starting — such as in primary reactor, primary resistor, soft starting or autotransformer starting; those contactors must be sized for the starting current, as must the wye contactor with wye-delta and autotransformer starting.

10.4.5.1.2 Starting contactors shall be sized for both the locked rotor current and the acceleration (starting) encountered.

> The sizing criteria for the selection of running and starting contactors in a fire pump controller are provided in 10.4.5.1.1 and 10.4.5.1.2. Across-the-line starting fire pump controllers generally employ one contactor that needs to meet the sizing criteria specified for a running contactor. Reduced-voltage starting fire pump controllers employ multiple contactors and the sizing criteria that an individual contactor needs to meet is based on its function in the controller.

10.4.5.2 Timed Acceleration.

10.4.5.2.1 For electrical operation of reduced-voltage controllers, timed automatic acceleration of the motor shall be provided.

10.4.5.2.2 The period of motor acceleration shall not exceed 10 seconds.

> The requirement in 10.4.5.2.2 ensures that the pump is running at full speed within 10 seconds of the signal to start.

10.4.5.3 Starting Resistors. Starting resistors shall be designed to permit one 5-second starting operation every 80 seconds for a period of not less than 1 hour.

> Starting resistors that are not sized properly based on duty cycle are subject to overheating and burnout. Duty cycle in this context is the proportion of time during which the motor-starting resistor is dissipating its maximum rated power. Duty cycle can be expressed as a ratio or as a percentage. The higher the duty cycle, the shorter the useful life is, if all other things are equal. To reduce costs, the resistors are allowed to use a 5 second start cycle rather than the usual 10. Also, a substantial cooling time is required, since these resistors typically get very hot after use.

10.4.5.4 Starting Reactors and Autotransformers.

10.4.5.4.1 Starting reactors and autotransformers shall comply with the requirements of ANSI/UL 508, *Standard for Industrial Control Equipment*, Table 92.1.

10.4.5.4.2 Starting reactors and autotransformers over 200 hp shall be permitted to be designed to Part 3 of ANSI/UL 508, *Standard for Industrial Control Equipment*, Table 92.1, in lieu of Part 4.

> Starting reactors and autotransformers over 200 hp are not required to meet the more stringent performance criteria of ANSI/UL 508, *Standard for Industrial Control Equipment*, Table 92.1, Part 4, because of the requirement in 10.4.5.2.2 limiting motor acceleration time. The criteria in Part 4 of ANSI/UL 508 include a higher duty cycle. In any case, the reactors and autotransformers easily handle multiple 10-second start cycles; although, a 3-second start cycle is normally enough to get the motor up to full speed, due to the high torque delivered by these two starting methods.

10.4.5.5 Soft Start Units.

10.4.5.5.1 Soft start units shall be horsepower rated or specifically designed for the service.

> ▶ **FAQ** Do soft start units for fire pump service need to be rated in horsepower?

Soft start units that are rated in amperes must have a rating at least equivalent to the motor full-load current.

10.4.5.5.2 The bypass contactor shall comply with 10.4.5.1.

> The bypass contactor must comply with 10.4.5.1 and its subsection 10.4.5.1.1 since it is a running contactor. This contactor is energized after the accelerate time (usually 10 seconds) to bypass the soft-start unit, which helps avoid the heat buildup which would otherwise occur if the soft-start unit ran the motor continuously. For example, a 60 hp, 208V ac unit would dissipate over 500 watts.
>
> The starting contactor must comply with 10.4.5.1.2. Note that the start contactor is kept energized with designs that provide a soft-start function. However, no substantial current flows through this contactor once the bypass contactor closes, especially since the soft-start unit does have some voltage drop across it when current flows through it. For the ramp-down (soft-stop) function, the controller de-energizes the bypass contactor, causing motor current to flow through the soft-start unit again. The controller also signals the soft starter to ramp down the motor voltage, normally over a 10-second period. At the end of this time, the controller also de-energizes the start contactor. Some designs also employ an isolating contactor that keeps the soft starter unit off-line and immune to line voltage transients during the standby periods.

10.4.5.5.3 Soft start units shall comply with the duty cycle requirements in accordance with 10.4.5.4.1 and 10.4.5.4.2.

> ▶ **FAQ** Are soft start units considered to be reduced voltage starting controllers?

Soft start units are categorized as reduced voltage starting controllers, similar to resistor, reactor, and autotransformer controllers and, therefore, must meet the same performance criteria as these controllers. They are often specified for hydraulic reasons. They gradually ramp up the motor voltage over a 10-second start cycle, which reduces the water surge that can otherwise occur, especially when the system pressure has dropped significantly or when the water demand is large. Most controllers also ramp the motor speed down prior to shutting down the motor. This reduces the sudden pressure fall-off and can greatly reduce the banging of the main check valve, especially in high-rise applications. Note that the accelerate and ramp-down are factory set (non-adjustable) with many of these types of controllers.

10.4.5.6 Operating Coils. For controllers of 600 V or less, the operating coil(s) for any motor contactor(s) and any bypass contactor(s), if provided, shall be supplied directly from the main power voltage and not through a transformer.

▶ **FAQ** Why is the operating coil for the motor contactor not allowed to be supplied through a transformer?

Supplying the operating coil directly from the main power voltage maximizes the reliability of contactor operation since this also reduces the number of items in the critical starting path. Also note that a transformer supplying the coil (or coils, in reduced voltage controllers) would have to be substantial in size. Failure of the transformer due to line spikes or surges could cause significant damage to the controller, making it more difficult and time consuming to repair.

10.4.5.7* Single-Phase Sensors in Controller.

Single phase sensors detect when one of the legs (lines) of a three phase power source is lost or open. Some controllers provide single phase start protection for all three single phasing conditions. Section 10.4.5.7 governs this function. Note, however, that all controllers inherently provide single phase start protection when either of the two legs providing control power are lost. This is because, regardless of the state of the control circuitry, the motor contactor(s) cannot pick up (close) if control power is lost.

A.10.4.5.7 The signal should incorporate local visible indication and contacts for remote indication. The signal can be incorporated as part of the power available indication and loss of phase signal. *(See 10.4.6.1 and 10.4.7.2.2.)*

This is the normal means by which a motor running under single phase conditions uses the power failure (phase loss) contacts to signal the remote indication (such as an alarm set). The result will be the simultaneous occurrence of both a motor running and a power failure indication.

10.4.5.7.1 Sensors shall be permitted to prevent a three-phase motor from starting under single-phase condition.

Three-phase motors at rest cannot be started with single-phase power. All three phases are needed to create the rotating magnetic vector needed to get a standing motor rotor to develop enough torque to turn the motor shaft. Trying to do so can cause damage to or destruction of the motor windings, and often to the controller as well.

10.4.5.7.2 Such sensors shall not cause disconnection of the motor if it is running at the time of single-phase occurrence.

However, a motor running at or near full speed (above the breakdown speed) can continue to run if one of the three legs is lost (due to an open circuit). It will draw between 173 percent and 200 percent more current than it was when it was running under three phase power. If the motor was running under light load and drawing less than 57 percent of FLA (115% / 2), it can run continuously with only a slight drop in running speed. The winding current will still be less than 115 percent of rated motor FLA, and will not overheat the windings. This is the usual case when fighting a fire, since normally only a few heads are open then. If the motor was at full load or into its service factor, most motors can still run at close to normal speed, if the voltage doesn't drop too far. They will draw up to 230 percent of FLA (115% × 2) and eventually overheat the windings until the motor fails. This will take some time but will provide fire water until the motor runs to destruction.

10.4.5.7.3 Such sensors shall be monitored to provide a local visible signal in the event of malfunction of the sensors.

This requirement is intended to address the possibility that false detection of a single phase event will prevent a motor from starting when needed, and when there is full three phase power.

10.4.5.8 No ground fault protection (tripping) shall be allowed.

10.4.5.9 A ground fault alarm shall be permitted.

While this section permits the use of an alarm to sound when a ground fault occurs, tripping is not permitted by section 10.4.5.8.

10.4.6* Signal Devices on Controller.

Sections 4.6 and 4.7 along with 10.8.3.12.1, 10.8.3.12.2, and 10.8.3.14, indicate which functions need to be monitored in fire pump controllers and fire pump power transfer switches. Some controllers also include a "failure to start" alarm and a "motor overload" alarm. The former is valuable to indicate that there is a demand for the fire pump but that the pump is not working. The latter is important since the controller will run the motor to destruction, and this alarm warns of an impending failure. An overload alarm is important if the pump is not fighting a fire since it provides notification that the pump is in danger of being damaged, and the alarm is important if there is a fire since it indicates impending failure of the motor. In either case, the alarm should serve to provide warning to determine the motor current.

A.10.4.6 The pilot lamp for signal service should have operating voltage less than the rated voltage of the lamp to ensure long operating life. When necessary, a suitable resistor or potential transformer should be used to reduce the voltage for operating the lamp.

10.4.6.1 Power Available Visible Indicator.

10.4.6.1.1 A visible indicator shall monitor the availability of power in all phases at the line terminals of the motor contactor or of the bypass contactor, if provided.

Sensing at the motor contactor line side terminals allows detection of any tripping or failure in either the controller isolating switch or the circuit breaker, as well as any failure of the power source. Some controller designs trip at around 85 percent of nominal voltage, with a time delay to avoid tripping on transients or momentary line voltage dips.

10.4.6.1.2 If the visible indicator is a pilot lamp, it shall be accessible for replacement.

The contacts required by 10.4.8 are needed to operate circuits for the three conditions specified in 10.4.7.2.1 thru 10.4.7.2.3, which apply to all controllers; namely motor running, phase loss (power loss) and phase reversal. There are two other controller contacts for remote indication which indicate that the transfer switches, are being supplied by the emergency system; or 10.8.3.13 (see 10.8.3.12) which indicate that the transfer switch emergency side isolating switch is open. Commentary Tables II.10.4 and II.10.5 summarize the required alarms.

Commentary Table II.10.4 *Required Alarms – Single Source Controllers*

Alarm	Contacts	Visible	Audible
Power Available	Yes	Yes	Optional
Phase Loss (Power Failure)	Yes	N/A	Optional
Motor (Pump) Running	Yes	Optional	Optional
Phase Sequence Reversed*	Yes	Yes	Optional

*Phase sequence circuitry is omitted on single phase (limited service) controllers.

Commentary Table II.10.5 *Required Alarms – Dual Source (Transfer Switch) Controllers*

Alarm	Contacts	Visible	Audible
Power Available	Yes	Yes	Optional
Phase Loss (Power Failure)	Yes	N/A	Optional
Motor (Pump) Running	Yes	Optional	Optional
Phase Sequence Reversed*	Yes	Yes	Optional
Transfer Switch in Normal Position	Optional	Yes	Optional
Transfer Switch in Emergency Position	Yes	Yes	Optional**
Emergency Isolating Switch Open	Yes	Yes	Yes

**Some controllers provide this audible alarm as standard in addition to the "emergency isolating switch open" audible alarm.

10.5 Starting and Control

10.5.1* Automatic and Nonautomatic.

A.10.5.1 The following definitions are derived from *NFPA 70, National Electrical Code*:

(1) *Automatic*. Self-acting, operating by its own mechanism when actuated by some impersonal influence, as, for example, a change in current strength, pressure, temperature, or mechanical configuration.
(2) *Nonautomatic*. Action requiring intervention for its control. As applied to an electric controller, nonautomatic control does not necessarily imply a manual controller, but only that personal intervention is necessary.

▶ **FAQ** What's the difference between "nonautomatic" and "manual" operation?

Manual controllers require manual intervention for all operations, including both starting and reduced voltage starting, if so equipped. An example would be a manual motor starter such as are used on machine tools, or manual start wound rotor fire pump controllers used in some high-rise buildings in New York City for pressure modulation.

> A nonautomatic controller requires manual intervention to start and/or stop the motor, but acceleration may be by means of automatic circuitry. An example would be a shop exhaust blower which employs multi-step reduced inrush starting means. The pressure switch in modern controllers is a pressure transducer and sensing circuitry.
>
> The standard was revised in 2003 to allow an automatic controller to be actuated by means other than a pressure switch. An example of a nonpressure switch–actuated controller would be one actuated by a flow switch, smoke detector, or heat sensor, such as in deluge systems for airplane hangars.

10.5.1.1 An automatic controller shall be self-acting to start, run, and protect a motor.

10.5.1.2 An automatic controller shall be arranged to start the driver upon actuation of a pressure switch or nonpressure switch actuated in accordance with 10.5.2.1 or 10.5.2.2.

10.5.1.3 An automatic controller shall be operable also as a nonautomatic controller.

▶ **FAQ** What is a "nonautomatic" controller?

Nonautomatic controllers have no pressure sensing components. They may be started locally by way of the local start pushbutton, or a wired remote pushbutton. Examples of nonautomatic controllers are a pump supplying water to water cannons (monitor nozzles) on an oil loading jetty, and manual-start wound-rotor variable-speed controllers in high rise buildings. See also the commentary to A.10.5.1(2).

10.5.1.4 A nonautomatic controller shall be actuated by manually initiated electrical means and by manually initiated mechanical means.

> The electrical means is the local start pushbutton (switch) discussed in 10.5.3.1. The mechanical means is the emergency-run mechanical operator covered in 10.5.3.2.

10.5.2 Automatic Controller.

10.5.2.1* Water Pressure Control.

> Most fire protection sprinkler systems are pressurized. Normally a pressure maintenance (jockey) pump maintains the required system pressure against leaks. When the water demand exceeds the capacity of the jockey pump, the preset start pressure set-point of the fire pump controller(s) is finally reached, and that controller starts the fire pump(s). In most instances, the controller stops the pump, after a minimum running period, when the preset stop pressure set-point is reached.
>
> If the fire pump is the sole source of water supply to the fire protection system, the code prohibits stopping it automatically per 10.5.4.2(3). See Exhibit II.10.10 for an example of a controller with pressure switch.

▶ **FAQ** What is a set-point?

A set-point is the setting of an adjustable or programmable means where an action is supposed to occur. It can be a turn-on or a turn-off point, or a desired parametric value such as a room thermostat temperature setting or the desired system pressure in a variable speed fire pump controller.

EXHIBIT II.10.10 *Controller with Pressure Switch.*

A.10.5.2.1 Installation of the pressure sensing line between the discharge check valve and the control valve is necessary to facilitate isolation of the jockey pump controller (and sensing line) for maintenance without having to drain the entire system. *[See Figure A.4.30(a) and Figure A.4.30(b).]*

10.5.2.1.1 Pressure-Actuated Switches.

10.5.2.1.1.1 There shall be provided a pressure-actuated switch or electronic pressure sensor having adjustable high- and low-calibrated set-points as part of the controller.

> In the 2003 edition of NFPA 20 this requirement read "... having independent high- and low-calibrated adjustments ...". This was directly applicable to the once ubiquitous Mercoid pressure switches where in universal use for this purpose. The language was revised for the 2007 edition to clarify that the requirement is for the pressure-actuated switch to have individual, adjustable high- and low-pressure set points. This was done to accommodate the possibility of using adjustable differential pressure switches when mercury switches were being outlawed in various states in the U.S. A non-mercury type switch, which did have independent high and low set-point adjustments was used by some manufacturers for a brief period of time. All these were replaced with the advent, of pressure transducers. In the 2010 edition of NFPA 20 this requirement was revised to reflect current technology and practice in permitting the use of said pressure transducers.

10.5.2.1.1.2 The requirements of 10.5.2.1.1.1 shall not apply in a nonpressure-actuated controller, where the pressure-actuated switch shall not be required.

10.5.2.1.2 There shall be no pressure snubber or restrictive orifice employed within the pressure switch or pressure responsive means.

> While two of these are required to be installed in the pressure sensing line upstream of the controller (4.30.4), they are prohibited within the controller to avoid restrictions and/or obstructions which may prevent proper operation of the pressure switch (sensor). Restrictions of any type are not permitted within the controller ahead of the pressure switch or pressure transducer as they are an unnecessary item in the critical starting path of fire pump controllers. Their presence would reduce the reliability of the fire pump controller.

10.5.2.1.3 There shall be no valve or other restrictions within the controller ahead of the pressure switch or pressure responsive means.

> Even more critical is avoiding any sort of valve, including solenoid valves in the path to the pressure sensor, either manual or automatic. These have the potential to inhibit or prevent operation of the pressure sensor, or switch, during a fire demand.

10.5.2.1.4 This switch shall be responsive to water pressure in the fire protection system.

10.5.2.1.5 The pressure sensing element of the switch shall be capable of withstanding a momentary surge pressure of 400 psi (27.6 bar) or 133 percent of fire pump controller rated operating pressure, whichever is higher, without losing its accuracy.

> Approval agency testing is necessary to verify compliance with this requirement. This requirement is important to prevent damage or alteration of the calibration of the pressure sensor or switch. This is particularly true when the system pressure is close to the full scale pressure rating of the switch or sensor, since there are always pressure surges or pressure spikes which occur in the normal operation of a sprinkler system.

10.5.2.1.6 Suitable provision shall be made for relieving pressure to the pressure-actuated switch to allow testing of the operation of the controller and the pumping unit. *[See Figure A.4.30(a) and Figure A.4.30(b).]*

▶ **FAQ** | How is a controller tested for responsiveness to a pressure drop in a nonfire situation?

> Testing the controller for operation through the pressure-actuated switch is normally accomplished by opening a solenoid-operated or manual drain valve in the water supply line to the controller pressure switch. The flow through this valve drops the pressure across either one or both of the upstream restrictive orifices in the pressure sensing line below the lower set point of the pressure switch or sensor, which initiates the start-and-run sequence for the fire pump motor. This avoids using the system test drain valve which is problematic in some cases and some climates at certain times. The valves upstream of the ¼" tee allow attaching a precision pressure gage. The valve downstream of the ¼" tee allows modulating the flow in the pressure sensing line to modulate the pressure seen (sensed) by the controller. This is needed in order to accurately test the start and stop pressure settings of the controller.

10.5.2.1.7 Water pressure control shall be in accordance with 10.5.2.1.7.1 through 10.5.2.1.7.6.

10.5.2.1.7.1 Pressure switch actuation at the low adjustment setting shall initiate pump starting sequence (if pump is not already in operation).

10.5.2.1.7.2* A pressure recording device shall record the pressure in each fire pump controller pressure-sensing line at the input to the controller.

A.10.5.2.1.7.2 The pressure recorder should be able to record a pressure at least 150 percent of the pump discharge pressure under no-flow conditions. In a high-rise building, this requirement can exceed 400 psi (27.6 bar). This pressure recorder should be readable without opening the fire pump controller enclosure. This requirement does not mandate a separate recording device for each controller. A single multichannel recording device can serve multiple sensors. If the pressure recording device is integrated into the pressure controller, the pressure sensing element should be used to record system pressure.

The change to this paragraph clarifies the need for the pressure recorder to record the output from the controllers start/stop pressure sensor. The extant requirement for pressure range protection is the same as in 10.5.2.1.5.

▶ **FAQ** Why is a pressure recorder required?

The recorder data is useful for verifying that the required periodic pump testing has been occurring. More importantly this provides a record of the performance of system pressure before, during, and after a fire event. This information is very valuable to those conducting a fire investigation.

10.5.2.1.7.3 The pressure recorder shall be listed as part of the controller or shall be a separately listed unit installed to sense the pressure at the input of the controller.

This new clause clarifies the need for the internal or external pressure recorder to be listed.

10.5.2.1.7.4 The recorder shall be capable of operating for at least 7 days without being reset or rewound.

▶ **FAQ** What is meant by the phrase *capable of operating for at least 7 days?*

The recorder must be able to record or store at least 7 days' worth of data. This clause is often misinterpreted to mean the recorder must be able to operate off a backup source of power for at least 7 days, which is not the case.

This clause is historic regarding the use of paper chart or strip chart recorders. However, modern controllers typically have enough on-board pressure recording memory to hold at least one week of pressure data. Certain recorders respond to changes in pressure, rather than fixed time interval recording. As a result, they provide far better detail during rapidly changing pressure events, such as during a fire.

10.5.2.1.7.5 The pressure sensing element of the recorder shall be capable of withstanding a momentary surge pressure of at least 400 psi (27.6 bar) or 133 percent of fire pump controller rated operating pressure, whichever is greater, without losing its accuracy.

See the commentary to 10.5.2.1.5 for more information on this topic.

10.5.2.1.7.6 For variable speed pressure limiting control, a ½ in. (12.7 mm) nominal size inside diameter pressure line shall be connected to the discharge piping at a point recommended by the variable speed control manufacturer. The connection shall be between the discharge check valve and the discharge control valve.

10.5.2.2 Nonpressure Switch–Actuated Automatic Controller.

EXHIBIT II.10.11 Pressure Recorder (top) and USB Flashdrive Log Recorder (bottom). [(top) Courtesy of Clarke Fire Protection Products, Inc.; (bottom) Courtesy of Master Control Systems, Inc.]

10.5.2.2.1 Nonpressure switch–actuated automatic fire pump controllers shall commence the controller's starting sequence by the automatic opening of a remote contact(s).

> The device described in 10.5.2.2.1 is a "closed loop" control. The controller starts the pump when the remote contacts open, or if there is a break in either of the two wires connecting the remote contact to the controller. See the commentary to 10.5.1.2 and 10.5.1.3 for more detailed discussion.

10.5.2.2.2 The pressure switch shall not be required.

10.5.2.2.3 There shall be no means capable of stopping the fire pump motor except those on the fire pump controller.

> The standard was revised in the 2003 edition to allow an automatic controller to be actuated by means other than a pressure switch, although this had been common practice in non-pressurized systems such as deluge systems. In those systems, any deluge valve which opens causes electrical contacts to open to signal the fire pump controllers to start. Typical deluge systems consist of multiple fire pumps, (commonly four to eight), and usually agent pumps such as foam pumps. These are all signaled to start when any set of deluge valve contacts open.
>
> The reason for this requirement is to prohibit any external remote stop switches. This is done to avoid anyone shutting down a running fire pump inadvertently. There is also danger that the wiring to such a switch could either break or short, and permanently disable the fire pump. If the stopping can only be done at the controller, it can also be re-started manually, if needed.

10.5.2.3 Fire Protection Equipment Control.

10.5.2.3.1 Where the pump supplies special water control equipment (deluge valves, dry pipe valves, etc.), it shall be permitted to start the motor before the pressure-actuated switch(es) would do so.

> The control circuit of this special fire protection equipment is allowed to bypass the pressure-actuated switch of the controller. Note that many fire pumps supplying deluge systems are of the non-pressure actuated type. When they are pressure actuated, it is usually because there are both wet and dry pipe sprinkler systems in the facility in addition to the deluge system area(s).

10.5.2.3.2 Under such conditions the controller shall be equipped to start the motor upon operation of the fire protection equipment.

10.5.2.3.3 Starting of the motor shall be initiated by the opening of the control circuit loop containing this fire protection equipment.

> Initiating the controller's starting sequence by opening the control circuit containing this special fire protection equipment allows the fire pump to start in the event that breakage or disconnection of conductors occurs in this control circuit. See also the commentary to 10.5.2.2.1.

10.5.2.4 Manual Electric Control at Remote Station. Where additional control stations for causing nonautomatic continuous operation of the pumping unit, independent of the pressure-actuated switch, are provided at locations remote from the controller, such stations shall not be operable to stop the motor.

> When additional control stations (start pushbuttons or switches) are used to initiate the controller's starting sequence, the controller circuitry must be arranged to latch (lock onto) the start signal(s) from these stations so that they cannot cause the controller to shut down the fire pump motor. Stopping the motor from a remote location is undesirable because it is difficult to determine if all fires have been extinguished. See also the commentary to 10.5.2.2.3.

10.5.2.5 Sequence Starting of Pumps.

> See Exhibit II.10.12 for a diagram of fire pumps in series.

10.5.2.5.1 The controller for each unit of multiple pump units shall incorporate a sequential timing device to prevent any one driver from starting simultaneously with any other driver.

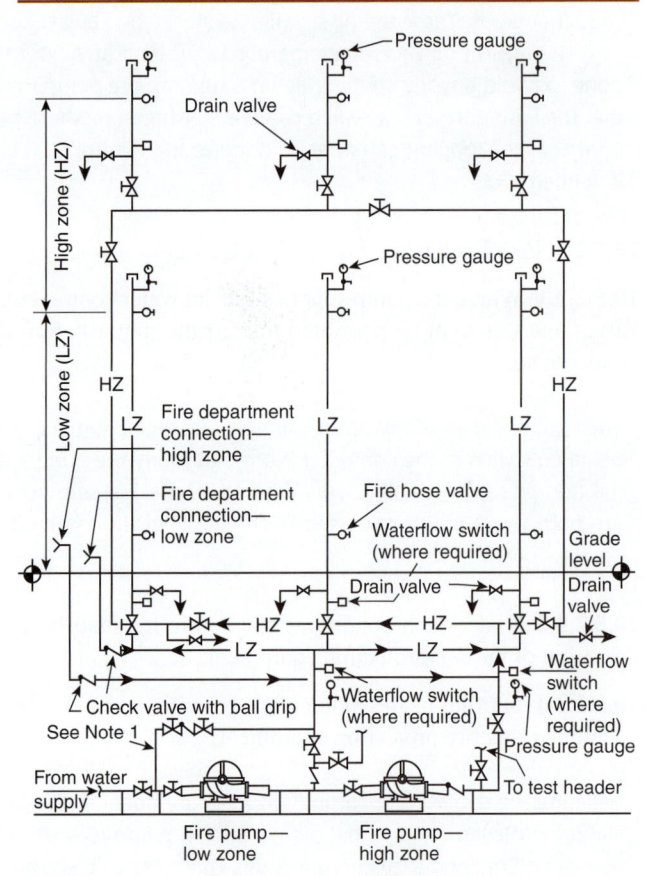

Notes:
1. Bypass in accordance with NFPA 20, *Standard for the Installation of Stationary Pumps for Fire Protection.*
2. High zone pump can be arranged to take suction directly from source of supply.

EXHIBIT II.10.12 *Fire Pumps in Series. [Source: NFPA 14, 2007, Figure A.7.1(b)]*

> There are two main types of multiple pump systems: pumps in series and pumps in parallel. (A third type is parallel pumps in series.) For pumps in parallel, two common reasons requiring sequential starting of the pumps are:
>
> 1. To avoid excessive starting current demand and the resultant voltage drop
> 2. To reduce the hydraulic stress on the system
>
> This second point is important for either deluge or other non-pressurized systems, or for systems where the pressure has dropped to or near zero and which have either a large demand or a significant amount of air in the system. Any of these three conditions results in a large flow (surge) of water — with its inherently large inertia — when that surge hits elbows and/or tees. The result is a large stress on the piping system.
>
> Some controller designs allow subsequent controllers to cancel the pressure demand signal if previous pumps have satisfied the flow demand and raised the system pressure enough to reset the pressure switch or sensor. Running pumps will continue to run until they are either manually stopped, in the case of manual stop arrangement, or after the minimum running timer has timed out in a particular controller, if arranged for automatic stop.

10.5.2.5.2 Each pump supplying suction pressure to another pump shall be arranged to start within 10 seconds before the pump it supplies.

> Pumps in series require sequential starting to avoid starting and running a high or mid-zone controller without adequate suction pressure from the pump or pumps supplying said pump. If suction pressure is too low, the higher zone pump can cavitate. If there is no suction pressure, the pump will run dry and ruin it in short order. Hence, when a higher demand pump receives a demand, it must signal the pump feeding it to start, and etc. Some controller designs apply this "high zone delay" to all start demand signals, not just pressure.
>
> The 2010 edition of NFPA 20 added the maximum time interval for starting pumps in series in order to ensure adequate water flow — especially in high-rise buildings, where the ability of one pump to feed another is critical. The latest change specifies that higher zone controllers must signal lower zone controllers electrically with a normally open control loop arrangement. The requirements for control wiring are in 695.14 of the NEC®.

10.5.2.5.2.1 Starting of the motor shall be initiated by the opening of the control circuit loop containing this fire protection equipment.

10.5.2.5.3 If water requirements call for more than one pumping unit to operate, the units shall start at intervals of 5 to 10 seconds.

> ▶ **FAQ** What is the starting sequence for a system of multiple pumps?
>
> Some installations require multiple pumps, such as in high-rise buildings, in order to deliver adequate water pressure to the sprinklers. Starting such pumps simultaneously is undesirable. There is no reason to start a pump unless adequate water supply is available on its suction side. The 5- to 10-second delay between pumping unit starts allows each pump supplying suction pressure to another pump to operate near or at full speed to provide an adequate water supply. If a leading pump driver fails to start, the operation of subsequent pumping units is imperative to deliver water to the fire-fighting system. See the commentary to 10.1.2.7 for more details on the starting sequence.

10.5.2.5.4 Failure of a leading driver to start shall not prevent subsequent pumping units from starting.

Subsection 10.5.2.5.4 provides another requirement which essentially sacrifices the pump in an effort to put out a fire, running the pump to destruction if necessary to avoid stopping the entire unit. While having an adequate suction supply is highly desirable, lack of suction is not allowed to prevent higher zones from starting, but may simply delay the start. In most cases, most of the pumps in a high-rise building are in the basement, and most are booster pumps; if a lower zone pump fails, there will still be some supply of water to the higher zone pump(s), especially since these pumps are required to have a bypass path around the pump in most cases per 4.14.4. The sprinkler system response to most fires consists of a small number of heads flowing water, resulting in reduced pressure and limited flow capacity, but maintaining the possibility of effectively fighting a fire.

10.5.2.6 External Circuits Connected to Controllers.

10.5.2.6.1 External control circuits that extend outside the fire pump room shall be arranged so that failure of any external circuit (open, ground-fault, or short circuit) shall not prevent operation of pump(s) from all other internal or external means.

10.5.2.6.2 Breakage, disconnecting, shorting of the wires, ground fault, or loss of power to these circuits shall be permitted to cause continuous running of the fire pump but shall not prevent the controller(s) from starting the fire pump(s) due to causes other than these external circuits.

External control circuits that extend outside the fire pump room are vulnerable to damage. It is critical that any such damage not prevent the controller from starting the fire pump when the pump start sequence is initiated by other means, such as the pressure-actuated switch. Any failure of these external control circuits must be allowed to start and run the fire pump motor continuously.

10.5.2.6.3 All control conductors within the fire pump room that are not fault tolerant as described in 10.5.2.6.1 and 10.5.2.6.2 shall be protected against mechanical injury.

▶ **FAQ** What is the meaning of the term *fault tolerant*?

Control conductors within the fire pump room are susceptible to damage in the same manner as control conductors outside the fire pump room. *Fault tolerant external control circuit*, as defined in 3.3.7.2, means protected against ground-fault or short circuit. Installation of these conductors in one of the wiring methods permitted by 695.14(F) of the 2011 edition of the *NEC* is one means of providing protection against mechanical injury and fire risk.

10.5.3 Nonautomatic Controller.

10.5.3.1 Manual Electric Control at Controller.

10.5.3.1.1 There shall be a manually operated switch on the control panel so arranged that, when the motor is started manually, its operation cannot be affected by the pressure-actuated switch.

The manually operated switch referred to in 10.5.3.1.1 is normally a start pushbutton.

10.5.3.1.2 The arrangement shall also provide that the unit will remain in operation until manually shut down.

The manually operated switch is one method of testing the controller for proper operation. The switch can also be used to start the fire pump motor in the event that a failure of the automatic starting circuitry occurs. The requirement "until manually shut down" means that if a controller is equipped or arranged for automatic stop operation, starting it this way will not shut down the motor.

10.5.3.2* Emergency-Run Mechanical Control at Controller.

Should all else fail, the emergency-run mechanical operator will get most motors running despite failures within the controller. Only a closed isolating switch and circuit breaker is needed in addition to the ability to mechanically close the contactor coil. The only manual operation that might be needed is manual electrical or manual mechanical switching of a transfer switch, if used per 10.8.3.5.

A.10.5.3.2 The emergency-run mechanical control provides means for externally and manually closing the motor contactor across-the-line to start and run the fire pump motor. It is intended for emergency use when normal electric/magnetic operation of the contactor is not possible.

When so used on controllers designed for reduced-voltage starting, the 15 percent voltage drop limitation in Section 9.4 is not applicable.

▶ **FAQ** Is compliance with the 15 percent voltage drop limitation in Section 9.4 required when the fire pump motor is started using the emergency-run mechanical control?

The 15 percent voltage drop limitation under motor-starting conditions is waived for reduced-voltage starting controllers because they are normally used in applications where the power supply cannot deliver locked rotor current when the motor is started at full line voltage. The emergency-run mechanical control bypasses the reduced-voltage starting components of the controller to apply full line voltage to the motor. Although the 15 percent voltage drop limitation is waived, reduced-voltage starting controllers are required to start and run the motor through the use of the emergency-run mechanical control.

10.5.3.2.1 The controller shall be equipped with an emergency-run handle or lever that operates to mechanically close the motor-circuit switching mechanism.

The "manual mechanical operator," as it is often called, pushes directly against the plunger of the motor contactor(s) to close the circuit to the motor, which causes the motor to run, if it is able. In part winding and both of the wye-delta starting type of controllers, the mechanical operator has a "balance beam" or similar mechanism for closing both of the two main motor contactors. All other schemes require only the one main motor contactor to be energized.

To minimize damage to the contactors, most modern controllers have an "electrical assist" function to energize the contactor. This consists of a limit switch that activates partway through the operating travel of the mechanical operator handle. This is to provide the normal high speed closing (and opening) of the contactors to avoid excessive arcing, pitting, corrosion, or even welding that can occur when the operator is used slowly.

While medium voltage controllers incorporate either over-center (toggle) or spring loaded mechanisms to ensure rapid actuation, low voltage controllers typically use a simple mechanism to push the contactor plunger. This is why the handle should always be used rapidly and purposely when either starting or stopping the motor.

See also the commentary to 9.4.2.

Stationary Fire Pumps Handbook   2013

10.5.3.2.1.1 This handle or lever shall provide for nonautomatic continuous running operation of the motor(s), independent of any electric control circuits, magnets, or equivalent devices and independent of the pressure-activated control switch.

10.5.3.2.1.2 Means shall be incorporated for mechanically latching or holding the handle or lever for manual operation in the actuated position.

> This function of the requirement in 10.5.3.2.1.2 is to keep the motor running without having to hold the handle in position. Note that some fire departments routinely engage this mechanism when they are fighting a fire in a building, to minimize the possibility of an interruption of the water supply, especially since hose lines are usually involved. It would be extremely dangerous if, using a hose line on the roof of a high rise building, the water pressure was lost, then re-applied; this has resulted in at least one fatality in the past when someone was thrown off of a roof as a result.

10.5.3.2.1.3 The mechanical latching shall be designed to be automatic or manual.

> The requirement on latching was revised in the 2013 edition because it is safer to use the controller's circuit breaker (disconnecting means) to stop the motor. When a controller has a burned out coil, or other problems which prevent the electrical assist from operating properly, some might choose, since the controller is out of service anyway, use of the circuit breaker for both starting and stopping the motor, and might accomplish this by leaving the mechanical operator engaged or by engaging it before closing the breaker. However, doing so would leave the building without automatic fire protection, so a fire watch would be required, and prompt repair of the controller would of course be urgent.

10.5.3.2.2 The handle or lever shall be arranged to move in one direction only from the off position to the final position.

> The requirement in 10.5.3.2.2 is intended to avoid handle or lever designs which use a gear shift type operation.

10.5.3.2.3 The motor starter shall return automatically to the off position in case the operator releases the starter handle or lever in any position but the full running position.

> The starter handle mechanism must be free moving to allow the mechanism to reset itself back to the off position. (This is only correct for medium voltage designs.)

10.5.3.2.4 The operating handle shall be marked or labeled as to function and operation.

10.5.4 Methods of Stopping. Shutdown shall be accomplished by the methods in 10.5.4.1 and 10.5.4.2.

10.5.4.1 Manual. Manual shutdown shall be accomplished by operation of a pushbutton on the outside of the controller enclosure that, in the case of automatic controllers, shall return the controller to the full automatic position.

> The manual shutdown means described in 10.5.4.1 is the stop pushbutton. In most designs, this button will stop the motor, regardless of the start demand. If the motor was started by the start pushbutton, the stop pushbutton will stop the motor. If there is another demand, most designs will stop the motor while the pushbutton is pressed, but the controller will re-start the motor when the pushbutton is released. In other types of designs, the controller will pause for the duration of the two to three second "restart delay."

10.5.4.2 Automatic Shutdown After Automatic Start. Where provided, automatic shutdown after automatic start shall comply with the following:

(1) Unless the requirements of 10.5.4.2(3) are met, automatic shutdown shall be permitted only where the controller is arranged for automatic shutdown after all starting and running causes have returned to normal.
(2) A running period timer set for at least 10 minutes running time shall be permitted to commence at initial operation.

> Setting the running period timer runs the pump for at least 10 minutes after an automatic start per 10.5.4.2(2). The running period prevents rapid stops and starts during small demands where the pump output pressure is high enough to reach the controller stop pressure set-point which would otherwise occur. This also allows the motor fan to cool the motor windings which warm up during every starting cycle.

(3) The requirements of 10.5.4.2(1) shall not apply and automatic shutdown shall not be permitted where the pump constitutes the sole supply of a fire sprinkler or standpipe system or where the authority having jurisdiction has required manual shutdown.

▶ **FAQ** Why is automatic stopping not allowed with sole source pumps?

> Keeping the motor running is the safest way to maintain the flow of fire water. This is because starting the motor often involves more circuitry than simply keeping the pump running. More importantly, the motor might be running under a single phase condition, in which case it can stay running for a long time, including indefinitely under a light load, such as when fighting a fire with only a few sprinkler heads open. Even worse, it is possible that the current may be well over 115 percent, whether running under a heavier load during single phase running, or due to an incipient fault in the motor windings. It may well be that, even under "run it to destruction" philosophy, the controller is now too degraded or damaged to start the motor even one more time.

10.6 Controllers Rated in Excess of 600 V

> Note that these controllers used to be commonly called "high voltage" controllers (see 9.5.1.2). In modern terminology, these are considered to be "medium voltage" controllers. Listed units are available in voltages ranging from 1200V ac up to 7200V ac (1.2 Kv to 7.2 Kv). Using these controllers often eliminates the need for a separate transformer for the fire pump, along with the transformer primary overcurrent protection. Also, these controllers are available in much higher horsepower than low voltage units. These controllers are mostly full voltage A-T-L (D-O-L) starting types, although listed reduced voltage primary or neutral reactor starting and autotransformer starting units are available. Some designs allow reduced voltage starting even when using the emergency-run mechanical means.

10.6.1 Control Equipment. Controllers rated in excess of 600 V shall comply with the requirements of Chapter 10, except as provided in 10.6.2 through 10.6.8.

10.6.2 Provisions for Testing.

10.6.2.1 The provisions of 10.3.4.3 shall not apply.

Subsection 10.3.4.3 contains the provision for metering all three phases voltages. Older designs had metering for only one phase voltage and one line current. Full three phase metering was introduced in 1993. Most modern controllers now provide full three phase metering.

10.6.2.2 An ammeter(s) shall be provided on the controller with a suitable means for reading the current in each phase.

10.6.2.3 An indicating voltmeter(s), deriving power of not more than 125 V from a transformer(s) connected to the high-voltage supply, shall also be provided with a suitable means for reading each phase voltage.

▶ **FAQ** How is voltage and current measured in these controllers without risk of electric shock?

Measuring voltage and current within equipment operating in excess of 600 V poses a high risk of electric shock. Meters on the outside of the controller enclosure enable measuring voltage and current without such risk. The voltage at the voltmeter terminals must be provided through a stepdown transformer to isolate voltages in excess of 600 V from the controller enclosure and operator.

10.6.3 Disconnecting Under Load.

10.6.3.1 Provisions shall be made to prevent the isolating switch from being opened under load.

Controllers in excess of 600 V are generally provided with an isolating switch that is not rated to interrupt current. In such cases the isolating switch must be prevented from being opened or closed when the contactor is closed, by interlocking the contactor with the isolating switch. Although not rated to interrupt current, the isolating switch is generally tested to determine whether it is capable of making and breaking the current of the primary winding of the control power transformer. See 10.6.5.2. Note that the interlocking must be two way to be effective; the isolating switch must not be allowed to open a load to avoid the arcing that would otherwise occur, or to close either on a fault or on the motor locked rotor current to avoid the arcing and possible welding that would otherwise occur if the main contactor is closed by such means as the emergency-run mechanical operator.

10.6.3.2 A load-break disconnecting means shall be permitted to be used in lieu of the isolating switch if the fault closing and interrupting ratings equal or exceed the requirements of the installation.

▶ **FAQ** When is interlocking between the isolating switch and contactor in a controller rated over 600 V not required?

If a switch is used that can safely close on the available fault current and interrupt the locked rotor current of the motor, no interlocking is needed.

10.6.4 Pressure-Actuated Switch Location.
Special precautions shall be taken in locating the pressure-actuated switch called for in 10.5.2.1 to prevent any water leakage from coming in contact with high-voltage components.

Water leakage on high-voltage components in the controller can cause a risk of electric shock for someone coming in contact with the controller enclosure. The pressure-actuated switch

EXHIBIT II.10.13 *Pressure-Actuated Switch.*

> (see Exhibit II.10.13) is normally located at the bottom of the enclosure and in a barriered compartment for controllers rated in excess of 600 V. Some designs put the pressure switch and/or pressure transducers in a separate enclosure with a sealed pass-through into the low voltage compartment of the controller, to minimize the chance of water mixing with electricity. Most controllers use "E" rated fuses for the transformer primary protection.

10.6.5 Low-Voltage Control Circuit.

10.6.5.1 The low-voltage control circuit shall be supplied from the high-voltage source through a stepdown transformer(s) protected by high-voltage fuses in each primary line.

10.6.5.2 The transformer power supply shall be interrupted when the isolating switch is in the open position.

> De-energizing the control circuit power prevents the contactor contacts from being closed by the operating coil of the contactor with the isolating switch open. This requirement supplements the interlocking mentioned in the commentary to 10.6.3.1 to prevent the isolating switch from operating under load. The contacts of the isolating switch are generally used to make and break only the current of the primary winding of the control power transformer or transformers.

10.6.5.3 The secondary of the transformer and control circuitry shall otherwise comply with 10.3.5.

10.6.5.4 One secondary line of the high voltage transformer or transformers shall be grounded unless all control and operator devices are rated for use at the high (primary) voltage.

Low voltage in this case means the control circuit operates at 600 V or less. Operating the control circuit at 600 V or less reduces the risk of electric shock for operators of equipment and maintenance personnel. One secondary line of the transformer is normally grounded as control and operator devices are seldom rated for use at the high (primary) voltage. In the 2010 edition of NFPA 20, 10.6.5.4 was changed to clarify that the requirement applies to high (medium) voltage transformers.

10.6.5.5 Current Transformers. Unless rated at the incoming line voltage, the secondaries of all current transformers used in the high voltage path shall be grounded.

Unless the current transformer is rated for use at the incoming line (medium) voltage, the secondaries of all current transformers should be grounded to reduce the risk of electric shock to anyone who may come in contact with the fire pump controller. See also the commentary to 10.6.5.4.

10.6.6 Indicators on Controller.

10.6.6.1 Specifications for controllers rated in excess of 600 V shall differ from those in 10.4.6.

While it has historically been true that controllers rated in excess of 600 V are subject to different specifications, most modern controllers are subject to basically the same specifications as low voltage controllers.

10.6.6.2 A visible indicator shall be provided to indicate that power is available.

10.6.6.3 The current supply for the visible indicator shall come from the secondary of the control circuit transformer through resistors, if found necessary, or from a small-capacity stepdown transformer, which shall reduce the control transformer secondary voltage to that required for the visible indicator.

The voltage at the visible indicator must be provided through a stepdown transformer or similar means to isolate voltages in excess of 600 V from the controller enclosure and operator. Most modern controllers use low voltage dc indicators, including LEDs and/or display panels similar to or the same as their low voltage equivalent.

10.6.6.4 If the visible indicator is a pilot lamp, it shall be accessible for replacement.

10.6.7 Protection of Personnel from High Voltage.
Necessary provisions shall be made, including such interlocks as might be needed, to protect personnel from accidental contact with high voltage.

Contact with equipment operating in excess of 600 V poses a high risk of electric shock. The equipment must be constructed such that high-voltage circuits are de-energized when the equipment enclosure is opened. Intentional grounding of the power circuits may be necessary to discharge any charge which remains on the motor circuitry, and as extra protection against electrical shock or electrocution. Protection from contact with high-voltage circuits during operation of the equipment is also critical. Such protection includes the necessary means to isolate voltages in excess of 600 V from the controller enclosure and operator. For further information on electrical safety work practices, see NFPA 70E, Electrical Safety in the Workplace.

10.6.8 Disconnecting Means. A contactor in combination with current-limiting motor circuit fuses shall be permitted to be used in lieu of the circuit breaker (disconnecting means) required in 10.4.3.1 if all of the following requirements are met:

> ▶ **FAQ** Why is a contactor permitted to be used in lieu of a circuit breaker in a controller operating in excess of 600 V?

> All modern controllers in excess of 600 V, unlike those rated 600 V or less, may be supplied with fuses that provide short-circuit protection and minimal locked rotor current protection. For this construction, closing on a short circuit will always be performed by the high-voltage contactor that is rated for such duty. While medium voltage circuit breakers do exist, they are expensive and very large in size. Used in combination of properly coordinated fuses and the contactor, these controllers are both practical and more readily available.

(1) Current-limiting motor circuit fuses shall be mounted in the enclosure between the isolating switch and the contactor and shall interrupt the short-circuit current available at the controller input terminals.

> Listed controllers make use of "R" rated fuses for this purpose. These are "silver-sand" fuses. The elements are some number of perforated silver ribbons with the fuse body filled with special sand to absorb the arc and quickly extinguish them. The location of the fuses allows them to be replaced with de-energized fuseholders.

(2) These fuses shall have an adequate interrupting rating to provide the suitability rating *(see 10.1.2.2)* of the controller.

> These controllers are typically rated in MVA (megavolt-ampers), as well as short circuit current rating. A typical rating is 50 Ka (50,000 amps). See Commentary Table II.10.2.

(3) The current-limiting fuses shall be sized to hold 600 percent of the full-load current rating of the motor for at least 100 seconds.

> The sizing of the fuses to hold 600 percent of the full-load current rating of the motor for at least 100 seconds is necessary to prevent opening of the fuses due to motor overload, and to coordinate with the motor locked rotor tripping circuit curve.

(4) A spare set of fuses of the correct rating shall be kept readily available in a compartment or rack within the controller enclosure.

> Having a spare set of fuses is necessary in the event of the opening of one or more fuses, and it minimizes the amount of time the fire pump is out of service.

> ▶ **FAQ** Who supplies the spare fuses that are required by 10.6.8(4)?

> The controller manufacturers are required to supply a means of storing the spare fuses, such as a rack in the low voltage compartment of the controller. However, the fuses themselves are not supplied as standard. This is due to their cost, and also due to the fact that some customers would rather supply the brand of fuses which they use for the other medium voltage equipment on the premises or property.

10.6.9 Locked Rotor Overcurrent Protection.

10.6.9.1 Tripping of the locked rotor overcurrent device required by 10.4.4 shall be permitted to be accomplished by opening the motor contactor coil circuit(s) to drop out the contactor.

> ▶ **FAQ** If high-voltage fuses are provided in lieu of a circuit breaker in a controller operating in excess of 600 V, how are motor overload currents interrupted?

This paragraph is a companion requirement to 10.6.8. Since all listed controllers are provided with fuses in lieu of a circuit breaker, the only means to de-energize the motor under overload conditions is through the opening of the contactor that is rated for such duty.

10.6.9.2 Means shall be provided to restore the controller to normal operation by an external manually reset device.

The requirement in 10.6.9.2 is a companion requirement to 10.6.8, which permits a contactor in combination with current-limiting motor circuit fuses to be used in lieu of the circuit breaker under certain circumstances. With controllers in excess of 600 V, a separate device to reset the locked rotor overcurrent protection must be provided external to the controller enclosure. The typical means is a latching relay or circuit, which is reset with an external pushbutton.

> ▶ **FAQ** Why does NFPA 20 require a "manually reset device"?

Once the motor protection is tripped (off), it has to be kept off to prevent re-cycling power to the motor and re-tripping. This simulates the tripping of the circuit breaker in a low voltage controller, which also must be manually reset.

10.6.10 Emergency-Run Mechanical Control at Controller.

10.6.10.1 The controller shall comply with 10.5.3.2.1 and 10.5.3.2.2, except that the mechanical latching can be automatic.

10.6.10.2 Where the contactor is latched in, the locked rotor overcurrent protection of 10.4.4 shall not be required.

> ▶ **FAQ** When the motor is started by the emergency-run mechanical control, the motor overload protection is rendered inoperative. In this situation, what protects the motor and the controller from overcurrent?

The provision for latching in the contactor in 10.6.10.2 is a companion requirement to 10.6.9.1, which allows opening of the motor contactor coil circuit as a means of tripping the locked rotor OCPD. With currently listed controllers, when the motor contactor is mechanically latched in by the emergency-run mechanical operator, the motor overload protection is rendered inoperative. If an overload condition occurs at this time, the motor can only be de-energized by the opening of the fuses, which is allowed after 100 seconds at 600 percent of the full-load current rating of the motor. See 10.6.8(3).

10.7 Limited Service Controllers

10.7.1 Limitations. Limited service controllers consisting of automatic controllers for across-the-line starting of squirrel-cage motors of 30 hp or less, 600 V or less, shall be permitted to be installed where such use is acceptable to the authority having jurisdiction.

> **▶ FAQ** How is a limited service controller different from a full service fire pump controller?

> The use of limited service controllers was restricted by their maximum allowable horsepower rating (30 hp). Limited service controllers were introduced many years ago as an economical alternative for premises that would otherwise have no fixed fire suppression equipment. Although these controllers have a number of differences compared to regular (often called "full service") fire pump controllers, they do provide a degree of protection against the spread of fire when applied properly. They should not be confused with pressure maintenance (jockey or make-up) pump controllers, which are used to maintain system pressure against leaks.

10.7.2 Requirements. The provisions of Sections 10.1 through 10.5 shall apply, unless specifically addressed in 10.8.2.1 through 10.8.2.3.

10.7.2.1 In lieu of 10.1.2.5.1, each controller shall be marked "Limited Service Controller" and shall show plainly the name of the manufacturer, the identifying designation, the maximum operating pressure, the enclosure type designation, and the complete electrical rating.

> Previous editions of NFPA 20 allowed the use of an inverse time nonadjustable circuit breaker having a standard rating between 150 percent and 250 percent of the motor full-load current in lieu of 10.4.3.3.1(2) and 10.4.4. The breakers used were typically thermal-magnetic circuit breakers. The 2012 edition of this standard eliminates this exemption; limited service controllers must now have the same circuit breaker and overcurrent characteristics as do "full service" controllers. As a result, the main difference between limited service units and full service controllers is the absence of the separate isolating switch. This change was made to better match the motor protection provided with full service controllers. The motor protection characteristics (current-time curve) of limited service controllers had wider variation than full service units.
>
> In this edition, two new marking items are added; namely, the "maximum operating pressure" and the "enclosure type designation." Although it was common practice for these controllers to be marked by the manufacturer, this addition ensures that the markings will be provided.

10.7.2.2 The controller shall have a short-circuit current rating not less than 10,000 A.

10.7.2.3 The manually operated isolating switch specified in 10.4.2 shall not be required.

10.8* Power Transfer for Alternate Power Supply

A.10.8 Typical fire pump controller and transfer switch arrangements are shown in Figure A.10.8. Other configurations can also be acceptable.

> Figure A.10.8 was revised for the 2012 edition; one of the changes was to add a schematic diagram showing the emergency source of the combination fire pump controller and power transfer switch unit being fed by an upstream transfer switch. This combination unit, often used in hospitals and high-rise buildings in U.S. cities, is commonly referred to as an "emergency bus" installation, since the emergency bus is normally powered by a utility supply. This feed may be from the buildings normal utility supply or from a separate utility supply depending on the project specifications, local codes or the AHJ. As a result, the combination fire pump controller emergency side circuitry will normally be energized, as if it was a "dual utility" application. However, the generator's engine start contacts should still be wired to the generator, to start it upon failure of the normal power source independent of any other such start signals on

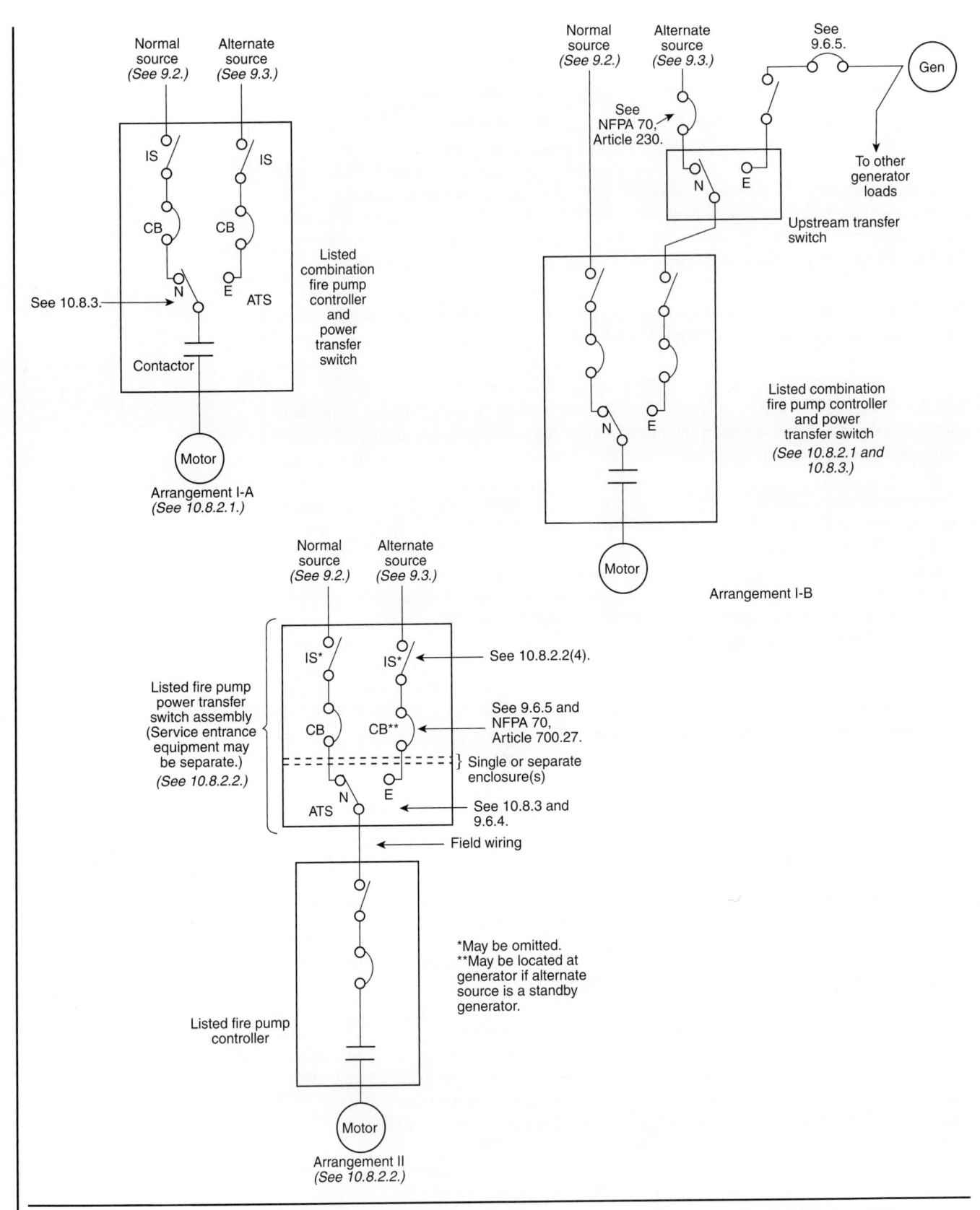

270 Part II • NFPA 20 • Chapter 10 Electric-Drive Controllers and Accessories

FIGURE A.10.8 *Typical Fire Pump Controller and Transfer Switch Arrangements.*

the premises. The change to Item #3 revises the "Arrangement II" schematic diagram to more closely reflect a listed fire pump power transfer switch, including optional isolating switches and emergency (alternate) side circuit breaker.

10.8.1 General.

10.8.1.1 Where required by the authority having jurisdiction or to meet the requirements of 9.3.2 where an on-site electrical power transfer device is used for power source selection, such switch shall comply with the provisions of Section 10.8 as well as Sections 10.1, 10.2, 10.3 and 10.4.1.

A transfer switch is used when the normal power source is not reliable (see Exhibit II.10.14). The function of the transfer switch is to transfer the fire pump motor to an alternate source of power when the power from the normal source is interrupted.

10.8.1.2 Manual transfer switches shall not be used to transfer power between the normal supply and the alternate supply to the fire pump controller.

All transfer switches used in fire pump applications must be of the automatic type.

10.8.1.3 No remote device(s) shall be installed that will prevent automatic operation of the transfer switch.

The installation of any device (remote sensing or other) that inhibits the operation of the transfer switch reduces the reliability of the transfer operation from the normal to the alternate source of power.

EXHIBIT II.10.14 *Transfer Switch.*

Stationary Fire Pumps Handbook 2013

10.8.2* Fire Pump Controller and Transfer Switch Arrangements.

A.10.8.2 The compartmentalization or separation is to prevent propagation of a fault in one compartment to the source in the other compartment.

10.8.2.1 Arrangement I (Listed Combination Fire Pump Controller and Power Transfer Switch).

10.8.2.1.1 Self-Contained Power Switching Assembly. Where the power transfer switch consists of a self-contained power switching assembly, such assembly shall be housed in a barriered compartment of the fire pump controller or in a separate enclosure attached to the controller and marked "Fire Pump Power Transfer Switch."

> This type of arrangement is listed as one assembly, a fire pump controller, although the enclosure contains both a fire pump controller and a transfer switch. The compartment or enclosure housing the fire pump controller is marked "Fire Pump Controller," and the portion containing the transfer switch is marked accordingly.

10.8.2.1.2 Isolating Switch.

10.8.2.1.2.1 An isolating switch, complying with 10.4.2, located within the power transfer switch enclosure or compartment shall be provided ahead of the alternate input terminals of the transfer switch.

> The isolating switch connected to the alternate power source terminals of the transfer switch provides for disconnection of the alternate power source from the transfer switch and components in the fire pump controller for maintenance and repair. This alternate power could be from one of the following sources:
>
> 1. a standby generator
> 2. an emergency bus
> 3. a second utility
>
> When the generator engine start terminals are wired, this switch should always be opened ahead of either the normal source circuit breaker or the normal source isolating switch, to keep the controller from signaling the generator to start.

10.8.2.1.2.2 The isolating switch shall be suitable for the available short circuit of the alternate source.

> Transfer switches may have different short-circuit ratings on the normal and alternate sources. This situation is typical when the transfer switch alternate source is a premises-owned generator. The normal source is usually a utility source, and the available short-circuit current at the normal source terminals of the transfer switch would be somewhat or significantly higher than the available short-circuit current at the alternate source terminals.

10.8.2.1.3 Circuit Breaker. The transfer switch emergency side shall be provided with a circuit breaker complying with 10.4.3 and 10.4.4.

> The 2010 edition of NFPA 20 added the requirement for the transfer switch emergency source side to be equipped with a circuit breaker as an isolation switch. Previous editions of NFPA 20 included a requirement for this circuit breaker if the alternate source was supplied by a

generator with a capacity that exceeded 225 percent of the fire pump motor's rated full-load current, or by an emergency bus or by a second utility source.

There have been numerous cases of units being supplied by a second utility, but this was unknown to the manufacturer, since the units were ordered as standard single circuit breaker types. Note that the second (emergency source) circuit breaker used to be a fairly expensive option. Several controllers have been severely damaged by installation faults when they were connected to a second utility source, but missing the necessary second circuit breaker. See also the commentary to 10.4.2.1.3 regarding short circuit ratings.

10.8.2.1.4 Cautionary Marking. The fire pump controller and transfer switch (see 10.8.2.1) shall each have a cautionary marking to indicate that the isolation switch for both the controller and the transfer switch is opened before servicing the controller, transfer switch, or motor.

▶ **FAQ** Why is an isolating switch required for both the normal and alternate input terminals of the transfer switch?

When a transfer switch is used, both the normal and alternate source must be disconnected for servicing the controller, its circuit breakers, the transfer switch, or motor. The disconnection is accomplished through the isolating switches required for both normal and alternate sources. As above, the emergency source isolating switch should be opened first, to avoid initiating an unintended start signal being sent to the generator's engine by the controller.

10.8.2.1.5 Turning off the normal source isolating switch or the normal source circuit breaker shall not inhibit the transfer switch from operating as required by 10.8.3.6.1 through 10.8.3.6.4.

10.8.2.2 Arrangement II (Individually Listed Fire Pump Controller and Power Transfer Switch). The following shall be provided:

(1) A fire pump controller power transfer switch complying with Sections 9.6 and 10.8 and a fire pump controller shall be provided.

▶ **FAQ** What is meant by the term *separately mounted fire pump transfer switch*?

The "power transfer switch" described in this arrangement is commonly referred to as a *separately mounted fire pump transfer switch* and is not part of the fire pump controller. These are installed upstream electrically from a "single source" (non-transfer switch combination) controller.

(2) The transfer switch overcurrent protection for both the normal and alternate sources shall comply with 9.2.3.4 or 9.2.3.4.1.

This clause used to require an isolating switch – or service disconnect where required – ahead of the normal input terminals of the transfer switch. This was changed to align with the revisions to "Arrangement I" by adding the requirement for overcurrent protection in both the normal and the emergency (alternate) sources. Currently, these may or may not be part of the power transfer switch unit.

(3) The transfer switch overcurrent protection shall be selected or set to indefinitely carry the locked rotor current of the fire pump motor where the alternate source is supplied by a second utility.

Stationary Fire Pumps Handbook   2013

> **▶ FAQ** How is overcurrent protection provided for a separately mounted fire pump transfer switch?

The overcurrent protection specified in 10.8.2.2(3) is the same as that required for a transfer switch required by the *NEC*. Because it is not the overcurrent protection provided by the circuit breaker in the fire pump controller, its selection is critical in that it must not trip unless the circuit breaker in the fire pump controller trips.

(4) An isolating switch ahead of the alternate source input terminals of the transfer switch shall meet the following requirements:
   (a) The isolating switch shall be lockable in the on position.
   (b) A placard shall be externally installed on the isolating switch stating "Fire Pump Isolating Switch," with letters at least 1 in. (25 mm) in height.
   (c) A placard shall be placed adjacent to the fire pump controller stating the location of the isolating switch and the location of the key (if the isolating switch is locked).
   (d) The isolating switch shall be supervised to indicate when it is not closed, by one of the following methods:
      i. Central station, proprietary, or remote station signal service
      ii. Local signaling service that will cause the sounding of an audible signal at a constantly attended point
      iii. Locking the isolating switch closed
      iv. Sealing of isolating switches and approved weekly recorded inspections where isolating switches are located within fenced enclosures or in buildings under the control of the owner
   (e) This supervision shall operate an audible and visible signal on the transfer switch and permit monitoring at a remote point where required.

> **▶ FAQ** Are special requirements given for the isolating switch ahead of the alternate source input terminals on a separately mounted fire pump transfer switch?

The requirements in 10.8.2.2(4) are similar to those applied to the remote disconnecting means permitted in 9.2.3. These additional requirements are necessary for the alternate source isolating switch provided for the separately mounted transfer switch, as this isolating switch may be remote from the fire pump controller and not readily accessible.

10.8.2.3 Transfer Switch. Each fire pump shall have its own dedicated transfer switch(es) where a transfer switch(es) is required.

> **▶ FAQ** What is meant by a "dedicated transfer switch" in a fire pump circuit?

A transfer switch is not allowed to supply more than one fire pump unless it is through one or more downstream transfer switches — that is, a fire pump circuit may have additional transfer switches upstream from the dedicated transfer switch, and those transfer switches may supply other fire pumps and building loads. Only the dedicated transfer switch is subject to the requirements of NFPA 20. All other transfer switches upstream of the dedicated transfer switch are outside the scope of NFPA 20.

> **FAQ** What are the differences between a listed transfer switch and a transfer switch listed for fire pump service?

Not all listed transfer switches are suitable for use in a fire pump branch circuit. Only those that are listed and marked as being suitable for fire pump service are acceptable in this application. Additional requirements are applied to transfer switches that are intended to be used in a fire pump application. Those additional requirements are specified in 10.8.3.3 to 10.8.3.14.

10.8.3 Power Transfer Switch Requirements.

10.8.3.1 Listing. The power transfer switch shall be specifically listed for fire pump service.

10.8.3.2 Suitability. The power transfer switch shall be suitable for the available short-circuit currents at the transfer switch normal and alternate input terminals.

The suitability requirement in 10.8.3.2 is crucial for both personnel safety and fire protection reliability. Since the upstream protection, per 10.8.2.2(3), must hold a motor locked rotor indefinitely, this affects the size of the transfer switch itself. The minimum size (rating) of the transfer switch is given in 10.8.3.4.1 (horsepower) or 10.8.4.2 (115 percent of motor FLA). However, the listing requirements for short circuit current rating of transfer switches is a function of the overcurrent protection (OCP) installed ahead of both sides (inputs) to the switch. Since the OCP must be sized to at least 600 percent of motor FLA, the switch must almost always be increased in size to coordinate with this OCP size (rating).

10.8.3.3 Electrically Operated and Mechanically Held. The power transfer switch shall be electrically operated and mechanically held.

This eliminates contactor type transfer switches or schemes that do not include a mechanical latching mechanism.

10.8.3.4 Horsepower or Ampere Rating.

As stated in the commentary to 10.8.3.2, coordinating with the minimum 600 percent FLA OCP may require a larger transfer switch than the minimum size mandated under this section.

10.8.3.4.1 Where rated in horsepower, the power transfer switch shall have a horsepower rating at least equal to the motor horsepower.

10.8.3.4.2 Where rated in amperes, the power transfer switch shall have an ampere rating not less than 115 percent of the motor full-load current and also be suitable for switching the motor locked rotor current.

10.8.3.5 Manual Means of Operation.

The safest means to manually transfer power in a pump controller is to always de-energize both sides of the transfer switch before opening the door(s) to manually transfer the switch, then close the door(s) before re-energizing the switch. Depending on which source is available and what the voltage is, the transfer switch could immediately switch back to its former position, which is why the doors should first be closed.

10.8.3.5.1 A means for safe manual (nonelectrical) operation of the power transfer switch shall be provided.

10.8.3.5.2 This manual means shall not be required to be externally operable.

10.8.3.6 Undervoltage- and Phase-Sensing Devices.

10.8.3.6.1 The power transfer switch shall be provided with undervoltage-sensing devices to monitor all ungrounded lines of the normal power source.

10.8.3.6.2 Where the voltage on any phase of the normal source falls below 85 percent of motor-rated voltage, the power transfer switch shall automatically initiate starting of the standby generator, if provided and not running, and initiate transfer to the alternate source.

> The phrase "of the normal source" replaced the requirement for sensing at the motor contactor to accommodate Arrangement II type installations. The extant 85 percent value allows adequate voltage for the motor while avoiding excessive running of an on-site generator. Note that the generator must be started and run anytime normal power is lost, whether there is a demand for the pump or not, because the controller needs its nominal power in order to monitor the pressure sensor and any other starting demand signals.

10.8.3.6.3 Where the voltage on all phases of the normal source returns to within acceptable limits, the fire pump controller shall be permitted to be retransferred to the normal source.

10.8.3.6.4 Phase reversal of the normal source power *(see 10.4.6.2)* shall cause a simulated normal source power failure upon sensing phase reversal.

> This requirement also applies to an Arrangement II type installation, and is often overlooked.

10.8.3.6.5 For Arrangement II Units, the sensing of voltage described in 10.8.3.6.2 shall be permitted at the input to the power transfer switch instead of at the load terminals of the fire pump controller circuit breaker.

> This paragraph was rewritten to clarify and simplify it. Simply put, the voltage sensing for Arrangement II transfer switches may occur in or above the transfer switches, rather than within the downstream fire pump controller.

10.8.3.7 Voltage- and Frequency-Sensing Devices. Unless the requirements of 10.8.3.7.3 are met, the requirements of 10.8.3.7.1 and 10.8.3.7.2 shall apply.

10.8.3.7.1 Voltage- and frequency-sensing devices shall be provided to monitor at least one ungrounded conductor of the alternate power source.

> Note that this requirement for frequency sensing devices also applies to installations where the emergency (alternate) source is provided by an emergency bus, since the transfer switch still has to signal the generator engine to start, regardless of any other start signals that may be wired to the generator. The presumption is that these other start signals could be inoperative, defective, or disabled, and the fire pump system must stand on its own.

10.8.3.7.2 Transfer to the alternate source shall be inhibited until there is adequate voltage and frequency to serve the fire pump load.

> The ability to transfer to the alternate source is important for at least the following two reasons:
>
> 1. The generator needs to come up to speed and voltage before the fire pump motor load is applied. In most installations, the fire pump or pumps are the largest load on the generator.

> 2. The voltage and frequency need to be reasonably close to the nominal (expected) values, to provide adequate starting torque and not draw excessive running current due to inadequate voltage. Also, the pump running speed is directly proportional to the frequency supplied by the generator.

10.8.3.7.3 Where the fire pump controller is marked to indicate that the alternate source is provided by a second utility power source, the requirements of 10.8.3.7.1 and 10.8.3.7.2 shall not apply, and undervoltage-sensing devices shall monitor all ungrounded conductors in lieu of a frequency-sensing device.

10.8.3.8 Visible Indicators. Two visible indicators shall be provided to externally indicate the power source to which the fire pump controller is connected.

10.8.3.9 Retransfer.

10.8.3.9.1 Means shall be provided to delay retransfer from the alternate power source to the normal source until the normal source is stabilized.

> ▶ **FAQ** Why is the delay of retransfer from the alternate power source to the normal source necessary until the normal source is stabilized?

> Damage could occur to the fire pump motor if the voltage and frequency to serve the fire pump load is not adequate. The delay in retransfer allows the normal source to achieve rated and consistent voltage and frequency.

10.8.3.9.2 This time delay shall be automatically bypassed if the alternate source fails.

> If the alternate source fails during the period in which the transfer switch is attempting to return to the normal power source, it is important to continue driving the fire pump even if the normal power source does not have adequate voltage and frequency to serve the fire pump load.

10.8.3.10 In-Rush Currents. Means shall be provided to prevent higher than normal in-rush currents when transferring the fire pump motor from one source to the other.

10.8.3.10.1 The use of an "in-phase monitor" or an intentional delay via an open neutral position of the transfer switch to comply with the requirements of 10.8.3.10 shall be prohibited.

> A new second sentence was added to this section on in-rush currents, to prohibit the use of "in-phase monitors" to achieve the required in-rush current control. The following are two reasons for this prohibition:
>   No such monitor can keep up with the rapid spin-down of a centrifugal pump due to its very low inertia and its minimum horsepower demand.
>   The resulting time delay simply prolongs the time when there is no fire water. In most cases, the in-phase monitor will time out and unlock the action of the switch at some point, but this is only a supposition; the situation becomes even more complex when multiple generators are involved.

10.8.3.11* Overcurrent Protection. The power transfer switch shall not have short circuit or overcurrent protection as part of the switching mechanism of the transfer switch.

Stationary Fire Pumps Handbook 2013

> **FAQ** Why are fire pump transfer switches not allowed to have integral overcurrent protection?

If an overcurrent condition exists in the fire pump branch circuit, only the circuit breaker in the fire pump controller is permitted to operate to interrupt the overcurrent. Because of the special characteristics of this circuit breaker per 10.4.3.3 and associated locked rotor overcurrent protection per 10.4.4, any short-circuit or overcurrent protection integral to the transfer switch could render the fire pump inoperable in conditions where its operation is critical and it otherwise should be allowed to operate.

A.10.8.3.11 Internal protection refers to any tripping elements contained within the switching mechanism of the transfer switch. This is to prevent a switching mechanism from inhibiting transfer of power.

10.8.3.12 Additional Requirements. The following shall be provided:

(1) A device to delay starting of the alternate source generator to prevent nuisance starting in the event of momentary dips and interruptions of the normal source
(2) A circuit loop to the alternate source generator whereby either the opening or closing of the circuit will start the alternate source generator (when commanded by the power transfer switch) *(see 10.8.3.6)*
(3) A means to prevent sending of the signal for starting of the alternate source generator when commanded by the power transfer switch, if the alternate isolating switch or the alternate circuit breaker is in the open or tripped position

10.8.3.12.1 The alternate isolating switch and the alternate circuit breaker shall be monitored to indicate when one of them is in the open or tripped position, as specified in 10.8.3.12(3).

If either of these devices is open, there is no emergency power available. Previous editions of NFPA 20 included an exception to the requirement in 10.8.3.12.1 that monitoring of both the alternate isolating switch and alternate circuit breaker was not required if they were interlocked. In the 2010 edition of NFPA 20 this exception was deleted, since either or both the switch and the breaker may be tripped instead of simply being turned off (opened). Hence, there would be no warning of either a tripped or a manually opened breaker on the alternate side without this monitoring.

10.8.3.12.2 Supervision shall operate an audible and visible signal on the fire pump controller/automatic transfer switch combination and permit monitoring at a remote point where required.

10.8.3.13 Momentary Test Switch. A momentary test switch, externally operable, shall be provided on the enclosure that will simulate a normal power source failure.

Note that some of the pushbutton switches referred to in this section need to be held for a few seconds until the transfer occurs.

10.8.3.14 Remote Indication. Auxiliary open or closed contacts mechanically operated by the fire pump power transfer switch mechanism shall be provided for remote indication in accordance with 10.4.8.

The remote indication requirement in 10.4.8 states that where two sources of power are supplied to meet the requirements of 9.3.2, controllers with transfer switches or separately mounted power transfer switches must be equipped with contacts (open or closed) to operate circuits for the condition where the alternate source is the source supplying power to the controller.

10.9 Controllers for Additive Pump Motors

▶ FAQ Do the requirements of Section 10.9 apply to controllers for foam concentrate pumps?

In previous editions of NFPA 20, this section was titled "Controllers for Foam Concentrate Pump Motors." The title was revised to clarify that the requirements apply to all controllers used in systems where additives are provided to aid in fire suppression.

10.9.1 Control Equipment. Controllers for additive pump motors shall comply with the requirements of Sections 10.1 through 10.5 (and Section 10.8, where required) unless specifically addressed in 10.9.2 through 10.9.5.

10.9.2 Automatic Starting. In lieu of the pressure-actuated switch described in 10.5.2.1, automatic starting shall be capable of being accomplished by the automatic opening of a closed circuit loop containing this fire protection equipment.

Controllers for additive pump motors are generally started by remote means and conditions other than a drop in water pressure due to a sprinkler system activation. The closed circuit loop ensures that the pump will start even if a failure of the starting circuit and/or starting circuit components exists.

10.9.3 Methods of Stopping.

10.9.3.1 Manual shutdown shall be provided.

10.9.3.2 Automatic shutdown shall not be permitted.

In fire suppression systems where an additive is employed, manual shutdown is the required method of stopping the pump. No automatic detection and signaling means are available to determine whether a need for this type of fire suppression system to operate still exists. Automatic shutdown of fire suppression systems is not permitted because malfunction of an automatic shutdown function can occur.

10.9.4 Lockout.

10.9.4.1 Where required, the controller shall contain a lockout feature where used in a duty-standby application.

10.9.4.2 Where supplied, this lockout shall be indicated by a visible indicator and provisions for annunciating the condition at a remote location.

▶ FAQ Is there a means available to stop the pump in controllers for additive pump motors should the additive tank run dry?

The lockout feature allows for stopping the pump in the event the source of additive has been depleted. Generally, the operation of the fire pump in the event that the additive is no longer available is undesirable. For example, attempting to suppress a chemical-based fire with water only could result in spreading or increasing the size of the fire.

10.9.5 Marking. The controller shall be marked "Additive Pump Controller."

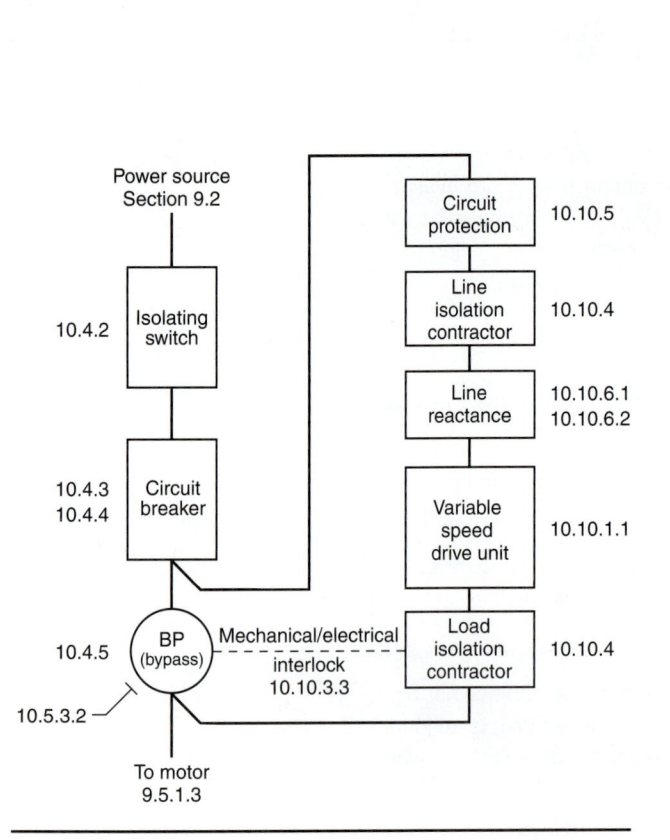

10.10* Controllers with Variable Speed Pressure Limiting Control or Variable Speed Suction Limiting Control

Section 10.10 was added to NFPA 20 in the 2007 edition. These added requirements reflect the latest technology in fire suppression systems. In the 2010 edition of NFPA 20, this section was expanded to include variable speed suction-limiting control systems. These speed control systems are used to maintain a minimum positive suction pressure at the pump inlet by reducing the pump driver speed from rated speed while monitoring pressure in the suction piping through a sensing line. These controllers are, in reality, two controllers; namely, a full service full speed controller (the bypass path) and a variable speed controller, both of which are rated for continuous duty.

A.10.10 See Figure A.10.10.

10.10.1 Control Equipment.

See Exhibit II.10.15 for a typical variable speed pressure limiting fire pump controller with key components identified.

10.10.1.1 Controllers equipped with variable speed pressure limiting control or variable speed suction limiting control shall comply with the requirements of Chapter 10, except as provided in 10.10.1 through 10.10.11.

FIGURE A.10.10 *Variable Speed Pressure Limiting Control.*

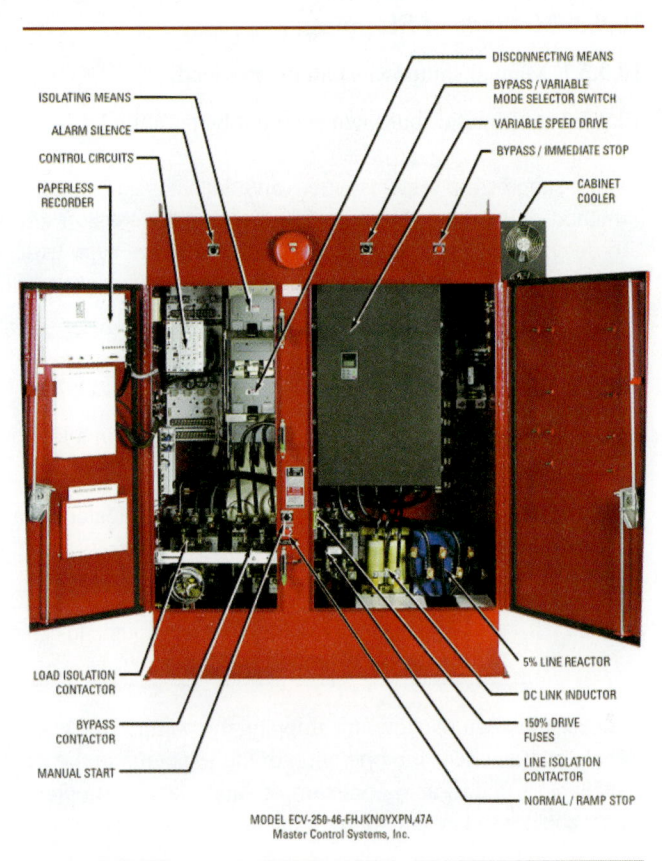

EXHIBIT II.10.15 *Variable Speed Pressure Limiting Fire Pump Controller. (Photograph courtesy of Master Control Systems, Inc.)*

Section 10.10 • Controllers with Variable Speed Pressure Limiting Control or Variable Speed Suction Limiting Control   **281**

> ▶ **FAQ** When are controllers equipped with variable speed pressure limiting control used?
>
> Controllers equipped with variable speed pressure limiting control are used in systems where overpressurizing the piping, fittings, connection points, and/or sprinklers could result in damage to one or more of these components. See Exhibit II.10.16 for a controller equipped with variable speed pressure limiting control.

10.10.1.2 Controllers with variable speed pressure limiting control or variable speed suction limiting control shall be listed for fire service.

10.10.1.3 The variable speed pressure limiting control or variable speed suction limiting control shall have a horsepower rating at least equal to the motor horsepower or, where rated in amperes, shall have an ampere rating not less than the motor full-load current.

> Some variable speed drives specify their ratings in amperes. Therefore, an allowance is made for those devices. Note that these variable speed units are variously known as variable speed drives (VFD), adjustable speed drives (ASD), and variations such as AFD and VSD. Most manufacturers make use of listed units in their variable speed controllers.

10.10.2 Additional Marking. In addition to the markings required in 10.1.2.5.1, the controller shall be marked with the maximum ambient temperature rating.

EXHIBIT II.10.16 *Controller Equipped with Variable Speed Pressure Limiting Control. (Photograph courtesy of Master Control Systems, Inc.)*

Stationary Fire Pumps Handbook 2013

> Marking with ambient temperature is vitally important, since all VFDs give off heat, and all will shut down when their ambient temperature (the internal temperature of the controller) is exceeded. The common maximum ambient temperature rating is +30°C (86°F). Others are rated at +40°C (104°F) with optional ratings available to +50°C (122°F).

10.10.3* Bypass Operation.

A.10.10.3 The bypass path constitutes all of the characteristics of a non-variable speed fire pump controller.

> ▶ **FAQ** Why are controllers equipped with variable speed pressure limiting control provided with a bypass path?
>
> In the event of failure of the variable speed pressure limiting control, a means to energize and operate the pump motor needs to be provided, and the bypass path serves that purpose. As such, all components in the bypass path must comply with all of the starting, stopping, and control requirements of a standard fire pump controller.

10.10.3.1* Upon failure of the variable speed pressure limiting control to keep the system pressure at or above the set pressure of the variable speed pressure limiting control system, the controller shall bypass and isolate the variable speed pressure limiting control system and operate the pump at rated speed.

> In the event of failure of the variable speed pressure limiting control, the controller is expected to operate as a standard across-the-line fire pump controller. In this mode the possibility that the system could be overpressurized exists. However, the continued operation of the fire pump to suppress the fire is critical even if water is discharged through the main relief valve, when supplied. These are mandatory whenever the pump discharge at churn – with the highest anticipated suction pressure and at full speed – can exceed the pressure rating of the system (system components). See 4.18.1.3. Most designs assume that manual stop mode will be used under this condition, even if the controller is otherwise arranged for automatic stop. Manual stop is assumed because the source of the failure cannot be anticipated, and restarting the motor for a third time is considered unduly risky. This full speed mode is normally called the *bypass mode*.

A.10.10.3.1 The bypass contactor should be energized only when there is a pump demand to run and the variable speed pressure limiting control or variable speed suction limiting control is in the fault condition.

> The bypass contactor should not be used to test the system until and unless the full discharge pressure has been measured with the pump discharge control valve closed and – for pumps which can exceed the system pressure ratings – the main relief valve settings and operation has been verified. Note that the pump itself is always tested in the full speed (bypass) mode. This often requires blocking the relief valve closed, meaning that the pump discharge (system) valve must also be closed. As a result, the limited amount of piping still pressurized needs to be rated for this full speed pump churn (shutoff) pressure.

10.10.3.1.1 Low Pressure. If the system pressure remains below the set pressure for more than 15 seconds, the bypass operation shall occur.

10.10.3.1.2* Drive Not Operational. If the variable speed drive indicates that it is not operational within 5 seconds, the bypass operation shall occur.

> All VFDs used in this service have either a "drive ready" or a "drive not ready" signal. Some have additional drive failure signals. Some designs monitor all of these signals and switch to the bypass mode if any single one of these events is abnormal after the five second trial period. Note that most VFDs undergo a host of self tests, and some also test the condition of the motor.

A.10.10.3.1.2 Variable speed drive units (VSDs) should have a positive means of indicating that the drive is operational within a few seconds after power application. If the VSD fails, there is no need to wait for the low pressure bypass time of 10.10.3.1.1.

> The term a few seconds was chosen as a reasonable amount of time to allow the variable speed pressure limiting control to either begin operation or signal a fault. If the variable speed pressure limiting control is not operational within a few seconds, allowing further time would only delay starting the pump in the full speed mode.

10.10.3.1.3* Means shall be provided to prevent higher than normal in-rush currents when transferring the fire pump motor from the variable speed mode to the bypass mode.

> A restart delay of two to three seconds is the most common method used for this purpose, which is sufficient for the types of motors used in this service. See the commentary to 10.4.3.3.1(8) for more on this.

A.10.10.3.1.3 A motor running at a reduced frequency cannot be connected immediately to a source at line frequency without creating high transient currents that can cause tripping of the fire pump circuit breaker. It is also important to take extra care not to connect (back feed) line frequency power to the VSD since this will damage the VSD and, more important, can cause the fire pump circuit breaker to trip, which takes the pump out of service.

> Reducing the speed of the fire pump motor or stopping it altogether before transferring the fire pump motor from the variable speed mode to the bypass mode is essential. The speed and phase angle differences cannot be easily determined. A time delay is normally incorporated to allow the motor magnetic flux, and hence, it's back EMF to decay to safe values. This time delay is fixed by the manufacturer in some designs to avoid altering it or defeating it in the field.

10.10.3.2 When the variable speed pressure limiting control is bypassed, the unit shall remain bypassed until manually restored.

> If a variable speed fire pump controller is running in the bypass mode, it is doing so either by way of manual intervention (mode switch) or by way of automatic switching to bypass upon failure of the variable frequency drive to maintain the required pressure. If manually set to bypass, the controller should stay in that mode. If switched automatically, it should also stay there, since switching back to automatic will likely result in cycling and needless interruption of fire water. Allowing automatic stopping in either case can leave a property or premises unprotected.

10.10.3.3 The bypass contactors shall be operable using the emergency-run handle or lever defined in 10.5.3.2.

> **▶ FAQ** What performs the function of the emergency-run mechanical control in a variable speed pressure limiting fire pump controller?

In the bypass mode, the controller acts as a "full service" full speed controller and must meet all applicable requirements of one. As such, the bypass contactor(s) need to be operated in the same manner as that provided for the emergency-run mechanical control. This operation is accomplished through the emergency-run handle or lever specified in 10.5.3.2.

10.10.3.4 Automatic Shutdown. When the variable speed pressure limiting control is bypassed, automatic shutdown of the controller shall be as permitted by 10.5.4.2.

This clause was added to the 2013 edition to accommodate those controller designs that provide for automatic stop in the bypass mode.

10.10.3.5 When the manual selection means required in 10.10.7.3 is used to initiate a switch over from variable speed to bypass mode, if the pump is running in the variable speed mode and none of the conditions in 10.10.3 that require the controller to initiate the bypass operation exist, the controller shall be arranged to provide a restart delay to allow the motor to be de-energized before it is re-energized in the bypass mode.

This new clause was added to cover the case of a controller being switched manually from the variable speed mode to the bypass (full speed, full voltage) mode. See the commentary to 10.10.3.1.3 for more.

10.10.4 Isolation.

10.10.4.1 The variable speed drive shall be line and load isolated when not in operation.

The variable speed pressure limiting control is generally a solid-state device that is sensitive to electrical impulses and voltage surges from the connected source(s). Therefore, isolating them from the connected source when not in use is important. This feature extends the life of the components, minimizes their failure, and increases their reliability. Further, all such drives have a capacitor overvoltage shutdown mode. With many VFD designs, this shutdown can be triggered by line transients or surges. This shutdown mode requires manual resetting (such as cycling power to the controller) and forces the controller into the bypass mode when started to supply fire water.

10.10.4.2 The variable speed drive load isolation contactor and the bypass contactor shall be mechanically and electrically interlocked to prevent simultaneous closure.

This mechanical interlocking is important to prevent inadvertent feeding VFD output voltage to the ac power supply, and vice versa. Either will result in damage and a tripped circuit breaker, further resulting in no fire protection. Although these are always electrically interlocked, the presumption is that since a failure in the variable speed path is of unknown origin, the electrical interlocking may also be defective.

10.10.5* Circuit Protection.

A.10.10.5 The intent is to prevent tripping of the fire pump controller circuit breaker due to a variable speed drive failure and thus maintain the integrity of the bypass circuit.

Section 10.10 • Controllers with Variable Speed Pressure Limiting Control or Variable Speed Suction Limiting Control

10.10.5.1 Separate variable speed drive circuit protection shall be provided between the line side of the variable speed drive and the load side of the circuit breaker required in 10.4.3.

> Refer to Figure A.10.10 for the location of this circuit protection. This overcurrent circuit protection could be a circuit breaker; most, if not all, controllers employ fuses. While this OCP can provide protection for the VFD, the main reason is to prevent a fault in the VFD (short circuit or excessive current draw) from tripping the fire pump controller circuit breaker, because a trip would disable the fire pump. The requirement in 10.10.5.2 and it's associated Annex A text spell this out definitively.

10.10.5.2 The circuit protection required in 10.10.5.1 shall be coordinated such that the circuit breaker in 10.4.3 does not trip due to a fault condition in the variable speed circuitry.

> ▶ **FAQ** Are there special requirements for the overcurrent protection of the variable speed pressure limiting control?
>
> The selection of the overcurrent circuit protective devices for the VFD is critical, in that the device should operate to protect the VFD but not allow high enough current for long enough to trip the fire pump circuit breaker. Generally, fast-acting or semiconductor fuses can be selected which satisfy this condition. The curves of the fuses must be analyzed in conjunction with the controller's circuit breaker curve. The fuses must be large enough to handle the VFD overload compatibility (150 percent of FLA in certain designs) and meet the VFD manufacturer's recommendations and minimums where possible. However, the coordination with the controller's circuit breaker is the primary coordination and selection priority.

10.10.6 Power Quality.

10.10.6.1 Power quality correction equipment shall be located in the variable speed circuit.

10.10.6.1.1 As a minimum, 5 percent line reactance shall be provided.

10.10.6.2 Coordination shall not be required where the system voltage does not exceed 480 V and cable lengths between the motor and controller do not exceed 100 ft (30.5 m) *(see 10.10.6.3)*.

> ▶ **FAQ** Is cable length an issue when a variable speed pressure limiting control is used?
>
> Voltage transients are generally not a problem when the supply voltage does not exceed 480 V and cable lengths between the motor and controller do not exceed 100 ft (30.5 m). Otherwise the requirements of 10.10.6.3 apply.

10.10.6.3* Where higher system voltages or longer cable lengths exist, the cable length and motor requirements shall be coordinated.

> All modern VFDs supply a modulated waveform to the motor, typically at a frequency of 2.0 kHz or higher, which creates voltage spikes in the motor windings due to their inductance. This is one of the main ways in which "inverter duty motors" differ from standard motors. See the commentary to 9.5.1.4.2 for more on the differences between inverter duty and standard motors. Both higher voltages and longer motor lead (cable) lengths tend to increase the magnitude of these voltage spikes. The use of either an output (motor) filter reactor, or an output filter module can accomplish the coordination required to protect the motor.

Stationary Fire Pumps Handbook 2013

A.10.10.6.3 As the motor cable length between the controller and motor increases, the VSD high frequency switching voltage transients at the motor will increase. To prevent the transients from exceeding the motor insulation ratings, the motor manufacturer's recommended cable lengths must be followed.

10.10.7 Local Control.

10.10.7.1 All control devices required to keep the controller in automatic operation shall be within lockable enclosures.

> Restricting access to setting or adjustment of control devices required to keep the controller in automatic operation is necessary, because in the event of a fire no automatic means to start and run the fire pump are available. Modern VFD units are controlled by microprocessors or specialty devices known as digital signal processors. As a result, modern VFDs have numerous programmable parameters. All of these need to be in the range that results in successful and robust motor operation and pressure control. If any of these parameters is set wrong, an inoperative or incorrectly operating variable speed control occurs. This is one reason locking the unit against tampering is vital. Some controllers also include a means for downloading and storing the collection of parameters with the ability to load them back into the VFD.

10.10.7.2 Except as provided in 10.10.7.2.1, the variable speed pressure sensing element connected in accordance with 10.5.2.1.7.6 shall only be used to control the variable speed drive.

> The requirement in 10.10.7.2 was revised to add the exception of 10.10.7.2.1. The primary intent of 10.10.7.2 is to reserve the pressure transducer used for the VFD process control loop to be used for that purpose only, to separate the controller's start/stop control pressure transducer from the speed control transducer, thereby minimizing damage to one by inhibiting the function of the other. Since the controller supplies the power to the transducers, inadequate protection from line transients can and will damage the transducer. In a specific scenario where the VFD transducer fails or is damaged, an ineffective variable speed function can result; however, the other transducer would sense the absence of sufficient pressure and initiate the changeover (automatic switchover) to bypass full speed mode. If that other transducer fails, either the unit wants to run all the time (in which case it is invariably turned off) or the unit fails to start on pressure demand. In either case, the manual start pushbutton can be used to start the controller in the variable speed mode or the bypass mode, depending on the specifics of the failure.

10.10.7.2.1 Where redundant pressure sensing elements are provided as part of a water mist positive displacement pumping unit, they shall be permitted for other system functions.

> This requirement was added to the 2012 edition to accommodate a particular water mist central control scheme.

10.10.7.3 Means shall be provided to manually select between variable speed and bypass mode.

> Manual selection between variable speed and bypass mode is necessary in the event that a failure of the circuitry to automatically close the bypass contactor occurs. This "mode switch" is also used for testing the pump and for other maintenance and/or for repair purposes.

10.10.7.4 Except as provided in 10.10.7.4.2, common pressure control shall not be used for multiple pump installations.

> This requirement was revised for the 2012 edition to add the exception of 10.10.7.4.1. The purpose of this clause is to keep the controllers (and hence the pumps) in multiple pump installations completely independent of one another. While this is common practice with installations having fire pumps in parallel, it is not the case for multiple variable speed domestic pumps, which typically make use of a common controller to stage pumps on and off and often to send the speed control set-points to the individual VFDs. Since fire pumps are used to supply domestic water systems, this requirement is included to ensure that no such common control is involved, since this is a point of common failure which could disable all pumps and result in no fire water.

10.10.7.4.1 Each controller pressure sensing control circuit shall operate independently.

10.10.7.4.2 A common pressure control shall be permitted to be used for a water mist positive displacement pumping unit controller.

10.10.8 Indicating Devices on Controller.

10.10.8.1 Drive Failure. A visible indicator shall be provided to indicate when the variable speed drive has failed.

10.10.8.2 Bypass Mode. A visible indicator shall be provided to indicate when the controller is in bypass mode.

10.10.8.3 Variable Speed Pressure Limiting Control Overpressure. Visible indication shall be provided on all controllers equipped with variable speed pressure limiting control to actuate at 115 percent of set pressure.

> The visible indicator alerts personnel that the system is experiencing an overpressure condition and some precautions may need to be taken if this condition persists. Under such circumstances shutting down the fire pump is not desirable. Preventing the spread of fire is more critical than preventing the system from being over-pressured. The main result of such an occurrence will be water flow from the main relief valve. This can diminish the amount of fire water available to the sprinkler system and/or hose lines.

10.10.9 Controller Contacts for Remote Indication. Controllers shall be equipped with contacts (open or closed) to operate circuits for the conditions in 10.10.8.

> Subsection 10.10.9 requires open or closed alarm contacts operated by the variable speed pressure limiting control to be provided for remote indication. The function of these contacts is similar to those required in 10.4.8.

10.10.10 System Performance.

10.10.10.1* The controller shall be provided with suitable adjusting means to account for various field conditions.

> ▶ FAQ What types of field adjustments are available for variable speed pressure limiting controllers?

> The variable speed drive used for the pressure limiting control generally has many adjustments that affect its performance. These adjustments allow the controller to be coordinated with the motor to provide maximum motor efficiency. These adjustments cover a myriad of performance factors such as voltage ramping for acceleration and deceleration, frequency control, and voltage sag.

A.10.10.10.1 This allows for field adjustments to reduce hunting, excessive overshooting, or oscillating.

10.10.10.2 Operation at reduced speed shall not result in motor overheating.

> ▶ **FAQ** Why do motors overheat when operated at less than their rated frequency, and how is this avoided when using variable speed controllers?
>
> Any centrifugal pump motor operating at less than rated speed (RPM) gets less cooling due to the fact that the motor's cooling fan is also running at a lower speed, and thus providing less cooling air flow over the motor windings. However, the motor is also delivering less mechanical (shaft) horsepower and, as a result, is drawing less electrical power from the VFD. Both the voltage and current are reduced reducing the heating in the windings. Hence there is a trade-off between less heating and less cooling at any given speed.

10.10.10.3 The maximum operating frequency shall not exceed line frequency.

> Many modern VFDs have the ability to run the motor above its normal synchronous speed, through the use of a "field weakening" function. While this function is designed to protect the motor, no fire pump is listed to operate above its rated speed (RPM). Thus, running a pump above its rated speed may void its listing. Running a pump above its listed speed is also dangerous, since the pressure gain of the pump, and hence its discharge pressure, is proportional to the square of the speed increase.

10.10.11 Critical Settings. Means shall be provided and permanently attached to the inside of the controller enclosure to record the following settings:

(1) Variable speed pressure limiting set point setting
(2) Pump start pressure
(3) Pump stop pressure

> ▶ **FAQ** What are the critical pressure settings that may be field adjustable in variable speed pressure limiting controllers?
>
> Recording the settings specified in 10.10.11 is critical as the need for future adjustment of these settings is common for controllers with variable speed pressure limiting control. The maintenance of these settings for proper operation of the system is important, especially if the VFD needs to be either repaired or replaced.

10.10.12 Variable Speed Drives for Vertical Pumps.

> In the 2010 edition of NFPA 20, this new subsection was added to recognize that the application of variable speed drives to vertical turbine pumps differs somewhat from other centrifugal pumps. While most vertical turbine pumps are used for below ground storage reservoirs, this section is primarily intended for well pumps due to the greater shaft and column lengths involved.

10.10.12.1 The pump supplier shall inform the controller manufacturer of any and all critical resonant speeds within the operating speed range of the pump, which is from zero speed up to full speed.

Section 10.10 • Controllers with Variable Speed Pressure Limiting Control or Variable Speed Suction Limiting Control

With increasing shaft lengths, the range of shaft (pump) resonance gets closer to the range of intended operating speeds of the pump. These critical resonant speeds must be avoided due to their destructive nature. If there are, indeed, such critical speeds below the maximum pump speed, the VFD must be programmed to avoid these speeds by rapidly skipping over these speeds.

10.10.12.1.1 The controller shall avoid operating at or ramping through these speeds.

10.10.12.1.2 The controller shall make use of skip frequencies with sufficient bandwidth to avoid exciting the pump into resonance.

10.10.12.2 When water-lubricated pumps with line shaft bearings are installed, the pump manufacturer shall inform the controller manufacturer of the maximum allowed time for water to reach the top bearing under the condition of the lowest anticipated water level of the well or reservoir.

Most vertical turbine pumps have oil lubricated shaft bearings. However, if the pump is also pumping water for domestic use (potable water), the bearings are more likely to be water lubricated instead. In this case, the bearings are dry (not being wetted) when the pump is not running, since there is no water in the pump column pipe. Hence, when starting these types of pumps, the VFD must be programmed to ramp up the pump speed rapidly enough to wet the bearings within the maximum time specified by the pump manufacturer, including when the well water level is at its lowest allowed level. It is the upper bearing or bearings which are of the greatest concern. This is usually not a problem since there is only the static water height head during this time.

10.10.12.2.1 The controller shall provide a ramp up speed within this time period.

10.10.12.3 The ramp down time shall be approved or agreed to by the pump manufacturer.

The requirement for ramp up speed in 10.10.12.3 also applies when the pump is being shut down. The ramp down time cannot be so long as to run the upper bearings dry for too long a time.

10.10.12.4 Any skip frequencies employed and their bandwidth shall be included along with the information required in 10.10.11.

This information is vital should the VFD need to be re-programmed due to either repair or replacement.

10.10.12.5 Ramp up and ramp down times for water-lubricated pumps shall be included along with the information required in 10.10.11.

References Cited in Commentary

National Fire Protection Association, 1 Batterymarch Park, Quincy, MA 02169-7471.

Earley, M. W., et al., eds., *National Electrical Code® Handbook*, 2011 edition.
NFPA 14, *Standard for the Installation of Standpipe and Hose Systems*, 2010 edition.
NFPA 70®, *National Electrical Code®*, 2011 edition.
National Electrical Manufacturers Association, 1300 North 17th Street, Suite 1752, Rosslyn, VA 22209.

NEMA ICS 14, *Application Guide for Electric Pump Controllers*, 2001 edition.

NEMA ICS 15 *Instructions for the Handling, Installation, Operation, and Maintenance of Electric Fire Pump Controllers Rated Not More Than 600 V.*
NEMA 250, *Enclosures for Electrical Equipment,* 1991 edition.

Underwriters Laboratories Inc., 333 Pfingsten Road, Northbrook, IL 60062-2096.

ANSI/UL 508, *Standard for Industrial Control Equipment,* 2006 edition.
ANSI/UL 50 Enclosures for Electrical Equipment, Non-Environmental Considerations
ANSI/UL 50E *"Enclosures for Electrical Equipment, Environmental Considerations"*
UL 347 — *High Voltage Industrial Control Equipment*

CHAPTER 11

Diesel Engine Drive

> Chapter 11 describes the requirements for diesel engine drives. Included in this chapter are requirements for diesel engines, fire pump and engine protection, diesel engine fuel supply and arrangement, engine exhaust, and driver system operation.
>
> Diesel engine–driven fire pumps are often used where there is insufficient or unreliable electrical power for an electric-driven fire pump. They are used in conjunction with, or in addition to, motor-driven fire pumps as the best combination for reliable pumping systems. When combined with on-site water storage — such as a ground storage tank, an underground reservoir, a tower, a pond, or a well — a diesel-driven fire pump is completely self-contained. Since diesel engine–driven fire pumps are completely independent of external utilities, diesel drive units are well suited for remote locations or areas without a reliable power grid. Properly designed and installed units can survive extended periods in the absence of ac power. Some installations can remain in service indefinitely when the controller, engine, and other battery loads are low enough for the weekly run function to replenish the batteries. As with all fire protection equipment, the functionality of the fire pump system is only as good as the performance of the periodic maintenance and inspection programs. Refer to Chapter 8 of NFPA 25, *Standard for the Inspection, Testing, and Maintenance of Water-Based Fire Protection Systems*, and the equipment manufacturers' OEM manuals, instructions, and requirements for the full requirements for periodic care and maintenance schedules.

11.1 General

11.1.1 This chapter provides requirements for minimum performance of diesel engine drivers.

11.1.2 Accessory devices, such as monitoring and signaling means, are included where necessary to ensure minimum performance of the aforementioned equipment.

11.1.3* Engine Type.

A.11.1.3 The compression ignition diesel engine has proved to be the most dependable of the internal combustion engines for driving fire pumps.

11.1.3.1 Diesel engines for fire pump drive shall be of the compression ignition type.

> The compression ignition diesel engine accomplishes combustion by spraying atomized fuel into a cylinder of compressed, heated air. The fuel spray is provided by an injector(s) controlled by the electronic fuel management control [electronic control module (ECM)] system at a precise time and for a specific volume. The cylinder air is compressed by the upstroke of the piston, which provides high compression that heats the air 2°F (1.1°C) for each pound per square inch (psi) of compression. This very simple ignition process has proven to be extremely reliable.
>
> Diesel engines are designed with the necessary strength to accommodate high compression ratios and to provide long life of the fire pump. A typical diesel engine drive is shown in Exhibit II.11.1.

EXHIBIT II.11.1 Diesel Engine Drive. (Photograph courtesy of Mechanical Designs Ltd.)

11.1.3.2 Spark-ignited internal combustion engines shall not be used.

There are still spark ignition engines operating throughout the world. The spark engines for gasoline and natural gas were dropped from the 1974 edition of NFPA 20, *Standard for the Installation of Stationary Pumps for Fire Protection*. There are no equipment listings for ignition spark internal combustion engines and controls. Owing to the safety and environmental concerns about gasoline engines, it is advisable to replace these units with up-to-date diesel-driven units and controllers.

11.2 Engines

11.2.1 Listing. Engines shall be listed for fire pump service.

▶ **FAQ** What information is provided on the engine's nameplate?

Listed engines carry the performance ratings and the listing laboratory's mark. This information is displayed on a nameplate mounted on the engine to aid the owner and authority having jurisdiction (AHJ) when reviewing the system components.

11.2.2 Engine Ratings.

11.2.2.1 Engines shall have a nameplate indicating the listed horsepower rating available to drive the pump.

Exhibit II.11.2 shows an example of a listed engine nameplate for which the engine is listed for a speed range with a lower limit (1760 rpm) and an upper limit (2100 rpm). The engine may be operated at any rpm between the lower and upper limits. For example, if the engine were operated at 2000 rpm, the maximum corresponding horsepower generated would be 329.

11.2.2.2* The horsepower capability of the engine, when equipped for fire pump service, shall have a 4-hour minimum horsepower rating not less than 10 percent greater than the listed horsepower on the engine nameplate.

Section 11.2 • Engines 293

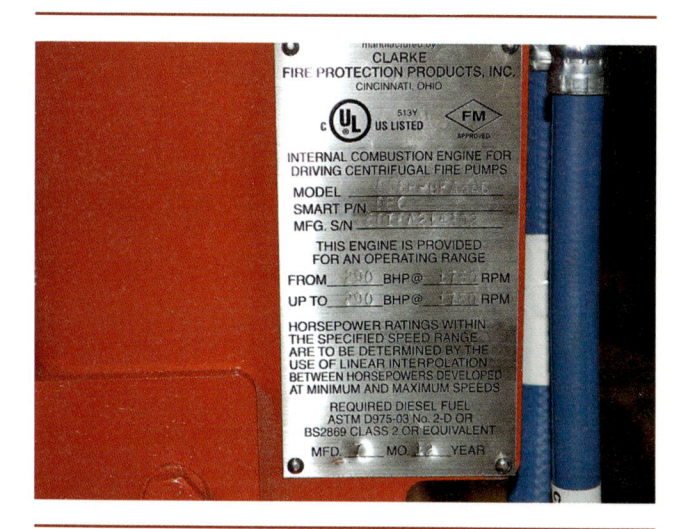

EXHIBIT II.11.2 Listed Engine Nameplate. (Photograph courtesy of Clarke Fire Protection Products, Inc.)

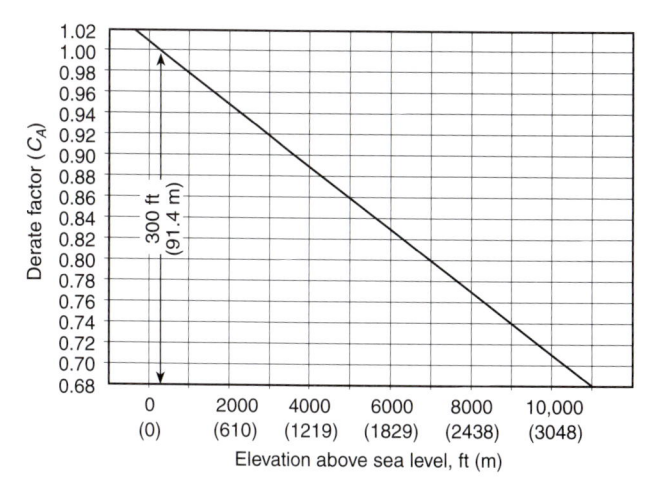

FIGURE A.11.2.2.4 Elevation Derate Curve.

Note: The correction equation is as follows:

Corrected engine horsepower = $(C_A + C_T - 1) \times$ listed engine horsepower
where:
C_A = derate factor for elevation
C_T = derate factor for temperature

A.11.2.2.2 For more information, see SAE J-1349, *Engine Power Test Code — Spark Ignition and Compression Engine*. The 4-hour minimum power requirement in NFPA 20 has been tested and witnessed during the engine listing process.

11.2.2.3 Engines shall be acceptable for horsepower ratings listed by the testing laboratory for standard SAE conditions.

11.2.2.4* A deduction of 3 percent from engine horsepower rating at standard SAE conditions shall be made for each 1000 ft (300 m) of altitude above 300 ft (91 m).

> Each engine is rated for a certain horsepower at standard SAE conditions. The performance of any internal combustion engine is affected by the quantity of oxygen molecules in the combustion chamber. The oxygen is needed to support the burning of the fuel, which creates expanding gases that are then converted to mechanical energy. Higher altitude and/or ambient temperature causes the expansion of air, resulting in fewer molecules of oxygen per volume of air.

A.11.2.2.4 See Figure A.11.2.2.4.

11.2.2.5* A deduction of 1 percent from engine horsepower rating as corrected to standard SAE conditions shall be made for every 10°F (5.6°C) above 77°F (25°C) ambient temperature.

▶ **FAQ** How is the engine horsepower rating corrected to the installation conditions, and who provides this information?

Because engine performance is affected by the combustion air temperature, the power rating of the engine must be corrected to the actual installation conditions, as in the example that follows. The example provides the derating process needed to choose the correct engine for a particular installation. This information is required prior to selecting the engine; this information should be provided by project engineering, and the installing contractor to the pump manufacturer, to ensure that the correct equipment is chosen.

Stationary Fire Pumps Handbook 2013

CALCULATION ▶ An engine rated at 95 hp is used to drive a fire pump in Yellowstone National Park, where the elevation is 7000 ft (2134 m) above sea level and the ambient temperature could be 105°F (40.5°C). What is the adjusted rating of this engine?

Solution: See Figure A.11.2.2.4 and Figure A.11.2.2.5, and use the following formula:

$$EH_{Corr} = [(C_A + C_T) - 1] \times \text{listed engine horsepower}$$

where:

EH_{Corr} = corrected engine horsepower
C_A = derate factor for elevation
C_T = derate factor for temperature

From Figure A.11.2.2.4, the derate factor for altitude (C_A) is approximately 0.80 for an elevation of 7000 ft (2134 m). From Figure A.11.2.2.5, the derate factor for temperature (C_T) is approximately 0.973 for an ambient temperature of 105°F (40.5°C). Using the following correction equation, as indicated by the figures, the adjusted horsepower rating can be determined.

$$EH_{Corr} = [(C_A + C_T) - 1] \times \text{listed engine horsepower}$$
$$= [(0.80 + 0.973) - 1] \times 302$$
$$= 233.5$$

Therefore, an engine rated at 302 bhp at sea level is derated to 233 bhp at an elevation of 7,000 ft (2134 m) and a temperature of 105°F (49.5°C).

A.11.2.2.5 Pump room temperature rise should be considered when determining the maximum ambient temperature specified. *(See Figure A.11.2.2.5.)*

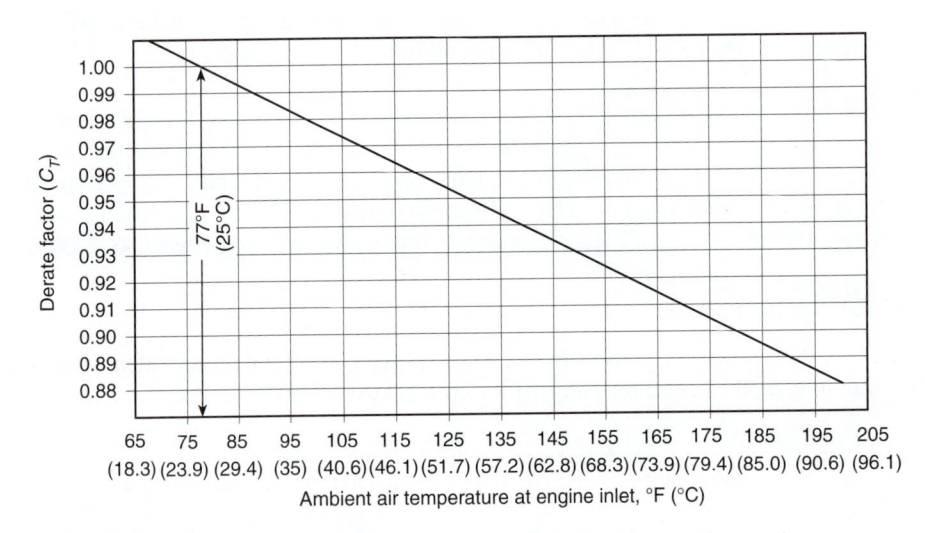

Note: The correction equation is as follows:

Corrected engine horsepower = $(C_A + C_T - 1) \times$ listed engine horsepower

where:

C_A = derate factor for elevation
C_T = derate factor for temperature

FIGURE A.11.2.2.5 *Temperature Derate Curve.*

11.2.2.6 Where right-angle gear drives *(see 11.2.3.2)* are used between the vertical turbine pump and its driver, the horsepower requirement of the pump shall be increased to allow for power loss in the gear drive.

> To ensure that the engine has adequate horsepower, the power loss through the right-angle gear must be considered as additional load and added to the maximum pump load. The gear manufacturer provides the power loss of the right-angle gear drive.

11.2.2.7 After the requirements of 11.2.2.1 through 11.2.2.6 have been complied with, engines shall have a 4-hour minimum horsepower rating equal to or greater than the brake horsepower required to drive the pump at its rated speed under any conditions listed for environmental conditions under pump load.

> Typically, the shorter the performance duration, the higher the manufacturer rates the engine. Ratings of less than 4 hours continuous are not acceptable for fire pump drivers.

11.2.3 Engine Power Connection to Pump.

> The fire pump and diesel engine are both rendered useless if the connecting drive fails. A flexible coupling or flexible-connecting shaft (universal driveshaft) that has been tested and listed is required by NFPA 20 for connecting a fire pump to a diesel engine.
>
> The drive system between the pump and the engine should be as simple as possible, with no means of disconnecting the pump and engine shafts without dismantling the system. This type of drive arrangement is considered to provide the greatest reliability. The manufacturer's requirements for the installation and alignment must be used to ensure proper operation of drive couplings and/or driveshafts.

11.2.3.1 Horizontal Shaft Pumps.

> The installation and setup of the engine and (right-angle) gear drive must follow the instructions and requirements set forth in the installation data sheets and/or O&M manual provided by the manufacturers of the connecting shafts, the engine, and the gear drive. Special installation requirements must be followed to ensure a proper installation.
>
> Caution should be exercised during the servicing of the connecting shaft or the diesel driver. Lockout and tag-out procedures should be used. If the engine attempts to start without warning, the shaft can cause serious injury or damage.

11.2.3.1.1 Engines shall be connected to horizontal shaft pumps by means of a flexible coupling or flexible connecting shaft listed for this service.

11.2.3.1.2 The flexible coupling or flexible connecting shaft shall be directly attached to the engine flywheel adapter or stub shaft. *(See Section 6.5.)*

11.2.3.2 Vertical Shaft Turbine–Type Pumps.

11.2.3.2.1 Unless the requirements of 11.2.3.2.2 are met, engines shall be connected to vertical shaft pumps by means of a right-angle gear drive with a listed flexible connecting shaft that will prevent undue strain on either the engine or gear drive. *(See Section 7.5.)*

11.2.3.2.2 The requirements of 11.2.3.2.1 shall not apply to diesel engines and steam turbines designed and listed for vertical installation with vertical shaft turbine–type pumps, which shall be permitted to employ solid shafts and shall not require a right-angle drive but shall require a nonreverse ratchet.

11.2.4 Engine Speed Controls.

11.2.4.1 Speed Control Governor.

11.2.4.1.1 Engines shall be provided with a governor capable of regulating engine speed within a range of 10 percent between shutoff and maximum load condition of the pump.

11.2.4.1.2 The governor shall be field adjustable and set and secured to maintain rated pump speed at maximum pump load.

11.2.4.1.3 Engines shall accelerate to rated output speed within 20 seconds.

11.2.4.2* Electronic Fuel Management Control.

> ECM-equipped diesel engines require periodic and annual testing, inspection, and service. Refer to NFPA 25, 8.3.3.8 and A.8.3.3.8, for further guidance in the performance of the required ECM test and inspection.

A.11.2.4.2 Traditionally, engines have been built with mechanical devices to control the injection of fuel into the combustion chamber. To comply with requirements for reduced exhaust emissions, many engine manufacturers have incorporated electronics to control the fuel injection process, thus eliminating levers and linkages. Many of the mechanically controlled engines are no longer manufactured.

11.2.4.2.1 Alternate Electronic Control Module. Engines that incorporate an electronic control module (ECM) to accomplish and control the fuel injection process shall have an alternate ECM permanently mounted and wired so the engine can produce its full rated power output in the event of a failure of the primary ECM.

11.2.4.2.2 ECM Voltage Protection. ECMs shall be protected from transient voltage spikes and reverse dc current.

11.2.4.2.3 ECM Selector Switch.

11.2.4.2.3.1 Operation.

(A) The transition from the primary ECM to the alternate or alternate to primary shall be controlled by a hand/automatic switch without an off position.
(B) When the switch required in 11.2.4.2.3.1(A) is in the automatic position, the transition from the primary ECM to the alternate or alternate to primary shall be accomplished automatically upon failure of the either ECM.
(C) When the switch required in 11.2.4.2.3.1(A) is in the hand position, the transition from the primary ECM to the alternate or from alternate to primary shall be accomplished manually.

11.2.4.2.3.2 Supervision. A visual indicator shall be provided on the engine instrument panel, and a supervisory signal shall be provided to the controller when the ECM selector switch is positioned to the alternate ECM.

11.2.4.2.3.3 Contacts.

(A) The contacts for each circuit shall be rated for both the minimum and maximum current and voltage.
(B) The total resistance of each ECM circuit through the selector switch shall be approved by the engine manufacturer.

11.2.4.2.3.4 Enclosure.

(A) The selector switch shall be enclosed in a NEMA Type 2 dripproof enclosure.
(B) Where special environments exist, suitably rated enclosures shall be used.

11.2.4.2.3.5 Mounting.

(A) The selector switch and enclosure shall be engine mounted.
(B) The selector switch enclosure and/or the selector switch inside shall be isolated from engine vibration to prevent any deterioration of contact operation.

11.2.4.2.4* Engine Power Output. The ECM (or its connected sensors) shall not, for any reason, intentionally cause a reduction in the engine's ability to produce rated power output.

A.11.2.4.2.4 ECMs can be designed by engine manufacturers to monitor various aspects of engine performance. A stressed engine condition (such as high cooling water temperature) is usually monitored by the ECM and is built into the ECM control logic to reduce the horsepower output of the engine, thus providing a safeguard for the engine. Such engine safeguards are not permitted for ECMs in fire pump engine applications.

11.2.4.2.5 ECM Sensors. Any sensor necessary for the function of the ECM that affects the engine's ability to produce its rated power output shall have a redundant sensor that shall operate automatically in the event of a failure of the primary sensor.

11.2.4.2.6 ECM Engine Supervision. A common supervisory signal shall be provided to the controller in the event of any of the following:

(1) Fuel injection failure
(2) Low fuel pressure
(3) Any primary sensor failure

11.2.4.2.7 ECM and Engine Power Supply.

11.2.4.2.7.1* In the standby mode, the engine batteries or battery chargers shall be used to power the ECM.

A.11.2.4.2.7.1 When engines are in standby and the battery chargers have the batteries in float, it is actually the chargers that are providing the current to support the engine, controller, and pump room as defined in 11.2.7.2.3.2.

11.2.4.2.7.2 Engines shall not require more than 0.5 ampere from the battery or battery charger while the engine is not running.

11.2.4.3 Variable Speed Pressure Limiting Control or Variable Speed Suction Limiting Control (Optional).

> The addition of the variable speed pressure limiting control (VSPLC) feature for a diesel engine has increased the tools available to the fire protection professional in the design of fire protection systems. The pressure limiting feature can be used to limit the available pressure to more workable levels when a variable pressure supply is used to provide suction to the fire pump, or when the fire protection system needs large quantities of water at high pressures because the discharge pressure curve from the fire pump drops off quickly as the flow increases. Similar results are available from electric motor–driven fire pumps. See Chapters 9 and 10 for details.

11.2.4.3.1 Variable speed pressure limiting control or variable speed suction limiting control systems used on diesel engines for fire pump drive shall be listed for fire pump service and be capable of limiting the pump output total rated head (pressure) or suction pressure by reducing pump speed.

11.2.4.3.2 Variable speed control systems shall not replace the engine governor as defined in 11.2.4.1.

11.2.4.3.3 In the event of a failure of the variable speeds control system, the engine shall be fully functional with the governor defined in 11.2.4.1.

11.2.4.3.4 Pressure Sensing Line.

> The pressure sensing line required for operation of the variable speed feature outlined in 11.2.4.3.4.1 is needed in addition to that specified in Section 4.30. The pressure sensing line required by Section 4.30 is intended to control the starting and running time for the fire pump, while the pressure sensing line required in 11.2.4.3.4 is used to control the VSPLC function in the diesel engine driver on board controls.

11.2.4.3.4.1 A pressure sensing line shall be provided to the engine with a ½ in. (12.7 mm) nominal size inside diameter line.

11.2.4.3.4.2 For pressure limiting control, a sensing line shall be installed from a connection between the pump discharge flange and the discharge check valve to the engine.

11.2.4.3.4.3 For suction limiting control, a sensing line shall be installed from a connection at the pump inlet flange to the engine.

11.2.4.4 Engine Overspeed Shutdown Control, Low Oil Pressure Signal, and High and Low Coolant Temperature Signals.

11.2.4.4.1 Engines shall be provided with an overspeed shutdown device.

11.2.4.4.2 The overspeed device shall be arranged to shut down the engine in a speed range of 10 to 20 percent above rated engine speed and to be manually reset.

11.2.4.4.3 A means shall be provided to indicate an overspeed trouble signal to the automatic engine controller such that the controller cannot be reset until the overspeed shutdown device is manually reset to normal operating position.

11.2.4.4.4 Means shall be provided for verifying overspeed switch and circuitry shutdown function.

11.2.4.4.5 Means shall be provided for signaling critically low oil pressure in the engine lubrication system to the controller.

11.2.4.4.5.1 Means shall be provided on the engine for testing the operation of the oil pressure signal to the controller resulting in visible and common audible alarm on the controller as required in 12.4.1.3.

11.2.4.4.5.2 Instructions for performing the test in 11.2.4.4.5.1 shall be included in the engine manual.

11.2.4.4.6 Means shall be provided for signaling high engine temperature to the controller.

11.2.4.4.6.1 Means shall be provided on the engine for testing the operation of the high engine temperature signal to the controller, resulting in visible and common audible alarm on the controller as required in 12.4.1.3.

11.2.4.4.6.2 Instructions for performing the test in 11.2.4.4.6.1 shall be included in the engine manual.

11.2.4.4.7 Means shall be provided for signaling low engine temperature to the controller.

11.2.4.4.7.1 Means shall be provided on the engine for testing the operation of the low engine temperature signal to the controller, resulting in visible and common audible alarm on the controller as required in 12.4.1.3.

11.2.4.4.7.2 Instructions for performing the test in 11.2.4.4.7.1 shall be included in the engine manual.

11.2.4.5 Engine Running and Crank Termination Control.

11.2.4.5.1 Engines shall be provided with a speed-sensitive switch to signal engine running and crank termination.

11.2.4.5.2 Power for this signal shall be taken from a source other than the engine generator or alternator.

11.2.5 Instrumentation.

11.2.5.1 Instrument Panel.

11.2.5.1.1 All engine instruments shall be placed on a panel secured to the engine or inside an engine base plate–mounted controller.

11.2.5.1.2 The engine instrument panel shall not be used as a junction box or conduit for any ac supply.

11.2.5.2 Engine Speed.

11.2.5.2.1 A tachometer or other means shall be provided to indicate revolutions per minute of the engine, including zero, at all times.

11.2.5.2.2 The tachometer shall be the totalizing type, or an hour meter or other means shall be provided to record total time of engine operation.

11.2.5.2.3 Tachometers with digital display shall be permitted to be blank when the engine is not running.

11.2.5.3 Oil Pressure. Engines shall be provided with an oil pressure gauge or other means to indicate lubricating oil pressure.

11.2.5.4 Temperature. Engines shall be provided with a temperature gauge or other means to indicate engine coolant temperature at all times.

11.2.6 Wiring Elements.

11.2.6.1 Automatic Controller Wiring in Factory.

11.2.6.1.1* All connecting wires for automatic controllers shall be harnessed or flexibly enclosed, mounted on the engine, and connected in an engine junction box to terminals numbered to correspond with numbered terminals in the controller.

> The interconnecting wiring between the engine control panel termination strip and the matching field connection strip in the controller should be wired as required by the equipment manufacturers' data sheets or NFPA electrical codes and standards. The power-carrying conductors should be sized to meet the requirements for the current and voltage between the engine and the automatic controller.
>
> The minimum wire size is normally No. 10 AWG CHHN stranded wire for up to 25 ft (7.62 m) runs for the battery charging circuits. Undersized wiring components can cause failure of the circuit and possible failure of the starting system. Wiring schematics are included on the inside of the controller to assist in the installation of the wiring.

A.11.2.6.1.1 A harness on the enclosure will ensure ready wiring in the field between the two sets of terminals.

11.2.6.1.2 All wiring on the engine, including starting circuitry, shall be sized on a continuous-duty basis.

11.2.6.2* Automatic Control Wiring in the Field.

A.11.2.6.2 Terminations should be made using insulated ring-type compression connectors for post-type terminal blocks. Saddle-type terminal blocks should have the wire stripped with about 1/16 in. (1.6 mm) of bare wire showing after insertion in the saddle, to ensure that no insulation is below the saddle. Wires should be tugged to ensure adequate tightness of the termination.

11.2.6.2.1 Interconnections between the automatic controller and the engine junction box shall be made using stranded wire sized on a continuous-duty basis.

11.2.6.2.1.1 Interconnection wire size shall be based on length as recommended for each terminal by the controller manufacturer.

11.2.6.2.2 The dc interconnections between the automatic controller and engine junction box and any ac power supply to the engine shall be routed in separate conduits.

11.2.6.3 Battery Cables.

11.2.6.3.1 Battery cables shall be sized in accordance with the engine manufacturer's recommendations considering the cable length required for the specific battery location.

11.2.7 Starting Methods.

11.2.7.1 Starting Devices. Engines shall be equipped with a reliable starting device.

> The main battery contactors are normally open relays used for connecting and disconnecting one of the two battery units to one starter. To ensure that each battery is maintained at full charge, the two battery units must be isolated from each other. A main battery contactor is installed in the cable of each battery to the starter.
>
> The controller provides a voltage signal to a selected main battery contactor, energizing its coil, which causes the relay to close and send battery potential to the starter, thus providing cranking power to start the engine. The main battery contactor is also equipped with a snap-action lever to allow the operator to manually crank the engine in the event of a circuit failure within the controller. Some engines are equipped with two starters, one for each battery unit. Therefore, main battery contactors are not required on the engines.
>
> Main battery contactors or dual starter systems are intended to utilize only one battery unit at a time to crank an engine. However, in the event of an emergency, and where cranking the engine is difficult, either main battery contactors or both starters might need to be used simultaneously.

11.2.7.2 Electric Starting. Where electric starting is used, the electric starting device shall take current from a storage battery(ies).

11.2.7.2.1 Batteries.

11.2.7.2.1.1 Each engine shall be provided with two storage battery units.

> ▶ **FAQ** Are all batteries within a battery set to be changed when a problem occurs in one of the units?

Because it is important to have sufficient power available during the starting sequence, it is advisable to keep the batteries as equal in power as possible. If one battery becomes weak or fails a battery test, the other battery is close to failing as well. The battery must always be replaced with the same class and amp-hour rating recommended by the engine manufacturer.

11.2.7.2.1.2 Lead-acid batteries shall be furnished in a dry charge condition with electrolyte liquid in a separate container.

> To ensure that the batteries are ready for in-service testing of the pump, battery electrolyte must be added a minimum of 24 hours prior to putting the pump in service so that the batteries have time to charge before testing begins.

11.2.7.2.1.3 Nickel-cadmium or other kinds of batteries shall be permitted to be installed in lieu of lead-acid batteries, provided they meet the engine manufacturer's requirements.

11.2.7.2.1.4 At 40°F (4°C), each battery unit shall have twice the capacity sufficient to maintain the cranking speed recommended by the engine manufacturer through a 3-minute attempt-to-start cycle, which is six consecutive cycles of 15 seconds of cranking and 15 seconds of rest.

▶ **FAQ** Why is the battery sized for twice the capacity?

> Because battery output goes down with temperature, batteries must be sized to meet the engine requirements at 40°F (4.5°C), or lower for extreme conditions. Exhibit II.11.3 shows how a battery's performance is affected by temperature. By sizing the battery for twice the capacity, it has remaining capacity to start the engine, even if it has gone through a full attempt-to-start cycle and shut down for overcrank. See 12.8.2 for requirements concerning starting equipment arrangement.

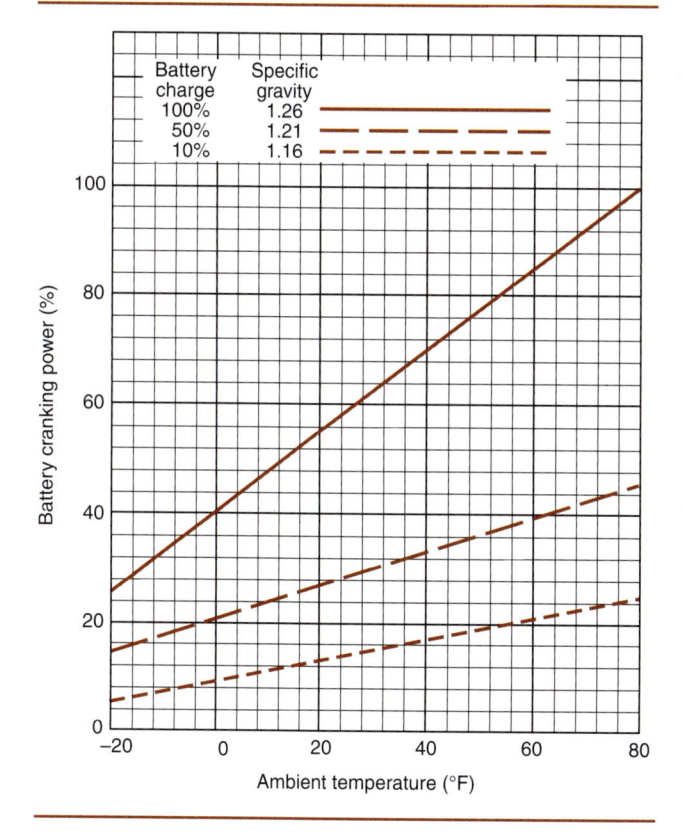

EXHIBIT II.11.3 *Battery Performance versus Temperature.*

11.2.7.2.1.5* Batteries shall be sized, based on calculations, to have capacity to carry the loads defined in 11.2.7.2.3 for 72 hours of standby power followed by three 15-second attempt-to-start cycles per battery unit as defined in 11.2.7.2.1.4, without ac power being available for battery charging.

A.11.2.7.2.1.5 The 72-hour requirement is intended to apply when batteries are new. Some degradation is expected as batteries age.

11.2.7.2.2* Battery Isolation.

A.11.2.7.2.2 Manual mechanical operation of the main battery contactor will bypass all of the control circuit wiring within the controller.

> In the event of a failure of control circuits within the controller or the interconnecting wires between the controller and engine, the main battery contactor provides a means of locally cranking the engine for emergency operation. Engines equipped with two starters, in lieu of main battery contactors, provide control switches on the engine instrument panel for local cranking during this type of emergency.

11.2.7.2.2.1 Engines with only one cranking motor shall include a main battery contactor installed between each battery and the cranking motor for battery isolation.

(A) Main battery contactors shall be listed for fire pump driver service.
(B) Main battery contactors shall be rated for the cranking motor current.
(C) Main battery contactors shall be capable of manual mechanical operation, including positive methods such as spring-loaded, over-center operator to energize the starting motor in the event of controller circuit failure.

11.2.7.2.2.2 Engines with two cranking motors shall have one cranking motor dedicated to each battery.

(A) Each cranking motor shall meet the cranking requirements of a single cranking motor system.
(B) To activate cranking, each cranking motor shall have an integral solenoid relay to be operated by the pump set controller.
(C) Each cranking motor integral solenoid relay shall be capable of being energized from a manual operator listed and rated for the cranking motor solenoid relay and include a mechanical switch on the engine panel to energize the starting motor in the event of controller circuit failure.

11.2.7.2.3 Battery Loads.

11.2.7.2.3.1 Nonessential loads shall not be powered from the engine starting batteries.

11.2.7.2.3.2 Essential loads, including the engine, controller, and all pump room equipment combined, shall not exceed 0.5 ampere each for a total of 1.5 amperes, on a continuous basis.

11.2.7.2.4* Battery Location.

DESIGN ALERT ▶

> When calculating cable length for sizing purposes, all cable for the circuit must be included (positive cable, negative cable, battery jumper, cable from main battery contactor to starter, ground, and so forth). A diesel engine creates a tremendous amount of vibration that can cause component failure; the base of the battery compartment should be rigid to control this vibration. The frame must first be aligned with the pump and drivers and then securely anchored to the concrete foundation. To eliminate further concern about vibration, the base frame should be grouted under the beams and then filled with non-shrink grout or concrete.

A.11.2.7.2.4 Location at the side of and level with the engine is recommended to minimize lead length from battery to starter.

11.2.7.2.4.1 Storage batteries shall be rack supported above the floor, secured against displacement, and located where they will not be subject to excessive temperature, vibration, mechanical injury, or flooding with water.

11.2.7.2.4.2 Current-carrying parts shall not be less than 12 in. (305 mm) above the floor level.

11.2.7.2.4.3 Storage batteries shall be readily accessible for servicing.

11.2.7.2.4.4 Storage batteries shall not be located in front of the engine-mounted instruments and controls.

11.2.7.2.4.5 Storage battery racks and their location shall meet the requirements of *NFPA 70, National Electrical Code*.

11.2.7.3 Hydraulic Starting.

11.2.7.3.1 Where hydraulic starting is used, the accumulators and other accessories shall be enclosed or so protected that they are not subject to mechanical injury.

11.2.7.3.2 The enclosure shall be installed as close to the engine as practical so as to prevent serious pressure drop between the engine and the enclosure.

11.2.7.3.3 The diesel engine as installed shall be without starting aid except that as required in 11.2.8.2.

11.2.7.3.4 The diesel as installed shall be capable of carrying its full rated load within 20 seconds after cranking is initiated with the intake air, room ambient temperature, and all starting equipment at 32°F (0°C).

11.2.7.3.5 Hydraulic starting means shall comply with the following conditions:

(1) The hydraulic cranking device shall be a self-contained system that will provide the required cranking forces and engine starting revolutions per minute (rpm) as recommended by the engine manufacturer.
(2) Electrically operated means shall automatically recharge and maintain the stored hydraulic pressure within the predetermined pressure limits.
(3) The means of automatically maintaining the hydraulic system within the predetermined pressure limits shall be energized from the main bus and the final emergency bus if one is provided.
(4) Engine driven means shall be provided to recharge the hydraulic system when the engine is running.
(5) Means shall be provided to manually recharge the hydraulic system.
(6) The capacity of the hydraulic cranking system shall provide not fewer than six cranking cycles of not less than 15 seconds each.
(7) Each cranking cycle — the first three to be automatic from the signaling source — shall provide the necessary number of revolutions at the required rpm to permit the diesel engine to meet the requirements of carrying its full rated load within 20 seconds after cranking is initiated with intake air, room ambient temperature, and hydraulic cranking system at 32°F (0°C).
(8) The capacity of the hydraulic cranking system sufficient for three starts under conditions described in 11.2.7.3.5(6) shall be held in reserve and arranged so that the operation of a single control by one person will permit the reserve capacity to be employed.

(9) All controls for engine shutdown in the event of overspeed shall be 12 V dc or 24 V dc source to accommodate controls supplied on the engine, and the following also shall apply:
   (a) In the event of such failure, the hydraulic cranking system shall provide an interlock to prevent the engine from recranking.
   (b) The interlock shall be manually reset for automatic starting when engine failure is corrected.

11.2.7.4 Air Starting.

11.2.7.4.1 In addition to the requirements of Section 11.1 through 11.2.7, 11.2.8.1, 11.2.8 through 11.6.2, 11.6.4, and 11.6.6, the requirements of 11.2.7.4 shall apply.

11.2.7.4.2 Automatic Controller Connections in Factory.

11.2.7.4.2.1 All conductors for automatic controllers shall be harnessed or flexibly enclosed, mounted on the engine, and connected in an engine junction box to terminals numbered to correspond with numbered terminals in the controller.

11.2.7.4.2.2 These requirements shall ensure ready connection in the field between the two sets of terminals.

11.2.7.4.3 Signal for Engine Running and Crank Termination.

11.2.7.4.3.1 Engines shall be provided with a speed-sensitive switch to signal running and crank termination.

11.2.7.4.3.2 Power for this signal shall be taken from a source other than the engine compressor.

11.2.7.4.4* Air Starting Supply.

A.11.2.7.4.4 Automatic maintenance of air pressure is preferable.

11.2.7.4.4.1 The air supply container shall be sized for 180 seconds of continuous cranking without recharging.

11.2.7.4.4.2 There shall be a separate, suitably powered automatic air compressor or means of obtaining air from some other system, independent of the compressor driven by the fire pump engine.

11.2.7.4.4.3 Suitable supervisory service shall be maintained to indicate high and low air pressure conditions.

11.2.7.4.4.4 A bypass conductor with a manual valve or switch shall be installed for direct application of air from the air container to the engine starter in the event of control circuit failure.

11.2.8 Engine Cooling System.

> The cooling system for modern engines is very complex. The system not only must remove waste heat from the internal parts of the engine but must also accomplish the following:
>
> 1. Remove entrapped air from the coolant
> 2. Allow for coolant expansion
> 3. Allow for a specific volume of coolant loss
> 4. Maintain minimum flow under varying conditions
>
> Therefore, the cooling system is installed and provided as part of the listed engine.

11.2.8.1 The engine cooling system shall be included as part of the engine assembly and shall be one of the following closed-circuit types:

(1) A heat exchanger type that includes a circulating pump driven by the engine, a heat exchanger, and an engine jacket temperature regulating device

> A heat exchanger is the type of cooling system traditionally used for fire pump drivers. It takes water from the discharge piping of the fire pump and uses it to cool the engine. The water from the pump is passed through the engine heat exchanger, where it absorbs heat from the engine coolant. The water is then piped to a drain. In installations where the water is potable, the cooling water line is required to be equipped with listed backflow devices before being discharged to a drain or the building exterior. This is required because of the inhibitors that are part of the antifreeze solution.

(2) A radiator type that includes a circulating pump driven by the engine, a radiator, an engine jacket temperature regulating device, and an engine-driven fan for providing positive movement of air through the radiator

> Engine-mounted radiator cooling does not require water from the fire pump. It does, however, require a considerable volume of air to pass through the pump room. Provision for sufficient airflow for radiator cooling through the pump room requires special attention. See 11.3.2 for more information on pump room ventilation.
>
> A radiator-cooled engine eliminates the normal cooling water outlet from the pump nozzle. In addition to providing cooling for the engine, the water used also keeps the pump cool during static or churn operations. A circulating or casing relief valve should be added to provide cooling to the pump casing.

11.2.8.2 A means shall be provided to maintain 120°F (49°C) at the combustion chamber.

▶ **FAQ** Why must the temperature be maintained at 120°F (49°C)?

In a diesel engine, combustion of the fuel is caused by the rapid compression of air in the combustion chamber or cylinder. When an engine starts, the temperature in the cylinder is greatly affected by the temperature of the block and cylinder head. Therefore, maintaining the engine at 120°F (49°C), as required in 11.2.8.2, greatly improves the ability of the engine to start quickly. All listed and approved diesel engines require the coolant system to operate under pressure. This pressure is typically achieved by using a preset pressure relief cap on the coolant and fill openings. (See 11.2.8.3.)

The coolant provides a way to remove heat from the engine and to lubricate seals and prevent corrosion. Therefore, the coolant must be compatible with engine seals and gaskets. The engine manufacturer's recommendations for antifreeze and additives must be followed.

When air bubbles form in the engine coolant, the coolant's ability to remove heat is reduced. The pressure cap and antifreeze raise the boiling point of the coolant, provide the heat transfer necessary, and increase the engine's reliability. Exhibit II.11.4 shows how antifreeze not only lowers the freezing point but also raises the boiling point of water.

The installer or mechanic should always mix the coolant and the water prior to filling the engine water jacket. The power switch to the electric jacket heater or block heaters must be in the "off" or "open" position until the coolant mix is completely installed. Keeping the power off the block heater will ensure the heater is not destroyed. Energizing the heater without completely filling the cooling system with the coolant and water mix will cause immediate failure of the electric heater. Premixing of the coolant with water greatly reduces problems with burned out block heaters. The block heater is a continuous operation heater. The block heater is not

EXHIBIT II.11.4 Coolant Freezing and Boiling Temperature versus Antifreeze Concentration (Sea Level).

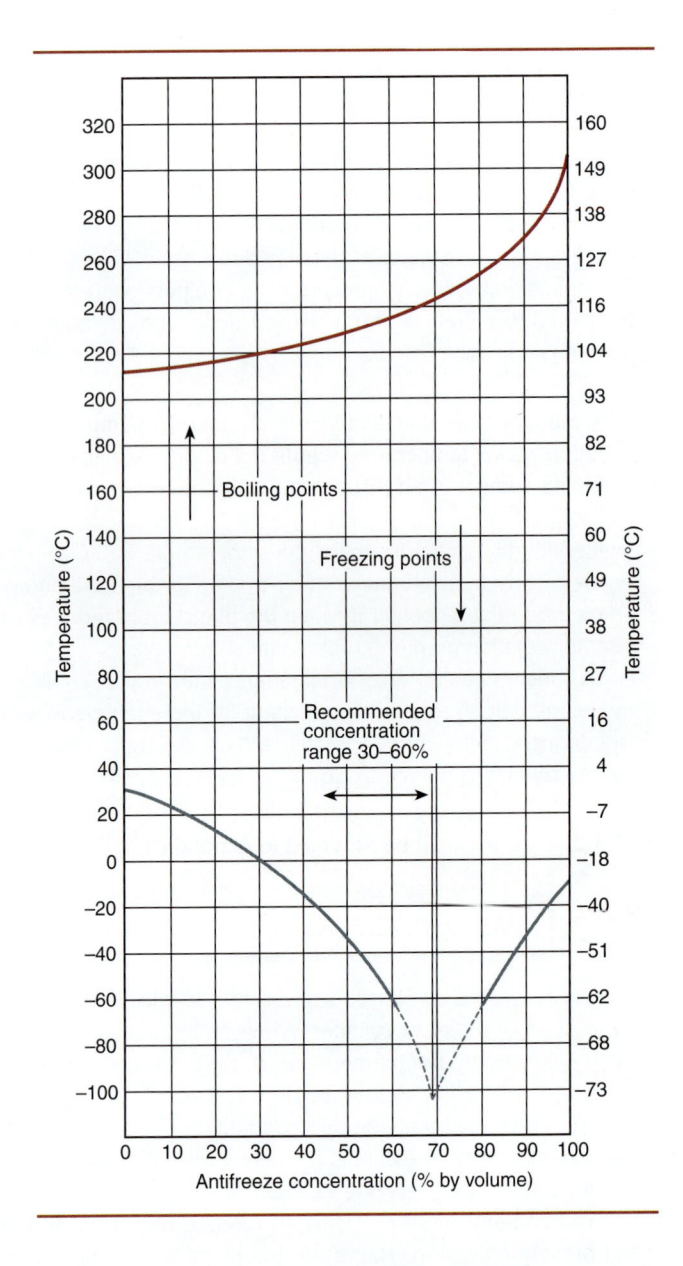

intended to prevent the engine from freezing temperatures. The coolant should be checked periodically to determine the protection level, and the solution should be checked for cleanliness. When changes are made to the coolant, a run test must be performed based on NFPA 25, Table 8.6.1, Summary of Components Replacement Testing Requirements.

11.2.8.3 An opening shall be provided in the circuit for filling the system, checking coolant level, and adding make-up coolant when required.

11.2.8.4 The coolant shall comply with the recommendation of the engine manufacturer.

11.2.8.5* Heat Exchanger Water Supply.

A.11.2.8.5 See Figure A.11.2.8.5. Water supplied for cooling the heat exchanger is sometimes circulated directly through water-jacketed exhaust manifolds or engine aftercoolers, or both, in addition to the heat exchangers.

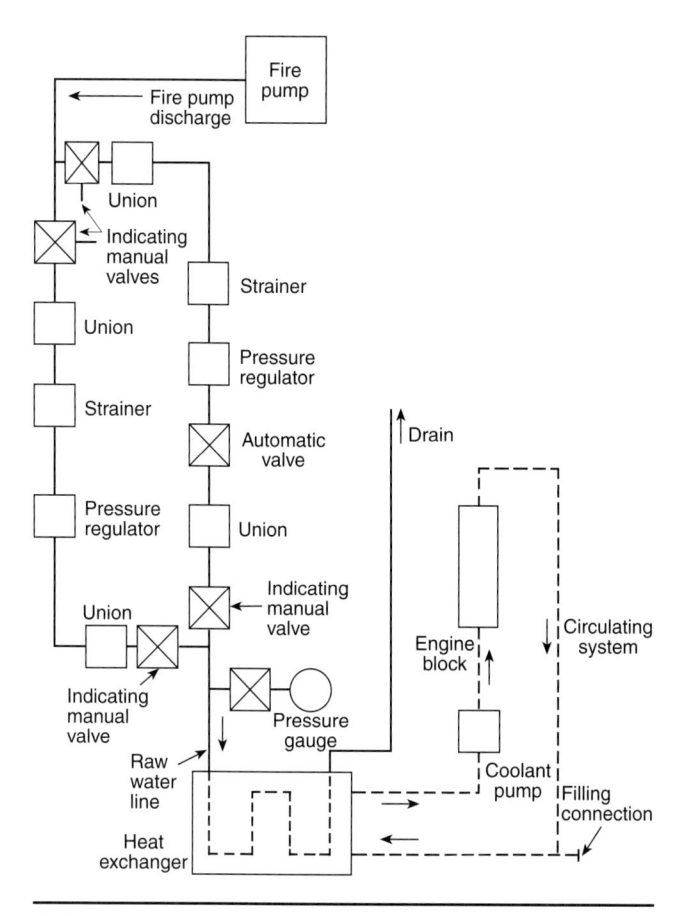

FIGURE A.11.2.8.5 *Cooling Water Line with Bypass.*

11.2.8.5.1 The cooling water supply for a heat exchanger–type system shall be from the discharge of the pump taken off prior to the pump discharge check valve.

11.2.8.5.2 The cooling water flow required shall be set based on the maximum ambient cooling water.

> The water supply for heat exchanger–cooled engines must provide an adequate volume (up to 120 percent of the cooling water required when the engine is operating at maximum brake horsepower) to sufficiently cool the engine. The efficiency of the heat exchanger depends on the temperature of the cooling water that enters the exchanger. Refer to the engine manufacturer's instructions for the recommended waterflow through the heat exchanger, based on the cooling water inlet temperature. See 11.2.8.5.3.4 for more information.

11.2.8.5.3 Heat Exchanger Water Supply Components.

> Engine heat exchangers have low-pressure limits 15 psi to 60 psi (1 bar to 5 bars). The pressure regulator in the water supply and the bypass piping is provided to reduce the pump outlet pressure to a level below the heat exchanger limit.

11.2.8.5.3.1 Threaded rigid piping shall be used for this connection.

11.2.8.5.3.2 Nonmetallic flexible sections shall be allowed between the pump discharge and cooling water supply assembly inlet, and between the cooling water supply assembly discharge and engine inlet, provided they have at least 2 times the fire pump discharge rated pressure and have a 30-minute fire resistance rating equal to ISO 15540, *Fire Resistance of Hose Assemblies.*

11.2.8.5.3.3 The pipe connection in the direction of flow shall include an indicating manual shutoff valve, an approved flushing-type strainer in addition to the one that can be a part of the pressure regulator, a pressure regulator, an automatic valve, and a second indicating manual shutoff valve or a spring-loaded check valve.

11.2.8.5.3.4 The indicating manual shutoff valves shall have permanent labeling with minimum ½ in. (12.7 mm) text that indicates the following: For the valve in the heat exchanger water supply, "Normal/Open" for the normal open position when the controller is in the automatic position and "Caution: Nonautomatic/Closed" for the emergency or manual position.

11.2.8.5.3.5 The pressure regulator shall be of such size and type that it is capable of and adjusted for passing approximately 120 percent of the cooling water required when the engine is operating at maximum brake horsepower and when the regulator is supplied with water at the pressure of the pump when it is pumping at 150 percent of its rated capacity.

11.2.8.5.3.6 Automatic Valve.

(A) An automatic valve listed for fire protection service shall permit flow of cooling water to the engine when it is running.
(B) Energy to operate the automatic valve shall come from the diesel driver or its batteries and shall not come from the building.
(C) The automatic valve shall be normally closed.
(D) The automatic valve shall not be required on a vertical shaft turbine–type pump or any other pump when there is no pressure in the discharge when the pump is idle.

11.2.8.5.3.7 A pressure gauge shall be installed in the cooling water supply system on the engine side of the last valve in the heat exchanger water supply and bypass supply.

11.2.8.5.3.8 Potable Water Separation (Optional.) Where two levels of separation for possible contaminants to the ground or potable water source are required by the authority having jurisdiction, dual spring-loaded check valves or backflow preventers shall be installed.

(A)* The spring-loaded check valve(s) shall replace the second indicating manual shutoff valve(s) in the cooling loop assembly as stated in 11.2.8.5.3.4.
(B)* If backflow preventers are used, the devices shall be listed for fire protection service and installed in parallel in the water supply and water supply bypass assembly.
(C) Where the authority having jurisdiction requires the installation of backflow prevention devices in connection with the engine, special consideration shall be given to the increased pressure loss, which will require that the cooling loop pipe size be evaluated and documented by engineering calculations to demonstrate compliance with the engine manufacturer's recommendation.

A.11.2.8.5.3.8(A) See Figure A.11.2.8.5.3.8(A).

A.11.2.8.5.3.8(B) See Figure A.11.2.8.5.3.8(B).

11.2.8.6* Heat Exchanger Water Supply Bypass.

A.11.2.8.6 Where the water supply can be expected to contain foreign materials, such as wood chips, leaves, lint, and so forth, the strainers required in 11.2.8.5 should be of the duplex filter type. Each filter (clean) element should be of sufficient filtering capacity to permit full water flow for a 3-hour period. In addition, a duplex filter of the same size should be installed in the bypass line. *(See Figure A.11.2.8.5.)*

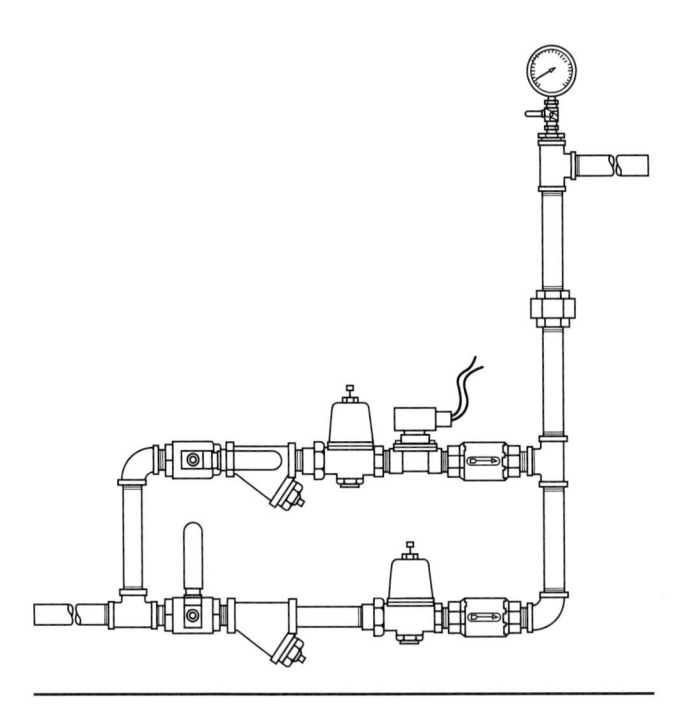

FIGURE A.11.2.8.5.3.8(A) Spring-Loaded Check Valve Arrangement.

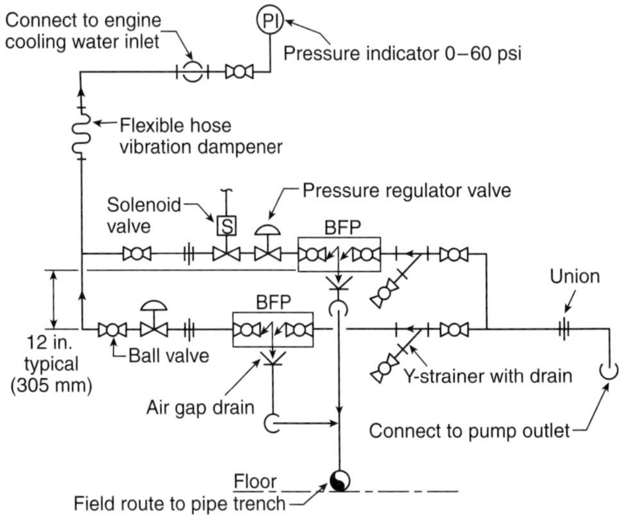

FIGURE A.11.2.8.5.3.8(B) Backflow Preventer Arrangement.

11.2.8.6.1 A threaded rigid pipe bypass line shall be installed around the heat exchanger water supply.

11.2.8.6.2 The pipe connection in the direction of flow shall include an indicating manual shutoff valve, an approved flushing-type strainer in addition to the one that can be a part of the pressure regulator, a pressure regulator, and an indicating manual shutoff valve or a spring-loaded check valve.

11.2.8.6.3 The indicating manual shutoff valves shall have permanent labeling with minimum ½ in. (12.7 mm) text that indicates the following: For the valve in the heat exchanger water supply bypass, "Normal/Closed" for the normal closed position when the controller is in the automatic position and "Emergency/Open" for manual operation or when the engine is overheating.

> The water supply bypass is required to ensure the engine always has an adequate flow of cooling water, even during maintenance of the cooling water line. If the flow through the water supply begins to deteriorate, the reading on the pressure gauge goes down, indicating a potential overheating condition.

11.2.8.7 Heat Exchanger Waste Outlet.

11.2.8.7.1 An outlet shall be provided for the wastewater line from the heat exchanger, and the discharge line shall not be less than one size larger than the inlet line.

> The wastewater line piping must be sized to handle the flow of cooling water leaving the discharge outlet. The size of the diesel engine drive will determine the size of the piping and drain. Small diesel engines may require only 12 gpm to 15 gpm (45 L/min to 57 L/min) to keep the engine temperature at the correct level. However, a large-tier three-engine arrangement may

> require as much as 30 gpm (114 L/min) to control the engine temperature at an acceptable level. Contact the engine manufacturer's data sheet for the required flow needed to cool the engine. Local environmental rules should also be followed for the disposal of the wastewater.

11.2.8.7.2 The outlet line shall be as short as practical, shall provide discharge into a visible open waste cone, and shall have no valves in it.

11.2.8.7.3 The outlet shall be permitted to discharge to a suction reservoir, provided a visual flow indicator and temperature indicator are installed.

11.2.8.7.4 When the waste outlet piping is longer than 15 ft (4.6 m) or its outlet discharges are more than 4 ft (1.2 m) higher than the heat exchanger, or both, the pipe size shall be increased by at least one size.

11.2.8.8 Radiators.

11.2.8.8.1 The heat from the primary circuit of a radiator shall be dissipated by air movement through the radiator created by a fan included with, and driven by, the engine.

11.2.8.8.2 The radiator shall be designed to limit maximum engine operating temperature with an inlet air temperature of 120°F (49°C) at the combustion air cleaner inlet.

11.2.8.8.3 The radiator shall include the plumbing to the engine and a flange on the air discharge side for the connection of a flexible duct from the discharge side to the discharge air ventilator.

11.2.8.8.4 Fan.

11.2.8.8.4.1 The fan shall push the air through the radiator to be exhausted from the room via the air discharge ventilator.

11.2.8.8.4.2 To ensure adequate airflow through the room and the radiator, the fan shall be capable of a 0.5 in. water column (13 mm water column) restriction created by the combination of the air supply and the discharge ventilators in addition to the radiator, fan guard, and other engine component obstructions.

11.2.8.8.4.3 The fan shall be guarded for personnel protection.

11.2.9 Engine Lubrication.

11.2.9.1 The engine manufacturer's recommendations for oil heaters shall be followed.

> Certain countries require the employment of an oil heater in the lube oil reservoir. The manufacturer's selection and installation instructions must be followed in the application of a lube oil heater. A thermo-well installation may be required to prevent the "cooking" of the lubrication oil. If the crankcase oil heater damages the oil, the engine may fail during emergency operation.

11.3* Pump Room

A.11.3 The engine-driven pump can be located with an electric-driven fire pump(s) in a pump house or pump room that should be entirely cut off from the main structure by noncombustible construction. The fire pump house or pump room can contain facility pumps and/or equipment as determined by the authority having jurisdiction.

11.3.1 The floor or surface around the pump and engine shall be pitched for adequate drainage of escaping water away from critical equipment, such as pump, engine, controller, fuel tank, and so forth.

11.3.2* Ventilation.

A.11.3.2 For optimum room ventilation, the air supply ventilator and air discharge should be located on opposite walls.

When calculating the maximum temperature of the pump room, the radiated heat from the engine, the radiated heat from the exhaust piping, and all other heat-contributing sources should be considered.

If the pump room is to be ventilated by a power ventilator, reliability of the power source during a fire should be considered. If the power source is unreliable, the temperature rise calculation should assume the ventilator is not operable.

Air consumed by the engine for combustion should be considered as part of the air changes in the room.

Pump rooms with heat exchanger–cooled engines typically require more air changes than engine air consumption can provide. To control the temperature rise of the room, additional air flow through the room is normally required. *[See Figure A.11.3.2(a).]*

Pump rooms with radiator-cooled engines could have sufficient air changes due to the radiator discharge and engine consumption. *[See Figure A.11.3.2(b).]*

11.3.2.1 Ventilation shall be provided for the following functions:

(1) To control the maximum temperature to 120°F (49°C) at the combustion air cleaner inlet with engine running at rated load
(2) To supply air for engine combustion
(3) To remove any hazardous vapors
(4) To supply and exhaust air as necessary for radiator cooling of the engine when required

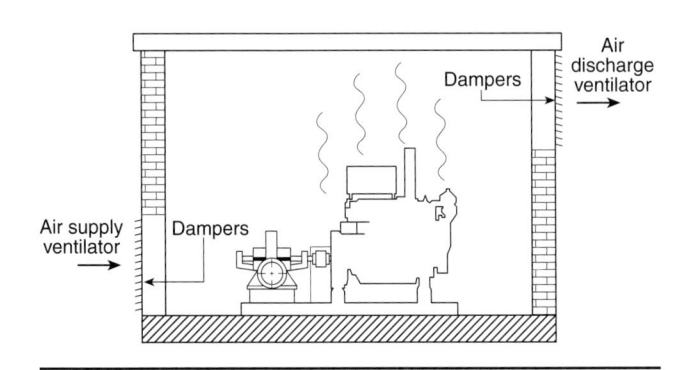

FIGURE A.11.3.2(a) *Typical Ventilation System for a Heat Exchanger–Cooled Diesel-Driven Pump.*

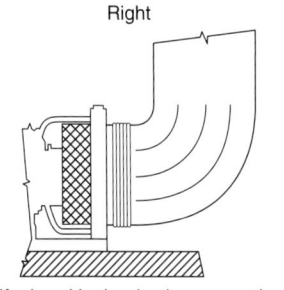

If a bend in the ducting cannot be avoided, it should be radiused and should include turning vanes to prevent turbulence and flow restriction.

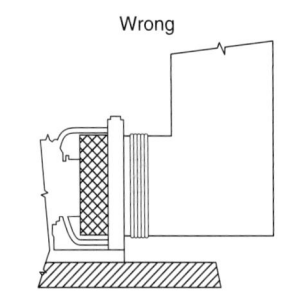

This configuration should not be used; turbulence will not allow adequate air flow.

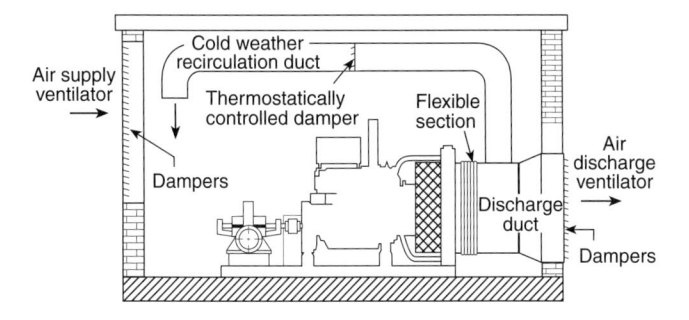

FIGURE A.11.3.2(b) *Typical Ventilation System for a Radiator-Cooled Diesel-Driven Pump.*

11.3.2.2 The ventilation system components shall be coordinated with the engine operation.

11.3.2.3* Air Supply Ventilator.

A.11.3.2.3 When motor-operated dampers are used in the air supply path, they should be spring operated to the open position and motored closed. Motor-operated dampers should be signaled to open when or before the engine begins cranking to start.

It is necessary that the maximum air flow restriction limit for the air supply ventilator be compatible with listed engines to ensure adequate air flow for cooling and combustion. This restriction typically includes louvers, bird screens, dampers, ducts, and anything else in the air supply path between the pump room and the outdoors.

Motor-operated dampers are recommended for the heat exchanger–cooled engines to enhance convection circulation.

Gravity-operated dampers are recommended for use with radiator-cooled engines to simplify their coordination with the air flow of the fan.

Another method of designing the air supply ventilator in lieu of dampers is to use a vent duct (with rain cap), the top of which extends through the roof or outside wall of a pump house and the bottom of which is approximately 6 in. (152.4 mm) off the floor of the pump house. This passive method reduces heat loss in the winter. Sizing of this duct must meet the requirements of 11.3.2.1.

11.3.2.3.1 The air supply ventilator shall be considered to include anything in the air supply path to the room.

11.3.2.3.2 The total air supply path to the pump room shall not restrict the flow of the air more than 0.2 in. water column (5.1 mm water column).

11.3.2.4* Air Discharge Ventilator.

A.11.3.2.4 When motor-operated dampers are used in the air discharge path, they should be spring operated to the open position, motored closed, and signaled to open when or before the engine begins cranking to start.

Prevailing winds can work against the air discharge ventilator. Therefore, the winds should be considered in determining the location for the air discharge ventilator. *(See Figure A.11.3.2.4 for the recommended wind wall design.)*

For heat exchanger–cooled engines, an air discharge ventilator with motor-driven dampers designed for convection circulation is preferred in lieu of a power ventilator. This arrangement requires the size of the ventilator to be larger, but it is not dependent on a power source that might not be available during the pump operation.

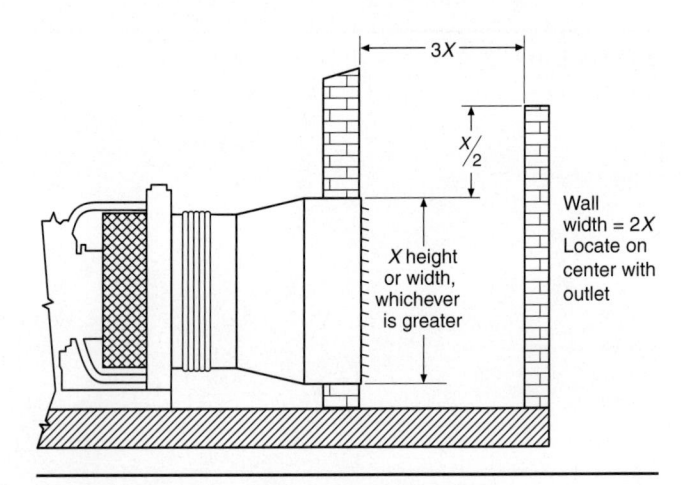

FIGURE A.11.3.2.4 *Typical Wind Wall.*

For radiator-cooled engines, gravity-operated dampers are recommended. Louvers and motor-operated dampers are not recommended due to the restriction to air flow they create and the air pressure against which they must operate.

The maximum air flow restriction limit for the air discharge ventilator is necessary to be compatible with listed engines to ensure adequate air flow cooling.

11.3.2.4.1 The air discharge ventilator shall be considered to include anything in the air discharge path from the engine to the outdoors.

11.3.2.4.2 The air discharge ventilator shall allow sufficient air to exit the pump room to satisfy 11.3.2.

11.3.2.4.3 Radiator-Cooled Engines.

11.3.2.4.3.1 For radiator-cooled engines, the radiator discharge shall be ducted outdoors in a manner that will prevent recirculation.

11.3.2.4.3.2 The duct shall be attached to the radiator via a flexible section.

11.3.2.4.3.3 The air discharge path for radiator-cooled engines shall not restrict the flow of air more than 0.3 in. water column (7.6 mm water column).

11.3.2.4.3.4* A recirculation duct shall be permitted for cold weather operation provided that the following requirements are met:

(1) The recirculation airflow shall be regulated by a thermostatically controlled damper.
(2) The control damper shall fully close in a failure mode.
(3) The recirculated air shall be ducted to prevent direct recirculation to the radiator.
(4) The recirculation duct shall not cause the temperature at the combustion air cleaner inlet to rise above 120°F (49°C).
(5) The bypass shall be installed in such a way as to supply air to the room when needed and when the control damper is open, and not exhaust air from the room.

A.11.3.2.4.3.4 If not properly installed, the bypass duct can draw air rather than supply it, due to venturi effect.

11.3.3 The entire pump room shall be protected with fire sprinklers in accordance with NFPA 13, *Standard for the Installation of Sprinkler Systems*, as an Extra Hazard Group 2 space.

> **▼ AHJ FAQ**
>
> **Sprinkler plans show the protection criteria for the pump room with a diesel engine–driven pump based on Ordinary Hazard Group 2. Is this protection criteria acceptable?**
>
> *ANSWER:* No. The fire pump room that houses a diesel engine–driven fire pump will also include a diesel fuel supply. Since the presence of this fuel is considered storage of a combustible liquid, the protection criteria is required to be Extra Hazard Group 2.

11.4 Fuel Supply and Arrangement

11.4.1 General.

11.4.1.1 Plan Review. Before any fuel system is installed, plans shall be prepared and submitted to the authority having jurisdiction for agreement on suitability of the system for prevailing conditions.

11.4.1.2 Tank Construction.

11.4.1.2.1 Tanks shall be single wall or double wall and shall be designed and constructed in accordance with recognized engineering standards such as ANSI/UL 142, *Standard for Steel Aboveground Tanks for Flammable and Combustible Liquids*.

11.4.1.2.2 Tanks shall be securely mounted on noncombustible supports.

11.4.1.2.3 Tanks used in accordance with the rules of this standard shall be limited in size to 1320 gal (4996 L).

11.4.1.2.3.1 For situations where fuel tanks in excess of 1320 gal (4996 L) are being used, the rules of NFPA 37, *Standard for the Installation and Use of Stationary Combustion Engines and Gas Turbines*, shall apply.

11.4.1.2.4 Single wall fuel tanks shall be enclosed with a wall, curb, or dike sufficient to hold the entire capacity of the tank.

> When calculating the size of the containment wall or dike, the engineer must consider the overhead flow from the required sprinkler system. Sprinkler systems for pump rooms are calculated at 0.25 gpm/ft² (10.19 L/min · m²) over the entire area. The sprinkler flow duration for the hazard must be used when calculating the size of the containment. In most applications, a listed double-wall fuel tank requires less space within the pump room and is much safer. The minimum oversizing of the containment should be at least 110 percent of the total tank capacity.

11.4.1.2.5 Each tank shall have suitable fill, drain, and vent connections.

11.4.1.2.6 Fill pipes that enter the top of the tank shall terminate within 6 in. (152 mm) of the bottom of the tank and shall be installed or arranged so that vibration is minimized.

11.4.1.2.7 The fuel tank shall have one 2 in. (50.8 mm) NPT threaded port in the top, near the center, of the tank to accommodate the low fuel level switch.

11.4.1.2.8 Tank Venting. If a double-wall tank is installed, the interstitial space between the shells of the diesel fuel storage tank shall be monitored for leakage and annunciated by the engine drive controller. The signal shall be of the supervisory type.

> The signal from the initiating device will be transmitted from the engine controller through to a constantly attended area utilizing a central station monitoring system or a remote alarm panel.

11.4.1.2.8.1 Normal vents shall be sized in accordance with ANSI/UL 142, *Standard for Steel Aboveground Tanks for Flammable and Combustible Liquids*, or other approved standards.

11.4.1.2.8.2 As an alternate to the requirement in 11.4.1.2.8.1, the normal vent shall be at least as large as the largest filling or withdrawal connection, but in no case shall it be less than 1¼ in. (32 mm) nominal inside diameter.

11.4.1.2.8.3 Vent Piping.

(A) Vent piping shall be arranged so that the vapors are discharged upward or horizontally away from adjacent walls and so that vapors will not be trapped by eaves or other obstructions.

(B) Outlets shall terminate at least 5 ft (1.5 m) from building openings.

11.4.1.2.9 Engine Supply Connection.

11.4.1.2.9.1 The fuel supply pipe connection shall be located on a side of the tank.

11.4.1.2.9.2 The engine fuel supply (suction) pipe connection shall be located on the tank so that 5 percent of the tank volume provides a sump volume not usable by the engine.

11.4.2* Fuel Supply Tank and Capacity.

A.11.4.2 The quantity 1 gal per hp (5.07 L per kW) is equivalent to 1 pint per hp (0.634 L per kW) per hour for 8 hours. Where prompt replenishment of fuel supply is unlikely, a reserve supply should be provided along with facilities for transfer to the main tanks.

11.4.2.1* Fuel supply tank(s) shall have a capacity at least equal to 1 gal per hp (5.07 L per kW), plus 5 percent volume for expansion and 5 percent volume for sump.

> ▶ **FAQ** How is a fuel storage tank sized?
>
> An example of how to size a fuel storage tank follows.
>
> ▶ **EXAMPLE**
>
> What is the required fuel tank capacity for a 120 hp (89.5 kW) engine driving a fire pump?   ◀ **CALCULATION**
>
> *Solution:* The initial calculations based on 11.4.2.1 are 1 gal/hp × 120 hp = 120 gal (5.07 L/kW × 89.5 kW × 454 L).
>
> Total capacity is determined by adding the allowances for fuel expansion and sump to the answer obtained in the preceding equation:
>
> 120 gal + [120 gal × (5% + 5%)] = 120 gal + 12 gal = 132 gal
> Or 454 L + [454 L × (5% + 5%)] = 454 L + 45.4 L = 499.4 L
>
> Therefore, the minimum size fuel tank for a 120 hp (89.5 kW) engine driving a fire pump is 132 gal (499.4 L).
>
> The 5 percent sump in the fuel tank is provided for any sediment or condensation that may be in the tank. By positioning the supply connection to the engine on the side of the tank at an elevation no lower than the engine fuel pump, the plumbing to the engine is always under a small head pressure. In the event of a leak, fuel will escape and be observed. This method is preferable to air being drawn in, which could result in the possible loss of engine prime, if the supply line is under negative pressure.

A.11.4.2.1 Where the authority having jurisdiction approves the start of the fire pump on loss of ac power supply, provisions should be made to accommodate the additional fuel needed for this purpose.

11.4.2.2 Whether larger-capacity fuel supply tanks are required shall be determined by prevailing conditions, such as refill cycle and fuel heating due to recirculation, and shall be subject to special conditions in each case.

11.4.2.3 The fuel supply tank and fuel shall be reserved exclusively for the fire pump diesel engine.

11.4.2.4 There shall be a separate fuel supply tank for each engine.

11.4.2.5 There shall be a separate fuel supply and return line for each engine.

11.4.2.6 Tank Level Indication.

11.4.2.6.1 Means other than sight tubes for continuous indicating of the amount of fuel in each storage tank shall be provided.

11.4.2.6.2 A fuel level indicator shall be provided to activate at the two-thirds tank level.

11.4.2.6.3 The low fuel level condition shall initiate a supervisory signal.

11.4.3* Fuel Tank Supply Location.

A.11.4.3 Diesel fuel storage tanks preferably should be located inside the pump room or pump house, if permitted by local regulations. Fill and vent lines in such case should be extended to outdoors. The fill pipe can be used for a gauging well where practical.

11.4.3.1 Diesel fuel supply tanks shall be located above ground in accordance with municipal or other ordinances and in accordance with requirements of the authority having jurisdiction and shall not be buried.

11.4.3.2 In zones where freezing temperatures [32°F (0°C)] are possible, the fuel supply tanks shall be located in the pump room.

11.4.3.3 The supply tank shall be located so the fuel supply pipe connection to the engine is no lower than the level of the engine fuel transfer pump.

11.4.3.4 The engine manufacturer's fuel pump static head pressure limits shall not be exceeded when the level of fuel in the tank is at a maximum.

11.4.4* Fuel Piping.

A.11.4.4 NFPA 31, *Standard for the Installation of Oil-Burning Equipment*, can be used as a guide for diesel fuel piping. Figure A.11.4.4 shows a suggested diesel engine fuel system.

11.4.4.1 Flame-resistant reinforced flexible hose with a 30–minute fire resistance rating equal to ISO 15540, *Fire Resistance of Hose Assemblies*, and a pressure rating no less than 2 times the fuel supply and return working pressure with threaded connections shall be provided at the engine for connection to fuel system piping.

11.4.4.2 Fuel piping shall not be galvanized steel or copper.

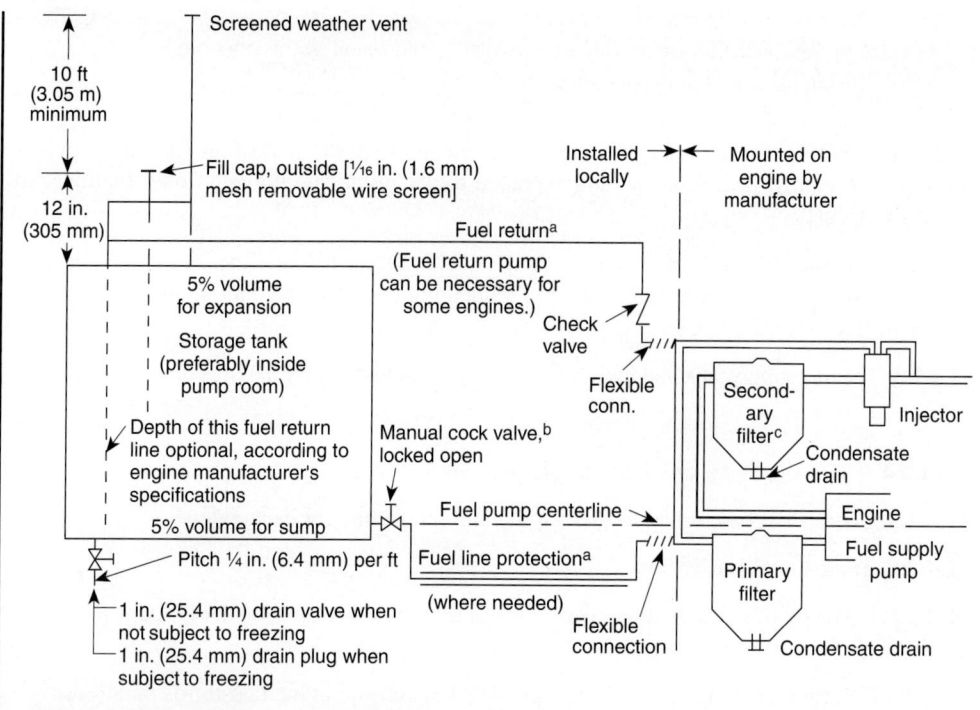

[a]Size fuel piping according to engine manufacturer's specifications.
[b]Excess fuel can be returned to fuel supply pump suction, if recommended by engine manufacturer.
[c]Secondary filter behind or before engine fuel pump, according to engine manufacturer's specifications.

FIGURE A.11.4.4 *Fuel System for Diesel Engine–Driven Fire Pump.*

> It is very important to follow the requirements set forth in two NFPA publications – NFPA 37, *Standard for the Installation and Use of Stationary Combustion Engines and Gas Turbines*, and NFPA 30, *Flammable and Combustible Liquids Code*.
>
> All piping materials must be black steel or stainless steel pipe and fittings. The use of galvanized piping creates a problem with scaling of the coating and possible clogging of the fuel lines or injectors. Copper materials cause microbiologically influenced corrosion (MIC) growth in the fuel system and storage tank and can cause failure of the fuel system.
>
> The fuel supply and return lines shall be installed using the correct materials to prevent fouling of the stored fuel. The lines are required to be protected from damage. The lines are required to be installed without any of the traps that often occur when block or concrete walls are used for containment.

11.4.4.3 The fuel return line shall be installed according to the engine manufacturer's recommendation.

11.4.4.4 There shall be no shutoff valve in the fuel return line to the tank.

11.4.4.5 A manual shutoff valve shall be provided within the tank fuel supply line.

11.4.4.5.1 The valve shall be locked in the open position.

11.4.4.5.2 No other valve than a manual locked open valve shall be put in the fuel line from the fuel tank to the engine.

11.4.4.6* Fuel Line Protection. A guard, pipe protection, or approved double-walled pipe shall be provided for all exposed fuel lines.

A.11.4.4.6 A means, such as covered floor trough, angle, channel steel, or other adequate protection cover(s) (mechanical or nonmechanical), should be used on all fuel line piping "exposed to traffic," to prevent damage to the fuel supply and return lines between the fuel tank and diesel driver.

11.4.4.7 Fuel Solenoid Valve. Where an electric solenoid valve is used to control the engine fuel supply, it shall be capable of manual mechanical operation or of being manually bypassed in the event of a control circuit failure.

11.4.5* Fuel Type.

A.11.4.5 The pour point and cloud point should be at least 10°F (5.6°C) below the lowest expected fuel temperature. *(See 4.12.2 and 11.4.3.)*

11.4.5.1* The type and grade of diesel fuel shall be as specified by the engine manufacturer.

A.11.4.5.1 Biodiesel and other alternative fuels are not recommended for diesel engines used for fire protection because of the unknown storage life issues. It is recommended that these engines use only petroleum fuels.

11.4.5.2 In areas where local air quality management regulations allow only the use of DF #1 fuel and no diesel fire pump driver is available listed for use with DF #1 fuel, an engine listed for use with DF #2 shall be permitted to be used but shall have the nameplate rated horsepower derated 10 percent, provided the engine manufacturer approves the use of DF #1 fuel.

11.4.5.3 The grade of fuel shall be indicated on the engine nameplate required in 11.2.2.1.

11.4.5.4 The grade of fuel oil shall be indicated on the fuel tank by letters that are a minimum of 6 in. (152 mm) in height and in contrasting color to the tank.

11.4.5.5 Residual fuels, domestic heating furnace oils, and drained lubrication oils shall not be used.

11.4.6* Static Electricity.

A.11.4.6 The prevention of electrostatic ignition in equipment is a complex subject. Refer to NFPA 77, *Recommended Practice on Static Electricity,* for more guidance.

11.4.6.1 The tank, pump, and piping shall be designed and operated to prevent electrostatic ignitions.

11.4.6.2 The tank, pump, and piping shall be bonded and grounded.

11.4.6.3 The bond and ground shall be physically applied or shall be inherently present by the nature of the installation.

11.4.6.4 Any electrically isolated section of metallic piping or equipment shall be bonded and grounded to prevent hazardous accumulation of static electricity.

> **AHJ FAQ**
>
> The fire pump room for a diesel engine–driven fire pump is not constantly attended; the concern is that a low fuel indication on the controller will not be heard. How should this problem be addressed?
>
> *ANSWER:* Section 11.4.6.2.3 now requires that the low fuel indication also be a supervisory signal. This signal ensures that a low fuel (below ⅔) warning is audible where it will be heard.

11.5 Engine Exhaust

11.5.1 Exhaust Manifold. Exhaust manifolds and turbochargers shall incorporate provisions to avoid hazard to the operator or to flammable material adjacent to the engine.

11.5.2* Exhaust Piping.

A.11.5.2 A conservative guideline is that, if the exhaust system exceeds 15 ft (4.5 m) in length, the pipe size should be increased one pipe size larger than the engine exhaust outlet size for each 5 ft (1.5 m) in added length.

11.5.2.1 Each pump engine shall have an independent exhaust system.

11.5.2.2 A flexible connection with a section of stainless steel, seamless or welded corrugated (not interlocked), not less than 12 in. (305 mm) in length shall be made between the engine exhaust outlet and exhaust pipe.

11.5.2.3 The exhaust pipe shall not be any smaller in diameter than the engine exhaust outlet and shall be as short as possible.

11.5.2.4 The exhaust pipe shall be covered with high-temperature insulation or otherwise guarded to protect personnel from injury.

11.5.2.5 The exhaust pipe and muffler, if used, shall be suitable for the use intended, and the exhaust back pressure shall not exceed the engine manufacturer's recommendations.

11.5.2.6 Exhaust pipes shall be installed with clearances of at least 9 in. (229 mm) to combustible materials.

11.5.2.7 Exhaust pipes passing directly through combustible roofs shall be guarded at the point of passage by ventilated metal thimbles that extend not less than 9 in. (229 mm) above and 9 in. (229 mm) below roof construction and are at least 6 in. (152 mm) larger in diameter than the exhaust pipe.

11.5.2.8 Exhaust pipes passing directly through combustible walls or partitions shall be guarded at the point of passage by one of the following methods:

(1) Metal ventilated thimbles not less than 12 in. (305 mm) larger in diameter than the exhaust pipe
(2) Metal or burned clay thimbles built in brickwork or other approved materials providing not less than 8 in. (203 mm) of insulation between the thimble and construction material

11.5.2.9* Exhaust emission after treatment devices that have the potential to adversely impact the performance and reliability of the engine shall not be permitted.

A.11.5.2.9 Exhaust emission after treatment devices are typically dependent upon high exhaust temperature to burn away collected materials to prevent clogging. Due to the lower exhaust temperatures produced by the engine when operating at pump shutoff during weekly operation, there is a high possibility the after treatment device will accumulate collected material and will not be capable of flowing the volume of exhaust in the event the engine is required to produce full rated power for an emergency.

11.5.2.10 Where required by the authority having jurisdiction, the installation of an exhaust emission after treatment device shall be of the active regeneration type with a pressure limiting device that permits the engine exhaust to bypass the after treatment device when the engine manufacturer's maximum allowed exhaust backpressure is exceeded.

11.5.3 Exhaust Discharge Location.

11.5.3.1 Exhaust from the engine shall be piped to a safe point outside the pump room and arranged to exclude water.

11.5.3.2 Exhaust gases shall not be discharged where they will affect persons or endanger buildings.

11.5.3.3 Exhaust systems shall terminate outside the structure at a point where hot gases, sparks, or products of combustion will discharge to a safe location. [**37**:8.2.3.1]

11.5.3.4 Exhaust system terminations shall not be directed toward combustible material or structures, or into atmospheres containing flammable gases, flammable vapors, or combustible dusts. [**37**:8.2.3.2]

> **AHJ FAQ**
>
> **The plans for a fire pump installation show the exhaust for a diesel engine fire pump terminating below the outside deck of a restaurant. Is this arrangement acceptable?**
>
> *ANSWER:* There are two problems with this proposed installation. First, the exhaust gases pose a risk to the occupants on the deck. Second, if the deck is made of combustible material, the hot gases need to be directed away from that area. While this language appears in NFPA 20, it is actually material extracted from NFPA 37. While occupants of the deck would be evacuated during a fire, the pump would need to be run weekly, creating an ongoing problem with regard to deck occupants.

11.5.3.5 Exhaust systems equipped with spark-arresting mufflers shall be permitted to terminate in Division 2 locations as defined in Article 500 of *NFPA 70, National Electrical Code*. [**37**:8.2.3.3]

11.6* Diesel Engine Driver System Operation

A.11.6 Internal combustion engines necessarily embody moving parts of such design and in such number that the engines cannot give reliable service unless given diligent care. The manufacturer's instruction book covering care and operation should be readily available, and pump operators should be familiar with its contents. All of its provisions should be observed in detail.

11.6.1 Weekly Run.

11.6.1.1 Engines shall be designed and installed so that they can be started no less than once a week and run for no less than 30 minutes to attain normal running temperature.

> Running the engine for a minimum of 30 minutes provides sufficient time for the engine to reach its full operating temperature and for condensation to evaporate from the crankcase. Any irregularities in the engine operation should be investigated and repaired immediately. Running the engine for this length of time also gives the operator sufficient time to check out the pump operations, the cooling loop operation, the battery chargers, and all other pump room components requiring an inspection. Subsection 8.1.2 and Table 8.1.1.2 of NFPA 25 describe the schedules and the specific requirements for periodically required maintenance items. Additionally, 8.6.1 of NFPA 25 provides the necessary testing required after repairs or upgrades are completed on any pump room component.

11.6.1.2 Engines shall run smoothly at rated speed, except for engines addressed in 11.6.1.3.

11.6.1.3 Engines equipped with variable speed pressure limiting control shall be permitted to run at reduced speeds provided factory-set pressure is maintained and they run smoothly.

11.6.2* Engine Maintenance. Engines shall be designed and installed so that they can be kept clean, dry, and well lubricated to ensure adequate performance.

A.11.6.2 See NFPA 25, *Standard for the Inspection, Testing, and Maintenance of Water-Based Fire Protection Systems*, for proper maintenance of engine(s), batteries, fuel supply, and environmental conditions.

11.6.3 Battery Maintenance.

11.6.3.1 Storage batteries shall be designed and installed so that they can be kept charged at all times.

11.6.3.2 Storage batteries shall be designed and installed so that they can be tested frequently to determine the condition of the battery cells and the amount of charge in the battery.

> Lead-acid batteries last between 24 months and 30 months of service. When replacements of batteries are required, it is best to replace both units where two batteries are required to make a starting set. This is particularly true in a system requiring two batteries to make a set (24 V dc). See NFPA 25 for the recommended replacement schedule.

11.6.3.3 Only distilled water shall be used in battery cells.

11.6.3.4 Battery plates shall be kept submerged at all times.

> If the battery requires an excessive amount of replacement water, the battery should be checked to make sure it is not overcharging. Also, the battery plates need to be covered with water to help eliminate the possibility of battery explosions. Exposed plates may create an arc or spark that can ignite the hydrogen gases that are present during the starting and charging cycles.

11.6.3.5 The automatic feature of a battery charger shall not be a substitute for proper maintenance of battery and charger.

11.6.3.6 The battery and charger shall be designed and installed so that periodic inspection of both battery and charger is physically possible.

11.6.3.6.1 This inspection shall determine that the charger is operating correctly, the water level in the battery is correct, and the battery is holding its proper charge.

11.6.4* Fuel Supply Maintenance.

A.11.6.4 Active systems that are permanently added to fuel tanks for removing water and particulates from the fuel can be acceptable, provided the following apply:

(1) All connections are made directly to the tank and are not interconnected with the engine or its fuel supply and return piping in any way.
(2) There are no valves or other devices added to the engine or its fuel supply and return piping in any way.

11.6.4.1 The fuel storage tanks shall be designed and installed so that they can be kept as full and maintained as practical at all times but never below 66 percent (two-thirds) of tank capacity.

11.6.4.2 The tanks shall be designed and installed so that they can always be filled by means that will ensure removal of all water and foreign material.

11.6.5* Temperature Maintenance.

A.11.6.5 Proper engine temperature, in accordance with 11.2.8.2 and 11.6.5.1, maintained through the use of a supplemental heater has many benefits, as follows:

(1) Quick starting (a fire pump engine might have to carry a full load as soon as it is started)
(2) Reduced engine wear
(3) Reduced drain on batteries
(4) Reduced oil dilution
(5) Reduced carbon deposits, so that the engine is far more likely to start every time

11.6.5.1 The temperature of the pump room, pump house, or area where engines are installed shall be designed so that the temperature is maintained at the minimum recommended by the engine manufacturer and is never less than the minimum recommended by the engine manufacturer.

> Fuel tanks installed outdoors generate a greater volume of water from condensation than tanks installed inside the fire pump room. Maintenance for outdoor tanks should be increased to ensure that water in the tank is kept to a minimum. Drain test valves should be added to the fuel tank to allow for removal of sediment and water. The valves should be plugged at all times, except during a drain test.

11.6.6 Emergency Starting and Stopping.

> All personnel should be factory trained and experienced in starting and stopping the engine in all modes of operation. The manufacturer's operating requirements for starting and stopping should be followed. In addition to the manufacturer's instructions and qualification, the personnel performing the inspection, the test, or other service function should be experienced in all phases of the fire protection system. The fire pump installation interfaces with the alarm system, all sprinkler systems, and other features of the total fire protection system; therefore, the entire fire protection system is impacted by the operation of the fire pump.

11.6.6.1 The sequence for emergency manual operation, arranged in a step-by-step manner, shall be posted on the fire pump engine.

11.6.6.2 It shall be the engine manufacturer's responsibility to list any specific instructions pertaining to the operation of this equipment during the emergency operation.

References Cited in Commentary

National Fire Protection Association, 1 Batterymarch Park, Quincy, MA 02169-7471.

NFPA 25, *Standard for the Inspection, Testing, and Maintenance of Water-Based Fire Protection Systems*, 2011 edition.

NFPA 30, *Flammable and Combustible Liquids Code*, 2012 edition.

NFPA 37, *Standard for the Installation and Use of Stationary Combustion Engines and Gas Turbines*, 2010 edition.

Engine Drive Controllers

CHAPTER 12

Diesel engine–driven fire pumps are often used when independent self-contained fire protection is required. Diesel-driven fire pump controllers maximize this self-sufficiency in two ways: the engine start is powered by batteries, and the controller is independent of the ac power supply mains. Furthermore, these controllers contain extensive monitoring, supervision, and annunciation provisions, as well as contacts for remote alarm operation. Annunciation options might include pump room or pump house conditions, such as temperature, reservoir level, or valve positions, among others. Substantial redundant circuitry is employed, as well as a means for manual intervention if required.

This chapter specifies the requirements for the application, location, construction, and internal components of engine drive controllers and also delineates their relevant operating requirements, including starting and stopping of the engine. Chapter 12 also provides requirements that clarify battery failure, high cooling water temperature as a condition to be monitored and signaled, allow for silencing of the audible alarm and signal for all trouble conditions, and allow for a pressure-responsive means as an alternative to a pressure-actuated switch.

The commentary in this chapter is intended to help explain the meaning of, and often the rationale behind, the chapter's requirements. A more thorough discussion of the theory, parameters, and practical application considerations of fire pump controllers can be found in the National Fire Protection Association publication titled *Pumps for Fire Protection Systems*.

12.1 Application

12.1.1 This chapter provides requirements for minimum performance of automatic/nonautomatic diesel engine controllers for diesel engine–driven fire pumps.

Although not available for many years, a nonautomatic (manual only) engine fire pump controller, which was a small, simple switch box that energized the fuel and cranking circuits, was once utilized. Most modern engines still incorporate this switch (in addition to the manual devices on the engine itself) and are called combined automatic/nonautomatic controllers. These controllers are the type commonly used for automatic fire protection and, while equipped to be used either in manual or in automatic (standby) mode, the automatic method is normally used. Exhibit II.12.1 shows a typical solid-state automatic/nonautomatic diesel engine–driven fire pump controller.

12.1.2 Accessory devices, such as fire pump alarm and signaling means, are included where necessary to ensure minimum performance of the equipment mentioned in 12.1.1.

This provision addresses items separate from the controller itself. One example is remote alarm equipment.

12.1.3 General.

12.1.3.1 All controllers shall be specifically listed for diesel engine–driven fire pump service.

EXHIBIT II.12.1 Diesel Engine–Driven Fire Pump Controller. (Courtesy of Master Control Systems, Inc.)

323

> See 3.2.3 and A.3.2.3 for the definition and explanation of the term *listed*. "Specifically listed" means that the equipment has been examined for compliance with these requirements. Note that most, but not all, fire pump controllers are both Underwriters Laboratories (UL) listed *and* Factory Mutual (FM) approved. Furthermore, some options or modifications specified on a purchase order form may void one or another listing.

12.1.3.2 All controllers shall be completely assembled, wired, and tested by the manufacturer before shipment from the factory.

> ▶ **FAQ** How can the buyer be certain the listed equipment has been adequately tested?
>
> The amount of factory testing can vary substantially among the various listed manufacturers. Also, most, if not all, manufacturers can provide copies of test data sheets, and many can also accommodate witness testing, although this is typically an extra cost item.

12.1.3.3 Markings.

12.1.3.3.1 All controllers shall be marked "Diesel Engine Fire Pump Controller" and shall show plainly the name of the manufacturer, the identifying designation, rated operating pressure, enclosure type designation, and complete electrical rating.

> ▶ **FAQ** What is the best way to confirm that the controller received is appropriate?
>
> It is extremely important to verify that the controller is a listed and approved controller in the proper category. The words "Diesel Engine Fire Pump Controller" should be on the same nameplate as the listing mark. There have been pump controllers sold and marked as a "Fire Pump Controller" that have only general industrial or control panel listing. While this violates listing agency procedures or rules, the violation has to be reported before action can occur. See Exhibit II.12.2 for an example of a suitable fire pump controller nameplate.

EXHIBIT II.12.2 *Diesel Drive Controller Nameplate. (Courtesy of Master Control Systems, Inc.)*

12.1.3.3.2 Where multiple pumps serving different areas or portions of the facility are provided, an appropriate sign shall be conspicuously attached to each controller indicating the area, zone, or portion of the system served by that pump or pump controller.

> Fire pumps are provided in different locations, and for a number of reasons, including the following:
>
> 1. Plant expansion
> 2. Fire water drawn from different sources or water mains
> 3. Independent zones in a high-rise building, such as an upper zone fed by a separate fire water reservoir (e.g., a swimming pool)
> 4. Some large processing plants
> 5. Some multi-building campuses, especially where a fire pump may not supply the campus fire water loop (main)
>
> It is important that each controller have its own independent pressure sensing line in accordance with Section 4.30. It is specifically required that jockey pumps (pressure maintenance pumps) also have their own independent pressure sensing line. When pressure sensing lines are combined, the system can, and usually does, malfunction until the required separate pressure sensing lines are installed. See Figure A.4.30(b) and 12.7.2.1.

12.1.4 It shall be the responsibility of the pump manufacturer or its designated representative to make necessary arrangements for the services of a controller manufacturer's representative, where needed, for services and adjustment of the equipment during the installation, testing, and warranty periods.

> Typically, the installing fire pump contractor arranges for the services identified in 12.1.4.

12.2 Location

12.2.1* Controllers shall be located as close as is practical to the engines they control and shall be within sight of the engines.

A.12.2.1 If the controller must be located outside the pump room, a glazed opening should be provided in the pump room wall for observation of the motor and pump during starting. The pressure control pipeline should be protected against freezing and mechanical injury.

> It is important to provide the ability to see and hear whether an engine and pump are functioning as expected.

12.2.2 Controllers shall be so located or so protected that they will not be damaged by water escaping from pumps or pump connections.

> ▶ **FAQ** How can controllers be protected from water damage?
>
> Fire pump installations are sometimes located on lower floor levels or below grade. In the event of fire, water from fire-fighting operations can collect at these low points. By raising the current-carrying components above the floor, the likelihood of an electric hazard caused by water coming into contact with "hot" components is reduced. The electrical equipment associated with fire pump controllers is not waterproof or weatherproof, which is another reason that the equipment should be elevated above the floor.

Stationary Fire Pumps Handbook 2013

12.2.3 Current carrying parts of controllers shall not be less than 12 in. (305 mm) above the floor level.

> While the 12 in. (305 mm) requirement is adequate in most cases, it is recommended that a housekeeping pad be used for floor-mounted controllers to keep them out of puddles or casual water in the pump room. Extra protection may be needed when controllers are mounted in pits or where sump pumps are needed to keep water away from the pump installation.

12.2.4 Working clearances around controllers shall comply with *NFPA 70, National Electrical Code*, Article 110.

12.3 Construction

12.3.1 Equipment.

12.3.1.1* All equipment shall be suitable for use in locations subject to a moderate degree of moisture, such as a damp basement.

> Where conditions warrant, a more robust and protective enclosure, such as Type 4, may be appropriate. This is also true for humidistatically controlled cabinet space heaters.

A.12.3.1.1 In areas affected by excessive moisture, heat can be useful in reducing the dampness.

> This paragraph refers to cabinet (enclosure) space heaters. See the commentary following 12.3.3.1.2 for more information on cabinet heaters.

12.3.1.2 Reliability of operation shall not be adversely affected by normal dust accumulations.

> Note that Type 12 or better enclosures provide dust protection.

12.3.2 Mounting. All equipment not mounted on the engine shall be mounted in a substantial manner on a single noncombustible supporting structure.

12.3.3 Enclosures.

12.3.3.1* Mounting.

A.12.3.3.1 For more information, see NEMA 250, *Enclosures for Electrical Equipment*.

> Also refer to ANSI/UL 50, *Standard for Enclosures for Electrical Equipment, Non-Environmental Considerations*.

12.3.3.1.1 The structure or panel shall be securely mounted in, as a minimum, a NEMA Type 2 dripproof enclosure(s) or an enclosure(s) with an ingress protection (IP) rating of IP 31.

> Commentary Table II.12.1 lists common enclosure types.
> Note that the NEMA type designation is self-certified by the manufacturer and, thus, is not necessarily a reliable assurance of compliance. However, when a controller is marked with a UL type designation, such as "Type 4 Enclosure," the construction of the enclosure is included in the UL investigation (conformity assessment) of the product and is also part of the periodic

COMMENTARY TABLE II.12.1 *Fire Pump Controller Enclosure Types*

Enclosure Type*	Location	Protection Provided Against:
Type 2	Indoor	Falling water and dirt
Type 12	Indoor	Dripping water and dust (previously splash and oil resistant)
Type 13	Indoor	Spraying water (and oil) and dust
Type 3	Outdoor	Rain, dust, and ice
Type 3R	Outdoor	Falling rain (allows water, dust, and dirt to enter)
Type 4	Outdoor	Rain, dust, ice, splashing and hose-directed water (hose-tight)
Type 4XA	Outdoor	Same as NEMA 4 but corrosion resistant (special paint finish)
Type 4XB	Outdoor	Same as NEMA 4 but also of corrosion-resistant Type 304 stainless steel
Type 4XC	Outdoor	Same as NEMA 4 but also of corrosion-resistant Type 316 stainless steel
Type 4XCL	Outdoor	Same as NEMA 4XC but of Type 316 low carbon stainless steel

Note: The NEMA 4X, 4XA, 4XB, and 4XC designations are unofficial and are for reference only.

follow-up inspection. This designation does give reasonable assurance of the compliance of the enclosure construction. For example, NEMA "X" is not the same as UL Type X with respect to assurance or confidence in construction.

The IP 31 rating was new to the 2010 edition of NFPA 20, *Standard for the Installation of Stationary Pumps for Fire Protection*. It is an IEC rating that provides the same protection from dripping (vertical) water and slightly smaller openings than NEMA 2 enclosures.

12.3.3.1.2 Where the equipment is located outside or special environments exist, suitably rated enclosures shall be used.

For outdoor installations, the enclosure must be an outdoor-rated type, such as Type 3, Type 4, or Type 4X. Moreover, a sunscreen or roof is highly recommended to keep the controller internal temperature from exceeding the rating of the components, especially since the battery charger gives off heat when charging the batteries. For corrosive environments, including salt air, one of the Type 4X enclosures should be employed.

▶ **FAQ** What is the most efficient way to protect outdoor equipment where temperatures drop significantly at night?

To protect outdoor equipment from cooler temperatures at night, an optional enclosure space heater should be employed, preferably one that is thermostatically controlled, or, better yet, one that is controlled by a humidistat. This minimizes running the heater(s) when unnecessary. Better-built controllers run the heater at half voltage to lower the surface temperature of the heater. This is easily accomplished, for example, by using 240 Vac rated heaters at 120 Vac. The power is one-fourth of the rated power and, thus, the surface temperature is substantially lower.

12.3.3.2 Grounding. The enclosures shall be grounded in accordance with *NFPA 70, National Electrical Code*, Article 250.

> Grounding is important to protect the equipment from lightning and/or other surges, but it is also an essential safety feature for personnel. Local codes must be followed. Most enclosures are grounded by way of conduits or raceways and the pressure sensing line.

12.3.4 Locked Lockable Cabinet. All switches required to keep the controller in the automatic position shall be within locked enclosures having breakable glass panels.

> In an emergency, the operator can break the glass window and switch the controller from automatic to manual mode and start the diesel engine manually. Likewise, an operator can break the glass window and switch the control switch to the "off" position to perform an emergency shutdown. Some modern controllers make use of glass rods or similar devices as an alternative to the usual break glass window. Note that the old phrase *locked cabinet* has been changed to the more correct *lockable enclosure.*

12.3.5 Connections and Wiring.

12.3.5.1 Field Wiring.

12.3.5.1.1 All wiring between the controller and the diesel engine shall be stranded and sized to carry the charging or control currents as required by the controller manufacturer.

> Article 695 in *NFPA 70®*, *National Electrical Code®*, governs wiring methods for fire pump controllers. However, note that some of the material in Article 695 of *NFPA 70* is text extracted from NFPA 20. See Part 4, Supplement 3 of this handbook for the entire text of *NFPA 70,* Article 695.
>
> ▶ **FAQ** Why does 12.3.5.1.1 specify stranded wire when most wiring between the controller and the diesel engine is solid wire?
>
> Note that 12.3.5.1.1 specifies stranded wire so as to better withstand engine vibration, and it allows for the use of ring lugs. Ring lugs are used because, if a screw should come loose, as happens occasionally, the connection may still be good enough for emergency running of the pump.

12.3.5.1.2 Such wiring shall be protected against mechanical injury.

> The best wiring practice makes use of high-quality crimped ring lugs for connection to the engine junction terminal box terminal strip and the controller terminal strip(s).

12.3.5.1.3 Controller manufacturer's specifications for distance and wire size shall be followed.

> In some cases, the wire size recommended by the controller manufacturer will be larger than the minimum size specified in *NFPA 70.* Use of the larger wire size is intended to reduce the voltage drop in the wiring from the battery charger to the batteries.

12.3.5.2 Wiring Elements. Wiring elements of the controller shall be designed on a continuous-duty basis.

12.3.5.3 Field Connections.

12.3.5.3.1 A diesel engine fire pump controller shall not be used as a junction box to supply other equipment.

Only wiring directly involved with the fire pump engine and fire pump controller, and associated alarm and signals, is allowed in the fire pump controller — no other circuitry is allowed. This requirement is to both minimize the need for access to the controller, which should be locked, and to minimize any hazard or threat posed to the controller by foreign signals, voltages, or circuitry.

12.3.5.3.2 No external contacts or changes to the controller that interfere with the operation of the controller shall be installed.

In the 2013 edition of NFPA 20, the wording has been changed to make the requirement appropriate for diesel engine controllers. The previous wording was more applicable to electric drive controllers. Modifying the controller in any way could void the listing(s) and likely void the manufacturer's warranty.

12.3.5.3.3 Electrical supply conductors for pressure maintenance (jockey or make-up) pump(s) shall not be connected to the diesel engine fire pump controller.

This paragraph reinforces the requirement of 12.3.5.3.1. Specifically, anything connected to the controller ac power (mains) circuit or circuitry is capable of tripping the upstream branch circuit breaker (or fuse), leaving the controller without ac power and the batteries without a charger.

12.3.5.3.4 Diesel engine fire pump controllers shall be permitted to supply essential and necessary ac or dc power, or both, to operate pump room dampers and engine oil heaters and other associated required engine equipment only when provided with factory-equipped dedicated field terminals and overcurrent protection.

▶ **FAQ** Why is the diesel engine fire pump controller allowed to provide power to pump room dampers and engine oil heaters in the situation described in 12.3.5.3.4?

This allowance appears to be an exception to 12.3.5.3.1. However, the difference is a factory-equipped arrangement that has gone through the approval process with the listing agency. Done properly, the controller's overcurrent protection (fuses or circuit breakers) will be coordinated with upstream protection and will prevent tripping that protection if a short circuit or fault occurs in the auxiliary equipment or circuit.

12.3.6 Electrical Diagrams and Instructions.

12.3.6.1 A field connection diagram shall be provided and permanently attached to the inside of the enclosure.

The diagram required in 12.3.6.1 is used by the owner and contractor to identify internal circuits of the controller. See Exhibit II.12.3 and Exhibit II.12.4 for typical engine-to-controller wiring diagrams.

12.3.6.2 The field connection terminals shall be plainly marked to correspond with the field connection diagram furnished.

All required field connection terminals must be shown on the drawing(s) and permanently attached to the inside of the controller in addition to being shown in the instruction manual.

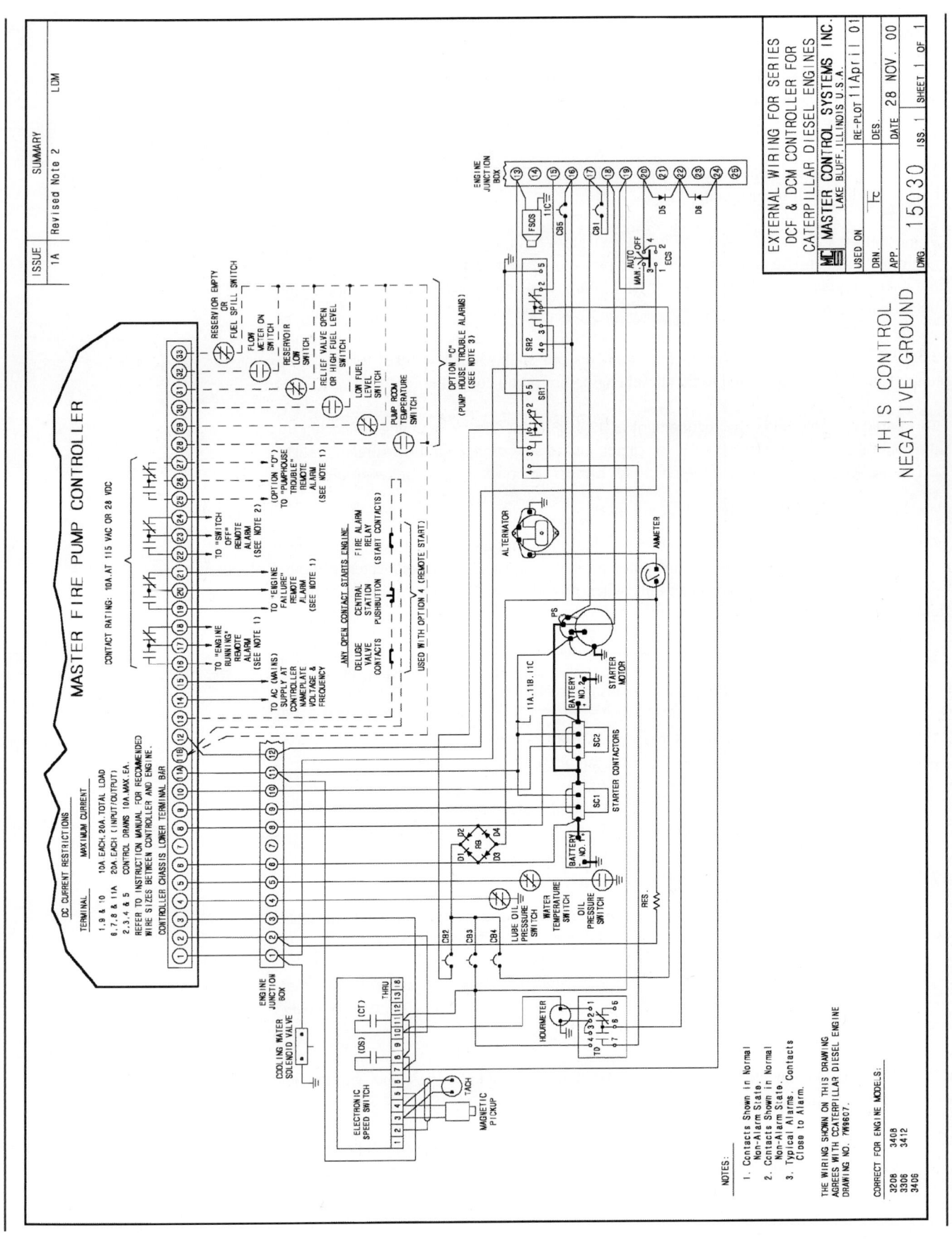

EXHIBIT II.12.3 External Wiring Diagram for Caterpillar Engines to Stop Diesel Engine. (Courtesy of Master Control Systems, Inc.)

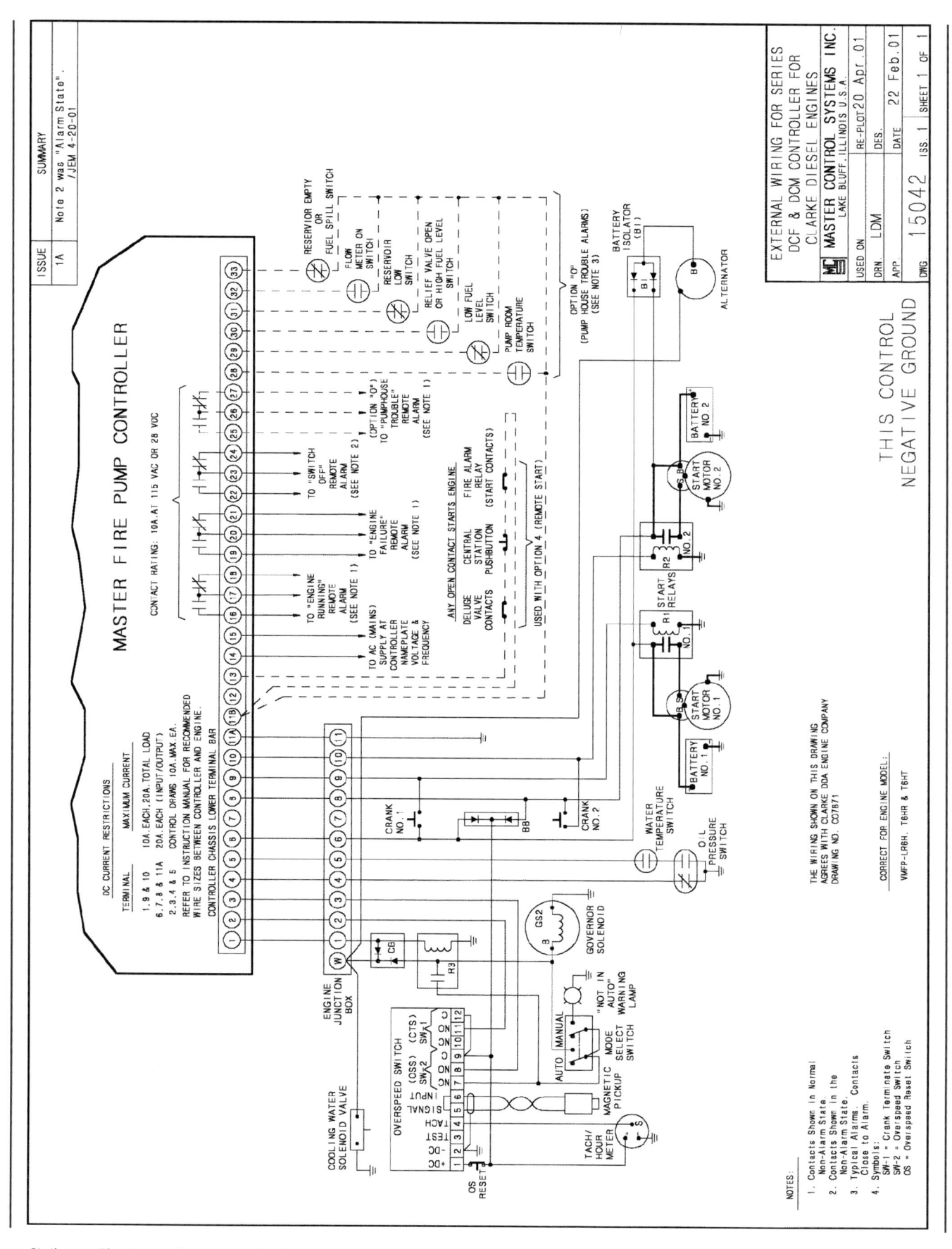

EXHIBIT II.12.4 External Wiring Diagram for Clarke Dual Starter Motor Diesel Engine. (Courtesy of Master Control Systems, Inc.)

12.3.6.3 For external engine connections, the field connection terminals shall be commonly numbered between the controller and the engine terminals.

12.3.6.4 The installation instructions of the manufacturer of the fire pump controller shall be followed.

> ▶ **FAQ** Why is it important to follow the manufacturer's instructions when installing a fire pump controller?

Too often, equipment is not installed according to the manufacturer's instructions or is installed without regard to the manufacturer's instructions. These practices lead to noncompliant installations. Noncompliant installations are sometimes, but not always, corrected. The requirement of 12.3.6.4 provides inspectors with a definitive mandated reference when they are asked questions such as the following: "Where does it say I am prohibited from doing that?"; "Where does it say I am required to do that?"; or "Where does it say I am required to do it that way?"

12.3.7 Marking.

12.3.7.1 Each operating component of the controller shall be plainly marked with the identification symbol that appears on the electrical schematic diagram.

> This marking is to make field replacement of components and assemblies easier and reduce the likelihood of mistakes.

12.3.7.2 The markings shall be located so as to be visible after installation.

> Labels and designations must not be covered over by other components or assemblies.

12.3.8* Instructions. Complete instructions covering the operation of the controller shall be provided and conspicuously mounted on the controller.

> ▶ **FAQ** What steps should the instructions include to be considered "complete"?

The instruction plate should include clear and concise instructions for emergency starting and stopping and also how to set the controller's selector, control, and/or mode switch(es) for automatic (standby) fire protection service mode.

A.12.3.8 Pump operators should be familiar with instructions provided for controllers and should observe in detail all their recommendations.

12.4 Components

> Exhibit II.12.5 shows the components of a relay-style diesel drive fire pump controller, while Exhibit II.12.18 shows components for a modern solid-state controller that employs hard-wired logic modules and relays for control, reporting, and communication.

12.4.1 Indicators on Controller.

12.4.1.1 All visible indicators shall be plainly visible.

12.4.1.2* Visible indication shall be provided to indicate that the controller is in the automatic position. If the visible indicator is a pilot lamp, it shall be accessible for replacement.

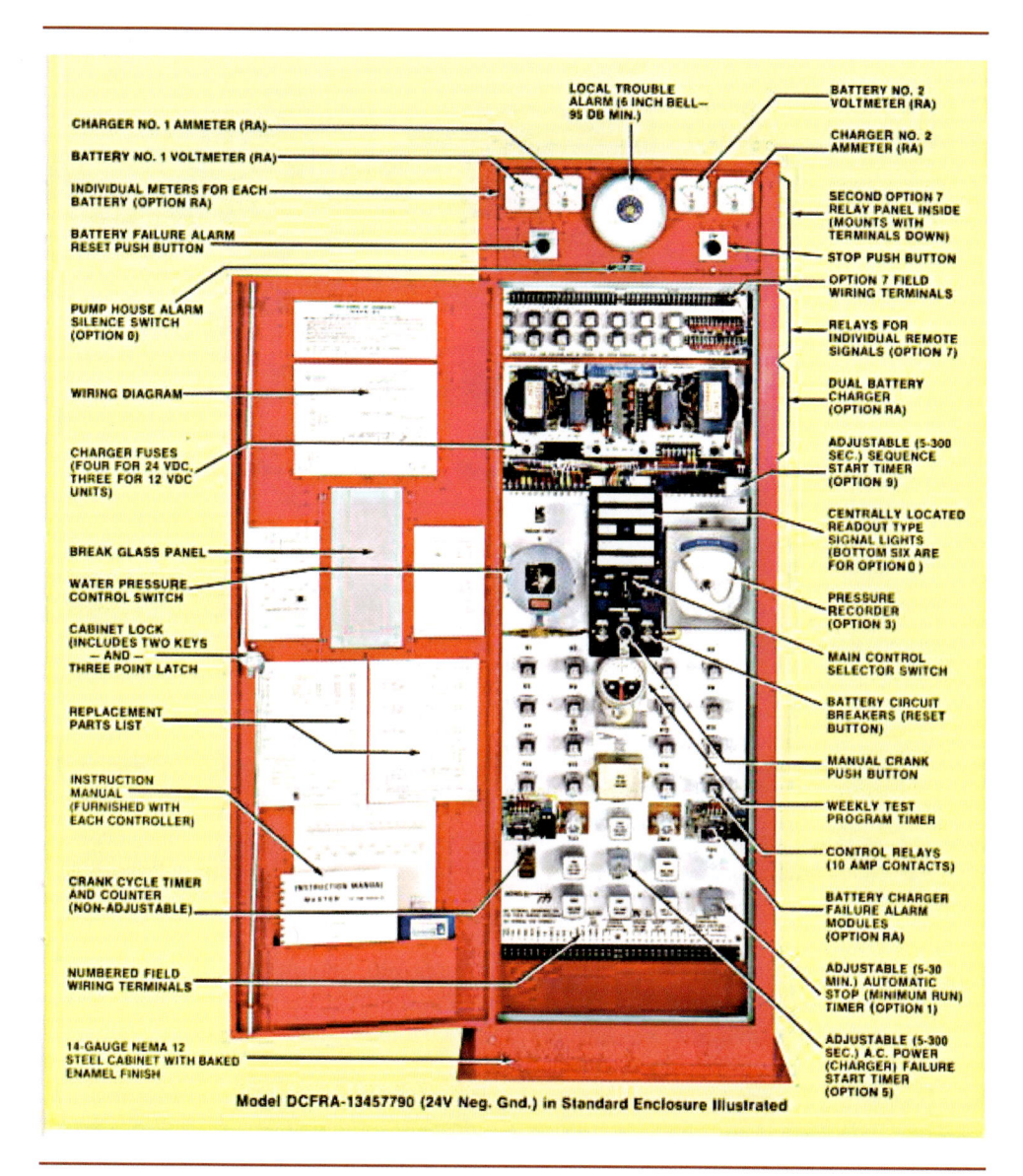

EXHIBIT II.12.5 Components of a Relay Logic–Type Diesel Drive Controller. (Courtesy of Master Control Systems, Inc.)

> Visible indication is necessary whether the controller is located in a dark room or in a brightly lit pump house or is located outdoors. This requirement applies to alarm, trouble, and status indicators, as well as digital displays.
>
> Legacy, or traditional, controllers commonly used incandescent lamp– (pilot light–) type indicators. Modern controllers use LED-type indicators and/or digital displays in one form or another. See Exhibit II.12.6 for an example of such a display.

A.12.4.1.2 It is recommended that the pilot lamp for signal service have operating voltage less than the rated voltage of the lamp to ensure long operating life. When necessary, a suitable resistor should be used to reduce the voltage for operating the lamp.

EXHIBIT II.12.6 Diesel Controller Door Digital Display Panel. (Courtesy of Master Control Systems, Inc.)

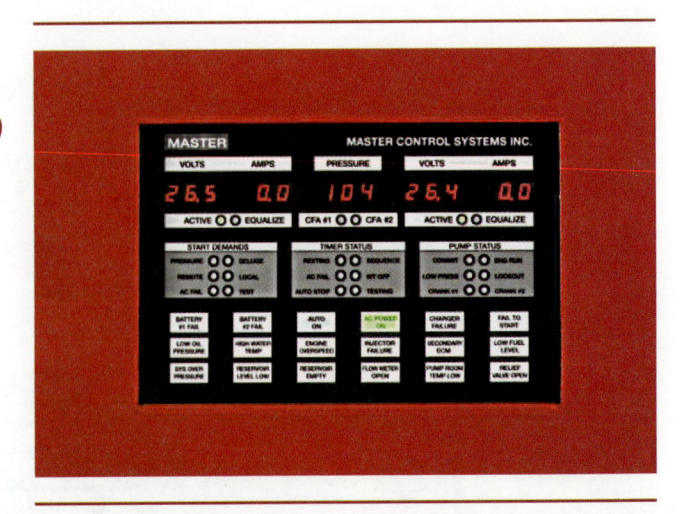

This recommendation was important for legacy controllers, which made use of filament-type pilot lights. Better-designed controllers used 14 volt and 28 volt rated lamps for 12 Vdc and 24 Vdc rated systems, respectively.

▶ **FAQ** How can the pilot lamp be kept operable as long as possible?

The life of filament (incandescent) lamps is inversely proportional to the 12th power of the voltage. Hence, operating a 14 volt lamp at 13.2 Vdc (typical battery charger float voltage) is a 9.1 percent reduction in voltage, which results in approximately a 314 percent increase in lamp life.

12.4.1.3 Separate visible indicators and a common audible fire pump alarm capable of being heard while the engine is running and operable in all positions of the main switch except the off position shall be provided to immediately indicate the following conditions:

The audible device must be capable of being heard. Traditionally, controllers used alarm bells or horns, which could usually be heard over the noise of a running engine. The exception was 3000 rpm engines, which are not as prevalent as they once were. It is important to note that solid-state devices, such as piezoelectric beepers, are often not audible over the sound of a running engine, and the rating type of the enclosure is affected by the audible device that is used. For example, NEMA Type 4 alarm devices are not as common as lower-rated types.

(1) Critically low oil pressure in the lubrication system

The text of 12.4.1.3(1) has been revised in the 2013 edition to specifically require testing of the low oil pressure alarm and circuit. This testing is to be performed in the same manner as testing of other engine monitoring and sensing circuits.

The oil pressure referred to is lube oil pressure and not fuel oil pressure. In most, if not all, cases, the testing is accomplished by switching the controller to the manual mode or one of the two manual modes (MAN1 or MAN2) and not by cranking the engine. The low oil pressure switch should be closed in manual mode, and the controller visible indicator should be lit. However, the audible device and remote failure contacts should remain in their "normal" (non-alarm) state. Typically, a delay time of a few seconds is used after the controller receives the cranking terminating (engine running) signal from the engine speed switch prior to actuating the audible device and alarm contacts. This delay is intended to prevent false alarms. See also 12.7.5.2(4).

(2) High engine temperature

Similar to the low lube oil pressure alarm, the controller will not shut down the engine due to a high water alarm condition if it is running under a true demand, such as pressure loss, remote start, and so forth. The controller will shut down the engine if it is running under test. Note that some engines add an oil pressure switch in series with the high temperature switch to avoid false alarms after a high horsepower demand running. See also 12.7.5.2(4) for further requirements and information.

(3) Failure of engine to start automatically

This condition occurs if the engine fails to start after six cranking attempts or if the engine (speed switch) fails to notify the controller that the engine is running. See 12.7.4 for more information. The controller treats this condition as a shutdown requirement, even though the engine may not be running. The reason for this shutdown is to kill the water and fuel solenoids to avoid further battery drain and also to avoid cooling water drain, which can be significant over an extended period, such as a 3-day weekend.

(4) Shutdown from overspeed

Typically, the engine overspeed switch is set to trip at approximately 20 percent of the engine's rated running speed.

Although the general philosophy is that the fire pump and the engine are sacrificial if an overspeed condition is detected by the engine speed switch and transmitted to the controller, the controller will immediately shut down the engine. The reason is that a true engine overspeed condition is considered an imminent destruction condition. It is also a personnel safety consideration, since engines sometimes throw parts in the process of overspeed destruction. By shutting down the engine, there may also be a possibility of manual intervention to correct or temporarily remedy the situation.

12.4.1.3.1 The controller shall provide means for testing the low oil pressure alarms and circuit in conjunction with the engine circuit testing method.

12.4.1.3.2 Instructions shall be provided on how to test the operation of the signals in 12.4.1.3.

▶ **FAQ** Why is it important to provide instructions for testing alarm circuits?

Paragraph 12.4.1.3.1 has been added to the 2013 edition to mandate that instructions be provided for testing the alarm circuits specified in 12.4.1.3, since the test method for a specific engine controller was not always evident, which resulted in inadequate or no testing of alarm circuits.

12.4.1.4 Separate visible indicators and a common audible signal capable of being heard while the engine is running and operable in all positions of the main switch except the off position shall be provided to immediately indicate the following conditions:

(1)* Battery failure or missing battery. Each controller shall be provided with a separate visible indicator for each battery. The battery failure signal shall initiate at no lower than two-thirds of battery nominal voltage rating (8.0 V dc on a 12 V dc system). Sensing shall be delayed to prevent nuisance signals.

> **▶ FAQ** What is meant by "battery failure," and how is it determined?

The battery failure signal initiates when the engine cranking voltage is too low to effectively crank (turn over) the engine. In addition to initiating the signal, the controller must also switch to the other battery set. Text has been added in the 2013 edition to quantify the initiation of the battery failure signal by setting a voltage limit at which initiation must commence.

> **▶ FAQ** Why is the oldest battery in the engine controller suddenly stronger than the newer ones when cranking the engine?

As batteries age, their actual capacity (amp-hours) increases somewhat, then declines over the life of the battery. Once a battery's actual capacity returns to its initial capacity, the capacity curve will indicate that this capacity is declining rapidly and will soon be inadequate. Battery replacement is advised at this time.

(2) Battery charger failure. Each controller shall be provided with a separate visible indicator for battery charger failure and shall not require the audible signal for battery charger failure.

The controller is required to trip a signal if the battery charger either fails to maintain the charge in the battery or fails to recharge the battery as required. See Sections 12.5 and 12.6 for more requirements.

(3) Low air or hydraulic pressure. Where air or hydraulic starting is provided *(see 11.2.7 and 11.2.7.4)*, each pressure tank shall provide to the controller separate visible indicators to indicate low pressure.

A low air or low hydraulic pressure condition is analogous to the battery failure alarm and its need for a signal. Note that, while not commonly used, a controller for either hydraulically or pneumatically started engines will usually make use of two batteries for controller power, although these batteries may be fairly small. In any event, the controller usually will still be provided with the battery failure and charger failure alarms, since both the battery and charger remain critical to the operation of the fire pump.

(4) System overpressure, for engines equipped with variable speed pressure limiting controls, to actuate at 115 percent of set pressure.

Pressure limiting diesel drivers use feedback from the pump discharge pressure to modulate the speed governor setting. The system overpressure signal is used to indicate a problem or manual intervention resulting in pressure higher than the desired set point value. This signal also indicates that the main pressure relief valve may be open and flowing large quantities of water or that it may not be working at all. See Section 4.18 for more information on relief valves.

(5) ECM selector switch in alternate ECM position (for engines with ECM control only).

Note that this position should also indicate when the alternate electronic control module (ECM) is selected automatically by the engine.

(6) Fuel injection malfunction (for engines with ECM only).

This signal also occurs when the ECM detects an internal failure.

(7) Low fuel level. Signal at two-thirds tank capacity.

▶ **FAQ** | What is the purpose of a fuel maintenance system?

This requirement was added in the 2003 edition. Prior to that edition, signaling at two-thirds tank capacity was a popular, but not a mandatory, option. Exhibit II.12.7 shows a low fuel level switch.

(8) Low air pressure (air-starting engine controllers only). The air supply container shall be provided with a separate visible indicator to indicate low air pressure.

(9) Low engine temperature.

Paragraph 12.4.1.4(9) was new to the 2010 edition. The low engine temperature alarm picks up the low engine temperature sensor added to the engine as mentioned in Chapter 11. This signal indicates that there is a loss of ac (mains) power to the engine heater or a malfunction of the heater, that the heater is unplugged or disconnected, or that the heater has a kinked or plugged hose(s).

(10) Supervisory signal for interstitial space liquid intrusion.

This requirement has been added to the 2013 edition to correlate with 11.4.1.2.8, which states that the interstitial space between the shells of the diesel fuel storage tank be monitored for leakage and annunciated by the engine drive controller with a signal of the supervisory type.

EXHIBIT II.12.7 *Low Fuel Level Switch. (Courtesy of Master Control Systems, Inc.)*

(11) High cooling water temperature.

▶ **FAQ** | What is the purpose of the high cooling water temperature signal?

This requirement has been added to the 2013 edition to correlate with 4.18.7.2, which states:

> Where pump discharge water is piped back to pump suction, and the pump is driven by a diesel engine with heat exchanger cooling, a high cooling water temperature signal at 40°C (104°F) from the engine inlet of the heat exchanger water supply shall be sent to the fire pump controller and the controller shall stop the engine provided there are no active emergency requirements for the pump to run.

This prevents water in the heat exchanger from reaching a temperature that results in insufficient cooling of the engine during testing procedures.

Failure to promptly investigate and correct any of the conditions indicated in 12.4.1.4 is likely to cause unsatisfactory pump performance or failure of the fire protection system. Although not specifically required by this chapter, some installations also include monitoring of the following conditions:

1. Low pump room temperature
2. Relief valve discharge
3. Low reservoir level
4. Low city water pressure
5. Control valve closure

Stationary Fire Pumps Handbook 2013

A.12.4.1.4(1) The controller can set the signal trip point above the two-thirds level. But, higher than ¾ of nominal is not recommended to avoid false signals during normal battery aging.

12.4.1.5 A separate signal silencing switch or valve, other than the controller main switch, shall be provided for the conditions reflected in 12.4.1.3 and 12.4.1.4.

> ▶ **FAQ** Why and when is audible alarm and signal silencing allowed?

Paragraph 12.4.1.5 has been revised in the 2013 edition to eliminate the requirement that no audible signal silencing switch or valve, other than the controller main switch, be permitted for the conditions reflected in 12.4.1.3 and 12.4.1.4. It was replaced with a requirement for a separate signal silencing switch or valve. In concert with this change, new 12.4.1.5.1, 12.4.1.5.2, and 12.4.1.5.3 were added to specify the conditions under which any of the audible alarms in 12.4.1.3 or signals in 12.4.1.4 may be silenced. These new requirements address situations where conditions that are being alarmed cannot be corrected immediately (such as low engine temperature or fuel injection malfunction alarms). In such situations, it may take 24 hours or longer to correct the condition. Under such circumstances, it is not necessary to acknowledge the alarm until the condition is corrected. In addition, the alarm is unavailable for additional alarm conditions during this period. Field experience has demonstrated situations in which audible alarms were disabled, resulting in no audible alarm for a new alarm condition and in which the fire pump failed to respond when needed. The words "or valve" were added in the 2010 edition to address controllers that operate pneumatically or use fluidic logic rather than traditional electrical circuitry.

12.4.1.5.1 The switch or valve shall allow the audible device to be silenced for up to 4 hours and then re-sound repeatedly for the conditions in 12.4.1.3.

> The trouble conditions specified in 12.4.1.3 may exist for extended periods. Silencing of these alarms is needed in the case of attended pump rooms. Silencing is also needed to avoid unnecessary battery drain during power outages and to avoid the controller being taken out of service to quiet the alarm.

12.4.1.5.2 The switch or valve shall allow the audible device to be silenced for up to 24 hours and then re-sound repeatedly for the conditions in 12.4.1.4.

> The trouble conditions specified in 12.4.1.4 may exist for extended periods. Silencing of these alarms is needed in the case of attended pump rooms. Silencing is also needed to avoid unnecessary battery drain during power outages and to avoid the controller being taken out of service to quiet the alarm.

12.4.1.5.3 The audible device shall re-sound until the condition is corrected or the main switch is placed in the off position.

12.4.1.6* The controller shall automatically return to the nonsilenced state when the alarm(s) have cleared (returned to normal).

A.12.4.1.6 This automatic reset function can be accomplished by the use of a silence switch of the automatic reset type or of the self-supervising type.

12.4.1.6.1 This switch shall be clearly marked as to its function.

12.4.1.7 Where audible signals for the additional conditions listed in A.4.24 are incorporated with the engine fire pump alarms specified in 12.4.1.3, a silencing switch or valve for the additional A.4.24 audible signals shall be provided at the controller.

> The conditions mentioned in A.4.24 are pump house (pump room) trouble conditions and include the following:
>
> 1. Low pump room temperature (freeze alarm)
> 2. Relief valve discharge (relief valve open)
> 3. Flowmeter left on, bypassing the pump
> 4. Water level in suction supply below normal (reservoir low)
> 5. Water level in suction supply near depletion (reservoir empty)
> 6. Steam pressure below normal
>
> These trouble condition alarms are all optional signals that the controller can monitor when so equipped. The following are other often-used (monitored) signals:
>
> 1. Low suction pressure
> 2. Fuel spill (primarily used in California)
> 3. High fuel level
> 4. Reservoir temperature low (freeze alarm)
> 5. Pump room intrusion (pump room door open)
> 6. Low discharge pressure
>
> Exhibit II.12.8 shows pump room signal contacts for connection to the controller on the left, while contacts for remote annunciation are shown on the right.

12.4.1.8 The circuit shall be arranged so that the audible signal will be actuated if the silencing switch or valve is in the silent position when the supervised conditions are normal.

12.4.2 Signal Devices Remote from Controller.

12.4.2.1 Where the pump room is not constantly attended, audible or visible signals powered by a source other than the engine starting batteries and not exceeding 125 V shall be provided at a point of constant attendance.

> Paragraph 12.4.2.1 requires the fire pump controller to be constantly monitored or supervised. This is not a requirement of the controller, although the controller is required to provide the contacts necessary to fulfill this requirement.
>
> ▶ **FAQ** What is an example of a constantly attended location?
>
> A boiler room where operating personnel are required 24 hours a day is one type of a constantly attended location. Because most new installations are not able to satisfy the provisions for a constantly attended location, a remote alarm panel is often installed at a place of constant attendance, such as a guard house or a 24-hour phone switchboard. Although these units are termed *alarm sets* or *alarm panels,* they are usually not built or sold as alarm equipment in accordance with *NFPA 72®, National Fire Alarm and Signaling Code.*
>
> As an alternative, the authority having jurisdiction (AHJ) may require an off-site fire alarm monitoring station or alarm service. Note that a fire alarm monitoring station is often preferred. These signals are not considered a *fire alarm* as defined in *NFPA 72.* They would instead be considered supervisory or trouble signals. The monitoring station would normally notify the plant personnel, rather than the fire department, when these signals occur. A notable exception can be the engine running signal. This signal is often used to meet the required weekly line test function of *NFPA 72,* since this signal will, or should, occur during the weekly ½-hour engine test run. When an engine running signal occurs during any time other than during the expected weekly test run, the alarm company may signal the fire department of a potential fire.
>
> The 125 V maximum requirement is to avoid imposing higher voltages on the controller remote alarm contacts of 12.4.3. This voltage is often the maximum voltage rating of these contacts.

Stationary Fire Pumps Handbook 2013

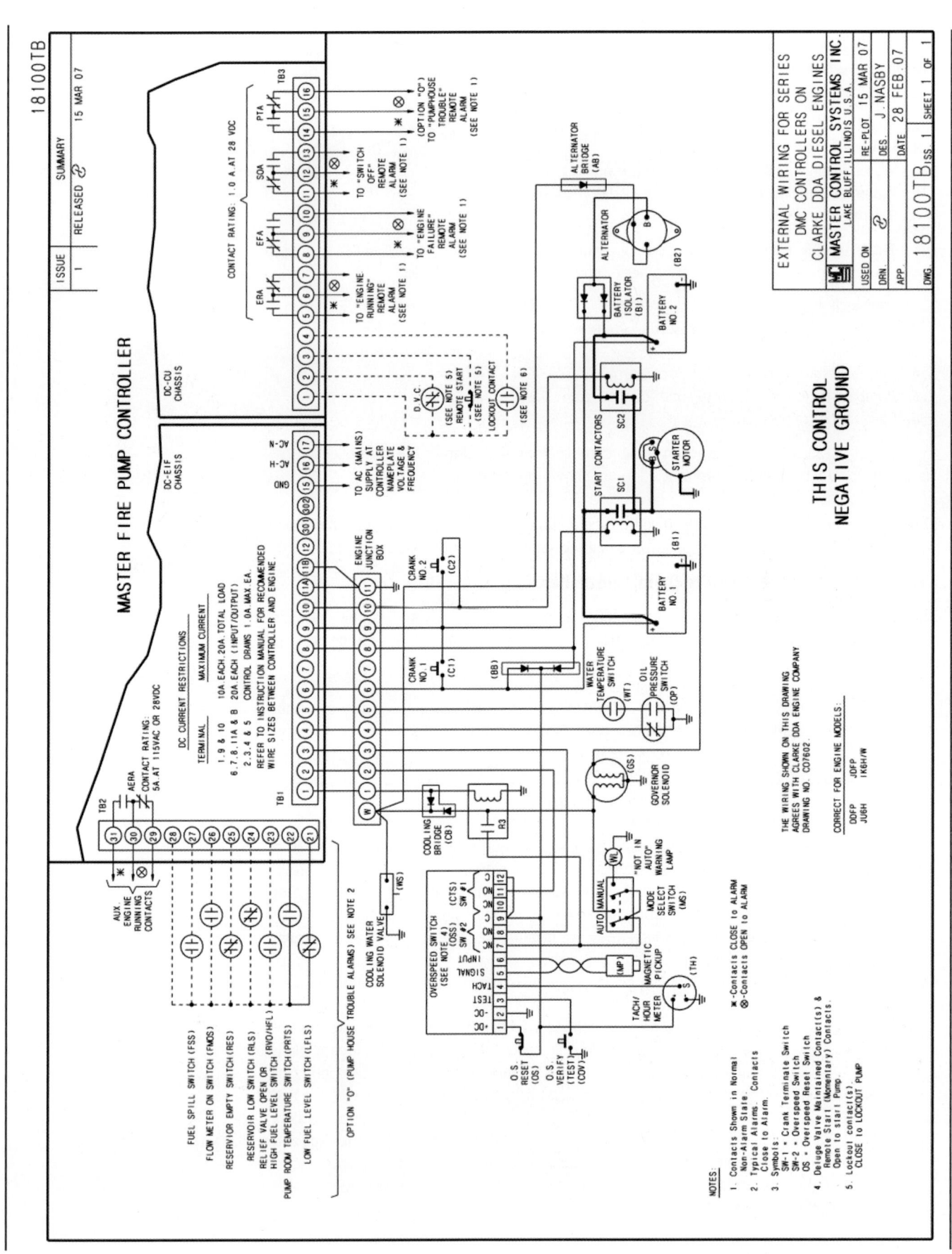

EXHIBIT II.12.8 Controller Wiring Diagram with Alarm Signals. (Courtesy of Master Control Systems, Inc.)

12.4.2.2 Remote Indication. Controllers shall be equipped to operate circuits for remote indication of the conditions covered in 12.4.1.3, 12.4.1.4, and 12.4.2.3.

> The controller is required to have at least three sets of contacts for engine running, switch-off, and engine (or controller) failure. Many controllers have more than these three sets as a standard, and all controllers are available with additional or individual alarm contacts for the various conditions supervised by the controller.

12.4.2.3 The remote panel shall indicate the following:

(1) The engine is running (separate signal).

> It is good practice for the weekly test run to be attended to monitor the starting and running conditions of the engine and controller. It is imperative that appropriate personnel be immediately dispatched to the pump room at any other time the engine is running to determine the cause and to monitor the engine and controller operation, since it is highly likely that the engine and controller are supplying, or attempting to supply, water for fire fighting.

(2) The controller main switch has been turned to the off or manual position (separate signal).

> If the controller control or selector switch is not in the auto (automatic) mode position, the controller and engine and, thus, the fire pump are out of service. For a sole source pump installation, which is a common situation, this means that the protected premises (building, plant, etc.) has no fire protection. This condition may require a fire watch or may even require a fire pump truck to be connected to the building until fire protection is restored.

(3)* There is trouble on the controller or engine (separate or common signals). *(See 12.4.1.4 and 12.4.1.5.)*

A.12.4.2.3(3) The following signals should be monitored remotely from the controller:

(1) A common signal can be used for the following trouble indications: the items in 12.4.1.4(1) through 12.4.1.4(7) and loss of output of battery charger on the load side of the dc overcurrent protective device.
(2) If there is no other way to supervise loss of power, the controller can be equipped with a power failure circuit, which should be time delayed to start the engine upon loss of current output of the battery charger.
(3) The arrangement specified in A.12.4.2.3(3)(2) is only permitted where approved by the authority having jurisdiction in accordance with Section 1.5 and allows, upon loss of the ac power supply, the batteries to maintain their charge, activates ventilation in case conditions require cooling the engine, and/or maintains engine temperature in case conditions require heating the engine. *(See also A.4.6.4 and A.11.4.2.1.)*

> It is suggested that, when the engine is allowed by the authority having jurisdiction (AHJ) to be run in order to charge the batteries or keep the pump room temperature from freezing, it should not be run any longer than necessary.
>
> In some cases, an AHJ requires ac power (mains) pump starting where hot processes are involved, such as with furnaces and salt baths. The purpose is to bring the fire water supply on line, since plantwide ac (mains) power loss will result in the loss of cooling fans, blowers, and/or pumps, which presents an increased likelihood of a fire.

12.4.3 Controller Contacts for Remote Indication. Controllers shall be equipped with open or closed contacts to operate circuits for the conditions covered in 12.4.2.

These contacts are part of the fire pump controller. Modern controllers employ contacts rated for 1.0 amp or less at 125 Vac or 28 Vdc. See also the commentary to 12.4.2.1 and 12.4.4.

12.4.4* Pressure Recorder.

The pressure recording device can be examined after operation of the fire pump. The recorder registers the time the pump was in operation and the output pressure during operation. This information can be used to help evaluate the performance of the pump.

▶ **FAQ** Why is a pressure recorder needed?

A pressure recorder is of paramount importance for two main reasons — first, to record the weekly test run of the engine and pump, and second, and most important, to record pressure events and excursions (abnormalities) during a fire for forensic analysis.

After a fire, forensic specialists can extract vital information regarding the operation or failure of equipment such as the pump and sprinklers, as well as the progress of the water supply during a fire. It also is important to provide definitive evidence that the engine, pump, and controller are being exercised (tested) weekly.

A.12.4.4 The pressure recorder should be able to record a pressure at least 150 percent of the pump discharge pressure under no-flow conditions. In a high-rise building, this requirement can exceed 400 psi (27.6 bar). This requirement does not mandate a separate recording device for each controller. A single multichannel recording device can serve multiple sensors.

12.4.4.1 A listed pressure recording device shall be installed to sense and record the pressure in each fire pump controller pressure-sensing line at the input to the controller.

Too often, a multiple fire pump installation has only a single pressure recorder. Fortunately, most modern controllers come with built-in pressure recorders as a standard. It is paramount that each controller have a dedicated recorder to verify weekly tests and provide forensic data if a fire occurs. Exhibit II.12.9 shows a traditional circular chart-type recorder, while Exhibit II.12.10 shows the control panel and 160-character display for a modern digital alarm, status, and pressure recorder.

EXHIBIT II.12.9 Circular Chart Pressure Recorder. (Courtesy of Clarke Fire Protection Products, Inc.)

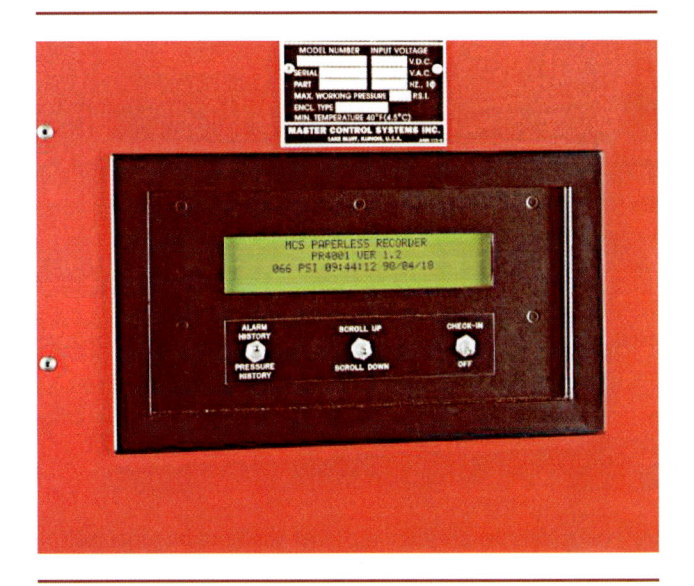

EXHIBIT II.12.10 *Digital Recorder Display and Control Panel. (Courtesy of Master Control Systems, Inc.)*

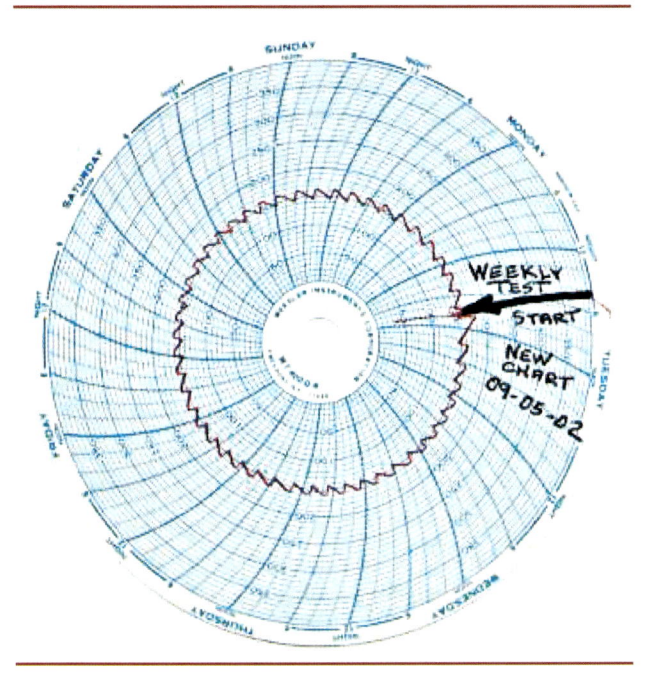

EXHIBIT II.12.11 *Circular Pressure Chart.*

12.4.4.2 The recorder shall be capable of operating for at least 7 days without being reset or rewound.

> Older recorders were of the spring-wound type. See Exhibit II.12.9, for an example. The circular charts were good for 7 days. Most modern recorders are either battery operated or draw their power from the engine batteries. Exhibit II.12.11 shows a typical chart from the recorder shown in Exhibit II.12.9. The saw tooth pattern is the cycling of the jockey (pressure maintenance) pump.

12.4.4.3 The pressure-sensing element of the recorder shall be capable of withstanding a momentary surge pressure of at least 400 psi (27.6 bar) or 133 percent of fire pump controller rated operating pressure, whichever is higher, without losing its accuracy.

> This requirement is important because severe hydraulic transients and surges often occur. These transients and surges can occur on pump shutdown, when the main check valve slams shut, or on start-up, when the engine rapidly accelerates to full speed and the pump attempts to accelerate tons of water. Severe transients also occur on vertical pumps, especially deep well pumps, when the water column hits the air relief valve.

12.4.4.4 The pressure recording device shall be spring wound mechanically or driven by reliable electrical means.

> After spring-wound recorders became obsolete, pressure recorders were often ac powered. Later, they were powered by dry cell batteries.

12.4.4.5 The pressure recording device shall not be solely dependent upon alternating current (ac) electric power as its primary power source.

The ac powered chart recorders were equipped with a spring backup to keep the clock motor running when ac power was lost.

> ▶ **FAQ** Why must a pressure recorder be designed to operate without ac power?
>
> A diesel engine drive is used because it is independent of ac power. Therefore, the pressure recorder must also be designed to operate entirely without ac power or without ac power for at least 24 hours.

12.4.4.6 Upon loss of ac electric power, the electric-driven recorder shall be capable of at least 24 hours of operation.

> In the past, the spring had to be operable for at least 1 full day of recording without ac (mains) power. Now almost all modern controllers make use of digital pressure recorders.

12.4.4.7 In a non-pressure-actuated controller, the pressure recorder shall not be required.

> An example of a non-pressure-actuated controller is a controller used for nonpressurized systems, such as deluge or monitor systems. These controllers have no pressure switch and no pressure sensing line and, thus, have no need for a pressure recorder. Most modern controllers, however, still have a digital recorder for the purpose of recording alarms and status change events.

12.4.5 Voltmeter. A voltmeter with an accuracy of ±5 percent shall be provided for each battery bank to indicate the voltage during cranking or to monitor the condition of batteries used with air-starting engine controllers.

12.5* Battery Recharging

A.12.5 A single charger that automatically alternates from one battery to another can be used on two battery installations.

> The type of charger referenced in A.12.5 was used until the 1970s, before controllers with built-in chargers were commonly available. These chargers had a relay that periodically connected the charger to one battery and then to the other.

12.5.1 Two means for recharging storage batteries shall be provided.

12.5.2 One method shall be the generator or alternator furnished with the engine.

> The engine-driven generator/alternator is completely independent of other energy sources. This equipment recharges the batteries, after cranking, while the engine is running much faster than the ac powered, automatically controlled charger.

12.5.3 The other method shall be an automatically controlled charger taking power from an ac power source.

> To recharge storage batteries, one means used is the static (ac powered) battery charger, and the other is the belt-driven battery-charging alternator (or generator on older engines). Engines have used alternators instead of generators since the 1970s, although some generators

are still in service. Controllers were equipped with optional battery chargers starting in the late 1960s or early 1970s. It was a significant cost item that slowly became popular. Currently, the charger is generally standard equipment in modern controllers.

12.5.4 If an ac power source is not available or is not reliable, another charging method in addition to the generator or alternator furnished with the engine shall be provided.

Although uncommon, an inverter powered by 48 Vdc telephone batteries or 120 Vdc or higher station batteries is an example of an inverter using alternative charging methods.

12.6 Battery Chargers

The requirements for battery chargers shall be as follows:

(1) Chargers shall be specifically listed for fire pump service and be part of the diesel fire pump controller.

Chargers were not always specifically listed for fire pump service, and older chargers included wide variations in charger quality, performance, and reliability.
   Exhibit II.12.12 shows a modern dual fire pump battery charger, while Exhibit II.12.13 shows the schematic diagram for such a charger.

(2) Additional chargers also listed for fire pump service shall be permitted to be installed external to the diesel fire pump controller for added capacity or redundancy.

▶ **FAQ**   Why would additional battery chargers be needed?

Section 12.6(2) is new to the 2013 edition. It addresses those situations where batteries or other accessory equipment requires additional charging capacity or an additional battery charger(s) is desired to increase system reliability.

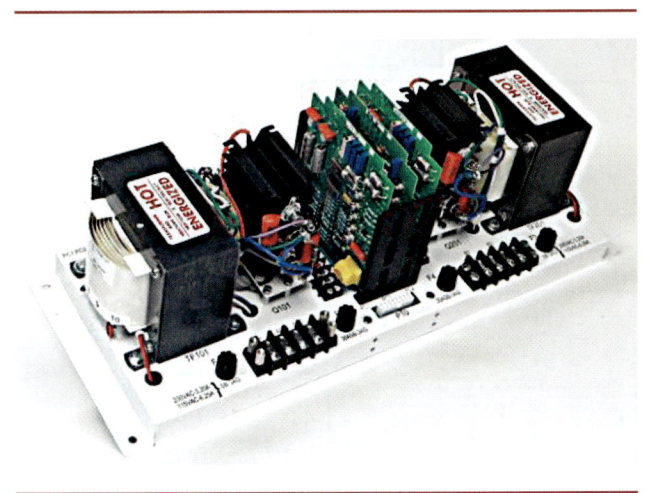

EXHIBIT II.12.12 *Dual 24-V dc 10-A dc Battery Charger. (Courtesy of Master Control Systems, Inc.)*

346 Part II • NFPA 20 • Chapter 12 Engine Drive Controllers

EXHIBIT II.12.13 Schematic Diagram of a Dual Battery Charger. (Courtesy of Master Control Systems, Inc.)

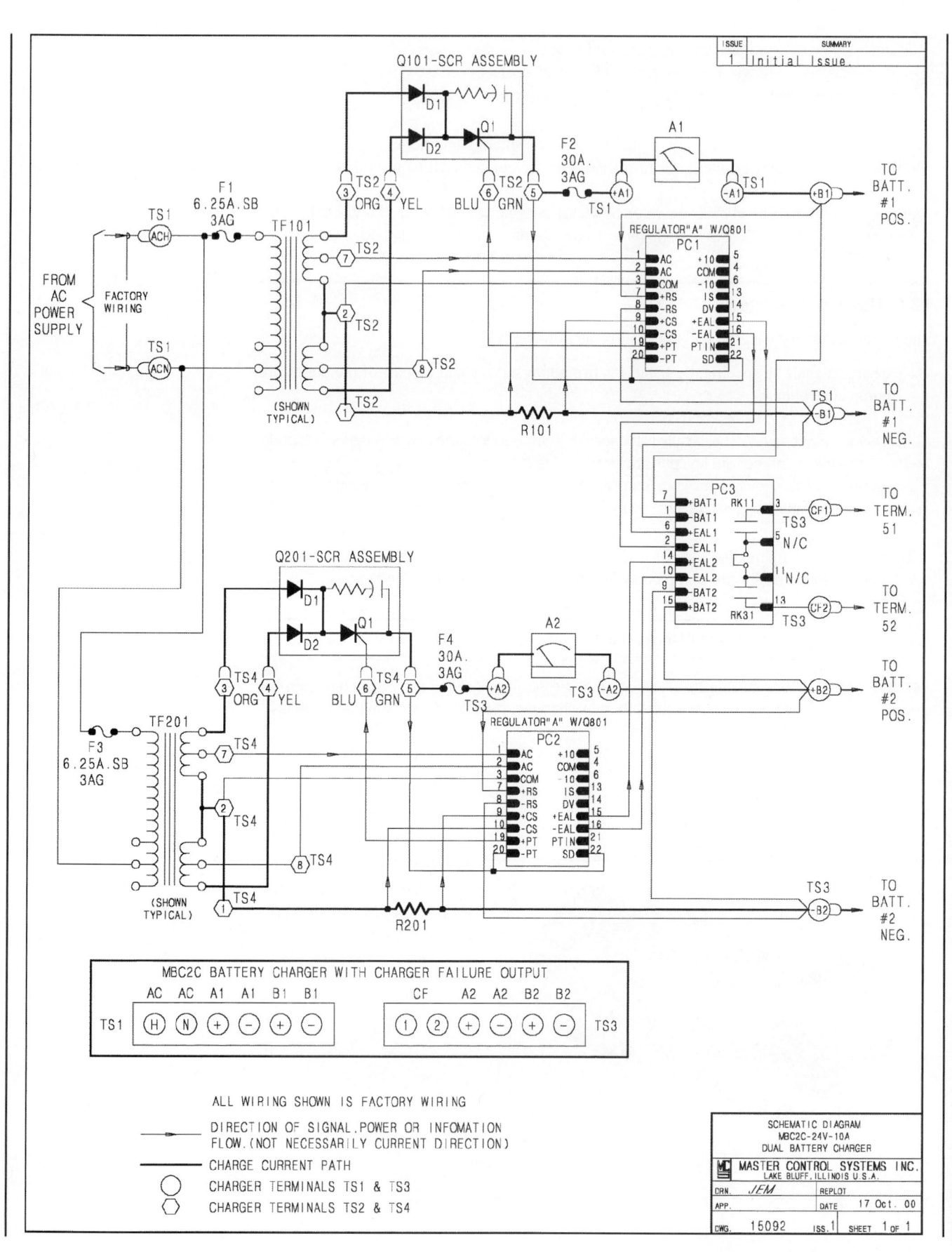

2013 Stationary Fire Pumps Handbook

(3) The rectifier shall be a semiconductor type.

A semiconductor-type rectifier is specified, as opposed to a vacuum tube–type rectifier, a type still in use when this requirement was added.

(4) The charger for a lead-acid battery shall be a type that automatically reduces the charging rate to less than 500 mA when the battery reaches a full charge condition.

A common problem with historic chargers was called "battery boiling." This phenomenon involves the dissolution of water in the electrolyte well past the time when the battery is fully charged. In the past, this problem resulted from the poor regulation of these chargers, or the complete lack of any voltage regulation, which was the norm in that era. Modern chargers are regulated well enough that this requirement is easily met if the charger is properly adjusted for the battery type involved. For this reason, field adjustment of battery charger settings is not recommended without the specific approval and guidance of the controller manufacturer. The amount of battery water consumption is directly proportional to final current flow.

(5) The battery charger at its rated voltage shall be capable of delivering energy into a fully discharged battery in such a manner that it will not damage the battery.

Modern battery chargers are of the current-limiting type to avoid excessive initial charging currents. This was not the case with historic unregulated chargers.

(6) The battery charger shall restore to the battery 100 percent of the battery's reserve capacity or ampere-hour rating within 24 hours.

In order to fulfill this requirement, all modern fire pump battery chargers are of the mode-switching type — that is, they operate in two different modes: (1) float mode (low rate) and (2) equalize mode (high rate). The float mode is a lower voltage-maintaining mode that allows the charger to meet the maximum 500 mA (½ amp) maintaining current. Better chargers reduce this maintaining current to 100 mA or less for batteries in good condition.

It is vitally important that the battery size and type match the battery charger. If the battery type does not match the type for which the charger is built and adjusted, one of the following two failure conditions will very likely occur: (1) the charger will not be able to meet the 24-hour recharge time requirement, or (2) sooner or later, the charger will wind up in the mode lock-up condition.

This requirement was added in the 1972 edition to avoid the case in which the batteries were depleted after acceptance testing, leaving the premises with questionable, unreliable, or inoperative fire protection.

(7) The charger shall be marked with the reserve capacity or ampere-hour rating of the largest capacity battery that it can recharge in compliance with 12.6(4).

A 10 amp charger can supply no more than approximately 240 amp-hours in a 24-hour period. This is sufficient for most SAE Group 4D or Group 8D size batteries but not for parallel battery banks. When batteries are connected in parallel, either Group 4D or Group 8D, the battery charger has to be a 20 amp rated unit. Marking the charger with the maximum battery size allows a field inspector to determine immediately if the charger is not the correct size for the installation involved. Chargers of insufficient size are highly vulnerable to the mode lock-up phenomenon described in the commentary for 12.6(6).

(8) An ammeter with an accuracy of ±5 percent of the normal charging rate shall be furnished to indicate the operation of the charger.

> While most modern controllers make use of digital ammeters, there is no assurance that the readings will be accurate within ½ amp for a 10 amp charger unless this is checked as part of the listing (conformity assessment) process.

(9) The charger shall be designed such that it will not be damaged or blow fuses during the cranking cycle of the engine when operated by an automatic or manual controller.
(10) The charger shall automatically charge at the maximum rate whenever required by the state of charge of the battery.
(11) The battery charger shall be arranged to indicate loss of current output on the load side of the direct current (dc) overcurrent protective device where not connected through a control panel. [See 12.4.1.4(2).]

> During starter motor breakaway, the battery voltage can drop to one-half of nominal voltage. This puts an immediate large demand on the battery charger. If the current-limiting circuitry is too slow or imprecise, or if the overcurrent protection (fuses or circuit breakers) is inadequate, blown fuses or tripped circuit breakers will result.
> The battery charger is required to be of the well-regulated type that will sense small changes in the battery's desired voltage set point. Such a charger will go to full output when it detects insufficient battery voltage.
> It is important to note that the battery charger is required to detect loss of ac (mains) power, malfunction of the charger, blown fuses or tripped circuit breakers, or any loss of connection between the battery and the charger.

(12) The charger(s) shall remain in float mode or switch from equalize to float mode while the batteries are under the loads in 12.5.2.

> This requirement helps ensure that the charger does not become stuck in the equalize (high rate) mode.
> For a more detailed discussion of battery charger theory and operation, refer to the NFPA publication *Pumps for Fire Protection Systems*.

12.7* Starting and Control

A.12.7 The following definitions are derived from *NFPA 70, National Electrical Code*:

(1) *Automatic.* Self-acting, operating by its own mechanism when actuated by some impersonal influence (e.g., a change in current strength, pressure, temperature, or mechanical configuration).
(2) *Nonautomatic.* The implied action requires personal intervention for its control. As applied to an electric controller, nonautomatic control does not necessarily imply a manual controller, but only that personal intervention is necessary.

12.7.1 Automatic and Nonautomatic.

12.7.1.1 An automatic controller shall be operable also as a nonautomatic controller.

> There are four main modes of operation for these controllers:
>
> 1. *Emergency control:* This mode is strictly manual operation (see 12.7.6).
> 2. *Test mode:* Starting is initiated manually by draining water through a test solenoid valve, but the controller initiates starting and running automatically.

3. *Automatic operation:* This mode is normal standby fire protection mode. It starts on pressure loss (drop) or other signal.
4. *Manual starting:* A manual local or remote switch initiates starting similar to the test mode, but the signal is electrical rather than hydraulic.

12.7.1.2 The controller's primary source of power shall not be ac electric power.

All power needed for controller operation, as well as engine operation, is derived from the engine batteries. The ac (mains) power is used for ancillary functions, such as the battery charger, cabinet heater, and engine heater; pump room loads (heat, light, sump pumps); and so forth. Older versions of controllers also used the ac power for timers, such as the weekly test timer and the minimum running (automatic stop) timer.

When both the controller dc and the engine dc standby power use (battery drain) are low enough, the weekly 30-minute run can be enough to replenish the charge in the battery. In this case, the fire pump can remain in service despite extended periods of ac power (mains) loss.

▶ **FAQ** Why does the controller need to be independent of ac (mains) power?

Diesel drive fire pumps are often specified due to the lack of reliable ac power needed for an electric motor–driven pump. Also, some diesel drive fire pump controllers are equipped with an ac power fail start option.

12.7.2 Automatic Operation of Controller.

12.7.2.1 Water Pressure Control.

Most controllers monitor the water pressure in a pressurized sprinkler system and start the pump (engine) when the pressure drops below a specific set point. Controllers make use of a pressure switch or pressure transducer with suitable circuitry to take these measurements and determine when to start the pump.

12.7.2.1.1 Pressure-Actuated Switch.

12.7.2.1.1.1 There shall be provided a pressure-actuated switch or electronic pressure sensor having adjustable high- and low-calibrated set-points as part of the controller.

These requirements have been revised in the 2013 edition to recognize the use of other types of devices that sense pressure (such as pressure transducers). The revised text also brings Chapter 12 into agreement with similar requirements that already exist in Chapter 10 for electric drive controllers.

Exhibit II.12.14, Exhibit II.12.15, and Exhibit II.12.16 are examples of pressure switches and a pressure transducer mounted inside of the controller.

At one time, all controllers used similar bourdon-type pressure switches. These switches had independent high and low set points and had a means for locking the settings. Exhibit II.12.15 shows one example. Although, traditionally, these switches used a mercury tilt bottle switch element for sensitivity, the switch shown uses a magnetic reed-type switch element instead.

12.7.2.1.1.2 The requirement of 12.7.2.1.1.1 shall not apply to a non-pressure-actuated controller, where the pressure-actuated switch or pressure responsive means shall not be required.

Stationary Fire Pumps Handbook 2013

EXHIBIT II.12.14 Pressure Switch Mounted Inside of Controller.

EXHIBIT II.12.15 Bourdon Tube–Type Pressure Switch.

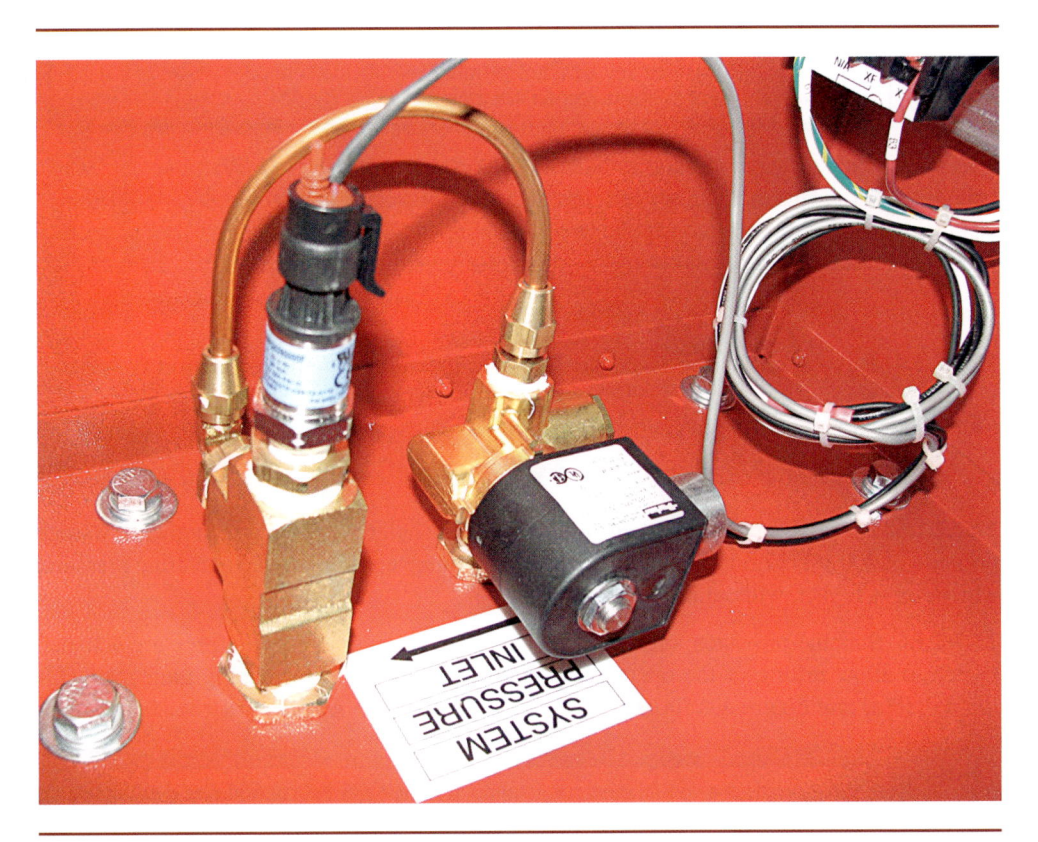

EXHIBIT II.12.16 *Pressure Transducer Mounted Inside of Controller.*

This requirement has been revised in the 2013 edition to recognize the use of other types of devices that respond to pressure. The revised text also brings Chapter 12 into agreement with similar requirements that already exist in Chapter 10 for electric drive controllers.

12.7.2.1.2 There shall be no pressure snubber or restrictive orifice employed within the pressure switch or pressure responsive means.

▶ **FAQ**   Why is a pressure snubber prohibited?

A pressure snubber was once an option with bourdon-type pressure switches but was a problem for fire water installations, since the switch could, under certain circumstances, trap pressure and thus prevent the pressure switch from responding to a drop in pressure. This situation would then prevent the pump from starting when it should. This requirement has been revised in the 2013 edition to recognize the use of other types of devices that respond to pressure. The revised text also brings Chapter 12 into agreement with similar requirements that already exist in Chapter 10 for electric drive controllers.

12.7.2.1.3 There shall be no valve or other restrictions within the controller ahead of the pressure switch or pressure responsive means.

> **FAQ** Why are restrictions of any type not permitted within the controller ahead of the pressure switch or pressure-responsive means?

This requirement has been added in the 2013 edition to bring Chapter 12 into agreement with similar requirements that already exist in Chapter 10 for electric drive controllers. Restrictions of any type are not permitted within the controller ahead of the pressure switch or pressure-responsive means, because they are an unnecessary item in the critical starting path of fire pump controllers. Their presence would reduce the reliability of the fire pump controller.

12.7.2.1.4 This switch shall be responsive to water pressure in the fire protection system.

As opposed to other types of sensing, such as flow switches, the switch is responsive only to water pressure and not to waterflow.

12.7.2.1.5 The pressure sensing element of the switch shall be capable of a momentary surge pressure of 400 psi (27.6 bar) or 133 percent of fire pump controller rated operating pressure, whichever is higher, without losing its accuracy.

12.7.2.1.6 Suitable provision shall be made for relieving pressure to the pressure-actuated switch to allow testing of the operation of the controller and the pumping unit. *[See Figure A.4.30(a) and Figure A.4.30(b).]*

> **FAQ** What provisions should be made to allow for testing of the controller and pumping unit?

The provisions consist of a solenoid valve that is actuated in the test mode of the controller and is also actuated by the weekly test timer. Note that modern controllers re-close (de-energize) the valve once the controller receives the signal that the engine is running. This avoids constant draining of the water. Better controllers have no valves or solenoids between the pressure sensing line connection and the pressure switch or transducer.

12.7.2.1.7 Water pressure control shall be as follows:

(1) There shall be no shutoff valve in the pressure sensing line.

This requirement is intended to prohibit valves of any type, including solenoid valves, in the pressure sensing line or in the path of the pressure switch or transducer, either outside of, or within, the controller.

(2) Pressure switch actuation at the low adjustment setting shall initiate the pump starting sequence if the pump is not already in operation.

When the system fire water pressure drops to the start pressure set point of the pressure-sensing means (either the pressure switch or pressure transducer circuitry), the controller needs to start the engine, either immediately or in accordance with permitted or mandated delays; that is, sequence start delay or high zone delay.

12.7.2.2 Fire Protection Equipment Control.

12.7.2.2.1 Where the pump supplies special water control equipment (e.g., deluge valves, dry-pipe valves), the engine shall be permitted to start before the pressure-actuated switch(es) would do so.

12.7.2.2.2 Under such conditions, the controller shall be equipped to start the engine upon operation of the fire protection equipment.

> ▶ **FAQ** What are examples of fire protection equipment control?
>
> This equipment includes remote manual start (push-button) stations; fire detection equipment, as in aircraft hangers; or deluge valve systems. Most, but not all, of these systems will be pressurized systems. Any circuits should be of the normally open (drop-out) type, so that opening the circuit results in starting the engine. See 10.5.2.3.3 for more information on open circuits.
>
> Since the inputs for external demand (start) signals have been, and often still are, optional, the controller must be ordered with the proper specified options where required.

12.7.2.2.3 Starting of the engine shall be initiated by the opening of the control circuit loop containing this fire protection equipment.

> Engine starting upon opening of the control circuit starting path provides for a fail-safe operation. This requirement applies to both diesel and electric drive fire pump installations.

12.7.2.3 Manual Electric Control at Remote Station. Where additional control stations for causing nonautomatic continuous operation of the pumping unit, independent of the pressure-actuated switch or control valve, are provided at locations remote from the controller, such stations shall not be operable to stop the engine.

> Once a pump is started from one of these remote start (push-button) stations, the controller is required to keep the pump running, regardless of external circuitry or signals. The reason is because shutdown of the pump is to occur only by operating the switch on the controller within the pump room or pump house.
>
> Some controllers provide a separate terminal for this input; some actuate on closing the remote start circuit; others actuate on opening the remote start circuit. Better controllers latch this signal, since it is often actuated from a momentary push-button station. If the sequence start delay is used, better controllers route this signal through the high zone delay.

12.7.2.4 Sequence Starting of Pumps.

> This paragraph applies to pumps operating in parallel or in series. Pumps operating in parallel are those that take suction directly from the suction supply and discharge into the same system (same header or manifold). Parallel pumps are used either to provide more capacity than a single pump can deliver or to increase the reliability of the system.
>
> The sequence starting requirement prevents excessive hydraulic stress to piping, control valves, and other system components during pump start-up. The time delay in each unit allows independent controller operation without the need for electrical interconnections. The lead (first to start) controller may not be equipped with sequential starting, especially if there are no electric motor–drive fire pumps in the system. When the lead controller is equipped with sequence starting, the time delay setting of the first pump is usually set to zero, unless there are electric motor–driven pumps required to start before the diesel-driven pumps. Better installations equip all controllers with sequential starting, since this allows adjustment of the lead and lagging pumps in the system. Many modern controllers come equipped with this optional mode of operation as standard equipment.
>
> The pressure switch (transducer) settings of all the pumps in parallel may be set to the same set point or to different set points. However, upon a sudden demand for fire water, all pressure switches will trip in unison. This is the reason sequential starting is absolutely

necessary for fire protection systems with multiple pumps. Without adherence to this rule, pump houses have been destroyed when the surge of tons of water breaks the headers or elbows in the underground piping. See Figure A.4.20.1.2(a) for a parallel pump arrangement.

▶ **FAQ** Is staggering the pressure switch start settings permitted?

Staggering of the pressure switches will result in sequential starting of the pumps *only* when a very low and slowly rising demand for fire water exists. With a sudden sharp drop in pressure, all the switches will trip at the same time.

Pumps operating in series have at least one pump that takes suction from the discharge of another pump. This arrangement is often used in high-rise installations to achieve the pressure required for the upper floors or a zone(s). The starting of the higher zone pump is delayed to allow the lower zone pump to start first. The delay prevents cavitation and/or dry running of the higher zone pump or pumps. Exhibit II.12.17 shows a diagram of two pumps in series.

In addition to provisions for delaying the starting of higher zone pumps, the lower zone controller(s) must be equipped with remote starting capabilities to receive the start signal sent by higher zone controllers. In three-zone systems, the mid zone controller must be equipped both ways — that is, with both high zone delay and with remote starting capabilities.

It should also be noted that sometimes high-rise buildings make use of multiple pumps in each zone for reliability, with one pump acting as a backup for the other. These cases require an additional special interlocking scheme to avoid false cycling or shutdown and restarting of pumps. While it is common for all pumps of a series pumping arrangement to be in the same room (basement), this is not always the case. Also, the highest zone in a high-rise building may be an independent pump with its own suction supply, such as a swimming pool.

12.7.2.4.1 The controller for each unit of multiple pump units shall incorporate a sequential timing device to prevent any one driver from starting simultaneously with any other driver.

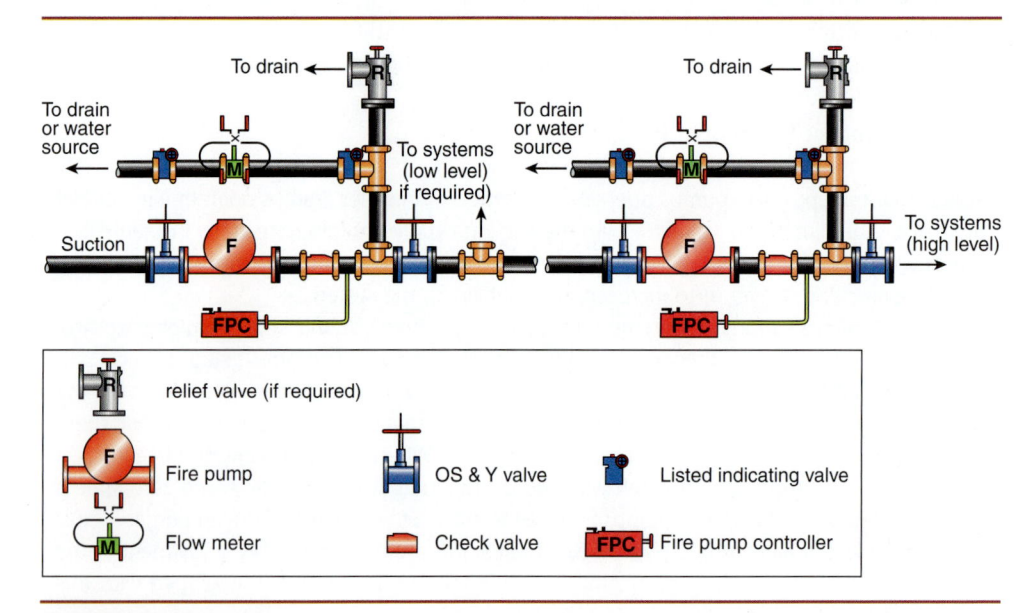

EXHIBIT II.12.17 Two Fire Pumps in Series. (Courtesy of Stephan Laforest, Summit Sprinkler Design Services, Inc.)

> The controller timing feature for pumps in parallel is usually called sequence start or sequential starting. The controller timing for pumps in series is usually called high zone delay. Better controllers apply these time delay functions differently with different demand signals.

12.7.2.4.2 Each pump supplying suction pressure to another pump shall be arranged to start within 10 seconds before the pump it supplies.

> Some controllers for high zone operation delay all starting demands, including local test start. This situation provides the signal to start a lower zone pump(s) and delays long enough for suction water to arrive. The only exception is manual (emergency) starting, since this should bypass all automatic circuitry. However, some controllers will still send the starting signal to the lower zone controller(s) to start the pump(s). Better controllers also latch in the starting demand signal and start the pump, even if the demand clears during the delay interval. The following example indicates the possible time delay sequencing of controllers in a high rise building where pumps are located at the low zone, midzone, and high zone of the high rise.
>
> ▶ EXAMPLE
>
> With the high zone pump delay set to 15 seconds and the midzone delay set to 7.5 seconds, and with a pump demand in the high zone, the sequence would be as follows:
>
> | Low zone start: | no delay (immediate starting) |
> | Midzone start: | 7.5-second total delay |
> | High zone start: | 15-second total delay |

12.7.2.4.2.1 The controllers for pumps arranged in series shall be interlocked to ensure the correct pump starting sequence.

12.7.2.4.3 If water requirements call for more than one pumping unit to operate, the units shall start at intervals of 5 to 10 seconds.

> Some controllers do not latch in a pressure demand until after the sequence time interval has elapsed. As a result, only the number of pumps actually needed will start and run, while the remaining pumps remain in standby mode. Some controllers apply this sequential starting delay only to so-called "static demands" such as pressure, deluge, or ac power loss start, among others. Remote manual start (push button) signals start the pump immediately to avoid an unexpected delay for fire water. The following example indicates the possible time delay sequencing of controllers where fire pumps are arranged in parallel.
>
> ▶ EXAMPLE
>
> For four pumps in parallel set at 0, 10, 20, and 30 seconds, with a sudden large pressure drop, the starting sequence is as follows:
>
> | Lead pump: | No delay (immediate starting); if the pressure remains low, then |
> | Second pump: | 10 seconds; if the pressure still remains low, then |
> | Third pump: | 20 seconds; if the pressure still remains low, then |
> | Fourth pump: | 30 seconds |

Stationary Fire Pumps Handbook 2013

> **▶ FAQ** Why wait out the sequence start delay instead of latching the demand signal right away?

> By waiting to latch the demand, these controllers start only as many pumps as are needed to supply the required pressure and flow. The process is as follows:
> If the pressure rises enough to reset the pressure switch before the sequence start delay times out, the controller allows the demand to expire at the end of the delay period if the demand no longer exists; therefore, the controller will not start a pump that is not needed.

12.7.2.4.4 Failure of a leading driver to start shall not prevent subsequent drivers from starting.

> Properly designed sequential start controllers have no interwiring and thus operate independently of one another. Therefore, failure of any other pumps (engines or controller) does not affect starting and running of the other units.
> For pumps in series, high zone units will start the pump, regardless of the state of lower zone units — that is, whether or not the lower zone units work.

12.7.2.5 External Circuits Connected to Controllers.

12.7.2.5.1 With pumping units operating singly or in parallel, the control conductors entering or leaving the fire pump controller and extending outside the fire pump room shall be so arranged as to prevent failure to start due to fault.

> Listed controllers are available with a lockout (interlock) option where required by an AHJ. This requirement is especially important in preventing defective external wiring from inhibiting the start of the pump when needed.

12.7.2.5.2 Breakage, disconnecting, shorting of the wires, or loss of power to these circuits shall be permitted to cause continuous running of the fire pump but shall not prevent the controller(s) from starting the fire pump(s) due to causes other than these external circuits.

> This requirement allows the controller to employ fail-safe-type design in order to comply with 12.7.2.5.1. Better controllers include circuitry that detects a failure in the controlling transducer. These failures include a short circuit, an open circuit, or a disconnection. Upon detecting any such fault, a fail-safe-type controller will start the pump. The only viable alternative is a signal and an alarm to monitor system pressure and report if the pump fails to respond. If the controller is the heart of a sprinkler system, the transducer is the heart of the controller.

12.7.2.5.3 All control conductors within the fire pump room that are not fault tolerant shall be protected against mechanical injury.

> Any such wiring will be best protected by running the wiring in threaded rigid metal conduit (RMC) or RMC raceway.

12.7.2.5.4 When a diesel driver is used in conjunction with a positive displacement pump, the diesel controller shall provide a circuit and timer to actuate and then close the dump valve after engine start is finished.

> This requirement mandates that the controller employ circuitry to actuate an "unloader" (dump) valve to allow the starter motor to crank the engine. A timer de-energizes the valve

a few seconds after receipt of the crank terminate (engine running) signal from the engine. Without this provision, the starter motor would attempt to crank against water-filled cylinders.

12.7.2.6 Sole Supply Pumps.

12.7.2.6.1 Shutdown shall be accomplished by manual or automatic means.

12.7.2.6.2 Automatic shutdown shall not be permitted where the pump constitutes the sole source of supply of a fire sprinkler or standpipe system or where the authority having jurisdiction has required manual shutdown.

Note that, at the present time, Factory Mutual does not allow automatic shutdown on their installations without special exemption.

12.7.2.7 Weekly Program Timer.

12.7.2.7.1 To ensure dependable operation of the engine and its controller, the controller equipment shall be arranged to automatically start and run the engine for at least 30 minutes once a week.

The 30-minute running time is normally sufficient to heat the lube oil sufficiently to drive out any accumulated moisture and to prevent wet stacking of the exhaust system. Wet stacking occurs when the engine does not run long enough for the entire exhaust piping arrangement to be heated above the boiling point. In this case, water will gradually accumulate in low points. This weekly run also consumes some of the stored fuel, which helps avoid excessive aging of the fuel and also works to establish routine fuel replacement as part of the maintenance program.

It is most important that the weekly test be attended by appropriate personnel to observe cranking times and voltages, water temperature, oil pressure, running speed, and all of the other required inspections and observations delineated in NFPA 25, *Standard for the Inspection, Testing, and Maintenance of Water-Based Fire Protection Systems*.

12.7.2.7.2 The controller shall use the opposite battery bank (every other bank) for cranking on subsequent weeks.

This requirement ensures that each battery unit (battery bank) is exercised every other week, assuming that the engine starts during the first cranking cycle. If it does not, corrective action should be taken.

12.7.2.7.3 Means shall be permitted within the controller to manually terminate the weekly test, provided a minimum of 30 minutes has expired.

12.7.2.7.4 A solenoid valve drain on the pressure control line shall be the initiating means.

This requirement ensures that the pump is started by actuation of the pressure-sensing means — for example, hydraulically.

12.7.2.7.5 Performance of this weekly program timer shall be recorded as a pressure drop indication on the pressure recorder. *(See 12.4.4.)*

12.7.2.7.6 In a non-pressure-actuated controller, the weekly test shall be permitted to be initiated by means other than a solenoid valve.

> This provision permits electric-only initiation of the cranking and starting in the controller when there are no hydraulic (pressure) components. For example, for nonpressurized applications, neither the pressure switch (transducer), nor the pressure recorder, nor the weekly test drain solenoid valve is needed.

12.7.3 Nonautomatic Operation of Controller.

12.7.3.1 Manual Control at Controller.

12.7.3.1.1 There shall be a manually operated switch or valve on the controller panel.

> This switch is usually the mode or control selector switch, which is behind a break glass window or other means if it is part of the automatic control switch. See 12.3.4.

12.7.3.1.2 This switch or valve shall be so arranged that operation of the engine, when manually started, cannot be affected by the pressure-actuated switch.

12.7.3.1.3 The arrangement shall also provide that the unit will remain in operation until manually shut down.

> All automatic circuitry should be completely bypassed when the controller is in the manual position(s) (MAN 1 or MAN 2 on some controllers). See 12.7.3.1.4.

12.7.3.1.4 Failure of any of the automatic circuits shall not affect the manual operation.

12.7.3.2 Manual Testing of Automatic Operation.

12.7.3.2.1 The controller shall be arranged to manually start the engine by opening the solenoid valve drain when so initiated by the operator.

12.7.3.2.2 In a non-pressure-actuated controller, the manual test shall be permitted to be initiated by means other than a solenoid valve.

> ▶ **FAQ** How are the manual test valves used?
>
> One use of manual test valves is to determine and/or verify the start and reset pressures of the pressure switch or pressure transducer circuitry. This is done by slowly opening one or the other of the test valves so as to slowly bleed down the pressure read by the controller until the start set point pressure is reached as read on a calibrated precision pressure gauge attached to the controller pressure sensing line. The valve is slowly closed to cause (allow) the system pressure to slowly rise until the stop (reset) pressure set point is reached.

12.7.4 Starting Equipment Arrangement. The requirements for starting equipment arrangement shall be as follows:

(1) Two storage battery units, each complying with the requirements of 11.2.7.2, shall be provided and so arranged that manual and automatic starting of the engine can be accomplished with either battery unit.
(2) The starting current shall be furnished by first one battery and then the other on successive operations of the starter.
(3) The battery changeover shall be made automatically, except for manual start.
(4) In the event that the engine does not start after completion of its attempt-to-start cycle, the controller shall stop all further cranking and operate a visible indicator and audible fire pump alarm on the controller.

(5) The attempt-to-start cycle shall be fixed and shall consist of six crank periods of approximately 15-second duration separated by five rest periods of approximately 15-second duration.

> NFPA 20 requires fire pump engines to be of the dual battery type, with the assumption that one of the two batteries, or its contactors, wiring, or other components, may be defective. As explained in Chapter 11, Section 11.2.7.2.1.4, each battery unit must have "twice the capacity" for the full "3-minute attempt-to-start cycle, which is six consecutive cycles of 15 seconds of cranking and 15 seconds of rest."
>
> Follow the manufacturer's instruction manual to perform the crank cycle test, and check the temperature of the starter motor(s) after each crank cycle to avoid overheating. The cranking is to be on alternate batteries for a maximum of three cycles on each battery. The controller will perform the alternating of the batteries.
>
> At the end of the attempt-to-start cycle, the controller de-energizes terminal #1 to de-energize the water and/or fuel solenoids. Besides the local visible and audible alarm, the controller will transfer the engine failure alarm contacts to notify the remote monitoring alarm set or alarm service.
>
> The six cranks and five rest periods add up to a total of 90 seconds of cranking and 75 seconds of resting, or 2¾ minutes total.

(6) In the event that one battery is inoperative or missing, the control shall lock in on the remaining battery unit during the cranking sequence.

> Since each battery unit must be sized (have capacity) for twelve 15-second cranks, one complete automatic crank cycle will consume around half of the cranking capacity of the battery. This leaves six more cranking attempts of 15 seconds available for manual intervention and starting attempts when one battery unit is inoperative, missing, or defective.

12.7.5 Methods of Stopping.

12.7.5.1 Manual Electric Shutdown. Manual shutdown shall be accomplished by either of the following:

(1) Operation of the main switch or stop valve inside the controller
(2) Operation of a stop button or stop valve on the outside of the controller enclosure as follows:

   (a) The stop button or stop valve shall cause engine shutdown through the automatic circuits only if all starting causes have been returned to normal.
   (b) The controller shall then return to the full automatic position.

> ▶ **FAQ**   *How is the controller shut down?*
>
> There are two methods for shutting down the controller. One method is an emergency stopping function in which the main selector control switch, located behind the break glass window or equivalent, is switched to the "off" position. A second method is to utilize the external "stop" push button on the outside of the controller. It stops the engine by releasing the running latched circuitry if, and only if, all start demands (call to start) have cleared to the reset (normal) condition.
>
> If any demands remain active, this external stop push button is inoperative, and the engine remains running without interruption. (Note that this operation is different from that of the stop push button on electric motor–driven fire pump controllers.)

EXHIBIT II.12.18
Components and Modules of a Diesel Drive Controller. (Courtesy of Master Control Systems, Inc.)

It is not permitted to press the stop push button to take the controller out of service, whether or not it is effective in shutting down the engine.

Exhibit II.12.18 shows an interior view of a modern diesel drive controller. Various plug-in modules in the right-hand chassis determine which options and type(s) of remote operation are installed and operative.

12.7.5.2* Automatic Shutdown After Automatic Start. The requirements for automatic shutdown after automatic start shall be as follows:

(1)* If the controller is set up for automatic engine shutdown, the controller shall shut down the engine only after all starting causes have returned to normal and a 30-minute minimum run time has elapsed.

▶ **FAQ** Why would an AHJ prohibit automatic stopping of the fire pump?

Note that automatic stop (shutdown) is not permitted for sole supply pumps, and, in some cases, is prohibited by an AHJ. See 12.7.2.6 for more information on sole supply pumps. Stopping a running engine presents a possibility, regardless of how remote, that the engine will not restart if it is needed to fight a fire. If the fire pump is fighting a fire and the water supply to the sprinkler system is interrupted for only a few seconds, the system will most likely continue to contain the fire after the water supply is restored. If the water supply is interrupted for more than a few seconds, it is likely that a fire plume or heat plume will fuse (open) more sprinklers. Also, the fire will resume growing during the interruption. A fire plume or heat plume can overwhelm the water supply or render the sprinkler system incapable of containing the fire, even if the fire is being contained. A running fire pump should never be deliberately shut down without absolute certainty that it is safe to do so.

(2) When the engine overspeed shutdown device operates, the controller shall remove power from the engine running devices, prevent further cranking, energize the overspeed fire pump alarm, and lock out until manually reset.

(3) Resetting of the overspeed circuit shall be required at the engine and by resetting the controller main switch to the off position.

> ▶ **FAQ** Why must the resetting of the overspeed circuit be done at the main controller switch?

The purpose of this requirement is for both personnel safety and the preservation of the start of the local and remote alarms to ensure that manual intervention or repair occurs, since the engine will be off-line (out of service) until both the engine device and the controller are reset.

(4) The engine shall not shut down automatically on high engine temperature, low oil pressure, or high cooling water temperature when any automatic starting or running cause exists, and the following also shall apply:

  (a) If no other starting or running cause exists during engine test, the engine shall shut down automatically on high engine temperature, low oil pressure, or high cooling water temperature.
  (b) If after shutdown a starting cause occurs, the controller shall restart the engine and override the high engine temperature, low oil pressure, or high cooling water temperature shutdowns for the remainder of the test period.

> ▶ **FAQ** Why will the engine not shut down due to a high engine temperature, a low oil pressure, or a high cooling water temperature condition?

If the engine is running under a true demand condition, such as a pressure loss start, the controller will not shut down the engine. The engine will be run to destruction if it is fighting a fire. The fire pump and driver are considered sacrificial if running under a true demand.

The controller will shut down the engine during a test run only. This preserves the engine rather than allowing it to destroy itself during a test run. This also leaves the pump in service, even if the engine is in need of attention. If there is a small demand for water, such as a few sprinkler heads for controlling a fire, the pump discharge pressure will normally reset the pressure switch (or transducer circuit), which will clear the demand. If the controller is set for automatic stop (shutdown) during the test, the controller will shut down the engine on high engine temperature, low oil pressure, or high cooling water temperature conditions. If a demand occurs (reoccurs) after shutdown on high engine temperature, low oil pressure, or high cooling water temperature conditions, it is likely that the pump is fighting a fire. Thus, the controller is required to change to the "run to destruction" mode to keep fire water flowing as long as possible.

The 2013 edition has been revised to include high cooling water temperature as another condition under which the engine is not allowed to shut down when any automatic starting or running cause exists. As is the case for the conditions of high engine temperature and low oil pressure, provisions that allow for engine shutdown and require restart apply to a high cooling water temperature condition as well. This revision correlates with 4.18.7.2, which states:

> Where pump discharge water is piped back to pump suction, and the pump is driven by a diesel engine with heat exchanger cooling, a high cooling water temperature signal at 40°C (104°F) from the engine inlet of the heat exchanger water supply shall be sent to the fire pump controller and the controller shall stop the engine provided there are no active emergency requirements for the pump to run.

(5) The controller shall not be capable of being reset until the engine overspeed shutdown device is manually reset.

> Modern engines supply the speed switch and its contacts directly from the batteries. Hence, if the controller selector switch is cycled to the "off" position (to reset the visible, audible, and/or remote alarms), the alarm will reactivate as soon as the controller's selector switch is set back to the auto position.

A.12.7.5.2 Manual shutdown of fire pumps is preferred. Automatic fire pump shutdown can occur during an actual fire condition when relatively low-flow conditions signal the controller that pressure requirements have been satisfied.

> Shutting down a running engine is not advisable due to the possibility of problems restarting the engine, especially while hot. If the engine is distressed in one way or another, it may not restart.

A.12.7.5.2(1) A run time of 30 to 45 minutes is usually long enough to dry out the exhaust system and bring the engine and oil up to normal operating temperatures to dry them out. Longer times could require larger fuel tanks.

> The text of A.12.7.5.2(1) has been added to 12.7.5.2(1) for the 2013 edition to provide substantiation for the 30-minute minimum run time. Since the 30-minute run time is a minimum, in some installations it may be desirable to have a longer run time before automatic engine shutdown. Coordinating the size of the fuel tank with the desired run time is critical in such installations.

12.7.6 Emergency Control. Automatic control circuits, the failure of which could prevent engine starting and running, shall be completely bypassed during manual start and run.

▶ **FAQ** Why are automatic controls bypassed during a manual start?

> This requirement is intended to ensure that any failed, removed, or unplugged component, including computer board assemblies or modules, does not affect the ability to start the engine using the manual control switch position(s). See also 12.7.3.1.4.

12.8 Air-Starting Engine Controllers

12.8.1 Existing Requirements. In addition to the requirements in Sections 12.1 through 12.7, the requirements in Section 12.8 shall apply.

12.8.2 Starting Equipment Arrangement. The requirements for starting equipment arrangement shall be as follows:

(1) The air supply container, complying with the requirements of 11.2.7.4.4, shall be provided and so arranged that manual and automatic starting of the engine can be accomplished.

> This provision is intended to match the requirements and ability of electrically (battery) cranked engines.

(2) In the event that the engine does not start after completion of its attempt-to-start cycle, the controller shall stop all further cranking and operate the audible and visible fire pump alarms.

(3) The attempt-to-start cycle shall be fixed and shall consist of one crank period of an approximately 90-second duration.

> This requirement replaces the six-time 15-second electrical starting scheme. One reason for intermittent cranking of electric-cranked engines is to prevent overheating of the starter motor. This is not applicable to air- (or hydraulic-) started (cranked) engines. The air or hydraulic motor does not generate significant heat while cranking.

12.8.3 Manual Shutdown. Manual shutdown shall be accomplished by either of the following:

(1) Operation of a stop valve or switch on the controller panel
(2) Operation of a stop valve or switch on the outside of the controller enclosure

12.8.3.1 The stop valve shall cause engine shutdown through the automatic circuits only after starting causes have been returned to normal.

12.8.3.2 This action shall return the controller to full automatic position.

References Cited in Commentary

National Fire Protection Association, 1 Batterymarch Park, Quincy, MA 02169-7471.

Isman, K. E. and M. T. Puchovsky, *Pumps for Fire Protection Systems*, 2002.
NFPA 25, *Standard for the Inspection, Testing, and Maintenance of Water-Based Fire Protection Systems*, 2011 edition.
NFPA 70®, *National Electrical Code*®, 2011 edition.
NFPA 72®, *National Fire Alarm and Signaling Code*, 2013 edition.

Underwriters Laboratories Inc., 333 Pfingsten Road, Northbrook, IL 60062-2096.

ANSI/UL 50, *Standard for Enclosures for Electrical Equipment, Non-Environmental Considerations*, 2007.

CHAPTER 13

Steam Turbine Drive

Steam-driven fire pump technology has evolved to its current state over many years. The application of steam as a driver for fire pumps is not as common as that of electric drive or diesel engine drive, because steam is not as readily available as it once was. Furthermore, steam generation is not very energy efficient, so other forms of heating have been developed for most modern buildings. In fact, the number of new installations of steam-driven fire pumps worldwide is decreasing steadily each year. Chapter 13 covers the minimum requirements for the use of steam-driven fire pumps, and the standard requirements and commentary in this chapter reflect the current technology for this type of pump.

13.1 General

▶ **FAQ** Does NFPA 20 permit the use of steam turbines to drive fire pumps?

Although the use of steam turbines to drive fire pumps is rare, they are permitted by NFPA 20 and are likely to be used where steam is produced for manufacturing or industrial uses.

When selecting a turbine, it is important to analyze the total steam consumption and the demand of the driver and compare these factors to the steam supply that is available. The supply must be verified as being greater than the demand. See Exhibit II.13.1 through Exhibit II.13.3 for illustrations of steam-driven fire pumps. The use of a steam turbine–driven fire pump would usually serve as a backup in most facilities. The operating durations required for fire pumps makes the steam-driven unit a cause for concern.

The installation of a steam turbine driver for a fire pump should be performed by a person knowledgeable in boilers and turbines as well as fire pumps.

13.1.1 Acceptability.

13.1.1.1 Steam turbines of adequate power are acceptable prime movers for driving fire pumps.

13.1.1.1.1 Reliability of the turbines shall have been proved in commercial work.

▶ **FAQ** Are any steam turbine drivers listed?

At the present time, no steam turbine drivers are listed. The approval of the individual driver must be made using sound engineering principles and must be presented to the authority having jurisdiction (AHJ) for review and disposition.

As with other drivers, the reliability of the power source is one of the main considerations when choosing a means of supplying power to the fire pump. Very often, where used, a steam turbine–driven fire pump is placed in a remote location where steam generates the electrical power that operates an entire facility.

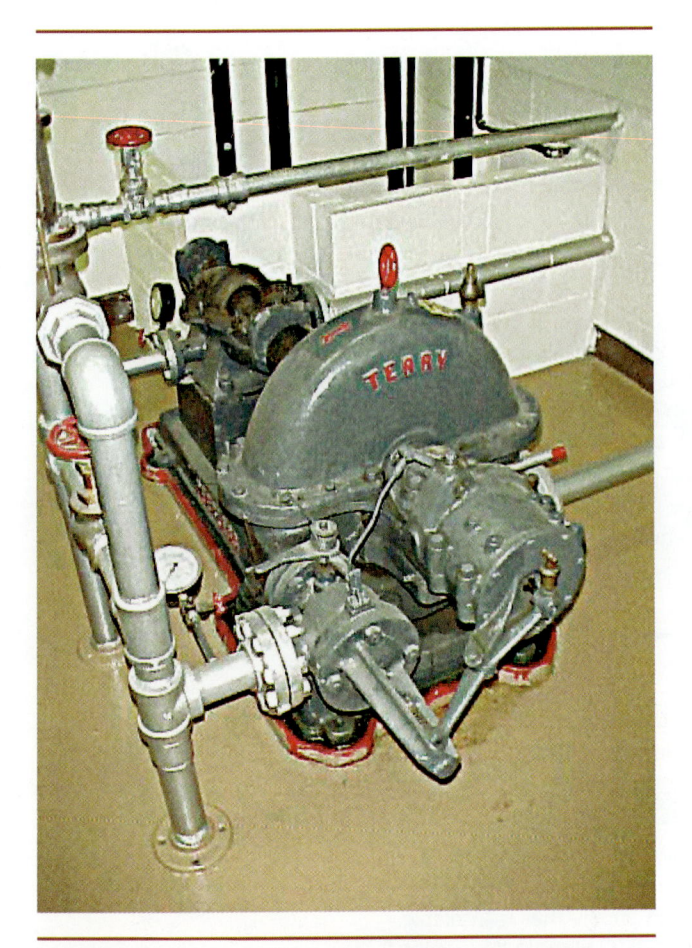

EXHIBIT II.13.1 Steam Turbine–Driven Fire Pump. (Photograph courtesy of Larry Wenzel)

EXHIBIT II.13.2 Fire Pump Information for Steam Turbine Pump. (Photograph courtesy of Larry Wenzel)

EXHIBIT II.13.3 Reciprocating Steam Engine–Driven Fire Pump. (Photograph courtesy of Larry Wenzel)

In most cases, a steam turbine–driven fire pump is not the primary automatic fire pump. Steam turbine–driven units normally are used as a secondary supply. The difference between today's evolved steam turbine drivers and the old reciprocating steam pumps is the difference between night and day. The newer equipment has automatic pressure regulating control valves that have improved the safety features of the turbine drive operations.

However, if for some reason the steam supply is shut down, the power to the steam turbine is also shut down. In such situations, an alternate source of steam or a pump with a different source of power (diesel) should be installed or provided as a backup.

13.1.1.2 The steam turbine shall be directly connected to the fire pump.

> Direct connection to a fire pump includes a flexible coupling that is used for alignment. It is not the intent of NFPA 20 to require the installation of a gear to reduce or increase transmission (the only approved gear drive is for vertical turbine fire pumps). The pump and the driver should run at the same rpm.

13.1.2 Turbine Capacity.

13.1.2.1 For steam boiler pressures not exceeding 120 psi (8.3 bar), the turbine shall be capable of driving the pump at its rated speed and maximum pump load with a pressure as low as 80 psi (5.5 bar) at the turbine throttle when exhausting against atmospheric back pressure with the hand valve open.

> The difference in maximum and minimum pressure identified in 13.1.2.1 represents the loss of steam due to friction in the steam piping system while the steam turbine is running.

13.1.2.2 For steam boiler pressures exceeding 120 psi (8.3 bar), where steam is continuously maintained, a pressure 70 percent of the usual boiler pressure shall take the place of the 80 psi (5.5 bar) pressure required in 13.1.2.1.

13.1.2.3 In ordering turbines for stationary fire pumps, the purchaser shall specify the rated and maximum pump loads at rated speed, the rated speed, the boiler pressure, the steam pressure at the turbine throttle (if possible), and the steam superheat.

13.1.3* Steam Consumption.

A.13.1.3 Single-stage turbines of maximum reliability and simplicity are recommended where the available steam supply will permit.

13.1.3.1 Prime consideration shall be given to the selection of a turbine having a total steam consumption commensurate with the steam supply available.

> As with other power supplies, the normal steam supply must be maintained for other on-line equipment while maintaining the quantity of steam needed to operate the steam turbine drive. The turbine must be able to come up to speed instantly when required by the fire water demand. If load shedding of other steam-driven equipment is required, the action must be completed automatically.
>
> The steam turbine should always be in a ready state. The turbine drive unit should always be kept warm to minimize the buildup of steam condensate when steam is injected into the unit. Manufacturers' requirements for static mode and operating mode functions should be followed. Full open steam pressure should not be injected into cold equipment. Only qualified personnel fully trained in the operation of steam-driven equipment should operate a steam-powered unit. Energizing a steam turbine too rapidly can result in a catastrophic event.
>
> The available steam supply should always be capable of supplying the steam at the minimum pressure required to operate the steam turbine for the minimum period of time required by the design of the fire protection system. A reserve supply of steam should be anticipated in the design to handle operation of the fire pump at 150 percent of its capacity.

13.1.3.2 Where multistage turbines are used, they shall be so designed that the pump can be brought up to speed without a warmup time requirement.

13.2 Turbine

▶ **FAQ** Where can more information regarding steam turbines be found?

A recognized standard for the manufacture of steam turbines is API Standard 611, *General-Purpose Steam Turbines for Petroleum, Chemical, and Gas Industry Services,* which outlines the minimum manufacturing tolerances that should be considered in approval of a turbine.

13.2.1 Casing and Other Parts.

13.2.1.1* The casing shall be designed to permit access with the least possible removal of parts or piping.

A.13.2.1.1 The casing can be of cast iron.

Some applications can require a turbine-driven fire pump to start automatically but not require the turbine to be on pressure control after starting. In such cases, a satisfactory quick-opening manual reset valve installed in a bypass of the steam feeder line around a manual control valve can be used.

Where the application requires the pump unit to start automatically and after starting continue to operate by means of a pressure signal, the use of a satisfactory pilot-type pressure control valve is recommended. This valve should be located in the bypass around the manual control valve in the steam feeder line. The turbine governor control valve, when set at approximately 5 percent above the normal full-load speed of the pump under automatic control, would act as a pre-emergency control.

In the arrangements set forth in the two preceding paragraphs, the automatic valve should be located in the bypass around the manual control valve, which would normally be kept in the closed position. In the event of failure of the automatic valve, this manual valve could be opened, allowing the turbine to come to speed and be controlled by the turbine governor control valve(s).

The use of a direct acting pressure regulator operating on the control valve(s) of a steam turbine is not recommended.

13.2.1.2 A safety valve shall be connected directly to the turbine casing to relieve high steam pressure in the casing.

13.2.1.3 Main Throttle Valve.

13.2.1.3.1 The main throttle valve shall be located in a horizontal run of pipe connected directly to the turbine.

13.2.1.3.2 There shall be a water leg on the supply side of the throttle valve.

13.2.1.3.3 This leg shall be connected to a suitable steam trap to automatically drain all condensate from the line supplying steam to the turbine.

13.2.1.3.4 Steam and exhaust chambers shall be equipped with suitable condensate drains.

13.2.1.3.5 Where the turbine is automatically controlled, these drains shall discharge through adequate traps.

13.2.1.3.6 In addition, if the exhaust pipe discharges vertically, there shall be an open drain at the bottom elbow.

13.2.1.3.7 This drain shall not be valved but shall discharge to a safe location.

13.2.1.4 The nozzle chamber, governor-valve body, pressure regulator, and other parts through which steam passes shall be made of a metal able to withstand the maximum temperatures involved.

13.2.2 Speed Governor.

13.2.2.1 The steam turbine shall be equipped with a speed governor set to maintain rated speed at maximum pump load.

13.2.2.2 The governor shall be capable of maintaining, at all loads, the rated speed within a total range of approximately 8 percent from no turbine load to full-rated turbine load by either of the following methods:

(1) With normal steam pressure and with hand valve closed
(2) With steam pressures down to 80 psi (5.5 bar) [or down to 70 percent of full pressure where this is in excess of 120 psi (8.3 bar)] and with hand valve open

13.2.2.3 While the turbine is running at rated pump load, the speed governor shall be capable of adjustment to secure speeds of approximately 5 percent above and 5 percent below the rated speed of the pump.

13.2.2.4 There shall also be provided an independent emergency governing device.

13.2.2.5 The independent emergency governing device shall be arranged to shut off the steam supply at a turbine speed approximately 20 percent higher than the rated pump speed.

13.2.3 Gauge and Gauge Connections.

13.2.3.1 A listed steam pressure gauge shall be provided on the entrance side of the speed governor.

13.2.3.2 A 0.25 in. (6 mm) pipe tap for a gauge connection shall be provided on the nozzle chamber of the turbine.

13.2.3.3 The gauge shall indicate pressures not less than one and one-half times the boiler pressure and in no case less than 240 psi (16.5 bar).

13.2.3.4 The gauge shall be marked "Steam."

13.2.4 Rotor.

13.2.4.1 The rotor of the turbine shall be of suitable material.

13.2.4.2 The first unit of a rotor design shall be type tested in the manufacturer's shop at 40 percent above rated speed.

13.2.4.3 All subsequent units of the same design shall be tested at 25 percent above rated speed.

13.2.5 Shaft.

13.2.5.1 The shaft of the turbine shall be of high-grade steel, such as open-hearth carbon steel or nickel steel.

13.2.5.2 Where the pump and turbine are assembled as independent units, a flexible coupling shall be provided between the two units.

13.2.5.3 Where an overhung rotor is used, the shaft for the combined unit shall be in one piece with only two bearings.

13.2.5.4 The critical speed of the shaft shall be well above the highest speed of the turbine so that the turbine will operate at all speeds up to 120 percent of rated speed without objectionable vibration.

13.2.6 Bearings.

13.2.6.1 Sleeve Bearings. Turbines having sleeve bearings shall have split-type bearing shells and caps.

13.2.6.2 Ball Bearings.

13.2.6.2.1 Turbines having ball bearings shall be acceptable after they have established a satisfactory record in the commercial field.

13.2.6.2.2 Means shall be provided to give visible indication of the oil level.

13.3* Installation

Details of steam supply, exhaust, and boiler feed shall be carefully planned to provide reliability and effective operation of a steam turbine–driven fire pump.

> Steam can be provided in a piping loop to the machine requiring the power. This piping should be designed with the same care that is required for fire sprinkler systems. The piping in earthquake-prone areas should be properly supported and braced. The boiler in earthquake-prone areas must be able to operate and resist the minimum horizontal and vertical force conditions as required by the AHJ.
>
> In addition to the requirements set forth in Chapter 13, regulatory requirements for the installation of steam piping control valves and special fabrication methods fall under the purview of ASME. Refer to ASME B31 series, *Standards of Pressure Piping Installation*, with specific reference to ASME B31.1, *Power Piping*, and ASME B31.9, *Building Services Piping*. All steam piping work is required to be performed by individuals or companies certified by ASME.

A.13.3 The following information should be taken into consideration when planning a steam supply, exhaust, and boiler feed for a steam turbine–driven fire pump.

The steam supply for the fire pump should preferably be an independent line from the boilers. It should be run so as not to be liable to damage in case of fire in any part of the property. The other steam lines from the boilers should be controlled by valves located in the boiler room. In an emergency, steam can be promptly shut off from these lines, leaving the steam supply entirely available for the fire pump. Strainers in steam lines to turbines are recommended.

The steam throttle at the pump should close against the steam pressure. It should preferably be of the globe pattern with a solid disc. If, however, the valve used has a removable composition ring, the disc should be of bronze and the ring made of sufficiently hard and durable material, and so held in place in the disc as to satisfactorily meet severe service conditions. Gate valves are undesirable for this service because they cannot readily be made leaktight, as is possible with the globe type of valve. The steam piping should be so arranged and trapped that the pipes can be kept free of condensed steam.

In general, a pressure-reducing valve should not be placed in the steam pipe supplying the fire pump. There is no difficulty in designing turbines for modern high-pressure steam, and this gives the simplest and most dependable unit. A pressure-reducing valve introduces a possible obstruction in the steam line in case it becomes deranged. In most cases, the turbines can be protected by making the safety valve required by 13.2.1.2 of such size that the pressure in the casing will not exceed 25 psi (1.7 bar). This valve should be piped outside of the pump room and, if possible, to some point where the discharge could be seen by the pump attendant. Where a pressure-reducing valve is used, the following points should be carefully considered:

(1) *Pressure-Reducing Valve.*

    (a) The pressure-reducing valve should not contain a stuffing box or a piston working in a cylinder.

    (b) The pressure-reducing valve should be provided with a bypass containing a globe valve to be opened in case of an emergency. The bypass and stop valves should be one pipe size smaller than the reducing valve, and they should be located so as to be

readily accessible. This bypass should be arranged to prevent the accumulation of condensate above the reducing valve.
 (c) The pressure-reducing valve should be smaller than the steam pipe required by the specifications for the turbine.
(2) *Exhaust Pipe.* The exhaust pipe should run directly to the atmosphere and should not contain valves of any type. It should not be connected with any condenser, heater, or other system of exhaust piping.
(3) *Emergency Boiler Feed.* A convenient method of ensuring a supply of steam for the fire pump unit, in case the usual boiler feed fails, is to provide an emergency connection from the discharge of the fire pump. This connection should have a controlling valve at the fire pump and also, if desired, an additional valve located in the boiler room. A check valve also should be located in this connection, preferably in the boiler room. This emergency connection should be about 2 in. (50 mm) in diameter.

This method should not be used when there is any danger of contaminating a potable water supply. In situations where the fire pump is handling salt or brackish water, it might also be undesirable to make this emergency boiler feed connection. In such situations, an effort should be made to secure some other secondary boiler feed supply that will always be available.

References Cited in Commentary

American Petroleum Institute, 1220 L Street, NW, Washington, DC 20005-4070.

ANSI/API Standard 611, *General-Purpose Steam Turbines for Petroleum, Chemical, and Gas Industry Services,* 2008.

American Society of Mechanical Engineers, Three Park Avenue, New York, NY 10016-5990

ASME B31 Series, *Standards of Pressure Piping Installation:*
ASME B31.1, *Power Piping,* 2012.
ASME B31.9, *Building Services Piping,* 2011.

Acceptance Testing, Performance, and Maintenance

CHAPTER 14

The final step in the design and installation of a fire pump system is to demonstrate that the fire pump functions as intended, the installation meets the requirements of NFPA 20, *Standard for the Installation of Stationary Pumps for Fire Protection,* and the work has been completed in a professional manner. The various tests and system checks required in Chapter 14 are intended to show that all of the critical components of the system are in place and the fire pump performs in accordance with design requirements. The fire pump should perform in accordance with the shop test curve as described in Chapter 6. If the water supply does not allow testing the fire pump at the full range on the shop test curve, as a minimum, the pump is expected to meet the fire protection system demand and 100 percent of rated flow.

This chapter also deals with maintenance of the pump and related equipment. For more detailed information regarding fire pump maintenance, see NFPA 25, *Standard for the Inspection, Testing, and Maintenance of Water-Based Fire Protection Systems,* and *Water-Based Fire Protection Systems Handbook.*

The acceptance test evaluates the pump's performance over a range of conditions in order to evaluate the performance of the installation so that the pump will perform as needed during an actual fire.

The acceptance test includes the following steps:

1. Demonstration of the adequacy of the pump and its ability to deliver water in accordance with the manufacturer's certified shop test characteristic curve, as described in 14.2.4
2. Operation of the pump driver under various conditions to confirm satisfactory performance under all expected conditions
3. Repeated operations of the primary and alternative power supply equipment (if supplied) to demonstrate satisfactory operation under all expected conditions using either automatic or manual operation of the controller (see 14.2.7)
4. Verification of the operation and pressure settings of the variable speed pressure limiting control (VSPLC) system (see 14.2.6.3)

Any defects, faults, or other performance problems discovered during the acceptance test must be corrected before final acceptance is granted.

Note that Annex B, Possible Causes of Pump Troubles, of NFPA 20 contains a partial list of pump problems and their possible causes and some suggested remedies for those problems. Annex B may be useful in identifying and correcting problems discovered during fire pump acceptance tests.

When defective components are replaced, Table 8.6.1 of NFPA 25, Summary of Component Replacement Testing Requirements, must be used to recommission the fire pump. If critical components are replaced, NFPA 20 requires that a complete acceptance test be performed.

14.1 Hydrostatic Tests and Flushing

14.1.1* Flushing.

Stones, gravel, blocks of wood, bottles, work gloves, burlap bags, and other foreign objects have been found in underground piping and piping taking suction from municipal or private water mains or water storage tanks when flushing and/or flow testing procedures are performed. Also, objects in underground piping that are quite remote from the fire pump installation and that would otherwise remain stationary during low-flow conditions may sometimes be carried into the pump during the high volume flows required during the fire pump acceptance test. Because many of these objects can cause damage to the fire pump, flushing of suction piping prior to connection to the fire pump minimizes potential damage from foreign objects during the fire pump acceptance flow tests. See Exhibit II.14.1 for sample objects removed from underground pipe.

For booster pumps taking suction from underground mains, the flushing flow rate should be at least equal to that in Table 14.1.1.1 or the maximum water demand, whichever is greater. The flushing flow rates given in Table 14.1.1.1 are based on 15 ft/sec (4.6 m/sec) waterflow velocity, which is the basis for the minimum suction inlet pipe sizes in Table 4.26(a) and Table 4.26(b). The minimum suction pipe inlet sizes in Table 4.26(a) and Table 4.26(b) are the same for pumps of several different capacities.

Experience has shown that the flow rates established in Table 14.1.1.1 are of sufficient velocity to clear the suction piping of debris. Additionally, most of these flow rates exceed the 150 percent point of the pump ratings where these pipe sizes are the minimum mandated in Table 4.26(a) and Table 4.26(b). For instance, the largest pump allowed to use a minimum 5 in. (125 mm) suction pipe is rated at 500 gpm (1892 L/min). The peak (150 percent) point for a 500 gpm (1892 L/min) pump is 750 gpm (2839 L/min). For 5 in. (125 mm) suction pipes, the flush rate is 920 gpm (3482 L/min). In another instance, the largest pump allowed to use an 8 in. (200 mm) suction pipe is rated at 1500 gpm (5677 L/min). The peak (150 percent) point for a 1500 gpm (5677 L/min) pump is 2250 gpm (8516 L/min). For 8 in. (200 mm) suction pipes, the flush rate is 2350 gpm (8895 L/min).

For fire pumps taking suction from a storage tank, reservoir, or wet pit, the following should be determined prior to the pump acceptance flow test:

1. No obstructing materials are in the water main, tank, reservoir, or pit.
2. Any piping to the pump suction inlet is free from obstructing material.

EXHIBIT II.14.1 Objects Found in a Fire Pump Impeller During Acceptance Testing. (Courtesy of Underwood Fire Equipment, Inc.)

> Although field experience is such that oftentimes actual flushing is done without a measurement of the flow rate, verification of the flow rate should be determined. The actual flow rate can be determined using methods such as pitot readings or flowmeters described in NFPA 291, *Recommended Practice for Fire Flow Testing and Marking of Hydrants* (see Part III, Section 1, of this handbook for the extract of NFPA 291). Flushing is normally accomplished at the maximum flow rate available from the water supply. In some jurisdictions, restrictions on the minimum suction pressures allowed may apply. In most cases, the flow at 150 percent of the fire pump rating should not result in unacceptable minimum residual suction pressures and should serve as an appropriate flushing flow rate. Where this scenario is not the case, actual flushing flow rates should at least be equal to the maximum water demand for the fire protection system. It is also important to know that the flushing rates required by NFPA 24, *Standard for the Installation of Private Fire Service Mains and Their Appurtenances*, are based on a velocity of 10 ft/sec (3.1 m/sec) and not the rates required by NFPA 20.
>
> Regardless of the flushing methods used, the installing contractor should never connect piping to the underground stub-up without obtaining a copy of the Contractor's Material and Test Certificate for Underground Piping from the site utilities contractor or the underground piping installing contractor. This certificate provides verification that the underground piping has been flushed in accordance with NFPA 24 and that no debris was present in the piping system at the time of the test.

A.14.1.1 The suction piping to a fire pump needs to be adequately flushed to make sure stones, silt, and other debris will not enter the pump or the fire protection system. The flow rates in Table 14.1.1.1 are the minimum recommended, which will produce a velocity of approximately 15 ft/sec (4.6 m/sec). If the flow rate cannot be achieved with the existing water supply, a supplemental source such as a fire department pumper could be necessary. The procedure is to be performed, witnessed, and signed off before connection to the suction piping is completed.

14.1.1.1 Suction piping shall be flushed at a flow rate not less than indicated in Table 14.1.1.1 or at the hydraulically calculated water demand rate of the system, whichever is greater.

TABLE 14.1.1.1 Minimum Flow Rates for Flushing Suction Piping

Nominal Pipe Size (in.)	Flow Rate (gpm)	Nominal Pipe Size (mm)	Flow Rate (L/min)
1½	85	38	345
2	150	50	540
3	330	75	1,380
4	590	100	2,160
5	920	125	3,490
6	1,360	150	4,850
8	2,350	200	8,900
10	3,670	250	13,900
12	5,290	300	20,030

14.1.1.2 Flushing shall occur prior to hydrostatic test.

> If the fire pump installer responsible for conducting the fire pump test does not have the contract responsibility for the underground pipe to the fire pump suction piping, the party responsible for the fire pump acceptance testing should obtain a flushing certificate from the installer of the underground pipe prior to conducting the fire pump acceptance test.

14.1.1.3 Where the maximum flow available from the water supply cannot provide the flow rate provided in Table 14.1.1.1, the flushing flow rate shall be the greater of 100 percent of rated flow of the connected fire pump or the maximum flow demand of the fire protection system(s).

14.1.1.3.1 This reduced flushing flow capacity shall constitute an acceptable test, provided that the flow rate exceeds the fire protection system design and flow rate.

14.1.2 Hydrostatic Test.

14.1.2.1 Suction and discharge piping shall be hydrostatically tested at not less than 200 psi (13.8 bar) pressure or at 50 psi (3.4 bar) in excess of the maximum pressure to be maintained in the system, whichever is greater.

> All new systems are required to be tested hydrostatically at a minimum pressure of either 200 psi (13.8 bar) or at 50 psi (3.4 bar) above the maximum discharge pressure, whichever is greater, at the fire pump discharge flange. Verification that the proper pressure rating for pipe and fittings is used in the installation is important. This value is set to ensure that all pipe joints and other equipment are installed properly to withstand that pressure without coming apart or leaking. Although the test is primarily a quality control test on the installation and not a materials performance test, damaged materials (e.g., cracked fittings, leaky valves, bad joints) are routinely discovered during the hydrostatic test.
>
> The measurement of the hydrostatic test pressure is taken at the lowest elevation within the system or portion of the system being tested. Testing at the high point of the system, which, due to static head, could increase the test pressure at lower points in the system to exceptionally high pressures, is not considered necessary.
>
> The hydrostatic test should include all suction and discharge piping between the suction pipe control valve and the discharge pipe control valve. Also, any bypass, meter, jockey pump, cooling water piping, and hose header piping should be included in the test. The normally closed control valve for the hose header should be open for this test. Suction piping on the upstream side of the suction piping control valve on booster pumps, and discharge piping on the system side of the discharge piping control valve, which feeds into underground mains, is required to be hydrostatically tested in accordance with NFPA 24. For discharge piping on the system side of the discharge piping control valves, which feeds directly into the building's fire protection systems, hydrostatic testing is required in accordance with other standards such as NFPA 13, *Standard for Installation of Sprinkler Systems,* and NFPA 14, *Standard for the Installation of Standpipe and Hose Systems.* The coordination of hydrostatic test procedures to facilitate simultaneous testing of piping for various portions of the entire fire protection system may be practical. However, when testing underground piping, NFPA 24 permits a minimal amount of leakage. As such, underground and aboveground piping should be isolated for the hydrostatic test. The actual leakage in underground piping can be closely approximated by measuring how much water is required to restore the initial pressure at the end of the test. The measured leakage should be compared to the leakage allowed in NFPA 24.
>
> Where the fire pump takes suction from a storage tank or reservoir, suction piping upstream of the suction piping control valve must be in accordance with the requirements of NFPA 22, *Standard for Water Tanks for Private Fire Protection.*
>
> Where a main pressure relief valve is provided on the discharge side of the fire pump, and the relief valve setting is less than the required hydrostatic test pressure, suitable means should be employed to allow full hydrostatic testing without discharging water through the relief valve. One method is to temporarily set the relief valve at or above the hydrostatic test pressure and then reset and verify the pressure setting of the relief valve during the fire pump acceptance flow test.

14.1.2.2 The pressure required in 14.1.2.1 shall be maintained for 2 hours.

> **▶ FAQ** What are the minimum acceptance criteria for the hydrostatic test?

> The minimum acceptance criteria for the hydrostatic test are that there be no visible leaks and no drop in the test pressure for above ground piping systems. The intention of this standard is to require the same procedure for hydrostatic testing as that of Chapter 25 of NFPA 13, 2013 edition.

14.1.3* The installing contractor shall furnish a certificate for flushing and hydrostatic test prior to the start of the fire pump field acceptance test.

A.14.1.3 See Figure A.14.1.3(a) for a sample of a contractor's material and test certificate for fire pumps and Figure A.14.1.3(b) for a sample certificate for private fire service mains.

14.2 Field Acceptance Tests

> The field acceptance test evaluates the pump performance over a range of conditions in order to assess the performance of the installation to ensure the pump will perform as needed during an actual fire.

14.2.1* The pump manufacturer, the engine manufacturer (when supplied), the controller manufacturer, and the transfer switch manufacturer (when supplied) or their factory-authorized representatives shall be present for the field acceptance test. *(See Section 4.4.)*

> As noted in 4.4.1 and A.4.4.1, NFPA 20 requires a single entity to have the responsibility for ensuring a properly completed, tested, and accepted fire pump installation. To this end, 14.2.1 requires that all key component manufacturers (or their representatives) be present at the field acceptance test. The presence of the manufacturer's representatives allows any problems with the quality of the installation, the equipment, and the performance of the completed fire pump installation to be effectively identified and corrected to the satisfaction of the owner, the design entity, the authority having jurisdiction (AHJ), the installing contractor, and any other involved parties.

A.14.2.1 In addition, representatives of the installing contractor, insurance company, and owner should be present.

14.2.2 The date, time, and location of the field acceptance test shall be coordinated with the authority having jurisdiction.

> **▶ FAQ** Who should be notified of the field acceptance test?

> In many cases, the local fire/building official ultimately has legal jurisdictional responsibility and may be needed to approve an installation or even issue a valid Certificate of Occupancy. As such, adequate notice needs to be given to arrange the testing. See Exhibit II.14.2 for a sample Certificate of Occupancy.
> While the AHJ can represent a number of different individuals, such as various code enforcement officers, it is best to have all involved parties, such as the owner, tenant, or insurance companies, notified and present rather than to have to repeat the acceptance test multiple times. Sometimes, the owner can assist in notifying other interested parties. The owner may also be needed to provide access to other important areas, such as electrical rooms.

Stationary Fire Pumps Handbook   2013

Contractor's Material and Test Certificate for Fire Pump Systems

PROCEDURE Upon completion of work, inspection and tests shall be made by the contractor's representative and witnessed by an owner's representative. All defects shall be corrected and system left in service before contractor's personnel finally leave the job.

A certificate shall be filled out and signed by both representatives. Copies shall be prepared for approving authorities, owners, and contractor. It is understood the owner's representative's signature in no way prejudices any claim against contractor for faulty material, poor workmanship, or failure to comply with approving authority's requirements or local ordinances.

PROPERTY NAME		DATE	
PROPERTY ADDRESS			

PLANS	ACCEPTED BY APPROVING AUTHORITIES (NAMES)		
	ADDRESS		
	INSTALLATION CONFORMS TO ACCEPTED PLANS	☐ YES	☐ NO
	ALL EQUIPMENT USED IS APPROVED FOR FIRE SYSTEM SERVICE IF NO, STATE DEVIATIONS	☐ YES	☐ NO
INSTRUCTIONS	HAS PERSON IN CHARGE OF FIRE PUMP EQUIPMENT BEEN INSTRUCTED AS TO LOCATION OF SYSTEM CONTROL VALVES AND CARE AND MAINTENANCE OF THIS NEW EQUIPMENT? IF NO, EXPLAIN	☐ YES	☐ NO
	HAVE COPIES OF APPROPRIATE INSTRUCTIONS AND CARE AND MAINTENANCE CHARTS BEEN LEFT ON PREMISES? IF NO, EXPLAIN	☐ YES	☐ NO
LOCATION	SUPPLIES BUILDING(S) (CAMPUS, WAREHOUSE, HIGH RISE) EXPLAIN		
PUMP ROOM EQUIPMENT	IS THE PUMP ROOM EQUIPMENT PER THE PLANS AND SPECS?	☐ YES	☐ NO
	IS THE FIRE PUMP PROPERLY MOUNTED AND ANCHORED TO THE FOUNDATION? IF NO, EXPLAIN	☐ YES	☐ NO
	IS THE FIRE PUMP BASE PROPERLY GROUTED? IF NO, EXPLAIN	☐ YES	☐ NO
	DOES THE PUMP ROOM HAVE THE PROPER FLOOR DRAINS? IF NO, EXPLAIN	☐ YES	☐ NO
	IS THE SUCTION AND DISCHARGE PIPING PROPERLY SUPPORTED?	☐ YES	☐ NO
	IS THE PUMP ROOM HEATED AND VENTILATED PER NFPA 20?	☐ YES	☐ NO
PIPES AND FITTINGS	PIPE TYPES AND CLASS		
	PIPE CONFORMS TO _____ STANDARD FITTINGS CONFORM TO _____ STANDARD IF NO, EXPLAIN	☐ YES ☐ YES	☐ NO ☐ NO
	SUCTION AND DISCHARGE PIPING ANCHORED OR RESTRAINED?:	☐ YES	☐ NO
PRE-PACKAGED PUMP HOUSE	IS THIS A PACKAGE OR SKID MOUNTED PUMP?	☐ YES	☐ NO
	IS THE PACKAGE/SKID PROPERLY ANCHORED TO A CONCRETE FOUNDATION? IF NO, EXPLAIN	☐ YES	☐ NO
	IS THE STRUCTURAL FOUNDATION FRAME FILLED WITH CONCRETE TO FORM A FINISHED FLOOR?	☐ YES	☐ NO
	IS THERE A FLOOR DRAIN INSTALLED?	☐ YES	☐ NO
TEST DESCRIPTION	HYDROSTATIC: Hydrostatic tests shall be made at not less than 200 psi (13.8 bar) for 2 hours or 50 psi (3.4 bar) above static pressure in excess of 200 psi (13.8 bar) for 2 hours.		
	HYDROSTATIC TEST: ALL NEW PIPING HYDROSTATICALLY TESTED AT: _____ PSI/BAR FOR _____ HOURS	NO LEAKAGE ALLOWED	
FLUSHING TESTS	FLUSHING: Flow the required rate until water is clear as indicated by no collection of foreign material in burlap bags at outlets such as hydrants and blowoffs. Flush at flows not less than 390 gpm (1476 L/min) for 4 in. pipe, 610 gpm (2309 L/min) for 5 in. pipe, 880 gpm (3331 L/min) for 6 in. pipe, 1560 gpm (5905 L/min) for 8 in. pipe, 2440 gpm (9235 L/min) for 10 in. pipe, and 3520 gpm (13,323 L/min) for 12 in. pipe. When supply cannot produce stipulated flow rates, obtain maximum available.		

© 2012 National Fire Protection Association (NFPA 20, 1 of 2)

FIGURE A.14.1.3(a) *Sample of Contractor's Material Test Certificate for Fire Pump Systems.*

FLUSHING TESTS (continued)	NEW PIPING FLUSHED ACCORDING TO _____ STANDARD ☐ YES ☐ NO BY (COMPANY) _____ IF NO, EXPLAIN		
	HOW FLUSHING FLOW WAS OBTAINED ☐ PUBLIC WATER ☐ TANK OR RESERVOIR ☐ OTHER (EXPLAIN)	THROUGH WHAT TYPE OPENING ☐ TEST HEADER ☐ OPEN PIPE	
	LEAD-INS FLUSHED ACCORDING TO _____ STANDARD ☐ YES ☐ NO BY (COMPANY) _____ IF NO, EXPLAIN		
	HOW FLUSHING FLOW WAS OBTAINED ☐ PUBLIC WATER ☐ TANK OR RESERVOIR ☐ OTHER (EXPLAIN)	THROUGH WHAT TYPE OPENING ☐ Y CONNECTION TO FLANGE & SPIGOT ☐ OPEN PIPE	
FIELD ACCEPTANCE TEST	ALL EQUIPMENT APPROVED?		☐ YES ☐ NO
	ALL REQUIRED REPRESENTATIVES PRESENT FOR TEST		☐ YES ☐ NO
	AHJ AND OWNER'S REPRESENTATIVE PRESENT FOR TEST IF NO, EXPLAIN		☐ YES ☐ NO
	ALL ELECTRICAL WIRING COMPLETE AND PER *NFPA 70* AND NFPA 20 IF NO, EXPLAIN		☐ YES ☐ NO
	CALIBRATE TEST EQUIPMENT USED CALIBRATION DATE _____		☐ YES ☐ NO
	FLOW TESTS PUMP DESIGN _____ GPM _____ PSI DOES THE PUMP MEET OR EXCEED THE CERTIFIED CURVE? PUMP TYPE ☐ HORIZONTAL ☐ VERTICAL TURBINE ☐ OTHER PUMP MAKE _____ MODEL # _____ SERIAL # _____ COMMENTS _____		☐ YES ☐ NO
	ELECTRIC DRIVER OPERATIONAL TEST SATISFACTORY ELEC. DRIVER _____ MODEL # _____ SERIAL # _____ VOLTAGE _____ VAC @ _____ HP _____ RPM _____ FLA		☐ YES ☐ NO
	ENGINE DRIVEN ENGINE MAKE _____ MODEL # _____ SERIAL # _____ _____ HP _____ RPM SPEED DIESEL DRIVER OPERATIONAL TEST SATISFACTORY? OTHER EXPLAIN		☐ YES ☐ NO ☐ YES ☐ NO

	CONTROLLER MAKE _____ MODEL # _____ SERIAL # _____
	VARIABLE SPEED PRESSURE LIMITING CONTROL
	TESTED AT MINIMUM, RATED, AND PEAK FLOW
	CONTROLLER TEST:
	    SIX AUTO STARTS
	    SIX MANUAL STARTS
	PHASE REVERSAL TEST PERFORMED (ELECTRIC ONLY)
	ALTERNATE POWER SOURCE TESTED (ELECTRIC ONLY)
	ELECTRONIC FUEL MANAGEMENT (ECM) FUNCTION TEST PERFORMED (DIESEL ONLY)
			☐ YES ☐ NO
			☐ YES ☐ NO
			☐ YES ☐ NO
			☐ YES ☐ NO
			☐ YES ☐ NO
			☐ YES ☐ NO
			☐ YES ☐ NO

CONTROL VALVES	SYSTEM CONTROL VALVES LEFT WIDE OPEN IF NO, STATE REASON		☐ YES ☐ NO
	HOSE THREADS OF FIRE DEPARTMENT CONNECTIONS AND HYDRANTS INTERCHANGEABLE WITH THOSE OF FIRE DEPARTMENT ANSWERING ALARM		☐ YES ☐ NO
REMARKS	DATE LEFT IN SERVICE _____ ADDITIONAL COMMENTS:		
SIGNATURES	NAME OF INSTALLING CONTRACTOR		
	TESTS WITNESSED BY FOR PROPERTY OWNER (SIGNED)	TITLE	DATE
	FOR INSTALLING CONTRACTOR (SIGNED)	TITLE	DATE
	ADDITIONAL COMMENTS AND NOTES:		

© 2012 National Fire Protection Association   (NFPA 20, 2 of 2)

FIGURE A.14.1.3(a) *Continued*

Contractor's Material and Test Certificate for Private Fire Service Mains

PROCEDURE Upon completion of work, inspection and tests shall be made by the contractor's representative and witnessed by an owner's representative. All defects shall be corrected and system left in service before contractor's personnel finally leave the job.

A certificate shall be filled out and signed by both representatives. Copies shall be prepared for approving authorities, owners, and contractor. It is understood the owner's representative's signature in no way prejudices any claim against contractor for faulty material, poor workmanship, or failure to comply with approving authority's requirements or local ordinances.

PROPERTY NAME		DATE	
PROPERTY ADDRESS			

PLANS	ACCEPTED BY APPROVING AUTHORITIES (NAMES)		
	ADDRESS		
	INSTALLATION CONFORMS TO ACCEPTED PLANS	☐ YES	☐ NO
	EQUIPMENT USED IS APPROVED IF NO, STATE DEVIATIONS	☐ YES	☐ NO
INSTRUCTIONS	HAS PERSON IN CHARGE OF FIRE EQUIPMENT BEEN INSTRUCTED AS TO LOCATION OF CONTROL VALVES AND CARE AND MAINTENANCE OF THIS NEW EQUIPMENT? IF NO, EXPLAIN	☐ YES	☐ NO
	HAVE COPIES OF APPROPRIATE INSTRUCTIONS AND CARE AND MAINTENANCE CHARTS BEEN LEFT ON PREMISES? IF NO, EXPLAIN	☐ YES	☐ NO
LOCATION	SUPPLIES BUILDINGS		
PIPES AND JOINTS	PIPE TYPES AND CLASS	TYPE JOINT	
	PIPE CONFORMS TO _____ STANDARD FITTINGS CONFORM TO _____ STANDARD IF NO, EXPLAIN	☐ YES ☐ YES	☐ NO ☐ NO
	BURIED JOINTS NEEDING ANCHORAGE CLAMPED, STRAPPED, OR BLOCKED IN ACCORDANCE WITH _____ STANDARD IF NO, EXPLAIN	☐ YES	☐ NO
TEST DESCRIPTION	FLUSHING: Flow the required rate until water is clear as indicated by no collection of foreign material in burlap bags at outlets such as hydrants and blowoffs. Flush at flows not less than 390 gpm (1476 L/min) for 4 in. pipe, 610 gpm (2309 L/min) for 5 in. pipe, 880 gpm (3331 L/min) for 6 in. pipe, 1560 gpm (5905 L/min) for 8 in. pipe, 2440 gpm (9235 L/min) for 10 in. pipe, and 3520 gpm (13323 L/min) for 12 in. pipe. When supply cannot produce stipulated flow rates, obtain maximum available. HYDROSTATIC: Hydrostatic tests shall be made at not less than 200 psi (13.8 bar) for 2 hours or 50 psi (3.4 bar) above static pressure in excess of 150 psi (10.3 bar) for 2 hours. LEAKAGE: New pipe laid with rubber gasketed joints shall, if the workmanship is satisfactory, have little or no leakage at the joints. The amount of leakage at the joints shall not exceed 2 qt/hr (1.89 L/hr) per 100 joints irrespective of pipe diameter. The amount of allowable leakage specified above can be increased by 1 fl oz per inch valve diameter per hour (30 mL/25 mm/hr) for each metal seated valve isolating the test section. If dry barrel hydrants are tested with the main valve open, so the hydrants are under pressure, an additional 5 oz per minute (150 mL/min) leakage is permitted for each hydrant.		
FLUSHING TESTS	NEW PIPING FLUSHED ACCORDING TO _____ STANDARD BY (COMPANY) IF NO, EXPLAIN	☐ YES	☐ NO
	HOW FLUSHING FLOW WAS OBTAINED ☐ PUBLIC WATER ☐ TANK OR RESERVOIR ☐ FIRE PUMP	THROUGH WHAT TYPE OPENING ☐ HYDRANT BUTT ☐ OPEN PIPE	
	LEAD-INS FLUSHED ACCORDING TO _____ STANDARD BY (COMPANY) IF NO, EXPLAIN	☐ YES	☐ NO
	HOW FLUSHING FLOW WAS OBTAINED ☐ PUBLIC WATER ☐ TANK OR RESERVOIR ☐ FIRE PUMP	THROUGH WHAT TYPE OPENING ☐ Y CONNECTION TO FLANGE & SPIGOT ☐ OPEN PIPE	

© 2012 National Fire Protection Association (NFPA 20, 1 of 2)

FIGURE A.14.1.3(b) *Sample of Contractor's Material and Test Certificate for Private Fire Service Mains.*

HYDROSTATIC TEST	ALL NEW PIPING HYDROSTATICALLY TESTED AT ———— PSI FOR ———— HOURS		BURIED JOINTS COVERED ☐ YES ☐ NO
LEAKAGE TEST	TOTAL AMOUNT OF LEAKAGE MEASURED ———— GALLONS	NO LEAKAGE ALLOWED FOR VISIBLE JOINTS ———— HOURS	
	ALLOWABLE LEAKAGE (BURIED) ———— GALLONS	NO LEAKAGE ALLOWED FOR VISIBLE JOINTS ———— HOURS	
HYDRANTS	NUMBER INSTALLED	TYPE AND MAKE	ALL OPERATE SATISFACTORILY ☐ YES ☐ NO
CONTROL VALVES	WATER CONTROL VALVES LEFT WIDE OPEN IF NO, STATE REASON		☐ YES ☐ NO
	HOSE THREADS OF FIRE DEPARTMENT CONNECTIONS AND HYDRANTS INTERCHANGEABLE WITH THOSE OF FIRE DEPARTMENT ANSWERING ALARM		☐ YES ☐ NO
REMARKS	DATE LEFT IN SERVICE ———— ADDITIONAL COMMENTS: ————		
SIGNATURES	NAME OF INSTALLING CONTRACTOR		
	TESTS WITNESSED BY		
	FOR PROPERTY OWNER (SIGNED)	TITLE	DATE
	FOR INSTALLING CONTRACTOR (SIGNED)	TITLE	DATE
ADDITIONAL EXPLANATION AND NOTES			

© 2012 National Fire Protection Association (NFPA 20, 2 of 2)

FIGURE A.14.1.3(b) *Continued*

14.2.3 Pump Room Electrical Wiring. All electric wiring to the fire pump motor(s), including control (multiple pumps) interwiring, normal power supply, alternate power supply where provided, and jockey pump, shall be completed and checked by the electrical contractor prior to the initial startup and acceptance test.

> The acceptance test is intended to evaluate the performance of a fully completed installation over a range of conditions, including a minimum flow rate that equals or exceeds the maximum fire protection demand, so that the pump will perform as needed during an actual fire. Therefore, it is important to verify that all wiring and connections are completed and tested prior to acceptance testing and the pump rotation direction verified so that any problems associated with incorrect or incomplete connections, unacceptable quality, incompatibilities due to mismatched components, or any other performance problems can be readily identified and corrected. Acceptance should not be granted on the assumption that an incomplete installation will be correctly completed at some future time.
>
> Most instances of driver failures during the acceptance testing of electric motor–driven fire pumps are usually traced back to noncompliant wiring methods, materials, or practices. By checking the entire installation prior to the acceptance test, unnecessary and time-consuming troubleshooting efforts can be avoided.

CERTIFICATE OF OCCUPANCY

Building Permit Number _____ Date _____

The undersigned hereby applies for a permit of occupancy in accordance with 780 CMR 120, sixth edition:

1. Location of building _____
                          Street Address   Unit Number
2. Applicant _____
3. Owner _____
   Address _____
4. Occupant _____
5. Use group _____ Occupancy _____
6. Construction type _____ Occupant load _____
7. Special stipulations or conditions _____
   _____
   _____
   _____
   _____
   _____
   _____
   _____

Plumbing/gas _____ Fire _____

Electrical _____ Water and sewer _____

Health _____ Public works _____

I hereby certify that the work specified by the above named building permit has been completed and is ready for occupancy.

Building Inspector _____

Inspection Director _____

Date _____

© 2005 National Fire Protection Association

EXHIBIT II.14.2 *Sample Certificate of Occupancy.*

14.2.4* Certified Pump Curve.

> The manufacturer's certified pump test characteristic curve provides a graphical representation of the pump's performance under controlled conditions, prior to any damage that occurs in transit and prior to the completion of the installation. The curve, therefore, provides a benchmark to which installed pumps can be compared (see Section 4.5).

A.14.2.4 If a complete fire pump submittal package is available, it should provide for comparison of the equipment specified. Such a package should include an approved copy of the fire pump room general arrangement drawings, including the electrical layout, the layout of the pump and water source, the layout of the pump room drainage details, the pump foundation layout, and the mechanical layout for heat and ventilation.

14.2.4.1 A copy of the manufacturer's certified pump test characteristic curve shall be available for comparison of the results of the field acceptance test.

14.2.4.1.1 For water mist positive displacement pumping units, a copy of the manufacturer's certified shop test data for both variable speed and non-variable speed operation shall be available for comparison of the results of the field acceptance test.

> The manufacturer must provide separate certified pump curves for water mist positive displacement pumping units, with the variable speed features both active and deactivated. The manufacturer also provides certified pump curves for each individual pump, with the variable speed features both active and deactivated. The individual curves are intended to assist in evaluating underperforming units and do not need to be verified during the acceptance test.

14.2.4.2 At all flow conditions, including those required to be tested in 14.2.6.2, the fire pump as installed shall equal the performance as indicated on the manufacturer's certified shop test characteristic curve within the accuracy limits of the test equipment.

> Exhibit II.14.3 shows fire pump test data obtained from a fire pump acceptance test, and Exhibit II.14.4 shows the plots of the test data with respect to the manufacturer's certified test curve.

14.2.4.2.1 For water mist positive displacement pumping units with variable speed features, the pump unit as installed shall equal the performance as indicated on the fire pump unit manufacturer's certified shop test data, with variable speed features deactivated within the accuracy limits of the test equipment.

> Water mist positive displacement pumping units are tested as a unit with variable speed features deactivated. The pumps do not need to be tested individually, unless it is necessary to investigate an underperforming unit. The unit test results are compared to the pump unit manufacturer's certified shop test data with variable speed features deactivated.

14.2.4.2.2 For water mist positive displacement pumping units, the pump unit as installed shall equal the performance as indicated on the fire pump unit manufacturer's certified shop test data, with variable speed features activated within the accuracy limits of the test equipment.

> Water mist positive displacement pumping units are tested as a unit with variable speed features active. The pumps do not need to be tested individually, unless it is necessary to investigate an underperforming unit. The unit test results are compared to the pump unit manufacturer's certified shop test data with variable speed features active.

384 Part II • NFPA 20 • Chapter 14 Acceptance Testing, Performance, and Maintenance

2011 ANNUAL FIRE PUMP TEST RECORD

Underwood Fire Equipment, Inc.
43000 W 9 Mile Rd., Suite 304
Novi, Michigan 48375
PH: 248-347-4975 FAX: 248-347-0843

MICHIGAN MECHANICAL LICENSE: 7115166
FIRE SUPPRESSION LICENSE: S-0487
SERVICE DATE: 4/14/2011
TECHNICIAN: J. GOOTEE - MI

PUMP RATING: GOOD

COMPANY NAME:
ADDRESS:
CITY/STATE/ZIP: DETROIT, MICHIGAN 48201
CONTACT NAME:
CONTACT PHONE:

MANUFACTURER: ITT A-C
MODEL NUMBER: 8x6x9F
SERIAL NUMBER:
PUMP ID:

PUMP TYPE: HSC - ELECTRIC
PUMP CAPACITY: 1000
DESIGN PRESSURE: 100
PUMP SPEED: 3550
SUCTION FLG. SIZE: 6   DISCHARGE FLG. SIZE: 5

FIRE PUMP PERFORMANCE RECORD

Flow Point	Start/Stop Time	Discharge (PSI)	Suction (PSI)	Speed (RPM)	Phase A Amps	Phase B Amps	Phase C Amps	Motor Overload (Y/N)	Phase A-B Volts	Phase A-C Volts	Phase B-C Volts	Over/Under Voltage (Y/N)	Fire Pump Casing Temp (°F)	Inboard Stuff Box Temp (°F)	Outboard Stuff Box Temp (°F)	Inboard Bearing Temp (°F)	Outboard Bearing Temp (°F)	Casing Relief Open (Y/N)	Actual Net Flow (GPM)	Actual Net Boost (PSI)	Speed Corrected Flow (GPM)	Speed Corrected Boost (PSI)	Velocity Head Corrected (PSI)
1	9:32	174	50	3576	69	66	65	NO	467	468	467	NO	73	72	72	79	77	YES	0	124.0	0	122.2	122.2
2	9:45	148	45	3551	105	102	102	NO	465	466	465	NO	48	56	57	100	80	NO	1001	103.0	1001	102.9	103.9
3	10:02	122	42	3545	111	108	105	YES	464	464	466	NO	41	50	52	98	88	NO	1480	80.0	1482	80.2	82.3
4																							
5																							
ALT. POWER																							

Flow Point	Nozzle Size (in)	Hose Req'd (ft)	Hose Stream Diffuser (Y/N)	Discharge Nozzle Coef.	Pitot #1 (PSI)	Pitot #2 (PSI)	Pitot #3 (PSI)	Pitot #4 (PSI)	Pitot #5 (PSI)	Pitot #6 (PSI)	Pitot #7 (PSI)	Pitot #8 (PSI)	Pitot #9 (PSI)	Pitot #10 (PSI)	Pitot #11 (PSI)	Pitot #12 (PSI)	Flow Meter (GPM)	Total Flow (GPM)	Deviation (%)	Fire Pump Performance Data
1	1.75	50	NO	0.99														0	0.2	CERTIFIED FACTORY TEST CURVE
2	1.75	50	NO	0.99	13	12	16											1001	3.7	FLOW (GPM): 0 / 1000 / 1500
3	1.75	50	NO	0.99	26	38	26											1480	6.4	BOOST (PSI): 122 / 100 / 76
4																				MOST RECENT FIELD TEST DATA
5																				FLOW (GPM): 0 / 989 / 1502
ALT. POWER																				BOOST (PSI): 124.5 / 106.7 / 87.3

INTERPRETATION OF FIELD TEST RESULTS AND SERVICE RECOMMENDATIONS PER NFPA #25 2011 EDITION SECTION 8.3.5.1.2

AS COMPARED TO THE CERTIFIED FACTORY TEST CURVE, THE FIRE PUMP TESTED GOOD AT ALL POINTS FLOWED. THE ONLY REMOTE ALARMS CONNECTED INSIDE THE CONTROL PANEL AT THE TIME OF OUR TEST WERE "PUMP RUN" AND "POWER AVAILABLE"; NOT "PHASE REVERSAL". THE MOTOR AMPS AT 150% CAPACITY EXCEEDS THE ALLOWABLE AMPERAGE OF THE MOTOR PER THE NAMEPLATE AND NEC FULL LOAD CURRENT TOLERANCES. THIS SHOULD BE INVESTIGATED FURTHER TO DETERMINE THE CAUSE.

EXHIBIT II.14.3 Fire Pump Test Data. (Courtesy of Underwood Fire Equipment, Inc.)

14.2.5 System Demand. The actual unadjusted fire pump discharge flows and pressures installed shall meet or exceed the fire protection system's demand.

> While speed adjustments are needed to indicate the condition of the pump itself, a pump that is in perfect condition may not deliver the required system demand if the driver is operating at substantially below the rated speed.

14.2.6* Field Acceptance Test Procedures.

A.14.2.6 The fire pump operation is as follows:

(1) *Motor-Driven Pump.* To start a motor-driven pump, the following steps should be taken in the following order:

  (a) See that pump is completely primed.
  (b) Close isolating switch and then close circuit breaker.
  (c) Automatic controller will start pump if system demand is not satisfied (e.g., pressure low, deluge tripped).

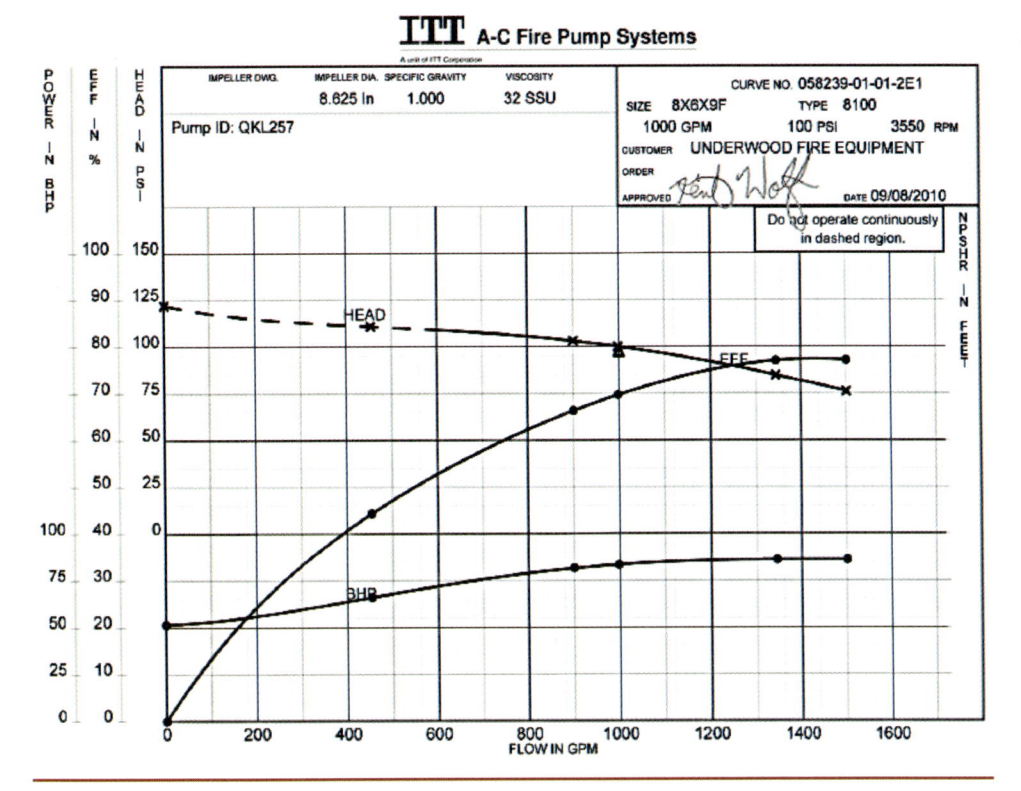

EXHIBIT II.14.4 Manufacturer Certified Test Curve. (Courtesy of Underwood Fire Equipment, Inc.)

(d) For manual operation, activate switch, pushbutton, or manual start handle. Circuit breaker tripping mechanism should be set so that it will not operate when current in circuit is excessively large.

(2) *Steam-Driven Pump.* A steam turbine driving a fire pump should always be kept warmed up to permit instant operation at full-rated speed. The automatic starting of the turbine should not be dependent on any manual valve operation or period of low-speed operation. If the pop safety valve on the casing blows, steam should be shut off and the exhaust piping examined for a possible closed valve or an obstructed portion of piping. Steam turbines are provided with governors to maintain speed at a predetermined point, with some adjustment for higher or lower speeds. Desired speeds below this range can be obtained by throttling the main throttle valve.

(3) *Diesel Engine–Driven Pump.* To start a diesel engine–driven pump, the operator should be familiar beforehand with the operation of this type of equipment. The instruction books issued by the engine and control manufacturer should be studied to this end. The storage batteries should always be maintained in good order to ensure prompt, satisfactory operation of this equipment (i.e., check electrolyte level and specific gravity, inspect cable conditions, corrosion, etc.).

(4) *Fire Pump Settings.* The fire pump system, when started by pressure drop, should be arranged as follows:

(a) The jockey pump stop point should equal the pump churn pressure plus the minimum static supply pressure.

(b) The jockey pump start point should be at least 10 psi (0.68 bar) less than the jockey pump stop point.

> Because the main fire pump start point is lower than the pressure maintenance pump start point, care should be taken to ensure that the settings given in A.14.2.6(4)(b) are not so low that a pressure shock (water hammer) is created when the main fire pump finally does come on.

    (c) The fire pump start point should be 5 psi (0.34 bar) less than the jockey pump start point. Use 10 psi (0.68 bar) increments for each additional pump.

> In the case where a booster pump from a municipal water supply is being used, the start point should not be set so low that the pressure of the water supply feeding the pump keeps the pump from starting, even at substantial flow rates.

    (d) Where minimum run times are provided, the pump will continue to operate after attaining these pressures. The final pressures should not exceed the pressure rating of the system.
    (e) Where the operating differential of pressure switches does not permit these settings, the settings should be as close as equipment will permit. The settings should be established by pressures observed on test gauges.

> An alternative to A.14.2.6(4)(e) that can be explored when the operating differential of a pressure switch (or transducer) does not permit close settings is to use separate, independent switches for start and stop settings.

    (f) Examples of fire pump settings follow (for SI units, 1 psi = 0.0689 bar):
        i. Pump: 1000 gpm, 100 psi pump with churn pressure of 115 psi
        ii. Suction supply: 50 psi from city — minimum static; 60 psi from city — maximum static
        iii. Jockey pump stop = 115 psi + 50 psi = 165 psi
        iv. Jockey pump start = 165 psi − 10 psi = 155 psi
        v. Fire pump stop = 115 psi + 50 psi = 165 psi
        vi. Fire pump start = 155 psi − 5 psi = 150 psi
        vii. Fire pump maximum churn = 115 psi + 60 psi = 175 psi

> Fire pumps should not be stopped automatically at the shutoff (or churn) pressure. In many cases, the AHJ requires that the pump be manually stopped (see 10.5.4.1 and 12.7.2.6.2).

    (g) Where minimum-run timers are provided, the pumps will continue to operate at churn pressure beyond the stop setting. The final pressures should not exceed the pressure rating of the system components.

(5) *Automatic Recorder.* The performance of all fire pumps should be automatically indicated on a pressure recorder to provide a record of pump operation and assistance in fire loss investigation.

14.2.6.1* Test Equipment.

A.14.2.6.1 The test equipment should be furnished by either the authority having jurisdiction, the installing contractor, or the pump manufacturer, depending upon the prevailing arrangements made between the aforementioned parties. The equipment should include, but not necessarily be limited to, the following:

(1) *Equipment for Use with Test Valve Header.* 50 ft (15 m) lengths of 2½ in. (65 mm) lined hose should be provided including Underwriters Laboratories' play pipe nozzles as needed to flow required volume of water. Where test meter is provided, however, these might not be needed.

(2) *Instrumentation.* The following test instruments should be of high quality, accurate, and in good repair:

  (a) Clamp-on volt/ammeter
  (b) Test gauges
  (c) Tachometer

> Exhibit II.14.5 shows the application of a typical tachometer for measuring pump speed.

EXHIBIT II.14.5
Tachometer Used to Measure Pump Engine Speed.

  (d) Pitot tube with gauge (for use with hose and nozzle)

◀ **CALCULATION**

> Exhibit II.14.6 (top) illustrates a pitot tube, which is used to measure the amount of water flowing out of an open nozzle. Exhibit II.14.6 (bottom) illustrates another type of flow measuring device that is commonly used for the purpose of measuring flow during hydrant flow tests for fire pump acceptance tests. The end with the orifice is placed vertically into the center of the stream of flowing water (with the orifice facing the nozzle opening) at a distance from the opening of the nozzle of approximately one-half the diameter of the nozzle. The pressure registered on the gauge corresponds to the amount of water flowing. The following equation is used to determine the waterflow in gpm:
>
> $$Q = 29.83 c d^2 \sqrt{P}$$
>
> where:
>
> Q = flow (gpm)
> c = coefficient of the nozzle opening (provided by the manufacturer)
> d = actual diameter of the nozzle opening (in.)
> P = pressure registered on the pitot tube gauge (psi)

(3) *Instrumentation Calibration.* All test instrumentation should be calibrated by an approved testing and calibration facility within the 12 months prior to the test. Calibration documentation should be available for review by the authority having jurisdiction.

Stationary Fire Pumps Handbook 2013

EXHIBIT II.14.6 *Pitot Tube (top) and Flow Measuring Device (bottom).*

> ▶ **FAQ** Why does NFPA 20 recommend that all test instrumentation be calibrated?
>
> It is a waste of time and effort to perform an acceptance test only to have the accuracy of the results questioned. The use of calibrated equipment can reduce the possibility of measurement errors and the need to repeat the test again at a later date.

A majority of the test equipment used for acceptance and annual testing has never been calibrated. This equipment can have errors of 15 to 30 percent in readings. The use of uncalibrated test equipment can lead to inaccurately reported test results.

While it is desirable to achieve a true churn condition (no flow) during the test for comparison to the manufacturer's certified pump test characteristic curve, it might not be possible in all circumstances. Pumps with circulation relief valves will discharge a small amount of water, even when no water is flowing into the fire protection system. The small discharge through the circulation relief valve should not be shut off during the test since it is necessary to keep the pump from overheating. For pumps with circulation relief valves, the minimum flow condition in the test is expected to be the situation where no water is flowing to the fire protection system but a small flow is present through the circulation relief valve. During a test on a pump with a pressure relief valve, the pressure relief valve should not open because these valves are installed purely as a safety precaution to prevent overpressurization during overspeed conditions.

Overspeed conditions should not be present during the test, so the pressure relief valve should not open. When pressure relief valves are installed on systems to relieve pressure under normal operating conditions, and if a true churn condition is desired during the acceptance test, the system discharge valve can be closed and the pressure relief valve can be adjusted to eliminate the flow. The pressure readings can be quickly noted and the pressure relief valve adjusted again to allow flow and relief of pressure. After this one-time test, a reference net pressure can be noted with the relief valve open so that the relief valve can remain open during subsequent annual tests with the comparison back to the reference residual net pressure rather than the manufacturer's curve.

> Testing of variable speed fire pumps requires adjusting the pressure relief valve (if provided) to prevent flow through the valve in order to test the pump performance at rated speed. The pressure relief valve should be reset and the setting retested as part of the acceptance test.

14.2.6.1.1 Calibrated test equipment shall be provided to determine net pump pressures, rate of flow through the pump, volts and amperes for electric motor–driven pumps, and speed.

14.2.6.1.2 Calibrated test gauges shall be used and bear a label with the latest date of calibration.

14.2.6.1.2.1 Gauges shall be calibrated a minimum of annually.

14.2.6.1.2.2 Calibration of test gauges shall be maintained at an accuracy level of ±1 percent.

> The intent of NFPA 20 is to require calibrated, listed, or approved test equipment to determine the performance of the fire pump system. Shop fabricated or untested equipment is not permitted.
>
> In many cases, the installing contractor is held responsible for an acceptable installation, including a satisfactory acceptance test. Therefore, the contractor should have the test equipment available, even when the AHJ has made arrangements to furnish the equipment. A good practice is to have a backup set of equipment available in the event of test equipment breakdown. This spare set can reduce the possibility of having to make arrangements for a future second test.

14.2.6.1.3 Requirements for personal protective equipment and procedures in accordance with *NFPA 70E, Standard for Electrical Safety in the Workplace*, shall be followed when working near energized electrical or rotating equipment.

14.2.6.2 Fire Pump Flow Testing(s).

14.2.6.2.1 For water mist positive displacement pumping units, each pump shall be operated manually a minimum of six times during the acceptance test.

14.2.6.2.2 For water mist positive displacement pumping units, each of the required automatic operations shall operate all pumps, except as provided in 14.2.6.2.2.1 and 14.2.6.2.2.2.

14.2.6.2.2.1 Where redundant pumps are provided, each of the automatic operations shall operate the number of pumps required to meet system demand.

14.2.6.2.2.2 Where redundant pumps are provided, each pump shall operate for a minimum of three automatic operations.

> Where a redundant standby fire pump(s) is included in a system designed for multiple fire pumps operating simultaneously, after the on-line pumps are tested, a different pump(s) should be placed in the standby mode and the redundant fire pump(s) tested for performance and automatic operation with the other pumps.

14.2.6.2.3 The fire pump shall perform at minimum, rated, and peak loads without objectionable overheating of any component.

> See Exhibit II.14.7 and Exhibit II.14.8 for examples of flow testing.
>
> ▶ **FAQ** What happens if overheating occurs during the acceptance test?
>
> Temperatures should be monitored at regular intervals during the entire test to ensure that any overheating can be detected and corrected before damage occurs to any component. If overheating does occur during the acceptance test, the test should be stopped until the problem is remedied. Once the problem is remedied, a retest should include a full 1-hour duration without overheating. Other successfully completed tests on the controller or other components do not need to be repeated.

14.2.6.2.4 Vibrations of the fire pump assembly shall not be of a magnitude to pose potential damage to any fire pump component.

14.2.6.2.5 The minimum, rated, and peak loads of the fire pump shall be determined by controlling the quantity of water discharged through approved test devices.

> Acceptance testing cannot be done using a closed loop flowmeter. Acceptance testing requires operating the pump at 100 percent and 150 percent without recirculating the water back to the pump suction.

EXHIBIT II.14.7 Determining Flow Using a Flowmeter. (Courtesy of Underwood Fire Equipment, Inc.)

EXHIBIT II.14.8 Determining Flow Using a Pitot Gauge. (Courtesy of Underwood Fire Equipment, Inc.)

14.2.6.2.5.1 Where simultaneous operation of multiple pumps is possible or required as part of a system design, the acceptance test shall include a flow test of all pumps operating simultaneously.

14.2.6.2.6 Where the maximum flow available from the water supply cannot provide a flow of 150 percent of the rated flow of the pump, the fire pump shall be operated at the greater of 100 percent of rated flow or the maximum flow demand of the fire protection system(s) maximum allowable discharge to determine its acceptance.

> ▶ **FAQ** Does NFPA 20 require pumps to flow at 150 percent of the rated capacity during tests?
>
> A test including a minimum flow rate of 150 percent of the rated pump capacity is required if it can be completed without creating a dangerously low suction pressure. If the 150 percent of rated pump capacity cannot be achieved at the initial acceptance test of a new installation, the pump may be oversized relative to the available water supply.
>
> Nevertheless, when pump flow tests are conducted at maximum rates of less than 150 percent of the pump's rated capacity, the maximum flow should at least exceed the greatest demand of all fire protection systems supplied by the respective pump. Based on 14.2.6.2.6, the inability to operate the fire pump at 150 percent of the pump's rated capacity should not result in an unacceptable installation. However, the pump will not have been properly tested for pump performance. The net pressure at 100 percent rated flow should be compared to the manufacturer's certified pump curve.

14.2.6.2.6.1 This reduced capacity shall constitute an acceptable test, provided that the pump discharge exceeds the fire protection system design and flow rate.

14.2.6.2.7 Where the suction to the fire pump is from a break tank, the tank refill rate shall be tested and recorded.

14.2.6.2.7.1 The refill device shall be operated a minimum of five times.

> The tank refill rate can be calculated by measuring the change in the water level in the tank over a fixed time period and taking into account the discharge from the fire pump over the same time period.

14.2.6.2.8 Water Level Detection. Water level detection shall be required for all vertical turbine pumps installed in wells to determine the suction pressure available at the shutoff, 100 percent flow, and 150 percent flow points, to determine if the pump is operating within its design conditions.

14.2.6.3 Variable Speed Pressure Limiting Control.

14.2.6.3.1 Pumps with variable speed pressure limiting control shall be tested at no-flow, 25 percent, 50 percent, 75 percent, 100 percent, 125 percent, and 150 percent of rated load in the variable speed mode.

14.2.6.3.1.1 They shall also be tested at minimum, rated, and peak loads, with the fire pump operating at rated speed.

> Paragraph 14.2.6.3.1 mandates a fire pump test with the simultaneous operation of the VSPLC and the fire pump at rated speed. This requirement is intended to ensure that both the pump and the VSPLC are operating correctly. When testing the performance of the pump without the VSPLC and with the pressure relief valve closed, the fire protection system components could be overpressurized, so the discharge isolation valve should be closed for these tests.

14.2.6.3.2 The fire protection system shall be isolated and the pressure relief valve closed for the rated speed tests required in 14.2.6.3.1.

14.2.6.3.3 The fire protection system shall be open and the relief valve set for the variable speed tests required in 14.2.6.3.1.

14.2.6.4* Measurement Procedure.

> ▶ **FAQ** At what flow rates should readings be recorded?
>
> In order to produce the most accurate curve, and to discover an occasional invalid reading, gathering readings at numerous flow rates, including at least 0 percent, 100 percent, and 150 percent of the rated capacity, is prudent. Flow testing at low-flow conditions for the pump may lead to false readings; therefore, flow testing should not be attempted below 50 percent of the rated flow. However, testing at 0 percent, or no flow, normally referred to as churn or shutoff, is necessary to verify the correct operating pressure of the fire pump.

A.14.2.6.4 The test procedure is as follows:

(1) Make a visual check of the unit. If hose and nozzles are used, see that they are securely tied down. See that the hose valves are closed. If a test meter is used, the valve on the discharge side of the meter should be closed.
(2) Start the pump.
(3) Partially open one or two hose valves, or slightly open the meter discharge valve.
(4) Check the general operation of the unit. Watch for vibration, leaks (oil or water), unusual noises, and general operation. Adjust packing glands.
(5) Measure water discharge. The steps to do so are as follows:
   (a) Where a test valve header is used, regulate the discharge by means of the hose valves and a selection of the nozzle tips. It will be noticed that the play pipe has a removable tip. This tip has a 1⅛ in. (28.6 mm) nozzle, and when the tip is removed, the play pipe has a 1¾ in. (44.4 mm) nozzle. Hose valves should be shut off before removing or putting on the 1⅛ in. (28.6 mm) tip.
   (b) Where a test meter is used, regulate the discharge valve to achieve various flow readings.
   (c) Important test points are at 150 percent rated capacity, rated capacity, and shutoff. Intermediate points can be taken if desired to help develop the performance curve.
(6) Record the following data at each test point *(see Figure A.14.2.6.4)*:
   (a) Pump rpm
   (b) Suction pressure
   (c) Discharge pressure
   (d) Number and size of hose nozzles, pitot pressure for each nozzle, and total gpm (L/min); for test meter, simply a record of gpm (L/min)
   (e) Amperes (each phase)
   (f) Volts (phase to phase)
(7) Calculation of test results is as follows:
   (a) *Rated Speed.* Determine that pump is operating at rated rpm.
   (b) *Capacity.* For hose valve header, using a fire stream table, determine the gpm (L/min) for each nozzle at each pitot reading. For example, 16 psi (1.1 bar) pitot pressure with 1¾ in. (44.4 mm) nozzle indicates 364 gpm (1378 L/min). Add the gpm for each hose line to determine total volume. For test meter, the total gpm (L/min) is read directly.

(c) *Total Head for Horizontal Pump.* Total head is the sum of the following:
   i. Pressure measured by the discharge gauge at pump discharge flange
   ii. Velocity head difference, pump discharge, and pump suction
   iii. Gauge elevation corrections to pump centerline (plus or minus)
   iv. Pressure measured by suction gauge at pump suction flange — negative value when pressure is above zero

(d) *Total Head for Vertical Pump.* Total head is the sum of the following:
   i. Pressure measured by the discharge gauge at pump discharge flange
   ii. Velocity head at the discharge
   iii. Distance to the supply water level
   iv. Discharge gauge elevation correction to centerline of discharge

(e) *Electrical Input.* Voltage and amperes are read directly from the volt/ammeter. This reading is compared to the motor nameplate full-load amperes. The only general calculation is to determine the maximum amperes allowed due to the motor service factor. In the case of 1.15 service factor, the maximum amperes are approximately 1.15 times motor amperes, because changes in power factor and efficiency are not considered. If the maximum amperes recorded on the test do not exceed this figure, the motor and pump will be judged satisfactory. It is most important to measure voltage and amperes accurately on each phase should the maximum amperes logged on the test exceed the calculated maximum amperes. This measurement is important because a poor power supply with low voltage will cause a high ampere reading. This condition can be corrected only by improvement in the power supply. There is nothing that can be done to the motor or the pump.

(f) *Correction to Rated Speed.* For purposes of plotting, the capacity, head, and power should be corrected from the test values at test speed to the rated speed of the pump. The corrections are made as follows.

Capacity:

$$Q_2 = \left(\frac{N_2}{N_1}\right) Q_1$$

where:

Q_1 = capacity at test speed in gpm (L/min)
Q_2 = capacity at rated speed in gpm (L/min)
N_1 = test speed in rpm
N_2 = rated speed in rpm

Head:

$$H_2 = \left(\frac{N_2}{N_1}\right)^2 H_1$$

where:

H_1 = head at test speed in ft (m)
H_2 = head at rated speed in ft (m)

Horsepower:

$$hp_2 = \left(\frac{N_2}{N_1}\right)^3 hp_1$$

where:

hp_1 = kW (horsepower) at test speed
hp_2 = kW (horsepower) at rated speed

The water supply to the fire pump should be tested before the fire pump test. Significant deviations between flow test results and the design analysis should also be investigated and resolved before the pump test. Partially closed valves in the underground piping can be identified prior to the pump test. In some jurisdictions, where a pump is fed from the public mains, the local code/fire official may want to see a hydrant test at a rate in excess of 150 percent of the pump's rated capacity. This test is conducted in order to eliminate any questions regarding the capability of the water supply prior to setting up the meters, gauges, and hoses. When the pump test begins, the capability of the water supply should already be known. In this instance, any deficiency will most likely be related to the fire pump, not the attached water supply. This testing should be limited to 165 percent of the rated flow because of the discharge curve of the pump.

An example of the plotted information obtained during a pump acceptance test is shown in Exhibit II.14.9.

Usually, test results with pump speed variations of 1 percent or 2 percent from rated speed are disregarded because the capacity, head, and horsepower differences are insignificant. The example that follows illustrates how to correct pump performance for its rated speed.

▶ **EXAMPLE**

A pump rated for 80 psi at 750 gpm (5.6 bar at 2839 L/min) when turning at 1780 rpm is operating at 1700 rpm during the acceptance test. The manufacturer's certified shop test indicates the following operational characteristics when the pump is operating at 1780 rpm:

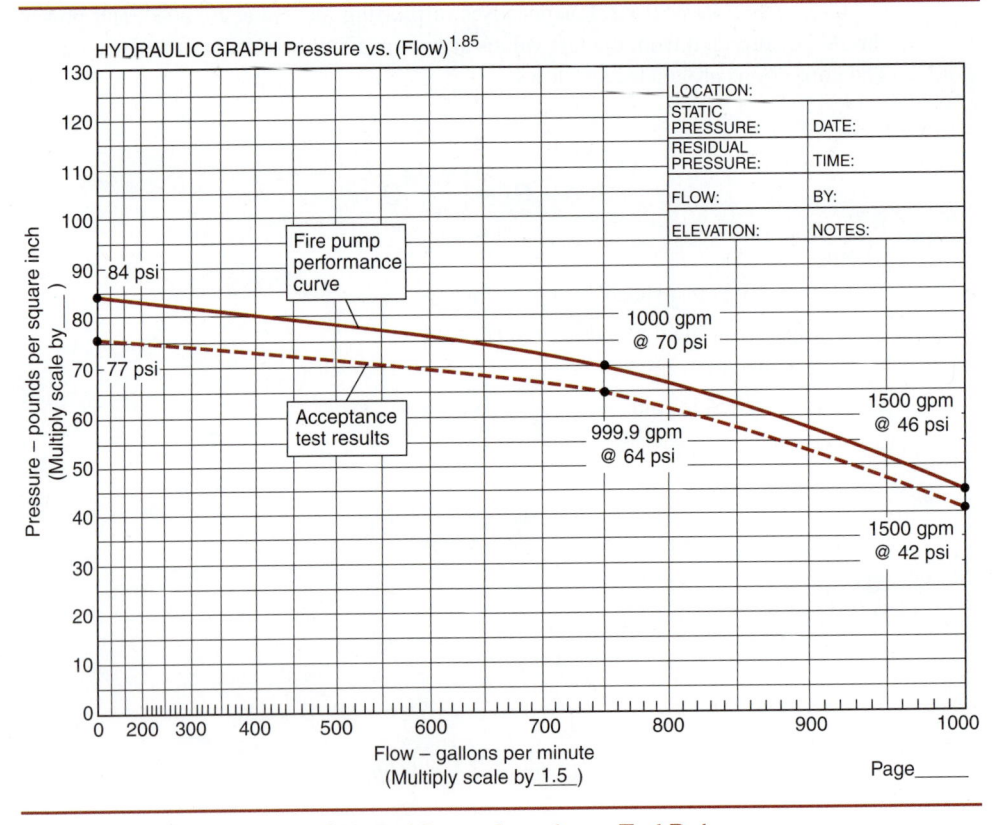

EXHIBIT II.14.9 Example of Plotted Pump Acceptance Test Data.

	Churn	Rated	Overload
Flow	0 gpm (0 L/min)	750 gpm (2839 L/min)	1125 gpm (4258 L/min)
Net pressure	96 psi (6.7 bar)	80 psi (5.6 bar)	52 psi (3.6 bar)

The acceptance test shows the following results when the pump is operating at 1700 rpm:

	Churn	Rated	Overload
Flow	0 gpm (0 L/min)	716 gpm (2710 L/min)	1075 gpm (4069 L/min)
Net pressure	88 psi (6.1 bar)	73 psi (5.0 bar)	47 psi (3.2 bar)

Solution: At first, the results when the pump is operating at 1700 rpm may seem to reflect a deficiency in the pump performance when compared to the shop curve. However, by adjusting for the proper speed, the corrected flow (Q_c) and pressure (H_c) results are obtained. Using the equations presented in A.14.2.6.4(f), the corrected flow and pressure can be obtained as follows:

$$Q_c = \left(\frac{N_2}{N_1}\right) \times Q_1$$

$$H_c = \left(\frac{N_2}{N_1}\right)^2 \times H_1$$

For churn,

$$Q_c = (1780/1700) \times 0 = 1.047 \times 0 = 0 \text{ gpm (0 L/min)}$$
$$H_c = (1780/1700)^2 \times 88 = (1.047)^2 \times 88 = 1.096 \times 88 = 96 \text{ psi (6.6 bar)}$$

For rated,

$$Q_c = (1780/1700) \times 716 = 1.047 \times 716 = 750 \text{ gpm (2839 L/min)}$$
$$H_c = (1780/1700)^2 \times 73 = (1.047)^2 \times 73 = 1.096 \times 73 = 80 \text{ psi (5.5 bar)}$$

For overload,

$$Q_c = (1780/1700) \times 1075 = 1.047 \times 1075 = 1125 \text{ gpm (4258 L/min)}$$
$$H_c = (1780/1700)^2 \times 47 = (1.047)^2 \times 47 = 1.096 \times 47 = 52 \text{ psi (3.6 bar)}$$

Exhibit II.14.10 shows the plots of these results. By plotting these results on hydraulic graph paper, overall pump performance can be better evaluated.

As a minimum, the pump as tested (without an rpm adjustment) must deliver the maximum flow and pressure required by any fire protection system it supplies. If the rpm adjusted curve indicates that the fire pump is performing satisfactorily, but the actual performance is below the fire protection requirement, the driver must be investigated to determine why it is operating below rated speed.

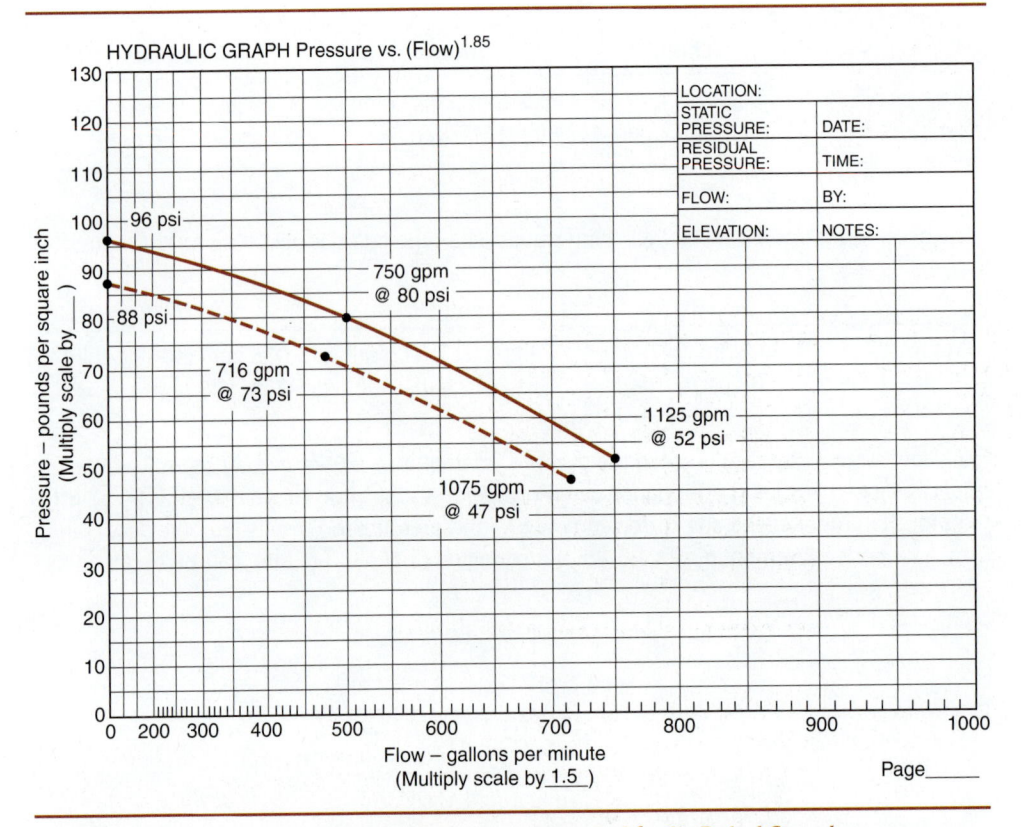

EXHIBIT II.14.10 *Plot of Pump Performance Corrected for Its Rated Speed.*

(g) *Conclusion.* The final step in the test calculation is generally a plot of test points. A head-capacity curve is plotted, and an ampere-capacity curve is plotted. A study of these curves will show the performance picture of the pump as it was tested.

14.2.6.4.1 The quantity of water discharging from the fire pump assembly shall be determined and stabilized.

14.2.6.4.2 Immediately thereafter, the operating conditions of the fire pump and driver shall be measured.

14.2.6.4.3 Positive Displacement Pumps.

14.2.6.4.3.1 The pump flow for positive displacement pumps shall be tested and determined to meet the specified rated performance criteria where only one performance point is required to establish positive displacement pump acceptability.

14.2.6.4.3.2 The pump flow test for positive displacement pumps shall be accomplished using a flowmeter or orifice plate installed in a test loop back to the supply tank, to the inlet side of a positive displacement water pump, or to drain.

14.2.6.4.3.3 The flowmeter reading or discharge pressure shall be recorded and shall be in accordance with the pump manufacturer's flow performance data.

14.2.6.4.3.4 If orifice plates are used, the orifice size and corresponding discharge pressure to be maintained on the upstream side of the orifice plate shall be made available to the authority having jurisdiction.

Centrifugal Fire Pump Acceptance Test Form

Information on this form covers the minimum requirements of NFPA 20-2013 for performing acceptance tests on pumps with electric motors or diesel engine drivers. Other forms are available for periodic inspection, testing and maintenance.

NFSA — The Voice of the Fire Sprinkler Industry

Owner: _____
Owner's Address: _____
_____
Property on which pump is installed: _____
_____
Property Address: _____
_____
Date of Test: _____
Demand(s) of Fire Protection Systems Supplied By Pump: _____
_____
Pump: ❏ Horizontal ❏ Vertical
Manufacturer: _____ Shop/Serial Number: _____
Model or Type: _____
Rated GPM _____ Rated Pressure _____ Rated RPM _____
Suction From _____ If Tank, Size and Height _____
Driver: ❏ Electric Motor ❏ Diesel Engine
Manufacturer: _____ Shop/Serial Number: _____
Model or Type: _____
Rated Horsepower: _____ Rated Speed: _____
If Electric Motor, Rated Voltage _____ Operating Voltage _____
Rated Amps _____ Phase Cycles _____ Service Factor _____
Controller Manufacturer: _____
Shop/Serial Number: _____ Model or Type: _____
Jockey Pump on System? ❏ Yes ❏ No   Settings: On ___ Off ___

All questions are to be answered Yes, No or Not Applicable.
All No answers are to be explained in the comments portion of this form.

I. Flush Test *(Conduct before Hydrostatic Test)*
Suction piping was flushed at ____ gpm? ❏ Yes ❏ No ❏ N/A
(See Table 14.1.1.1 of NFPA 20.)
Certificate presented showing flush test? ❏ Yes ❏ No ❏ N/A

II. Hydrostatic Test
Piping tested at ____ psi for 2 hours? ❏ Yes ❏ No ❏ N/A
(Note: NFPA 20 requires 200 psi or 50 psi above maximum system pressure whichever is greater.)
Piping passed test? ❏ Yes ❏ No ❏ N/A
Certificate presented showing test? ❏ Yes ❏ No ❏ N/A

III. People Present
Were the following present to witness the test:
A. Pump manufacturer/representative ❏ Yes ❏ No ❏ N/A
B. Engine manufacturer/representative ❏ Yes ❏ No ❏ N/A
C. Controller manufacturer/representative ❏ Yes ❏ No ❏ N/A
D. Transfer switch manufacturer/rep. ❏ Yes ❏ No ❏ N/A
E. Authority having jurisdiction/rep. ❏ Yes ❏ No ❏ N/A

IV. Electric Wiring
Was all electric wiring including control interwiring for multiple pumps, emergency power supply, and the jockey pump completed and checked by the electrical contractor prior to the initial start-up and acceptance test? ❏ Yes ❏ No ❏ N/A

V. Flow Test
Run the pump at no-load, rated load and peak load (usually 150% of rated load) conditions. For variable speed drivers, run the test with the pressure limiting control "on" at 25, 50, 75, 100, 125, and 150% of rated load and then again at rated speed with the pump isolated from the fire protection system and the relief valve closed.
A. Was the manufacturers' certified pump test curve available for comparison to the acceptance test? ❏ Yes ❏ No ❏ N/A
B. Equipment and gages calibrated? ❏ Yes ❏ No ❏ N/A
   Date of last calibration: _____
C. If dry charge batteries were supplied, was electrolyte added at least 24 hours prior to engine start and were batteries given a conditioning charge? ❏ Yes ❏ No ❏ N/A

D. For each test, record the following for each load condition:

Test	Driver Speed rpm	Suction Pressure psi	Discharge Pressure psi	Nozzle Size inch	Pitot Readings or Flow 1	2	3	4	5	6
0									×	×
100%										
150%										

E. For electric motor driven pumps also record:

Test	Voltage	Amperes
0		
100%		
150%		

F. Calculate Net Pressure and Total Flow:

$P_{Net} = P_{Discharge} - P_{Suction}$ $Q = 29.83\, cd^2 \sqrt{P}$

Test	Net Pressure	Flow (Q) (use formula above) 1	2	3	4	5	6	Total Flow
0								
100%								
150%								

G. For electric motors operating at rated voltage and frequency, is the ampere demand on each phase less than or equal to the product of the full load ampere rating times the allowable service factor as stamped on the motor nameplate? ❏ Yes ❏ No ❏ N/A

H. For electric motors operating under varying voltage:
   1. Was the product of the actual voltage and current demand on each phase less than or equal to the product of the rated full load current times the rated voltage times the allowable service factor? ❏ Yes ❏ No ❏ N/A
   2. Did the voltage stay within the range of 95 to 110% of the rated voltage during the test? ❏ Yes ❏ No ❏ N/A

I. Did engine-drive unit show no signs of overload or stress? ❏ Yes ❏ No ❏ N/A

J. Was the governor set to properly regulate the engine speed at rated pump speed? ❏ Yes ❏ No ❏ N/A

K. Did the gear drive assembly operate without excessive objectionable noise, vibration or heating? ❏ Yes ❏ No ❏ N/A

L. Was the unit started and brought up to rated speed without interruption under discharge conditions equal to peak load? ❏ Yes ❏ No ❏ N/A

M. Did the fire pump perform equal to the manufacturer's curve within the accuracy limits of the test equipment? ❏ Yes ❏ No ❏ N/A

N. Did the unadjusted performance discharge curve meet or exceed the fire protection system demand(s)? ❏ Yes ❏ No ❏ N/A

O. No vibrations that could potentially damage any fire pump component? ❏ Yes ❏ No ❏ N/A

P. The fire pump performed at all conditions without objectionable overheating of any component? ❏ Yes ❏ No ❏ N/A

Q. Electric motor pumps passed phase reversal test on normal and alternate (if provided) power? ❏ Yes ❏ No ❏ N/A

R. If a break tank is provided, was the refill mechanism tested 5 times and did it operate correctly? ❏ Yes ❏ No ❏ N/A
   Record refill rate: _____

S. For vertical turbine pumps taking suction from wells, is the water level capable of being recorded? ❏ Yes ❏ No ❏ N/A

© 2012 National Fire Sprinkler Association, 40 Jon Barrett Road, Patterson, NY 12563 (845) 878-4200   Form 20-A Sheet 1 of 2

Sample for Guidance

FIGURE A.14.2.6.4 *Centrifugal Fire Pump Acceptance Test Form. (Source: National Fire Sprinkler Association, Inc.)*

VI. Controller Test

A. Did the pump start at least 6 times from automatic sources? ❏ Yes ❏ No ❏ N/A

B. Was each automatic starting feature tested at least once? ❏ Yes ❏ No ❏ N/A

C. Did the pump start at least 6 times manually? ❏ Yes ❏ No ❏ N/A

D. Was the pump run for at least 5 minutes during each of the operations in Parts A, B and C above? ❏ Yes ❏ No ❏ N/A

(Note: An engine driver is not required to run for 5 minutes at full speed between successive starts until the cumulative cranking time of successive starts reaches 45 seconds.)

E. Were the starting operations divided between both sets of batteries for engine-driven controllers? ❏ Yes ❏ No ❏ N/A

F. Electric Driven Pump Controllers

1. Were all overcurrent protective devices (including the controller circuit-breaker) selected, sized and set in accordance with NFPA 20? ❏ Yes ❏ No ❏ N/A

2. Was the fire pump started at least once from each power service and run for at least 5 minutes? ❏ Yes ❏ No ❏ N/A

3. Upon simulation of a power failure, while the pump is operating at peak load, did the transfer switch transfer from the normal to the emergency source without opening overcurrent protection devices on either line? ❏ Yes ❏ No ❏ N/A

4. When normal power was restored, did retransfer from emergency to normal power occur without overcurrent protection devices opening on either line? ❏ Yes ❏ No ❏ N/A

5. Were at least half of the automatic and manual starts required by Parts A and C performed with the pump connected to the alternate source? ❏ Yes ❏ No ❏ N/A

G. Were all signal conditions simulated demonstrating satisfactory operation? ❏ Yes ❏ No ❏ N/A

H. Was the pump run for at least 1 hour total during all of the above tests? ❏ Yes ❏ No ❏ N/A

I. For engines with ECM fuel management systems, primary and alternate ECM passed function test? ❏ Yes ❏ No ❏ N/A

VII. Information For Owner

Was the owner given all of the following? ❏ Yes ❏ No ❏ N/A

A. A manual explaining the operation of all components.
B. Instructions for routine maintenance and repairs.
C. Parts list and parts identification.
D. List of recommended spare parts and lubricants to keep on hand.
E. Schematic electrical drawings of controller, transfer switch and alarm panels.
F. Manufacturer's certified shop test curve or acceptance test curve.

VIII. Tester Information

Tester: _____

Company: _____

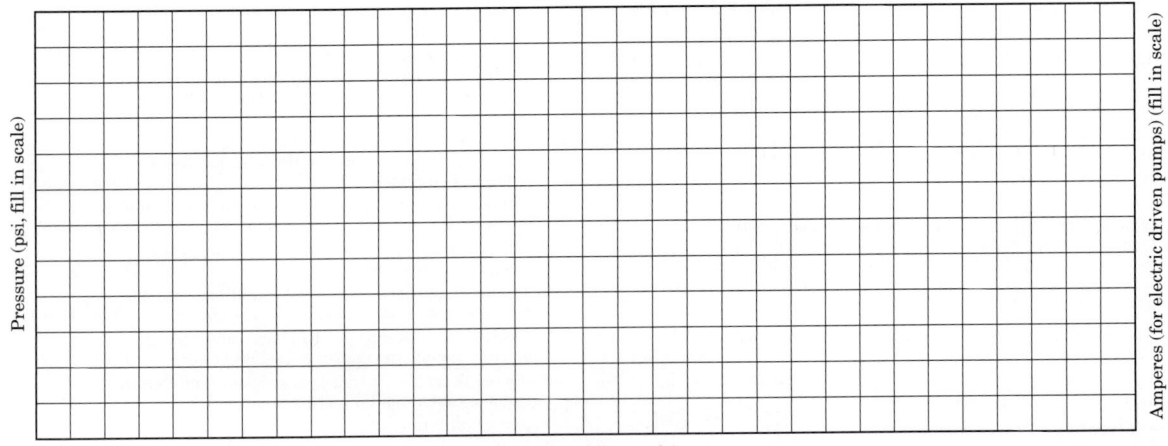

Company Address: _____

I state that the information on this form is correct at the time and place of my test, and that all equipment tested was left in operational condition upon completion of this test except as noted in the comments section below.

Signature of Tester: _____

Date: _____

License or Certification Number if Applicable: _____

IX. Comments

(Any "No" answers, test failures, or other problems must be explained here.)

_____

Pump Test Results

[Graph: Pressure (psi, fill in scale) vs. Flow (gpm, fill in scale); Amperes (for electric driven pumps) (fill in scale) on right axis]

© 2012 National Fire Sprinkler Association, 40 Jon Barrett Road, Patterson, NY 12563 (845) 878-4200

Form 20-A Sheet 2 of 2

FIGURE A.14.2.6.4 Continued

14.2.6.4.3.5 Flow rates shall be as specified while operating at the system design pressure. Tests shall be performed in accordance with HI 3.6, *Rotary Pump Tests*.

14.2.6.4.3.6 Positive displacement pumps intended to pump liquids other than water shall be permitted to be tested with water; however, the pump performance will be affected, and manufacturer's calculations shall be provided showing the difference in viscosity between water and the system liquid.

14.2.6.4.4 Electric Motor–Driven Units. For electric motors operating at rated voltage and frequency, the ampere demand on each phase shall not exceed the product of the full-load ampere rating times the allowable service factor as stamped on the motor nameplate.

14.2.6.4.5 For electric motors operating under varying voltage, the product of the actual voltage and current demand on each phase shall not exceed the product of the rated voltage and rated full-load current times the allowable service factor.

14.2.6.4.6 The voltage at the motor contacter output lugs shall not vary more than 5 percent below or 10 percent above rated (nameplate) voltage during the test. *(See Section 9.4.)*

> When recording the ampere reading, the difference in current between the phases should normally not be more than 10 percent. Regardless of the flow rate during the tests, the average current readings in the three phases should not exceed the nameplate current by more than 10 percent.

14.2.6.4.7 Engine-Driven Units.

14.2.6.4.7.1 When dry charge batteries have been supplied, electrolyte shall be added to the batteries a minimum of 24 hours prior to the time the engine is to be started and the batteries given a conditioning charge.

14.2.6.4.7.2 Engine-driven units shall not show signs of overload or stress.

14.2.6.4.7.3 The governor of such units shall be set at the time of the test to properly regulate the engine speed at rated pump speed. *(See 11.2.4.1.)*

> Many engines require special tools to reset speed in the field. While, in some cases, the fire pump installer may be able to accomplish this action, in other cases, the engine manufacturer may need to be present. For this reason, the governor must be set at the time of the test when all appropriate parties are present per the requirements of 14.2.6.

14.2.6.4.7.4 Engines equipped with a variable speed control shall have the variable speed control device nonfunctioning when the governor field adjustment in 11.2.4.1 is set and secured.

14.2.6.4.8 Steam Turbine–Driven Units. The steam turbine shall maintain its speed within the limits specified in 13.2.2.

14.2.6.4.9 Right Angle Gear Drive Units. The gear drive assembly shall operate without excessive objectionable noise, vibration, or heating.

14.2.6.5 Loads Start Test. The fire pump unit shall be started and brought up to rated speed without interruption under the conditions of a discharge equal to peak load.

> For centrifugal pumps, peak load (or maximum brake horsepower) usually occurs at approximately 145 percent of rated pump capacity. This test requires establishing the flow rate at the peak load with the pump running, shutting off the power to the pump with the nozzles still flowing. After the power is shut off, the flow rate out of the nozzles will reduce, and the motor will continue to turn while water flows through the pump impeller. The power is restored to the motor after the reduced flow and motor speed has stabilized.

Stationary Fire Pumps Handbook 2013

14.2.6.6* Phase Reversal Test. For electric motors, a test shall be performed to ensure that there is not a phase reversal condition in either the normal power supply configuration or from the alternate power supply (where provided).

> Phase reversal is a continuing problem that has not always been verified during pump acceptance testing when the alternate power supply may not yet be in service. Paragraph 14.2.6.6 requires a suitable test to be conducted for both supplies (when appropriate), even if performed at two separate times.

A.14.2.6.6 A simulated test of the phase reversal device is an acceptable test method.

14.2.7 Controller Acceptance Test for Electric and Diesel Driven Units.

14.2.7.1* Fire pump controllers shall be tested in accordance with the manufacturer's recommended test procedure.

> Some electric motor manufacturers suggest that starting and stopping large (over 250 hp) motors in rapid sequence could adversely affect motor life, especially motors installed prior to 1986. For proper cooling, the motor should be run for intervals of at least 10 minutes.

A.14.2.7.1 All controller starts required for tests described in 14.2.6 through 14.2.9 should accrue respectively to this number of tests.

> The minimum required number of starts is six. If an alternate power source is provided, three of the six starts should be from the primary power supply and three should be from the alternate power source (see 14.2.8). Tests required to satisfy 14.2.6.5, 14.2.6.6, 14.2.7.1, and 14.2.7.9 should be added to the six minimum required starts. Note that simultaneously satisfying several of these additional tests may be possible. For example, once the loads start test required by 14.2.6.5 is satisfied, the pump and motor may be left running in order to test the power transfer and retransfer between the primary and alternate power supplies as required by 14.2.8.

14.2.7.2 As a minimum, no fewer than six automatic and six manual operations shall be performed during the acceptance test.

14.2.7.3 An electric driven fire pump shall be operated for a period of at least 5 minutes at full speed during each of the operations required in 14.2.6.2.

14.2.7.4 An engine driver shall not be required to run for 5 minutes at full speed between successive starts until the cumulative cranking time of successive starts reaches 45 seconds.

14.2.7.5 The automatic operation sequence of the controller shall start the pump from all provided starting features.

14.2.7.6 This sequence shall include pressure switches or remote starting signals.

14.2.7.7 Tests of engine-driven controllers shall be divided between both sets of batteries.

14.2.7.8 The selection, size, and setting of all overcurrent protective devices, including fire pump controller circuit breaker, shall be confirmed to be in accordance with this standard.

14.2.7.9 The fire pump shall be started once from each power service and run for a minimum of 5 minutes.

> **CAUTION:** Manual emergency operation shall be accomplished by manual actuation of the emergency handle to the fully latched position in a continuous motion. The handle shall be latched for the duration of this test run.

If the emergency handle cannot be placed in the fully latched position in one continuous motion, single phasing and severe arcing can occur.

14.2.8 Alternate Power Supply.

14.2.8.1 On installations with an alternate source of power and an automatic transfer switch, loss of primary source shall be simulated and transfer shall occur while the pump is operating at peak load.

When the primary power source is interrupted while flowing at 150 percent of rated pump capacity, the transfer switch and the alternate power source should be achieved within 10 seconds, and the peak flow should be successfully redelivered within 20 seconds to 30 seconds.

14.2.8.2 Transfer from normal to alternate source and retransfer from alternate to normal source shall not cause opening of overcurrent protection devices in either line.

14.2.8.3 At least half of the manual and automatic operations of 14.2.7.2 shall be performed with the fire pump connected to the alternate source.

14.2.8.4 If the alternate power source is a generator set required by 9.3.2, installation acceptance shall be in accordance with NFPA 110, *Standard for Emergency and Standby Power Systems*.

14.2.9 Emergency Governor for Steam Driven Units.

14.2.9.1 Emergency governor valve for steam shall be operated to demonstrate satisfactory performance of the assembly.

14.2.9.2 Hand tripping shall be acceptable.

14.2.10 Simulated Conditions. Both local and remote signals and fire pump alarm conditions shall be simulated to demonstrate satisfactory operation.

14.2.11* Test Duration. The fire pump or foam concentrate pump shall be in operation for not less than 1 hour total time during all of the foregoing tests.

A.14.2.11 It is not the intent to discharge water for the full 1-hour test duration, provided all flow tests can be conducted in less time and efforts are taken to prevent the pump from overheating.

14.2.12* Electronic Fuel Management (ECM). For engines with electronic fuel management (ECM) control systems, a function test of both the primary and the alternate ECM shall be conducted.

A.14.2.12 To verify the operation of the alternate ECM, with the motor stopped, move the ECM selector switch to the alternate ECM position. Repositioning of the selector switch should cause a signal on the fire pump controller. Start the engine; it should operate normally with all functions. Shut engine down, switch back to the primary ECM, and restart the engine briefly to verify that correct switchback has been accomplished.

To verify the operation of the redundant sensor, with the engine running, disconnect the wires from the primary sensor. There should be no change in the engine operation. Reconnect the wires to the sensor. Next, disconnect the wires from the redundant sensor. There should be no change in the engine operation. Reconnect the wires to the sensor. Repeat this process for all primary and redundant sensors on the engines. *Note:* If desired, the disconnecting and reconnecting of wires to the sensors can be done while the engine is not running, then starting the engine after each disconnection and reconnection of the wires to verify engine operation.

14.3* Record Drawings, Test Reports, Manuals, Special Tools, and Spare Parts

A.14.3 It is the intent to retain the record drawing, equipment manual, and completed test report for the life of the fire pump system.

14.3.1 One set of record drawings shall be provided to the building owner.

14.3.2 One copy of the completed test report shall be provided to the building owner.

14.3.3* One set of instruction manuals for all major components of the fire pump system shall be supplied by the manufacturer of each major component.

A.14.3.3 Consideration should be given to stocking spare parts for critical items not readily available.

14.3.4 The manual shall contain the following:

(1) A detailed explanation of the operation of the component
(2) Instructions for routine maintenance
(3) Detailed instructions concerning repairs
(4) Parts list and parts identification
(5) Schematic electrical drawings of controller, transfer switch, and fire pump control panels
(6)* List of recommended spare parts and lubricants

> Ordinarily, the operation and maintenance manual is submitted by the installing contractor for review and approval by the building owner or his or her representative, who is usually the project engineer. Multiple copies of the manual need to be maintained on file for future reference. The manual should be made available at the time of the acceptance test for review of systems and components. Prior to final acceptance, the manual should be used for training of operating personnel.
>
> ▶ **FAQ** What information should the manual contain?
>
> In addition to the required information listed in 14.3.4, the manual should contain a complete list of parts and suppliers, including contact information. The manual should also list any recommended spare parts that should be kept on-site for immediate use if the need arises.

A.14.3.4(6) Recommended spare parts and lubricants should be stored on-site to minimize system impairment.

14.3.5 Any special tools and testing devices required for routine maintenance shall be available for inspection by the authority having jurisdiction at the time of the field acceptance test.

14.4 Periodic Inspection, Testing, and Maintenance

Fire pumps shall be inspected, tested, and maintained in accordance with NFPA 25, *Standard for the Inspection, Testing, and Maintenance of Water-Based Fire Protection Systems*.

> Nonflow testing, which consists of starting the pump and running it at full speed at the regular intervals required by NFPA 25, helps to ensure that the pump installation is always ready for service. The 2011 edition of NFPA 25 requires weekly nonflow testing for diesel engine–drive fire pumps and monthly nonflow testing for electric motor–drive fire pumps. In earlier editions of NFPA 25, weekly testing was required for electric motor–driven fire pumps. The appropriate test frequency for electric motor–driven fire pumps is still being reviewed, so the latest edition of NFPA 25 should be consulted for the test frequency.

> NFPA 25 requires an investigation whenever the net pump pressure is less than 95 percent of the original pump curve at 100 percent of rated flow. A more pronounced reduction in capacity may be due to one or more causes, as shown in Annex B. A plugged impeller, worn wear rings, and obstructed suction are among the most common causes. In some cases, the reading of the suction gauge indicates whether the trouble is due to an obstruction in the water supply or suction pipe. In other cases, the difference between the reading of the suction gauge and that of the discharge gauge is the indicator of a plugged impeller or overly worn wear rings.
>
> When testing a pump by discharging through a listed meter back to the pump suction, a hydraulic imbalance within the pump is possible. Additionally, this test method is not able to check the suction supply and piping upstream of the connection to the pump suction. If possible, the flowmeter should not be piped to the pump suction. Piping the flow back to the water storage tank (with an air gap) or to the pump house exterior results in accurate measurements.
>
> At least once a year, diesel engine–driven pumps should be checked for overheating by running them with the pump discharging at 150 percent or more of its rated capacity. Once the engine temperature has stabilized, the engine should be run for at least an additional 15 minutes. If the engine overheats, a blockage of the cooling system is probably the cause, such as an inadequate cooling waterflow caused by an obstruction in the cooling water system, a plugged strainer, or a partly closed valve.
>
> In addition to the requirements of NFPA 25 for engine-driven pumps, the engine should be kept clean and dry. The fuel tank should be kept at a level at least capable of running the engine for 8 hours at peak load. For instance, if tanks are not refilled and they drop to the two-thirds level, the tank size should be adequate for a 12-hour supply when full. The crankcase oil should be checked to see that it is at the proper level, that it has not become fouled, and that it has not lost its viscosity. Additionally, the strainers in the cooling water system should be cleaned, and the specific gravity of the battery electrolyte should be checked monthly.
>
> Even though NFPA 25 specifies the records that should be kept, it is also beneficial to maintain a specific record of the temperature and tightness of the glands, the readings of the suction and discharge gauges, and the condition of the suction supply.

14.5 Component Replacement

14.5.1 Positive Displacement Pumps.

14.5.1.1 Whenever a critical path component in a positive displacement fire pump is replaced, as defined in 14.5.2.5, a field test of the pump shall be performed.

14.5.1.2 If components that do not affect performance are replaced, such as shafts, then only a functional test shall be required to ensure proper installation and reassembly.

14.5.1.3 If components that affect performance are replaced, such as rotors, plungers, and so forth, then a retest shall be conducted by the pump manufacturer or designated representative or qualified persons acceptable to the authority having jurisdiction.

14.5.1.3.1 For water mist positive displacement pumping units, the retest shall include the pump unit as a whole.

14.5.1.4 Field Retest Results.

14.5.1.4.1 The field retest results shall be compared to the original pump performance as indicated by the fire pump manufacturer's original factory-certified test curve, whenever it is available.

14.5.1.4.2 The field retest results shall meet or exceed the performance characteristics as indicated on the pump nameplate, and the results shall be within the accuracy limits of field testing as stated elsewhere in this standard.

14.5.2 Centrifugal Pumps.

> The replacement of components in a listed fire pump, fire pump controller, or driver should be performed by the manufacturer's factory-trained representative. All replacement parts should be supplied by the original equipment manufacturer and should be listed for fire pump service. When the replacement parts are no longer available, the equipment should be replaced with new listed and approved components.
>
> The replacement of any of the components listed in 14.5.2.4 requires a complete performance retest, as defined in Table 8.6.1 of NFPA 25, by the manufacturer's representative or other qualified person. The field retest should equal or exceed the original factory-certified test curve, controller starting and trip curves, and material specifications. The tests should be within the accuracy limits established in 14.5.2.7.
>
> The major components of the fire pump system and fire pump room should be returned to service as quickly as possible when system impairment occurs. This work should be performed by a factory-trained and qualified technician who has established competency through training.
>
> After the initial acceptance test, NFPA 25 should be consulted for component replacement.

14.5.2.1 Whenever a critical path component in a piece of centrifugal pump equipment is replaced, changed, or modified, a field/on-site retest shall be performed.

14.5.2.2 The replacement of components in fire pumps, fire pump controllers, and drivers shall be performed by factory-authorized representatives or qualified persons acceptable to the authority having jurisdiction.

14.5.2.3* When an ECM on an electronic fuel management–controlled engine is replaced, the replacement ECM shall include the same software programming that was in the original ECM.

A.14.5.2.3 Fire pump engines have unique features compared to standard industrial engines. The standard industrial ECM programming can result in the reduction of power to self-protect the engine during a fire or the inability to accelerate the pump to rated speed in rated flow condition.

14.5.2.4 Component Replacement. The requirements of Table 8.6.1 of NFPA 25, *Standard for the Inspection, Testing, and Maintenance of Water-Based Fire Protection Systems*, shall be followed for component replacement testing.

14.5.2.4.1 Replacement parts shall be provided that will maintain the listing for the fire pump component whenever possible.

14.5.2.4.2 If it is not possible to maintain the listing of a component or if the component was not originally listed for fire protection use, the replacement parts shall meet or exceed the quality of the parts being replaced.

14.5.2.5 Critical path components include the following features of the pump equipment:

(1) Fire pumps
   (a) Impeller, casing
   (b) Gear drives
(2) Fire pump controllers (electric or diesel): total replacement
(3) Electric motor, steam turbines, or diesel engine drivers
   (a) Electric motor replacement
   (b) Steam turbine replacement or rebuild
   (c) Steam regulator or source upgrade
   (d) Engine replacement or engine rebuild

14.5.2.6 Whenever replacement, change, or modification to a critical path component is performed on a fire pump, driver, or controller, as described in 14.5.2.5, a new acceptance test shall be conducted by the pump manufacturer, factory-authorized representative, or qualified persons acceptable to the authority having jurisdiction.

14.5.2.7 Field Retests.

14.5.2.7.1 The field retest results shall be compared to the original pump performance as indicated by the original factory-certified test curve, whenever it is available.

14.5.2.7.2 The field retest results shall meet or exceed the performance characteristics as indicated on the pump nameplate, and they shall be within the accuracy limits of field testing as stated elsewhere in this standard.

References Cited in Commentary

National Fire Protection Association, 1 Batterymarch Park, Quincy, MA 02169-7471.

NFPA 13, *Standard for the Installation of Sprinkler Systems,* 2013 edition.
NFPA 14, *Standard for the Installation of Standpipe and Hose Systems,* 2010 edition.
NFPA 22, *Standard for Water Tanks for Private Fire Protection,* 2008 edition.
NFPA 24, *Standard for Installation of Private Fire Service Mains and Their Appurtenances,* 2013 edition.
NFPA 25, *Standard for the Inspection, Testing, and Maintenance of Water-Based Fire Protection Systems,* 2011 edition.
NFPA 291, *Recommended Practice for Fire Flow Testing and Marking of Hydrants,* 2013 edition.
Lake, J. D. and M. J. Klaus, eds., *Water-Based Fire Protection Systems Handbook,* 2011.

ANNEX A

Explanatory Material

The material contained in Annex A of NFPA 20 is included within the text of this handbook and therefore is not repeated here.

Possible Causes of Pump Troubles

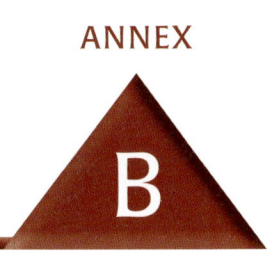

ANNEX B

This annex is not a part of the requirements of this NFPA document but is included for informational purposes only.

B.1 Causes of Pump Troubles

This annex contains a partial guide for locating pump troubles and their possible causes *(see Figure B.1)*. It also contains a partial list of suggested remedies. *(For other information on this subject, see Hydraulic Institute Standards for Centrifugal, Rotary and Reciprocating Pumps.)*

The causes listed here are in addition to possible mechanical breakage that would be obvious on visual inspection. In case of trouble, it is suggested that those troubles that can be checked easily should be corrected first or eliminated as possibilities.

B.1.1 Air Drawn into Suction Connection Through Leak(s). Air drawn into suction line through leaks causes a pump to lose suction or fail to maintain its discharge pressure. Uncover suction pipe and locate and repair leak(s).

B.1.2 Suction Connection Obstructed. Examine suction intake, screen, and suction pipe and remove obstruction. Repair or provide screens to prevent recurrence. *(See 4.14.8.)*

B.1.3 Air Pocket in Suction Pipe. Air pockets cause a reduction in delivery and pressure similar to an obstructed pipe. Uncover suction pipe and rearrange to eliminate pocket. *(See 4.14.6.)*

B.1.4 Well Collapsed or Serious Misalignment. Consult a reliable well drilling company and the pump manufacturer regarding recommended repairs.

B.1.5 Stuffing Box Too Tight or Packing Improperly Installed, Worn, Defective, Too Tight, or of Incorrect Type. Loosen gland swing bolts and remove stuffing box gland halves. Replace packing.

B.1.6 Water Seal or Pipe to Seal Obstructed. Loosen gland swing bolt and remove stuffing box gland halves along with the water seal ring and packing. Clean the water passage to and in the water seal ring. Replace water seal ring, packing gland, and packing in accordance with manufacturer's instructions.

B.1.7 Air Leak into Pump Through Stuffing Boxes. Same as the possible cause in B.1.6.

B.1.8 Impeller Obstructed. Does not show on any one instrument, but pressures fall off rapidly when an attempt is made to draw a large amount of water.

For horizontal split-case pumps, remove upper case of pump and remove obstruction from impeller. Repair or provide screens on suction intake to prevent recurrence.

For vertical shaft turbine–type pumps, lift out column pipe *(see Figure A.7.2.2.1 and Figure A.7.2.2.2)* and pump bowls from wet pit or well and disassemble pump bowl to remove obstruction from impeller.

For close-coupled, vertical in-line pumps, lift motor on top pull-out design and remove obstruction from impeller.

Fire pump Troubles	Suction				Pump																Driver and/or Pump					Driver						
	Air drawn into suction connection through leak(s)	Suction connection obstructed	Air pocket in suction pipe	Well collapsed or serious misalignment	Stuffing box too tight or packing improperly installed, worn, defective, too tight, or incorrect type	Water seal or pipe to seal obstructed	Air leak into pump through stuffing boxes	Impeller obstructed	Wearing rings worn	Impeller damaged	Wrong diameter impeller	Actual net head lower than rated	Casing gasket defective, permitting internal leakage (single-stage and multistage pumps)	Pressure gauge is on top of pump casing	Incorrect impeller adjustment (vertical shaft turbine-type pump only)	Impellers locked	Pump is frozen	Pump shaft or shaft sleeve scored, bent, or worn	Pump not primed	Seal ring improperly located in stuffing box, preventing water from entering space to form seal	Excess bearing friction due to lack of lubrication, wear, dirt, rusting, failure, or improper installation	Rotating element binds against stationary element	Pump and driver misaligned	Foundation not rigid	Engine-cooling system obstructed	Faulty driver	Lack of lubrication	Speed too low	Wrong direction of rotation	Speed too high	Rated motor voltage different from line voltage	Faulty electric circuit, obstructed fuel system, obstructed steam pipe, or dead battery
	1	2	3	4	5	6	7	8	9	10	11	12	13	14	15	16	17	18	19	20	21	22	23	24	25	26	27	28	29	30	31	32
Excessive leakage at stuffing box					X													X					X									
Pump or driver overheats				X	X	X		X			X				X			X	X	X	X	X	X	X	X	X			X	X	X	
Pump unit will not start				X	X										X	X	X				X					X	X					X
No water discharge	X	X	X					X									X															
Pump is noisy or vibrates				X	X			X	X									X			X	X	X	X		X						
Too much power required				X	X			X	X		X			X				X			X	X	X	X		X			X	X	X	
Discharge pressure not constant for same gpm	X				X	X	X																									
Pump loses suction after starting	X	X	X		X	X														X												
Insufficient water discharge	X	X	X		X	X	X	X	X	X	X		X															X	X		X	
Discharge pressure too low for gpm discharge	X	X	X		X	X	X	X	X	X	X	X																X	X		X	

FIGURE B.1 *Possible Causes of Fire Pump Troubles.*

B.1.9 Wearing Rings Worn. Remove upper case and insert feeler gauge between case wearing ring and impeller wearing ring. Clearance when new is 0.0075 in. (0.19 mm). Clearances of more than 0.015 in. (0.38 mm) are excessive.

B.1.10 Impeller Damaged. Make minor repairs or return to manufacturer for replacement. If defect is not too serious, order new impeller and use damaged one until replacement arrives.

B.1.11 Wrong Diameter Impeller. Replace with impeller of proper diameter.

B.1.12 Actual Net Head Lower than Rated. Check impeller diameter and number and pump model number to make sure correct head curve is being used.

B.1.13 Casing Gasket Defective, Permitting Internal Leakage (Single-Stage and Multistage Pumps). Replace defective gasket. Check manufacturer's drawing to see whether gasket is required.

B.1.14 Pressure Gauge Is on Top of Pump Casing. Place gauges in correct location. *[See Figure A.6.3.1(a).]*

B.1.15 Incorrect Impeller Adjustment (Vertical Shaft Turbine–Type Pump Only). Adjust impellers according to manufacturer's instructions.

B.1.16 Impellers Locked. For vertical shaft turbine–type pumps, raise and lower impellers by the top shaft adjusting nut. If this adjustment is not successful, follow the manufacturer's instructions.

For horizontal split-case pumps, remove upper case and locate and eliminate obstruction.

B.1.17 Pump Is Frozen. Provide heat in the pump room. Disassemble pump and remove ice as necessary. Examine parts carefully for damage.

B.1.18 Pump Shaft or Shaft Sleeve Scored, Bent, or Worn. Replace shaft or shaft sleeve.

B.1.19 Pump Not Primed. If a pump is operated without water in its casing, the wearing rings are likely to seize. The first warning is a change in pitch of the sound of the driver. Shut down the pump.

For vertical shaft turbine–type pumps, check water level to determine whether pump bowls have proper submergence.

B.1.20 Seal Ring Improperly Located in Stuffing Box, Preventing Water from Entering Space to Form Seal. Loosen gland swing bolt and remove stuffing box gland halves along with the water-seal ring and packing. Replace, putting seal ring in proper location.

B.1.21 Excess Bearing Friction Due to Lack of Lubrication, Wear, Dirt, Rusting, Failure, or Improper Installation. Remove bearings and clean, lubricate, or replace as necessary.

B.1.22 Rotating Element Binding Against Stationary Element. Check clearances and lubrication and replace or repair the defective part.

B.1.23 Pump and Driver Misaligned. Shaft running off center because of worn bearings or misalignment. Align pump and driver according to manufacturer's instructions. Replace bearings according to manufacturer's instructions. *(See Section 6.5.)*

B.1.24 Foundation Not Rigid. Tighten foundation bolts or replace foundation if necessary. *(See Section 6.4.)*

B.1.25 Engine Cooling System Obstructed. Heat exchanger or cooling water systems too small or cooling pump faulty. Remove thermostats. Open bypass around regulator valve and strainer. Check regulator valve operation. Check strainer. Clean and repair if necessary. Disconnect sections of cooling system to locate and remove possible obstruction. Adjust engine cooling water circulating pump belt to obtain proper speed without binding. Lubricate bearings of this pump.

If overheating still occurs at loads up to 150 percent of rated capacity, contact pump or engine manufacturer so that necessary steps can be taken to eliminate overheating.

B.1.26 Faulty Driver. Check electric motor, internal combustion engine, or steam turbine, in accordance with manufacturer's instructions, to locate reason for failure to start.

B.1.27 Lack of Lubrication. If parts have seized, replace damaged parts and provide proper lubrication. If not, stop pump and provide proper lubrication.

B.1.28 Speed Too Low. For electric motor drive, check that rated motor speed corresponds to rated speed of pump, voltage is correct, and starting equipment is operating properly.

Low frequency and low voltage in the electric power supply prevent a motor from running at rated speed. Low voltage can be due to excessive loads and inadequate feeder capacity or (with private generating plants) low generator voltage. The generator voltage of private generating plants can be corrected by changing the field excitation. When low voltage is from the other causes mentioned, it might be necessary to change transformer taps or increase feeder capacity.

Low frequency usually occurs with a private generating plant and should be corrected at the source. Low speed can result in older type squirrel-cage-type motors if fastenings of copper bars to end rings become loose. The remedy is to weld or braze these joints.

For steam turbine drive, check that valves in steam supply pipe are wide open; boiler steam pressure is adequate; steam pressure is adequate at the turbine; strainer in the steam supply pipe is not plugged; steam supply pipe is of adequate size; condensate is removed from steam supply pipe, trap, and turbine; turbine nozzles are not plugged; and setting of speed and emergency governor is correct.

For internal combustion engine drive, check that setting of speed governor is correct; hand throttle is opened wide; and there are no mechanical defects such as sticking valves, timing off, or spark plugs fouled, and so forth. These problems might require the services of a trained mechanic.

B.1.29 Wrong Direction of Rotation. Instances of an impeller turning backward are rare but are clearly recognizable by the extreme deficiency of pump delivery. Wrong direction of rotation can be determined by comparing the direction in which the flexible coupling is turning with the directional arrow on the pump casing.

With a polyphase electric motor drive, two wires must be reversed; with a dc driver, the armature connections must be reversed with respect to the field connections. Where two sources of electrical current are available, the direction of rotation produced by each should be checked.

B.1.30 Speed Too High. See that pump- and driver-rated speed correspond. Replace electric motor with one of correct rated speed. Set governors of drivers for correct speed. Frequency at private generating stations might be too high.

B.1.31 Rated Motor Voltage Different from Line Voltage. For example, a 220 V or 440 V motor on 208 V or 416 V line. Obtain motor of correct rated voltage or larger size motor. *(See Section 9.4.)*

B.1.32 Faulty Electric Circuit, Obstructed Fuel System, Obstructed Steam Pipe, or Dead Battery. Check for break in wiring open switch, open circuit breaker, or dead battery. If circuit breaker in controller trips for no apparent reason, make sure oil is in dash pots in accordance with manufacturer's specifications. Make sure fuel pipe is clear, strainers are clean, and control valves are open in fuel system to internal combustion engine. Make sure all valves are open and strainer is clean in steam line to turbine.

B.2 Warning

Chapters 9 and 10 include electrical requirements that discourage the installation of disconnect means in the power supply to electric motor–driven fire pumps. This requirement is intended to ensure the availability of power to the fire pumps. When equipment connected to those circuits is serviced or maintained, the employee can have unusual exposure to electrical and other hazards. It can be necessary to require special safe work practices and special safeguards, personal protective clothing, or both.

B.3 Maintenance of Fire Pump Controllers After a Fault Condition

B.3.1 Introduction. In a fire pump motor circuit that has been properly installed, coordinated, and in service prior to the fault, tripping of the circuit breaker or the isolating switch indicates a fault condition in excess of operating overload.

It is recommended that the following general procedures be observed by qualified personnel in the inspection and repair of the controller involved in the fault. These procedures are not intended to cover other elements of the circuit, such as wiring and motor, which can also require attention.

B.3.2 Caution. All inspections and tests are to be made on controllers that are de-energized at the line terminal, disconnected, locked out, and tagged so that accidental contact cannot be made with live parts and so that all plant safety procedures will be observed.

B.3.2.1 Enclosure. Where substantial damage to the enclosure, such as deformation, displacement of parts, or burning has occurred, replace the entire controller.

B.3.2.2 Circuit Breaker and Isolating Switch. Examine the enclosure interior, circuit breaker, and isolating switch for evidence of possible damage. If evidence of damage is not apparent, the circuit breaker and isolating switch can continue to be used after closing the door.

If there is any indication that the circuit breaker has opened several short-circuit faults, or if signs of possible deterioration appear within either the enclosure, circuit breaker, or isolating switch (e.g., deposits on surface, surface discoloration, insulation cracking, or unusual toggle operation), replace the components. Verify that the external operating handle is capable of opening and closing the circuit breaker and isolating switch. If the handle fails to operate the device, this would also indicate the need for adjustment or replacement.

B.3.2.3 Terminals and Internal Conductors. Where there are indications of arcing damage, overheating, or both, such as discoloration and melting of insulation, replace the damaged parts.

B.3.2.4 Contactor. Replace contacts showing heat damage, displacement of metal, or loss of adequate wear allowance of the contacts. Replace the contact springs where applicable. If deterioration extends beyond the contacts, such as binding in the guides or evidence of insulation damage, replace the damaged parts or the entire contactor.

B.3.2.5 Return to Service. Before returning the controller to service, check for the tightness of electrical connections and for the absence of short circuits, ground faults, and leakage current.

Close and secure the enclosure before the controller circuit breaker and isolating switch are energized. Follow operating procedures on the controller to bring it into standby condition.

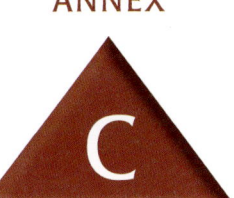

Informational References

ANNEX C

C.1 Referenced Publications

The documents or portions thereof listed in this annex are referenced within the informational sections of this standard and are not part of the requirements of this document unless also listed in Chapter 2 for other reasons.

C.1.1 NFPA Publications. National Fire Protection Association, 1 Batterymarch Park, Quincy, MA 02169-7471.

NFPA 1, *Fire Code,* 2012 edition.
NFPA 13, *Standard for the Installation of Sprinkler Systems,* 2013 edition.
NFPA 14, *Standard for the Installation of Standpipe and Hose Systems,* 2010 edition.
NFPA 15, *Standard for Water Spray Fixed Systems for Fire Protection,* 2012 edition.
NFPA 16, *Standard for the Installation of Foam-Water Sprinkler and Foam-Water Spray Systems,* 2011 edition.
NFPA 24, *Standard for the Installation of Private Fire Service Mains and Their Appurtenances,* 2013 edition.
NFPA 25, *Standard for the Inspection, Testing, and Maintenance of Water-Based Fire Protection Systems,* 2011 edition.
NFPA 31, *Standard for the Installation of Oil-Burning Equipment,* 2011 edition.
NFPA 70®, National Electrical Code®, 2011 edition.
NFPA 77, *Recommended Practice on Static Electricity,* 2007 edition.
NFPA 750, *Standard on Water Mist Fire Protection Systems,* 2010 edition.

C.1.2 Other Publications

C.1.2.1 ANSI Publications. American National Standards Institute, Inc., 25 West 43rd Street, 4th Floor, New York, NY 10036.

ANSI/IEEE C62.11, *IEEE Standard for Metal-Oxide Surge Arresters for AC Power Circuits,* 1987.

C.1.2.2 ASCE Publications. American Society of Civil Engineers, 1801 Alexander Bell Drive, Reston, VA 20190-4400.

SEI/ASCE 7, *Minimum Design Loads for Buildings and Other Structures,* 2010.

C.1.2.3 AWWA Publications. American Water Works Association, 6666 West Quincy Avenue, Denver, CO 80235.

AWWA C104, *Cement-Mortar Lining for Cast-Iron and Ductile-Iron Pipe and Fittings for Water,* 1990.

C.1.2.4 HI Publications. Hydraulic Institute, 6 Campus Drive, First Floor North, Parsippany, NJ 07054-4406.

Hydraulic Institute Standards for Centrifugal, Rotary and Reciprocating Pumps, 14th ed., 1983.
HI 3.5, *Standard for Rotary Pumps for Nomenclature, Design, Application and Operation,* 1994.
HI 3.6, *Rotary Pump Tests,* 1994.

C.1.2.5 IEEE Publications. Institute of Electrical and Electronics Engineers, Three Park Avenue, 17th Floor, New York, NY 10016-5997.

IEEE 141, *Electric Power Distribution for Industrial Plants,* 1986.
IEEE 241, *Electric Systems for Commercial Buildings,* 1990.

C.1.2.6 NEMA Publications. National Electrical Manufacturers Association, 1300 North 17th Street, Suite 1847, Rosslyn, VA 22209.

NEMA ICS 14, *Application Guide for Electric Fire Pump Controllers,* 2010.
NEMA 250, *Enclosures for Electrical Equipment,* 2008.

C.1.2.7 SAE Publications. Society of Automotive Engineers, 400 Commonwealth Drive, Warrendale, PA 15096.

SAE J-1349, *Engine Power Test Code — Spark Ignition and Compression Engine,* 1990.

C.1.2.8 UL Publications. Underwriters Laboratories Inc., 333 Pfingsten Road, Northbrook, IL 60062-2096.

ANSI/UL 508, *Standard for Industrial Control Equipment,* 1999, Revised 2010.
ANSI/UL 1008, *Standard for Transfer Switch Equipment,* 2011.

C.2 Informational References

The following documents or portions thereof are listed here as informational resources only. They are not a part of the requirements of this document.

C.2.1 UL Publications. Underwriters Laboratories Inc., 333 Pfingsten Road, Northbrook, IL 60062-2096.

ANSI/UL 80, *Standard for Steel Tanks for Oil Burner Fuels and Other Combustible Liquids,* 2007, Revised 2009.
UL 2080, *Standard for Fire Resistant Tanks for Flammable and Combustible Liquids,* 2000.
ANSI/UL 2085, *Standard for Protected Aboveground Tanks for Flammable and Combustible Liquids,* 1997, Revised 2010.

C.3 References for Extracts in Informational Sections

(Reserved)

ANNEX D

Material Extracted by NFPA 70, Article 695

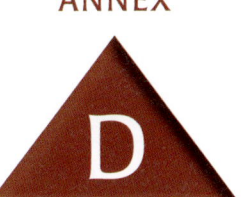

D.1 General

Table D.1 indicates corresponding sections of *NFPA 70*, Article 695.

TABLE D.1 *NFPA 70, National Electric Code, Extracted Material*

NFPA 20 2010 Edition	NFPA 20 2013 Edition	NFPA 70 Section 695 2008 Edition	NFPA 70 Section 695 2011 Edition	Section 695 Titles or Text
(Reference Only)		695.3	Same	Definitions
3.3.7.2	Same	695.2	Same	Fault Tolerant External Control Circuits
3.3.34	3.3.35	695.2	Same	On-Site Power Production Facility
3.3.35	3.3.36	695.2	Same	On-Site Standby Generator
9.2.1	Same	695.3	Same	Power Source(s) for Electric Motor-Driven Fire Pumps
9.2	Same	695.3(A)	Same	Individual Sources
9.2.2	Same	695.3(A)(1)	Same	Electric Utility Service Connection (2nd Sentence)
9.2.2	Same	695.3(A)(2)	Same	On-Site Power Production Facility
9.2.2(3)	Same	N/A	695.3(A)(3)	Dedicated Feeder
9.2.2	Same	695.3(B)	Same	Multiple Sources
9.2.2	Same	695.3(B)	695.3(B)(1)	Individual Sources
9.2.2	Same	695.3(B)	695.3(B)(2)	Individual Source and On-Site Standby Generator
9.3.4(1)	Same	695.3(B)(2)	695.3(C)	Multibuilding Campus-Style Complexes
9.3.4(1)	Same	695.3(B)(2)	695.3(C)(1)	Feeder Sources
9.3.4(1)	Same	695.3(B)(2)	695.3(C)(2)	Feeder and Alternate Sources
N/A	N/A	N/A	695.3(C)(3)	Selective Coordination
9.6.1	Same	695.3(B)(1)	695.3(D)	On-Site Standby Generator as Alternate Source
9.6.1	Same	695.3(B)(1)	695.3(D)(1)	Capacity
9.6.1	Same	695.3(B)(1)	695.3(D)(2)	Connection
9.6.1	Same	695.3(B)(1)	695.3(D)(3)	Adjacent Disconnections
9.3.2	Same	695.3(B)(3)	695.3(E)	Arrangement
N/A	N/A	N/A	695.3(F)	Phase Converters
9.2.1	Same	695.4	Same	Continuity of Power
9.2.2	Same	695.4(A)	Same	Direct Connection
9.2	Same	695.4(B)	Same	Connection Through Disconnecting Means and Overcurrent Device
9.2	Same	695.4(B)	695.4(B)(1)	Number of Disconnecting Means
9.2	Same	695.4(B)	695.4(B)(1)(a)	General
9.2	Same	695.4(B)-(1)	695.4(B)(1)(a)(1)	(listed fire pump controller)

(continued)

TABLE D.1 *Continued*

NFPA 20 2010 Edition	NFPA 20 2013 Edition	NFPA 70 Section 695 2008 Edition	NFPA 70 Section 695 2011 Edition	Section 695 Titles or Text
9.2	Same	695.4(B)-(2)	695.4(B)(1)(a)(2)	(listed fire pump power transfer switch)
9.2	Same	695.4(B)-(3)	695.4(B)(1)(a)(3)	(listed combo, fire pump controller/transfer switch)
9.2	Same	695.4(B)	695.4(B)(1)(b)	Feeder Sources
9.2	Same	695.4(B)	695.4(B)(1)(c)	On-Site Standby Generator
9.2.3.4	Same	695.4(B)(1)	695.4(B)(2)	Overcurrent Device Selection
9.2.3.4	Same	695.4(B)(1)	695.4(B)(2)(a)	Individual Sources
9.2	Same	695.4(B)	695.4(B)(2)(b)	On-Site Standby Generators
9.2.3.2	Same	695.4(B)(2)	695.4(B)(3)	Disconnecting Means
9.2.3.2	Same	695.4(B)(2)	695.4(B)(3)(a)	Features and Location – Normal Power Source
9.2.3.1	Same	695.4(B)(2)-(1)	695.4(B)(3)(a)(1)	Be identifiable as suitable for use as service equipment...
9.2.3.1(2)	Same	695.4(B)(2)-(2)	695.1(B)(3)(a)(2)	Be lockable in the closed position...
9.2.3.1(3)	Same	695.4(B)(2)-(3)	695.1(B)(3)(a)(3)	Not be located within equipment that feeds loads other than...
9.2.3.1(4)	Same	695.4(B)(2)-(4)	695.1(B)(3)(a)(4)	Be located sufficiently remote from...
N/A	N/A	N/A	695.4(B)(3)(b)	Features and Location – On – Site Standby Generator
9.2.3.1(5)	Same	695.4(B)(3)	695.4.(B)(3)(c)	Disconnect Marking
10.1.2.2	Same	695.4(B)(4)	695.4.(B)(3)(d)	Controller Marking
9.2.3.3	Same	695.4(B)(5)	695.4.(B)(3)(e)	Supervision
9.2.3.3(1)	Same	695.4(B)(5)-(1)	695.4.(B)(3)(e)(1)	Central station, proprietary, or remote station...
9.2.3.3(2)	Same	695.4(B)(5)-(2)	695.4.(B)(3)(e)(2)	Local signaling service...
9.2.3.3(3)	Same	695.4(B)(5)-(3)	695.4.(B)(3)(e)(3)	Locking the disconnecting means...
9.2.3.3(4)	Same	695.4(B)(5)-(4)	695.4.(B)(3)(e)(4)	Sealing of disconnecting means...
9.2.2(5)	Same	695.5	Same	Transformers
Reference Only		695.6	Same	Power Wiring
N/A	N/A	695.6(F)	695.6(E)	Loads Supplied by Controllers and Transfer Switches
9.8	Same	N/A	695.6(H)	Listed Electrical Circuit Protective System to Controller Wiring
9.8.1	Same	N/A	695.6(H)(1)	A junction box shall be installed ahead of the fire pump controller...
9.8.2	Same	N/A	695.6(H)(2)	... the raceway between the controller and junction box shall be sealed
9.8.3	Same	N/A	695.6(H)(3)	Standard wiring between the junction box and controller is permitted
9.7	Same	695.6(F)	695.6(I)	Junction boxes
9.7(1)	Same	N/A	695.6(I)(1)	The junction box shall be securely mounted
9.7(2)	Same	N/A	695.6(I)(2)	Mounting and Installing... shall not violate the enclosure type rating...
9.7(3)	Same	N/A	695.6(I)(3)	Mounting and Installing... shall not affect the short circuit rating...
9.7(4)	Same	N/A	695.6(I)(4)	...a Type 2... enclosure shall be used...
9.7(5)	Same	695.6(F)	695.6(I)(5)	Terminals, junction blocks, wire connectors, and splices shall be listed

Section D.1 • General

NFPA 20		NFPA 70 Section 695		
2010 Edition	*2013 Edition*	*2008 Edition*	*2011 Edition*	*Section 695 Titles or Text*
10.3.4.5.1, 10.3.4.6	Same	695.6(F)	695.6(I)(6)	A fire pump controller or . . . transfer switch shall not be used as a junction box . . .
9.9	Same	N/A	695.6(J)	Raceway Terminations
9.9.1	Same	N/A	695.6(J)(1)	Listed Conduit Hubs
9.9.2	Same	N/A	695.6(J)(2)	The type rating of conduit hubs
9.9.3	Same	N/A	695.6(J)(3)	Installation instructions
9.9.4	Same	N/A	695.6(J)(4)	Alterations to the controller
9.4	Same	695.7	Same	Voltage Drop
9.5.1.1	Same	695.10	Same	Listed Equipment
10.1.2.1	Same	695.10	Same	Listed Equipment
10.8.3.1	10.7.3.1	695.10	Same	Listed Equipment
12.1.3.1	Same	695.10	Same	Listed Equipment
(Reference Only)		695.12	Same	Equipment Location
10.2.1	Same	695.12(A)	Same	Controllers and Transfer Switches
12.2.1	Same	695.12(B)	Same	Engine-Drive Controllers
11.2.7.2.4	11.2.7.2.5	695.12(C)	Same	Storage Batteries
11.2.7.2.4.2	11.2.7.2.5.2	695.12(D)	Same	Energized Equipment
10.2.2	Same	695.12(E)	Same	Protection Against Pump Water
12.2.2	Same	695.12(E)	Same	Protection Against Pump Water
10.3.2	Same	695.12(F)	Same	Mounting
12.3.2	Same	695.12(F)	Same	Mounting
12.5.2.6	Same	695.14(A)	Same	Control Circuit Failures
12.7.2.5	Same	695.14(A)	Same	Control Circuit Failures
10.4.5.7	Same	695.14(B)	Same	Sensor Functioning
10.8.1.3	10.7.1.3	695.14(C)	Same	Remote Device(s)
12.3.5.1	Same	695.14(D)	Same	Engine-Drive Control Wiring
(Reference Only)	Same	695.14(E)	Same	Electric Fire Pump Control Wiring Methods

Additional References – Informational Only

A.9.3.2(3)	Same	695.6(A)	Same	Supply Conductors
9.6.4	Same	695.12(A)	Same	Controllers and Transfer Switches
A.9.3.2(3)	Same	695.14(F)	Same	Generator Control Wiring Methods

PART THREE

Water Supply Requirements, Water Demand, Hydrants, Tanks, and Piping

Part III of this handbook is a collection of standards, requirements, annexes, and associated commentary from several water supply documents and outlines the hydraulic calculation requirements for the most common types of fire protection systems: standpipe systems and sprinkler systems. The flow and pressure demands of standpipe and sprinkler systems determine the need for a fire pump. An important fact to remember is that a fire pump is a supplemental component of the water supply – a fire pump cannot create waterflow. However, a fire pump does transfer energy to the water, increasing the water pressure. In Part III, icons are located in the page margins to indicate material of special interest with respect to design and calculations.

Water supplies are a vital and integral part of every water-based fire suppression system. Before a water-based suppression system is designed, the system designer or registered design professional (RDP) should thoroughly analyze the available water supply to ensure that sufficient flow and pressure are available to meet system design parameters. Several scenarios that might result from analyzing the water supply include, but are not limited to, the following:

- Sufficient flow and pressure
- Insufficient pressure
- Insufficient flow and pressure
- Insufficient flow

If the flow and pressure are found to be sufficient, the need for a fire pump does not exist; however, for any of the other circumstances previously specified, a fire pump is a practical solution.

In preparation for the flow test, which is performed using the guidelines from NFPA 291, *Recommended Practice for Fire Flow Testing and Marking of Hydrants,* several aspects of the water supply system should be reviewed. To avoid what could be a costly mistake, a designer or an RDP should review civil underground piping plans, if available, and identify the correct main and hydrants to be used during the flow test. If civil plans are unavailable at the time of, or prior to, the performance of the flow test, the local water district should be contacted for details. Underground mains might be supplied from different sources. For example, one main may be supplied by a municipal pumping station, while another is supplied from a municipal water tank; hydrants might be on a dead-end main, or the mains may vary in pipe size. All of these circumstances can produce different flow test results. If the incorrect underground main and hydrants are used for flow test readings, any insufficient flows and pressures might remain undiscovered until late in the project or during final testing.

It is important to note that, in accordance with the requirements of NFPA 13, *Standard for the Installation of Sprinkler Systems* and NFPA 20, *Standard for the Installation of Stationary Pumps for Fire Protection,* waterflow tests must be performed within 12 months of submitting a working plan. Of course, it is possible that, due to restrictive waterflow conditions in several areas, the authority having jurisdiction (AHJ) may require or allow older, previously recorded flow test data to be used. Alternatively, the AHJ may allow the use of a waterflow analysis provided by the water district or approved consulting agency in lieu of a flow test. It is also a good practice to acquire previous flow test information from the local water district; this practice provides a means of comparing the current results to previous outcomes and assists in determining the accuracy of current results and/or highlights possible deficiencies. If substantial variations are observed, it might be a good indicator that more research into the water supply system is needed. One contributing factor could be the time of the flow test; ensuring that flow test results are taken during peak demand times can avoid shortfalls in system designs.

If the results of the flow test show a static pressure similar to that of the previous test, but show a large decrease in residual pressure and flow, several possible causes should be considered or investigated. A valve in the underground system might be partially closed, or a portion of a municipal grid system might be partially or completely shut down. The underground mains could have extreme degradation (corrosion and sedimentation buildup), depending on the age of the system, or there could be a large obstruction in the piping system. These are just some examples of the possible causes of reduced residual pressure and flow. Regardless of the cause, any problem needs to be resolved before the system designer or the RDP can proceed with the design of the system. Simply performing a flow test does not automatically ensure that accurate flow data is being used during design, and the results of using innacurate data could lead to expensive corrections at or near the end of the project.

The system designer or RDP should take into account the age of the project area and any future development that could affect the water supply over time. If an area is slated for a large industrial development, and initial flow test results are sufficient to meet the system demand requirements of the initial building, a very real possibility exists that supply will fall short of demand as additional structures are added to the water supply demand.

If flow test results show insufficient water supply, the use of a fire pump is a viable solution. Since a fire pump cannot create waterflow, a water source having a positive head is required. The supply can come from a municipal supply main to which an on-site fire supply line is designed and installed in accordance with NFPA 24, *Standard for the Installation of Private Fire Service Mains and Their Appurtenances.* In addition, if on-site fire hydrants are located on the on-site fire supply line, NFPA 1, *Fire Code,* provides fire flow requirements. Another possible water source could be a private fire supply, such as a tank designed and installed in accordance with NFPA 22, *Standard for Water Tanks for Private Fire Protection.*

As has been explained, different scenarios can exist that lead the system designer or the RDP to the conclusion that a specific project requires a fire pump. For example, the city supply might be capable of providing sufficient flow but could lack the pressure required to meet the demands of a standpipe system at the upper floors of a high-rise building; or the pressure, however adequate for ordinary systems, might be insufficient to meet the high-pressure demands of extra hazard occupancies or early suppression fast response (ESFR) sprinkler systems. Specific sections of NFPA 14, *Standard for the Installation of Standpipe and Hose Systems,* and NFPA 13, *Standard for the Installation of Sprinkler Systems,* are included in Part III to show how the water supply needs of sprinkler and standpipe systems are determined and to illustrate how the capacity of a fire pump is determined.

Complete Text of NFPA® 291, Recommended Practice for Fire Flow Testing and Marking of Hydrants, 2013 Edition, with Commentary

SECTION 1

> NFPA 291, *Recommended Practice for Fire Flow Testing and Marking of Hydrants*, is included in this handbook because water supply information is critical to the design of all types of water-based fire protection systems. Where the water for a fire protection system is to be supplied by a public or private water system, a waterflow test should be conducted prior to the commencement of system design. NFPA 291 outlines the correct procedure for obtaining water supply information and also provides the necessary information in the form of flow tables for calculating the flow test results.
>
> NFPA 291 is a recommended practice and, as such, contains none of the mandatory language that is normally found in an NFPA code or standard. Where codes and standards use the term *shall*, NFPA 291 uses the term *should*. The difference in the use of the terms is important, because a variety of methods and equipment can be employed to correctly perform a hydrant flow test. NFPA 291 also includes a color-coding system for hydrants that, if implemented by the water authority, can provide a quick visual indication of the flow capacity of each hydrant.

CHAPTER 1 ADMINISTRATION

1.1 Scope

The scope of this document is fire flow testing and marking of hydrants.

> NFPA 291 outlines the procedures necessary for conducting fire flow testing and color coding of hydrants.

1.2 Purpose

Fire flow tests are conducted on water distribution systems to determine the rate of flow available at various locations for fire-fighting purposes.

> In addition to its role in fire fighting, fire flow testing is also necessary for determining the available waterflow and pressure for fixed fire protection systems including sprinklers, standpipes, and fire pumps. In all cases, the available flow and pressure from a public or private water supply must be determined prior to the commencement of any system design. NFPA 13, *Standard for the Installation of Sprinkler Systems*; NFPA 14, *Standard for the Installation of Standpipe and Hose Systems*; and NFPA 20, *Standard for the Installation of Stationary Pumps for Fire Protection*, require that fire flow test information be less than 1 year old prior to the submission of working plans. This requirement is intended to ensure that the statistics on fire flow and pressures available are accurate and as up-to-date as possible. A best practice is to retest water supplies and verify available fire water supply data every 5 years.
>
> Whenever modifications to a water supply system are made, or when substantial development takes place in a given area, retesting the available water supply and verifying the available fire flow is recommended.

423

1.3 Application

A certain residual pressure in the mains is specified at which the rate of flow should be available. Additional benefit is derived from fire flow tests by the indication of possible deficiencies, such as tuberculation of piping or closed valves or both, which could be corrected to ensure adequate fire flows as needed.

1.4 Units

Metric units of measurement in this recommended practice are in accordance with the modernized metric system known as the International System of Units (SI). Two units (liter and bar), outside of but recognized by SI, are commonly used in international fire protection. These units are listed in Table 1.4 with conversion factors.

TABLE 1.4 SI Units and Conversion Factors

Unit Name	Unit Symbol	Conversion Factor
Liter	L	1 gal = 3.785 L
Liter per minute per square meter	(L/min)/m^2	1 gpm ft^2 = (40.746 L/min)/m^2
Cubic decimeter	dm^3	1 gal = 3.785 dm^3
Pascal	Pa	1 psi = 6894.757 Pa
Bar	bar	1 psi = 0.0689 bar
Bar	bar	1 bar = 10^5 Pa

Note: For additional conversions and information, see IEEE/ASTM-SI-10, *Standard for Use of the International System of Units (SI): The Modern Metric System*, 1992.

1.4.1 If a value for measurement as given in this recommended practice is followed by an equivalent value in other units, the first value stated is to be regarded as the recommendation. A given equivalent value might be approximate.

CHAPTER 2 REFERENCED PUBLICATIONS

2.1 General

The documents or portions thereof listed in this chapter are referenced within this recommended practice and shall be considered part of the recommendations of this document.

2.2 NFPA Publications

(Reserved)

2.3 Other Publications

2.3.1 IEEE Publications. Institute of Electrical and Electronics Engineers, Three Park Avenue, 17th Floor, New York, NY 10016-5997.

IEEE/ASTM-SI-10, *Standard for Use of the International System of Units (SI): The Modern Metric System*, 1992.

2.3.2 Other Publications.

Merriam-Webster's Collegiate Dictionary, 11th edition, Merriam-Webster, Inc., Springfield, MA, 2003.

2.4 References for Extracts in Recommendations Sections

(Reserved)

CHAPTER 3 DEFINITIONS

3.1 General

The definitions contained in this chapter apply to the terms used in this recommended practice. Where terms are not defined in this chapter or within another chapter, they should be defined using their ordinarily accepted meanings within the context in which they are used. *Merriam-Webster's Collegiate Dictionary*, 11th edition, is the source for the ordinarily accepted meaning.

3.2 NFPA Official Definitions

3.2.1* Authority Having Jurisdiction (AHJ). An organization, office, or individual responsible for enforcing the requirements of a code or standard, or for approving equipment, materials, an installation, or a procedure.

A.3.2.1 Authority Having Jurisdiction (AHJ). The phrase "authority having jurisdiction," or its acronym AHJ, is used in NFPA documents in a broad manner, since jurisdictions and approval agencies vary, as do their responsibilities. Where public safety is primary, the authority having jurisdiction may be a federal, state, local, or other regional department or individual such as a fire chief; fire marshal; chief of a fire prevention bureau, labor department, or health department; building official; electrical inspector; or others having statutory authority. For insurance purposes, an insurance inspection department, rating bureau, or other insurance company representative may be the authority having jurisdiction. In many circumstances, the property owner or his or her designated agent assumes the role of the authority having jurisdiction; at government installations, the commanding officer or departmental official may be the authority having jurisdiction.

3.2.2* Listed. Equipment, materials, or services included in a list published by an organization that is acceptable to the authority having jurisdiction and concerned with evaluation of products or services, that maintains periodic inspection of production of listed equipment or materials or periodic evaluation of services, and whose listing states that either the equipment, material, or service meets appropriate designated standards or has been tested and found suitable for a specified purpose.

A.3.2.2 Listed. The means for identifying listed equipment may vary for each organization concerned with product evaluation; some organizations do not recognize equipment as listed unless it is also labeled. The authority having jurisdiction should utilize the system employed by the listing organization to identify a listed product.

3.2.3 Should. Indicates a recommendation or that which is advised but not required.

3.3 General Definitions

3.3.1 Rated Capacity. The flow available from a hydrant at the designated residual pressure (rated pressure), either measured or calculated.

> The rated capacity of a hydrant also includes the flow capabilities of the underground main. The flow capacity of the underground main can be compared to the flow and pressure demand of an attached fire protection system. The required flow and pressure demand of the attached fire protection system must be less than that which can be supplied by the underground main.

3.3.2 Residual Pressure. The pressure that exists in the distribution system, measured at the residual hydrant at the time the flow readings are taken at the flow hydrants.

3.3.3 Static Pressure. The pressure that exists at a given point under normal distribution system conditions measured at the residual hydrant with no hydrants flowing.

CHAPTER 4 FLOW TESTING

4.1 Rating Pressure

4.1.1 For the purpose of uniform marking of fire hydrants, the ratings should be based on a residual pressure of 20 psi (1.4 bar) for all hydrants having a static pressure in excess of 40 psi (2.8 bar).

4.1.2 Hydrants having a static pressure of less than 40 psi (2.8 bar) should be rated at one-half of the static pressure.

4.1.3 It is generally recommended that a minimum residual pressure of 20 psi (1.4 bar) should be maintained at hydrants when delivering the fire flow. Fire department pumpers can be operated where hydrant pressures are less, but with difficulty.

> The minimum pressure required by many water authorities is 20 psi (1.4 bar). Pressures less than 20 psi (1.4 bar) may create vacuum conditions in the water system piping in areas of the water system that are located at higher elevations relative to the location of the flow. This can cause possible collapse of the piping or introduce groundwater contamination through leak points in the piping. Low pressures such as 20 psi (1.4 bar) can also cause a backflow of non-potable water through check valves, which can contaminate a potable source. While fire apparatus can operate at such low pressures, the minimum net positive suction head (NPSH) of the pump needs to be maintained, or the pump will not perform well, and damage due to cavitation can occur.

4.1.4 Where hydrants are well distributed and of the proper size and type (so that friction losses in the hydrant and suction line are not excessive), it might be possible to set a lesser pressure as the minimum pressure.

4.1.5 A primary concern should be the ability to maintain sufficient residual pressure to prevent developing a negative pressure at any point in the street mains, which could result in the collapse of the mains or other water system components or back-siphonage of polluted water from some other interconnected source.

4.1.6 It should be noted that the use of residual pressures of less than 20 psi (1.4 bar) is not permitted by many state health departments.

4.2 Procedure

4.2.1 Tests should be made during a period of ordinary demand.

> "A period of ordinary demand" should include testing during normal business hours and during seasons of high usage or low storage levels, such as summertime. Note that water supply test data obtained at times of high use and used to size a fire pump can result in overpressure problems. During periods of low demand, the available pressure can be considerably higher than at periods of high demand. If pressures fluctuate such that excessively high pressures (i.e., pressures that exceed the rated working pressure system components) risk causing system damage, measures must be taken to control these high pressures. Possible options for controlling excessive pressure include use of a break tank or variable speed fire pump. Chapter 4 of NFPA 20 addresses the proper applications of these options (see Part II of this handbook). NFPA 20 does allow the use of the fire pump main relief valve as a means of addressing excess fire pump suction pressure.

4.2.2 The procedure consists of discharging water at a measured rate of flow from the system at a given location and observing the corresponding pressure drop in the mains.

4.3 Layout of Test

4.3.1 After the location where the test is to be run has been determined, a group of test hydrants in the vicinity is selected.

4.3.2 Once selected, due consideration should be given to potential interference with traffic flow patterns, damage to surroundings (e.g., roadways, sidewalks, landscapes, vehicles, and pedestrians), and potential flooding problems both local and remote from the test site.

> In addition to the considerations specified in 4.3.2, the local environmental authority may need to be consulted regarding the disposal of chlorinated or non-potable test water.

4.3.3 One hydrant, designated the residual hydrant, is chosen to be the hydrant where the normal static pressure will be observed with the other hydrants in the group closed, and where the residual pressure will be observed with the other hydrants flowing.

> The exact effective point of the flow test will be located where the nonflowing water in the piping supplying the hydrant at which the residual pressures are being measured meets the flowing water in the underground piping. In all cases, as shown in Figure 4.3.4, this effective point is where the supply piping of the residual pressure hydrant, "R," meets the underground main.

4.3.4 This hydrant is chosen so it will be located between the hydrant to be flowed and the large mains that constitute the immediate sources of water supply in the area. In Figure 4.3.4, test layouts are indicated showing the residual hydrant designated with the letter R and hydrants to be flowed with the letter F.

> The residual hydrant should also be the hydrant closest to the property to be protected. This simplifies the hydraulic calculations needed to "move" the water test effective point from the location of the test to the base of the sprinkler or standpipe riser in the protected property.

4.3.5 The number of hydrants to be used in any test depends upon the strength of the distribution system in the vicinity of the test location.

4.3.6 To obtain satisfactory test results of theoretical calculation of expected flows or rated capacities, sufficient discharge should be achieved to cause a drop in pressure at the residual hydrant of at least 25 percent, or to flow the total demand necessary for fire-fighting purposes.

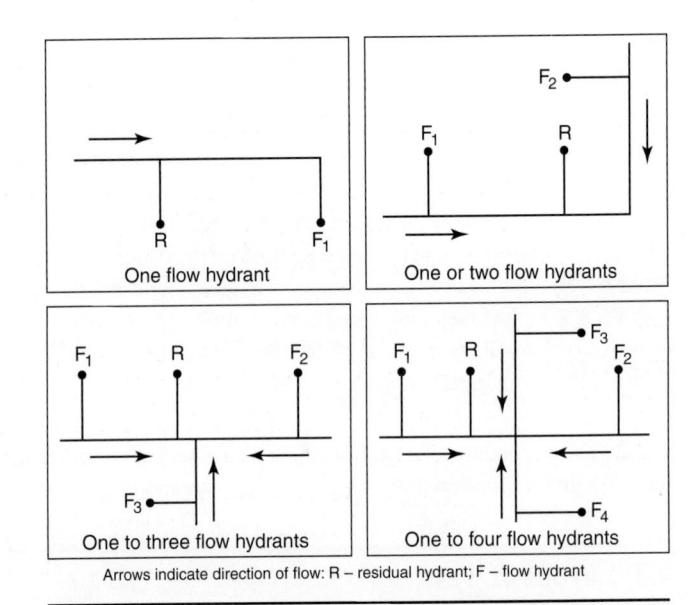

FIGURE 4.3.4 *Suggested Test Layout for Hydrants.*

> In order to produce a pressure drop of 25 percent, or the total fire demand, multiple outlets on the flowing hydrant may need to be opened during the test. In some cases, this may require that several hydrants be flowed simultaneously to produce a sufficient pressure drop and/or flow.

4.3.7 If the mains are small and the system weak, only one or two hydrants need to be flowed.

4.3.8 If, on the other hand, the mains are large and the system strong, it may be necessary to flow as many as seven or eight hydrants.

4.4 Equipment

4.4.1 The equipment necessary for field work consists of the following:

(1) A single 200 psi (14 bar) bourdon pressure gauge with 1 psi (0.0689 bar) graduations
(2) A number of pitot tubes
(3) Hydrant wrenches
(4) 50 or 60 psi (3.5 or 4.0 bar) bourdon pressure gauges with 1 psi (0.0689 bar) graduations, and scales with 1/16 in. (1.6 mm) graduations [one pitot tube, a 50 or 60 psi (3.5 or 4.0 bar) gauge, a hydrant wrench, a scale for each hydrant to be flowed]
(5) A special hydrant cap tapped with a hole into which is fitted a short length of 1/4 in. (6.35 mm) brass pipe provided with a T connection for the 200 psi (14 bar) gauge and a cock at the end for relieving air pressure

4.4.2 All pressure gauges should be calibrated at least every 12 months, or more frequently depending on use.

> Each calibrated gauge should be identified by a sticker from the company or person performing the calibration that indicates the date of the last calibration. Gauges can lose their calibration over time; therefore, it is recommended that recalibration be completed a minimum of every 12 months.

4.4.3 When more than one hydrant is flowed, it is desirable and could be necessary to use portable radios to facilitate communication between team members.

> Some hydrants may be located several hundred feet apart; therefore, portable radios are needed for proper communication. Testing personnel need to communicate so that measurements can be taken at the appropriate time. For example, when hydrants are fully opened, residual pressures should be taken.

4.4.4 It is preferred to use stream straightener with a known coefficient of discharge when testing hydrants due to a more streamlined flow and more accurate pitot reading.

> A stream straightener helps to improve the accuracy of the pressure reading but does not diffuse the energy of the water stream. In addition to improving the accuracy of test results, there are devices available that incorporate a diffuser that breaks up the energy of the water stream, improves the safety of the test personnel, and helps avoid property damage. Exhibit III.1 illustrates the use of a diffuser.

EXHIBIT III.1 *A Flow Diffuser in Use During a Hydrant Flow Test.*

4.5 Test Procedure

4.5.1 In a typical test, the 200 psi (14 bar) gauge is attached to one of the 2½ in. (6.4 cm) outlets of the residual hydrant using the special cap.

4.5.2 The cock on the gauge piping is opened, and the hydrant valve is opened full.

4.5.3 As soon as the air is exhausted from the barrel, the cock is closed.

4.5.4 A reading (static pressure) is taken when the needle comes to rest.

> Before attaching test equipment to any hydrant, the hydrant should be flushed to remove any debris within the hydrant barrel. Debris from the hydrant barrel can damage test gauges or pitot tubes. After flushing, the test equipment can be attached to the hydrant and the air bled off. At this point in the test, data such as date, time of day, weather conditions, and static pressure should be recorded. Exhibit III.2 illustrates flushing of the residual hydrant and shows the site for attachment of test equipment.

EXHIBIT III.2 *Flushing of Residual Hydrant (left) and Attachment of Test Equipment (right).*

4.5.5 At a given signal, each of the other hydrants is opened in succession, with discharge taking place directly from the open hydrant butts.

4.5.6 Hydrants should be opened one at a time.

4.5.7 With all hydrants flowing, water should be allowed to flow for a sufficient time to clear all debris and foreign substances from the stream(s).

4.5.8 At that time, a signal is given to the people at the hydrants to read the pitot pressure of the streams simultaneously while the residual pressure is being read.

4.5.9 The final magnitude of the pressure drop can be controlled by the number of hydrants used and the number of outlets opened on each.

4.5.10 After the readings have been taken, hydrants should be shut down slowly, one at a time, to prevent undue surges in the system.

> Closing a hydrant too quickly can cause damage in the system from pressure surges and can cause sprinkler system alarm valves or flow switches to operate, causing nuisance alarms.

4.6 Pitot Readings

4.6.1 When measuring discharge from open hydrant butts, it is always preferable from the standpoint of accuracy to use 2½ in. (6.4 cm) outlets rather than pumper outlets.

4.6.2 In practically all cases, the 2½ in. (6.4 cm) outlets are filled across the entire cross-section during flow, while in the case of the larger outlets there is very frequently a void near the bottom.

4.6.3 When measuring the pitot pressure of a stream of practically uniform velocity, the orifice in the pitot tube is held downstream approximately one-half the diameter of the hydrant outlet or nozzle opening, and in the center of the stream.

4.6.4 The center line of the orifice should be at right angles to the plane of the face of the hydrant outlet.

EXHIBIT III.3 *Diffuser – Slot for Pitot Tube Insertion.*

The pitot tube should be positioned at the one-half diameter position referenced in 4.6.3 and maintain the centerline of the pitot tube orifice at a right angle to the plane of the hydrant outlet. Exhibit III.3 illustrates the pitot tube opening or slot for correct positioning of the pitot tube in the water stream when using a diffuser.

4.6.5 The air chamber on the pitot tube should be kept elevated.

4.6.6 Pitot readings of less than 10 psi (0.7 bar) and more than 30 psi (2.0 bar) should be avoided, if possible.

Pitot readings of less than 10 psi (0.7 bar) and more than 30 psi (2.0 bar) can effect the accuracy of the readings and should be avoided.

4.6.7 Opening additional hydrant outlets will aid in controlling the pitot reading.

4.6.8 With dry barrel hydrants, the hydrant valve should be wide open to minimize problems with underground drain valves.

4.6.9 With wet barrel hydrants, the valve for the flowing outlet should be wide open to give a more streamlined flow and a more accurate pitot reading. *(See Figure 4.6.9.)*

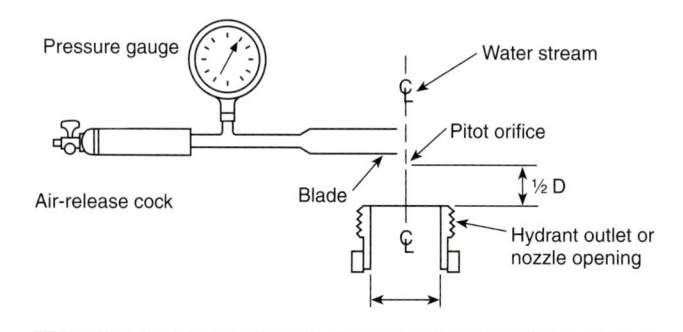

FIGURE 4.6.9 *Pitot Tube Position.*

4.7 Determination of Discharge

4.7.1 At the hydrants used for flow during the test, the discharges from the open butts are determined from measurements of the diameter of the outlets flowed, the pitot pressure (velocity head) of the streams as indicated by the pitot gauge readings, and the coefficient of the outlet being flowed as determined from Figure 4.7.1.

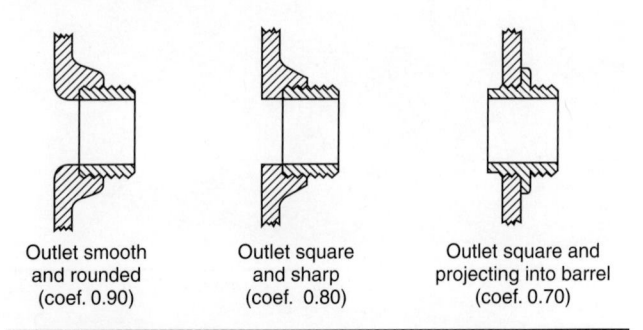

Outlet smooth and rounded (coef. 0.90)

Outlet square and sharp (coef. 0.80)

Outlet square and projecting into barrel (coef. 0.70)

FIGURE 4.7.1 *Three General Types of Hydrant Outlets and Their Coefficients of Discharge.*

4.7.2 If flow tubes (stream straighteners) are being utilized, a coefficient of 0.95 is suggested unless the coefficient of the tube is known.

4.7.3 The formula used to compute the discharge, Q, in gpm from these measurements is as follows:

$$Q = 29.84 cd^2 \sqrt{p} \qquad (4.7.3)$$

where:

c = coefficient of discharge *(see Figure 4.7.1)*

d = diameter of the outlet in inches

p = pitot pressure (velocity head) in psi

4.8 Use of Pumper Outlets

4.8.1 If it is necessary to use a pumper outlet, and flow tubes (stream straighteners) are not available, the best results are obtained with the pitot pressure (velocity head) maintained between 5 psi and 10 psi (0.3 bar and 0.7 bar).

4.8.2 For pumper outlets, the approximate discharge can be computed from Equation 4.7.3 using the pitot pressure (velocity head) at the center of the stream and multiplying the result by one of the coefficients in Table 4.8.2, depending upon the pitot pressure (velocity head).

4.8.3 These coefficients are applied in addition to the coefficient in Equation 4.7.3 and are for average-type hydrants.

4.9 Determination of Discharge Without a Pitot

4.9.1 If a pitot tube is not available for use to measure the hydrant discharge, a 50 or 60 psi (3.5 or 4.0 bar) gauge tapped into a hydrant cap can be used.

4.9.2 The hydrant cap with gauge attached is placed on one outlet, and the flow is allowed to take place through the other outlet at the same elevation.

TABLE 4.8.2 Pumper Outlet Coefficients

Pitot Pressure (Velocity Head)		
psi	bar	Coefficient
2	0.14	0.97
3	0.21	0.92
4	0.28	0.89
5	0.35	0.86
6	0.41	0.84
7 and over	0.48 and over	0.83

4.9.3 The readings obtained from a gauge so located, and the readings obtained from a gauge on a pitot tube held in the stream, are approximately the same.

4.10 Calculation Results

4.10.1 The discharge in gpm (L/min) for each outlet flowed is obtained from Table 4.10.1(a) and Table 4.10.1(b) or by the use of Equation 4.7.3.

◄ **DESIGN ALERT**

To determine the discharge flow, the pitot pressure, along with the outlet diameter and coefficient, must be known. The pitot pressure measured during the flow test in Exhibit III.4 is 50 psi (3.5 bar). From Table 4.10.1(a) and Table 4.10.1(b), a pitot pressure of 50 psi (3.5 bar) is 1319 gpm (4992 L/min) theoretical flow. After applying the outlet coefficient of 0.9, the actual flow measured during this flow test is 1187 gpm (4493 L/min).

4.10.1.1 If more than one outlet is used, the discharges from all are added to obtain the total discharge.

4.10.1.2 The formula that is generally used to compute the discharge at the specified residual pressure or for any desired pressure drop is Equation 4.10.1.2:

$$Q_R = Q_F \times \frac{h_r^{0.54}}{h_f^{0.54}} \qquad (4.10.1.2)$$

where:
Q_R = flow predicted at desired residual pressure
Q_F = total flow measured during test
h_r = pressure drop to desired residual pressure
h_f = pressure drop measured during test

◄ **DESIGN ALERT**

Applying the formula in 4.10.1.2, the total flow measured during the test is as previously calculated: 1187 gpm (4493 L/min). The pressure drop to the desired pressure in this case is 55 psi (3.8 bar) [75 psi − 20 psi = 55 psi (5.2 bar − 1.4 bar = 3.8 bar)]. The pressure drop measured during the test was 23 psi (1.6 bar) [75 psi − 52 psi = 23 psi (5.2 bar − 3.6 bar = 1.6 bar)]. The result of this calculation predicts a flow of 1901 gpm (7195.3 L/min) when the residual pressure is drawn down to the minimum recommended 20 psi (1.4 bar). As stated in 4.1.1, the ratings and uniform marking of hydrants should be based on a residual pressure of 20 psi (1.4 bar). Therefore, the rating for the hydrant in this example should be based on a flow of 1901 gpm (7195.3 L/min), not the 1187 gpm (4493 L/min) measured during the test.

Stationary Fire Pumps Handbook 2013

TABLE 4.10.1(a) Theoretical Discharge Through Circular Orifices (U.S. Gallons of Water per Minute)

Pitot Pressure* (psi)	Feet†	Velocity Discharge (ft/sec)	2	2.25	2.375	2.5	2.625	2.75	3	3.25	3.5	3.75	4	4.5
1	2.31	12.20	119	151	168	187	206	226	269	315	366	420	477	604
2	4.61	17.25	169	214	238	264	291	319	380	446	517	593	675	855
3	6.92	21.13	207	262	292	323	356	391	465	546	633	727	827	1047
4	9.23	24.39	239	302	337	373	411	451	537	630	731	839	955	1209
5	11.54	27.26	267	338	376	417	460	505	601	705	817	938	1068	1351
6	13.84	29.87	292	370	412	457	504	553	658	772	895	1028	1169	1480
7	16.15	32.26	316	400	445	493	544	597	711	834	967	1110	1263	1599
8	18.46	34.49	338	427	476	528	582	638	760	891	1034	1187	1350	1709
9	20.76	36.58	358	453	505	560	617	677	806	946	1097	1259	1432	1813
10	23.07	38.56	377	478	532	590	650	714	849	997	1156	1327	1510	1911
11	25.38	40.45	396	501	558	619	682	748	891	1045	1212	1392	1583	2004
12	27.68	42.24	413	523	583	646	712	782	930	1092	1266	1454	1654	2093
13	29.99	43.97	430	545	607	672	741	814	968	1136	1318	1513	1721	2179
14	32.30	45.63	447	565	630	698	769	844	1005	1179	1368	1570	1786	2261
15	34.61	47.22	462	585	652	722	796	874	1040	1221	1416	1625	1849	2340
16	36.91	48.78	477	604	673	746	822	903	1074	1261	1462	1679	1910	2417
17	39.22	50.28	492	623	694	769	848	930	1107	1300	1507	1730	1969	2491
18	41.53	51.73	506	641	714	791	872	957	1139	1337	1551	1780	2026	2564
19	43.83	53.15	520	658	734	813	896	984	1171	1374	1593	1829	2081	2634
20	46.14	54.54	534	676	753	834	920	1009	1201	1410	1635	1877	2135	2702
22	50.75	57.19	560	709	789	875	964	1058	1260	1478	1715	1968	2239	2834
24	55.37	59.74	585	740	825	914	1007	1106	1316	1544	1791	2056	2339	2960
26	59.98	62.18	609	770	858	951	1048	1151	1369	1607	1864	2140	2434	3081
28	64.60	64.52	632	799	891	987	1088	1194	1421	1668	1934	2220	2526	3197
30	69.21	66.79	654	827	922	1022	1126	1236	1471	1726	2002	2298	2615	3310
32	73.82	68.98	675	855	952	1055	1163	1277	1519	1783	2068	2374	2701	3418
34	78.44	71.10	696	881	981	1087	1199	1316	1566	1838	2131	2447	2784	3523
36	83.05	73.16	716	906	1010	1119	1234	1354	1611	1891	2193	2518	2865	3626
38	87.67	75.17	736	931	1038	1150	1268	1391	1656	1943	2253	2587	2943	3725
40	92.28	77.11	755	955	1065	1180	1300	1427	1699	1993	2312	2654	3020	3822
42	96.89	79.03	774	979	1091	1209	1333	1462	1740	2043	2369	2719	3094	3916
44	101.51	80.88	792	1002	1116	1237	1364	1497	1781	2091	2425	2783	3167	4008
46	106.12	82.70	810	1025	1142	1265	1395	1531	1821	2138	2479	2846	3238	4098
48	110.74	84.48	827	1047	1166	1292	1425	1563	1861	2184	2533	2907	3308	4186
50	115.35	86.22	844	1068	1190	1319	1454	1596	1899	2229	2585	2967	3376	4273
52	119.96	87.93	861	1089	1214	1345	1483	1627	1937	2273	2636	3026	3443	4357
54	124.58	89.61	877	1110	1237	1370	1511	1658	1974	2316	2686	3084	3508	4440
56	129.19	91.20	893	1130	1260	1396	1539	1689	2010	2359	2735	3140	3573	4522
58	133.81	92.87	909	1150	1282	1420	1566	1719	2045	2400	2784	3196	3636	4602
60	138.42	94.45	925	1170	1304	1445	1593	1748	2080	2441	2831	3250	3698	4681
62	143.03	96.01	940	1189	1325	1469	1619	1777	2115	2482	2878	3304	3759	4758
64	147.65	97.55	955	1209	1347	1492	1645	1805	2148	2521	2924	3357	3820	4834

TABLE 4.10.1(a) Continued

Pitot Pressure* (psi)	Feet[†]	Velocity Discharge (ft/sec)	2	2.25	2.375	2.5	2.625	2.75	3	3.25	3.5	3.75	4	4.5
66	152.26	99.07	970	1227	1367	1515	1670	1833	2182	2561	2970	3409	3879	4909
68	156.88	100.55	984	1246	1388	1538	1696	1861	2215	2599	3014	3460	3937	4983
70	161.49	102.03	999	1264	1408	1560	1720	1888	2247	2637	3058	3511	3995	5056
72	166.10	103.47	1013	1282	1428	1583	1745	1915	2279	2674	3102	3561	4051	5127
74	170.72	104.90	1027	1300	1448	1604	1769	1941	2310	2711	3144	3610	4107	5198
76	175.33	106.30	1041	1317	1467	1626	1793	1967	2341	2748	3187	3658	4162	5268
78	179.95	107.69	1054	1334	1487	1647	1816	1993	2372	2784	3228	3706	4217	5337
80	184.56	109.08	1068	1351	1505	1668	1839	2018	2402	2819	3269	3753	4270	5405
82	189.17	110.42	1081	1368	1524	1689	1862	2043	2432	2854	3310	3800	4323	5472
84	193.79	111.76	1094	1385	1543	1709	1885	2068	2461	2889	3350	3846	4376	5538
86	198.40	113.08	1107	1401	1561	1730	1907	2093	2491	2923	3390	3891	4428	5604
88	203.02	114.39	1120	1417	1579	1750	1929	2117	2519	2957	3429	3936	4479	5668
90	207.63	115.68	1132	1433	1597	1769	1951	2141	2548	2990	3468	3981	4529	5733
92	212.24	116.96	1145	1449	1614	1789	1972	2165	2576	3023	3506	4025	4579	5796
94	216.86	118.23	1157	1465	1632	1808	1994	2188	2604	3056	3544	4068	4629	5859
96	221.47	119.48	1169	1480	1649	1827	2015	2211	2631	3088	3582	4111	4678	5921
98	226.09	120.71	1182	1495	1666	1846	2035	2234	2659	3120	3619	4154	4726	5982
100	230.70	121.94	1194	1511	1683	1865	2056	2257	2686	3152	3655	4196	4774	6043
102	235.31	123.15	1205	1526	1700	1884	2077	2279	2712	3183	3692	4238	4822	6103
104	239.93	124.35	1217	1541	1716	1902	2097	2301	2739	3214	3728	4279	4869	6162
106	244.54	125.55	1229	1555	1733	1920	2117	2323	2765	3245	3763	4320	4916	6221
108	249.16	126.73	1240	1570	1749	1938	2137	2345	2791	3275	3799	4361	4962	6280
110	253.77	127.89	1252	1584	1765	1956	2157	2367	2817	3306	3834	4401	5007	6338
112	258.38	129.05	1263	1599	1781	1974	2176	2388	2842	3336	3869	4441	5053	6395
114	263.00	130.20	1274	1613	1797	1991	2195	2409	2867	3365	3903	4480	5098	6452
116	267.61	131.33	1286	1627	1813	2009	2215	2430	2892	3395	3937	4519	5142	6508
118	272.23	132.46	1297	1641	1828	2026	2234	2451	2917	3424	3971	4558	5186	6564
120	276.84	133.57	1308	1655	1844	2043	2252	2472	2942	3453	4004	4597	5230	6619
122	281.45	134.69	1318	1669	1859	2060	2271	2493	2966	3481	4038	4635	5273	6674
124	286.07	135.79	1329	1682	1874	2077	2290	2513	2991	3510	4070	4673	5317	6729
126	290.68	136.88	1340	1696	1889	2093	2308	2533	3015	3538	4103	4710	5359	6783
128	295.30	137.96	1350	1709	1904	2110	2326	2553	3038	3566	4136	4748	5402	6836
130	299.91	139.03	1361	1722	1919	2126	2344	2573	3062	3594	4168	4784	5444	6890
132	304.52	140.10	1371	1736	1934	2143	2362	2593	3086	3621	4200	4821	5485	6942
134	309.14	141.16	1382	1749	1948	2159	2380	2612	3109	3649	4231	4858	5527	6995
136	313.75	142.21	1392	1762	1963	2175	2398	2632	3132	3676	4263	4894	5568	7047

Notes:

(1) This table is computed from the formula $Q = 29.84cd^2\sqrt{p}$, with $c = 1.00$. The theoretical discharge of seawater, as from fireboat nozzles, can be found by subtracting 1 percent from the figures in Table 4.10.2.1, or from the formula $Q = 29.84cd^2\sqrt{p}$.

(2) Appropriate coefficient should be applied where it is read from hydrant outlet. Where more accurate results are required, a coefficient appropriate on the particular nozzle must be selected and applied to the figures of the table. The discharge from circular openings of sizes other than those in the table can readily be computed by applying the principle that quantity discharged under a given head varies as the square of the diameter of the opening.

*This pressure corresponds to velocity head.

[†]1 psi = 2.307 ft of water. For pressure in bars, multiply by 0.07.

TABLE 4.10.1(b) *Theoretical Discharge Through Circular Orifices (Liters of Water per Minute)*

Pitot Pressure* (kPa)	Meters†	Velocity Discharge (m/sec)	51	57	60	64	67	70	76	83	89	92	101	114
						Orifice Size (in.)								
6.89	0.70	3.72	455	568	629	716	785	857	1010	1204	1385	1578	1783	2272
13.8	1.41	5.26	644	804	891	1013	1111	1212	1429	1704	1960	2233	2524	3215
20.7	2.11	6.44	788	984	1091	1241	1360	1485	1750	2087	2400	2735	3091	3938
27.6	2.81	7.43	910	1137	1260	1433	1571	1714	2021	2410	2771	3158	3569	4547
34.5	3.52	8.31	1017	1271	1408	1602	1756	1917	2259	2695	3099	3530	3990	5084
41.4	4.22	9.10	1115	1392	1543	1755	1924	2100	2475	2952	3394	3867	4371	5569
48.3	4.92	9.83	1204	1504	1666	1896	2078	2268	2673	3189	3666	4177	4722	6015
55.2	5.63	10.51	1287	1608	1781	2027	2221	2425	2858	3409	3919	4466	5048	6431
62.0	6.33	11.15	1364	1704	1888	2148	2354	2570	3029	3613	4154	4733	5349	6815
68.9	7.03	11.75	1438	1796	1990	2264	2482	2709	3193	3808	4379	4989	5639	7184
75.8	7.73	12.33	1508	1884	2087	2375	2603	2841	3349	3995	4593	5233	5915	7536
82.7	8.44	12.87	1575	1968	2180	2481	2719	2968	3498	4172	4797	5466	6178	7871
89.6	9.14	13.40	1640	2048	2270	2582	2830	3089	3641	4343	4994	5690	6431	8193
96.5	9.84	13.91	1702	2126	2355	2680	2937	3206	3779	4507	5182	5905	6674	8503
103	10.55	14.39	1758	2196	2433	2769	3034	3312	3904	4656	5354	6100	6895	8784
110	11.25	14.87	1817	2269	2515	2861	3136	3423	4035	4812	5533	6304	7125	9078
117	11.95	15.33	1874	2341	2593	2951	3234	3530	4161	4963	5706	6502	7349	9362
124	12.66	15.77	1929	2410	2670	3038	3329	3634	4284	5109	5874	6693	7565	9638
131	13.36	16.20	1983	2477	2744	3122	3422	3735	4403	5251	6038	6880	7776	9906
138	14.06	16.62	2035	2542	2817	3205	3512	3834	4519	5390	6197	7061	7981	10168
152	15.47	17.43	2136	2668	2956	3363	3686	4023	4743	5657	6504	7410	8376	10671
165	16.88	18.21	2225	2779	3080	3504	3840	4192	4941	5893	6776	7721	8727	11118
179	18.28	18.95	2318	2895	3208	3650	4000	4366	5147	6138	7058	8042	9090	11580
193	19.69	19.67	2407	3006	3331	3790	4153	4534	5344	6374	7329	8350	9438	12024
207	21.10	20.36	2492	3113	3450	3925	4301	4695	5535	6601	7590	8648	9775	12453
221	22.50	21.03	2575	3217	3564	4055	4444	4851	5719	6821	7842	8935	10100	12867
234	23.91	21.67	2650	3310	3668	4173	4573	4992	5884	7018	8070	9195	10393	13240
248	25.31	22.30	2728	3408	3776	4296	4708	5139	6058	7225	8308	9466	10699	13630
262	26.72	22.91	2804	3502	3881	4416	4839	5282	6227	7426	8539	9729	10997	14010
276	28.13	23.50	2878	3595	3983	4532	4967	5422	6391	7622	8764	9986	11287	14379
290	29.53	24.09	2950	3685	4083	4646	5091	5557	6551	7813	8984	10236	11570	14740
303	30.94	24.65	3015	3767	4173	4748	5204	5681	6696	7986	9183	10463	11826	15066
317	32.35	25.21	3084	3853	4269	4857	5323	5810	6849	8169	9393	10702	12096	15410
331	33.75	25.75	3152	3937	4362	4963	5439	5937	6999	8347	9598	10935	12360	15747
345	35.16	26.28	3218	4019	4453	5067	5553	6061	7145	8522	9799	11164	12619	16077
358	36.57	26.80	3278	4094	4536	5161	5657	6175	7279	8681	9981	11373	12855	16377
372	37.97	27.31	3341	4173	4624	5261	5766	6294	7419	8849	10175	11593	13104	16694
386	39.38	27.80	3403	4251	4711	5360	5874	6412	7558	9014	10364	11809	13348	17005
400	40.78	28.31	3465	4328	4795	5456	5979	6527	7694	9176	10551	12021	13588	17311
414	42.19	28.79	3525	4403	4878	5551	6083	6640	7827	9335	10734	12230	13823	17611
427	43.60	29.26	3580	4471	4954	5637	6178	6743	7949	9481	10901	12420	14039	17885
441	45.00	29.73	3638	4544	5035	5729	6278	6853	8078	9635	11078	12622	14267	18176
455	46.41	30.20	3695	4616	5114	5819	6377	6961	8206	9787	11253	12821	14492	18462
469	47.82	30.65	3751	4686	5192	5908	6475	7067	8331	9936	11425	13017	14713	18744

TABLE 4.10.1(b) Continued

Pitot Pressure* (kPa)	Meters[†]	Velocity Discharge (m/sec)	51	57	60	64	67	70	76	83	89	92	101	114
												Orifice Size (in.)		
483	49.22	31.10	3807	4756	5269	5995	6570	7172	8454	10083	11594	13210	14931	19022
496	50.63	31.54	3858	4819	5340	6075	6658	7268	8567	10218	11749	13386	15131	19276
510	52.03	31.97	3912	4887	5415	6161	6752	7370	8687	10361	11913	13574	15343	19547
524	53.44	32.71	3965	4953	5488	6245	6844	7470	8806	10503	12076	13759	15552	19813
538	54.85	32.82	4018	5019	5561	6327	6934	7569	8923	10642	12236	13942	15758	20076
552	56.25	33.25	4070	5084	5633	6409	7024	7667	9038	10780	12394	14122	15962	20335
565	57.66	33.66	4118	5143	5699	6484	7106	7757	9144	10906	12539	14287	16149	20573
579	59.07	34.06	4168	5207	5769	6564	7194	7853	9256	11040	12694	14463	16348	20827
593	60.47	34.47	4218	5269	5839	6643	7280	7947	9368	11173	12846	14637	16544	21077
607	61.88	34.87	4268	5331	5907	6721	7366	8040	9478	11304	12997	14809	16738	21324
620	63.29	35.26	4313	5388	5970	6793	7444	8126	9578	11424	13136	14966	16917	21552
634	64.69	35.65	4362	5448	6037	6869	7528	8217	9686	11552	13283	15134	17107	21794
648	66.10	36.04	4410	5508	6103	6944	7610	8307	9792	11679	13429	15301	17294	22033
662	67.50	36.42	4457	5567	6169	7019	7692	8397	9898	11805	13573	15465	17480	22270
676	68.91	36.79	4504	5626	6234	7093	7773	8485	10002	11929	13716	15628	17664	22504
689	70.32	37.17	4547	5680	6293	7161	7848	8566	10097	12043	13847	15777	17833	22719
703	71.72	37.54	4593	5737	6357	7233	7927	8653	10200	12165	13987	15937	18013	22949
717	73.13	37.90	4638	5794	6420	7305	8005	8738	10301	12285	14126	16095	18192	23176
731	74.54	38.27	4684	5850	6482	7376	8083	8823	10401	12405	14263	16251	18369	23401
745	75.94	38.63	4728	5906	6544	7446	8160	8907	10500	12523	14399	16406	18544	23624
758	77.35	38.98	4769	5957	6601	7510	8231	8985	10591	12632	14524	16548	18705	23830
772	78.76	39.33	4813	6012	6662	7580	8307	9067	10688	12748	14658	16701	18877	24049
786	80.16	39.68	4857	6066	6722	7648	8382	9149	10785	12863	14790	16851	19047	24266
800	81.57	40.03	4900	6120	6781	7716	8456	9230	10880	12977	14921	17001	19216	24481
813	82.97	40.37	4939	6170	6836	7778	8525	9305	10968	13082	15042	17138	19371	24679
827	84.38	40.71	4982	6223	6895	7845	8598	9385	11063	13194	15171	17285	19538	24891
841	85.79	41.05	5024	6275	6953	7911	8670	9464	11156	13305	15299	17431	19702	25100
855	87.19	41.39	5065	6327	7011	7977	8742	9542	11248	13416	15425	17575	19866	25309
869	88.60	41.72	5107	6379	7068	8042	8813	9620	11340	13525	15551	17719	20028	25515
882	90.01	42.05	5145	6426	7121	8102	8879	9692	11424	13626	15667	17851	20177	25705
896	91.41	42.38	5185	6477	7177	8166	8949	9768	11515	13734	15791	17992	20336	25908
910	92.82	42.70	5226	6527	7233	8229	9019	9844	11604	13840	15914	18132	20495	26110
924	94.23	43.03	5266	6577	7288	8292	9088	9920	11693	13947	16036	18271	20652	26310
938	95.63	43.35	5305	6627	7343	8355	9156	9995	11782	14052	16157	18409	20807	26509

Notes:

(1) This table is computed from the formula $Q_m = 0.0666cd^2\sqrt{P_m}$, with $c = 1.00$. The theoretical discharge of seawater, as from fireboat nozzles, can be found by subtracting 1 percent from the figures in Table 4.10.2.1, or from the formula $Q_m = 0.065cd^2m\sqrt{P_m}$.

(2) Appropriate coefficient should be applied where it is read from the hydrant outlet. Where more accurate results are required, a coefficient appropriate on the particular nozzle must be selected and applied to the figures of the table. The discharge from circular openings of sizes other than those in the table can readily be computed by applying the principle that quantity discharged under a given head varies as the square of the diameter of the opening.

*This pressure corresponds to velocity head.

[†] 1 kPa = 0.102 m of water. For pressure in bars, multiply by 0.01.

FLOW TEST REPORT

Date: _6/2/10_
Inspector: _John Doe_

Residual hydrant location: _Main Street (see map on p. 2)_

Flow hydrant location: _"   "_

Static pressure (residual hydrant): _75_ psi

Residual pressure (residual hydrant): _52_ psi

Pitot pressure (flow hydrant): _50_ psi

Nozzle size (flow hydrant): _2.5_ in.
From Table 4.10.1(a) of NFPA 291, *Recommended Practice for Fire Flow Testing and Marking of Hydrants*, 2010 edition: _1319_ gpm

Nozzle coefficient (flow hydrant): _0.9_

_1319_ × _0.9_ = _1187_

Available water flow: _1187_ gpm at _52_ psi

Compute discharge:

$Q_R = Q_F \times h_r^{0.54}/h_f^{0.54}$

$Q_R =$ _1187_ × (_55_)$^{0.54}$ / (_23_)$^{0.54}$

$Q_R =$ _1901_ gpm at 20 psi

(p. 1 of 3)

EXHIBIT III.4 *Sample Flow Test Report.*

Section 4.10 • Calculation Results 439

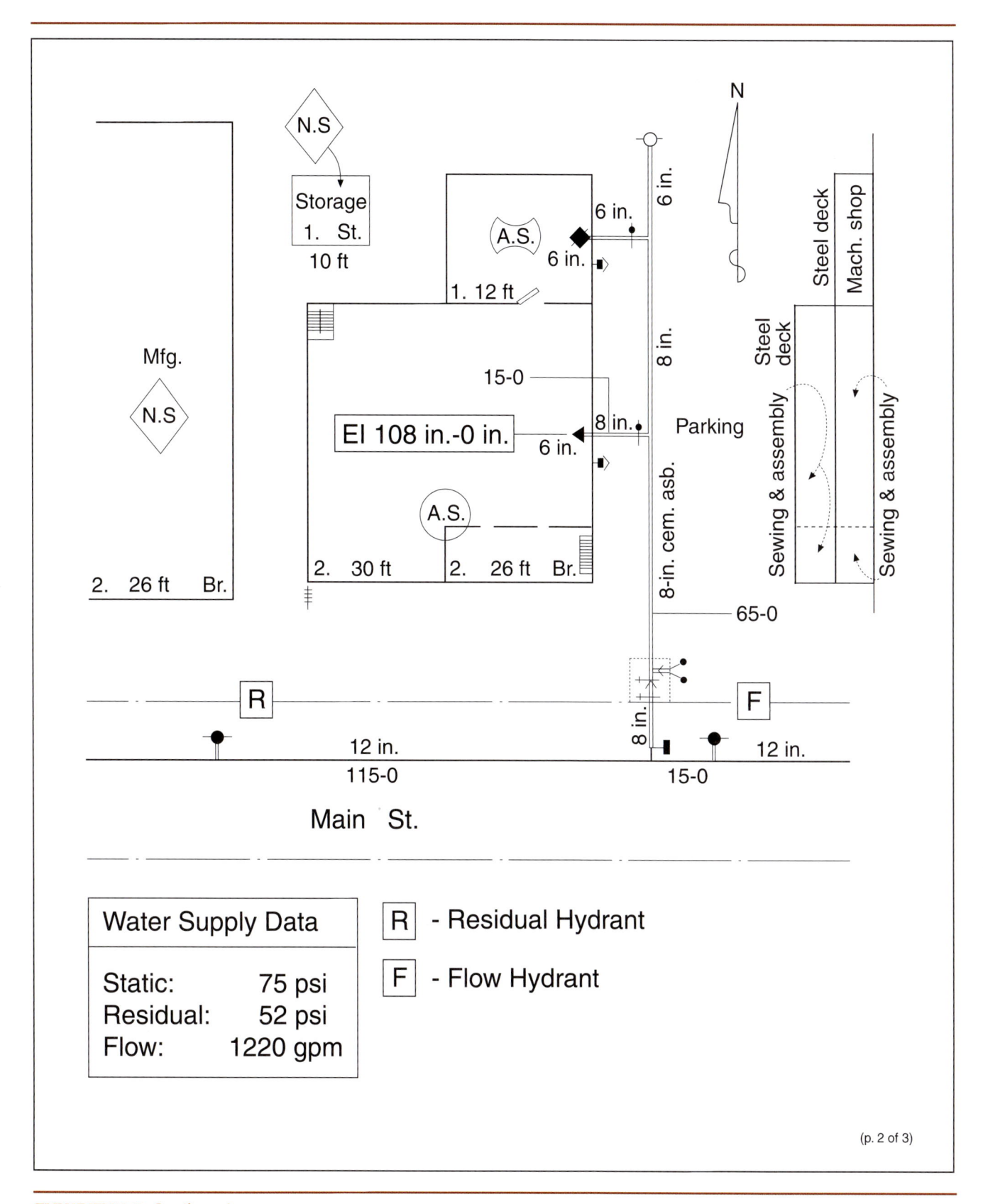

EXHIBIT III.4 Continued.

HYDRAULIC GRAPH Pressure vs. (Flow)$^{1.85}$

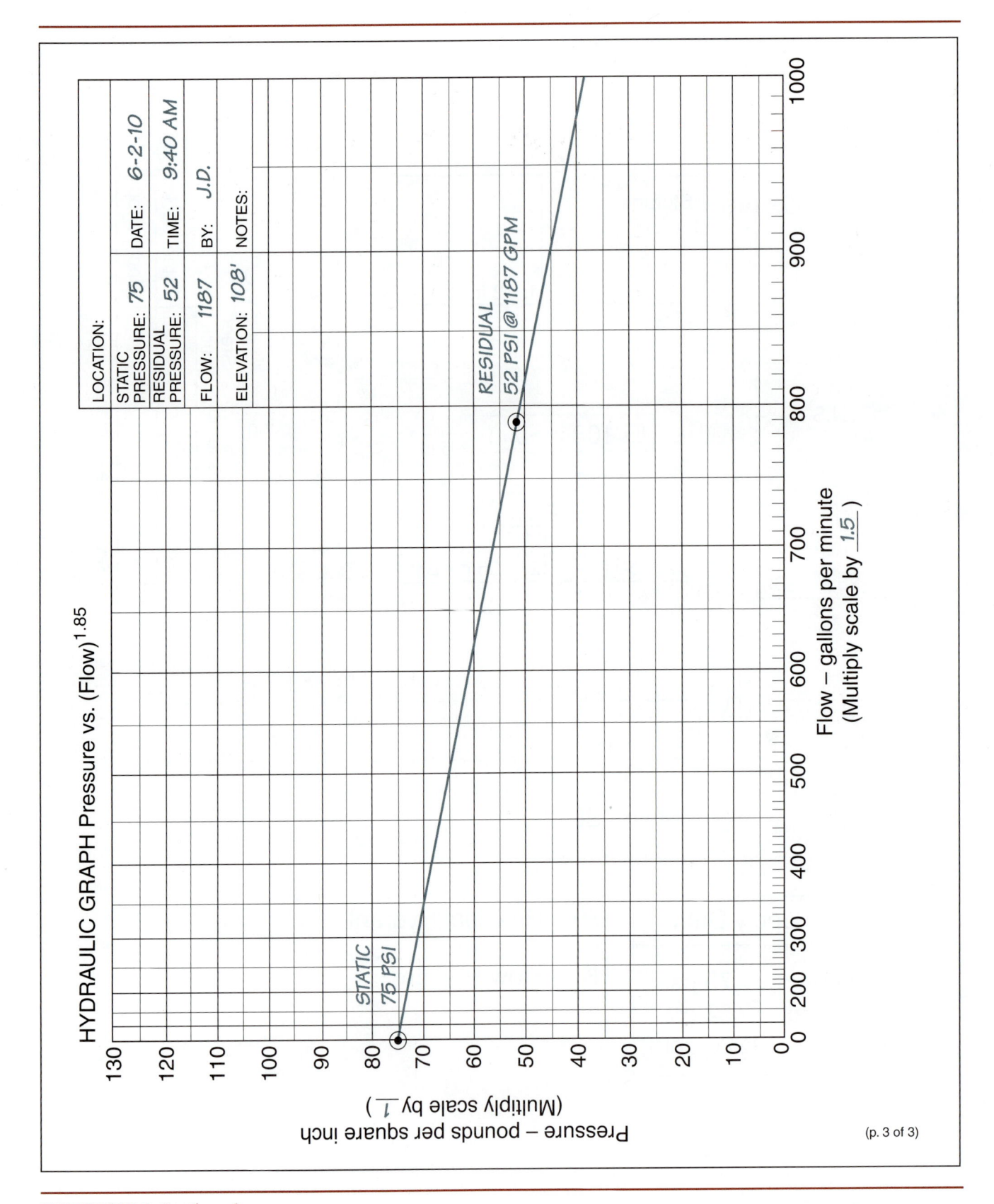

LOCATION:		
STATIC PRESSURE: 75	DATE: 6-2-10	
RESIDUAL PRESSURE: 52	TIME: 9:40 AM	
FLOW: 1187	BY: J.D.	
ELEVATION: 108'	NOTES:	

STATIC 75 PSI

RESIDUAL 52 PSI @ 1187 GPM

Pressure – pounds per square inch (Multiply scale by _1_)

Flow – gallons per minute (Multiply scale by _1.5_)

(p. 3 of 3)

EXHIBIT III.4 Continued.

4.10.1.3 In Equation 4.10.1.2, any units of discharge or pressure drop can be used as long as the same units are used for each value of the same variable.

4.10.1.4 In other words, if Q_R is expressed in gpm, Q_F must be in gpm, and if h_r is expressed in psi, h_f must be expressed in psi.

4.10.1.5 These are the units that are normally used in applying Equation 4.10.1.2 to fire flow test computations.

4.10.2 Discharge Calculations from Table.

4.10.2.1 One means of solving this equation without the use of logarithms is by using Table 4.10.2.1, which gives the values of the 0.54 power of the numbers from 1 to 175.

TABLE 4.10.2.1 Values of h *to the 0.54 Power*

h	$h^{0.54}$	h	$h^{0.54}$	h	$h^{0.54}$	h	$h^{0.54}$	h	$h^{0.54}$
1	1.00	36	6.93	71	9.99	106	12.41	141	14.47
2	1.45	37	7.03	72	10.07	107	12.47	142	14.53
3	1.81	38	7.13	73	10.14	108	12.53	143	14.58
4	2.11	39	7.23	74	10.22	109	12.60	144	14.64
5	2.39	40	7.33	75	10.29	110	12.66	145	14.69
6	2.63	41	7.43	76	10.37	111	12.72	146	14.75
7	2.86	42	7.53	77	10.44	112	12.78	147	14.80
8	3.07	43	7.62	78	10.51	113	12.84	148	14.86
9	3.28	44	7.72	79	10.59	114	12.90	149	14.91
10	3.47	45	7.81	80	10.66	115	12.96	150	14.97
11	3.65	46	7.91	81	10.73	116	13.03	151	15.02
12	3.83	47	8.00	82	10.80	117	13.09	152	15.07
13	4.00	48	8.09	83	10.87	118	13.15	153	15.13
14	4.16	49	8.18	84	10.94	119	13.21	154	15.18
15	4.32	50	8.27	85	11.01	120	13.27	155	15.23
16	4.48	51	8.36	86	11.08	121	13.33	156	15.29
17	4.62	52	8.44	87	11.15	122	13.39	157	15.34
18	4.76	53	8.53	88	11.22	123	13.44	158	15.39
19	4.90	54	8.62	89	11.29	124	13.50	159	15.44
20	5.04	55	8.71	90	11.36	125	13.56	160	15.50
21	5.18	56	8.79	91	11.43	126	13.62	161	15.55
22	5.31	57	8.88	92	11.49	127	13.68	162	15.60
23	5.44	58	8.96	93	11.56	128	13.74	163	15.65
24	5.56	59	9.04	94	11.63	129	13.80	164	15.70
25	5.69	60	9.12	95	11.69	130	13.85	165	15.76
26	5.81	61	9.21	96	11.76	131	13.91	166	15.81
27	5.93	62	9.29	97	11.83	132	13.97	167	15.86
28	6.05	63	9.37	98	11.89	133	14.02	168	15.91
29	6.16	64	9.45	99	11.96	134	14.08	169	15.96
30	6.28	65	9.53	100	12.02	135	14.14	170	16.01
31	6.39	66	9.61	101	12.09	136	14.19	171	16.06
32	6.50	67	9.69	102	12.15	137	14.25	172	16.11
33	6.61	68	9.76	103	12.22	138	14.31	173	16.16
34	6.71	69	9.84	104	12.28	139	14.36	174	16.21
35	6.82	70	9.92	105	12.34	140	14.42	175	16.26

4.10.2.2 If the values of h_f, h_r, and Q_F are known, the values of $h_f^{0.54}$ and $h_r^{0.54}$ can be read from Table 4.10.2.1 and Equation 4.10.1.2 solved for Q_R.

4.10.2.3 Results are usually carried to the nearest 100 gpm (380 L/min) for discharges of 1000 gpm (3800 L/min) or more, and to the nearest 50 gpm (190 L/min) for smaller discharges, which is as close as can be justified by the degree of accuracy of the field observations.

4.10.2.4 The values of $h_r^{0.54}$ and $h_f^{0.54}$ (determined from the table) and the value of Q_F are inserted in Equation 4.10.1.2, and the equation solved for Q_R.

4.11 Data Sheet

4.11.1 The data secured during the testing of hydrants for uniform marking can be valuable for other purposes.

4.11.2 With this in mind, it is suggested that the form shown in Figure 4.11.2 be used to record information that is taken.

4.11.3 The back of the form should include a location sketch.

4.11.4 Results of the flow test should be indicated on a hydraulic graph, such as the one shown in Figure 4.11.4.

4.11.5 When the tests are complete, the forms should be filed for future reference by interested parties.

Hydrant Flow Test Report

Location _____ Date _____
Test made by _____ Time _____
Representative of _____
Witness _____
State purpose of test _____
_____
Consumption rate during test _____
If pumps affect test, indicate pumps operating _____
Flow hydrants: ____A₁____A₂____A₃____A₄____
  Size nozzle   _____
  Pitot reading   _____
  Discharge coefficient _____ Total gpm
  gpm   _____
Static B _____ psi   Residual B _____ psi
Projected results @20 psi Residual ___ gpm; or @ ___ psi Residual ___ gpm
Remarks: _____
_____
_____
_____
Location map: Show line sizes and distance to next cross-connected line. Show valves and hydrant branch size. Indicate north. Show flowing hydrants – Label A₁, A₂, A₃, A₄. Show location of static and residual – Label B.
Indicate B Hydrant _____ Sprinkler _____ Other (identify) _____

FIGURE 4.11.2 *Sample Report of a Hydrant Flow Test.*

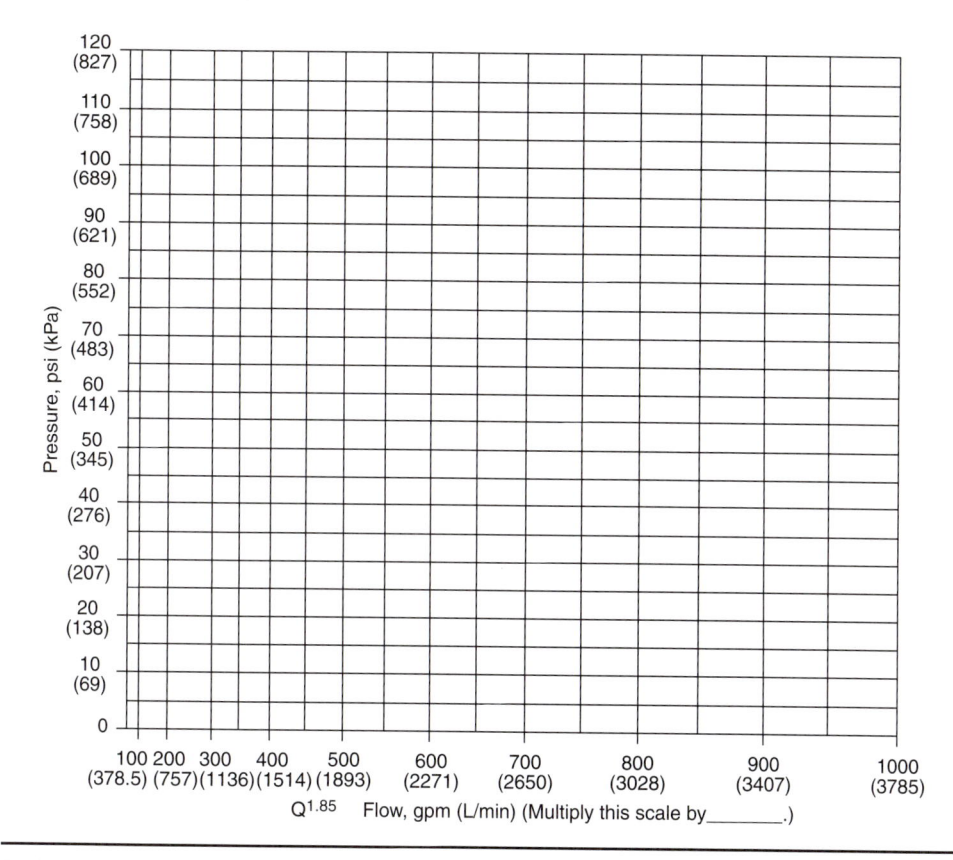

FIGURE 4.11.4 *Sample Graph Sheet.*

Due to changes in the water system, the water available for fire protection at a given location may change over time, so the flow test should be repeated. Waterflow test data should not be used for design purposes if the data are more than 1 year old. NFPA 25, *Standard for the Inspection, Testing, and Maintenance of Water-Based Fire Protection Systems,* requires a flow test of underground piping every 5 years to evaluate the internal condition of the system piping. Therefore, all subsequent testing should be documented and retained for comparison to previous results.

4.12 System Corrections

4.12.1 It must be remembered that flow test results show the strength of the distribution system and do not necessarily indicate the degree of adequacy of the entire water works system.

4.12.2 Consider a system supplied by pumps at one location and having no elevated storage.

4.12.3 If the pressure at the pump station drops during the test, it is an indication that the distribution system is capable of delivering more than the pumps can deliver at their normal operating pressure.

4.12.4 It is necessary to use a value for the drop in pressure for the test that is equal to the actual drop obtained in the field during the test, minus the drop in discharge pressure at the pumping station.

4.12.5 If sufficient pumping capacity is available at the station and the discharge pressure could be maintained by operating additional pumps, the water system as a whole could deliver the computed quantity.

4.12.6 If, however, additional pumping units are not available, the distribution system would be capable of delivering the computed quantity, but the water system as a whole would be limited by the pumping capacity.

4.12.7 The portion of the pressure drop for which a correction can be made for tests on systems with storage is generally estimated upon the basis of a study of all the tests made and the pressure drops observed on the recording gauge at the station for each.

4.12.8 The corrections may vary from very substantial portions of the observed pressure drops for tests near the pumping station, to zero for tests remote from the station.

4.13 Public Hydrant Testing and Flushing

4.13.1* Public fire hydrants should be flow tested every 5 years to verify capacity and marking of the hydrant.

A.4.13.1 When flow test data are needed, such data should not be more than 5 years old since conditions in the piping and system demands can change. It is not the intent of 4.13.1 to require routine 5-year testing of each hydrant if there is no immediate need for flow test data or if test data less than 5 years old are available from an adjacent hydrant on the same grid.

4.13.2 Public fire hydrants should be flushed at least annually to verify operation, address repairs, and verify reliability.

CHAPTER 5 MARKING OF HYDRANTS

5.1 Classification of Hydrants

Hydrants should be classified in accordance with their rated capacities [at 20 psi (1.4 bar) residual pressure or other designated value] as follows:

(1) Class AA — Rated capacity of 1500 gpm (5680 L/min) or greater
(2) Class A — Rated capacity of 1000–1499 gpm (3785–5675 L/min)
(3) Class B — Rated capacity of 500–999 gpm (1900–3780 L/min)
(4) Class C — Rated capacity of less than 500 gpm (1900 L/min)

> In accordance with Section 5.1, the hydrant in Exhibit III.4 should be rated as Class AA, since its capacity is greater than 1500 gpm (5680 L/min).

5.2 Marking of Hydrants

5.2.1 Public Hydrants.

5.2.1.1 All barrels are to be chrome yellow except in cases where another color has already been adopted.

5.2.1.2 The tops and nozzle caps should be painted with the following capacity-indicating color scheme to provide simplicity and consistency with colors used in signal work for safety, danger, and intermediate condition:

(1) Class AA — Light blue
(2) Class A — Green
(3) Class B — Orange
(4) Class C — Red

> In accordance with 5.2.1.2, the bonnet and nozzle caps of the hydrant in Exhibit III.4 should be painted light blue to indicate that the hydrant is rated as a Class AA hydrant.

5.2.1.3 For rapid identification at night, it is recommended that the capacity colors be of a reflective-type paint.

5.2.1.4 Hydrants rated at less than 20 psi (1.4 bar) should have the rated pressure stenciled in black on the hydrant top.

5.2.1.5 In addition to the painted top and nozzle caps, it can be advantageous to stencil the rated capacity of high-volume hydrants on the top.

5.2.1.6 The classification and marking of hydrants provided for in this chapter anticipate determination based on individual flow test.

5.2.1.7 Where a group of hydrants can be used at the time of a fire, some special marking designating group-flow capacity may be desirable.

5.2.1.8 Marking on private hydrants within private enclosures is to be done at the owner's discretion.

5.2.1.9 When private hydrants are located on public streets, they should be painted red or another color to distinguish them from public hydrants.

5.2.2 Permanently Inoperative Hydrants. Fire hydrants that are permanently inoperative or unusable should be removed.

5.2.3 Temporarily Inoperative Hydrants. Fire hydrants that are temporarily inoperative or unusable should be wrapped or otherwise provided with temporary indication of their condition.

5.2.4 Flush Hydrants. Location markers for flush hydrants should carry the same background color as stated above for class indication, with such other data stenciled thereon as deemed necessary.

5.2.5 Private Hydrants.

5.2.5.1 Marking on private hydrants within private enclosures is to be at the owner's discretion.

5.2.5.2 When private hydrants are located on public streets, they should be painted red or some other color to distinguish them from public hydrants.

Summary

Proper fire flow testing of hydrants provides the needed information for the design of any water-based fire protection system including fire pumps. By following the recommended practices of NFPA 291, the data obtained can be used for fire protection system design in addition to the proper color coding of hydrants.

Annex B Informational Referencess

(Reserved)

References Cited in Commentary

National Fire Protection Association, 1 Batterymarch Park, Quincy, MA 02169-7471.
NFPA 13, *Standard for the Installation of Sprinkler Systems,* 2013 edition.
NFPA 14, *Standard for the Installation of Standpipe and Hose Systems,* 2010 edition.
NFPA 20, *Standard for the Installation of Stationary Pumps for Fire Protection,* 2013 edition.
NFPA 25, *Standard for the Inspection, Testing, and Maintenance of Water-Based Fire Protection Systems,* 2011 edition.

Complete Text of NFPA® 24, Standard for the Installation of Private Fire Service Mains and Their Appurtenances, 2013 Edition, with Commentary

SECTION 2

> Because many fire pumps are supplied by a private fire service main, the requirements for the installation of private fire service mains must be well understood in order to properly install a fire pump with the water supply piping. In many cases, the installation of a new fire protection water supply includes the design and installation of a private fire service main and fire pump. For these reasons, NFPA 24, *Standard for the Installation of Private Fire Service Mains and Their Appurtenances,* is included in its entirety in this handbook to assist in the design, installation, and acceptance testing of a complete fire protection water supply.

CHAPTER 1 ADMINISTRATION

1.1 Scope

1.1.1 This standard shall cover the minimum requirements for the installation of private fire service mains and their appurtenances supplying the following:

(1) Automatic sprinkler systems
(2) Open sprinkler systems
(3) Water spray fixed systems
(4) Foam systems
(5) Private hydrants
(6) Monitor nozzles or standpipe systems with reference to water supplies
(7) Hose houses

> The scope of this standard establishes the requirements for the installation of private fire service mains, which distinguishes NFPA 24 from other standards that address public service mains, such as those published by the American Water Works Association (AWWA). An important distinction to remember is that NFPA 24 deals with the underground main once that pipe enters private property and becomes the responsibility of the building owner. See the definition of the term *private fire service main* in 3.3.11 and the accompanying annex text and commentary.

1.1.2 This standard shall apply to combined service mains used to carry water for fire service and other uses.

1.1.3 This standard shall not apply to the following situations:

(1) Mains under the control of a water utility
(2) Mains providing fire protection and/or domestic water that are privately owned but are operated as a water utility

1.1.4 This standard shall not apply to underground mains serving sprinkler systems designed and installed in accordance with NFPA 13R that are under 4 in. (102 mm) in size.

1.1.5 This standard shall not apply to underground mains serving sprinkler systems designed and installed in accordance with NFPA 13D.

> The 2010 edition of NFPA 24 was revised to clarify that the requirements of NFPA 24 do not apply to public mains or to mains owned by private companies that operate as water utilities. The technical committee believes that it is unrealistic and unnecessary to expect some utilities to meet the requirements for listed pipe and 150 psi (10 bar) rated piping. NFPA 13D, *Standard for the Installation of Sprinkler Systems in One- and Two-Family Dwellings and Manufactured Homes*, and NFPA 13R, *Standard for the Installation of Sprinkler Systems in Low-Rise Residential Occupancies*, address underground piping supplying these systems.

1.2 Purpose

The purpose of this standard shall be to provide a reasonable degree of protection for life and property from fire through installation requirements for private fire service main systems based on sound engineering principles, test data, and field experience.

1.3 Retroactivity

The provisions of this standard reflect a consensus of what is necessary to provide an acceptable degree of protection from the hazards addressed in this standard at the time the standard was issued.

1.3.1 Unless otherwise specified, the provisions of this standard shall not apply to facilities, equipment, structures, or installations that existed or were approved for construction or installation prior to the effective date of the standard. Where specified, the provisions of this standard shall be retroactive.

1.3.2 In those cases where the authority having jurisdiction determines that the existing situation presents an unacceptable degree of risk, the authority having jurisdiction shall be permitted to apply retroactively any portions of this standard deemed appropriate.

1.3.3 The retroactive requirements of this standard shall be permitted to be modified if their application clearly would be impractical in the judgment of the authority having jurisdiction and only where it is clearly evident that a reasonable degree of safety is provided.

1.4 Equivalency

Nothing in this standard is intended to prevent the use of systems, methods, or devices of equivalent or superior quality, strength, fire resistance, effectiveness, durability, and safety over those prescribed by this standard. Technical documentation shall be submitted to the authority having jurisdiction to demonstrate equivalency. The system, method, or device shall be approved for the intended purpose by the authority having jurisdiction.

1.5 Units

1.5.1 Metric units of measurement in this standard shall be in accordance with the modernized metric system known as the International System of Units (SI). Liter and bar units are not part of, but are recognized by, SI and are commonly used in international fire protection. These units are shown in Table 1.5.1 with conversion factors.

1.5.2 If a value for measurement as given in this standard is followed by an equivalent value in other units, the first stated is to be regarded as the requirement. A given equivalent value might be approximate.

TABLE 1.5.1 Conversion Table for SI Units

Name of Unit	Unit Symbol	Conversion Factor
Liter	L	1 gal = 3.785 L
Liter per minute per square meter	(L/min)/m^2	1 gpm/ft^2 = (40.746 L/min)/m^2
Cubic decimeter	dm^3	1 gal = 3.785 dm^3
Pascal	Pa	1 psi = 6894.757 Pa
Bar	bar	1 psi = 0.0689 bar
Bar	bar	1 bar = 10^5 Pa

Note: For additional conversions and information, see IEEE/ASTM-SI-10.

1.5.3 SI units have been converted by multiplying the quantity by the conversion factor and then rounding the result to the appropriate number of significant digits.

CHAPTER 2 REFERENCED PUBLICATIONS

2.1 General

The documents or portions thereof listed in this chapter are referenced within this standard and shall be considered part of the requirements of this document.

2.2 NFPA Publications. National Fire Protection Association, 1 Batterymarch Park, Quincy, MA 02169-7471.

NFPA 13, *Standard for the Installation of Sprinkler Systems*, 2013 edition.

NFPA 13D, *Standard for the Installation of Sprinkler Systems in One- and Two-Family Dwellings and Manufactured Homes*, 2013 edition.

NFPA 13R, *Standard for the Installation of Sprinkler Systems in Low-Rise Residential Occupancies*, 2013 edition.

NFPA 20, *Standard for the Installation of Stationary Pumps for Fire Protection*, 2013 edition.

NFPA 22, *Standard for Water Tanks for Private Fire Protection*, 2008 edition.

NFPA 25, *Standard for the Inspection, Testing, and Maintenance of Water-Based Fire Protection Systems*, 2011 edition.

NFPA 780, *Standard for the Installation of Lightning Protection Systems*, 2011 edition.

NFPA 1961, *Standard on Fire Hose*, 2013 edition.

NFPA 1963, *Standard for Fire Hose Connections*, 2009 edition.

2.3 Other Publications

2.3.1 ASME Publications. American Society of Mechanical Engineers, Three Park Avenue, New York, NY 10016-5990.

ASME B1.20.1, *Pipe Threads, General Purpose (Inch)*, 2001.

ASME B16.1, *Gray Iron Pipe Flanges and Flanged Fittings, Classes 25, 125, and 250*, 2005.

ASME B16.3, *Malleable Iron Threaded Fittings, Classes 150 and 300*, 2006.

ASME B16.4, *Gray Iron Threaded Fittings, Classes 125 and 250*, 2006.

ASME B16.5, *Pipe Flanges and Flanged Fittings NPS ½ through 24*, 2003.

ASME B16.9, *Factory-Made Wrought Steel Buttweld Fittings*, 2007.

ASME B16.11, *Forged Steel Fittings, Socket Welded and Threaded*, 2005.

ASME B16.18, *Cast Bronze Solder Joint Pressure Fittings*, 2001.

ASME B16.22, *Wrought Copper and Bronze Solder Joint Pressure Fittings*, 2001.

ASME B16.25, *Buttwelding Ends*, 2007.

2.3.2 ASTM Publications. ASTM International, 100 Barr Harbor Drive, P.O. Box C700, West Conshohocken, PA 19428-2959.

ASTM A 234, *Specification for Piping Fittings of Wrought Carbon Steel and Alloy Steel for Moderate and Elevated Temperatures*, 2007.

ASTM B 75, *Specification for Seamless Copper Tube*, 2002.

ASTM B 88, *Specification for Seamless Copper Water Tube*, 2003.

ASTM B 251, *Requirements for Wrought Seamless Copper and Copper-Alloy Tube*, 2002.

ASTM F 437, *Chlorinated Polyvinyl Chloride (CPVC) Specification for Schedule 80 CPVC Threaded Fittings*, 2006.

ASTM F 438, *Specification for Schedule 40 CPVC Socket-Type Fittings*, 2004.

ASTM F 439, *Specification for Schedule 80 CPVC Socket-Type Fittings*, 2006.

IEEE/ASTM-SI-10, *Standard for Use of the International System of Units (SI): The Modern Metric System*, 2002.

2.3.3 AWWA Publications. American Water Works Association, 6666 West Quincy Avenue, Denver, CO 80235.

AWWA C104, *Cement Mortar Lining for Ductile Iron Pipe and Fittings for Water*, 2008.

AWWA C105, *Polyethylene Encasement for Ductile Iron Pipe Systems*, 2005.

AWWA C110, *Ductile Iron and Gray Iron Fittings*, 2008.

AWWA C111, *Rubber-Gasket Joints for Ductile Iron Pressure Pipe and Fittings*, 2000.

AWWA C115, *Flanged Ductile Iron Pipe with Ductile Iron or Gray Iron Threaded Flanges*, 2005.

AWWA C116, *Protective Fusion-Bonded Epoxy Coatings for the Interior and Exterior Surfaces of Ductile-Iron and Gray-Iron Fittings for Water Supply Service*, 2003.

AWWA C150, *Thickness Design of Ductile Iron Pipe*, 2008.

AWWA C151, *Ductile Iron Pipe, Centrifugally Cast for Water*, 2002.

AWWA C153, *Ductile-Iron Compact Fittings for Water Service*, 2006.

AWWA C200, *Steel Water Pipe 6 in. and Larger*, 2005.

AWWA C203, *Coal-Tar Protective Coatings and Linings for Steel Water Pipelines Enamel and Tape — Hot Applied*, 2002.

AWWA C205, *Cement-Mortar Protective Lining and Coating for Steel Water Pipe 4 in. and Larger — Shop Applied*, 2007.

AWWA C206, *Field Welding of Steel Water Pipe*, 2003.

AWWA C207, *Steel Pipe Flanges for Waterworks Service — Sizes 4 in. Through 144 in.*, 2007.

AWWA C208, *Dimensions for Fabricated Steel Water Pipe Fittings*, 2007.

AWWA C300, *Reinforced Concrete Pressure Pipe, Steel-Cylinder Type*, 2004.

AWWA C301, *Prestressed Concrete Pressure Pipe, Steel-Cylinder Type*, 2007.

AWWA C302, *Reinforced Concrete Pressure Pipe, Non-Cylinder Type*, 2004.

AWWA C303, *Reinforced Concrete Pressure Pipe, Steel-Cylinder Type, Pretensioned*, 2002.

AWWA C400, *Standard for Asbestos-Cement Distribution Pipe, 4 in. Through 16 in. (100 mm through 400 mm), for Water Distribution Systems*, 2003.

AWWA C401, *Standard for the Selection of Asbestos-Cement Pressure Pipe 4 in. through 16 in. (100 mm through 400 mm)*, 2003.

AWWA C600, *Standard for the Installation of Ductile Iron Water Mains and Their Appurtenances*, 2005.

AWWA C602, *Cement-Mortar Lining of Water Pipe Lines 4 in. and Larger — in Place*, 2006.

AWWA C603, *Standard for the Installation of Asbestos-Cement Pressure Pipe*, 2005.

AWWA C900, *Polyvinyl Chloride (PVC) Pressure Pipe, 4 in. Through 12 in., for Water Distribution*, 2007.

AWWA C905, *AWWA Standard for Polyvinyl Chloride (PVC) Pressure Pipe and Fabricated Fittings, 14 in. Through 48 in. (350 mm Through 1,200 mm)*, 2010.

AWWA C906, *Polyethylene (PE) Pressure Pipe and Fittings, 4 in. (100 mm) Through 63 in. (1575 mm) for Water Distribution*, 2007.

AWWA M11, *A Guide for Steel Pipe Design and Installation, 4th edition*, 2004.

2.3.4 Other Publications.

Merriam-Webster's Collegiate Dictionary, 11th edition, Merriam-Webster, Inc., Springfield, MA, 2003.

2.4 References for Extracts in Mandatory Sections

NFPA 20, *Standard for the Installation of Stationary Pumps for Fire Protection*, 2013 edition.

CHAPTER 3 DEFINITIONS

3.1 General

The definitions contained in this chapter shall apply to the terms used in this standard. Where terms are not defined in this chapter or within another chapter, they shall be defined using their ordinarily accepted meanings within the context in which they are used. *Merriam-Webster's Collegiate Dictionary*, 11th edition, shall be the source for the ordinarily accepted meaning.

3.2 NFPA Official Definitions

3.2.1* Approved. Acceptable to the authority having jurisdiction.

A.3.2.1 Approved. The National Fire Protection Association does not approve, inspect, or certify any installations, procedures, equipment, or materials; nor does it approve or evaluate testing laboratories. In determining the acceptability of installations, procedures, equipment, or materials, the authority having jurisdiction may base acceptance on compliance with NFPA or other appropriate standards. In the absence of such standards, said authority may require evidence of proper installation, procedure, or use. The authority having jurisdiction may also refer to the listings or labeling practices of an organization that is concerned with product evaluations and is thus in a position to determine compliance with appropriate standards for the current production of listed items.

3.2.2* Authority Having Jurisdiction (AHJ). An organization, office, or individual responsible for enforcing the requirements of a code or standard, or for approving equipment, materials, an installation, or a procedure.

A.3.2.2 Authority Having Jurisdiction (AHJ). The phrase "authority having jurisdiction," or its acronym AHJ, is used in NFPA documents in a broad manner, since jurisdictions and approval agencies vary, as do their responsibilities. Where public safety is primary, the authority having jurisdiction may be a federal, state, local, or other regional department or individual such as a fire chief; fire marshal; chief of a fire prevention bureau, labor department, or health department; building official; electrical inspector; or others having statutory authority. For insurance purposes, an insurance inspection department, rating bureau, or other insurance company representative may be the authority having jurisdiction. In many circumstances, the property owner or his or her designated agent assumes the role of the authority having jurisdiction; at government installations, the commanding officer or departmental official may be the authority having jurisdiction.

3.2.3 Labeled. Equipment or materials to which has been attached a label, symbol, or other identifying mark of an organization that is acceptable to the authority having jurisdiction and concerned with product evaluation, that maintains periodic inspection of production of labeled equipment or materials, and by whose labeling the manufacturer indicates compliance with appropriate standards or performance in a specified manner.

3.2.4* Listed. Equipment, materials, or services included in a list published by an organization that is acceptable to the authority having jurisdiction and concerned with evaluation of products or services, that maintains periodic inspection of production of listed equipment or materials or periodic evaluation of services, and whose listing states that either the equipment, material, or service meets appropriate designated standards or has been tested and found suitable for a specified purpose.

A.3.2.4 Listed. The means for identifying listed equipment may vary for each organization concerned with product evaluation; some organizations do not recognize equipment as listed unless it is also labeled. The authority having jurisdiction should utilize the system employed by the listing organization to identify a listed product.

3.2.5 Shall. Indicates a mandatory requirement.

3.2.6 Should. Indicates a recommendation or that which is advised but not required.

3.2.7 Standard. A document, the main text of which contains only mandatory provisions using the word "shall" to indicate requirements and which is in a form generally suitable for mandatory reference by another standard or code or for adoption into law. Nonmandatory provisions are not to be considered a part of the requirements of a standard and shall be located in an appendix, annex, footnote, informational note, or other means as permitted in the *Manual of Style for NFPA Technical Committee Documents*.

3.3 General Definitions

3.3.1 Appurtenance. An accessory or attachment that enables the private fire service main to perform its intended function.

> As used in NFPA 24, an appurtenance can be a hydrant, valve, monitor nozzle (for master streams), or a fire department connection (FDC).

3.3.2 Corrosion-Resistant Piping. Piping that has the property of being able to withstand deterioration of its surface or its properties when exposed to its environment.

> Galvanized steel piping is corrosion resistant. However, this standard restricts the use of galvanized steel piping to FDCs. In constantly wet environments, such as those that exist in many underground installations, galvanized pipe can corrode and, therefore, is inappropriate for use. Piping intended for underground installation, such as polyvinyl chloride (PVC), high density polyethylene (HDPE), and fiber-reinforced epoxy pipe, is considered corrosion resistant.

3.3.3 Corrosion-Retardant Material. A lining or coating material that when applied to piping or appurtenances has the property of reducing or slowing the deterioration of the object's surface or properties when exposed to its environment.

> Ferrous piping used for underground applications is frequently coated or wrapped with corrosion-resistant materials such as cement mortar lining, paint, epoxy, asphalt, or bituminous coatings. The external coatings can increase the survivability of the piping when buried. Some soil conditions can create a corrosive environment that can exceed the corrosion resistance of these coatings, and alternative materials, such as PVC or HDPE, should be considered.

3.3.4 Fire Department Connection. A connection through which the fire department can pump supplemental water into the sprinkler system, standpipe, or other system, furnishing water for fire extinguishment to supplement existing water supplies.

3.3.5 Fire Pump. A pump that is a provider of liquid flow and pressure dedicated to fire protection. [20, 2010]

3.3.6 Hose House. An enclosure located over or adjacent to a hydrant or other water supply designed to contain the necessary hose nozzles, hose wrenches, gaskets, and spanners to be used in fire fighting in conjunction with and to provide aid to the local fire department.

3.3.7 Hydrant Butt. The hose connection outlet of a hydrant.

3.3.8 Hydraulically Calculated Water Demand Flow Rate. The waterflow rate for a system or hose stream that has been calculated using accepted engineering practices.

3.3.9 Pressure.

3.3.9.1 Residual Pressure. The pressure that exists in the distribution system, measured at the residual hydrant at the time the flow readings are taken at the flow hydrants.

> The residual pressure is measured when the flow hydrant is open and discharging through at least one outlet. This pressure is measured at the pressure or test hydrant when fire flow is occurring. See Exhibit III.5.

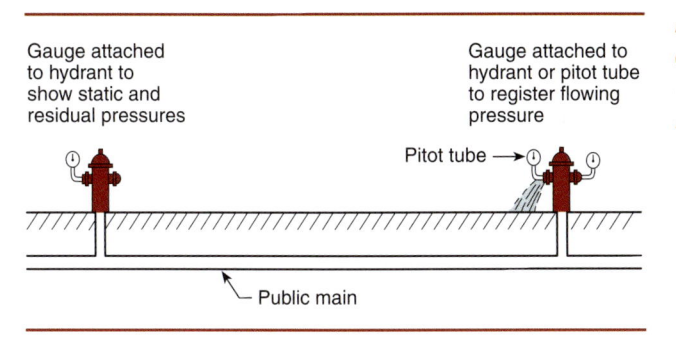

EXHIBIT III.5 *Method of Conducting Flow Tests. (Source: NFPA 13, 2013, Figure A.24.2.2)*

3.3.9.2 Static Pressure. The pressure that exists at a given point under normal distribution system conditions measured at the residual hydrant with no hydrants flowing.

> Static pressure is the pressure measured at the pressure or test hydrant when no flow is occurring.

3.3.10* Pressure-Regulating Device. A device designed for the purpose of reducing, regulating, controlling, or restricting water pressure.

> Pressure regulating devices are valves normally used only when pressures in the system exceed those for which the system components are rated [usually 175 psi (12.1 bar)]. The exposure of a private fire service main to such high pressures is unusual, unless the private fire service main is located on the discharge side of a fire pump.

A.3.3.10 Pressure-Regulating Device. Examples include pressure-reducing valves, pressure-control valves, and pressure-restricting devices.

3.3.11* Private Fire Service Main. Private fire service main, as used in this standard, is that pipe and its appurtenances on private property (1) between a source of water and the base of the system riser for water-based fire protection systems, (2) between a source of water and inlets to foam-making systems, (3) between a source of water and the base elbow of private hydrants or monitor nozzles, and (4) used as fire pump suction and discharge piping, (5) beginning at the inlet side of the check valve on a gravity or pressure tank.

A.3.3.11 Private Fire Service Main. See Figure A.3.3.11.

> Figure A.3.3.11 shows a typical private fire service main arrangement that is fed from a public water supply. As shown in the figure, the piping becomes the property of the building owner at the point where the pipe crosses from public to private land. The transition from public to private has implications for acceptance testing, because public mains and accessories are tested to a different standard and a different hydrostatic pressure. This difference in testing standards can create problems in the field when the private main needs to be tested and isolated from the public main. The two mains must be separated by a valve that holds the higher pressure, which is sometimes difficult, because the isolation valve on the public main may not be rated for the test pressure.
>
> Also, in Figure A.3.3.11, note that a fire pump is supplied by a private fire service main. The requirements in NFPA 24 for pipe, fittings, and valves are frequently used to design and install a water supply that serves a fire pump. NFPA 24 requirements are used for a variety of piping arrangements, including the following:
>
> - Direct connection from a public supply to a fire pump, as illustrated in Figure A.3.3.11
> - Piping between a private water tank and a fire pump
> - Piping from a fire pump to one or more buildings in a multi-building complex

3.3.12 Pumper Outlet. The hydrant outlet intended for use by fire departments for taking supply from the hydrant for pumpers.

3.3.13 Rated Capacity. The flow available from a hydrant at the designated residual pressure (rated pressure) either measured or calculated.

> The rated capacity and color code of a hydrant, as covered in Chapter 5 of NFPA 291, *Recommended Practice for Fire Flow Testing and Marking of Hydrants,* are based on the flow available at a residual pressure of 20 psi (1.4 bar). While not restricted in NFPA 24, some jurisdictions

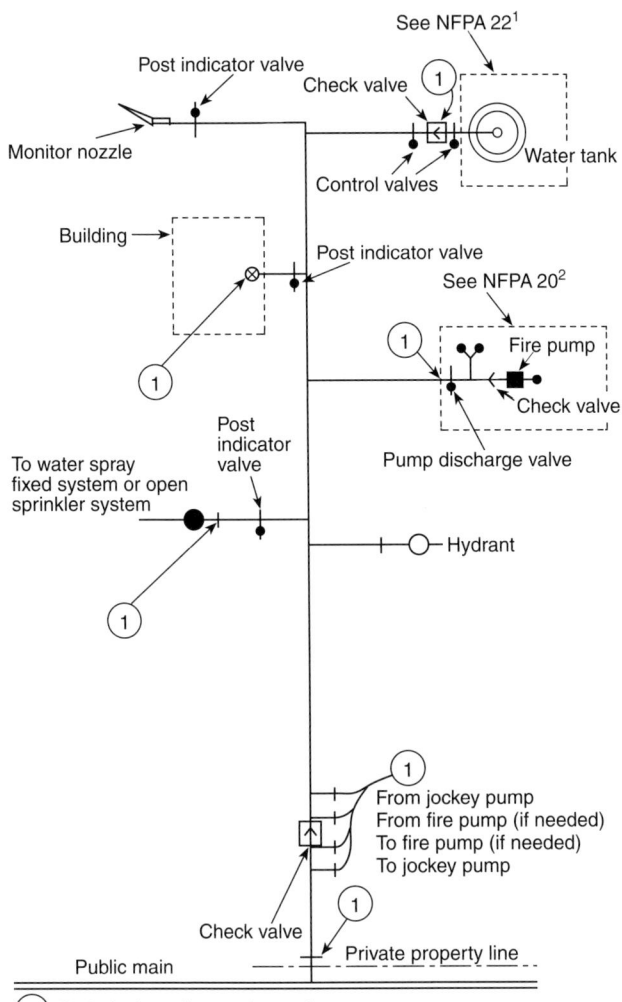

FIGURE A.3.3.11 *Typical Private Fire Service Main.*

restrict the use of pressures below 20 psi (1.4 bar), thus rating the hydrant at that pressure. By restricting the residual pressure to 20 psi (1.4 bar), problems such as drawing a vacuum on the main and inducing a backflow can be avoided. In some cases, a vacuum can cause the main to collapse, thus damaging the pipe.

3.3.14 Test.

3.3.14.1 Flow Test. A test performed by the flow and measurement of water from one hydrant and the static and residual pressures from an adjacent hydrant for the purpose of determining the available water supply at that location.

When water supply flow test data are supplied, consideration should be given to the occurrence of changes in the water system since the test was last conducted, as such changes could affect the available waterflow and pressure. Examples of such changes are modifications to the distribution system or increased development in the general area, which would place a higher demand on the system. Evidence of rapid deterioration of the system due to tuberculation, microbiologically influenced corrosion (MIC), or other potential obstruction sources is another condition that could impact the test data. NFPA 13, *Standard for the Installation of Sprinkler Systems*; NFPA 14, *Standard for the Installation of Standpipe and Hose Systems*; and NFPA 20, *Standard for the Installation of Stationary Pumps for Fire Protection*, require that the flow test data be no older than 1 year for the purposes of fire protection system design.

3.3.14.2 Flushing Test. A test of a piping system using high velocity flows to remove debris from the piping system prior to it being placed in service.

Dirt, rocks, and other obstructing material can enter piping during installation, creating a potential obstruction for sprinklers and valves. For this reason, the flushing test is intended to be conducted prior to connecting the underground system to any aboveground system.

3.3.14.3 Hydrostatic Test. A test of a closed piping system and its attached appurtenances consisting of subjecting the piping to an increased internal pressure for a specified period of duration to verify system integrity and leak rates.

The hydrostatic test is a "stress" test for the system to reveal leaks prior to placing the system in service.

3.3.15 Valve.

3.3.15.1 Check Valve. A valve that allows flow in one direction only.

3.3.15.2 Indicating Valve. A valve that has components that show if the valve is open or closed. Examples are outside screw and yoke (OS&Y) gate valves and underground gate valves with indicator posts.

All NFPA installation standards for fire protection systems require that all valves on fire protection systems be indicating valves. Because the leading cause of sprinkler system failure is due to a shut valve (over 35 percent according to the NFPA report, "U.S. Experience with Sprinklers and Other Automatic Fire Extinguishing Equipment"), indicating valves are required for all systems so that an inspector can easily determine whether the valve is open or closed.

3.4 Hydrant Definitions

3.4.1 Hydrant. An exterior valved connection to a water supply system that provides hose connections.

3.4.1.1* Dry Barrel Hydrant. This is the most common type of hydrant; it has a control valve below the frost line between the footpiece and the barrel.

A dry barrel hydrant is used in climates where freezing weather occurs. As the name suggests, the barrel or body of the hydrant does not contain water. When the operating nut on top of the hydrant is used to open the hydrant, the barrel is flooded with water. A small drain is located at the bottom of the hydrant, and, once the hydrant is closed, it should drain completely of water within 60 minutes.

A.3.4.1.1 Dry Barrel Hydrant. A drain is located at the bottom of the barrel above the control valve seat for proper drainage after operation.

3.4.1.2 Flow Hydrant. The hydrant that is used for the flow and flow measurement of water during a flow test.

3.4.1.3* Private Fire Hydrant. A valved connection on a water supply system having one or more outlets and that is used to supply hose and fire department pumpers with water on private property.

> A private hydrant may or may not have a 4 in. (100 mm) connection. For most installations, it is not desirable to have a 4 in. (100 mm) connection on a private hydrant, since a large waterflow drawn from hydrants located near a building could have a negative impact on the performance of the building fire protection systems.

A.3.4.1.3 Private Fire Hydrant. Where connected to a public water system, the private hydrants are supplied by a private service main that begins at the point of service designated by the authority having jurisdiction, usually at a manually operated valve near the property line.

3.4.1.4 Public Hydrant. A valved connection on a water supply system having one or more outlets and that is used to supply hose and fire department pumpers with water.

> Public hydrants are normally found on sidewalks or on the side of public streets or roads. These hydrants can be color coded to indicate their flow capacity at 20 psi (1.4 bar) residual pressure. See Exhibit III.6 for a typical dry barrel hydrant.

3.4.1.5 Residual Hydrant. The hydrant that is used for measuring static and residual pressures during a flow test.

3.4.1.6 Wet Barrel Hydrant. A type of hydrant that sometimes is used where there is no danger of freezing weather. Each outlet on a wet barrel hydrant is provided with a valved outlet threaded for fire hose.

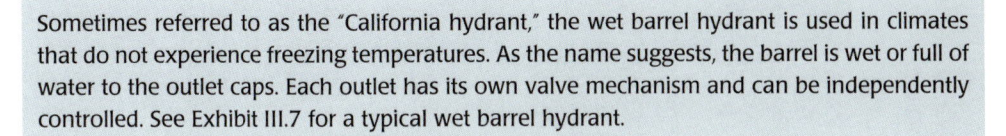

> Sometimes referred to as the "California hydrant," the wet barrel hydrant is used in climates that do not experience freezing temperatures. As the name suggests, the barrel is wet or full of water to the outlet caps. Each outlet has its own valve mechanism and can be independently controlled. See Exhibit III.7 for a typical wet barrel hydrant.

EXHIBIT III.6 Dry Barrel Hydrant.

CHAPTER 4 GENERAL REQUIREMENTS

4.1* Plans

A.4.1 Underground mains should be designed so that the system can be extended with a minimum of expense. Possible future plant expansion should also be considered and the piping designed so that it is not covered by buildings.

> When designing the piping layout, tees with blind flanges and isolation valves should be considered to accommodate future expansion. An increase in the pipe diameter should be considered in order to permit additional waterflow from future expansions.

4.1.1 Working plans shall be submitted for approval to the authority having jurisdiction before any equipment is installed or remodeled.

EXHIBIT III.7 Wet Barrel Hydrant.

> By submitting plans before installation, the designer and installing contractor avoid making later modifications to installed equipment based on the review by the authority having jurisdiction (AHJ). A permit to install such equipment is required in most jurisdictions.

4.1.2 Deviation from approved plans shall require permission of the authority having jurisdiction.

4.1.3 Working plans shall be drawn to an indicated scale on sheets of uniform size, with a plan of each floor as applicable, and shall include the following items that pertain to the design of the system:

(1) Name of owner
(2) Location, including street address
(3) Point of compass
(4) A graphic representation of the scale used on all plans
(5) Name and address of contractor
(6) Size and location of all water supplies
(7) Size and location of standpipe risers, hose outlets, hand hose, monitor nozzles, and related equipment
(8) The following items that pertain to private fire service mains:
   (a) Size
   (b) Length
   (c) Location
   (d) Weight
   (e) Material
   (f) Point of connection to city main
   (g) Sizes, types, and locations of valves, valve indicators, regulators, meters, and valve pits
   (h) Depth at which the top of the pipe is laid below grade
   (i) Method of restraint
(9) The following items that pertain to hydrants:
   (a) Size and location, including size and number of outlets and whether outlets are to be equipped with independent gate valves
   (b) Thread size and coupling adapter specifications if different from NFPA 1963
   (c) Whether hose houses and equipment are to be provided, and by whom
   (d) Static and residual hydrants used in flow
   (e) Method of restraint
(10) Size, location, and piping arrangement of fire department connections

4.1.4 The working plan submittal shall include the manufacturer's installation instructions for any specially listed equipment, including descriptions, applications, and limitations for any devices, piping, or fittings.

4.2 Installation Work

4.2.1 Installation work shall be performed by fully experienced and responsible persons.

4.2.2 The authority having jurisdiction shall always be consulted before the installation or remodeling of private fire service mains.

CHAPTER 5 WATER SUPPLIES

5.1* Connection to Waterworks Systems

A.5.1 If possible, dead-end mains should be avoided by arranging for mains to be supplied from both directions. Where private fire service mains are connected to dead-end public mains, each situation should be examined to determine if it is practical to request the water utility to loop the mains to obtain a more reliable supply.

5.1.1 A connection to a reliable waterworks system shall be an acceptable water supply source.

5.1.2* The volume and pressure of a public water supply shall be determined from water-flow test data or other approved method.

> Alternative methods of determining pressure at a public water supply could include calculations that modify the effective point of a nearby flow test or the use of modeling. Any alternative methods are required to be approved by the AHJ.

◀ **DESIGN ALERT**

A.5.1.2 An adjustment to the waterflow test data to account for the following should be made, as appropriate:

(1) Daily and seasonal fluctuations
(2) Possible interruption by flood or ice conditions
(3) Large simultaneous industrial use
(4) Future demand on the water supply system
(5) Other conditions that could affect the water supply

> In the 2013 edition of NFPA 24, the waterflow adjustments have been moved from the body of the document to Annex A for consistency with NFPA 13. Water supplies should be evaluated for all of the conditions listed in A.5.1.2. The water supply test should be conducted during normal business hours to include the effects of simultaneous business and industrial demand. The possibility of future expansion should be discussed with the AHJ and accounted for in the system demand, since expansion can be expected to create additional demand on the water supply.

5.2 Size of Fire Mains

5.2.1 Private Fire Service Mains. Pipe smaller than 6 in. (152 mm) in diameter shall not be installed as a private service main supplying hydrants.

> Pipes smaller than 6 in. (152 mm) do not permit the full flow capabilities for the hydrant.

5.2.2 Mains Not Supplying Hydrants. For mains that do not supply hydrants, sizes smaller than 6 in. (152 mm) shall be permitted to be used subject to the following restrictions:

(1) The main shall supply only the following types of systems:
    (a) Automatic sprinkler systems
    (b) Open sprinkler systems
    (c) Water spray fixed systems
    (d) Foam systems
    (e) Standpipe systems

(2) Hydraulic calculations shall show that the main is able to supply the total demand at the appropriate pressure.
(3) Systems that are not hydraulically calculated shall have a main at least as large as the riser.

> When private fire service mains supply water to a fire pump, the pipe size should be in accordance with Table 4.26(a) and Table 4.26(b) in NFPA 20 (see Part II of this handbook). The intent of pipe sizing in the aforementioned tables is to limit water velocity to not more than 15 ft/sec (4.57 m/sec).
>
> Private fire service main piping supplying standpipe systems should be sized in accordance with NFPA 14.

5.3 Pressure-Regulating Devices and Meters

5.3.1 No pressure-regulating valve shall be used in the water supply, except by special permission of the authority having jurisdiction.

DESIGN ALERT ▶

> Subsection 5.3.1 is intended to discourage the use of high pressures in a water supply piping system and permits the use of a pressure regulating device only with the approval of the AHJ. See Exhibit III.8 for an example of a pressure reducing valve.
>
> The piping and fittings for private fire service mains are required by 10.1.5 to be rated for a pressure of at least 150 psi (10 bar). Pressures in excess of this value in a private fire service main are unusual and should be avoided by the system designer. The 150 psi (10 bar) pressure limitation is allowed to be exceeded when a fire pump is supplying pressure to private fire service mains. In such a case, pipe and fittings rated for the higher pressure are necessary.

5.3.2 Where meters are required by other authorities, they shall be listed.

EXHIBIT III.8 *Pressure Reducing Valve.*

5.4* Connection from Waterworks Systems

A.5.4 Where connections are made from public waterworks systems, such systems should be guarded against possible contamination as follows *(see AWWA M14 or local plumbing code)*:

(1) For private fire service mains with direct connections from public waterworks mains only or with fire pumps installed in the connections from the street mains, no tanks or reservoirs, no physical connection from other water supplies, no antifreeze or other additives of any kind, and with all drains discharging to atmosphere, dry well, or other safe outlets, an approved double check valve assembly is recommended.

(2) For private fire service mains with direct connections from the public water supply main plus one or more elevated storage tanks or fire pumps taking suction from aboveground covered reservoirs or tanks (all storage facilities are filled or connected to public water only, and the water in the tanks is to be maintained in a potable condition), an approved double check valve assembly is recommended.

(3) For private fire service mains directly supplied from public mains with an auxiliary water supply, such as a pond or river on or available to the premises and dedicated to fire department use; or for systems supplied from public mains and interconnected with auxiliary supplies, such as pumps taking suction from reservoirs exposed to contamination or rivers and ponds; driven wells, mills, or other industrial water systems; or for systems or portions of systems where antifreeze or other solutions are used, an approved reduced-pressure zone-type backflow preventer is recommended.

5.4.1 The requirements of the public health authority having jurisdiction shall be determined and followed.

5.4.2 Where equipment is installed to guard against possible contamination of the public water system, such equipment and devices shall be listed for fire protection service.

The situations that follow illustrate the need for a backflow prevention device.

A backflow prevention device (double-check or reduced pressure type) is intended to protect the public water supply at the water user connection for the following situations:

1. The public water system serves water user premises where the fire protection system is directly supplied from the public water system and an auxiliary non-potable water supply is on the premises (tank) or is accessible to the premises (open reservoir) that is not part of the public water system.
2. The public water system serves water user premises where the fire protection system is supplied from the public water system and where either elevated, non-potable water storage tanks, or fire pumps that take suction from private reservoirs or tanks, are on the user's premises.

In situation 1, when the FDC is used, it is possible that a higher pressure exists in the system than is being supplied from the public water system through the FDC, and a backflow situation can occur. This can result in the fire protection water in the system flowing into the public water supply. Since fire protection systems are not sanitized for human consumption, protection against backflow conditions by means of a backflow preventer is often required. Exhibit III.9 illustrates the correct location for the installation of a backflow preventer.

In situation 2, either an elevated tank or a fire pump can pressurize a fire protection system so that a higher pressure exists in the system than in the water supply. Again, this case also creates the potential for a backflow and subsequent contamination of the public water supply.

EXHIBIT III.9 Typical Layout of a Water Supply System with Backflow Preventer for Automatic Sprinkler Systems. (Source: NFPA 13E, 2010, Figure A.4.2)

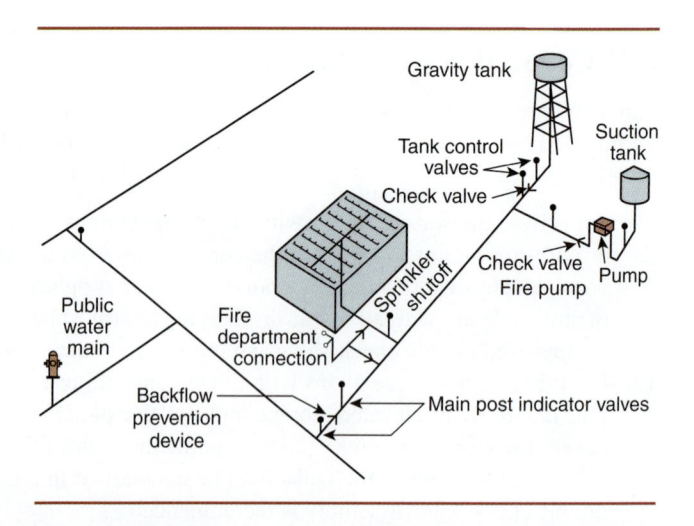

5.5 Connections to Public Water Systems

Connections to public water systems shall be arranged to be isolated by one of the methods permitted in 6.2.11.

> The 2010 edition of NFPA 24 was revised to recognize that there are many arrangements that satisfy the performance requirement of providing a safe way to turn off water to public water systems. See 6.2.11 for these methods.

5.6* Pumps

A single, automatically controlled fire pump installed in accordance with NFPA 20 shall be an acceptable water supply source.

A.5.6 A fire pump installation consisting of pump, driver, and suction supply, when of adequate capacity and reliability and properly located, makes a good supply. An automatically controlled fire pump taking water from a water main of adequate capacity, or taking draft under a head from a reliable storage of adequate capacity, is permitted to be, under certain conditions, accepted by the authority having jurisdiction as a single supply.

5.7 Tanks

Tanks shall be installed in accordance with NFPA 22.

5.8 Penstocks, Flumes, Rivers, Lakes, or Reservoirs

Water supply connections from penstocks, flumes, rivers, lakes, or reservoirs shall be arranged to avoid mud and sediment and shall be provided with approved, double, removable screens or approved strainers installed in an approved manner.

> A proper evaluation should be conducted before selecting a natural body of water to supply a fire protection system. Screens and strainers may remove larger obstructing material but do not remove mud or sediment. Fire protection strainers typically have ⅛ in. (3 mm) mesh size openings in the strainer basket, which are too large to remove fine particulate matter. See

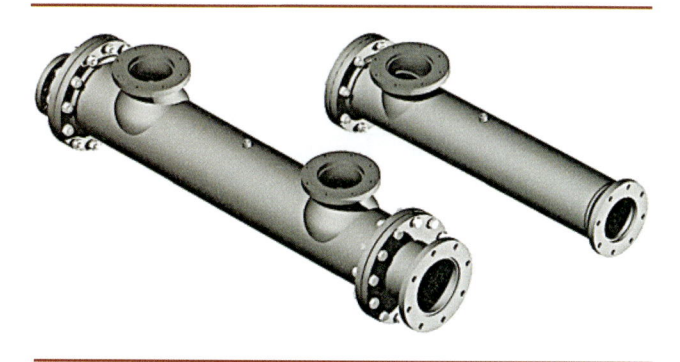

EXHIBIT III.10 *Fire Protection Strainer.*

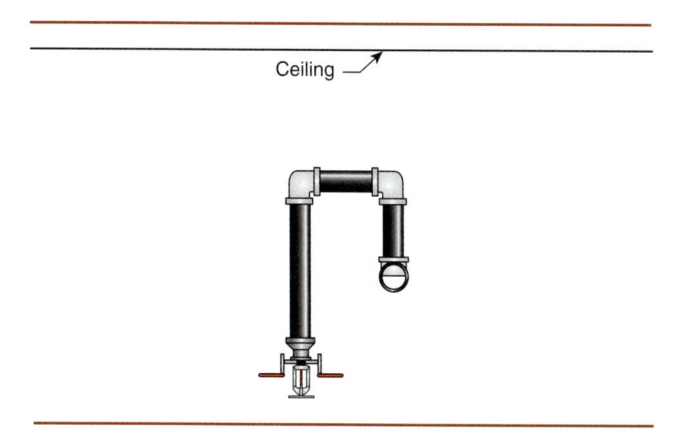

EXHIBIT III.11 *Return Bend Arrangement. (Courtesy of Stephan Laforest, Summit Sprinkler Design Systems, Inc.)*

> Exhibit III.10 for an illustration of a typical fire protection strainer. Sprinkler systems that are supplied from natural sources of water must be installed with return bends or goosenecks to prevent the settlement of mud or other sediment in the bottom of the sprinkler drop, which obstructs the sprinkler orifice. See Exhibit III.11 for an illustration of a return bend.

5.9* Fire Department Connections

A.5.9 The fire department connection should be located not less than 18 in. (457 mm) and not more than 4 ft (1.2 m) above the level of the adjacent grade or access level. Typical fire department connections are shown in Figure A.5.9(a) and Figure A.5.9(b). Where a hydrant is not available, other water supply sources such as a natural body of water, a tank, or a reservoir should be utilized. The water authority should be consulted when a nonpotable water supply is proposed as a suction source for the fire department.

5.9.1 General. Where the authority having jurisdiction requires a remote fire department connection, for systems requiring one by another standard, a fire department connection shall be provided as described in Section 5.9.

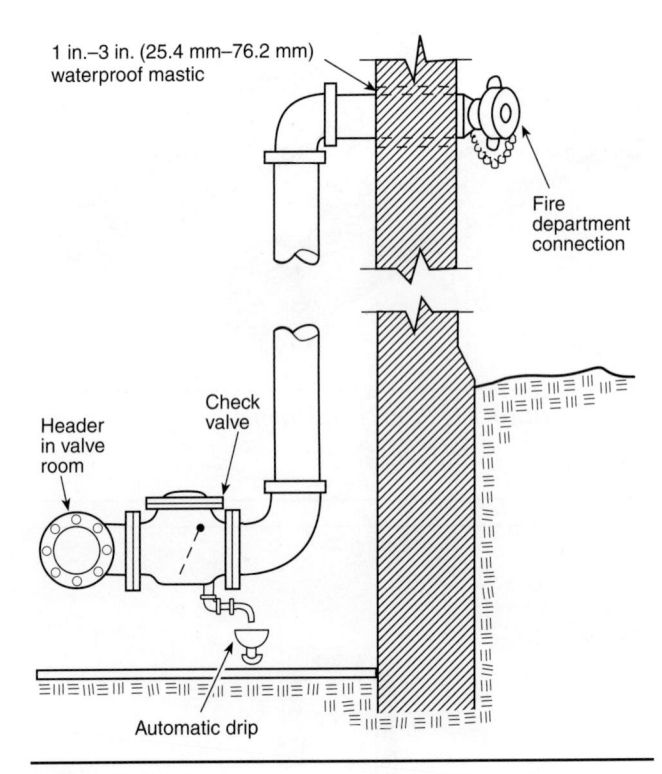

FIGURE A.5.9(a) *Typical Fire Department Connection.*

> The FDC may serve more than one building. The underground piping, as illustrated in Figure A.3.3.11, can be pressurized through the FDC, and, therefore, some guidelines on the design and installation of the FDC are provided in Section 5.9. See Exhibit III.12 for an illustration of a freestanding FDC supplying an underground private fire service main.

5.9.1.1 Fire department connections shall not be required where approved by the authority having jurisdiction.

> The AHJ in this case is typically the local fire department. Since the local fire department is the end user, the fire marshal or the equivalent authority should be consulted for the location and number of FDCs needed.

5.9.1.2 Fire department connections shall be supported.

5.9.1.3 Fire department connections shall be of an approved type.

5.9.1.4 Fire department connections shall be equipped with listed plugs or caps that are secured and arranged for easy removal by fire departments.

5.9.1.5 Fire department connections shall be protected where subject to mechanical damage.

5.9.2 Couplings.

5.9.2.1 The fire department connection(s) shall use an NH internal threaded swivel fitting(s) with an NH standard thread(s).

5.9.2.2 At least one of the connections shall be the 2.5 to 7.5 NH standard thread specified in NFPA 1963.

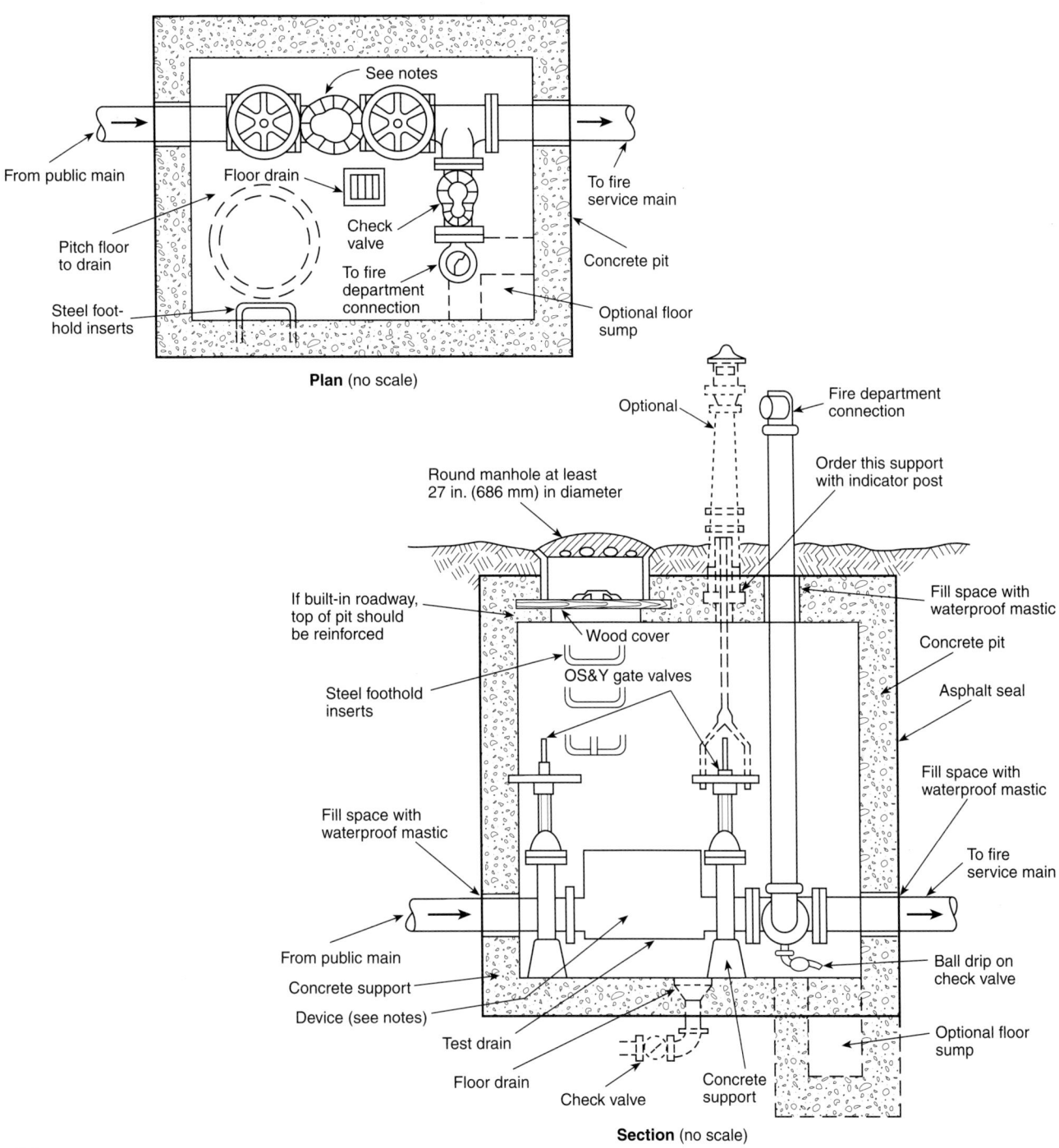

Section 5.9 • Fire Department Connections   **465**

Notes:
1. Various backflow prevention regulations accept different devices at the connection between public water mains and private fire service mains.
2. The device shown in the pit could be any or a combination of the following:
   (a) Gravity check valve
   (b) Detector check valve
   (c) Double check valve assembly
   (d) Reduced-pressure zone (RPZ) device
   (e) Vacuum breaker
3. Some backflow prevention regulations prohibit these devices from being installed in a pit.
4. In all cases, the device(s) in the pit should be approved or listed as necessary. The requirements of the local or municipal water department should be reviewed prior to design or installation of the connection.
5. Pressure drop should be considered prior to the installation of any backflow prevention device.

FIGURE A.5.9(b) Typical City Water Pit — Valve Arrangement.

Stationary Fire Pumps Handbook   2013

EXHIBIT III.12
Freestanding FDC.

5.9.2.3 Where local fire department connections do not conform to NFPA 1963, the authority having jurisdiction shall designate the connection to be used.

5.9.2.4 The use of threadless couplings shall be permitted where required by the authority having jurisdiction and where listed for such use.

> It is important to observe the requirement that the thread type must match that of the local fire department. Ideally, all fire departments would use the same hose threads, but this is not the case in all areas of the United States or in certain countries. Paragraph 5.9.2.4 provides some flexibility for jurisdictions that use threads or connections other than the national standard. See Exhibit III.13 for an illustration of a nonthreaded hose connection.

5.9.3 Valves.

5.9.3.1 A listed check valve shall be installed in each fire department connection.

5.9.3.2 No shutoff valve shall be permitted in the piping from the fire department connection piping to the point that the fire department connection piping connects to the system piping.

> Due to the presence of the check valve in the FDC, a shutoff valve is not needed. Installing a shutoff valve where it is not needed could increase the likelihood that the valve will be closed (rendering the FDC unusable) when needed.

5.9.4 Drainage.

5.9.4.1 The pipe between the check valve and the outside hose coupling shall be equipped with an approved automatic drip.

> For freestanding FDCs, the requirement in 5.9.4.1 presents a problem. Usually the check valve in a freestanding FDC is buried, or it might be installed in a valve pit. If installed in a valve pit, the usual method of draining, which is a ball-drip valve, can be used. See Exhibit III.14 for a typical freestanding FDC pit arrangement.

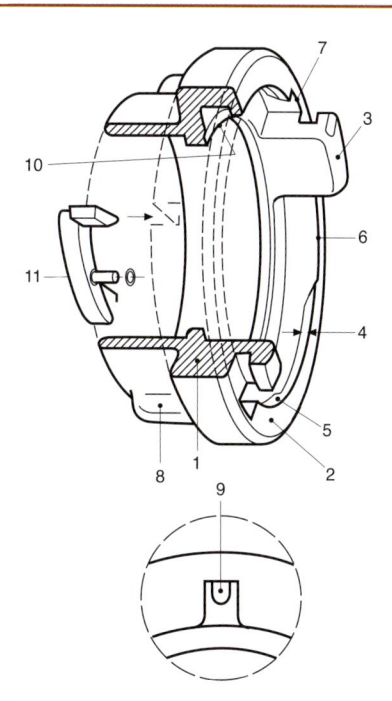

EXHIBIT III.13 *Diagram of a Nonthreaded Hose Connection. (Source: NFPA 1963, 2009, Figure A.6.1)*

Legend:
1. Circumferential O.D.: The largest outer diameter of connection that protects the connection from damage.
2. Coupling face: The front part of the connection from which dimensions are developed.
3. Internal lug: The two internal lugs with recesses that fit on the ramp under the face of the cam head.
4. Ramp: The inclined plane under the face of the cam head that, when turned clockwise, increases pressure on lip seals.
5. Lug stop: The stop at the end of the ramp that the internal lug comes against.
6. Lug recess: Recessed area where opposite internal lugs enter the ramp.
7. Cleaning port: Area on end of connection face where dirt is pushed in by mating lug.
8. External wrenching lug: The external ribs or lugs on back diameter of connection head.
9. External wrenching lug indicator: The identification on rib or lug that, when lined up together, indicates the connection is fully engaged.
10. Tail place recess: The recess counterbore on the interface of the cam head that the tail piece rides in.
11. Lock: To keep the connection from becoming unintentionally disengaged.

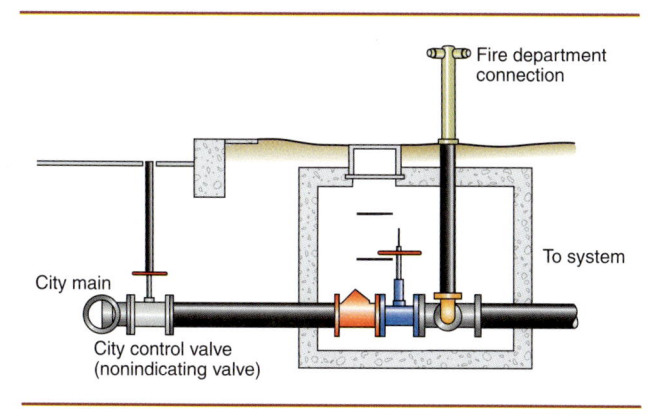

EXHIBIT III.14 *Pit for Gate Valve, Check Valve, and Fire Department Connection. (Courtesy of Stephan Laforest, Summit Sprinkler Design Systems, Inc.)*

Stationary Fire Pumps Handbook 2013

5.9.4.2 The automatic drip shall be installed in a location that permits inspection and testing as required by NFPA 25 and reduces the likelihood of freezing.

5.9.4.3 An automatic drip shall not be required in areas not subject to freezing.

5.9.5 Location and Signage.

5.9.5.1* Fire department connections shall be located at the nearest point of fire department apparatus accessibility or at a location approved by the authority having jurisdiction.

DESIGN ALERT ▶

> Careful consideration should be given to the location of FDCs that supply a standpipe system or combined sprinkler/standpipe system. NFPA 14 requires that an FDC be located within 100 ft (30.5 m) of a hydrant. When designing a private fire service main, the designer should consider this requirement when determining the relative location of FDCs and the hydrants intended to supply them.

A.5.9.5.1 The requirement in 5.9.5.1 applies to fire department connections attached to underground piping. If the fire department connection is attached directly to a system riser, the requirements of the appropriate installation standard apply.

5.9.5.2* Fire department connections shall be located and arranged so that hose lines can be attached to the inlets without interference.

> A clearance of 4 ft (1.2 m) is recommended to be maintained around all FDCs in order to permit fire fighters to connect and use FDCs. Protective bollards should also not be located within the recommended 4 ft (1.2 m) radius. See Exhibit III.15 for a recommended FDC arrangement.

EXHIBIT III.15
Recommended Arrangement for an FDC.

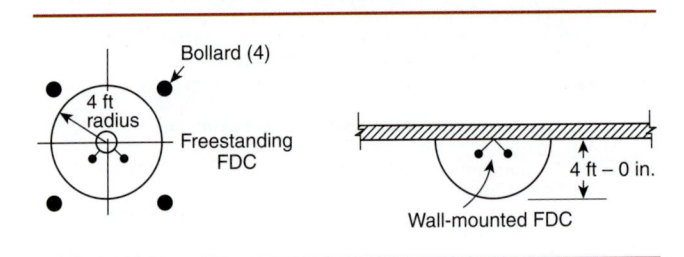

A.5.9.5.2 Obstructions to fire department connections include, but are not limited to, buildings, fences, posts, landscaping, other fire department connections, gas meters, and electrical equipment.

5.9.5.3* Each fire department connection shall be designated by a sign as follows:

(1) The sign shall have raised or engraved letters at least 1 in. (25.4 mm) in height on a plate or fitting.
(2)* The sign shall indicate the type of system for which the connection is intended.

A.5.9.5.3 Where a fire department connection services multiple buildings, structures, or locations, a sign should be provided indicating the buildings, structures, or locations served.

A.5.9.5.3(2) Examples for wording of signs are:

AUTOSPKR
OPEN SPKR
AND STANDPIPE

It is recommended that the sign required in 5.9.5.3 include a graphic representation of the type of FDC provided in NFPA 170, *Standard for Fire Safety and Emergency Symbols*, as follows:

FDC – Automatic Sprinkler System

FDC – Standpipe System Fire Department

FDC – Combined Automatic Sprinkler/Standpipe System

5.9.5.4 Where the system demand pressure exceeds 150 psi (10.3 bar), the sign required by 5.9.5.3 shall indicate the required design pressure.

5.9.5.5 Where a fire department connection only supplies a portion(s) of the building, a sign shall be attached to indicate the portion(s) of the building supplied.

CHAPTER 6 VALVES

6.1 Types of Valves

6.1.1 All valves controlling connections to water supplies and to supply pipes to sprinklers shall be listed indicating valves.

> All valves controlling fire protection systems must be listed indicating valves. As mentioned in the commentary to 3.3.15.2, a shut valve is the leading cause of system failure. The status of an indicating valve (open or shut) can be determined quickly and easily by visual observation of the indicator.

6.1.2 Indicating valves shall not close in less than 5 seconds when operated at maximum possible speed from the fully open position.

> The requirement in 6.1.2 is intended to reduce the possibility of water hammer or pressure surge in the system when water is flowing. A water hammer condition sounds like knocking in the pipes and is caused by quickly stopping the flow of water, which results in a sharp increase in pressure. Water hammer can briefly raise the pressure in the system beyond the pressure rating of the system components, which may result in damage or leaks.

6.1.3 A listed underground gate valve equipped with a listed indicator post shall be permitted.

6.1.4 A listed water control valve assembly with a reliable position indication connected to a remote supervisory station shall be permitted.

> Although the valves specified in 6.1.3 and 6.1.4 are permitted to be nonindicating valves, attachments are needed to provide an indication of their status. A post indicator valve (PIV) indicates its "open" or "shut" status through a window in the body of the post. See Exhibit III.16 for an illustration of a PIV.

6.1.5* A nonindicating valve, such as an underground gate valve with approved roadway box, complete with T-wrench, and accepted by the authority having jurisdiction, shall be permitted.

EXHIBIT III.16 *Post Indicator Valve (PIV).*

DESIGN ALERT ▶ Although a nonindicating valve with a roadway box is permitted by 6.1.5, this design should be avoided, since it offers no easy means of inspection or indication of the valve status. Additionally, access to these valves can be blocked by vehicles, or the box can get paved over or filled with sand. In areas subject to freezing, the roadway box can fill with water and freeze, rendering it inoperable.

A.6.1.5 A valve wrench with a long handle should be provided at a convenient location on the premises.

6.2 Valves Controlling Water Supplies

6.2.1 At least one valve shall be installed in each source of water supply.

The requirement for listed indicating valves has been eliminated in the 2013 edition of NFPA 24 to allow tapping sleeve and nonindicating valve assemblies to qualify as meeting the intent of 6.2.1.

6.2.2 No shutoff valve shall be permitted in the piping from the fire department connection to the point that the fire department connection piping connects to the system piping.

6.2.3 Where more than one source of water supply exists, a check valve shall be installed in each connection.

6.2.4 Where break tanks are used with automatic fire pumps, a check valve shall not be required in the break tank connection.

6.2.5* In a connection serving as one source of supply, listed indicating valves or post indicator valves shall be installed on both sides of all check valves required in 6.2.3.

The installation of two valves on each side of a check valve allows the check valve to be isolated for internal inspection, maintenance, or repair while the second water source remains in service.

A.6.2.5 See Figure A.6.2.5. For additional information on controlling valves, see NFPA 22.

6.2.6 In the discharge pipe from a pressure tank or a gravity tank of less than 15,000 gal (56.78 m³) capacity, a control valve shall not be required to be installed on the tank side of the check valve.

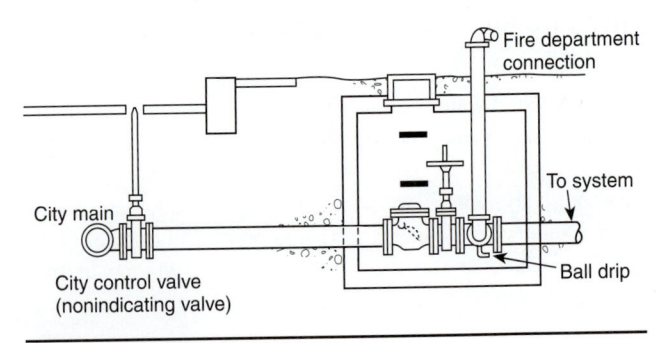

FIGURE A.6.2.5 *Pit for Gate Valve, Check Valve, and Fire Department Connection.*

6.2.7* The following requirements shall apply where a gravity tank is located on a tower in the yard:

(1) The control valve on the tank side of the check valve shall be an outside screw and yoke or a listed indicating valve.
(2) The other control valve shall be either an outside screw and yoke, a listed indicating valve, or a listed valve having a post-type indicator.

A.6.2.7 For additional information on controlling valves, see NFPA 22.

6.2.8* The following requirements shall apply where a gravity tank is located on a building:

(1) Both control valves shall be outside screw and yoke or listed indicating valves.
(2) All fittings inside the building, except the drain tee and heater connections, shall be under the control of a listed valve.

A.6.2.8 For additional information on controlling valves, see NFPA 22.

6.2.9 One of the following requirements shall be met where a pump is located in a combustible pump house or exposed to danger from fire or falling walls, or where a tank discharges into a private fire service main fed by another supply:

(1)* The check valve in the connection shall be located in a pit.
(2) The control valve shall be of the post indicator type and located a safe distance outside buildings.

A.6.2.9(1) Where located underground, check valves on tank or pump connections can be placed inside of buildings and at a safe distance from the tank riser or pump, except in cases where the building is entirely of one fire area. Where the building is one fire area, it is ordinarily considered satisfactory to locate the check valve overhead in the lowest level.

6.2.10* All control valves shall be located where accessible and free of obstructions.

A.6.2.10 It might be necessary to provide valves located in pits with an indicator post extending above grade or other means so that the valve can be operated without entering the pit.

6.2.11 All connections to private fire service mains for fire protection systems shall be arranged in accordance with one of the following so that they can be isolated:

(1)* A post indicator valve installed not less than 40 ft (12 m) from the building
   (a) For buildings less than 40 ft (12 m) in height, a post indicator valve shall be permitted to be installed closer than 40 ft (12 m) but at least as far from the building as the height of the wall facing the post indicator valve.
(2) A wall post indicator valve
(3) An indicating valve in a pit, installed in accordance with Section 6.4
(4)* A backflow preventer with at least one indicating valve not less than 40 ft (12 m) from the building
   (a) For buildings less than 40 ft (12 m) in height, a backflow preventer with at least one indicating valve shall be permitted to be installed closer than 40 ft (12 m) but at least as far from the building as the height of the wall facing the backflow preventer.
(5)* A nonindicating valve, such as an underground gate valve with an approved roadway box, complete with T-wrench, located not less than 40 ft (12 m) from the building
   (a) For buildings less than 40 ft (12 m) in height, a nonindicating valve, such as an underground gate valve with an approved roadway box, complete with T-wrench, shall be permitted to be installed closer than 40 ft (12 m) but at least as far from the building as the height of the wall facing the backflow preventer.

(6) Control valves installed in a fire-rated room accessible from the exterior
(7) Control valves in a fire-rated stair enclosure accessible from the exterior as permitted by the authority having jurisdiction

A.6.2.11(1) Distances greater than 40 ft (12 m) are not required but can be permitted regardless of the building height.

A.6.2.11(4) Distances greater than 40 ft (12 m) are not required but can be permitted regardless of the building height.

A.6.2.11(5) Distances greater than 40 ft (12 m) are not required but can be permitted regardless of the building height.

6.3 Post Indicator Valves

6.3.1 Where post indicator valves are used, they shall be set so that the top of the post is 32 in. to 40 in. (0.8 m to 1.0 m) above the final grade.

6.3.2 Where post indicator valves are used, they shall be protected against mechanical damage where needed.

> Ideally, the PIV should be installed in a location where it can be surrounded by a 4 ft (1.2 m) radius for proper access and to allow for the rotation of the handle. Bollards for mechanical protection should fall outside of the 4 ft (1.2 m) radius.

6.4 Valves in Pits

6.4.1 Valve pits located at or near the base of the riser of an elevated tank shall be designed in accordance with Chapter 14 of NFPA 22.

6.4.2 Where used, valve pits shall be of adequate size and accessible for inspection, operation, testing, maintenance, and removal of equipment contained therein.

6.4.3 Valve pits shall be constructed and arranged to properly protect the installed equipment from movement of earth, freezing, and accumulation of water.

6.4.3.1 Depending on soil conditions and the size of the pit, valve pits shall be permitted to be constructed of any of the following materials:

(1) Poured-in-place or precast concrete, with or without reinforcement
(2) Brick
(3) Other approved materials

6.4.3.2 Where the water table is low and the soil is porous, crushed stone or gravel shall be permitted to be used for the floor of the pit.

6.4.4 The location of the valve shall be marked, and the cover of the pit shall be kept free of obstructions.

6.5 Backflow Prevention Assemblies

6.5.1 Where used in accordance with 6.2.11(4), backflow prevention assemblies shall be installed in accordance with their installation instructions.

6.5.2 Where backflow prevention assemblies are used, they shall be protected against mechanical damage where needed.

6.6 Sectional Valves

6.6.1* Sectional valves shall be provided at appropriate points within piping sections such that the number of fire protection connections between sectional valves does not exceed six.

> The 2013 edition of NFPA 24 has been modified to include specific language to address the maximum number of connections allowed before a sectional valve is required. It was common practice to include these valves at increments of four to six connections; therefore, the technical committee established six connections as the maximum number. A connection can be a hydrant or a fire protection lead-in to a building. The purpose of a sectional control valve is to ensure the ability to isolate a segment of the underground main for maintenance or repair while impacting as few systems as possible. For locations with an underground loop that is provided with sectional valves, a common practice when sizing the underground pipe is to assume the shorter leg of the loop is out of service. This ensures that the system is capable of supplying the fire protection demand even when a portion of the underground loop is out of service. Exhibit III.17 is an example of a private fire service main that has been divided into segments by valves A, B, C, E, and F.

A.6.6.1 Sectional valves are necessary to allow isolation of piping sections to limit the number of fire protection connections impaired in event of a break or to make repairs or extensions to the system. Fire protection connections can consist of sprinkler system lead-ins, hydrants, or other fire protection connections.

6.6.2 A sectional valve shall be provided at the following locations:

(1) On each bank where a main crosses water
(2) Outside the building foundation(s) where a main or a section of a main runs under a building

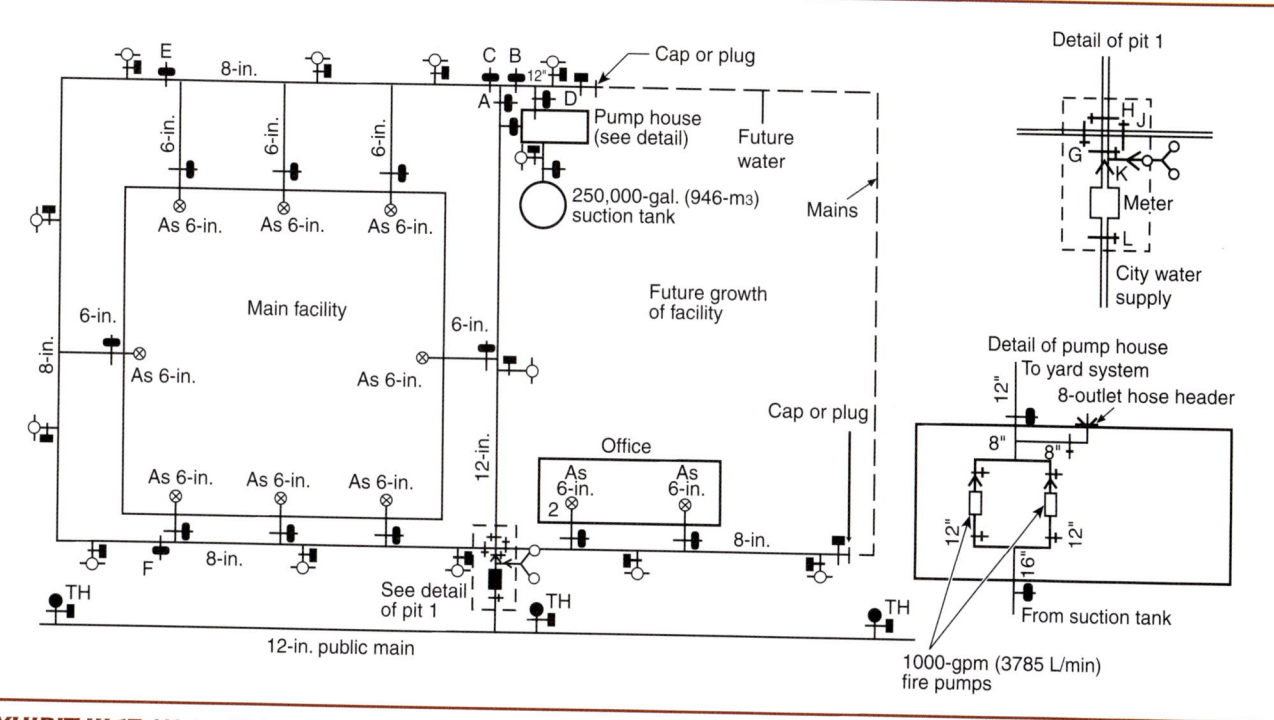

EXHIBIT III.17 Water Piping for Fire Protection of an Industrial Site. (Source: Fire Protection Handbook®, NFPA, 2003, Figure 10.3.30)

6.7 Identifying and Securing Valves

6.7.1 Identification signs shall be provided at each valve to indicate its function and what it controls.

> The graphical symbol for the valve function, which follows, is from NFPA 170 and can be used to fulfill this sign requirement.
> Automatic Sprinkler Control Valve

6.7.1.1 Identification signs in 6.7.1 shall not be required for underground gate valves with roadway boxes.

6.7.2* Control valves shall be supervised by one of the following methods:

(1) Central station, proprietary, or remote station signaling service
(2) Local signaling service that causes the sounding of an audible signal at a constantly attended location
(3) An approved procedure to ensure that valves are locked in the correct position
(4) An approved procedure to ensure that valves are located within fenced enclosures under the control of the owner, sealed in the open position, and inspected weekly

A.6.7.2 See Annex B.

6.7.3 Supervision of underground gate valves with roadway boxes shall not be required.

6.8 Check Valves

Check valves shall be installed in a vertical or horizontal position in accordance with their listing.

CHAPTER 7 HYDRANTS

7.1* General

A.7.1 For information regarding identification and marking of hydrants, see Annex D.

7.1.1 Hydrants shall be of an approved type and have not less than a 6 in. (152 mm) diameter connection with the mains.

7.1.1.1 A valve shall be installed in the hydrant connection.

7.1.1.1.1 Valves in the hydrant connection shall be installed within 20 ft (6.1 m) of the hydrant.

7.1.1.1.2 Where valves cannot be located in accordance with 7.1.1.1.1, valve locations shall be permitted where approved by the authority having jurisdiction.

7.1.1.2* The number, size, and arrangement of outlets; the size of the main valve opening; and the size of the barrel shall be suitable for the protection to be provided and shall be approved by the authority having jurisdiction.

A.7.1.1.2 The flows required for private fire protection service mains are determined by system installation standards or fire codes. The impact of the number and size of hydrant outlets on the fire protection system demand is not addressed in this standard. The appropriate code or standard should be consulted for the requirements for calculating system demand.

7.1.1.3 Independent gate valves on 2½ in. (64 mm) outlets shall be permitted.

7.1.2 Hydrant outlet threads shall have NHS external threads for the size outlet(s) supplied as specified in NFPA 1963.

7.1.3 Where local fire department connections do not conform to NFPA 1963, the authority having jurisdiction shall designate the connection to be used.

7.2 Number and Location

7.2.1* Hydrants shall be provided and spaced in accordance with the requirements of the authority having jurisdiction.

> Section 18.4 and Annex E of NFPA 1, *Fire Code*, provide guidance on the number and location of hydrants (see Part III, Section 3, of this handbook). The AHJ may require that specific local issues regarding the location of hydrants must be met.

A.7.2.1 Fire department pumpers will normally be required to augment the pressure available from public hydrants.

7.2.2 Public hydrants shall be permitted to be recognized as meeting all or part of the requirements of Section 7.2.

7.2.3* Hydrants shall be located not less than 40 ft (12 m) from the buildings to be protected.

> Hydrants in parking areas or in any area in which they are exposed to potential damage should be protected by bollards on all sides. The bollards should not be closer than 4 ft (1.2 m) to the protected hydrant. When located near a roadway, the hydrant should be placed within 6 ft (1.8 m) of the pavement, unless the AHJ determines another location is more acceptable.

A.7.2.3 Where wall hydrants are used, the authority having jurisdiction should be consulted regarding the necessary water supply and arrangement of control valves at the point of supply in each individual case. *(See Figure A.7.2.3.)*

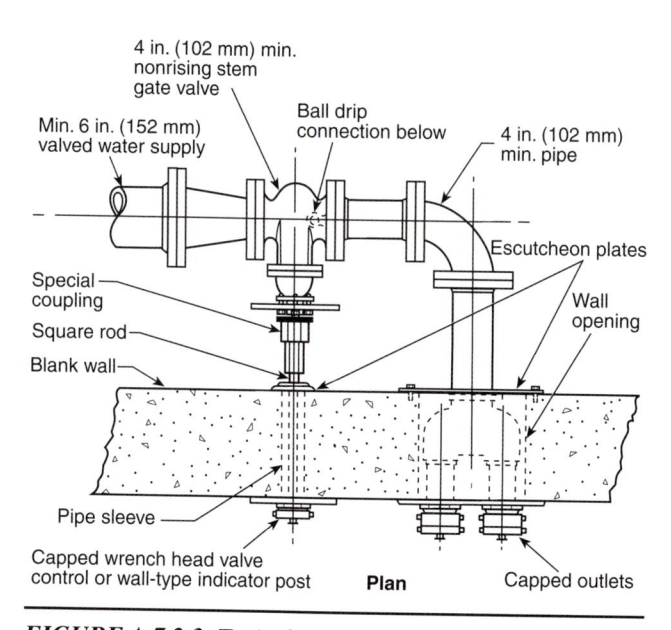

FIGURE A.7.2.3 Typical Wall Fire Hydrant Installation.

7.2.4 Where hydrants cannot be located in accordance with 7.2.3, locations closer than 40 ft (12.2 m) from the building or wall hydrants shall be permitted to be used where approved by the authority having jurisdiction.

7.2.5 Hydrants shall not be installed at less than the equivalent depth of burial from retaining walls where there is danger of frost through the walls.

7.3 Installation

7.3.1* Hydrants shall be set on flat stones or concrete slabs and shall be provided with small stones (or the equivalent) placed about the drain to ensure drainage.

> In cold climates, the hydrant barrel is expected to drain within 60 minutes. Where soil conditions do not permit this drainage, or where groundwater is above the hydrant drain, some jurisdictions have allowed the hydrant drain to be plugged and provisions made for pumping out the barrel after each use.

A.7.3.1 See Figure A.7.3.1(a) and Figure A.7.3.1(b).

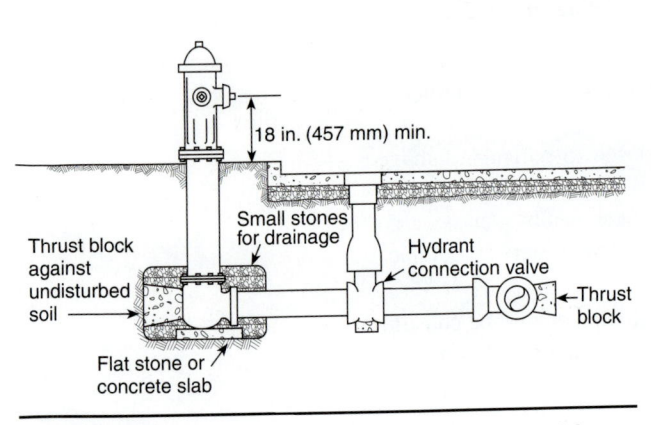

FIGURE A.7.3.1(a) *Typical Hydrant Connection with Minimum Height Requirement.*

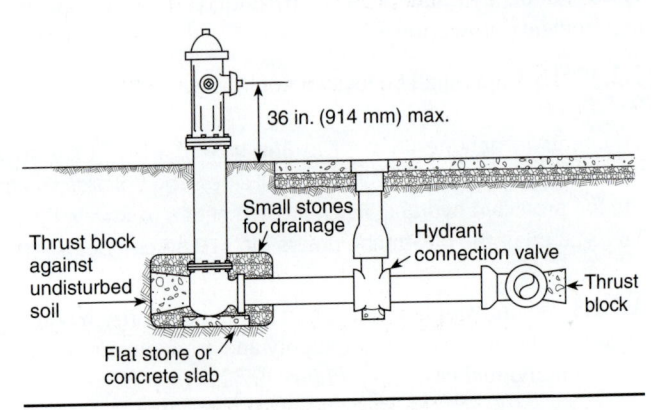

FIGURE A.7.3.1(b) *Typical Hydrant Connection with Maximum Height Requirement.*

7.3.2 Where soil is of such a nature that the hydrants will not drain properly with the arrangement specified in 7.3.1, or where groundwater stands at levels above that of the drain, the hydrant drain shall be plugged at the time of installation.

7.3.2.1 If the drain is plugged, hydrants in service in cold climates shall be pumped out after usage.

7.3.2.2 Such hydrants shall be marked to indicate the need for pumping out after usage.

7.3.3* The center of a hose outlet shall be not less than 18 in. (457 mm) above final grade or, where located in a hose house, 12 in. (305 mm) above the floor.

> It is recommended that the center of the hose outlet not be more than 36 in. (914.4 mm) above final grade [see Figure A.7.3.1(b)].

A.7.3.3 When setting hydrants, due regard should be given to the final grade line.

7.3.4 Hydrants shall be fastened to piping and anchored in accordance with the requirements of Chapter 10.

7.3.5 Hydrants shall be protected if subject to mechanical damage.

7.3.6 The means of hydrant protection shall be arranged in a manner that does not interfere with the connection to, or operation of, hydrants.

7.3.7 The following shall not be installed in the service stub between a fire hydrant and private water supply piping:

(1) Check valves
(2) Detector check valves
(3) Backflow prevention valves
(4) Other similar appurtenances

CHAPTER 8 HOSE HOUSES AND EQUIPMENT

8.1 General

8.1.1* A supply of hose and equipment shall be provided where hydrants are intended for use by plant personnel or a fire brigade.

> It is intended that this equipment be provided only where a trained industrial fire brigade is present. Industrial fire brigades should be trained in accordance with NFPA 600, *Standard on Industrial Fire Brigades*.

A.8.1.1 All hose should not be removed from a hose house for testing at the same time, since the time taken to return the hose in case of fire could allow a fire to spread beyond control. *(See NFPA 1962.)*

8.1.1.1 The quantity and type of hose and equipment shall depend on the following:

(1) Number and location of hydrants relative to the protected property
(2) Extent of the hazard
(3) Fire-fighting capabilities of potential users

8.1.1.2 The authority having jurisdiction shall be consulted regarding quantity and type of hose.

8.1.2 Hose shall be stored so it is accessible and is protected from the weather by storing in hose houses or by placing hose reels or hose carriers in weatherproof enclosures.

8.1.3* Hose shall conform to NFPA 1961.

A.8.1.3 Where hose will be subjected to acids, acid fumes, or other corrosive materials, as in chemical plants, the purchase of approved rubber-covered, rubber-lined hose is advised. For hose used in plant yards containing rough surfaces that cause heavy wear or used where working pressures are above 150 psi (10.3 bar), double-jacketed hose should be considered.

8.1.4 Hose Connections.

8.1.4.1 Hose connections shall have external national hose standard (NHS) threads, for the valve size specified, in accordance with NFPA 1963.

8.1.4.2 Hose connections shall be equipped with caps to protect the hose threads.

8.1.4.3 Where local fire department hose threads do not conform to NFPA 1963, the authority having jurisdiction shall designate the hose threads to be used.

8.2 Location

8.2.1 Where hose houses are utilized, they shall be located over, or immediately adjacent to, the hydrant.

8.2.2 Hydrants within hose houses shall be as close to the front of the house as possible and still allow sufficient room in back of the doors for the hose gates and the attached hose.

8.2.3 Where hose reels or hose carriers are utilized, they shall be located so that the hose can be brought into use at a hydrant.

8.3 Construction

8.3.1 Hose houses shall be of substantial construction on foundations.

8.3.2 The construction shall protect the hose from weather and vermin and shall be designed so that hose lines can be brought into use.

8.3.3 Clearance shall be provided for operation of the hydrant wrench.

8.3.4 Ventilation shall be provided.

8.3.5 The exterior shall be painted or otherwise protected against deterioration.

8.4* Size and Arrangement

Hose houses shall be of a size and arrangement that provide shelves or racks for the hose and equipment.

A.8.4 Typical hose houses are shown in Figure A.8.4(a) through Figure A.8.4(c).

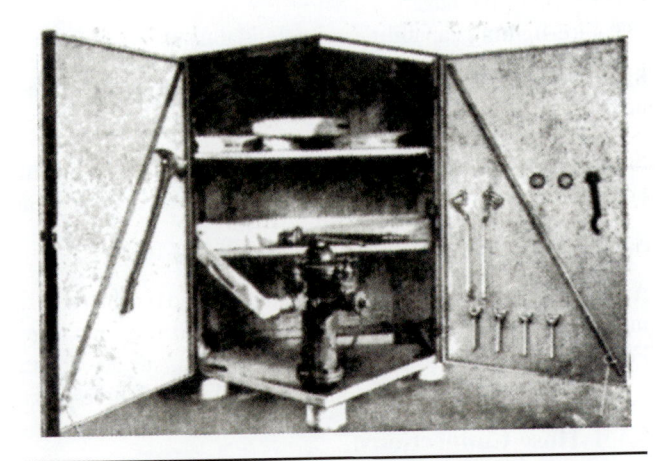

FIGURE A.8.4(a) Hose House of Five-Sided Design for Installation over Private Hydrant.

FIGURE A.8.4(b) Closed Steel Hose House of Compact Dimensions for Installation over Private Hydrant, in Which Top Lifts Up and Doors on Front Open for Complete Accessibility.

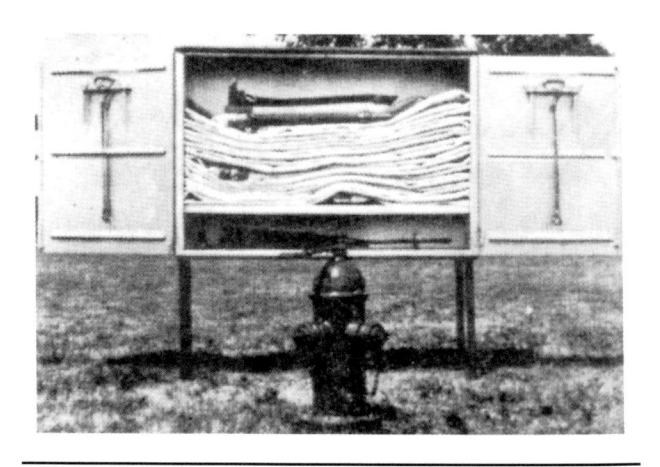

FIGURE A.8.4(c) Hose House That Can Be Installed on Legs, or Installed on Wall Near, but Not Directly Over, Private Hydrant.

8.5 Marking

Hose houses shall be plainly identified.

8.6 General Equipment

8.6.1* Where hose houses are used in addition to the hose, each shall be equipped with the following:

(1) Two approved adjustable spray–solid stream nozzles equipped with shutoffs for each size of hose provided
(2) One hydrant wrench (in addition to wrench on hydrant)
(3) Four coupling spanners for each size hose provided
(4) Two hose coupling gaskets for each size hose

A.8.6.1 All hose should not be removed from a hose house for testing at the same time, since the time taken to return the hose in case of fire could allow a fire to spread beyond control. *(See NFPA 1962.)*

8.6.2 Where two sizes of hose and nozzles are provided, reducers or gated wyes shall be included in the hose house equipment.

8.7 Domestic Service Use Prohibited

The use of hydrants and hose for purposes other than fire-related services shall be prohibited.

CHAPTER 9 MASTER STREAMS

9.1* Master Streams

Master streams shall be delivered by monitor nozzles, hydrant-mounted monitor nozzles, and similar master stream equipment capable of delivering more than 250 gpm (946 L/min).

EXHIBIT III.18 Master Stream Equipment. (Courtesy of Potter Electric Signal Company, LLC)

Some standards may specify higher flows for master streams. Monitor nozzles are available that can provide a master stream flow of 250 gpm to 2000 gpm (946 L to 7570 L) or more. See Exhibit III.18 for typical master stream equipment.

A.9.1 For typical master stream devices, see Figure A.9.1(a) and Figure A.9.1(b). Gear control nozzles are acceptable for use as monitor nozzles.

9.2 Application and Special Considerations

Master streams shall be provided as protection for the following:

(1) Large amounts of combustible materials located in yards
(2) Average amounts of combustible materials in inaccessible locations
(3) Occupancies presenting special hazards, as required by the authority having jurisdiction

CHAPTER 10 UNDERGROUND PIPING

10.1* Piping Materials

A.10.1 The term *underground* is intended to mean direct buried piping. For example, piping installed in trenches and tunnels but exposed should be treated as aboveground piping. Loop systems for yard piping are recommended for increased reliability and improved hydraulics. Loop systems should be sectionalized by placing valves at branches and at strategic locations to minimize the extent of impairments.

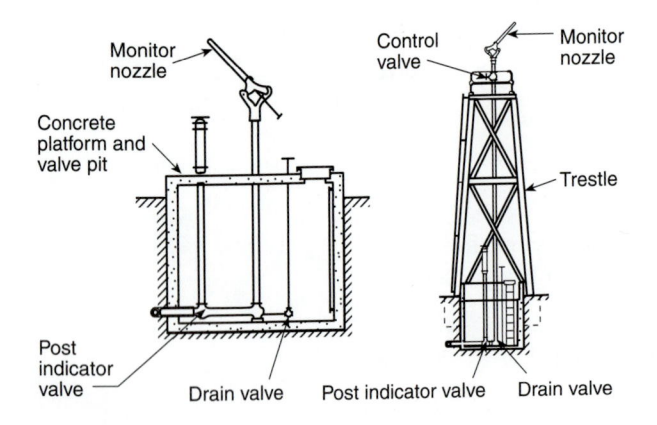

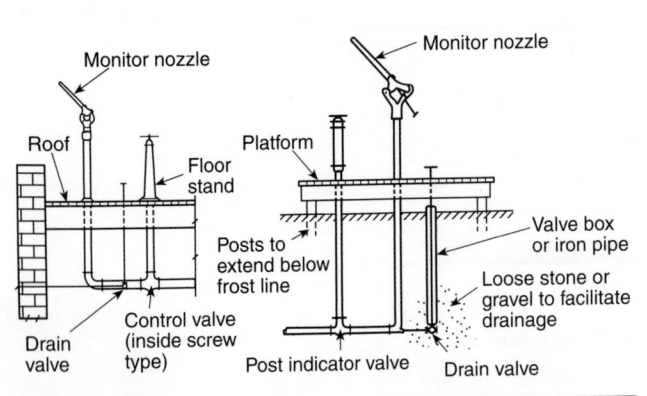

FIGURE A.9.1(a) Standard Monitor Nozzles.

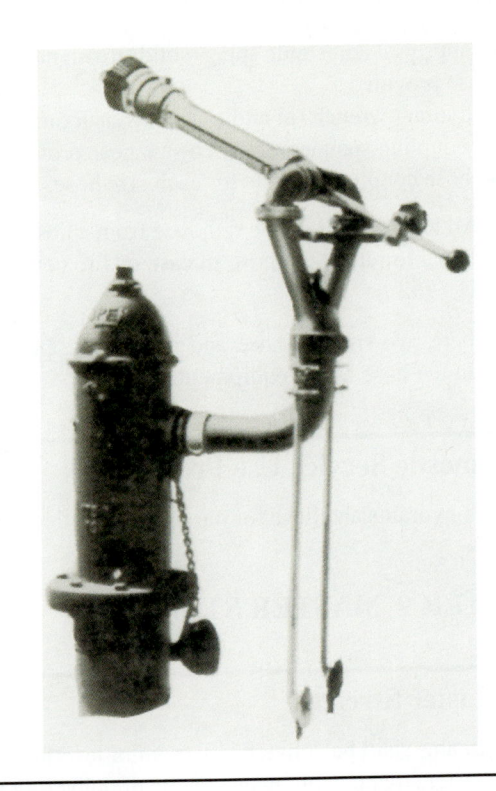

FIGURE A.9.1(b) Typical Hydrant-Mounted Monitor Nozzle.

10.1.1* Listing. Piping shall be listed for fire protection service or shall comply with the standards in Table 10.1.1.

TABLE 10.1.1 Manufacturing Standards for Underground Pipe

Materials and Dimensions	Standard
Ductile Iron	
Cement Mortar Lining for Ductile Iron Pipe and Fittings for Water	AWWA C104
Polyethylene Encasement for Ductile Iron Pipe Systems	AWWA C105
Ductile Iron and Gray Iron Fittings, 3 in. Through 48 in., for Water and Other Liquids	AWWA C110
Rubber-Gasket Joints for Ductile Iron Pressure Pipe and Fittings	AWWA C111
Flanged Ductile Iron Pipe with Ductile Iron or Gray Iron Threaded Flanges	AWWA C115
Protective Fusion-Bonded Epoxy Coatings for the Interior and Exterior Surfaces of Ductile-Iron and Gray-Iron Fittings for Water Supply Service	AWWA C116
Thickness Design of Ductile Iron Pipe	AWWA C150
Ductile Iron Pipe, Centrifugally Cast for Water	AWWA C151
Ductile-Iron Compact Fittings for Water Service	AWWA C153
Standard for the Installation of Ductile Iron Water Mains and Their Appurtenances	AWWA C600
Steel	
Steel Water Pipe 6 in. and Larger	AWWA C200
Coal-Tar Protective Coatings and Linings for Steel Water Pipelines Enamel and Tape — Hot Applied	AWWA C203
Cement-Mortar Protective Lining and Coating for Steel Water Pipe 4 in. and Larger — Shop Applied	AWWA C205
Field Welding of Steel Water Pipe	AWWA C206
Steel Pipe Flanges for Waterworks Service — Sizes 4 in. Through 144 in.	AWWA C207
Dimensions for Fabricated Steel Water Pipe Fittings	AWWA C208
A Guide for Steel Pipe Design and Installation	AWWA M11
Concrete	
Reinforced Concrete Pressure Pipe, Steel-Cylinder Type	AWWA C300
Prestressed Concrete Pressure Pipe, Steel-Cylinder Type	AWWA C301
Reinforced Concrete Pressure Pipe, Non-Cylinder Type	AWWA C302
Reinforced Concrete Pressure Pipe, Steel-Cylinder Type, Pretensioned	AWWA C303
Standard for Asbestos-Cement Distribution Pipe, 4 in. Through 16 in., for Water Distribution Systems	AWWA C400
Standard for the Selection of Asbestos-Cement Pressure Pipe	AWWA C401
Cement-Mortar Lining of Water Pipe Lines 4 in. and Larger — in Place	AWWA C602
Standard for the Installation of Asbestos-Cement Water Pipe	AWWA C603
Plastic	
Polyvinyl Chloride (PVC) Pressure Pipe, 4 in. Through 12 in., for Water Distribution	AWWA C900
Polyvinyl Chloride (PVC) Pressure Pipe, 14 in. Through 48 in., for Water Distribution	AWWA C905
Polyethylene (PE) Pressure Pipe and Fittings, 4 in. (100 mm) Through 63 in. (1575 mm) for Water Distribution	AWWA C906
Copper	
Specification for Seamless Copper Tube	ASTM B 75
Specification for Seamless Copper Water Tube	ASTM B 88
Requirements for Wrought Seamless Copper and Copper-Alloy Tube	ASTM B 251

A.10.1.1 Copper tubing (Type K) with brazed joints conforming to Table 10.1.1 and Table 10.2.2.1 is acceptable for underground service. Listing and labeling information, along with applicable publications for reference, is as follows:

(1) *Listing and Labeling.* Testing laboratories list or label the following:
   (a) Cast iron and ductile iron pipe (cement-lined and unlined, coated and uncoated)
   (b) Asbestos-cement pipe and couplings
   (c) Steel pipe
   (d) Copper pipe
   (e) Fiberglass filament-wound epoxy pipe and couplings
   (f) Polyethylene pipe
   (g) Polyvinyl chloride (PVC) pipe and couplings
   (h) Underwriters Laboratories Inc. lists, under re-examination service, reinforced concrete pipe (cylinder pipe, nonprestressed and prestressed)

(2) *Pipe Standards.* The various types of pipe are usually manufactured to one of the following standards:
   (a) ASTM C 296, *Standard Specification for Asbestos-Cement Pressure Pipe*
   (b) AWWA C151, *Ductile Iron Pipe, Centrifugally Cast for Water*
   (c) AWWA C300, *Reinforced Concrete Pressure Pipe, Steel-Cylinder Type*
   (d) AWWA C301, *Prestressed Concrete Pressure Pipe, Steel-Cylinder Type*
   (e) AWWA C302, *Reinforced Concrete Pressure Pipe, Non-Cylinder Type*
   (f) AWWA C303, *Reinforced Concrete Pressure Pipe, Steel-Cylinder Type, Pretensioned*
   (g) AWWA C400, *Standard for Asbestos-Cement Distribution Pipe, 4 in. Through 16 in. (100 mm through 400 mm), for Water Distribution Systems*
   (h) AWWA C900, *Polyvinyl Chloride (PVC) Pressure Pipe, 4 in. Through 12 in., for Water Distribution*

Underground pipe must meet either of the following criteria:

1. It must be listed for fire protection service.
2. It must comply with the American Water Works Association (AWWA) or ASTM International standards specified in Table 10.1.1.

The issue of listing or compliance with AWWA or ASTM had caused some confusion but was resolved in the 2010 edition following the issuance of a formal interpretation in 2007 clarifying that certain types of products, such as cross-linked polyethylene, chlorinated polyvinyl chloride (CPVC), and fiberglass, are acceptable based on listings alone.

Although several of the standards listed in Table 10.1.1 address steel pipe, flanges, welding, and installation, steel pipe is not permitted to be used in underground fire protection service, with the exception of FDCs or unless specially listed for such service. (See 10.1.2 and 10.1.3.)

Although the only plastic pipe included in Table 10.1.1 is PVC pipe manufactured in accordance with ANSI/AWWA C900-07, *Standard for Polyvinyl Chloride (PVC) Pressure Pipe and Fabricated Fittings, 4 in. Through 12 in. (100 mm Through 300 mm), for Water Transmission and Distribution*, listings are available for PVC with nominal diameters up to 30 in. (750 mm). Listings are also available for CPVC in sizes ¾ in. to 3 in. (20 mm to 75 mm) and for fiberglass-wound epoxy pipe in sizes 2 in. through 20 in. (50 mm through 500 mm) nominal diameter.

An often overlooked materials issue is that most nonmetallic underground piping brought up through the floor of a building could be vulnerable to fire exposure or spills of corrosive liquids. Regardless of material type, a section of underground pipe is allowed to extend above the floor level up to a maximum of 24 in. (600 mm) (6.3.1.1.1). Section 24.1.6.1, NFPA 13, 2013 edition, provides additional guidance on the transition from underground to aboveground piping and the need to protect the transition piece from damage. Possible methods for protecting the exposed section of underground pipe are to encase it in concrete, provide curbing, or provide a barrier.

10.1.2 Steel Pipe. Steel piping shall not be used for general underground service unless specifically listed for such service.

> Past experience and loss data indicate that a high potential exists for failure of buried steel pipe, even where the pipe is externally coated and wrapped and is internally galvanized. The integrity of the coating, wrapping, and galvanizing can be easily compromised in many underground applications. Accordingly, steel pipe can no longer be installed for general underground service, unless it is specially listed for the purpose. At this time, no such listings are available.

10.1.3 Steel Pipe Used with Fire Department Connections. Where externally coated and wrapped and internally galvanized, steel pipe shall be permitted to be used between the check valve and the outside hose coupling for the fire department connection.

> Steel pipe can be used for piping to an FDC because the FDC is an auxiliary, rather than a primary, water supply for the sprinkler system. Both coating and wrapping are required when internally galvanized steel pipe is used underground between an exterior FDC and the check valve where it connects to the system. Coating is also required for clamps, tie rods, and all bolted joint accessories. (See 10.3.6.2 and 10.8.3.5.)
>
> ANSI/AWWA C203-08, *Standard for Coal-Tar Protective Coatings and Linings for Steel Water Pipelines – Enamel and Tape – Hot Applied*, is a standard for coating materials referenced in Table 10.1.1, but other bituminous (coal-tar) coatings are available in the marketplace.
>
> ANSI/AWWA C105-10, *Standard for Polyethylene Encasement for Ductile-Iron Pipe Systems*, another of the standards referenced in Table 10.1.1, provides useful installation guidance where wrapping is required. The standard basically requires that high density polyethylene tubes or sheets with a minimum nominal thickness of 0.004 in. (0.1 mm), or low density polyethylene tubes or sheets with a minimum nominal thickness of 0.008 in. (0.2 mm), be used for the wrapping.

10.1.4* Pipe Type and Class. The type and class of pipe for a particular underground installation shall be determined through consideration of the following factors:

(1) Fire resistance of the pipe
(2) Maximum system working pressure
(3) Depth at which the pipe is to be installed
(4) Soil conditions
(5) Corrosion
(6) Susceptibility of pipe to other external loads, including earth loads, installation beneath buildings, and traffic or vehicle loads

A.10.1.4 The following pipe design manuals and standards can be used as guides:

(1) AWWA C150, *Thickness Design of Ductile Iron Pipe*
(2) AWWA C401, *Standard Practice for the Selection of Asbestos-Cement Water Pipe*
(3) AWWA C900, *Polyvinyl Chloride (PVC) Pressure Pipe, 4 in. Through 12 in. for Water Distribution*
(4) AWWA C905, *AWWA Standard for Polyvinyl Chloride (PVC) Pressure Pipe and Fabricated Fittings, 14 in. through 48 in. (350 mm through 1,200 mm)*
(5) AWWA C906, *Standard for Polyethylene (PE) Pressure Pipe and Fittings, 4 in. (100 mm) through 68 in. (1,600 mm), for Water Distribution and Transmission*
(6) AWWA M41, *Ductile Iron Pipe and Fittings*
(7) *Concrete Pipe Handbook*, American Concrete Pipe Association

10.1.5* Working Pressure. Piping, fittings, and other system components shall be rated for the maximum system working pressure to which they are exposed but shall not be rated at less than 150 psi (10 bar).

> Underground mains can and should have rated working pressures in excess of 150 psi (10.3 bar) where the system underground is expected to experience pressures in excess of 150 psi (10.3 bar). The 2010 edition of NFPA 24 was revised to recognize that an underground main may have working pressures in excess of 150 psi (10.3bar). Higher-rated piping may be needed on the discharge side of the pump for the private fire service mains.

A.10.1.5 For underground system components, a minimum system pressure rating of 150 psi (10 bar) is specified in 10.1.5, based on satisfactory historical performance. Also, this pressure rating reflects that of the components typically used underground, such as piping, valves, and fittings. Where system pressures are expected to exceed pressures of 150 psi (10.3 bar), system components and materials manufactured and listed for higher pressures should be used. Systems that do not incorporate a fire pump or are not part of a combined standpipe system do not typically experience pressures exceeding 150 psi (10.3 bar) in underground piping. However, each system should be evaluated on an individual basis, because the presence of a fire department connection introduces the possibility of high pressures being applied by fire department apparatus. It is not the intent of this section to include the pressures generated through fire department connections as part of the maximum working pressure.

10.1.6* Lining of Buried Pipe.

> Nominal diameters of available, listed, lined ferrous metal pipe range from 3 in. to 24 in. (75 mm to 600 mm). Ductile iron pipe is generally cement mortar–lined in accordance with ANSI/AWWA C104-08, *Standard for Cement–Mortar Lining for Ductile-Iron Pipe and Fittings.* Cement mortar linings were first used with cast-iron pipe in 1922 and help prevent tuberculation of ferrous pipe through the creation of a high pH at the pipe wall, as well as by providing a physical barrier to the water. The minimum thickness of the lining is $1/16$ in. (1.6 mm) for pipe sizes 3 in. to 12 in. (75 mm to 300 mm) nominal diameter and is greater for larger-diameter pipe.
>
> Care should be taken in using hydraulic calculations of underground piping, since manufacturers generally do not include the thickness of linings when providing information on the internal diameters of their piping products. Therefore, when calculating underground lined piping, it is imperative that the actual internal diameter, accounting for any lining material, be utilized throughout the calculations.

A.10.1.6 The following standards apply to the application of coating and linings:

(1) AWWA C104, *Cement Mortar Lining for Ductile Iron Pipe and Fittings for Water*
(2) AWWA C105, *Polyethylene Encasement for Ductile Iron Pipe Systems*
(3) AWWA C203, *Coal-Tar Protective Coatings and Linings for Steel Water Pipelines Enamel and Tape — Hot Applied*
(4) AWWA C205, *Cement-Mortar Protective Lining and Coating for Steel Water Pipe 4 in. and Larger — Shop Applied*
(5) AWWA C602, *Cement-Mortar Lining of Water Pipe Lines 4 in. and Larger — in Place*
(6) AWWA C116, *Protective Fusion-Bonded Epoxy Coatings for the Interior and Exterior Surfaces of Ductile-Iron and Gray-Iron Fittings for Water Supply Service.*

For internal diameters of cement-lined ductile iron pipe, see Table A.10.1.6.

10.1.6.1 Unless the requirements of 10.1.6.2 are met, all ferrous metal pipe shall be lined in accordance with the applicable standards in Table 10.1.1.

TABLE A.10.1.6 *Internal Diameters (IDs) for Cement-Lined Ductile Iron Pipe*

Pipe Size (in.)	OD (in.)	Pressure Class	Thickness Class	Wall Thickness	Minimum Lining Thickness*	ID (in.) with Lining
3	3.96	350		0.25	1/16	3.34
3	3.96		51	0.25	1/16	3.34
3	3.96		52	0.28	1/16	3.28
3	3.96		53	0.31	1/16	3.22
3	3.96		54	0.34	1/16	3.16
3	3.96		55	0.37	1/16	3.10
3	3.96		56	0.40	1/16	3.04
4	4.80	350		0.25	1/16	4.18
4	4.80		51	0.26	1/16	4.16
4	4.80		52	0.29	1/16	4.10
4	4.80		53	0.32	1/16	4.04
4	4.80		54	0.35	1/16	3.98
4	4.80		55	0.38	1/16	3.92
4	4.80		56	0.41	1/16	3.86
6	6.90	350		0.25	1/16	6.28
6	6.90		50	0.25	1/16	6.28
6	6.90		51	0.28	1/16	6.22
6	6.90		52	0.31	1/16	6.16
6	6.90		53	0.34	1/16	6.10
6	6.90		54	0.37	1/16	6.04
6	6.90		55	0.40	1/16	5.98
6	6.90		56	0.43	1/16	5.92
8	9.05	350		0.25	1/16	8.43
8	9.05		50	0.27	1/16	8.39
8	9.05		51	0.30	1/16	8.33
8	9.05		52	0.33	1/16	8.27
8	9.05		53	0.36	1/16	8.21
8	9.05		54	0.39	1/16	8.15
8	9.05		55	0.42	1/16	8.09
8	9.05		56	0.45	1/16	8.03
10	11.10	350		0.26	1/16	10.46
10	11.10		50	0.29	1/16	10.40
10	11.10		51	0.32	1/16	10.34
10	11.10		52	0.35	1/16	10.28
10	11.10		53	0.38	1/16	10.22
10	11.10		54	0.41	1/16	10.16
10	11.10		55	0.44	1/16	10.10
10	11.10		56	0.47	1/16	10.04
12	13.20	350		0.28	1/16	12.52
12	13.20		50	0.31	1/16	12.46
12	13.20		51	0.34	1/16	12.40
12	13.20		52	0.37	1/16	12.34
12	13.20		53	0.40	1/16	12.28
12	13.20		54	0.43	1/16	12.22
12	13.20		55	0.46	1/16	12.16
12	13.20		56	0.49	1/16	12.10

(continued)

TABLE A.10.1.6 Continued

Pipe Size (in.)	OD (in.)	Pressure Class	Thickness Class	Wall Thickness	Minimum Lining Thickness*	ID (in.) with Lining
14	15.30	250		0.28	3/32	14.55
14	15.30	300		0.30	3/32	14.51
14	15.30	350		0.31	3/32	14.49
14	15.30		50	0.33	3/32	14.45
14	15.30		51	0.36	3/32	14.39
14	15.30		52	0.39	3/32	14.33
14	15.30		53	0.42	3/32	14.27
14	15.30		54	0.45	3/32	14.21
14	15.30		55	0.48	3/32	14.15
14	15.30		56	0.51	3/32	14.09
16	17.40	250		0.30	3/32	16.61
16	17.40	300		0.32	3/32	16.57
16	17.40	350		0.34	3/32	16.53
16	17.40		50	0.34	3/32	16.53
16	17.40		51	0.37	3/32	16.47
16	17.40		52	0.40	3/32	16.41
16	17.40		53	0.43	3/32	16.35
16	17.40		54	0.46	3/32	16.29
16	17.40		55	0.49	3/32	16.23
16	17.40		56	0.52	3/32	16.17
18	19.50	250		0.31	3/32	18.69
18	19.50	300		0.34	3/32	18.63
18	19.50	350		0.36	3/32	18.59
18	19.50		50	0.35	3/32	18.61
18	19.50		51	0.35	3/32	18.61
18	19.50		52	0.41	3/32	18.49
18	19.50		53	0.44	3/32	18.43
18	19.50		54	0.47	3/32	18.37
18	19.50		55	0.50	3/32	18.31
18	19.50		56	0.53	3/32	18.25
20	21.60	250		0.33	3/32	20.75
20	21.60	300		0.36	3/32	20.69
20	21.60	350		0.38	3/32	20.65
20	21.60		50	0.36	3/32	20.69
20	21.60		51	0.39	3/32	20.63
20	21.60		52	0.42	3/32	20.57
20	21.60		53	0.45	3/32	20.51
20	21.60		54	0.48	3/32	20.45
20	21.60		55	0.51	3/32	20.39
20	21.60		56	0.54	3/32	20.33
24	25.80	200		0.33	3/32	24.95
24	25.80	250		0.37	3/32	24.87
24	25.80	300		0.40	3/32	24.81
24	25.80	350		0.43	3/32	24.75
24	25.80		50	0.38	3/32	24.85
24	25.80		51	0.41	3/32	24.79

TABLE A.10.1.6 Continued

Pipe Size (in.)	OD (in.)	Pressure Class	Thickness Class	Wall Thickness	Minimum Lining Thickness*	ID (in.) with Lining
24	25.80		52	0.44	3/32	24.73
24	25.80		53	0.47	3/32	24.67
24	25.80		54	0.50	3/32	24.61
24	25.80		55	0.53	3/32	24.55
24	25.80		56	0.56	3/32	24.49

ID: Internal diameter; OD: Outside diameter.

*Note: This table is appropriate for single lining thickness only. The actual lining thickness should be obtained from the manufacturer.

10.1.6.2 Steel pipe utilized in fire department connections and protected in accordance with the requirements of 10.1.3 shall not be required to be internally lined.

10.2 Fittings

10.2.1* Buried Fittings. Fittings shall be of an approved type with joints and pressure class ratings compatible with the pipe used.

A.10.2.1 Fittings generally used are cast iron with joints made to the specifications of the manufacturer of the particular type of pipe *(see the standards listed in A.10.3.1)*. Steel fittings also have some applications. The following standards apply to fittings:

(1) ASME B16.1, *Cast Iron Pipe Flanges and Flanged Fittings*
(2) AWWA C110, *Ductile Iron and Gray Iron Fittings, 3-in. Through 48-in., for Water and Other Liquids*
(3) AWWA C153, *Ductile Iron Compact Fittings, 3 in. through 24 in. and 54 in. through 64 in. for Water Service*
(4) AWWA C208, *Dimensions for Fabricated Steel Water Pipe Fittings*

10.2.2 Standard Fittings.

10.2.2.1 Fittings shall meet the standards in Table 10.2.2.1 or shall be in accordance with 10.2.3.

10.2.2.2 In addition to the standards in Table 10.2.2.2, CPVC fittings shall also be in accordance with 10.2.3 and with the portions of the ASTM standards specified in Table 10.2.2.2 that apply to fire protection service.

10.2.3 Special Listed Fittings. Other types of fittings investigated for suitability in automatic sprinkler installations and listed for this service, including, but not limited to, polybutylene, CPVC, and steel differing from that provided in Table 10.2.2.1, shall be permitted when installed in accordance with their listing limitations, including installation instructions.

10.2.4 Pressure Limits. Listed fittings shall be permitted for the system pressures as specified in their listings, but not less than 150 psi (10 bar).

TABLE 10.2.2.1 Fittings Materials and Dimensions

Materials and Dimensions	Standard
Cast Iron	
Gray Iron Threaded Fittings, Classes 125 and 250	ASME B16.4
Gray Iron Pipe Flanges and Flanged Fittings, Classes 12, 125, and 250	ASME B16.1
Malleable Iron	
Malleable Iron Threaded Fittings, Class 150 and 300	ASME B16.3
Steel	
Factory-Made Wrought Steel Buttweld Fittings	ASME B16.9
Buttwelding Ends	ASME B16.25
Specification for Piping Fittings of Wrought Carbon Steel and Alloy Steel for Moderate and Elevated Temperatures	ASTM A 234
Pipe Flanges and Flanged Fittings, NPS ½ Through 24	ASME B16.5
Forged Steel Fittings, Socket Welded and Threaded	ASME B16.11
Copper	
Wrought Copper and Bronze Solder Joint Pressure Fittings	ASME B16.22
Cast Bronze Solder Joint Pressure Fittings	ASME B16.18

TABLE 10.2.2.2 Specially Listed Fittings Materials and Dimensions

Materials and Dimensions	Standard
Chlorinated Polyvinyl Chloride (CPVC) Specification for Schedule 80 CPVC Threaded Fittings	ASTM F 437
Specification for Schedule 40 CPVC Socket-Type Fittings	ASTM F 438
Specification for Schedule 80 CPVC Socket-Type Fittings	ASTM F 439

10.3 Joining of Pipe and Fittings

Although 10.2.1 and 10.3.1 simply require that buried joints be approved and that buried fittings be of an approved type, acceptable joining methods for listed piping are generally controlled as part of the pipe listing. Both ductile iron and PVC pipe are generally joined by bell and spigot ends in conjunction with rubber gaskets or by listed mechanical joints. A simple rubber gasket joint is termed a *push-on joint*. In the case of a push-on joint, a single rubber ring gasket is fitted into the recess of a bell end of a length of pipe and is compressed by the plain end of the entering pipe, forming a seal. The gasket and the annular recess are specially shaped to lock the gasket in place against displacement. ANSI/AWWA C111-06, *Standard for Rubber-Gasket Joints for Ductile-Iron Pressure Pipe and Fittings,* one of the standards referenced in Table 10.1.1, requires that lubricants used in conjunction with push-on joints be labeled with the trade name or trademark and the pipe manufacturer's name to ensure compatibility with gasket material.

10.3.1* Buried Joints. Joints shall be approved.

A.10.3.1 The following standards apply to joints used with the various types of pipe:

(1) ASME B16.1, *Cast Iron Pipe Flanges and Flanged Fittings*
(2) AWWA C111, *Rubber-Gasket Joints for Ductile Iron Pressure Pipe and Fittings*
(3) AWWA C115, *Flanged Ductile Iron Pipe with Ductile Iron or Gray Iron Threaded Flanges*
(4) AWWA C206, *Field Welding of Steel Water Pipe*
(5) AWWA C606, *Grooved and Shouldered Joints*

10.3.2 Threaded Pipe and Fittings. All threaded steel pipe and fittings shall have threads cut in accordance with ASME B1.20.1.

10.3.3* Groove Joining Methods. Pipes joined with grooved fittings shall be joined by a listed combination of fittings, gaskets, and grooves.

A.10.3.3 Fittings and couplings are listed for specific pipe materials that can be installed underground. Fittings and couplings do not necessarily indicate that they are listed specifically for underground use.

10.3.4 Brazed and Pressure Fitting Methods. Joints for the connection of copper tube shall be brazed or joined using pressure fittings as specified in Table 10.2.2.1.

10.3.5 Other Joining Methods. Other joining methods listed for this service shall be permitted where installed in accordance with their listing limitations.

10.3.6 Pipe Joint Assembly.

10.3.6.1 Joints shall be assembled by persons familiar with the particular materials being used and in accordance with the manufacturer's instructions and specifications.

10.3.6.2* All bolted joint accessories shall be cleaned and thoroughly coated with asphalt or other corrosion-retarding material after installation.

A.10.3.6.2 It is not necessary to coat mechanical joint fittings or epoxy-coated valves and glands.

10.4 Depth of Cover

Buried piping must be located below the frost line to prevent freezing during the winter months. The piping must also be located deep enough to be protected from other surface loads that could cause mechanical damage. Piping located under driveways, roads, and railroad tracks must be buried deeper to prevent undue loads on the sprinkler piping.

10.4.1* The depth of cover over water pipes shall be determined by the maximum depth of frost penetration in the locality where the pipe is laid.

A.10.4.1 The following documents apply to the installation of pipe and fittings:

(1) AWWA C603, *Standard for the Installation of Asbestos-Cement Pressure Pipe*
(2) AWWA C600, *Standard for the Installation of Ductile-Iron Water Mains and Their Appurtenances*
(3) AWWA M11, *A Guide for Steel Pipe Design and Installation*
(4) AWWA M41, *Ductile Iron Pipe and Fittings*
(5) *Concrete Pipe Handbook*, American Concrete Pipe Association
(6) *Handbook of PVC Pipe*, Uni-Bell PVC Pipe Association
(7) *Installation Guide for Ductile Iron Pipe*, Ductile Iron Pipe Research Association
(8) *Thrust Restraint Design for Ductile Iron Pipe*, Ductile Iron Pipe Research Association

As there is normally no circulation of water in private fire mains, they require greater depth of covering than do public mains. Greater depth is required in a loose gravelly soil (or in rock) than in compact soil containing large quantities of clay. The recommended depth of cover above the top of underground yard mains is shown in Figure A.10.4.1.

10.4.2 The top of the pipe shall be buried not less than 1 ft (0.3 m) below the frost line for the locality.

10.4.3 In those locations where frost is not a factor, the depth of cover shall be not less than 2½ ft (0.8 m) to prevent mechanical damage.

10.4.4 Pipe under driveways shall be buried at a minimum depth of 3 ft (0.9 m).

10.4.5 Pipe under railroad tracks shall be buried at a minimum depth of 4 ft (1.2 m).

10.4.6 The depth of cover shall be measured from the top of the pipe to finished grade, and due consideration shall always be given to future or final grade and nature of soil.

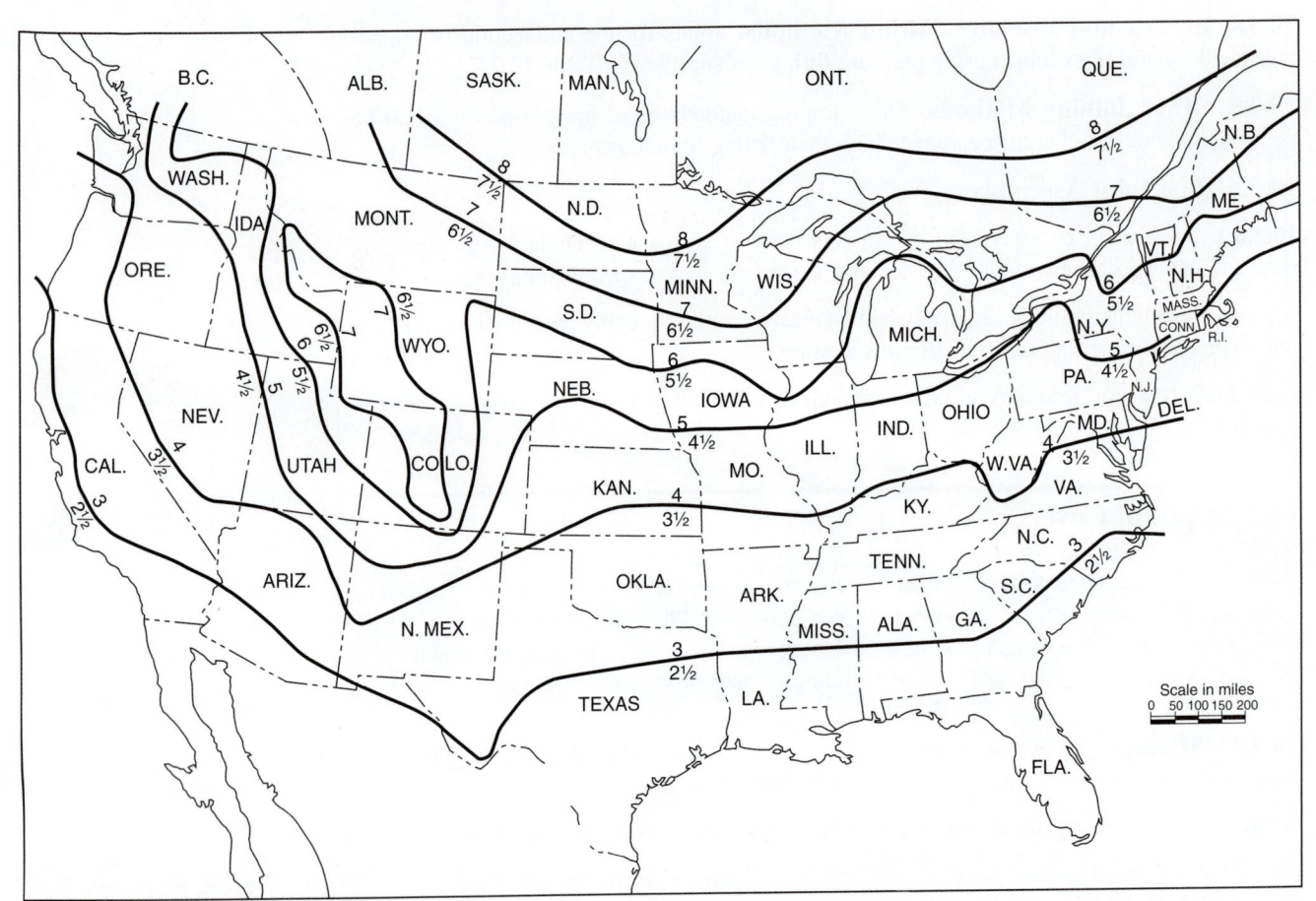

Notes:
1. For SI Units, 1 in. = 25.4 mm; 1 ft = 0.304 m.
2. Where frost penetration is a factor, the depth of cover shown averages 6 in. greater than that usually provided by the municipal waterworks. Greater depth is needed because of the absence of flow in yard mains.

FIGURE A.10.4.1 Recommended Depth of Cover (in feet) Above Top of Underground Yard Mains.

10.5 Protection Against Freezing

10.5.1* Where it is impracticable to bury pipe, pipe shall be permitted to be laid aboveground, provided that the pipe is protected against freezing and mechanical damage.

A.10.5.1 In determining the need to protect aboveground piping from freezing, the lowest mean temperature should be considered as shown in Figure A.10.5.1.

10.5.2 Pipe shall be buried below the frost line where entering streams and other bodies of water.

10.5.3 Where pipe is laid in water raceways or shallow streams, care shall be taken that there will be sufficient depth of running water between the pipe and the frost line during all seasons of frost; a safer method is to bury the pipe 1 ft (0.3 m) or more under the bed of the waterway.

10.5.4 Pipe shall be located at a distance from stream banks and embankment walls that prevents danger of freezing through the side of the bank.

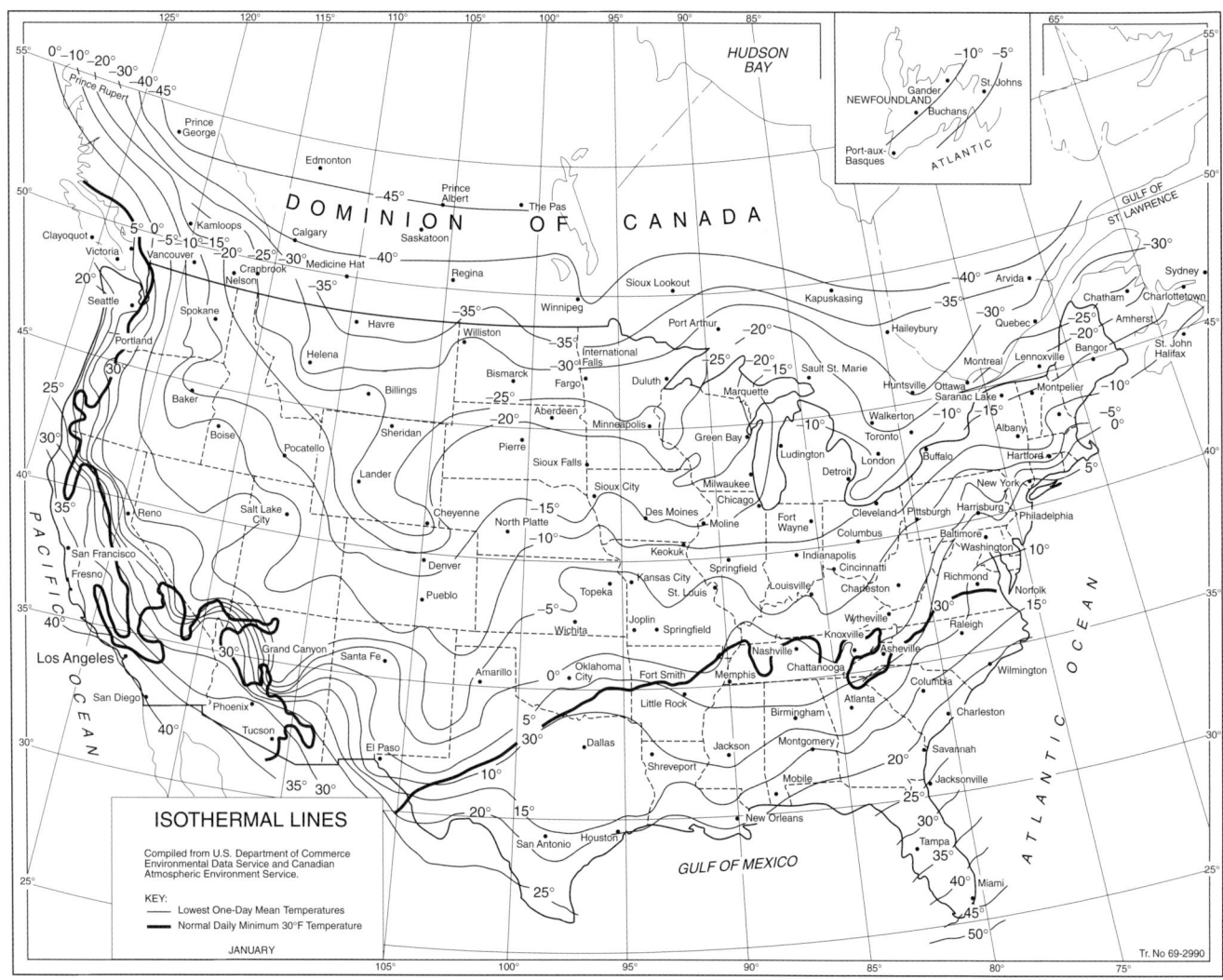

Source: Compiled from United States Weather Bureau records.
For SI units, °C = 5/9 (°F −32); 1 mi = 1.609 km.

FIGURE A.10.5.1 *Isothermal Lines — Lowest One-Day Mean Temperature (°F).*

10.6 Protection Against Damage

10.6.1 Pipe shall not be run under the building except where permitted in 10.6.2 and 10.6.3.

> When buildings are built over existing underground piping, the piping should be rerouted around the new building for the following reasons, among others:
>
> - Piping located under buildings is extremely difficult to repair, which is one of the reasons that 10.6.1 limits the installation of pipe under buildings. When piping under buildings requires repair, operations in the building must be curtailed, equipment may need to be moved, and the floor must be excavated.
> - Leaks in the buried piping underneath the building can go undetected for long periods, water can surface inside and damage building contents, and leaks can also undermine the building support.
> - System control valves need to be located in the center of a building, which is highly undesirable during a fire.

10.6.2 Where approved, pipe shall be permitted to be run under buildings, and special precautions shall be taken, including the following:

(1) Arching the foundation walls over the pipe
(2) Running pipe in covered trenches
(3) Providing valves to isolate sections of pipe under buildings

> Subsection 10.6.2 outlines some of the measures that can be taken to minimize, but not entirely eliminate, exposure to damage. Where the main passes beneath the foundation wall, arching of the foundation wall over the main should be accomplished, or another means of ensuring system and structural integrity should be provided.

10.6.3 Fire service mains shall be permitted to enter the building adjacent to the foundation.

> Subsection 10.6.3 permits a riser lead-in main to run underneath the building's footing or to penetrate the foundation wall if the main rises through the floor slab within 10ft (3 m) of the building exterior wall.

10.6.3.1* The requirements of 10.6.2(2) and 10.6.2(3) shall not apply where fire service mains enter under the building no more than 10 ft (3 m) as measured from the outside edge of the building to the center of the vertical pipe.

A.10.6.3.1 Items such as sidewalks or patios should not be included as they are no different from roadways. See Figure A.10.6.3.1.

> Even where locating risers adjacent to exterior walls is not practical, 10.6.3 nevertheless requires pipe runs to be minimized. NFPA 24 does not require a fire main of 10 ft (3 m) or less located under a building that connects to a system riser to be accessible through trench plates or other means of access, but, in accordance with the requirements of 10.6.2, this access is considered good practice. The alternative offered in 10.6.3.1 is provided to allow for the installation of sprinklers when the preferred options in 10.6.1 through 10.6.3 are not practical.

10.6.4* Pipe joints shall not be located under foundation footings.

A.10.6.4 The individual piping standards should be followed for load and bury depth, accounting for the load and stresses imposed by the building foundation.

Figure A.10.6.4 shows location where pipe joints would be prohibited.

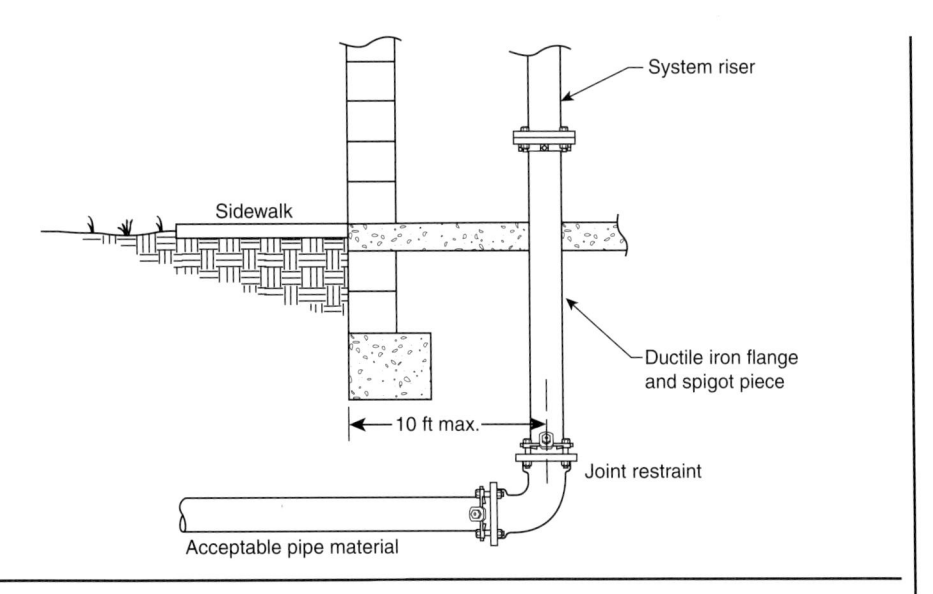

FIGURE A.10.6.3.1 *Riser Entrance Location.*

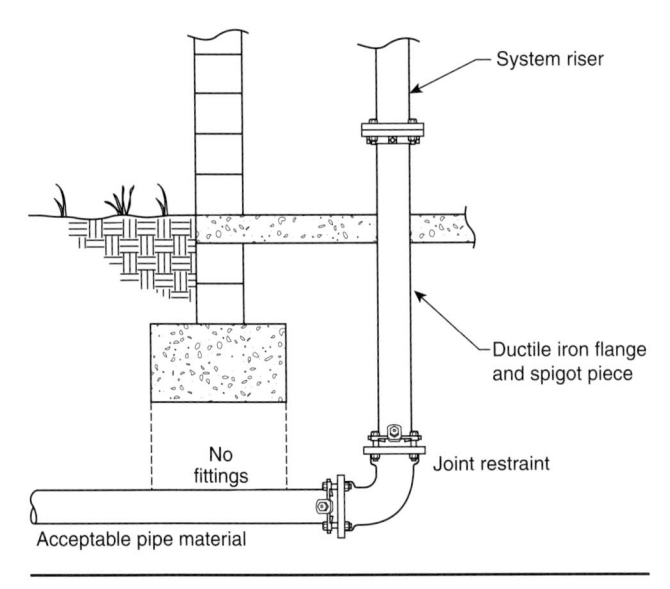

FIGURE A.10.6.4 *Pipe Joint Location in Relation to Foundation Footings.*

10.6.5* Piping shall be run at least 1 ft (305 mm) below the bottom of foundations/footers.

A.10.6.5 Sufficient clearance should be provided when piping passes beneath foundations or footers. See Figure A.10.6.5.

10.6.5.1 The requirements of 10.6.6 shall not apply when piping is sleeved.

10.6.6 Mains shall be subjected to an evaluation of the following specific loading conditions and protected, if necessary:

(1) Mains running under railroads carrying heavy cargo
(2) Mains running under large piles of heavy commodities
(3) Mains located in areas that subject the main to heavy shock and vibrations

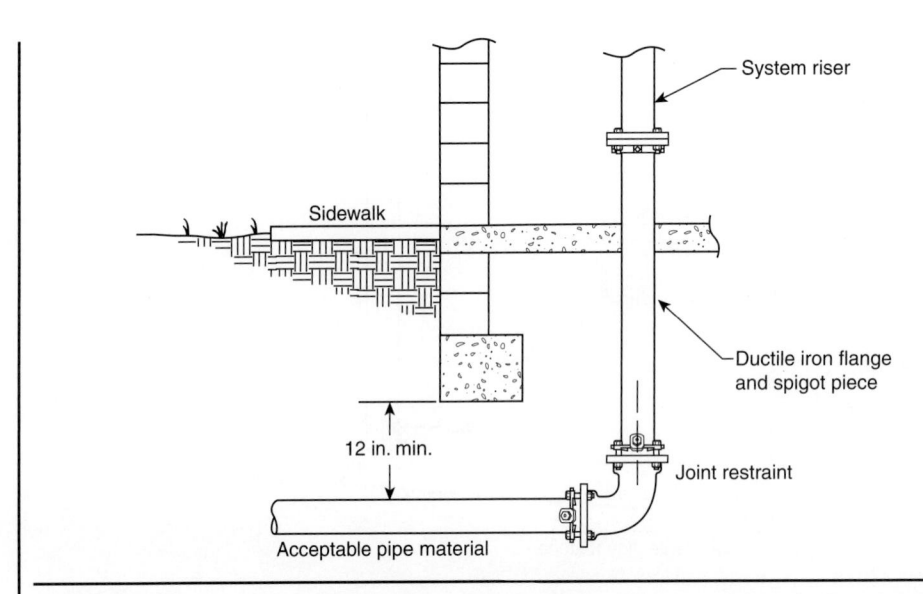

FIGURE A.10.6.5 *Piping Clearance from Foundation.*

10.6.7* Where it is necessary to join metal pipe with pipe of dissimilar metal, the joint shall be insulated against the passage of an electric current using an approved method.

A.10.6.7 Gray cast iron is not considered galvanically dissimilar to ductile iron. Rubber gasket joints (unrestrained push-on or mechanical joints) are not considered connected electrically. Metal thickness should not be considered a protection against corrosive environments. In the case of cast iron or ductile iron pipe for soil evaluation and external protection systems, see Appendix A of AWWA C105.

10.6.8* In no case shall the underground piping be used as a grounding electrode for electrical systems.

> The use of underground fire protection piping for electrical grounding increases the potential for stray ground currents and increased galvanic corrosion, which is why such use is prohibited by 10.6.8. Grounding to piping systems that may have nonconductive piping or joints is especially dangerous, since it may not provide the expected ground. In no case should the underground piping be used as a grounding electrode for electrical systems. Electrical equipment should be grounded in accordance with *NFPA 70®, National Electrical Code®*.

A.10.6.8 Where lightning protection is provided for a structure, NFPA 780, Section 4.14, requires that all grounding media, including underground metallic piping systems, be interconnected to provide common ground potential. These underground piping systems are not permitted to be substituted for grounding electrodes but must be bonded to the lightning protection grounding system. Where galvanic corrosion is of concern, this bond can be made via a spark gap or gas discharge tube.

10.6.8.1* The requirement of 10.6.8 shall not preclude the bonding of the underground piping to the lightning protection grounding system as required by NFPA 780 in those cases where lightning protection is provided for the structure.

A.10.6.8.1 While the use of the underground fire protection piping as the grounding electrode for the building is prohibited, *NFPA 70* requires that all metallic piping systems be bonded and grounded to disperse stray electrical currents. Therefore, the fire protection piping will be bonded to other metallic systems and grounded, but the electrical system will need an additional ground for its operation.

10.7 Requirement for Laying Pipe

The precautions that must be taken to minimize damage to underground piping, eliminate stresses, and ensure a long service life are outlined in Section 10.7.

In the United States, the federal Occupational Safety and Health Administration (OSHA) has a number of requirements relating to safety issues involved in underground piping installation. These OSHA regulations can be found in 29 CFR 1926, Subpart P, on excavations and 29 CFR 1910.146 on confined spaces.

For trenching operations, sloping, shoring, or shielding of the trench may be required. Soil classification is needed to select the correct options. In the absence of soil classification, sloping of the excavation wall can be performed, with the slope limited to 1½ horizontal to 1 vertical, which is a maximum 34 degrees from horizontal.

Confined space entry procedures often must be employed when access to control valves located in pits is needed. For this reason, control valves are brought above ground in areas where freezing is not a concern, and special prefabricated heated enclosures (see Exhibit III.19) are available to allow aboveground valve installations in areas subject to freezing.

EXHIBIT III.19 *Heated Enclosure. (Courtesy of AquaSHIELD)*

10.7.1 Pipes, valves, hydrants, gaskets, and fittings shall be inspected for damage when received and shall be inspected prior to installation. *(See Figure 10.10.1.)*

10.7.2 The torquing of bolted joints shall be checked.

10.7.3 Pipe, valves, hydrants, and fittings shall be clean inside.

10.7.4 When work is stopped, the open ends of pipe, valves, hydrants, and fittings shall be plugged to prevent stones and foreign materials from entering.

Precautions must be taken to prevent rocks and other foreign materials from entering piping during installation. These materials can be carried into fire protection system piping and cause obstructions and system failure during a fire. The precautions taken during installation are important but do not eliminate the need to flush the piping as part of the acceptance testing. Exhibit III.20 shows debris flushed out of underground piping.

EXHIBIT III.20 Debris Flushed Out of Underground Piping. (Courtesy of FM Global)

10.7.5 All pipe, fittings, valves, and hydrants shall be carefully lowered into the trench using appropriate equipment and carefully examined for cracks or other defects while suspended above the trench.

10.7.6 Plain ends shall be inspected for signs of damage prior to installation.

10.7.7 Under no circumstances shall water main materials be dropped or dumped.

10.7.8 Pipe shall not be rolled or skidded against other pipe materials.

10.7.9 Pipes shall bear throughout their full length and shall not be supported by the bell ends only or by blocks.

10.7.10 If the ground is soft or of a quicksand nature, special provisions shall be made for supporting pipe.

10.7.11 Valves and fittings used with nonmetallic pipe shall be supported and restrained in accordance with the manufacturer's specifications.

10.8 Joint Restraint

Section 10.8 addresses the need to restrain pipe movement caused by hydraulic pressure or dynamic thrust forces created by water changing direction as it flows through the pipe. Thrust blocks are provided at piping direction changes (elbow, bend, and tee fittings) to prevent unrestrained pipe joints from separating, and to prevent pipe movement in soft soil.

10.8.1 General.

10.8.1.1* All tees, plugs, caps, bends, reducers, valves, and hydrant branches shall be restrained against movement by using thrust blocks in accordance with 10.8.2 or restrained joint systems in accordance with 10.8.3.

A.10.8.1.1 It is a fundamental design principle of fluid mechanics that dynamic and static pressures, acting at changes in size or direction of a pipe, produce unbalanced thrust forces at locations such as bends, tees, wyes, dead ends, and reducer offsets. This design principle includes consideration of lateral soil pressure and pipe/soil friction, variables that can be reliably determined using current soil engineering knowledge. Refer to A.10.8.3 for a list of references for use in calculating and determining joint restraint systems.

Except for the case of welded joints and approved special restrained joints, such as is provided by approved mechanical joint retainer glands or locked mechanical and push-on joints, the usual joints for underground pipe are expected to be held in place by the soil in which the pipe is buried. Gasketed push-on and mechanical joints without special locking devices have limited ability to resist separation due to movement of the pipe.

10.8.1.2* Piping with fused, threaded, grooved, or welded joints shall not require additional restraining, provided that such joints can pass the hydrostatic test of 10.10.2.2 without shifting of piping or leakage in excess of permitted amounts.

A.10.8.1.2 Solvent-cemented and heat-fused joints such as those used with CPVC piping and fittings are considered restrained. They do not require thrust blocks.

10.8.1.3 Steep Grades. On steep grades, mains shall be additionally restrained to prevent slipping.

10.8.1.3.1 Pipe shall be restrained at the bottom of a hill and at any turns (lateral or vertical).

10.8.1.3.2 The restraint specified in 10.8.1.3.1 shall be to natural rock or to suitable piers built on the downhill side of the bell.

10.8.1.3.3 Bell ends shall be installed facing uphill.

10.8.1.3.4 Straight runs on hills shall be restrained as determined by the design engineer.

10.8.2* Thrust Blocks.

> Thrust blocks are covered in 10.8.2. Exhibit III.21 illustrates a typical thrust block arrangement.

A.10.8.2 Concrete thrust blocks are one of the methods of restraint now in use, provided that stable soil conditions prevail and space requirements permit placement. Successful blocking is dependent upon factors such as location, availability and placement of concrete, and possibility of disturbance by future excavations.

Resistance is provided by transferring the thrust force to the soil through the larger bearing area of the block such that the resultant pressure against the soil does not exceed the horizontal bearing strength of the soil. The design of thrust blocks consists of determining

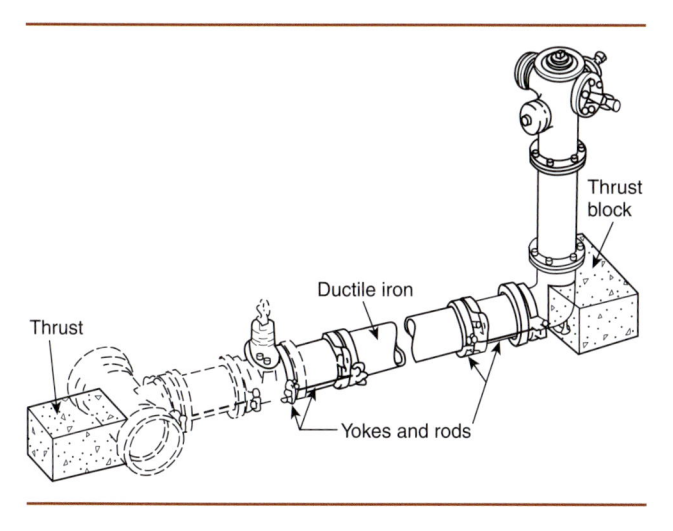

EXHIBIT III.21 Thrust Block Arrangement. (Courtesy of Los Angeles Department of Water and Power)

TABLE A.10.8.2(a) *Thrust at Fittings at 100 psi (6.9 bar) Water Pressure for Ductile Iron and PVC Pipe*

Nominal Pipe Diameter (in.)	Total Pounds					
	Dead End	90-Degree Bend	45-Degree Bend	22½-Degree Bend	11¼-Degree Bend	5⅛-Degree Bend
4	1,810	2,559	1,385	706	355	162
6	3,739	5,288	2,862	1,459	733	334
8	6,433	9,097	4,923	2,510	1,261	575
10	9,677	13,685	7,406	3,776	1,897	865
12	13,685	19,353	10,474	5,340	2,683	1,224
14	18,385	26,001	14,072	7,174	3,604	1,644
16	23,779	33,628	18,199	9,278	4,661	2,126
18	29,865	42,235	22,858	11,653	5,855	2,670
20	36,644	51,822	28,046	14,298	7,183	3,277
24	52,279	73,934	40,013	20,398	10,249	4,675
30	80,425	113,738	61,554	31,380	15,766	7,191
36	115,209	162,931	88,177	44,952	22,585	10,302
42	155,528	219,950	119,036	60,684	30,489	13,907
48	202,683	286,637	155,127	79,083	39,733	18,124

Notes:

(1) For SI units, 1 lb = 0.454 kg; 1 in. = 25.4 mm.

(2) To determine thrust at pressure other than 100 psi (6.9 bar), multiply the thrust obtained in the table by the ratio of the pressure to 100 psi (6.9 bar). For example, the thrust on a 12 in. (305 mm), 90-degree bend at 125 psi (8.6 bar) is 19,353 × 125/100 = 24,191 lb (10,973 kg).

the appropriate bearing area of the block for a particular set of conditions. The parameters involved in the design include pipe size, design pressure, angle of the bend (or configuration of the fitting involved), and the horizontal bearing strength of the soil.

Table A.10.8.2(a) gives the nominal thrust at fittings for various sizes of ductile iron and PVC piping. Figure A.10.8.2(a) shows an example of how thrust forces act on a piping bend.

Thrust blocks are generally categorized into two groups — bearing and gravity blocks. Figure A.10.8.2(b) depicts a typical bearing thrust block on a horizontal bend.

The following are general criteria for bearing block design:

(1) The bearing surface should, where possible, be placed against undisturbed soil.
(2) Where it is not possible to place the bearing surface against undisturbed soil, the fill between the bearing surface and undisturbed soil must be compacted to at least 90 percent Standard Proctor density.
(3) Block height (h) should be equal to or less than one-half the total depth to the bottom of the block (H_t) but not less than the pipe diameter (D).
(4) Block height (h) should be chosen such that the calculated block width (b) varies between one and two times the height.
(5) Gravity thrust blocks can be used to resist thrust at vertical down bends. In a gravity thrust block, the weight of the block is the force providing equilibrium with the thrust force. The design problem is then to calculate the required volume of the thrust block of a known density. The vertical component of the thrust force in Figure A.10.8.2(c) is balanced by the weight of the block. For required horizontal bearing block areas, see Table A.10.8.2(b).

TABLE A.10.8.2(b) Required Horizontal Bearing Block Area

Nominal Pipe Diameter (in.)	Bearing Block Area (ft²)	Nominal Pipe Diameter (in.)	Bearing Block Area (ft²)	Nominal Pipe Diameter (in.)	Bearing Block Area (ft²)
3	2.6	12	29.0	24	110.9
4	3.8	14	39.0	30	170.6
6	7.9	16	50.4	36	244.4
8	13.6	18	63.3	42	329.9
10	20.5	20	77.7	48	430.0

Notes:

(1) Although the bearing strength values in this table have been used successfully in the design of thrust blocks and are considered to be conservative, their accuracy is totally dependent on accurate soil identification and evaluation. The ultimate responsibility for selecting the proper bearing strength of a particular soil type must rest with the design engineer.

(2) Values listed are based on a 90-degree horizontal bend, an internal pressure of 100 psi, a soil horizontal bearing strength of 1000 lb/ft², a safety factor of 1.5, and ductile iron pipe outside diameters.

(a) For other horizontal bends, multiply by the following coefficients: for 45 degrees, 0.541; for 22½ degrees, 0.276; for 11¼ degrees, 0.139.

(b) For other internal pressures, multiply by ratio to 100 psi.

(c) For other soil horizontal bearing strengths, divide by ratio to 1000 lb/ft².

(d) For other safety factors, multiply by ratio to 1.5.

Example: Using Table A.10.8.2(b), find the horizontal bearing block area for a 6 in. diameter, 45–degree bend with an internal pressure of 150 psi. The soil bearing strength is 3000 lb/ft², and the safety factor is 1.5.

From Table A.10.8.2(b), the required bearing block area for a 6 in. diameter, 90-degree bend with an internal pressure of 100 psi and a soil horizontal bearing strength of 1000 psi is 7.9 ft².

For example:

$$\text{Area} = \frac{7.9 \text{ ft}^2 (0.541)\left(\frac{150}{100}\right)}{\left(\frac{3000}{1000}\right)} = 2.1 \text{ ft}^2$$

TABLE A.10.8.2(c) Horizontal Bearing Strengths

Soil	Bearing Strength (S_b) lb/ft²	Bearing Strength (S_b) kN/m²
Muck	0	0
Soft clay	1000	47.9
Silt	1500	71.8
Sandy silt	3000	143.6
Sand	4000	191.5
Sand clay	6000	287.3
Hard clay	9000	430.9

Note: Although the bearing strength values in this table have been used successfully in the design of thrust blocks and are considered to be conservative, their accuracy is totally dependent on accurate soil identification and evaluation. The ultimate responsibility for selecting the proper bearing strength of a particular soil type must rest with the design engineer.

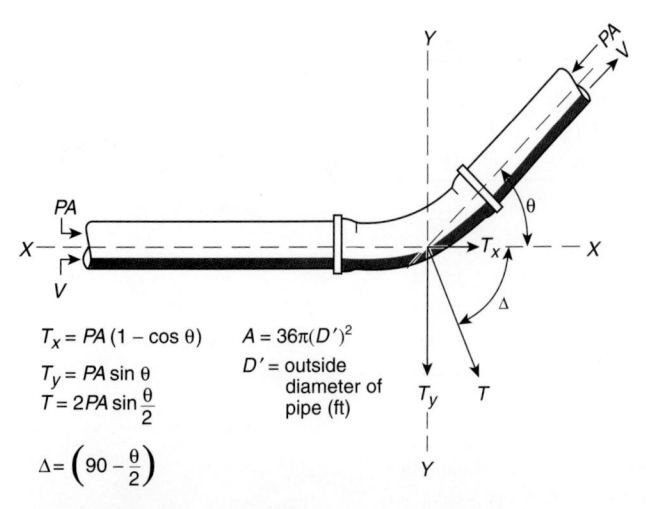

$T_x = PA(1 - \cos \theta)$ $A = 36\pi(D')^2$
$T_y = PA \sin \theta$
$T = 2PA \sin \dfrac{\theta}{2}$ D' = outside diameter of pipe (ft)

$\Delta = \left(90 - \dfrac{\theta}{2}\right)$

T = thrust force resulting from change in direction of flow (lbf)
T_x = component of thrust force acting parallel to original direction of flow (lbf)
T_y = component of thrust force acting perpendicular to original direction of flow (lbf)
P = water pressure (psi^2)
A = cross-sectional area of pipe based on outside diameter (in.^{2})
V = velocity in direction of flow

FIGURE A.10.8.2(a) *Thrust Forces Acting on Bend.*

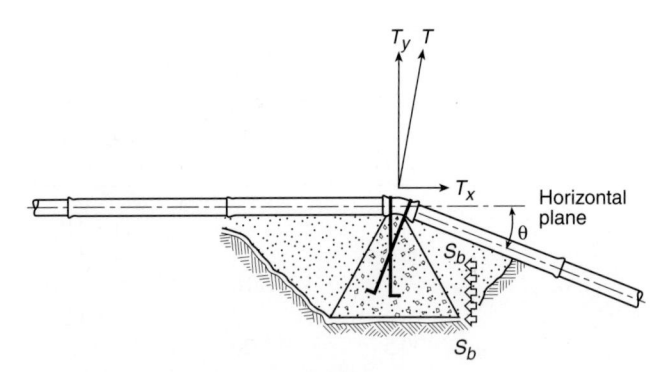

T = thrust force resulting from change of direction of flow
T_x = horizontal component of thrust force
T_y = vertical component of thrust force
S_b = horizontal bearing strength of soil

FIGURE A.10.8.2(c) *Gravity Thrust Block.*

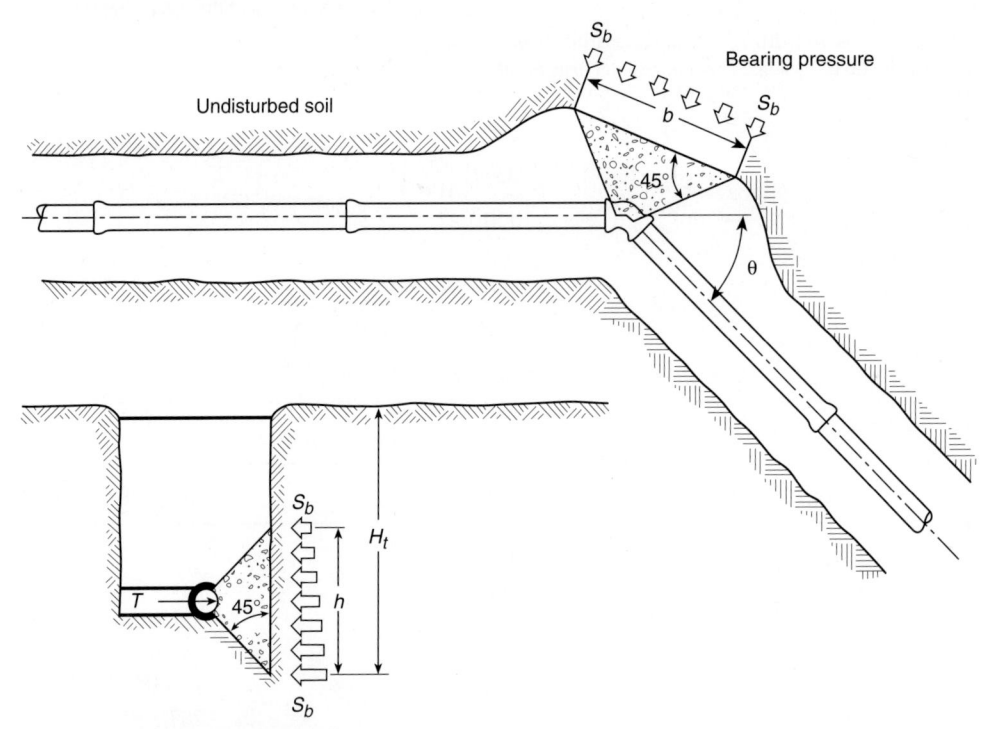

T = thrust force resulting from change in direction of flow
S_b = horizontal bearing strength of soil
h = block height
H_t = total depth to bottom of block

FIGURE A.10.8.2(b) *Bearing Thrust Block.*

The required block area (Ab) is as follows:

$$A_b = (h)(b) = \frac{T(S_f)}{S_b}$$

where:
A_b = required block area (ft²)
h = block height (ft)
b = calculated block width (ft)
T = thrust force (lbf)
S_f = safety factor (usually 1.5)
S_b = bearing strength (lb/ft²)

Then, for a horizontal bend, the following formula is used:

$$b = \frac{2(S_f)(P)(A)\sin\left(\frac{\theta}{2}\right)}{(h)(S_b)}$$

where:
b = calculated block width (ft)
S_f = safety factor (usually 1.5 for thrust block design)
P = water pressure (lb/in.²)
A = cross-sectional area of pipe based on outside diameter
h = block height (ft)
S_b = horizontal bearing strength of soil (lb/ft²)(in.²)

A similar approach can be used to design bearing blocks to resist the thrust forces at locations such as tees and dead ends. Typical values for conservative horizontal bearing strengths of various soil types are listed in Table A.10.8.2(c).

In lieu of the values for soil bearing strength shown in Table A.10.8.2(c), a designer might choose to use calculated Rankine passive pressure (P_p) or other determination of soil bearing strength based on actual soil properties.

It can be easily shown that $T_y = PA \sin \theta$. The required volume of the block is as follows:

$$V_g = \frac{S_f PA \sin \theta}{W_m}$$

where:
V_g = block volume (ft³)
S_f = safety factor
P = water pressure (psi)
A = cross-sectional area of pipe interior
W_m = density of block material (lb/ft³)

In a case such as the one shown, the horizontal component of thrust force is calculated as follows:

$$T_x = PA(1 - \cos \theta)$$

where:
T_x = horizontal component of thrust force
P = water pressure (psi)
A = cross-sectional area of pipe interior

The horizontal component of thrust force must be resisted by the bearing of the right side of the block against the soil. Analysis of this aspect follows the same principles as the previous section on bearing blocks.

The soil bearing strength value to be used for sizing thrust blocks is the horizontal bearing strength, which is shown in Table A.10.8.2(b). Vertical bearing strength values are often determined for purposes of structural support but are not necessarily the same as vertical bearing strength values.

Table A.10.8.2(c) provides a simplified approach to calculating the required bearing area. The tabular areas are based on the simple conditions indicated in the note to the table, and ratios can simply be applied for conditions other than 90-degree bends, 100 psi maximum pressure, and soil horizontal bearing strength of 1000 psi/ft², as indicated in the example following the notes to the table.

CALCULATION ▶

The following is a sample calculation to determine the size of a thrust block using the formula in A.10.8.2(5), Figure A.10.8.2(b), and Table A.10.8.2(b):

Design parameters: 12 in. ductile iron pipe (A) = 136.8 in.²
45-degree bend (θ)
85 psi water pressure at bend (P)
Soil type = sandy silt
Safety factor (S_f) = 1.5
Block height (h) = 1.5 ft

The cross-sectional area of the pipe is determined by the following formula:

$$A = 36\pi(D')^2$$

where:

D' = outside diameter of the pipe (ft)

Therefore:

$$A = 36\pi(1.1)^2$$
$$A = 136.8 \text{ in.}^2$$

Using the formula in A.10.8.2(5):

$$b = \frac{2(1.5)(85)(136.8)\sin(45/2)}{(1.5)(3000)}$$

$$b = \frac{13325.688}{4500}$$

$$b = 2.96 \text{ ft}$$

Note that the result of 2.96 ft is consistent with A.10.8.2(4) in that the block base (b = 2.96 ft) is between one to two times the height (h = 1.5 ft). This method is the correct method for selecting the size and shape of a thrust block.

10.8.2.1 Thrust blocks shall be considered satisfactory where soil is suitable for their use.

10.8.2.2 Thrust blocks shall be of a concrete mix not leaner than one part cement, two and one-half parts sand, and five parts stone.

10.8.2.3 Thrust blocks shall be placed between undisturbed earth and the fitting to be restrained and shall be capable of resisting the calculated thrust forces.

10.8.2.4 Wherever possible, thrust blocks shall be placed so that the joints are accessible for repair.

10.8.3* Restrained Joint Systems. Fire mains utilizing restrained joint systems shall include one or more of the following:

(1) Locking mechanical or push-on joints
(2) Mechanical joints utilizing setscrew retainer glands

(3) Bolted flange joints
(4) Heat-fused or welded joints
(5) Pipe clamps and tie rods
(6) Threaded or grooved joints

> Any one of the restraining systems listed in 10.8.3 is considered an acceptable means of restraining a joint. Flange adapter fittings and other mechanical joints do not necessarily provide joint restraint. For example, flange adapter fittings are listed by Underwriters Laboratories as "fittings, retainer type" in UL 194, *Standard for Gasketed Joints for Ductile-Iron Pipe and Fittings for Fire Protection Service*. UL 194 further categorizes such fittings as either "gasketed joints with self-restraining feature" or "gasketed joints without self-restraining feature." The latter includes gasketed joints consisting of merely a pipe or fitting bell and a spigot end without a feature for joint restraint. Under the provisions of UL 194, all gasketed joints are provided with a pressure test at twice the rated working pressure, with their joints deflected to the maximum angle specified by the manufacturer. Joints for which the manufacturer claims a self-restraining feature are not provided with external restraint during the test. The manufacturer's literature should specify whether external restraint is required. An example of a restrained joint system is shown in Exhibit III.22.
>
> Rods and clamps can be used as an alternative to the restrained joint systems shown in Figure A.10.8.3.

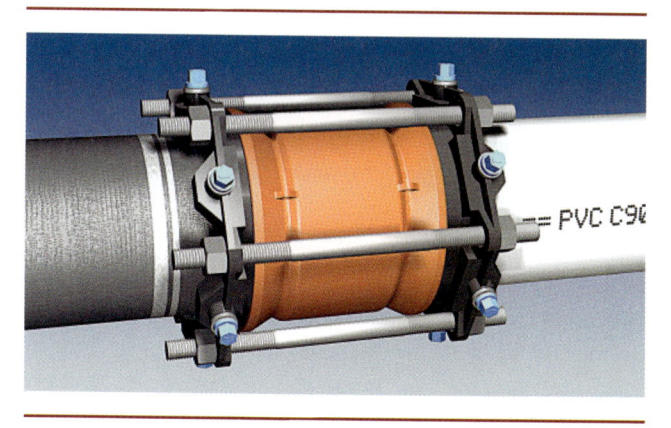

EXHIBIT III.22 *Restrained Joint System. (Courtesy of EBAA Iron, Inc.)*

(7) Other approved methods or devices

A.10.8.3 A method for providing thrust restraint is the use of restrained joints. A restrained joint is a special type of joint that is designed to provide longitudinal restraint. Restrained joint systems function in a manner similar to that of thrust blocks, insofar as the reaction of the entire restrained unit of piping with the soil balances the thrust forces.

The objective in designing a restrained joint thrust restraint system is to determine the length of pipe that must be restrained on each side of the focus of the thrust force. This will be a function of the pipe size, the internal pressure, the depth of cover, and the characteristics of the solid surrounding the pipe.

The following documents apply to the design, calculation, and determination of restrained joint systems:

(1) *Thrust Restraint Design for Ductile Iron Pipe*, Ductile Iron Pipe Research Association
(2) AWWA M41, *Ductile Iron Pipe and Fittings*
(3) AWWA M9, *Concrete Pressure Pipe*

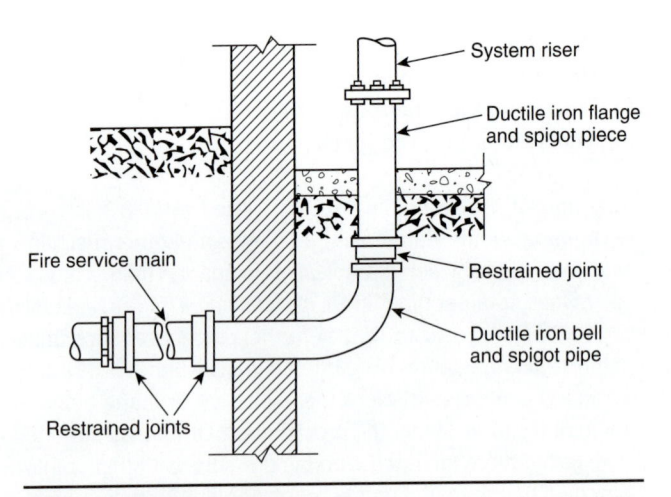

FIGURE A.10.8.3 Typical Connection to Fire Protection System Riser Illustrating Restrained Joints.

(4) AWWA M11, *A Guide for Steel Pipe Design and Installation*
(5) *Thrust Restraint Design Equations and Tables for Ductile Iron and PVC Pipe*, EBAA Iron, Inc.

Figure A.10.8.3 shows an example of a typical connection to a fire protection system riser utilizing restrained joint pipe.

10.8.3.1 Sizing Clamps, Rods, Bolts, and Washers.

10.8.3.1.1 Clamps.

10.8.3.1.1.1 Clamps shall have the following dimensions:

(1) ½ in. × 2 in. (12.7 mm × 50.8 mm) for 4 in. (102 mm) to 6 in. (152 mm) pipe
(2) ⅝ in. × 2½ in. (15.9 mm × 63.5 mm) for 8 in. (204 mm) to 10 in. (254 mm) pipe
(3) ⅝ in. × 3 in. (15.9 mm × 76.2 mm) for 12 in. (305 mm) pipe

10.8.3.1.1.2 The diameter of a bolt hole shall be ¹⁄₁₆ in. (1.6 mm) larger than that of the corresponding bolt.

10.8.3.1.2 Rods.

10.8.3.1.2.1 Rods shall be not less than ⅝ in. (15.9 mm) in diameter.

10.8.3.1.2.2 Table 10.8.3.1.2.2 provides the numbers of various diameter rods that shall be used for a given pipe size.

10.8.3.1.2.3 Where using bolting rods, the diameter of mechanical joint bolts shall limit the diameter of rods to ¾ in. (19.1 mm).

10.8.3.1.2.4 Threaded sections of rods shall not be formed or bent.

10.8.3.1.2.5 Where using clamps, rods shall be used in pairs for each clamp.

10.8.3.1.2.6 Assemblies in which a restraint is made by means of two clamps canted on the barrel of the pipe shall be permitted to use one rod per clamp if approved for the specific installation by the authority having jurisdiction.

TABLE 10.8.3.1.2.2 *Rod Number — Diameter Combinations*

Nominal Pipe Size (in.)	⅝ in. (15.9 mm)	¾ in. (19.1 mm)	⅞ in. (22.2 mm)	1 in. (25.4 mm)
4	2	—	—	—
6	2	—	—	—
8	3	2	—	—
10	4	3	2	—
12	6	4	3	2
14	8	5	4	3
16	10	7	5	4

Note: This table has been derived using pressure of 225 psi (15.5 bar) and design stress of 25,000 psi (172.4 MPa).

10.8.3.1.2.7 Where using combinations of rods, the rods shall be symmetrically spaced.

10.8.3.1.3 Clamp Bolts. Clamp bolts shall have the following diameters:

(1) ⅝ in. (15.9 mm) for pipe 4 in. (102 mm), 6 in. (152 mm), and 8 in. (204 mm)
(2) ¾ in. (19.1 mm) for 10 in. (254 mm) pipe
(3) ⅞ in. (22.2 mm) for 12 in. (305 mm) pipe

10.8.3.1.4 Washers.

10.8.3.1.4.1 Washers shall be permitted to be cast iron or steel and round or square.

10.8.3.1.4.2 Cast iron washers shall have the following dimensions:

(1) ⅝ in. × 3 in. (15.9 mm × 76.2 mm) for 4 in. (102 mm), 6 in. (152 mm), 8 in. (204 mm), and 10 in. (254 mm) pipe
(2) ¾ in. × 3½ in. (19.1 mm × 88.9 mm) for 12 in. (305 mm) pipe

10.8.3.1.4.3 Steel washers shall have the following dimensions:

(1) ½ in. × 3 in. (12.7 mm × 76.2 mm) for 4 in. (102 mm), 6 in. (152 mm), 8 in. (204 mm), and 10 in. (254 mm) pipe
(2) ½ in. × 3½ in. (12.7 mm × 88.9 mm) for 12 in. (305 mm) pipe

10.8.3.1.4.4 The diameter of holes shall be ⅛ in. (3.2 mm) larger than that of rods.

10.8.3.2 Sizes of Restraint Straps for Tees.

10.8.3.2.1 Restraint straps for tees shall have the following dimensions:

(1) ⅝ in. (15.9 mm) thick and 2½ in. (63.5 mm) wide for 4 in. (102 mm), 6 in. (152 mm), 8 in. (204 mm), and 10 in. (254 mm) pipe
(2) ⅝ in. (15.9 mm) thick and 3 in. (76.2 mm) wide for 12 in. (305 mm) pipe

10.8.3.2.2 The diameter of rod holes shall be ¹⁄₁₆ in. (1.6 mm) larger than that of rods.

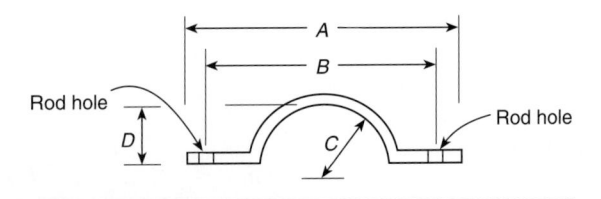

FIGURE 10.8.3.2.3 *Restraint Strap for Tees*

TABLE 10.8.3.2.3 *Restraint Straps for Tees*

Nominal Pipe Size (in.)	A in.	A mm	B in.	B mm	C in.	C mm	D in.	D mm
4	12½	318	10⅛	257	2½	64	1¾	44
6	14½	368	12⅛	308	3 9/16	90	2 13/16	71
8	16¾	425	14⅜	365	4 21/32	118	3 29/32	99
10	19 1/16	484	16 11/16	424	5¾	146	5	127
12	22 5/16	567	19 3/16	487	6¾	171	5⅞	149

10.8.3.2.3 Figure 10.8.3.2.3 and Table 10.8.3.2.3 shall be used in sizing the restraint straps for both mechanical and push-on joint tee fittings.

10.8.3.3 Sizes of Plug Strap for Bell End of Pipe.

10.8.3.3.1 The strap shall be ¾ in. (19.1 mm) thick and 2½ in. (63.5 mm) wide.

10.8.3.3.2 The strap length shall be the same as dimension *A* for tee straps as shown in Figure 10.8.3.2.3.

10.8.3.3.3 The distance between the centers of rod holes shall be the same as dimension *B* for tee straps as shown in Figure 10.8.3.2.3.

10.8.3.4 Material. Clamps, rods, rod couplings or turnbuckles, bolts, washers, restraint straps, and plug straps shall be of a material that has physical and chemical characteristics that indicate its deterioration under stress can be predicted with reliability.

10.8.3.5* Corrosion Resistance. After installation, rods, nuts, bolts, washers, clamps, and other restraining devices shall be cleaned and thoroughly coated with a bituminous or other acceptable corrosion-retarding material.

A.10.8.3.5 Examples of materials and the standards covering these materials are as follows:

(1) Clamps, steel *(see discussion on steel in the following paragraph)*
(2) Rods, steel *(see discussion on steel in the following paragraph)*
(3) Bolts, steel (ASTM A 307)
(4) Washers, steel *(see discussion on steel in the following paragraph)*; cast iron (Class A cast iron as defined by ASTM A 126)
(5) Anchor straps and plug straps, steel *(see discussion on steel in the following paragraph)*
(6) Rod couplings or turnbuckles, malleable iron (ASTM A 197)

Steel of modified range merchant quality as defined in U.S. Federal Standard No. 66C, April 18, 1967, change notice No. 2, April 16, 1970, as promulgated by the U.S. Federal Government General Services Administration.

The materials specified in A.10.8.3.5(1) through (6) do not preclude the use of other materials that also satisfy the requirements of this section.

10.9 Backfilling

10.9.1 Backfill shall be tamped in layers or puddled under and around pipes to prevent settlement or lateral movement and shall contain no ashes, cinders, refuse, organic matter, or other corrosive materials.

> Improper backfill is a major cause of underground piping failures. Proper consolidation of backfill can prevent voids that eventually place stress on the piping and joints.
>
> Underground piping should be laid on a firm bed of earth for its entire length, with the earth scooped out at the joints. Clean earth or screened gravel should be tamped under and around the pipe and above the pipe to a level of 1 ft (0.3 m). The excavation should then be backfilled to grade, compacting the fill layers in the trench as they are added. Puddling, which involves the use of water to help consolidate soil around and below the piping, is occasionally used to assist the backfill compaction effort.

10.9.2 Rocks shall not be placed in trenches.

> Subsection 10.9.2 prohibits the placement of rocks in trenches. Clean fill cushions the pipe and evenly distributes the load to the surrounding earth. The presence of cinders, refuse, and other organic matter can create points of accelerated corrosion that can reduce the life of underground piping.

10.9.3 Frozen earth shall not be used for backfilling.

10.9.4 In trenches cut through rock, tamped backfill shall be used for at least 6 in. (150 mm) under and around the pipe and for at least 2 ft (0.6 m) above the pipe.

10.10 Testing and Acceptance

10.10.1 Approval of Underground Piping. The installing contractor shall be responsible for the following:

(1) Notifying the authority having jurisdiction and the owner's representative of the time and date testing is to be performed
(2) Performing all required acceptance tests
(3) Completing and signing the contractor's material and test certificate(s) shown in Figure 10.10.1

10.10.2 Acceptance Requirements.

10.10.2.1* Flushing of Piping.

A.10.10.2.1 Underground mains and lead-in connections to system risers should be flushed through hydrants at dead ends of the system or through accessible aboveground flushing outlets

Contractor's Material and Test Certificate for Underground Piping

PROCEDURE
Upon completion of work, inspection and tests shall be made by the contractor's representative and witnessed by an owner's representative. All defects shall be corrected and system left in service before contractor's personnel finally leave the job.

A certificate shall be filled out and signed by both representatives. Copies shall be prepared for approving authorities, owners, and contractor. It is understood the owner's representative's signature in no way prejudices any claim against contractor for faulty material, poor workmanship, or failure to comply with approving authority's requirements or local ordinances.

Property name	Date
Property address	

Plans	Accepted by approving authorities (names)	
	Address	
	Installation conforms to accepted plans	☐ Yes ☐ No
	Equipment used is approved If no, state deviations	☐ Yes ☐ No
Instructions	Has person in charge of fire equipment been instructed as to location of control valves and care and maintenance of this new equipment? If no, explain	☐ Yes ☐ No
	Have copies of appropriate instructions and care and maintenance charts been left on premises? If no, explain	☐ Yes ☐ No
Location	Supplies buildings	

Underground pipes and joints	Pipe types and class	Type joint
	Pipe conforms to _____ standard	☐ Yes ☐ No
	Fittings conform to _____ standard If no, explain	☐ Yes ☐ No
	Joints needing anchorage clamped, strapped, or blocked in accordance with _____ standard If no, explain	☐ Yes ☐ No

Test description

Flushing: Flow the required rate until water is clear as indicated by no collection of foreign material in burlap bags at outlets such as hydrants and blow-offs. Flush at one of the flow rates as specified in 10.10.2.1.3.

Hydrostatic: All piping and attached appurtenances subjected to system working pressure shall be hydrostatically tested at 200 psi (13.8 bar) or 50 psi (3.5 bar) in excess of the system working pressure, whichever is greater, and shall maintain that pressure ±5 psi (0.35 bar) for 2 hours.

Hydrostatic Testing Allowance: Where additional water is added to the system to maintain the test pressures required by 10.10.2.2.1, the amount of water shall be measured and shall not exceed the limits of the following equation (for metric equation, see 10.10.2.2.6):

$$L = \frac{SD\sqrt{P}}{148{,}000}$$

L = testing allowance (makeup water), in gallons per hour
S = length of pipe tested, in feet
D = nominal diameter of the pipe, in inches
P = average test pressure during the hydrostatic test, in pounds per square inch (gauge)

Flushing tests

New underground piping flushed according to _____ standard by (company) If no, explain	☐ Yes ☐ No

How flushing flow was obtained
☐ Public water   ☐ Tank or reservoir   ☐ Fire pump

Through what type opening
☐ Hydrant butt   ☐ Open pipe

Lead-ins flushed according to _____ standard by (company)
If no, explain   ☐ Yes ☐ No

How flushing flow was obtained
☐ Public water   ☐ Tank or reservoir   ☐ Fire pump

Through what type opening
☐ Y connection to flange and spigot   ☐ Open pipe

© 2012 National Fire Protection Association   NFPA 24 (p. 1 of 2)

FIGURE 10.10.1 *Sample of Contractor's Material and Test Certificate for Underground Piping.*

Hydrostatic test	All new underground piping hydrostatically tested at _____ psi for _____ hours		Joints covered ☐ Yes ☐ No
Leakage test	Total amount of leakage measured _____ gallons _____ hours		
	Allowable leakage _____ gallons _____ hours		
Forward flow test of backflow preventer	Foward flow test performed in accordance with 10.10.2.5.2:		☐ Yes ☐ No
Hydrants	Number installed	Type and make	All operate satisfactorily ☐ Yes ☐ No
Control valves	Water control valves left wide open If no, state reason		☐ Yes ☐ No
	Hose threads of fire department connections and hydrants interchangeable with those of fire department answering alarm		☐ Yes ☐ No
Remarks	Date left in service		
Signatures	Name of installing contractor		
	Tests witnessed by		
	For property owner (signed)	Title	Date
	For installing contractor (signed)	Title	Date

Additional explanation and notes

© 2012 National Fire Protection Association   NFPA 24 (p. 2 of 2)

FIGURE 10.10.1 *Continued*

allowing the water to run until clear. Figure A.10.10.2.1 shows acceptable examples of flushing the system. If water is supplied from more than one source or from a looped system, divisional valves should be closed to produce a high-velocity flow through each single line. The flows specified in Table 10.10.2.1.3 will produce a velocity of at least 10 ft/sec (3 m/sec), which is necessary for cleaning the pipe and for lifting foreign material to an aboveground flushing outlet.

10.10.2.1.1 Underground piping, from the water supply to the system riser, and lead-in connections to the system riser shall be completely flushed before the connection is made to downstream fire protection system piping.

10.10.2.1.2 The flushing operation shall be continued for a sufficient time to ensure thorough cleaning.

10.10.2.1.3 The minimum rate of flow shall be not less than one of the following:

(1) Hydraulically calculated water demand flow rate of the system, including any hose requirements
(2)* Flow in accordance with Table 10.10.2.1.3
(3) Maximum flow rate available to the system under fire conditions

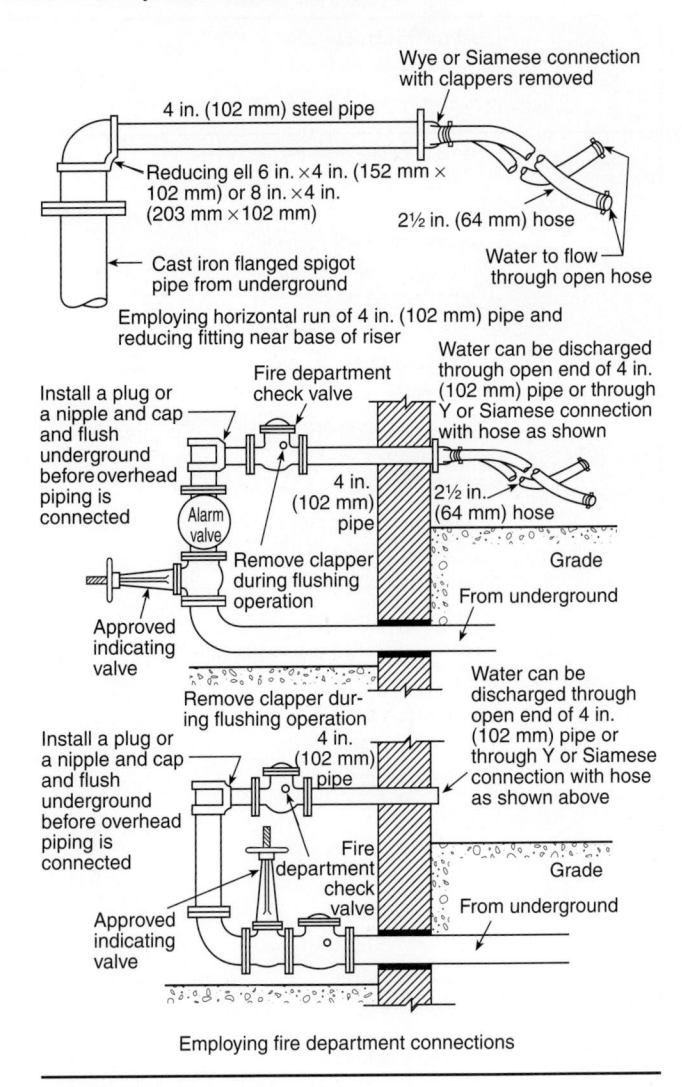

FIGURE A.10.10.2.1 Methods of Flushing Water Supply Connections.

TABLE 10.10.2.1.3 *Flow Required to Produce Velocity of 10 ft/sec (3 m/sec) in Pipes*

Nominal Pipe Size		Flow Rate	
in.	mm	gpm	L/min
2	51	100	379
2½	63	150	568
3	76	220	833
4	102	390	1,476
5	127	610	2,309
6	152	880	3,331
8	204	1,560	5,905
10	254	2,440	9,235
12	305	3,520	13,323

In some cases, the fire protection water supply may not be capable of producing the flows listed in Table 10.10.2.1.3. If the flow is not available, a flow equal to or greater than the fire protection system's maximum demand should be used.

Stones, gravel, blocks of wood, bottles, work tools, work clothes, and other objects have been found in piping when flushing procedures were performed. Objects in underground piping that are quite remote from a sprinkler installation — and that would otherwise remain stationary — can sometimes be carried into sprinkler system piping when sprinkler systems operate. Sprinkler systems can draw greater flows than most domestic or process uses. Fire department pumpers taking suction from hydrants for pumping into sprinkler systems and normal fire-fighting operations further increase flow rates and velocities and can dislodge other materials in the piping network, forcing them into sprinkler system piping.

Because of the inherent nature of sprinkler system design in which pipe sizes usually decrease beginning at the point of connection to the underground piping, objects that move from the underground piping into the sprinkler system can become lodged at a point in the system where they may obstruct the passage of water. Exhibit III.23 illustrates material captured while flushing an underground main.

Studies by FM Global concluded that the size of objects that move upward in piped water streams can be determined if the density of the object and the velocity of the water stream are known. For example, granite that is 2 in. (51 mm) in diameter will move upward in piping if the water stream velocity is 5.64 ft/sec (1.72 m/sec). The flow rates in Table 10.10.2.1.3 reflect velocities of approximately 10 ft/sec (3.1 m/sec), which is generally agreed upon as a reasonably fast flow capable of removing most obstructing debris.

A.10.10.2.1.3(2) The velocity of approximately 10 ft/sec (3.1 m/sec) was used to develop Table 10.10.2.1.3 because this velocity has been shown to be sufficient for moving obstructive material out of the pipes. It is not important that the velocity equal exactly 10 ft/sec (3.1 m/sec), so there is no reason to increase the flow during the test for slightly different internal pipe dimensions. Note that where underground pipe serves as suction pipe for a fire pump, NFPA 20 requires greater flows for flushing the pipe.

10.10.2.1.4 Provision shall be made for the proper disposal of water used for flushing or testing.

10.10.2.2 Hydrostatic Test.

EXHIBIT III.23 *Pipe Scale (Rust) and Work Gloves Found When Flushing an Underground Main. (Courtesy of John Jensen)*

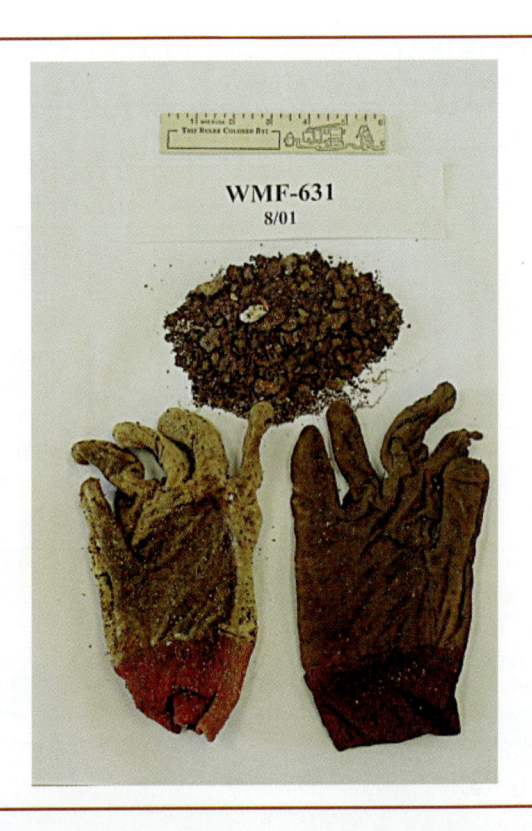

The trench must be backfilled prior to hydrostatic testing to prevent movement of underground piping. The backfilling can take place between joints if it is desired to observe the joints for leakage and if the backfill depth is sufficient to prevent movement. As an alternative, the joints can also be covered, but the contractor remains responsible for locating and correcting excessive leakage.

A 2-hour hydrostatic test is required at not less than 200 psi (13.8 bar) and at not less than 50 psi (3.5 bar) above the maximum expected static pressure. The piping between an exterior fire department connection and the check valve in the connection's inlet pipe must also be hydrostatically tested. All thrust blocks should be hardened before testing takes place.

Allowable leakage rates during hydrostatic testing are based on the total length and diameter of the underground piping. In the past, the allowable leakage had been based on the number of gaskets or joints, regardless of pipe diameter.

When underground piping is repaired, it should be tested as a new installation is tested, but it is not required that all the underground piping be subjected to the hydrostatic test if it can be isolated. Blind flanges or skillets can be used for this purpose. Procedures must be implemented to ensure that these devices are removed following hydrostatic testing.

10.10.2.2.1* All piping and attached appurtenances subjected to system working pressure shall be hydrostatically tested at 200 psi (13.8 bar) or 50 psi (3.5 bar) in excess of the system working pressure, whichever is greater, and shall maintain that pressure at ±5 psi (0.35 bar) for 2 hours.

The removal of all air from the system piping when conducting the hydrostatic test is extremely important. Air in the piping causes a fluctuation in the test pressure. Any time the test pressure drops below the minimum specified [200 psi (13.8 bar), or 50 psi (3.5 bar) in excess of the system working pressure], the test must be repeated and the correct pressure held for the 2-hour duration.

A.10.10.2.2.1 A sprinkler system has for its water supply a connection to a public water service main. A 100 psi (6.9 bar) rated pump is installed in the connection. With a maximum normal public water supply of 70 psi (4.8 bar), at the low elevation point of the individual system or portion of the system being tested and a 120 psi (8.3 bar) pump (churn) pressure, the hydrostatic test pressure is 70 psi (4.8 bar) + 120 psi (8.3 bar) + 50 psi (3.5 bar), or 240 psi (16.5 bar).

To reduce the possibility of serious water damage in case of a break, pressure can be maintained by a small pump, the main controlling gate meanwhile being kept shut during the test.

Polybutylene pipe will undergo expansion during initial pressurization. In this case, a reduction in gauge pressure might not necessarily indicate a leak. The pressure reduction should not exceed the manufacturer's specifications and listing criteria.

When systems having rigid thermoplastic piping such as CPVC are pressure tested, the sprinkler system should be filled with water. The air should be bled from the highest and farthest sprinklers. Compressed air or compressed gas should never be used to test systems with rigid thermoplastic pipe.

A recommended test procedure is as follows: The water pressure is to be increased in 50 psi (3.5 bar) increments until the test pressure described in 10.10.2.2.1 is attained. After each increase in pressure, observations are to be made of the stability of the joints. These observations are to include such items as protrusion or extrusion of the gasket, leakage, or other factors likely to affect the continued use of a pipe in service. During the test, the pressure is not to be increased by the next increment until the joint has become stable. This applies particularly to movement of the gasket. After the pressure has been increased to the required maximum value and held for 1 hour, the pressure is to be decreased to 0 psi while observations are made for leakage. The pressure is again to be slowly increased to the value specified in 10.10.2.2.1 and held for 1 more hour while observations are made for leakage and the leakage measurement is made.

10.10.2.2.2 Pressure loss shall be determined by a drop in gauge pressure or visual leakage.

10.10.2.2.3 The test pressure shall be read from one of the following, located at the lowest elevation of the system or the portion of the system being tested:

(1) A gauge located at one of the hydrant outlets
(2) A gauge located at the lowest point where no hydrants are provided

> Note that the gauge is required to be located at the lowest portion of the system. If located at the highest portion of the system, in some cases, an excessively high test pressure would result in the lower portions of the system. These high pressures could cause failure of some system components.

10.10.2.2.4* The trench shall be backfilled between joints before testing to prevent movement of pipe.

> Backfilling the trench between the pipe joints helps keep the pipe in place while the trench (or the piping) is being filled and pressurized.

A.10.10.2.2.4 Hydrostatic tests should be made before the joints are covered, so that any leaks can be detected. Thrust blocks should be sufficiently hardened before hydrostatic testing is begun. If the joints are covered with backfill prior to testing, the contractor remains responsible for locating and correcting any leakage in excess of that permitted.

10.10.2.2.5 Where required for safety measures presented by the hazards of open trenches, the pipe and joints shall be permitted to be backfilled, provided the installing contractor takes the responsibility for locating and correcting leakage.

10.10.2.2.6* Hydrostatic Testing Allowance. Where additional water is added to the system to maintain the test pressures required by 10.10.2.2.1, the amount of water shall be measured and shall not exceed the limits of Table 10.10.2.2.6, which are based upon the following equations:

U.S. Customary Units:

$$L = \frac{SD\sqrt{P}}{148,000} \qquad [10.10.2.2.6(a)]$$

where:
L = testing allowance (makeup water) [gph (gal/hr)]
S = length of pipe tested (ft)
D = nominal diameter of pipe (in.)
P = average test pressure during hydrostatic test (gauge psi)

Metric Units:

$$L = \frac{SD\sqrt{P}}{794,797} \qquad [10.10.2.2.6(b)]$$

where:
L = testing allowance (makeup water) (L/hr)
S = length of pipe tested (m)
D = nominal diameter of pipe (mm)
P = average test pressure during hydrostatic test (kPa)

TABLE 10.10.2.2.6 Hydrostatic Testing Allowance at 200 psi (gph/100 ft of Pipe)

Nominal Pipe Diameter (in.)	Testing Allowance
2	0.019
4	0.038
6	0.057
8	0.076
10	0.096
12	0.115
14	0.134
16	0.153
18	0.172
20	0.191
24	0.229

Notes:
(1) For other length, diameters, and pressures, utilize Equation 10.10.2.2.6(a) or 10.10.2.2.6(b) to determine the appropriate testing allowance.
(2) For test sections that contain various sizes and sections of pipe, the testing allowance is the sum of the testing allowances for each size and section.

> **◀ CALCULATION**
>
> Using equation 10.10.2.2.6(a), if testing 350 ft of a 10 in. main at an average of 203 psi during the duration of the test,
>
> $$L = \frac{(350)(10)\sqrt{203}}{148,000}$$
>
> $L = 0.3369$ gph (from Table 10.10.2.2.6: 0.096 gph loss is permitted per 100 ft of pipe.)
>
> $0.096 \times 3.5 = 0.336$ gph
>
> When testing 350 ft of 10 in. pipe at an average test pressure of 203 psi a total of 0.67 gal of water is permitted to be lost.

A.10.10.2.2.6 One acceptable means of completing this test is to utilize a pressure pump that draws its water supply from a full container. At the completion of the 2-hour test, the amount of water to refill the container can be measured to determine the amount of makeup water. In order to minimize pressure loss, the piping should be flushed to remove any trapped air. Additionally, the piping should be pressurized for 1 day prior to the hydrostatic test to account for expansion, absorption, entrapped air, and so on.

The use of a blind flange or skillet is preferred for hydrostatically testing segments of new work. Metal-seated valves are susceptible to developing slight imperfections during transport, installation, and operation and thus can be likely to leak more than 1 fl oz/in. (1.2 mL/mm) of valve diameter per hour. For this reason, the blind flange should be used when hydrostatically testing.

10.10.2.3 Other Means of Hydrostatic Tests. Where required by the authority having jurisdiction, hydrostatic tests shall be permitted to be completed in accordance with the requirements of AWWA C600, AWWA C602, AWWA C603, and AWWA C900.

10.10.2.4 Operating Test.

10.10.2.4.1 Each hydrant shall be fully opened and closed under system water pressure.

> Exercising the hydrant in this fashion should be done annually after the hydrant is placed into service. Threaded outlets should be lubricated and checked for burrs or potential cross-threading at this time.

10.10.2.4.2 Dry barrel hydrants shall be checked for proper drainage.

> The hydrant should completely drain within 60 minutes.

10.10.2.4.3 All control valves shall be fully closed and opened under system water pressure to ensure proper operation.

> PIVs should be opened completely until pressure is felt in the connecting rod. This pressure is an indication that the rod is still connected to the valve. Additional pressure should be exerted on the handle at this time and then suddenly released to determine if the handle "springs" back. The spring indicates that the valve is fully open and verifies that the gate is still attached to the handle. If the gate were jammed, it would be unlikely for the handle to spring back. If the gate were loose or detached from the handle, the handle would continue to turn with little or no resistance.
>
> Post indicator and gate valves should be turned back one-quarter turn from the fully open position to prevent jamming.

10.10.2.4.4 Where fire pumps are available, the operating tests required by 10.10.2.4 shall be completed with the pumps running.

10.10.2.5 Backflow Prevention Assemblies.

10.10.2.5.1 The backflow prevention assembly shall be forward flow tested to ensure proper operation.

10.10.2.5.2 The minimum flow rate tested in 10.10.2.5.1 shall be the system demand, including hose stream demand where applicable.

> In addition to the forward flow test, local regulations may require a backflow test at the time of the forward flow test. The backflow test should also be documented and submitted to the building owner along with the contractor's material and test certificate.

CHAPTER 11 HYDRAULIC CALCULATIONS

CALCULATION ▶

When calculating friction loss through underground piping, the Hazen–Williams formula must be used. This formula is easily converted to table format to simplify the calculation process. The C value used must correspond to those listed in Commentary Table III.3.

COMMENTARY TABLE III.3 Hazen–Williams C Values*

Pipe or Tube	C Value*
Unlined cast iron or ductile iron	100
Black steel (dry systems including preaction)	100
Black steel (wet systems including deluge)	120
Galvanized (all)	120
Plastic (listed) all	150
Cement-lined cast iron or ductile iron	140
Copper tube or stainless steel	150
Asbestos cement	140
Concrete	140

*The authority having jurisdiction is permitted to consider other C values.

Source: NFPA 13, 2013, Table 23.4.4.7.1

11.1* Calculations in U.S. Customary Units

Pipe friction losses shall be determined based on the Hazen–Williams formula, as follows:

$$p = \frac{4.52 Q^{1.85}}{C^{1.85} d^{4.87}}$$

where:

p = frictional resistance (psi/ft of pipe)

Q = flow (gpm)

C = friction loss coefficient

d = actual internal diameter of pipe (in.)

A.11.1 When calculating the actual inside diameter of cement mortar–lined pipe, twice the thickness of the pipe wall and twice the thickness of the lining need to be subtracted from the outside diameter of the pipe. The actual lining thickness should be obtained from the manufacturer.

Table A.11.1(a) and Table A.11.1(b) indicate the minimum lining thickness.

TABLE A.11.1(a) Minimum Thickness of Lining for Ductile Iron Pipe and Fittings

Pipe and Fitting Size		Thickness of Lining	
in.	mm	in.	mm
3–12	76–305	1/16	1.6
14–24	356–610	3/32	2.4
30–64	762–1600	1/8	3.2

Source: AWWA C104.

TABLE A.11.1(b) Minimum Thickness of Lining for Steel Pipe

Nominal Pipe Size		Thickness of Lining		Tolerance	
in.	mm	in.	mm	in.	mm
4–10	100–250	1/4	6	−1/16, +1/8	−1.6, +3.2
11–23	280–580	5/16	8	−1/16, +1/8	−1.6, +3.2
24–36	600–900	3/8	10	−1/16, +1/8	−1.6, +3.2
>36	>900	1/2	13	−1/16, +3/16	−1.6, +4.8

Source: AWWA C205.

11.2 Calculations in SI Units

Pipe friction losses shall be determined based on the Hazen–Williams formula in SI units, as follows:

$$p_m = 6.05 \left(\frac{Q_m^{1.85}}{C^{1.85} d_m^{4.87}} \right) 10^5$$

where:

p_m = frictional resistance (bar/m of pipe)
Q_m = flow (L/min)
C = friction loss coefficient
d_m = actual internal diameter of pipe (mm)

CHAPTER 12 ABOVEGROUND PIPE AND FITTINGS

12.1 General

Aboveground pipe and fittings shall comply with the applicable sections of Chapters 6 and 8 of NFPA 13 that address pipe, fittings, joining methods, hangers, and installation.

12.2 Protection of Piping

12.2.1 Aboveground piping for private fire service mains shall not pass through hazardous areas and shall be located so that it is protected from mechanical and fire damage.

12.2.2 Aboveground piping shall be permitted to be located in hazardous areas protected by an automatic sprinkler system.

12.2.3 Where aboveground water-filled supply pipes, risers, system risers, or feed mains pass through open areas, cold rooms, passageways, or other areas exposed to freezing temperatures, the pipe shall be protected against freezing by the following:

(1) Insulating coverings
(2) Frostproof casings
(3) Other reliable means capable of maintaining a minimum temperature between 40°F and 120°F (4°C and 48.9°C)

12.2.4 Where corrosive conditions exist or piping is exposed to the weather, corrosion-resistant types of pipe, fittings, and hangers or protective corrosion-resistant coatings shall be used.

12.2.5 To minimize or prevent pipe breakage where subject to earthquakes, aboveground pipe shall be protected in accordance with the seismic requirements of NFPA 13.

12.2.6 Mains that pass through walls, floors, and ceilings shall be provided with clearances in accordance with NFPA 13.

CHAPTER 13 SIZES OF ABOVEGROUND AND BURIED PIPE

13.1 Private Service Mains

Pipe smaller than 6 in. (152 mm) in diameter shall not be installed as a private service main supplying hydrants.

13.2 Mains Not Supplying Hydrants

For mains that do not supply hydrants, sizes smaller than 6 in. (152 mm) shall be permitted to be used, subject to the following restrictions:

(1) The main shall supply only the following types of systems:
   (a) Automatic sprinkler systems
   (b) Open sprinkler systems
   (c) Water spray fixed systems
   (d) Foam systems
   (e) Class II standpipe systems
(2) Hydraulic calculations shall show that the main is able to supply the total demand at the appropriate pressure.
(3) Systems that are not hydraulically calculated shall have a main at least as large as the riser.

13.3 Mains Supplying Fire Protection Systems

The size of private fire service mains supplying fire protection systems shall be approved by the authority having jurisdiction, and the following factors shall be considered:

(1) Construction and occupancy of the plant
(2) Fire flow and pressure of the water required
(3) Adequacy of the water supply

CHAPTER 14 SYSTEM INSPECTION, TESTING, AND MAINTENANCE

14.1 General

A private fire service main and its appurtenances installed in accordance with this standard shall be properly inspected, tested, and maintained in accordance with NFPA 25 to provide at least the same level of performance and protection as designed.

ANNEX B VALVE SUPERVISION ISSUES

This annex is not a part of the requirements of this NFPA document but is included for informational purposes only.

B.1 Responsibility

The management is responsible for the supervision of valves controlling the water supply for fire protection and should exert every effort to see that the valves are maintained in the normally open position. This effort includes special precautions to ensure that protection is promptly restored by completely opening valves that are necessarily closed during repairs or alterations. The precautions apply equally to the following:

(1) Valves controlling sprinklers and other fixed water-based fire suppression systems
(2) Hydrants
(3) Tanks
(4) Standpipes
(5) Pumps
(6) Street connections
(7) Sectional valves

Central station supervisory service systems or proprietary supervisory service systems, or a combination of these methods of valve supervision, as described in the following paragraphs, are considered essential to ensure that the valves controlling fire protection systems are in the normally open position. The methods described are intended as an aid to the person responsible for developing a systematic method of determining that the valves controlling sprinkler systems and other fire protection devices are open.

Continual vigilance is necessary if valves are to be kept in the open position. Responsible day and night employees should be familiar with the location of all valves and their proper use.

The authority having jurisdiction should be consulted as to the type of valve supervision required. Contracts for equipment should specify that all details are to be subject to the approval of the authority having jurisdiction.

B.2 Central Station Supervisory Service Systems

Central station supervisory service systems involve complete, constant, and automatic supervision of valves by electrically operated devices and circuits. The devices and circuits are continually under test and operate through an approved outside central station in compliance with *NFPA 72*. It is understood that only the portions of *NFPA 72* that relate to valve supervision should apply.

B.3 Proprietary Supervisory Service Systems

Proprietary supervisory service systems include systems in which the operation of a valve produces some form of signal and record at a common point by electrically operated devices and circuits. The device and circuits are continually under test and operate through a central supervising station at the protected property in compliance with the standards for the installation, maintenance, and use of local protective, auxiliary protective, remote-station protective, and proprietary signaling systems. It is understood that only the portions of the standards that relate to valve supervision should apply.

B.4 Locking and Sealing

The standard method of locking, sealing, and tagging valves to prevent, as far as possible, their unnecessary closing, to obtain notification of such closing, and to aid in restoring the valve to normal condition is a satisfactory alternative to valve supervision. The authority having jurisdiction should be consulted for details for specific cases.

Where electrical supervision is not provided, locks or seals should be provided on all valves and should be of a type acceptable to the authority having jurisdiction.

Seals can be marked to indicate the organization under whose jurisdiction the sealing is conducted. All seals should be attached to the valve in such a manner that the valves cannot be operated without breaking the seals. Seals should be of a character that prevents injury in handling and that prevents reassembly when broken. Where seals are used, valves should be inspected weekly. The authority having jurisdiction can require a valve tag to be used in conjunction with the sealing.

A padlock, with a chain where necessary, is especially desirable to prevent unauthorized closing of valves in areas where valves are subject to tampering. Where such locks are employed, valves should be inspected monthly.

If valves are locked, any distribution of keys should be restricted to only those directly responsible for the fire protection system. Multiple valves should not be locked together; they should be individually locked.

The individual performing inspections should determine that each valve is in the normal position and properly locked or sealed, and so noted on an appropriate record form while still at the valve. The authority having jurisdiction should be consulted for assistance in preparing a suitable report form for this activity.

Identification signs should be provided at each valve to indicate its function and what it controls.

The position of the spindle of OS&Y valves or the target on the indicator valves cannot be accepted as conclusive proof that the valve is fully open. The opening of the valve should be followed by a test to determine that the operating parts have functioned properly.

The test consists of opening the main drain valve and allowing a free flow of water until the gauge reading becomes stationary. If the pressure drop is excessive for the water supply involved, the cause should be determined immediately and the proper remedies taken. Where sectional valves or other special conditions are encountered, other methods of testing should be used.

If it becomes necessary to break a seal for emergency reasons, the valve, following the emergency, should be opened by the individual responsible for the fire protection of the plant or his or her designated representative. The responsible individual should apply a seal at the time of the valve opening. The seal should be maintained in place until such time as the authority having jurisdiction can replace it with a seal of its own.

Seals or locks should not be applied to valves that have been reopened after closure until such time as the inspection procedure is carried out.

Where water is shut off to the sprinkler or other fixed water-based fire suppression systems, a guard or other qualified person should be placed on duty and required to continuously patrol the affected sections of the premises until such time as protection is restored.

During specific critical situations, a responsible individual should be stationed at the valve so that the valve can be reopened promptly if necessary. It is the intent of this recommendation that the individual remain within sight of the valve and have no additional duties. This recommendation is considered imperative when fire protection is shut off immediately following a fire.

An inspection of all other fire protection equipment should be made prior to shutting off water in order to ensure that it is in operative condition.

Where changes to fire protection equipment are to be made, as much work as possible should be done in advance of shutting off the water, so that final connections can be made quickly and protection restored promptly. With careful planning, open outlets often can be plugged and protection can be restored on a portion of the equipment while the alterations are being made.

Where changes are to be made in underground piping, as much piping as possible should be laid before shutting off the water for final connections. Where possible, temporary feed lines, such as temporary piping for reconnection of risers by hose lines, should be used to afford maximum protection. The plant, public fire department, and other authorities having jurisdiction should be notified of all impairments to fire protection equipment.

ANNEX C RECOMMENDED PRACTICE FOR FIRE FLOW TESTING

Annex C of NFPA 24 is based on Chapter 4 of NFPA 291 and is, therefore, not reprinted here. See Part III, Section 1, of this handbook for NFPA 291, Chapter 4, and associated commentary.

ANNEX D RECOMMENDED PRACTICE FOR MARKING OF HYDRANTS

Annex D of NFPA 24 is based on Chapter 5 of NFPA 291 and is, therefore, not reprinted here. See Part III, Section 1, of this handbook for NFPA 291, Chapter 5, and associated commentary.

ANNEX E INFORMATIONAL REFERENCES

E.1 Referenced Publications

The documents or portions thereof listed in this annex are referenced within the informational sections of this standard and are not part of the requirements of this document unless also listed in Chapter 2 for other reasons.

E.1.1 NFPA Publications. National Fire Protection Association, 1 Batterymarch Park, Quincy, MA 02169-7471.

NFPA 20, *Standard for the Installation of Stationary Pumps for Fire Protection,* 2013 edition.

NFPA 22, *Standard for Water Tanks for Private Fire Protection,* 2008 edition.

NFPA 70®, National Electrical Code®, 2011 edition.

NFPA 72®, National Fire Alarm and Signaling Code, 2013 edition.

NFPA 291, *Recommended Practice for Fire Flow Testing and Marking of Hydrants,* 2010 edition.

NFPA 780, *Standard for the Installation of Lightning Protection Systems,* 2011 edition.

NFPA 1962, *Standard for the Inspection, Care, and Use of Fire Hose, Couplings, and Nozzles and the Service Testing of Fire Hose,* 2008 edition.

E.1.2 Other Publications.

E.1.2.1 ACPA Publications. American Concrete Pipe Association, 1303 West Walnut Hill Lane, Suite 305, Irving, TX 75038-3008.

Concrete Pipe Handbook.

E.1.2.2 ASME Publications. American Society of Mechanical Engineers, Three Park Avenue, New York, NY 10016-5990.

ASME B16.1, *Cast Iron Pipe Flanges and Flanged Fittings*, 1989.

E.1.2.3 ASTM Publications. ASTM International, 100 Barr Harbor Drive, P.O. Box C700, West Conshohocken, PA 19428-2959.

ASTM A 126, *Standard Specification for Gray Iron Castings for Valves, Flanges and Pipe Fittings*, 1993.

ASTM A 197, *Standard Specification for Cupola Malleable Iron*, 1987.

ASTM A 307, *Standard Specification for Carbon Steel Bolts and Studs*, 1994.

ASTM C 296, *Standard Specification for Asbestos-Cement Pressure Pipe*, 1988.

IEEE/ASTM-SI-10, *Standard for Use of the International System of Units (SI): The Modern Metric System*, 1997.

E.1.2.4 AWWA Publications. American Water Works Association, 6666 West Quincy Avenue, Denver, CO 80235.

AWWA C104, *Cement Mortar Lining for Ductile Iron Pipe and Fittings for Water*, 2008.

AWWA C105, *Polyethylene Encasement for Ductile Iron Pipe Systems*, 2005.

AWWA C110, *Ductile Iron and Gray Iron Fittings*, 2008.

AWWA C111, *Rubber-Gasket Joints for Ductile Iron Pressure Pipe and Fittings*, 2000.

AWWA C115, *Flanged Ductile Iron Pipe with Ductile Iron or Gray Iron Threaded Flanges*, 2005.

AWWA C116, *Protective Fusion-Bonded Epoxy Coatings for the Interior and Exterior Surfaces of Ductile-Iron and Gray-Iron Fittings for Water Supply Service*, 2003.

AWWA C150, *Thickness Design of Ductile Iron Pipe*, 2008.

AWWA C151, *Ductile Iron Pipe, Centrifugally Cast for Water*, 2002.

AWWA C153, *Ductile Iron Compact Fittings, 3 in. through 24 in. and 54 in. through 64 in. for Water Service*, 2006.

AWWA C203, *Coal-Tar Protective Coatings and Linings for Steel Water Pipelines Enamel and Tape — Hot Applied*, 2002.

AWWA C205, *Cement-Mortar Protective Lining and Coating for Steel Water Pipe 4 in. and Larger — Shop Applied*, 2007.

AWWA C206, *Field Welding of Steel Water Pipe*, 2003.

AWWA C208, *Dimensions for Fabricated Steel Water Pipe Fittings*, 2007.

AWWA C300, *Reinforced Concrete Pressure Pipe, Steel-Cylinder Type*, 2004.

AWWA C301, *Prestressed Concrete Pressure Pipe, Steel-Cylinder Type*, 2007.

AWWA C302, *Reinforced Concrete Pressure Pipe, Non-Cylinder Type*, 2004.

AWWA C303, *Reinforced Concrete Pressure Pipe, Steel-Cylinder Type, Pretensioned*, 2002.

AWWA C400, *Standard for Asbestos-Cement Distribution Pipe, 4 in. Through 16 in. (100 mm through 400 mm) for Water Distribution Systems*, 2003.

AWWA C401, *Standard for the Selection of Asbestos-Cement Pressure Pipe 4 in. through 16 in. (100 mm through 400 mm)*, 2003.

AWWA C600, *Standard for the Installation of Ductile-Iron Water Mains and Their Appurtenances*, 2005.

AWWA C602, *Cement-Mortar Lining of Water Pipe Lines 4 in. and Larger — in Place*, 2006.

AWWA C603, *Standard for the Installation of Asbestos-Cement Pressure Pipe*, 2005.

AWWA C606, *Grooved and Shouldered Joints*, 1997.

AWWA C900, *Polyvinyl Chloride (PVC) Pressure Pipe, 4 in. Through 12 in., for Water Distribution*, 2007.

AWWA C905, *AWWA Standard for Polyvinyl Chloride (PVC) Pressure Pipe and Fabricated Fittings 14 in. Through 48 in. (350 mm Through 1,200 mm)*, 2010.

AWWA C906, *Standard for Polyethylene (PE) Pressure Pipe and Fittings, 4 in. (100 mm) Through 63 in. (1,600 mm), for Water Distribution and Transmission*, 2007.

AWWA M9, *Concrete Pressure Pipe*, 2008.

AWWA M11, *A Guide for Steel Pipe Design and Installation*, 4th edition, 2004.

AWWA M14, *Recommended Practice for Backflow Prevention and Cross Connection Control*, 2004.

AWWA M41, *Ductile Iron Pipe and Fittings*, 2003.

E.1.2.5 DIPRA Publications. Ductile Iron Pipe Research Association, 245 Riverchase Parkway East, Suite O, Birmingham, AL 35244.

Installation Guide for Ductile Iron Pipe.

Thrust Restraint Design for Ductile Iron Pipe.

E.1.2.6 EBAA Iron Publications. EBAA Iron, Inc., P.O. Box 857, Eastland, TX 76448.

Thrust Restraint Design Equations and Tables for Ductile Iron and PVC Pipe.

E.1.2.7 UBPPA Publications. Uni-Bell PVC Pipe Association, 2655 Villa Creek Drive, Dallas, TX 75234.

Handbook of PVC Pipe.

E.1.2.8 U.S. Government Publications. U.S. Government Printing Office, Washington, DC 20402.

U.S. Federal Standard No. 66C, *Standard for Steel Chemical Composition and Harden Ability*, April 18, 1967 change notice No. 2, April 16, 1970, as promulgated by the U.S. Federal Government General Services Administration.

E.2 Informational References

The following documents or portions thereof are listed here as informational resources only. They are not a part of the requirements of this document.

AWWA M17, *Installation, Field Testing and Maintenance of Fire Hydrants*, 1989.

E.3 References for Extracts in Informational Sections

(Reserved)

Summary

NFPA 24 provides the complete information needed to design and install a private fire service water supply. In many cases, a private fire service main is used to supply water to a fire pump and must be properly interfaced with other systems designed and installed by other standards.

References Cited in Commentary

National Fire Protection Association, 1 Batterymarch Park, Quincy, MA 02169-7471.

NFPA 1, *Fire Code,* 2012 edition.
NFPA 13, *Standard for the Installation of Sprinkler Systems,* 2013 edition.
NFPA 13D, *Standard for the Installation of Sprinkler Systems in One- and Two-Family Dwellings and Manufactured Homes,* 2013 edition.
NFPA 13E, *Recommended Practice for Fire Department Operations in Properties Protected by Sprinkler and Standpipe Systems,* 2010 edition.
NFPA 13R, *Standard for the Installation of Sprinkler Systems in Low-Rise Residential Occupancies,* 2013 edition.
NFPA 14, *Standard for the Installation of Standpipe and Hose Systems,* 2010 edition.
NFPA 20, *Standard for the Installation of Stationary Pumps for Fire Protection,* 2013 edition.
NFPA 70®, *National Electrical Code®,* 2011 edition.
NFPA 170, *Standard for Fire Safety and Emergency Symbols,* 2012 edition.
NFPA 291, *Recommended Practice for Fire Flow Testing and Marking of Hydrants,* 2013 edition.
NFPA 600, *Standard on Industrial Fire Brigades,* 2010 edition.
NFPA 1963, *Standard for Fire Hose Connections,* 2009 edition.
Cote, A. E., ed., *Fire Protection Handbook®,* 20th edition, 2008.
Hall, John R., Jr., "U.S. Experience with Sprinklers and Other Automatic Fire Extinguishing Equipment," Fire Analysis and Research Division, February 2010.

American Water Works Association, 6666 West Quincy Avenue, Denver, CO 80235.

ANSI/AWWA C104-08, *Standard for Cement–Mortar Lining for Ductile-Iron Pipe and Fittings,* 2008.
ANSI/AWWA C105-10, *Standard for Polyethylene Encasement for Ductile-Iron Pipe Systems,* 2010.
ANSI/AWWA C111-06, *Standard for Rubber-Gasket Joints for Ductile-Iron Pressure Pipe and Fittings,* 2006.
ANSI/AWWA C203-08, *Standard for Coal-Tar Protective Coatings and Linings for Steel Water Pipelines – Enamel and Tape – Hot Applied,* 2008.
ANSI/AWWA C900-07, *Standard for Polyvinyl Chloride (PVC) Pressure Pipe and Fabricated Fittings, 4 in. Through 12 in. (100 mm Through 300 mm), for Water Transmission and Distribution,* 2007.

Underwriters Laboratories Inc., 333 Pfingsten Road, Northbrook, IL 60062-2096.

UL 194, *Standard for Gasketed Joints for Ductile-Iron Pipe and Fittings for Fire Protection Service,* 2005.

U.S. Government Printing Office, Washington, DC 20402.

Title 29, *Code of Federal Regulations,* Part 1910.146, Appendix A.
Title 29, *Code of Federal Regulations,* Part 1926, Subpart P, Appendix F.

Section 18.4 and Annex E
NFPA® 1, *Fire Code*, 2012 Edition with Commentary

SECTION 3

> In some cases, fire pumps may be needed to provide flow and pressure to an entire industrial complex, including private fire hydrants, as well as to fire protection systems. Section 18.4 of NFPA 1, *Fire Code*, provides the authority having jurisdiction (AHJ) and the design engineer with one method for determining the required fire flow for manual fire-fighting purposes. Annex E of NFPA 1 provides useful information in determining the location and distribution of hydrants. The method referenced in Section 18.4 does not take into consideration any fire flow from the fire suppression system or hose stream requirements from system installation standards such as NFPA 13, *Standard for the Installation of Sprinkler Systems,* and NFPA 14, *Standard for the Installation of Standpipe and Hose Systems*. However, the method does take into consideration the installation of an automatic sprinkler system. This method is used solely for fire department fire flow requirements. When fire pumps are used to supply flow and pressure to a system of fire hydrants, this method can be used to determine the correct size and capacity of the fire pump. Annex E provides valuable information regarding the number of hydrants and their distribution. Once the total flow requirements are determined from Section 18.4, Annex E can then be used to determine how many hydrants are needed to achieve that flow. For installations involving multiple hydrants, Annex E provides the average spacing between hydrants for uniform distribution.

18.4 Fire Flow Requirements for Buildings

18.4.1* Scope.

A.18.4.1 Section 18.4 and the associated tables are only applicable for determining minimum water supplies for manual fire suppression efforts. Water supplies for fire protection systems are not addressed by this section. It is not the intent to add the minimum fire protection water supplies, such as for a fire sprinkler system, to the minimum fire flow for manual fire suppression purposes required by this section.

18.4.1.1* The procedure determining fire flow requirements for buildings hereafter constructed or moved into the jurisdiction shall be in accordance with Section 18.4.

A.18.4.1.1 For the purpose of this section, a building subdivided by fire walls constructed in accordance with the building code is considered to be a separate building.

18.4.1.2 Section 18.4 shall not apply to structures other than buildings.

18.4.2 Definitions.
See definitions 3.3.14.5, Fire Flow Area, and 3.3.120, Fire Flow.

> The definition of the term *fire flow area* is as follows:
>
> **3.3.14.5 Fire Flow Area.** The floor area, in square feet, used to determine the required fire flow.
>
> The fire area should be determined based on the area within the surrounding exterior walls and 4-hour-rated fire walls, exclusive of courts. Areas of the building without surrounding

> exterior walls should be included in the fire area, if such areas are within the horizontal projection of the roof or floor above.
>
> The definition of the term *fire flow* is as follows:
>
> > **3.3.120 Fire Flow.** The flow rate of a water supply, measured at 20 psi (137.9 kPa) residual pressure, that is available for fire fighting.
>
> According to NFPA 291, *Recommended Practice for Fire Flow Testing and Marking of Hydrants* (see Part III, Section 1, of this handbook), hydrant output should be identified by means of a color code that is based on the available flow at a residual pressure of 20 psi (1.4 bar). Hydrants on particularly weak systems — that is, those producing less than 40 psi (2.8 bar) static pressure — should be identified based on the available flow at one-half of the static pressure.
>
> Generally, a minimum residual pressure of 20 psi (1.4 bar) should be maintained within the water supply system at all times. Fire department pumpers can operate at pressures of less than 20 psi (1.4 bar) but do so with difficulty. Operating at a minimum pressure of 20 psi (1.4 bar) also prevents the development of a negative pressure at any point in the system. Negative pressures in an underground piping system can cause the collapse of the water main and can create a backflow of contaminated water from sources that may be cross-connected with the potable water supply system. Paragraph 4.14.3.1 of NFPA 20, *Standard for the Installation of Stationary Pumps for Fire Protection* (see Part II of this handbook), permits the suction pressure in the fire pump suction pipe to drop as low as 0 psi (0 bar); however, many water purveyors will not permit pressures below 20 psi (1.4 bar) in their piping system.

18.4.3 Modifications.

18.4.3.1 Decreases. Fire flow requirements shall be permitted to be decreased by the AHJ for isolated buildings or a group of buildings in rural areas or suburban areas where the development of full fire flow requirements is impractical as determined by the AHJ.

18.4.3.1.1 The AHJ shall be authorized to establish conditions on fire flow reductions approved in accordance with 18.4.3.1 including, but not limited to, fire sprinkler protection, type of construction of the building, occupancy, and setbacks.

18.4.3.2 Increases. Fire flow shall be permitted to be increased by the AHJ where conditions indicate an unusual susceptibility to group fires or conflagrations. An upward modification shall not be more than twice that required for the building under consideration.

> An evaluation of issues such as separation distances and other means of protection should be completed before making a determination of required fire flow. NFPA 80A, *Recommended Practice for Protection of Buildings from Exterior Fire Exposures,* should be used to determine the appropriate separation distances and other means of protection, such as clear space between buildings, blank walls, or walls with protected openings and/or water curtain protection.

18.4.4 Fire Flow Area.

18.4.4.1 General. The fire flow area shall be the total floor area of all floor levels of a building except as modified in 18.4.4.1.1.

18.4.4.1.1 Type I (443), Type I (332), and Type II (222) Construction. The fire flow area of a building constructed of Type I (443), Type I (332), and Type II (222) construction shall be the area of the three largest successive floors.

18.4.5 Fire Flow Requirements for Buildings.

18.4.5.1 One- and Two-Family Dwellings.

18.4.5.1.1 The minimum fire flow and flow duration requirements for one- and two-family dwellings having a fire flow area that does not exceed 5000 ft² (334.5 m²) shall be 1000 gpm (3785 L/min) for 1 hour.

18.4.5.1.1.1 A reduction in required fire flow of 50 percent shall be permitted when the building is provided with an approved automatic sprinkler system.

18.4.5.1.1.2 A reduction in the required fire flow of 25 percent shall be permitted when the building is separated from other buildings by a minimum of 30 ft (9.1 m).

18.4.5.1.1.3 The reduction in 18.4.5.1.1.1 and 18.4.5.1.1.2 shall not reduce the required fire flow to less than 500 gpm (1900 L/min).

18.4.5.1.2 Fire flow and flow duration for dwellings having a fire flow area in excess of 5000 ft² (334.5 m²) shall not be less than that specified in Table 18.4.5.1.2.

TABLE 18.4.5.1.2 Minimum Required Fire Flow and Flow Duration for Buildings

\multicolumn{5}{c}{Fire Flow Area ft² (× 0.0929 for m²)}	Fire Flow gpm† (× 3.785 for L/min)	Flow Duration (hours)
I(443), I(332), II(222)*	II(111), III(211)*	IV(2HH), V(111)*	II(000), III(200)*	V(000)*		
0–22,700	0–12,700	0–8200	0–5900	0–3600	1500	
22,701–30,200	12,701–17,000	8201–10,900	5901–7900	3601–4800	1750	
30,201–38,700	17,001–21,800	10,901–12,900	7901–9800	4801–6200	2000	
38,701–48,300	21,801–24,200	12,901–17,400	9801–12,600	6201–7700	2250	2
48,301–59,000	24,201–33,200	17,401–21,300	12,601–15,400	7701–9400	2500	
59,001–70,900	33,201–39,700	21,301–25,500	15,401–18,400	9401–11,300	2750	
70,901–83,700	39,701–47,100	25,501–30,100	18,401–21,800	11,301–13,400	3000	
83,701–97,700	47,101–54,900	30,101–35,200	21,801–25,900	13,401–15,600	3250	
97,701–112,700	54,901–63,400	35,201–40,600	25,901–29,300	15,601–18,000	3500	3
112,701–128,700	63,401–72,400	40,601–46,400	29,301–33,500	18,001–20,600	3750	
128,701–145,900	72,401–82,100	46,401–52,500	33,501–37,900	20,601–23,300	4000	
145,901–164,200	82,101–92,400	52,501–59,100	37,901–42,700	23,301–26,300	4250	
164,201–183,400	92,401–103,100	59,101–66,000	42,701–47,700	26,301–29,300	4500	
183,401–203,700	103,101–114,600	66,001–73,300	47,701–53,000	29,301–32,600	4750	
203,701–225,200	114,601–126,700	73,301–81,100	53,001–58,600	32,601–36,000	5000	
225,201–247,700	126,701–139,400	81,101–89,200	58,601–65,400	36,001–39,600	5250	
247,701–271,200	139,401–152,600	89,201–97,700	65,401–70,600	39,601–43,400	5500	
271,201–295,900	152,601–166,500	97,701–106,500	70,601–77,000	43,401–47,400	5750	
Greater than 295,900	Greater than 166,500	106,501–115,800	77,001–83,700	47,401–51,500	6000	4
		115,801–125,500	83,701–90,600	51,501–55,700	6250	
		125,501–135,500	90,601–97,900	55,701–60,200	6500	
		135,501–145,800	97,901–106,800	60,201–64,800	6750	
		145,801–156,700	106,801–113,200	64,801–69,600	7000	
		156,701–167,900	113,201–121,300	69,601–74,600	7250	
		167,901–179,400	121,301–129,600	74,601–79,800	7500	
		179,401–191,400	129,601–138,300	79,801–85,100	7750	
		Greater than 191,400	Greater than 138,300	Greater than 85,100	8000	

*Types of construction are based on NFPA 220.
†Measured at 20 psi (139.9 kPa).

18.4.5.1.2.1 Required fire flow shall be reduced by 50 percent when the building is provided with an approved automatic sprinkler system.

18.4.5.2 Buildings Other Than One- and Two-Family Dwellings. The minimum fire flow and flow duration for buildings other than one- and two-family dwellings shall be as specified in Table 18.4.5.1.2.

18.4.5.2.1 Required fire flow shall be reduced by 75 percent when the building is protected throughout by an approved automatic sprinkler system. The resulting fire flow shall not be less than 1000 gpm (3785 L/min).

18.4.5.2.2 Required fire flow shall be reduced by 75 percent when the building is protected throughout by an approved automatic sprinkler system, which utilizes quick response sprinklers throughout. The resulting fire flow shall not be less than 600 gpm (2270 L/min).

DESIGN ALERT ▶

The following example illustrates the method used to calculate the required fire flow for a building other than a one- or two-family dwelling, including the allowed reduction for the installation of an automatic sprinkler system.

CALCULATION ▶

▶EXAMPLE

Referring to Exhibit III.24, the building is a six-story mill, Type V(000) construction (structural elements, walls, arches, floors, and roofs are entirely or partially of wood or other approved material). The main building measures 75 ft by 370 ft, for a total floor area per floor of 27,750 ft².

27,750 ft² × 6 floors = 166,500 ft² (per 18.4.4.1)

8000 gpm (from Table 18.4.5.1.2)

8000 gpm × 0.25 = 2000 gpm (per 18.4.5.2.1)

For metric equivalents, the main building measures 22.9 m by 112.8 m, for a total floor area per floor of 2583.12 m².

2583.12 m² × 6 floors = 15,498.72 m² (per 18.4.4.1)

30,280 L/min (converted from Table 18.4.5.1.2)

30,280 L/min × 0.25 = 7570 L/min (per 18.4.5.2.1)

Based on this approach, the minimum number of fire hydrants would be two, and the average spacing between them would be 450 ft (137 m).

This example uses the maximum credit for the installation of automatic sprinklers. In this particular example, since 18.4.5.2.1 of NFPA 1 permits a reduction of up to 75 percent, a smaller reduction can be expected due to a number of factors, including the judgment of the AHJ. The building in Exhibit III.24 is an existing mill building that has been converted to office space. The required fire flow is considerably higher than that for the typical modern office building, because the mill is still of Type V(000) construction. However, taking into consideration the previous use of the building, which was textile manufacturing, the credit for sprinklers should be considerably less, perhaps on the order of 25 percent to 50 percent. Applying a reduction of only 25 percent produces the following result:

8000 gpm × 0.75 = 6000 gpm (per 18.4.5.2.1)

30,278 L/min × 0.75 = 22,709 L/min

Based on these results, the minimum number of hydrants per Table E.3 would be six, and the average spacing between each would be 250 ft (76 m). Given the nature of the hazard, even considering the building's present use, the higher number of hydrants may be more appropriate for the hazard.

> In cases where a fire pump is used to supply the hydrants for a facility, the pump capacity would be selected based on the flow of 6000 gpm (22,712 L/min) and sized sufficiently to produce at least 20 psi (1.4 bar) at the most hydraulically demanding hydrant.
>
> Determination of the available water supply using this method should be conducted only by those having extensive experience in fire fighting and/or fire protection system design and knowledge of the construction of the building and its intended use.

18.4.5.3* For a building with an approved fire sprinkler system, the fire flow demand and the fire sprinkler system demand shall not be required to be added together. The water supply shall be capable of delivering the larger of the individual demands.

A.18.4.5.3 The fire sprinkler system demand is generally significantly less than the demands in Table 18.4.5.1.2, even after hose stream demands are applied. The sprinkler system demand can be a part of the overall flow available to a building site. There is no need to add these flow demands together, which would penalize the building owner that has decided to put fire sprinkler systems in place.

ANNEX E FIRE HYDRANT LOCATIONS AND DISTRIBUTION

This annex is not a part of the requirements of this NFPA document unless specifically adopted by the jurisdiction.

E.1 Scope

Fire hydrants shall be provided in accordance with Annex E for the protection of buildings, or portions of buildings, hereafter constructed.

E.2 Location

Fire hydrants shall be provided along required fire apparatus access roads and adjacent public streets.

> To be usable, fire hydrants must be reasonably accessible to the fire department in accordance with Section E.2. Hydrants should be located a minimum of 40 ft (12.1 m) away from the building to provide some protection for the end user. Hydrants should also be sufficient in number and location such that they are within 100 ft (30.5 m) of the fire department connection (FDC) serving the building's sprinkler system. Where hydrants cannot be located at the minimum distance from protected buildings, the application of wall hydrants should be considered. A wall hydrant should be used only where approved by the AHJ.

E.3 Number of Fire Hydrants

The minimum number of fire hydrants available to a building shall not be less than that listed in Table E.3. The number of fire hydrants available to a complex or subdivision shall not be less than that determined by spacing requirements listed in Table E.3 when applied to fire apparatus access roads and perimeter public streets from which fire operations could be conducted.

> Section E.3 and Table E.3 do not require hydrants to be within a minimum distance of the building. The minimum distance to a building is determined by other issues, such as fire

TABLE E.3 Number and Distribution of Fire Hydrants

Fire Flow Requirements		Number and Distribution of Fire Hydrants			Maximum Distance from Any Point on Street or Road Frontage to a Hydrant[4,5]	
		Minimum Number of Hydrants	Average Spacing Between Hydrants[1,2,3,4]			
gpm	L/min		ft	m	ft	m
1750 or less	6650 or less	1	500	152	250	76
2000–2250	7600–8550	2	450	137	225	69
2500	9500	3	450	137	225	69
3000	11,400	3	400	122	225	69
3500–4000	13,300–15,200	4	350	107	210	64
4500–5000	17,100–19,000	5	300	91	180	55
5500	20,900	6	300	91	180	55
6000	22,800	6	250	76	150	46
6500–7000	24,700–26,500	7	250	76	150	46
7500 or more	28,500 or more	8 or more[6]	200	61	120	37

Note: 1 gpm = 3.8 L/min; 1 ft = 0.3 m.

[1] Reduce by 100 ft (30.5 m) for dead-end streets or roads.

[2] Where streets are provided with median dividers that can be crossed by fire fighters pulling hose lines, or arterial streets are provided with four or more traffic lanes and have a traffic count of more than 30,000 vehicles per day, hydrant spacing shall average 500 ft (152.4 m) on each side of the street and be arranged on an alternating basis up to a fire flow requirement of 7000 gpm (26,500 L/min) and 400 ft (122 m) or higher fire flow requirements.

[3] Where new water mains are extended along streets where hydrants are not needed for protection of structures or similar fire problems, fire hydrants shall be provided at spacing not to exceed 1000 ft (305 m) to provide for transportation hazards.

[4] For detached one- and two-family dwellings protected by an automatic sprinkler system, with a required fire flow of not greater than 1750 gpm, the average distance between hydrants shall not exceed 800 ft, and the maximum distance from a street or road frontage to a hydrant shall not exceed 400 ft. Notes 1 and 5 shall also apply to detached one- and two-family dwellings constructed on a dead-end street or road.

[5] Reduce by 50 ft (15.2 m) for dead-end streets or roads.

[6] One hydrant for each 1000 gpm (3785 L/min) or fraction thereof.

> department access road design and the proximity to an FDC. When the layout of the fire department access road and the separation between protected buildings are determined, the number of hydrants can be selected from Table E.3, based on the average spacing requirement between hydrants. In addition to this number of hydrants, each building should be provided with the minimum number of hydrants as stated in Table E.3.

E.4 Consideration of Existing Fire Hydrants

Existing fire hydrants on public streets shall be permitted to be considered as available. Existing fire hydrants on adjacent properties shall not be considered available unless fire apparatus access roads extend between properties and easements are established to prevent obstruction of such roads.

> As illustrated in Exhibit III.24, only four private hydrants are provided. However, four public hydrants are located in the immediate vicinity and would be available for fire fighting if needed. Section E.4 permits the use of the public hydrants in Exhibit III.24.

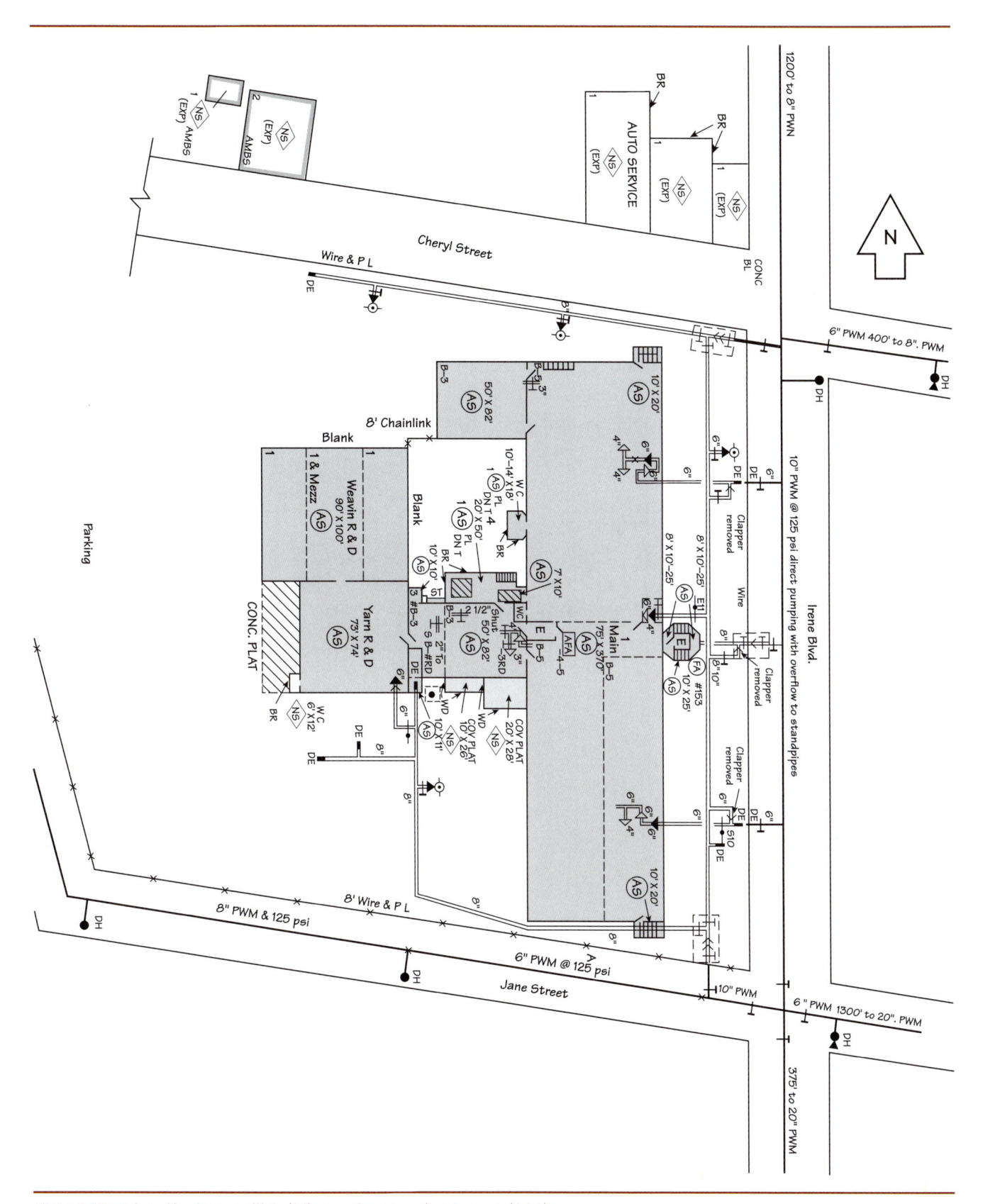

Annex E • Fire Hydrant Locations and Distribution 531

EXHIBIT III.24 Six-Story Mill Building — Construction Type V(000).

Stationary Fire Pumps Handbook 2013

E.5 Distribution of Fire Hydrants

The average spacing between fire hydrants shall not exceed that listed in Table E.3.

Exception: The AHJ shall be permitted to accept a deficiency of up to 10 percent where existing fire hydrants provide all or a portion of the required fire hydrant service. Regardless of the average spacing, fire hydrants shall be located such that all points on streets and access roads adjacent to a building are within the distances listed in Table E.3.

Summary

Section 18.4 and Annex E of NFPA 1 are included to assist the designer in determining the correct number of hydrants needed to properly design a private fire service system. Hydrants are a critical component of any water supply, and their proper number and distribution are necessary to provide adequate fire protection. The information provided by Section 18.4 and Annex E is needed to coordinate the design and installation of underground piping and other aspects of a complete water supply system.

References Cited in Commentary

National Fire Protection Association, 1 Batterymarch Park, Quincy, MA 02169-7471.

NFPA 13, *Standard for the Installation of Sprinkler Systems,* 2013 edition.
NFPA 14, *Standard for the Installation of Standpipe and Hose Systems,* 2010 edition.
NFPA 20, *Standard for the Installation of Stationary Pumps for Fire Protection,* 2013 edition.
NFPA 80A, *Recommended Practice for Protection of Buildings from Exterior Fire Exposures,* 2012 edition.
NFPA 291, *Recommended Practice for Fire Flow Testing and Marking of Hydrants,* 2013 edition.

Extracts with Commentary from NFPA® 22, *Standard for Water Tanks for Private Fire Protection*, 2008 Edition

SECTION 4

> Requirements extracted from NFPA 22, *Standard for Water Tanks for Private Fire Protection,* are included in this section of the handbook, since water tanks are frequently used to supply water to fire pumps. Only requirements that directly relate to fire pumps are included in this section. Paragraphs from the standard that have been omitted are indicated by ellipses. The user should note that this section of the handbook is not intended to be a design guide for water tanks.

CHAPTER 1 INTRODUCTION

1.1 Scope

This standard provides the minimum requirements for the design, construction, installation, and maintenance of tanks and accessory equipment that supply water for private fire protection, including the following:

(1) Gravity tanks, suction tanks, pressure tanks, and embankment-supported coated fabric suction tanks
(2) Towers
(3) Foundations
(4) Pipe connections and fittings
(5) Valve enclosures
(6) Tank filling
(7) Protection against freezing

> By specifying water tanks for private fire protection in its scope, NFPA 22 clarifies that it is intended for use as a standard for water tanks that serve fire protection systems only. The standard is not intended to provide water for domestic consumption and, therefore, does not include chlorination or disinfection requirements to achieve drinking water quality. Use of water tanks for other than fire protection is recognized by the standard but is limited to situations that are unavoidable and do not affect the quantity of fire protection water.

. . .

1.5 Types of Tanks

This standard addresses elevated tanks on towers or building structures, water storage tanks that are at grade or below grade, and pressure tanks.

> Section 1.5 describes, in general terms, the types of tanks normally used for fire protection. Of these types of tanks, those that are installed at or below grade are normally used in series with fire pumps.

. . .

CHAPTER 4 GENERAL INFORMATION

4.1 Capacity and Elevation

4.1.1* The size and elevation of the tank shall be determined by conditions at each individual property after due consideration of all factors involved.

A.4.1.1 Where tanks are to supply sprinklers, see separately published NFPA standards; also see NFPA 13.

> As discussed in Part II of this handbook, other systems, such as standpipe systems or private water supplies (including master streams), may require higher flow rates for longer durations than most sprinkler systems. When serving more than one system, the size and capacity of the water tank should be determined based on the highest system demand for the longest required duration of flow. The installation standard for each type of system (i.e., sprinkler, standpipe, or master stream) specifies the water supply duration. The water supply duration multiplied by the flow in gallons per minute (liters per minute) determines the water tank capacity.

4.1.2 Wherever possible, standard sizes of tanks and heights of towers shall be as specified in 5.1.3, 6.1.2, 8.1.3, and Section 9.2.

4.1.3 For suction tanks, the net capacity shall be the number of U.S. gallons (cubic meters) between the inlet of the overflow and the level of the vortex plate.

CALCULATION ▶
> The amount of water between the overflow and the anti-vortex plate constitutes the amount of usable water in the storage tank. The tank size as determined in 4.1.2 may be slightly larger than the quantity needed by the fire protection system. For example, a standpipe system that requires 1000 gpm (3785 L/min) for the required 30-minute duration needs a total quantity of 30,000 gal (113.55 m³) of fire protection water. In this case, the next size tank may be needed [40,000 gal (151.40 m³)] in order to provide the required 30,000 gal (113.55 m³) of fire protection water. This requirement applies to all types of tanks except break tanks. For the required size of break tanks, see 4.31 of NFPA 20, *Standard for the Installation of Stationary Pumps for Fire Protection*.

. . .

4.2 Location of Tanks

4.2.1 The location of tanks shall be such that the tank and structure are not subject to fire exposure.

> As is the case with fire pumps, water tanks should be protected from exposure to fire. For ground-mounted suction tanks or elevated gravity tanks, this protection usually involves physical separation from buildings or other exposures. Tanks that are installed inside the protected building should be located in the fire pump room or enclosed by sufficient fire-rated construction to prevent exposure. Buried tanks are considered to be protected.

4.2.1.1 If lack of yard room makes this impracticable, the exposed steel work shall be suitably fireproofed or shall be protected by open sprinklers *(see A.13.1.1)*.

4.2.1.2 Fireproofing, where necessary, shall be provided for steelwork within 20 ft (6.1 m) of exposures, combustible buildings or windows, and doors from which fire might issue.

4.2.1.3 Where used for supports near combustible construction or occupancy inside the building, steel or iron shall be fireproofed 6 ft (1.8 m) above combustible roof coverings, and within 20 ft (6.1 m) of windows and doors from which fire might issue.

4.2.1.4 Steel beams or braces that join two building columns that support a tank structure shall also be suitably fireproofed where near combustible construction or occupancy.

4.2.1.5 Interior timber shall not be used to support or brace tank structures.

. . .

4.3 Tank Materials

4.3.1 Materials shall be limited to steel, wood, concrete, coated fabrics, and fiberglass-reinforced plastic tanks.

> The 2008 edition of NFPA 22 recognizes that fiberglass-reinforced tanks are allowed by the standard, so 4.3.1 has been revised to include fiberglass-reinforced plastic (FRP) tanks in the list of permitted materials. A new chapter has also been developed on FRP; see the extracts from Chapter 11.

4.3.2 The elevated wood and steel tanks shall be supported on steel towers or reinforced concrete towers.

. . .

4.5 Plans

4.5.1 The contractor shall furnish stress sheets and plans required by the purchaser and the authority having jurisdiction for approval or for obtaining building permits and licenses for the erection of the structure.

4.5.2 Approval of Layouts.

> Construction should not proceed until plans and calculations have been reviewed and approved and all permits and fees have been applied for and paid. This documentation should be modified by the installing contractor to reflect field conditions, and should be submitted to the authority having jurisdiction (AHJ) and building owner for final review upon the completion of work. In addition, the manufacturer's inspection, testing, and maintenance instructions and a copy of NFPA 25, *Standard for the Inspection, Testing, and Maintenance of Water-Based Fire Protection Systems,* should be provided in the final submission package.
>
> Complete information regarding the tank piping on the tank side of the connection to the yard or sprinkler system is required to be submitted by the installing contractor to the AHJ for approval. This submittal should be made before work begins and should accompany the submittal requirements referenced in NFPA 22. No work should begin until these plans have been accepted by each AHJ involved in the project. In addition to these items, complete details of the heating system (where required) should be provided and approved prior to construction.

4.5.2.1 Complete information regarding the tank piping on the tank side of the connection to the yard or sprinkler system shall be submitted to the authority having jurisdiction for approval.

4.5.2.2 The information submitted shall include the following:

(1) Size and arrangement of all pipes

(2) Size, location, and type of all valves, tank heater, and other accessories
(3) Steam pressures available at the heater
(4) Arrangement of, and full information regarding, the steam supply and return system together with pipe sizes
(5) Details of construction of the frostproof casing
(6) Where heating is required, heat loss calculations
(7) Structural drawings and calculations
(8) Seismic bracing details and calculations
(9) Operational settings and sequence of operation
(10) Monitoring equipment and connections
(11) Underground details including foundations, compaction, and backfill details and calculations
(12) Buoyancy calculations for buried tanks

4.6 Tank Contractor Responsibility

4.6.1 Any necessary work shall be handled by experienced contractors.

4.6.1.1 Careful workmanship and expert supervision shall be employed.

4.6.1.2 The manufacturer shall warranty the tank for at least 1 year from the date of completion and final customer acceptance.

4.6.2 Upon completion of the tank construction contract, and after the contractor has tested the tank and made it watertight, the tank contractor shall notify the authority having jurisdiction so that the tank can be inspected and approved.

> A test report should be completed and signed by the installing contractor, the registered design professional (RDP), and the AHJ. This test report should include a description of the installation and tests conducted and should be included in the final submission package.

4.6.3 Cleaning Up.

4.6.3.1 During and upon completion of the work, the contractor shall remove or dispose of all rubbish and other unsightly material in accordance with NFPA 241.

4.6.3.2 The condition of the premises shall be as it was before tank construction.

> In addition to construction debris, NFPA 241, *Standard for Safeguarding Construction, Alteration, and Demolition Operations,* addresses issues such as temporary construction; equipment and storage; process and hazards; utilities; fire protection; safeguarding construction and alteration operations; and safeguarding roofing, demolition, and underground operations. A problem within any one of these areas can cause exposure of the tank installation to fire and should be remedied before demobilization.

. . .

CHAPTER 11 FIBERGLASS-REINFORCED PLASTIC TANKS

11.1 General

Fiberglass-reinforced plastic tanks shall be permitted to be used for fire protection systems when installed in accordance with this standard.

11.2* Application

Fiberglass-reinforced plastic tanks shall be permitted only for storage of water at atmospheric pressure.

> FRP tanks are permitted only for storage of water at atmospheric pressure and should not be used as pressure tanks.

A.11.2 See Figure A.11.2 for an example of a fiberglass tank being used under ground as a cistern to supply fire flow for fire department apparatus in a rural area.

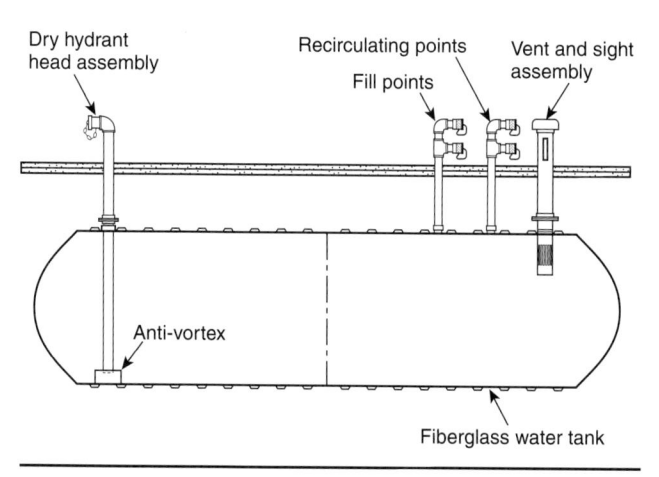

FIGURE A.11.2 *Fiberglass Tank as an Underground Cistern.*

> FRP tanks are permitted to be used only when buried to protect the tank against mechanical and fire damage. NFPA 22 does not address the use of tanks inside buildings. However, when tanks are proposed for use inside buildings, if the methods used for the fire pumps listed in Section 4.12 of NFPA 20 as protection from mechanical and fire damage are applied, and the proposed installation is approved by the AHJ, NFPA 22 does permit FRP tanks to be used inside a building.

11.3* Tank Specification

Fiberglass-reinforced plastic tanks shall meet the requirements of AWWA D120.

A.11.3 The standard capacities shall be from 2,000 gal to 50,000 gal (7.6 m³ to 190 m³). Tanks of other capacities are permitted.

11.4 Monolithic Tanks

Monolithic tanks shall be tested for leakage by the manufacturer prior to shipment.

11.4.1 Tanks that are assembled on site shall be tested for leakage by the manufacturer.

11.5 Protection of Buried Tanks

11.5.1 Tanks shall be designed to resist the pressure of earth against them.

11.5.2 Tanks shall meet local building code requirements for resisting earthquake damage.

11.5.3 Tanks shall be installed in accordance with the manufacturer's instructions and 11.5.4 through 11.5.13.

11.5.4 Bedding and backfill shall be noncorrosive inert material, of a type recommended by the tank manufacturer, such as crushed stone or pea gravel that is properly compacted.

11.5.5 Tanks shall be set on the minimum depth of bedding, as recommended by the tank manufacturer, that extends 1 ft (0.3 m) beyond the end and sides of the tank.

11.5.6 Tanks shall be located completely below the frost line to protect against freezing.

> Tank installation below the frost line prevents freezing and negates the need to install a heating system and temperature alarm.

11.5.7 Where tanks are buried below railroad tracks, the minimum depth of cover shall be 4 ft (1.2 m).

11.5.8 Where the tanks are not subjected to traffic, tanks shall be covered with not less than 12 in. (305 mm) of compacted backfill and topped with up to 18 in. (457 mm) of compacted backfill or with not less than 12 in. (305 mm) of compacted backfill, on top of which shall be placed a slab of reinforced concrete not less than 4 in. (100 mm) thick.

11.5.9 Where tanks are, or are likely to be, subjected to traffic, they shall be protected from vehicles passing over them by at least 36 in. (914 mm) of backfill, or 18 in. (457 mm) of compacted backfill, of a type recommended by the tank manufacturer, plus either 6 in. (152 mm) of reinforced concrete or 9 in. (229 mm) of asphaltic concrete or greater where specified by the tank manufacturer.

11.5.10 Where asphaltic or reinforced concrete paving is used as part of the protection, it shall extend at least 12 in. (305 mm) horizontally beyond the outline of the tank in all directions.

11.5.11 Tanks shall be safeguarded against movement when exposed to high groundwater or floodwater by anchoring with non-metallic straps to a bottom hold-down pad or deadman anchors with fittings built up or protected to prevent corrosion failure over the life of the tank or by securing by other equivalent means using recognized engineering standards.

11.5.12 The depth of cover shall be measured from the top of the tank to the finished grade, and due consideration shall be given to future or final grade and the nature of the soil.

11.5.13 Maximum burial depths, measured from the top of the tank, are established by underground tank manufacturers and independent testing laboratories. Maximum burial depth shall be specified by the tank manufacturer and shall be marked on the tank.

> The requirements in 11.5.7 through 11.5.10 and 11.5.12 are based on Section 10.4 of NFPA 24, *Standard for the Installation of Private Fire Service Mains and Their Appurtenances,* and are consistent with the burial depths for underground piping systems. These measures are meant to protect the tank from damage when heavy objects roll over the surface above the tank location. It is preferable that tanks be located where railroad tracks and roadways do not pass over them.
>
> This information is provided to standardize the burial depth for the tank, considering finished grade and the potential for soil settlement.

11.6 Protection of Aboveground Tanks

11.6.1 Tanks shall meet local building code requirements for resisting earthquake damage.

11.6.2 Tanks shall be installed in accordance with the manufacturer's instructions and 11.6.3 through 11.5.13.

11.6.3 Fiberglass-reinforced plastic (FRP) tanks located inside a building shall be protected by automatic sprinklers in accordance with ordinary hazard Group 2 occupancies.

> Where FRP tanks are used to supply fire pumps and/or are used as break tanks and are installed in the immediate vicinity of fire pumps, the tank should be protected in the same fashion as the fire pump. Section 4.12 of NFPA 20 provides guidance on the type and extent of protection needed for the pump (see Part II of this handbook). This protection should include separation from the hazard by means of a barrier with at least a 1-hour fire resistance rating for buildings that are completely sprinklered and a barrier with a 2-hour fire resistance rating for buildings that are only partially sprinklered. Where the hazard is greater than ordinary hazard Group II, tank protection should be in accordance with NFPA 13, *Standard for the Installation of Sprinkler Systems.*

11.6.3.1 Where the hazard is greater than OH2, protection shall be in accordance with NFPA 13.

11.6.4 Horizontal fiberglass-reinforced plastic tanks that are greater than 4 ft (1.2 m) in diameter and are positioned 18 in. (457 mm) or greater above finished floor shall be protected in accordance with the obstruction rules of NFPA 13.

> When the size of the tank exceeds 4 ft (1.2 m) and the tank is installed more than 18 in. (457 mm) above the finished floor, the tank becomes an obstruction to the sprinkler spray pattern. To solve this problem, a line of sprinklers must be installed below the tank in order to comply with NFPA 13.

11.6.5 Fiberglass tanks installed outdoors shall be protected from freezing and mechanical and UV damage.

> Outdoor use of FRP tanks is permitted, provided that the tanks are protected from freezing and mechanical and ultraviolet (UV) damage. For outdoor installation, the tank should be placed in a protective enclosure to prevent UV exposure, and access should be limited to prevent mechanical damage.
>
> Electronic supervision of the tank water temperature should be provided and can be accomplished by the installation of a temperature supervisory switch. This switch must be installed in accordance with *NFPA 72®, National Fire Alarm and Signaling Code,* and should produce two separate and distinct signals. One signal should indicate a decrease in water temperature to 40°F (4.4°C), and the other signal should indicate a return to water temperature above 40°F (4.4°C). A single switch can be used to accomplish this task. See Exhibit III.25 for a typical tank water temperature supervisory switch.

11.7 Tank Connections

11.7.1 Tanks shall have a vent that extends above the ground to prevent against pressurization of the tank during filling and creation of a vacuum during use. Tank venting systems shall be provided with a minimum 2.0 in. (50 mm) nominal inside diameter.

EXHIBIT III.25 *Tank Water Temperature Supervisory Switch. (Courtesy of Potter Electric Signal Company, LLC)*

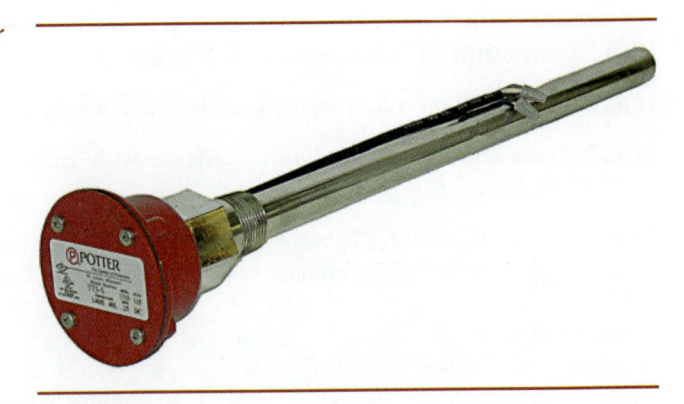

11.7.2* For underground tanks, water level monitoring required by 14.1.8 shall be capable of being read above ground.

A.11.7.2 See Figure A.11.7.2 for an example of a combination vent and sight assembly, which allows the tank to stay at atmospheric pressure while allowing the user to know the water level in the tank. While these two devices are not required to be combined, it is convenient since they are both required to be above ground.

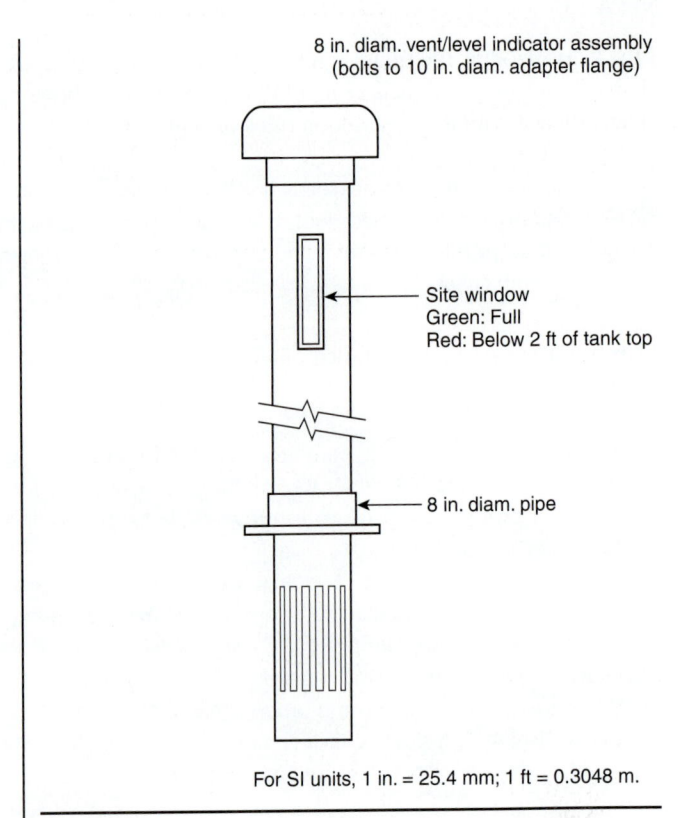

FIGURE A.11.7.2 *Typical Combination Vent and Sight Assembly.*

> Water level monitoring should be capable of being read above ground. The water level in a tank must be monitored for obvious reasons. According to NFPA 25, the water level should

be inspected monthly when not supervised by an alarm connected to a constantly attended location, and it should be inspected quarterly when supervised by an alarm that is connected to a constantly attended location. It is important to note that the quantity of water required in the tank is determined by the required flow rate times a specified duration; any quantity less than that determined by this calculation constitutes insufficient water for fire fighting and is not in compliance with NFPA 22. For tanks with an automatic fill connection, the flow rate of the fill connection is frequently included in the tank capacity calculation, thus reducing the overall size of the tank. If a tank is reduced to a size below the required capacity for the allotted duration due to an automatic flow rate, the requirements for break tanks apply in accordance with 4.31 of NFPA 20.

• • •

CHAPTER 14 PIPE CONNECTIONS AND FITTINGS

14.1* General Information

A.14.1 For embankment-supported coated fabric suction tanks, see Section 9.6.

14.1.1 Watertight Intersections at Roofs and Floors.

14.1.1.1 The intersections of all tank pipes with roofs and concrete or waterproof floors of buildings shall be watertight.

14.1.1.2 Where tank pipes pass through concrete roofs, a watertight intersection shall be obtained by using fittings that are caulked with oakum or by pouring the concrete solidly around the pipes, which first shall be wrapped with two or three thicknesses of building paper.

14.1.1.3 Where concrete is used, the upper side of the intersection shall be well flashed with a suitable, firm, waterproof material that is noncracking and that retains its adhesion and flexibility.

14.1.1.4 Wood roofs also shall be built tightly around the pipes and shall be made watertight by means of fittings that are caulked with oakum or by using flashing.

14.1.1.5 Where tank pipes pass through a concrete or waterproof floor, a watertight intersection, as described in 14.1.1.1, shall be obtained so that water from above cannot follow down the pipe to the lower floors or to the basement.

14.1.2 Rigid connections to steel tanks shall be made by means of a welded joint with approval by the authority having jurisdiction.

14.1.2.1 A rigid connection to a wood tank shall be made by means of a running nipple or by means of threaded flanges, one inside the tank and one outside the tank, bolted together through the wood with movable nuts outside.

14.1.3* Placing Tank in Service. All tank piping shall be installed immediately after completion of the tank and tower construction so that the tank can be filled and placed in service promptly.

A.14.1.3 Wood tanks can be extensively damaged by shrinkage if left empty after they are erected.

A water storage tank is a critical component of a fire protection system. If fire protection is needed for the construction site, the tank should be scheduled for early completion so it can be placed in service early on during construction.

14.1.4 The Contract. To ensure the installation of equipment, the contract shall specify that the finished work shall conform with this standard in all respects.

14.1.5 Precautions During Repairs. The authority having jurisdiction shall be notified well in advance when the tank is to be drained. The precautions required by 14.1.5.1 through 14.1.5.5 shall be observed.

> Rarely is a facility more exposed to a potentially catastrophic fire loss than when a fire protection system is impaired or out of service, particularly when the water supply, such as a water tank, is out of service. Experience has shown that, if a proper impairment program had been in place and followed, many devastating fires that have occurred could have been mitigated.
>
> The cause of an impairment, whether it is preplanned or occurs during an emergency, seems to have little bearing on the propensity for catastrophe. The following measures have been identified as major factors in preventing large fire loss:
>
> 1. Recognize the exposures created by the impairment.
> 2. Establish a fire watch and provide backup fire protection.
> 3. Control ignition sources and/or discontinue hazardous processes.
> 4. Expedite repairs.
> 5. Verify that fire protection systems are properly placed back into service.
>
> Chapter 15 of NFPA 25 should be followed any time a water storage tank is taken out of service for maintenance or repair. Note that 15.5.2(4) of NFPA 25 requires a comprehensive impairment procedure any time a fire protection system is removed from service for more than 10 hours in a 24-hour period. The extract of 15.5.2 is as follows:
>
> **15.5.2** Before authorization is given, the impairment coordinator shall be responsible for verifying that the following procedures have been implemented:
>
> (1) The extent and expected duration of the impairment have been determined.
> (2) The areas or buildings involved have been inspected and the increased risks determined.
> (3) Recommendations have been submitted to management or the property owner or designated representative.
> (4) Where a required fire protection system is out of service for more than 10 hours in a 24-hour period, the impairment coordinator shall arrange for one of the following:
>   (a) Evacuation of the building or portion of the building affected by the system out of service
>   (b)* An approved fire watch
>   (c)* Establishment of a temporary water supply
>   (d)* Establishment and implementation of an approved program to eliminate potential ignition sources and limit the amount of fuel available to the fire
> (5) The fire department has been notified.
> (6) The insurance carrier, the alarm company, property owner or designated representative, and other authorities having jurisdiction have been notified.
> (7) The supervisors in the areas to be affected have been notified.
> (8) A tag impairment system has been implemented. *(See Section 15.3.)*
> (9) All necessary tools and materials have been assembled on the impairment site.
> [**NFPA 25**, 15.5.2]

14.1.5.1 Work shall be planned carefully to enable its completion in the shortest possible time.

14.1.5.2 Where available, a second, reasonably reliable water supply with constant suitable pressure and volume, usually public water, shall be connected to the system.

14.1.5.3 Where such a supply is not available, the fire pump shall be started and kept running to maintain suitable pressure in the system.

14.1.5.4 Additional portable fire extinguishers shall be placed in buildings where protection is impaired, and extra, well-instructed watch personnel shall be continuously on duty.

14.1.5.5 The members of the private fire brigade, as well as the public fire department, shall be familiar with conditions that affect repairs.

14.1.6* Heater Thermometer.

A.14.1.6 One of the chief advantages of the gravity circulation system of heating tanks is that it enables convenient observation of the temperature of the coldest water at a thermometer located in the cold-water return pipe near the heater. Failure to provide an accurate thermometer at this point or failure to observe it daily and ensure that it registers the proper temperature forfeits this advantage and can result in the freezing of the equipment. *[See Figure 16.1.7.5.5(a) and Figure 16.1.7.5.5(b), Figure B.1(i) through Figure B.1(k), and Figure B.1(s), Figure B.1(t), and Figure B.1(v).]*

> Note that, for the purposes of this handbook, only Figure 16.1.7.5.5(a) and Figure 16.1.7.5.5(b) are reprinted here.

For SI units, 1 in. = 25.4 mm; 1 ft = 0.3048 m; 1 psi = 0.0689 bar; °C = 5/9 (°F−32).

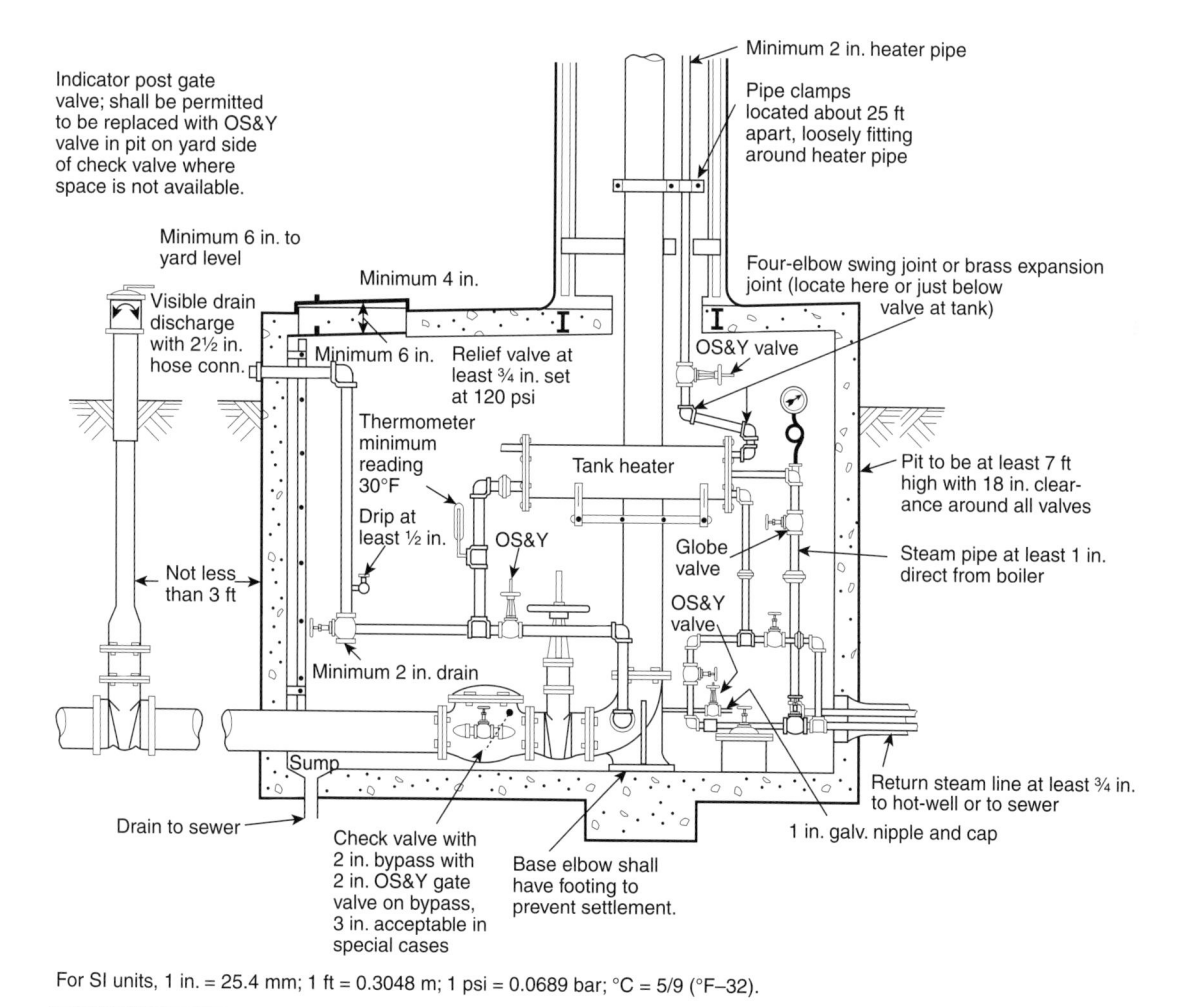

FIGURE 16.1.7.5.5(a) Tank Heater Drain Arrangement at Base of Riser.

FIGURE 16.1.7.5.5(b) Tank Heater Drain Arrangement.

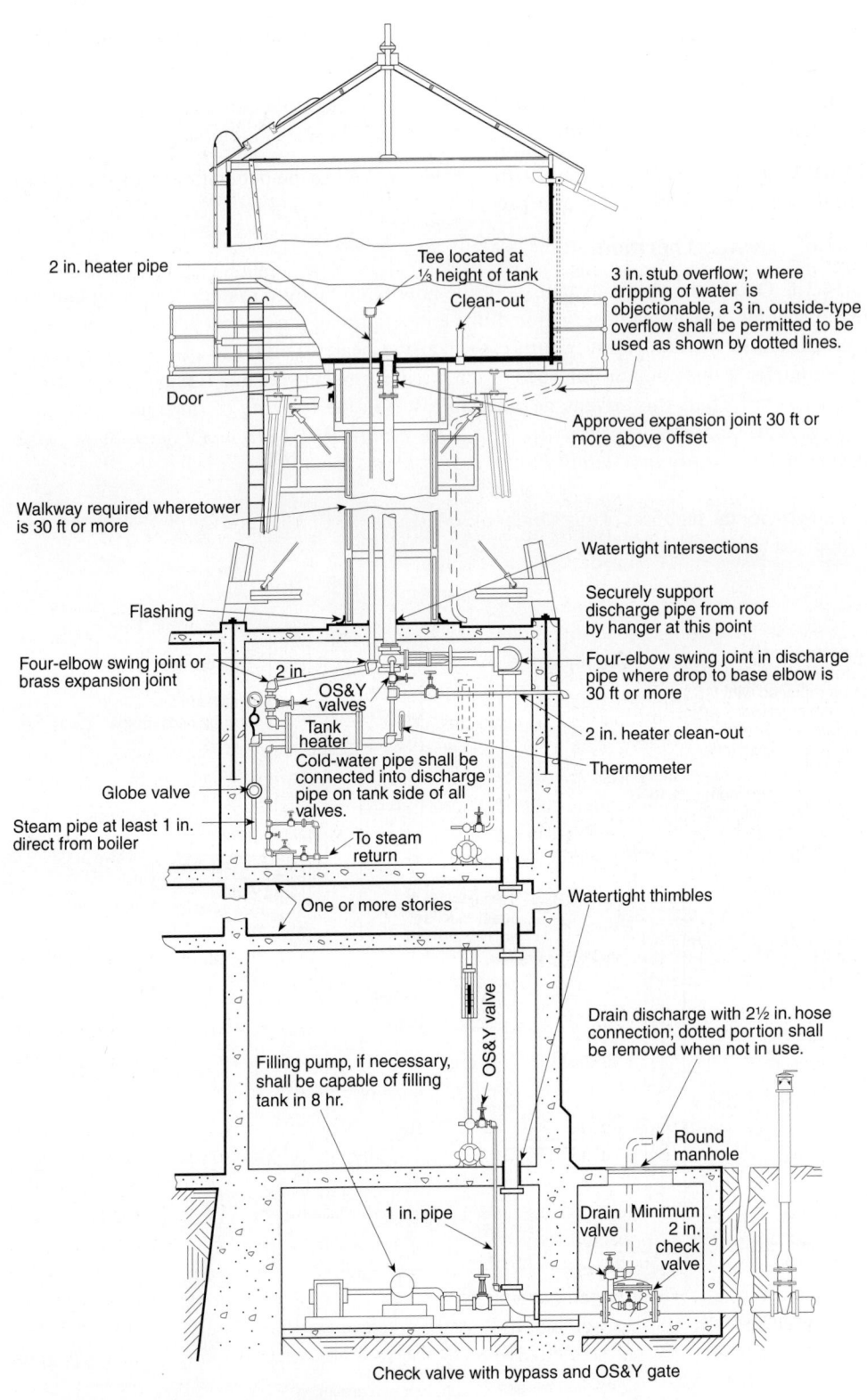

14.1.6.1 In the case of a gravity circulating heating system, an accurate thermometer shall be located as specified in 16.1.7.5.

> Paragraph 16.1.7.5.2 of NFPA 22 requires an accurate thermometer that is graduated at least as low as 30°F (−1.1°C) to be placed in the cold-water pipe at a point where it registers the temperature of the coldest water in the system. This requirement is clearly illustrated in Figure 16.1.7.5.5(a) and Figure 16.1.7.5.5(b).

14.1.6.2 Where a tank contains a radiator steam heater, an accurate socket thermometer shall be located as specified in 16.3.6.

> Subsection 16.3.6 of NFPA 22 requires an accurate angle socket thermometer that has at least a 6 in. (152 mm) stem and that is calibrated as low as 30°F (−1.1°C) to be permanently inserted through the plate or standpipe and located as far from the heating unit as possible. An angle socket thermometer is not required for suction tanks with a maximum height of 25 ft (7.6 m).

14.1.7* Connections for Use Other Than for Fire Protection. The authority having jurisdiction shall be consulted before the tank is designed where water for other than fire protection purposes is to be drawn from the tank.

A.14.1.7 The circulation of water through the tank causes an accumulation of sediment that can obstruct the piping or sprinklers. A leak or break in a pipe for use other than fire protection may seriously impair the fire protection by partly or completely draining the elevated tank.

> When a buildup of sediment occurs, the tank must be drained and cleaned; this creates an impaired system and leads to a potentially expensive cleaning and refilling process. A buildup of sediment can be avoided by dedicating the water tank to fire protection use only. Any multiple-use water tank should be closely monitored for increases in the domestic or industrial consumption to prevent drawing too much water from the tank. The connections for use other than for fire protection should be made above the full level for the fire protection system to prevent depleting the stored fire protection supply.

14.1.8* Water-Level Gauge. A water-level gauge of suitable design shall be provided. It shall be carefully installed, adjusted, and properly maintained.

A.14.1.8 Water-Level Gauges. The following information is provided for existing installations where mercury gauges are in use. Mercury gauges are no longer permitted for new installations.

(1) *Mercury Gauge Materials.* Pipe and fittings that contain mercury should be iron or steel. Brass, copper, or galvanized parts, if in contact with mercury, are amalgamated, and leaks result.

(2) *Water Pipe.* The water pipe to the mercury gauge should be 1 in. (25 mm) galvanized throughout and connected into the discharge pipe on the tank side of the check valve. Where possible, the pipe should be short, should be run with a continual upward pitch toward the tank piping, and should be without air pockets to avoid false readings. The pipe should be buried well below the frost line or located in a heated conduit.

(3) *Valves.* The valve at the mercury gauge should be a listed OS&Y gate valve. An additional listed OS&Y gate valve should be installed close to the discharge pipe where the distance to the mercury gauge exceeds 50 ft (15.2 m).

(4) *Mercury Catcher.* Occasionally, fluctuating water pressures require a mercury catcher at the top of the gauge glass to prevent loss of mercury. The catcher is not a standard part of the equipment and is not furnished by the gauge manufacturer unless specially ordered.

(5) *Extension Piece.* Where the mercury catcher is not needed, it can be replaced by approximately a 3 ft (0.91 m) extension of ⅛ in. (3 mm) pipe, vented at the top.

(6) *Water-Drain Plug.* A plugged tee should be provided in the mercury pipe between the mercury pot and the gauge glass to allow water that sometimes accumulates on top of the mercury column to drain off.

(7) *Location.* The gauge should be installed in a heated room such as a boiler room, engine room, or office, where it is readily accessible for reading, testing, and maintenance. It should be so located that it is not liable to break or to be damaged.

The column of mercury, extending from the mercury pot to the top, is roughly 1/13 the height from the mercury pot to the top of the tank. This fact should be considered when planning a location for the instrument.

(8) *Cleaning.* Before installing the gauge, all grease, dirt, and moisture should be removed from the pot and piping that are to contain mercury, and it should be ensured that the mercury itself is clean. Warm water that contains a small amount of washing soda is a good cleaning agent.

(9) *Installing.* The gauge should be accurately installed so that when the tank is filled to the level of the overflow, the mercury level is opposite the FULL mark on the gauge board.

(10) *Testing.* To determine that it is accurate, the instrument should be tested occasionally as follows:

(a) Overflow the tank.

(b) Close the OS&Y valve. Open the test cock. The mercury should quickly drop into the mercury pot. If it does not, there is an obstruction that must be removed from the pipe or pot between the test cock and the gauge glass.

(c) If the mercury lowers at once, as expected, close the test cock and open the OS&Y valve. If the mercury responds immediately and comes to rest promptly opposite the FULL mark on the gauge board, the instrument is operating properly.

(d) If the mercury column does not respond promptly and read correctly during the test specified in A.14.1.8(10)(c), there are probably air pockets or possibly obstructions in the water-connecting pipe. Open the test cock. Water should flow out forcefully. Allow water to flow through the test cock until all air is expelled and rusty water from the tank riser appears. Then close the test cock. The gauge should now read correctly. If air separates from the water in the 1 in. (25 mm) pipe due to being enclosed in a buried tile conduit with steam pipes, the air can be automatically removed by installing a ¾ in. (19 mm) air trap at the high point of the piping. The air trap can usually be best installed in a tee connected by a short piece of pipe at a location between the OS&Y valve and the test cock, using a plug in the top of the tee, so that mercury can be added in the future, if necessary, without removing the trap. If there are inaccessible pockets in the piping, such as locations below grade or under concrete floors, the air can be removed only through the test cock.

(e) If, in the test specified in A.14.1.8(10)(d), the water does not flow forcefully through the test cock, there is an obstruction that must be removed from the outlet of the test cock or from the waterpipe between the test cock and the tank riser.

(f) If there is water on top of the mercury column in the gauge glass, it will cause inaccurate readings and must be removed. First lower the mercury into the pot as in the test specified in A.14.1.8(10)(b). Close the test cock and remove the plug at the base of the mercury gauge. Open the OS&Y valve very slowly, causing mercury to rise slowly and the water above it to drain through the plug at the base of the mercury gauge. Close the OS&Y valve quickly when mercury appears at the outlet at the base of the mercury gauge, but have a receptacle ready to catch any mercury that drains out. Replace the plug. Replace any escaped mercury in the pot by removing the plug between the OS&Y valve and the test cock, and with the OS&Y valve closed, fill the pot with mercury to the mark on the cover corresponding to the height above the pot that indicates the full water level in the tank. Replace the plug.

(g) After testing leave the OS&Y valve open, except as noted in A.14.1.8(11).

(11) *Excessive Water Pressures.* If found necessary, to prevent forcing mercury and water into the mercury catcher, the controlling OS&Y valve may be closed when filling the tank but should be left open after the tank is filled, except when the gauge is subjected to continual fluctuation of pressure, when it may be necessary to keep the gauge shut off, except when it is being read. Otherwise it may be necessary to frequently remove water from the top of the mercury column as in A.14.18(10)

> Since mercury is a hazardous substance, it is no longer used in liquid level gauges for new installations. The text of A.14.1.8 remains in the standard for those installations that still use this technology. New installations rely on other means to indicate water level, such as the use of an altitude gauge or float mechanism in the tank. Exhibit III.26 illustrates a liquid level gauge on a tank.

14.1.8.1 Where an altitude gauge is used, it shall be at least 6 in. (152 mm) in diameter and shall be of noncorrodable construction.

14.1.8.2 The gauge shall be located to prevent it from freezing.

EXHIBIT III.26 *Liquid Level Gauge on a Tank.*

EXHIBIT III.27 Tank Water Level Supervisory Switch. (Courtesy of Potter Electric Signal Company, LLC)

14.1.8.2.1 If necessary, it shall be located in a heated building or enclosure.

14.1.8.2.2 A blow-off cock shall be located between the gauge and the connection to the tank.

14.1.8.3 A listed, closed-circuit, high-water and low-water level electric alarm shall be permitted to be used in place of the gauge where acceptable to the authority having jurisdiction.

> When a water level switch is used, the requirements of *NFPA 72* state that the switch must provide two separate and distinct signals — one indicating that the required water level has been raised or lowered (off-normal), and the other indicating restoration of the normal water level. See Exhibit III.27 for an example of a water level supervisory switch.

14.1.8.3.1 Provisions shall be made for the attachment of a calibrated test gauge.

14.1.8.4 For underground tanks, water-level monitoring shall be capable of being read and/or supervised above ground.

14.1.9* Frostproof Casing. The frostproof casing shall be maintained in good repair and shall be weathertight throughout.

A.14.1.9 The insulating qualities of frostproof casing are seriously impaired if joints spring open, if the casing settles away from the tank, or if rotting occurs around the base.

14.1.10 Tanks with Large Risers.

14.1.10.1* Large steel-plate riser pipes of 3 ft (0.91 m) or more in diameter and without frostproof casing shall be acceptable where properly heated.

A.14.1.10.1 By heating the large steel-plate riser pipes, the fire hazard and upkeep of the frostproof casing and the provision of an expansion joint or walkway are avoided. However, painting and heating the larger riser and building the stronger and larger valve pit cost more than the equipment for smaller risers.

A blow-off valve is sometimes furnished near the base of the larger riser.

A check valve and gates in the discharge pipe, filling arrangement, overflow, and drain are generally provided.

14.1.10.2 A manhole at least 12 in. × 16 in. (305 mm × 406 mm) shall be provided, and its lower edge shall be level with the discharge piping protection specified.

14.1.11 Discharge Piping Protection.

14.1.11.1 In the case of tanks with a large steel-plate riser [3 ft (0.91 m) diameter or larger], the inlet to the vertical discharge pipe that is located within the large riser shall be protected against the entry of foreign material.

14.1.11.2 The inlet can be done with an American National Standards Institute 125 lb/in. (8.6 bar) flanged tee, with the "run" of the reducing tee placed horizontally and with horizontal outlets one pipe size smaller than the discharge pipe, or with a fabricated plate extending at least 4 in. (102 mm) beyond the outside diameter of the pipe.

14.1.11.3 The plate shall be supported by at least three supporting bars 1½ in. × ¼ in. (38.1 mm × 6.4 mm), by ⅝ in. (15.9 mm) round rods, or by the equivalent, that elevate all portions of the plate at a height at least equal to the pipe diameter located above the discharge pipe inlet.

14.1.11.4 The attachment of the supports to the discharge pipe shall be made directly by welding or bolting or by means of a ¼ in. (6.4 mm) thick tightly fitting sectional clamp or collar that has ⅝ in. (15.9 mm) bolts in the outstanding legs of the clamps or collar.

14.1.11.5 A clearance of at least 6 in. (152 mm) shall be provided between all portions of the flanges of a tee or fabricated plate and the large riser plate.

14.1.12 Steel Pipe.

14.1.12.1 Steel pipe shall conform to ASTM A 53, Type E, Type F, Type S, Grade A, or Grade B, manufactured by the open-hearth, electric furnace, or basic oxygen process, or it shall conform to ASTM A 106, Grade A or Grade B.

14.1.12.2 Paragraphs 14.1.12.2.1 through 14.1.12.2.3 shall apply to steel pipe that is in contact with storage water.

14.1.12.2.1 Steel pipe smaller than 2 in. (50 mm) shall not be used.

14.1.12.2.2 Steel pipe of 2 in. to 5 in. (50 mm to 125 mm) to 5 in. (125 mm) shall be extra-strong weight.

> Piping in fire protection systems is ordinarily Schedule 10 (light wall) or Schedule 40 (standard weight). Occasionally, extra strong (Schedule 80) pipe is used. In this application, due to continual contact with storage water, pipe with a thicker wall is required. A comparison of wall thickness for Schedules 10, 40, and 80 pipe is shown in Commentary Table III.1.
>
> **COMMENTARY TABLE III.1** Comparison of Wall Thickness for Schedules 10, 40, and 80 Pipe
>
> | Pipe Size | | Schedule | Outside Diameter | | Inside Diameter | | Wall Thickness | |
> |-----------|----|----------|------------------|------|-----------------|------|----------------|-----|
> | in. | mm | Number | in. | mm | in. | mm | in. | mm |
> | 2 | 50 | 10 | 2.375 | 60.3 | 2.157 | 54.8 | 0.109 | 2.8 |
> | 2 | 50 | 40 | 2.375 | 60.3 | 2.067 | 52.5 | 0.154 | 3.9 |
> | 2 | 50 | 80 | 2.375 | 60.3 | 1.939 | 49.3 | 0.218 | 5.5 |

14.1.12.2.3 All steel pipe 6 in. (150 mm) and larger shall be standard weight.

14.2 Discharge Pipe

14.2.1 At Roofs and Floors. The intersection of discharge pipes, as well as the intersection of all other tank pipes, with roofs or with waterproof or concrete floors shall be watertight.

14.2.2 Size.

14.2.2.1 The conditions at each plant shall determine the size of the discharge pipe that is needed.

> The size of the discharge pipe is determined by fire protection system demand. In the case of sprinkler systems, the discharge density, system size, and distance from the tank determine the diameter of the discharge pipe. Where elevated tanks provide the sole supply of water, the hydraulic calculations should terminate at the base of the tank. The calculations must include flow through the discharge pipe and the elevation difference. The elevation to the top of the full water level is not considered, because this level decreases during discharge. Sizes other than those specified can be used, provided that they are approved by the AHJ. However, selection of the discharge pipe diameter should also take into consideration possible future demand due to expansion.

14.2.2.2 The size shall not be less than 6 in. (150 mm) for tanks up to and including a 25,000 gal (94.63 m^3) capacity and shall not be less than 8 in. (200 mm) for capacities of 30,000 gal to 100,000 gal (113.55 m^3 to 378.50 m^3) inclusive, or 10 in. (250 mm) for greater capacities.

> Where tanks supply fire pumps directly, the size of the discharge pipe, when it is located within 10 pipe diameters of the pump suction flange, should not be smaller than that specified in Table 4.26(a) and Table 4.26(b) of NFPA 20 (see Part II of this handbook).

14.2.2.3 Pipe that is smaller than specified in 14.2.2.2 [not less than 6 in. (150 mm)] shall be permitted in some cases where conditions are favorable and large flows of water are not needed.

14.2.2.3.1 Larger pipe shall be required where deemed necessary because of the location and arrangement of piping, height of buildings, or other conditions.

14.2.2.3.2 In all cases, approval of the pipe sizes shall be obtained from the authority having jurisdiction.

14.2.3 Pipe Material.

14.2.3.1 Underground Pipe Material. Piping shall be in accordance with NFPA 24.

14.2.3.2 Aboveground Pipe Material. Aboveground pipe material shall be in accordance with NFPA 13 and NFPA 20.

14.2.4 Braces.

14.2.4.1 Either the pipe or the large steel-plate riser pipe, or both, shall be braced laterally by rods of not less than ⅝ in. (15.9 mm) in diameter and shall be connected to the tower columns near each panel point.

14.2.4.2 The end connection of braces shall be by means of eyes or shackles; open hooks shall not be permitted.

14.2.5 Support.

14.2.5.1 The discharge pipe shall be supported at its base by a double-flanged base elbow that rests on a concrete or masonry foundation.

14.2.5.1.1 The base elbow of tanks with steel-plate risers, of suction tanks, or of standpipes shall have bell ends.

14.2.5.2 The joint at the connection of yard piping to the base elbow shall be strapped, or the base elbow shall be backed up by concrete.

14.2.5.2.1 If the discharge pipe is offset inside a building, it shall be supported at the offset by suitable hangers that extend from the roof or floors, in which case the base elbow might not be required.

14.2.5.2.2 Large steel riser pipes shall be supported on a reinforced concrete pier that is designed to support the load specified in Section 13.3.

14.2.5.2.3 Concrete grout shall be provided beneath the large riser to furnish uniform bearing when the tank is empty.

14.2.6 Offsets.

14.2.6.1 The discharge pipe outside of buildings shall extend vertically to the base elbow or building roof without offsets where possible.

> Offsets in the discharge pipe increase the need for additional bracing. They also add to the friction loss in the hydraulic calculations and should be avoided.

14.2.6.2 If an offset is unavoidable, it shall be supported at the offsetting elbows and at intermediate points not over 12 ft (3.7 m) apart, and it also shall be rigidly braced laterally.

14.2.6.3 The supports shall consist of steel beams that run across the tower struts or of steel rods from the tower columns arranged so that there is no slipping or loosening.

14.2.7 Expansion Joint.

14.2.7.1 Tanks with flanged or welded pipe risers [12 in. (250 mm) and under] shall have a listed expansion joint on the fire-service discharge pipe where the tank is on a tower that elevates the bottom 30 ft (9.1 m) or more above the base elbow or any offset in the discharge pipe.

> Due to the rigid connections of the discharge pipe at the top and bottom of the pipe, an expansion joint is needed. Any fluctuations in the length of the piping material due to temperature variations could cause stress at either connection that could result in connection failure.

14.2.7.2 Expansion joints shall be built to conform to Section 14.3.

14.2.8 Rigid Connection.

14.2.8.1 When the distance between the tank bottom and the base elbow or supporting hanger is less than 30 ft (9.1 m), the discharge pipe shall be connected by an expansion joint that is built to conform to Section 14.3 or shall be rigidly connected in accordance with 14.1.2.

14.2.8.2 The top of the pipe (or the fitting attached to the top) shall extend above the inside of the tank bottom or base of a steel-plate riser to form a settling basin.

14.2.8.2.1 The top of a steel-plate riser shall be connected rigidly to the suspended bottom of the tank.

14.2.8.2.2 The discharge pipe from a steel plate riser of a tank that is located over a building shall be connected rigidly to the base of the larger riser.

14.2.8.2.3 A rigid flanged connection or welded joint shall be permitted to be used between the discharge pipe and the bottom of a suction tank, a standpipe, or the base of a steel-plate riser of a tank that is located on an independent tower where special approval is obtained from the authority having jurisdiction.

14.2.8.2.4 When the base of a steel-plate riser is in its final position on a concrete support, it shall be grouted to obtain complete bearing.

14.2.9 Swing Joints. Where the vertical length of a discharge pipe that is located below an offset, either inside or outside a building, is 30 ft (9.1 m) or more, a four-elbow swing joint that is formed, in part, by the offset shall be provided in the pipe.

> Because the four elbows in an offset form a swing joint, the riser can tolerate movement in any direction without producing any stress on the end connections.

14.2.10 Settling Basin.

14.2.10.1 The depth of the settling basin in the tank bottom shall be 4 in. (102 mm) for a flat-bottom tank and 18 in. (457 mm) for a suspended-bottom tank.

14.2.10.2 The settling basin at the base of a large steel-plate riser shall be at least 3 ft (0.91 m) deep.

> The settling basin is intended to permit potentially obstructing material to settle on the tank bottom without entering the discharge pipe. Obstructing material can be in the form of paint chips or scale from steel tanks or steel piping, rocks and other debris from the fill connection, construction debris that has not been removed, or aquatic growth. See Exhibit III.28 for an example of potentially obstructing material.

EXHIBIT III.28 Material Found During an Interior Tank Inspection.

14.2.11 Check Valve.

A control valve is needed to isolate the tank or check valve for inspection and maintenance purposes. When tanks are used to supply water to a fire pump, devices in the suction pipe in close proximity to the fire pump suction flange are of concern. NFPA 20 permits the suction pressure to a fire pump to be as low as 0 psi (0 bar). When the water supply is a suction tank and the base of the tank is even with or above the elevation of the fire pump, the suction pressure is permitted to be as low as −3 psi (−0.2 bar). The friction loss caused by the devices required in NFPA 22, and any offsets in the tank discharge pipe, must be carefully calculated to verify that these pressure limitations are not exceeded. (See 4.14.3 of NFPA 20 in Part II of this handbook.)

14.2.11.1 A listed check valve shall be placed horizontally in the discharge pipe and shall be located in a pit under the tank where the tank is located on an independent tower.

The check valve prevents water from flowing back into the tank or discharge pipe when sprinklers are located above the elevation of the discharge pipe or tank bottom.

14.2.11.2 Where the tank is located over a building, the check valve shall ordinarily be placed in a pit, preferably outside the building.

14.2.11.3 Where yard room is not available, the check valve shall be located on the ground floor or in the basement of a building, provided it is protected against breakage.

14.2.12 Controlling Valves.

14.2.12.1 A listed gate valve shall be placed in the discharge pipe on the yard side of the check valve between the check valve and any connection of the tank discharge to other piping.

A gate valve enables the isolation of the tank for the purpose of filling where a separate fill pipe or fill pump is used. The valve should be equipped with a seal, lock, or tamper switch and should be inspected based on the inspection procedure and frequencies of NFPA 25.

14.2.12.1.1 The listed gate valve shall be permitted to be equipped with an indicating post.

14.2.12.2 Where yard room for an indicator post is not available, a listed outside screw and yoke (OS&Y) gate valve that is of similar arrangement, but that is located inside the valve pit or room, shall be used.

14.2.12.3 A listed indicating control valve shall be placed in the discharge pipe on the tank side of the check valve.

> An indicating control valve is needed to isolate the discharge check valve for maintenance or repair. Note that, according to NFPA 25, each system check valve must be inspected internally every 5 years, and isolation of the check valve is critical in order to comply with this requirement. Isolation of the check valve will probably also be needed at some point for replacement or repair.

14.2.12.3.1 Where the tank is on an independent tower, the valve shall be placed in the pit with the check valve, preferably on the yard side of the base elbow.

14.2.12.3.2 Where a tank is used as a suction source for a fire pump, the listed indicating control valve shall be of the OS&Y type.

> NFPA 20 requires any valve within 50 ft (15.3 m) of the pump suction flange to be an OS&Y-type valve.

14.2.12.3.3 Where the tank is located over a building, the valve shall be placed under the roof near the point where the discharge pipe enters the building.

14.2.12.3.4 For suction tanks, the valve shall be as close to the tank as possible.

14.2.13 Anti-Vortex Plate Assembly.

14.2.13.1 Where a tank is used as the suction source for a fire pump, the discharge outlet shall be equipped with an assembly that controls vortex flow.

14.2.13.2* The assembly shall consist of a horizontal steel plate that is at least twice the diameter of the outlet on an elbow fitting, where required, mounted at the outlet a distance above the bottom of the tank equal to one-half the diameter of the discharge pipe.

> As its name suggests, the anti-vortex plate assembly is intended to prevent the development of a vortex when the fire pump is in operation. A vortex entrains air and, if present, could cause the fire pump to cavitate. The inspection required by 17.1.1 of NFPA 22 should include inspection of the anti-vortex plate to verify that it is in place and has been installed correctly. The inspection report should indicate that an anti-vortex plate was installed and that it was found to be acceptable.

A.14.2.13.2 Large, standard size anti-vortex plates [48 in. × 48 in. (1219 mm × 1219 mm)] are desirable, as they are adequate for all sizes of pump suction pipes normally used. Smaller plates may be used; however, they should comply with 14.2.13.

14.2.13.3* The minimum distance above the bottom of the tank shall be 6 in. (152 mm).

A.14.2.13.3 See Figure B.1(o).

14.3 Expansion Joint

14.3.1 Connection to Tank.

14.3.1.1 A listed expansion joint shall be used where required by 14.2.7 and 14.2.8.1.

14.3.1.2 The expansion joint shall be placed immediately above the foot elbow or shall be connected to the tank bottom using welding for a steel tank and bolts or a special screw fitting for a wood tank.

Stationary Fire Pumps Handbook   2013

14.3.1.3 The movable nuts on bolts shall be located on the outside of the tank.

14.3.2 General Design.

14.3.2.1 The design shall be such that the joint operates reliably over years without attention and shall be of adequate strength to resist the stresses and corrosion to which it is subjected.

14.3.2.2 One or both of the two parts that slide, one on the other, shall be of brass or other noncorrodible metal of ample strength and resistance to wear.

14.3.3 Clearances. A minimum 1/16 in. (1.6 mm) clearance shall be provided around all movable parts to prevent binding, and at least 1/2 in. (12.7 mm) shall be provided between the cast-iron body and an iron or steel slip-tube.

14.3.4 Body.

14.3.4.1 The body shall be of steel or cast-iron and, if connected to the tank bottom, shall provide for a settling basin extension of proper length.

14.3.4.2 Provisions shall be made for a packing space of adequate size.

14.3.5 Gland. The adjustable gland shall be of brass or iron and shall be connected to the body casting, preferably with four standard bolts of at least 5/8 in. (15.9 mm) and of sufficient length to allow full adjustment.

14.3.6 Slip-Tube.

14.3.6.1 The sliding tube at the top of the discharge pipe shall be of brass or iron.

14.3.6.2 If the gland is iron, the slip-tube shall have a triple-plated brass outer surface.

14.3.6.3 If the gland is brass, the slip-tube shall be of cast-iron or steel, and the top of the packing space shall be formed with brass and a clearance of at least 1/2 in. (12.7 mm) provided at all points between the cast-iron body and the slip-tube.

14.3.6.4 The upper part of the slip-tube shall be machined over a length such that the top of the gland can be dropped to 6 in. (152 mm) below the bottom of the body casting so as to enable repacking.

14.3.6.5 The top of the slip-tube shall be located approximately 5 in. (127 mm) below the top of 4 in. (102 mm) settling-basin extensions and 12 in. (305 mm) below the top of 18 in. (457 mm) settling-basin extensions.

14.3.7 Packing.

14.3.7.1 The packing shall consist of asbestos wicking that is saturated with rape oil and graphite or an equally suitable material.

14.3.7.2 Packing at least 2 in. (51 mm) deep and 1/2 in. (12.7 mm) thick shall be provided in the packing space.

14.3.8 Connections for Use Other Than for Fire Protection.

14.3.8.1 Connections for a use other than fire protection shall not be made.

14.3.8.2 Where unavoidable connections for other than fire protection shall be permitted, connections shall be rigidly made to the tank bottom, and a standard expansion joint, where needed, shall be provided in each such pipe that is located below, and entirely independent of, the tank.

14.4 Filling

14.4.1 A permanent pipe connected to a water supply shall be provided to fill the tank.

14.4.2 The means to fill the tank shall be sized to fill the tank in a maximum time of 8 hours.

14.4.3 The tank shall be kept filled, and the water level shall never be more than 4 in. (102 mm) below the designated fire service level.

14.4.4 The filling bypass shall be kept closed when not in use.

> Section 14.4 establishes the requirement for a permanently installed filling connection. The intent of this requirement is to prohibit the filling of a tank from either a hydrant and hose arrangement or from a filling truck.
>
> This section has been revised to clarify the requirement to design and install a fill connection that is capable of filling the tank in not more than 8 hours.
>
> A tolerance of 4 in. (102 mm) is permitted by the standard. Failure to keep this bypass closed results in wasting water by causing a constant discharge into the overflow pipe.

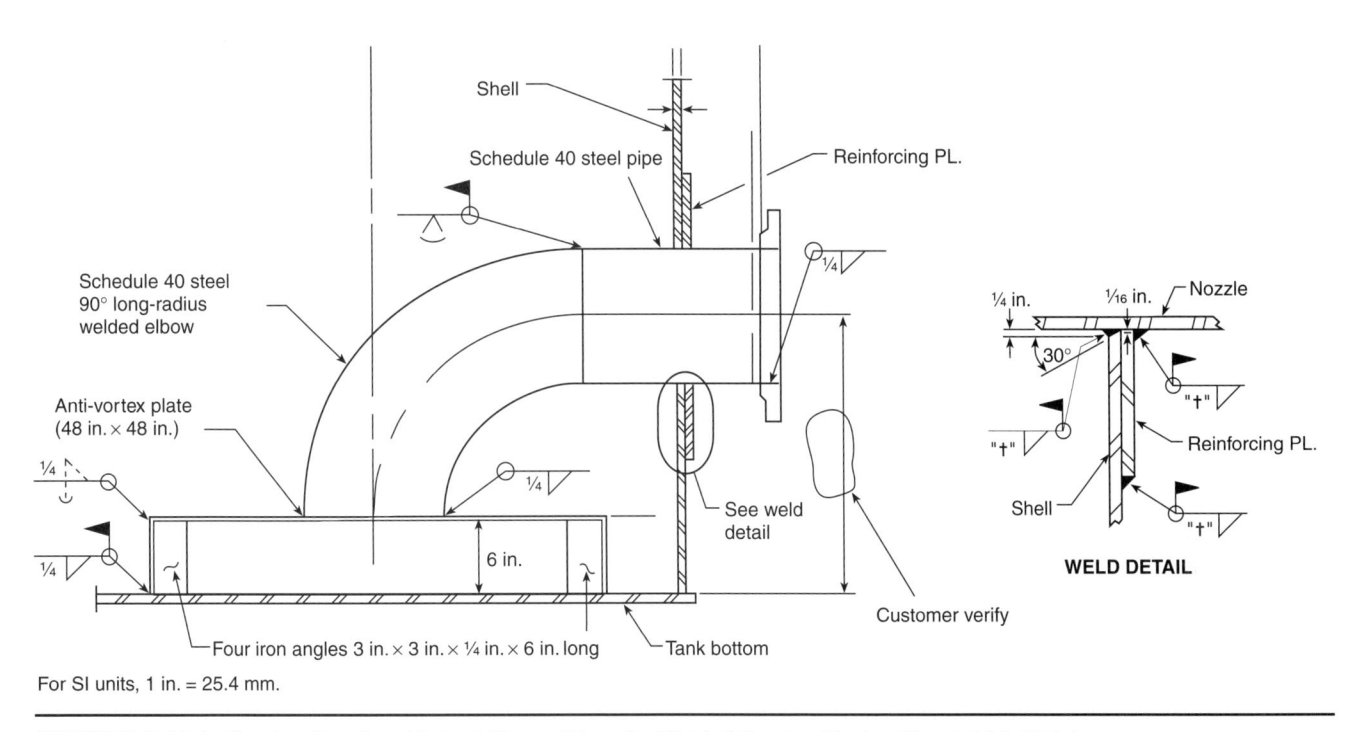

FIGURE B.1(o) Suction Nozzle with Anti-Vortex Plate for Welded Suction Tanks. (See A.14.2.13.2.)

14.4.5 Bypass Around Check Valve.

14.4.5.1 Where the tank is to be filled from the fire protection system under city or fire-pump pressure, the filling pipe shall be a bypass around the check.

> The filling arrangement specified in 14.4.5.1 is allowed only when the tank is used to supplement another water supply, such as a city supply, or when the tank connects into the fire protection main on the discharge side of the fire pump. With this arrangement, the control valve in the bypass is opened only for filling the tank and relies on city or fire pump pressure to fill the tank.

14.4.5.2 The bypass shall be connected into tapped bosses on the check valve or into the discharge pipe between the check valve and all other valves.

Most large bore [6 in. (150 mm) and larger] check valves can be tapped for up to 2 in. (50 mm) connections on each side of the clapper. The bypass is required to be connected by means of this tap in the check valve body. This method provides a convenient connection for the bypass. The bypass consists of a few short lengths of pipe, two 90-degree elbows, and a threaded gate valve. The gate valve is normally closed and is only opened to fill the tank. Exhibit III.29 illustrates a typical fill connection.

EXHIBIT III.29 Check Valve with Bypass.

14.4.5.3 The bypass shall be sized to fill the tank in accordance with 14.4.2 but shall not be smaller than 2 in. (50 mm).

14.4.5.4 A listed indicating control valve shall be placed in the bypass and shall be kept closed except when the tank is being filled.

14.4.6 Filling Pumps.

14.4.6.1 When the tank is to be filled by a filling pump, the pump and connections shall be of such size that the tank can be filled in accordance with 14.4.2.

The 8-hour requirement of 14.4.2 is a maximum fill time. Any arrangement that fills the tank sooner should be considered to be acceptable.

14.4.6.2 The filling pipe shall be of at least 2 in. (50 mm) and, except as noted in 14.4.7, shall be connected directly into the tank discharge pipe, in which case a listed indicating control valve and a check valve shall be placed in the filling pipe near the tank discharge pipe, with the check valve located on the pump side of the listed indicating valve.

14.4.6.3 The filling pump suction pipe shall not be connected to a fire service main that is supplied from the tank. The filling valve shall be open only when the tank is being filled.

14.4.7 Where a separate fill pipe is used, automatic filling shall be permitted.

For automatic filling, the tank water level is monitored, and the fill pump is turned on when the water level drops. With this arrangement, the fill valve is normally open. If the exact fill rate is known and the reliability of the fill pump actuation mechanism is verified, the fill rate can be considered when the tank capacity is calculated. This approach to filling must be presented to the AHJ for approval prior to the purchase and installation of the tank.

14.4.8 Filling from Drinking Water Supply. Where the water in the fire protection system is not suitable for drinking purposes and the tank is filled from a potable water supply, the filling pipe shall be installed in accordance with the regulations of the local health authority.

> When filling from a drinking water supply, the fill connection can create a cross-connection that could ultimately contaminate the potable water supply. When this arrangement is proposed, a backflow preventer should be installed to prevent contamination of the potable water supply. The friction loss through the backflow preventer must be considered when calculating the flow capacity of the fill connection in order to meet the 8-hour fill requirement.

◀ **DESIGN ALERT**

14.4.9 Filling Pipe at Roofs and Floors. The intersection of a separate filling pipe with a roof or a waterproof or concrete floor shall be watertight.

14.4.10 Suction Tanks.

14.4.10.1 Pipes for the automatic filling of suction tanks shall discharge into the opposite half of the tanks from the pump suction pipe.

14.4.10.2 Where an over-the-top fill line is used, the outlet shall be directed downward.

14.5 Overflow

14.5.1 Size. The overflow pipe shall be of adequate capacity for the operating conditions and shall be of not less than 3 in. (75 mm) throughout.

> In cases where the fill valve is left open or the automatic fill connection fails to open, an overflow connection is required to prevent overflowing of the tank and potential damage to the tank roof. Subsection 14.5.2 is intended to ensure that the overflow pipe is properly pitched to automatically drain and that the overflow piping capacity exceeds that of the fill connection.

14.5.2 Inlet.

14.5.2.1 The inlet of the overflow pipe shall be located at the top capacity line or high waterline.

14.5.2.2 The inlet also shall be located at least 1 in. (25 mm) below the bottom of the flat cover joists in a wood tank, but shall never be closer than 2 in. (50 mm) to the top of the tank.

14.5.2.3 Unless the maximum fill capacity is known and the overflow capacity is calculated to be at least equal to the fill capacity, the overflow pipe shall be at least one pipe size larger than the fill line and shall be equipped with an inlet such as a concentric reducer, or equivalent, that is at least 2 in. (50 mm) larger in diameter.

14.5.2.4 The inlet shall be arranged so that the flow of water is not retarded by any obstruction.

14.5.2.5 An overflow pipe that is cut with the opening to fit the roof shall be used on a steel tank, provided a suitable horizontal suction plate and vortex breaker are used to ensure full capacity flow for the overflow.

14.5.3* Stub Pipe.

A.14.5.3 On column-supported tanks with outside overflow, vertical extensions of the pipe that is located below the balcony are not recommended, as they can become plugged with ice.

14.5.3.1 Where dripping water or a small accumulation of ice is not objectionable, the overflow shall be permitted, at the discretion of the owner, to pass through the side of the tank near the top.

14.5.3.2 The pipe shall be extended with a slight downward pitch to discharge beyond the tank or balcony and away from the ladders and shall be adequately supported.

14.5.3.3 Overflows for pedestal tanks shall be extended to ground level within the access tube and pedestal.

14.5.4 Inside Pipe.

14.5.4.1 Where a stub pipe is undesirable, the overflow pipe shall extend down through the tank bottom and inside the frostproof casing or steel-plate riser and shall discharge through the casing near the ground or roof level.

14.5.4.2* The section of the pipe inside the tank shall be of brass, flanged cast-iron, or steel.

A.14.5.4.2 See 14.1.12.

14.5.4.2.1 Inside overflow pipes shall be braced by substantial clamps to tank and riser plates at points not over 25 ft (7.6 m) apart.

14.5.4.2.2 The discharge shall be visible, and the pipe shall be pitched to drain.

14.5.4.2.3 Where the discharge is exposed, the exposed length shall not exceed 4 ft (1.2 m) and shall avoid the entrance to the valve pit or house.

14.6 Clean-Out and Drain

14.6.1 Handhole. A standard handhole, with a minimum dimension of 3 in. (76 mm), or a manhole shall be provided in the saucer plate outside of the frostproof casing and at the bottom of an elevated steel tank with a suspended bottom unless the tank has a large riser pipe 3 ft (0.91 m) or more in diameter.

14.6.2 Manholes.

14.6.2.1 Two manholes shall be provided in the first ring of the steel suction tank shell at locations to be designated by the purchaser.

> The manholes required by 14.6.2 can serve the following two purposes:
>
> 1. They accommodate confined space entry requirements, thus providing a separate exit route.
> 2. They permit the temporary installation of an exhaust fan to introduce fresh air when the tank is drained and opened for maintenance.

14.6.2.1.1 The design of the manholes for steel tanks shall be in accordance with AWWA D100 for welded steel tanks, and AWWA D103 for bolted steel tanks.

14.6.3 For Elevated Flat-Bottom Tanks.

14.6.3.1 Where elevated, at least a 2 in. (50 mm) pipe clean-out also shall be provided outside of the frostproof casing in the bottom of a wood tank or a flat-bottom steel tank.

14.6.3.2 The clean-out connection for wood tanks shall consist of a special screw fitting with a gasket or a pair of 2 in. (50 mm) pipe flanges.

14.6.3.3 The connection for steel tanks shall consist of an extra-heavy coupling welded to the bottom plate.

14.6.3.4 The coupling shall be welded to both sides of the tank plates.

14.6.3.5 A piece of 2 in. (50 mm) brass pipe about 5 in. (127 mm) long that is capped at the top with a brass cap shall be screwed into the inner fitting or flange.

14.6.3.6* The clean-out shall be watertight.

A.14.6.3.6 See Figure B.1(k).

14.6.4 Riser Drain.

14.6.4.1 A drain pipe of at least 2 in. (50 mm) that is fitted with a controlling valve and a ½ in. (12 mm) drip valve shall be connected into the tank discharge pipe near its base and, where possible, on the tank side of all valves.

14.6.4.2 Where the outlet is an open end outlet, it shall be fitted with a 2½ in. (65 mm) hose connection unless it discharges into a funnel or cistern piped to a sewer.

14.6.4.3 Where the drain is piped directly to a sewer, a sight glass or a ¾ in. (19.1 mm) test valve on the underside of the pipe shall be provided.

14.6.4.4 Where the drain pipe is to be used for a hose stream, the controlling valve shall be a listed gate valve or angle valve.

14.6.4.5* Where a circulation-tank heater is located near the base of the tank riser, the drain pipe shall, if possible, be connected from the cold-water return pipe between the cold-water valve and the heater in order to permit flushing water from the tank through the hot-water pipe heater and drain for clean-out purposes.

A.14.6.4.5 See Figure B.1(i).

14.7 Connections for Other Than Fire Protection

14.7.1* Dual-Service Tanks. Where dual service is necessary, an adequate supply of water shall be constantly and automatically reserved in the tank for fire protection purposes.

A.14.7.1 The use of an elevated tank, in part, for purposes other than fire protection, is not advised. Frequent circulation of the water results in an accumulation of sediment that can obstruct the piping of sprinklers, and a fluctuating water level hastens decaying of wood and corrosion of steel.

14.7.2 Pipe for Other Than Fire Protection Purposes.

14.7.2.1 Pipe used for other than fire protection purposes shall be entirely separate from fire-service pipes and shall extend to an elevation inside the tank below which an adequate quantity of water is constantly available for fire protection.

14.7.2.2 Pipe inside the tank that is used for other than fire protection purposes shall be brass.

14.7.2.2.1 Steel pipe shall be permitted to be used where the pipe is larger than 3 in. (75 mm), or cast-iron shall be permitted where the pipe is 6 in. (150 mm) or larger.

14.7.2.3 Pipe inside the tank shall be braced near the top and at points not over 25 ft (7.6 m) apart.

14.7.2.4* Where an expansion joint exists, it shall be of the standard type, shall be located below the tank, and shall be without connection to the tank plates.

A.14.7.2.4 See 14.3.8.

14.7.3* At Roofs and Floors. Where a pipe used for other than fire protection purposes intersects with a building roof or a waterproof or concrete floor, the intersection shall be watertight.

A.14.7.3 See 14.1.1.

14.8* Sensors

A.14.8 It is not the intent of this standard to require the electronic supervision of tanks; however, where such supervision is required in accordance with *NFPA 72*, the following alarms should be required:

(1) Water temperature below 40°F (4.4°C)
(2) Return of water temperature to 40°F (4.4°C)
(3) Water level 3 in. (76.2 mm) (pressure tanks) or 5 in. (127 mm) (all other tanks) below normal
(4) Return of water level to normal
(5) Pressure in pressure tank 10 psi (0.48 kPa) below normal
(6) Pressure in pressure tank 10 psi (0.48 kPa) above normal

14.8.1 Provisions shall be made for the installation of sensors in accordance with *NFPA 72* for two critical water temperatures and two critical water levels.

14.8.2 Pressure Tanks. In addition to the requirements of 14.8.1, pressure tanks shall be provided with connections for the installation of high- and low-water pressure supervisory signals in accordance with *NFPA 72*.

> Wood tanks should be filled and tested for liquid tightness for 48 hours. The leakage test for wood tanks must last for 48 hours in order to allow the wood to become completely wet and swell. Dry wood, when first installed, will most likely leak. This test should be done under the supervision of a qualified wood tank specialist. The test should be in accordance with the National Wood Tank Institute Bulletin S82, "Specifications for Tanks and Pipes."
>
> For FRP tanks, a 4 in. (100 mm) standpipe should be attached to the tank that extends 4 ft (1.2 m) above the top of the tank. The tank and standpipe should be filled with water and allowed to stand for 24 hours. The tank should be examined for leakage or any drop in water elevation in the standpipe. The tank should show no visible signs of leakage, and the water level should not fall more than ½ in. (13 mm) within the 24-hour test period.
>
> The hydrostatic test for an FRP tank should be completed before backfill in order to inspect the tank for leaks. The tank should be in place and properly secured before being filled with water.

• • •

CHAPTER 17 ACCEPTANCE TEST REQUIREMENTS

17.1 Inspection of Completed Equipment

> The installing contractor should notify the local AHJ and encourage the AHJ's involvement in the inspection of the completed tank and accessories. In some jurisdictions, work or commissioning cannot proceed without a written authorization from the AHJ. On large projects, this inspection should also involve the general contractor and insurance company representative for their approval. Verification in writing that the inspection has been completed and the work has been accepted is important before concealment of any work or portion of the system and equipment, since once concealed, inspection is difficult.

17.1.1 Prior to placing the tank in service, a representative of the tank contractor and a representative of the owner shall conduct a joint inspection of the completed equipment.

> Immediately following construction of the tank, an inspection of the joints provides a training opportunity for the operating personnel to become familiar with the operation and maintenance of the tank and related accessories. Operation and maintenance personnel should become familiar with NFPA 25 at this time, and the critical inspection, test, and maintenance requirements for tanks and valves should be highlighted during this training period.

17.1.2* Written reports of completed equipment inspections shall be made in triplicate, and a copy that has been signed by the contractors and the owners shall be sent to the authority having jurisdiction.

> These final written inspection reports can be part of the test report previously mentioned and should be included in the final submission as an "as-built" record of completion of work. The tank installation specification and contract should include a clause that payment will be withheld until all of the aforementioned documentation has been submitted and approved by the AHJ and RDP.

A.17.1.2 This joint inspection reasonably ensures that there are no defects in the work of sufficient importance to prevent the system from being put into service immediately. The inspection also permits the owner's representatives to become more familiar with the system and equipment.

17.2 Testing

17.2.1 All coated steel tanks shall be tested for holidays and coating thickness.

17.2.2 Corrective action shall be completed prior to acceptance.

> The paint system on steel tanks must be inspected to verify that it has been applied correctly. Failure of the paint system results in excessive corrosion and potential failure of the tank.

17.3 Welded Steel Tanks

17.3.1 Flat Bottoms. Upon completion of the welding of the tank bottom, it shall be tested by one of the following methods and shall be made entirely tight:

(1) Air pressure or vacuum applied to the joints, using soap suds, linseed oil, or other suitable material for the detection of leaks
(2) Joints tested by the magnetic particle method

17.3.2 General. Upon completion of the tank construction, it shall be filled with water furnished at the tank site by the owner's representative using the pressure necessary to fill the tank to the maximum working water level.

> The fill test described in 17.3.2, although directed primarily towards welded-steel tanks, should be conducted on all tanks prior to placing them in service.

17.3.3 Any leaks in the shell, bottom, or roof (if the roof contains water) that are disclosed by the test shall be repaired by chipping or melting out any defective welds and then rewelding.

17.3.4 Repair work shall be done on joints only when the water in the tank is a minimum of 2 ft (0.6 m) below the point under repair.

17.3.5 The tank shall be tested as watertight to the satisfaction of the authority having jurisdiction and/or the owner's representative.

> Although documentation of the fill test is not required by NFPA 22, the fill test should be documented and signed by the owner's representative, RDP, and AHJ. A copy of this test report should be included in the final submission package.

17.4* Bolted Steel Tanks

The completed tank shall be tested by filling it with water, and any detected leaks shall be repaired in accordance with AWWA D103.

A.17.4 Care should be taken when retorquing bolts in leaking areas. Overtorqued bolts can cause linings to crack, to splinter, or to be otherwise damaged. Manufacturers' recommendations for the repair or replacement of panels should be followed.

17.5 Pressure Tanks

Tests shall be performed according to 17.5.1 through 17.5.4.

17.5.1 Each pressure tank shall be tested in accordance with the ASME *Boiler and Pressure Vessel Code*, "Rules for the Construction of Unfired Pressure Vessels," before painting.

17.5.1.1 The hydrostatic test pressure shall be a minimum of 150 lb/in.^{2} (10.3 bar).

17.5.2 In addition to the ASME tests, each pressure tank shall be filled to two-thirds of its capacity and tested at the normal working pressure with all valves closed and shall not lose more than ½ psi (0.03 bar) pressure in 24 hours.

17.5.3 A certificate signed by the manufacturer that certifies that the foregoing tests have been made shall be filed with the authority having jurisdiction.

> The certificate should also be signed by the owner's representative, RDP, and AHJ. The completed certificate or test report should be included in the final submission package.

17.5.4 A repetition of the tests specified in 17.5.1 through 17.5.3 shall be required after the tank has set in place and connected. Where conditions do not allow shipping the tank after it is assembled, these tests shall be conducted following its assembly in the presence of a representative of the authority having jurisdiction.

17.6 Embankment-Supported Coated Fabric Tanks

17.6.1 The tank shall be tested for leakage prior to shipment.

> The shop test report should accompany shipment of the tank and should be included in the final submission package. This test is conducted in the shop prior to shipment.

17.6.2 The tank also shall be tested for leakage after installation.

> The field test report should be signed by the owner's representative, RDP, and AHJ and should be included in the final submission package.

17.7 Concrete Tanks

17.7.1 Leakage Testing. On completion of the tank and prior to any specified backfill placement at the footing or wall, the test specified in 17.7.2 through 17.7.4 shall be applied to ensure watertightness.

17.7.2 Preparation. The tank shall be filled with water to the maximum level and left to stand for at least 24 hours.

17.7.3 Measurement. The drop in liquid level shall be measured over the next 72-hour period to determine the liquid volume loss. Evaporative losses shall be measured or calculated and shall be deducted from the measured loss to determine whether there is net leakage.

17.7.4 There shall be no measurable leakage after the tank is placed in service.

> A test report should be completed and signed by the installing contractor, owner's representative, RDP, and AHJ and should be included in the final submission package.

17.8 Wood Tanks

17.8.1 Wood tanks shall be filled and tested for liquid tightness for 48 hours.

17.8.2 Testing shall be done under the supervision of a qualified wood tank specialist.

17.8.3 Tests shall be in accordance with the National Wood Tank Institute Bulletin S82.

17.9 Fiberglass-Reinforced Plastic Tanks — Hydrostatic Test

17.9.1 After the excavation hole is backfilled to the bottom of the influent and effluent piping, influent and effluent piping shall be sealed off with watertight caps or plugs.

17.9.2 The tank shall be filled with water up to 3 in. (76 mm) into the access openings.

17.9.3 The water shall be allowed to stand in the tank for a minimum of 2 hours.

17.9.4 The tank shall be examined for leakage or drop in water elevation.

17.9.5 If the water level drops, plugs or caps sealing off piping shall be checked to see that they are tight.

17.9.6 If tightening is required, more water shall be added to fill air voids back to the standard testing level.

17.9.7 The tank shall show no visible signs of leakage, and the water level shall stabilize within a 2-hour test period.

17.10 Disposal of Test Water

The owner's representative shall provide a means for disposing of test water up to the tank inlet or drain pipe.

CHAPTER 18 INSPECTION, TESTING, AND MAINTENANCE OF WATER TANKS

18.1 General

Tanks shall be periodically inspected, tested, and maintained in accordance with NFPA 25.

. . .

Summary

Portions of NFPA 22 have been included in this handbook because water storage tanks are frequently used to supply water to fire pumps. Understanding the interplay of the water storage tank installation and the fire pump installation can be difficult if the installation requirements of each individual system are not fully understood. Water storage tanks are frequently used as break tanks to provide a separation between the attached water supply and the fire pump. This arrangement stabilizes the supply pressure to fire pumps and prevents the fire pump from overpressurizing the fire protection system due to varying pressures in the attached water supply. The intention of this handbook is to provide the proper guidance for the design, installation, and acceptance testing of both tanks and pumps to improve overall system performance.

References Cited in Commentary

National Fire Protection Association, 1 Batterymarch Park, Quincy, MA 02169-7471.

NFPA 13, *Standard for the Installation of Sprinkler Systems,* 2013 edition.
NFPA 20, *Standard for the Installation of Stationary Pumps for Fire Protection,* 2013 edition.
NFPA 24, *Standard for the Installation of Private Fire Service Mains and Their Appurtenances,* 2013 edition.
NFPA 25, *Standard for the Inspection, Testing, and Maintenance of Water-Based Fire Protection Systems,* 2011 edition.
NFPA 72®, *National Fire Alarm and Signaling Code,* 2013 edition.
NFPA 241, *Standard for Safeguarding Construction, Alteration, and Demolition Operations,* 2009 edition.

National Wood Tank Institute, 5500 North Water Street, P.O. Box 2755, Philadelphia, PA 19120.

NWTI Bulletin S82, "Specifications for Tanks and Pipes," 1982.

Extracts with Commentary from NFPA 14,
Standard for the Installation of Standpipe and Hose Systems, 2010 Edition

SECTION 5

Because their intended use is providing waterflow for manual fire fighting, standpipe systems present a substantial demand on water supplies. Historically, these systems have been designed to flow from 500 gpm at 50 psi (1893 L/min at 3.4 bar) — beginning with the 1915 edition of NFPA 14, *Standard for the Installation of Standpipe and Hose Systems* — to as much as 1250 gpm at 100 psi (4731 L/min at 6.9 bar) — in the 2010 edition. Several flow rates and pressures have been specified for standpipes between the first edition and current editions. The flow requirements, and particularly the pressure requirements, in the current edition are based on the type of equipment used by the fire department in what is commonly referred to as a *standpipe pack*.

In preparation for fire operations in buildings equipped with standpipes, NFPA 13E, *Recommended Practice for Fire Department Operations in Properties Protected by Sprinkler and Standpipe Systems*, states that fire departments should plan on providing "hose and nozzles of appropriate size and length along with proper accessory equipment for the anticipated fire conditions" (see NFPA 13E, 6.3.5). Following the tragic fire at One Meridian Plaza in Philadelphia, Pennsylvania, on February 23, 1991, discussions with several fire departments revealed that fire departments across the United States were using different tactics in fighting fires in buildings equipped with standpipe systems. In addition, they were using a variety of equipment in their standpipe packs. As a result of this fire, NFPA surveyed approximately 200 fire departments across the country and several fire departments in British Columbia, Canada. Those surveyed were asked to identify the type of standpipe equipment that was installed in their jurisdiction and the type of equipment that was employed in their standpipe packs. (The compiled results are shown in Table A.7.8 of NFPA 14.)

The results of the survey indicated a wide variety of equipment configurations and, most importantly, indicated the use of combination-type nozzles that require a minimum of 100 psi (6.9 bar) to achieve an effective hose stream. Because of the pressure requirements of this very popular nozzle, the 1993 edition required 100 psi (6.9 bar) of pressure at the topmost outlet of the standpipe system. If additional pressure is needed, the fire department or the stationary fire pump can supplement the needed pressure beyond the minimum requirement.

DESIGN

The following section on design is extracted from Chapter 7 of NFPA 14 and also provides commentary to explain the requirements.

7.1* General

The design of the standpipe system is governed by building height, area per floor occupancy classification, egress system design, required flow rate and residual pressure, and the distance of the hose connection from the source(s) of the water supply.

A.7.1 The building height determines the number of vertical zones. The area of a floor or fire area and exit locations, as well as the occupancy classification, determines the number and locations of hose connections. Local building codes influence types of systems, classes of systems, and locations of hose connections. Pipe sizing is dependent on the number of hose connections flowing, the quantity of water flowed, the required residual pressure, and the vertical distance and horizontal distance of those hose connections from the water supplies.

For typical elevation drawings, see Figure A.7.1(a), Figure A.7.1(b), Figure A.7.1(c), and Figure A.7.1(d). See Chapter 7 for general system requirements.

> Because building height is not limited or restricted, a standpipe system may be required to provide the needed flow and pressure to floors in high-rise buildings that are beyond the reach of the local fire department. The definition of the term *high-rise building* in NFPA 14 deals with fire department vehicle access and is "a building where the floor of an occupiable story is greater than 75 ft (23 m) above the lowest level of fire department vehicle access."
>
> A design scenario involving building heights beyond the reach of fire department access presents several problems that a designer must address. The first is how to move water to such heights, and the second is how to prevent dangerous and damaging pressures from reaching lower floors when high pressure is needed to reach upper floors. Another consideration is the reliability of critical systems, such as fuel and power, since loss of these systems on upper floors of a high-rise building leaves few fire-fighting options (see Part II of this handbook for more information on this topic).
>
> A fire pump in this case will solve the first pressure problem but will present the second problem, that of too much pressure at the lower points in the system. In lower levels of

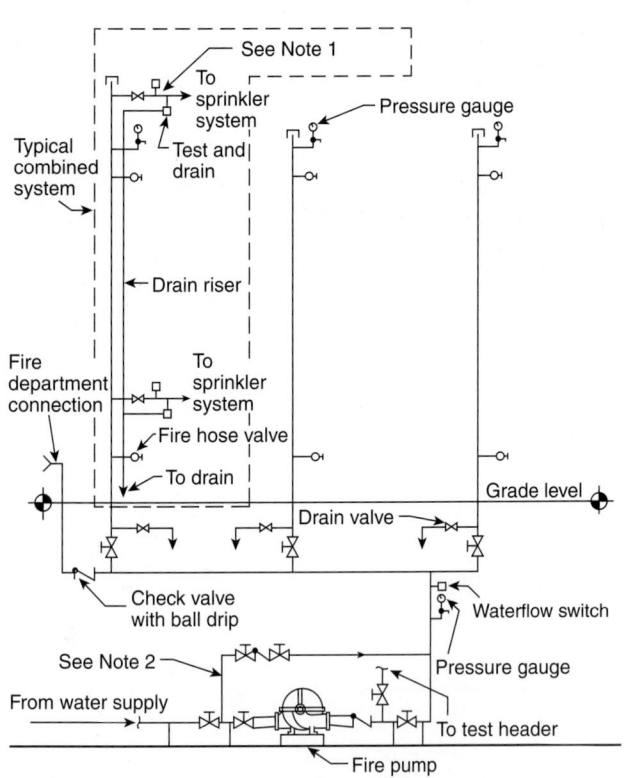

Notes:
1. Sprinkler floor assembly in accordance with NFPA 13, *Standard for the Installation of Sprinkler Systems*.
2. Bypass in accordance with NFPA 20, *Standard for the Installation of Stationary Pumps for Fire Protection*.

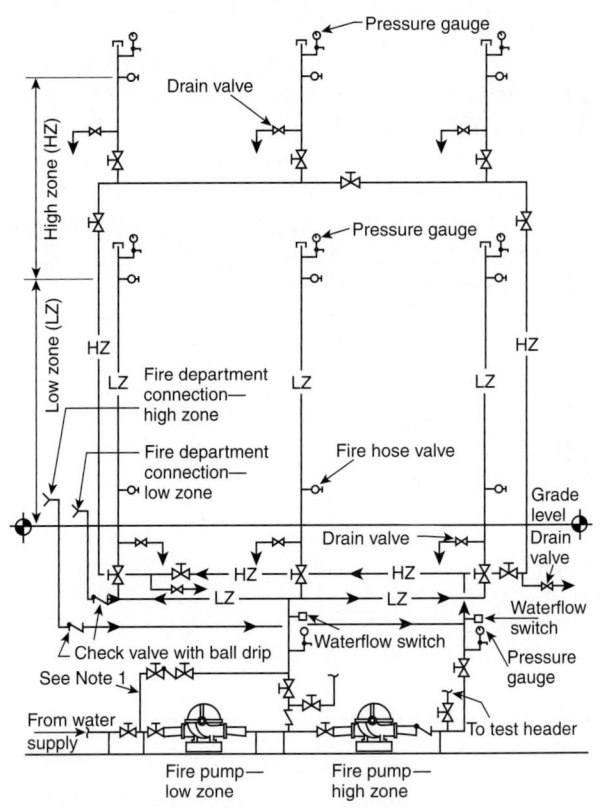

FIGURE A.7.1(a) *Typical Single-Zone System.*

Notes:
1. Bypass in accordance with NFPA 20, *Standard for the Installation of Stationary Pumps for Fire Protection*.
2. High zone pump can be arranged to take suction directly from source of supply.

FIGURE A.7.1(b) *Typical Two-Zone System.*

Section 7.1 • General **567**

high-rise buildings, excessively high pressures are detrimental to the system components and also present a safety issue to the end user — the fire fighter. In this instance, pressure regulating devices must be employed.

7.1.1* When pressure-regulating devices are used, they shall be approved for installation within the maximum and minimum anticipated flow conditions.

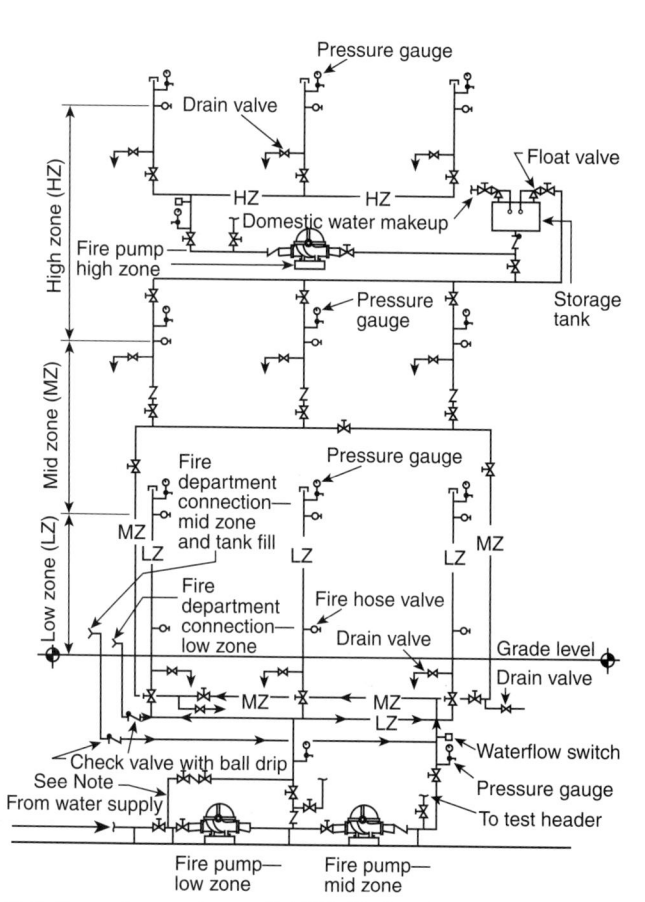

FIGURE A.7.1(c) Typical Multizone System.

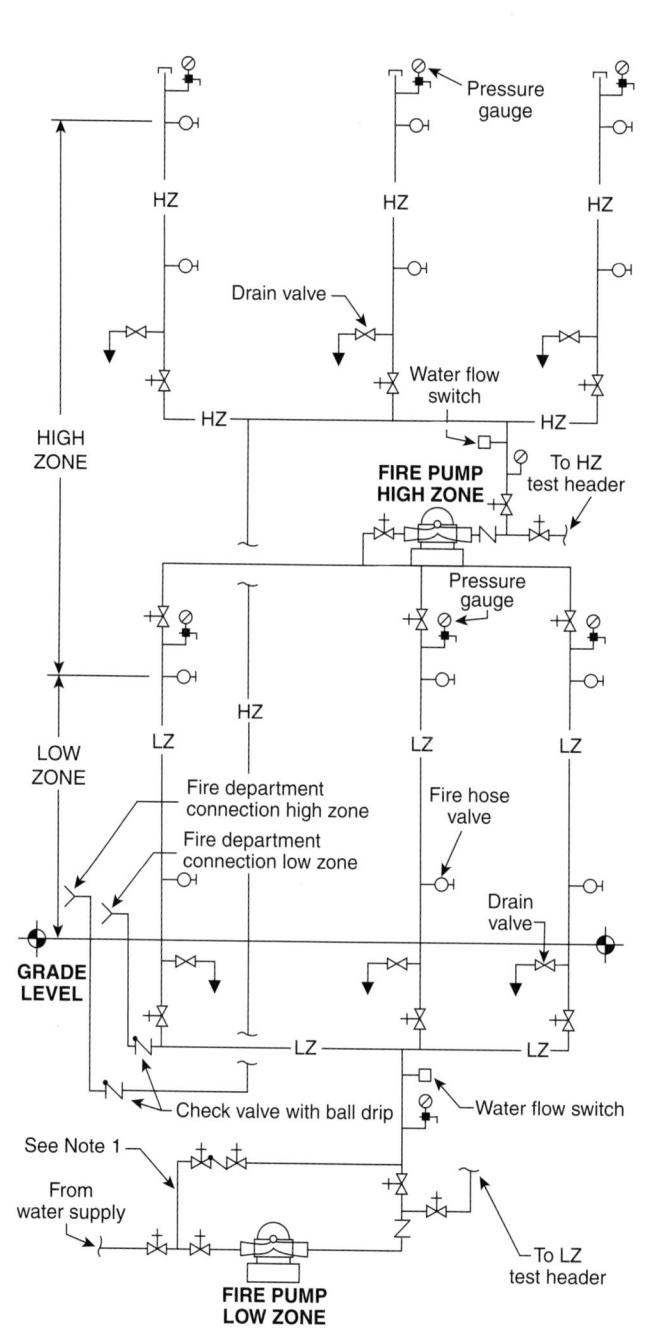

Note: Bypass in accordance with NFPA 20, *Standard for the Installation of Stationary Pumps for Fire Protection.*

FIGURE A.7.1(d) Vertically Staged Pumps for Two-Zone System.

Note: Bypass in accordance with NFPA 20, *Standard for the Installation of Stationary Pumps for Fire Protection.*

Stationary Fire Pumps Handbook   2013

A.7.1.1 It is important to determine the exact operating range to ensure that pressure-regulating devices function in accordance with the manufacturer's instructions for both maximum and minimum anticipated flow rates. Minimum flow can be from a single sprinkler for combined systems or flow from a 1½ in. (40 mm) hose connection on standpipe systems that do not supply sprinklers. This could require the use of two devices installed in parallel.

> When selecting the appropriate pressure regulating device, it is important to check the manufacturer's data and verify that the correct size device is selected to ensure that the flow and pressure settings are appropriate for the design condition. In cases where the pressure regulating device serves a combined sprinkler and standpipe system, both the upper and lower values of flow and pressure must be evaluated. The lower value can be as little as 15 gpm (56.7 L/min), which is the flow from a ½ in. orifice (K-5.6) sprinkler at 7 psi (48.3 kPa). The higher value can be as much as 1750 gpm (5678 L/min), which is the maximum anticipated flow from a combined system with partial sprinkler protection [1250 gpm (3785 L/min) is the maximum anticipated flow from a standpipe system plus 500 gpm (1893 L/min), which is the required flow from an ordinary hazard sprinkler system; see Section 7.10 of NFPA 14.]. The manufacturer's literature should provide the limits of flow and pressure for each device.

7.2* Pressure Limitation

A.7.2 The system pressure limits have been implemented to replace the former height units. Because the issue addressed by the height limits has always been maximum pressure, pressure limitations are a more direct method of regulation and allow flexibility in height units where pumps are used, because a pump curve with less excess pressure at churn yields lower maximum system pressures while achieving the required system demand.

The maximum system pressure normally is at pump churn. The measurement should include both the pump boost and city static pressures. The 350 psi (24 bar) limit was selected because it is the maximum pressure at which most system components are available, and it recognizes the need for a reasonable pressure unit.

> Design pressures for standpipe systems were initially addressed by limiting the height of a standpipe system zone. The previous height limitation was 275 ft (84 m), with an exception limiting height to a maximum of 400 ft (122 m). With this limitation and a pressure requirement of 100 psi (6.9 bar) at the topmost outlet, pressures at the bottom of the riser could be expected to approach approximately 220 psi (15.2 bar), requiring pressure regulation of the hose valves on lower floors and the use of extra heavy valves and fittings to cope with pressures in excess of the 175 psi (12.1 bar) limitation on standard weight valves and fittings.
>
> The 1993 edition of NFPA 14 eliminated the height limitation in favor of a logical pressure limitation of 350 psi (24 bar), based on the pressure ratings of available devices at the time. For the 2010 edition of NFPA 14, the Technical Committee on Standpipes opted for a performance-based approach for express mains supplying higher standpipe system zones. This performance-based approach referenced manufacturers' pressure listings for components rather than a prescriptive pressure limitation, thus permitting pressures in excess of 350 psi (24 bar). Regardless of the exact pressure requirement in the standard, most system components are listed based on 300 psi (21 bar).

7.2.1 The maximum pressure at any point in the system at any time shall not exceed 350 psi (24 bar).

7.2.2 Express mains supplying higher standpipe zones shall be permitted to be designed with pressures in excess of 350 psi (24 bar) in accordance with their materials listings or as approved by the AHJ.

7.2.2.1 Where express mains supply higher standpipe zones, there shall be no hose outlets on any portion of the system where the pressure exceeds 350 psi (24 bar).

7.2.3* Maximum Pressure at Hose Connections.

A.7.2.3 Due to the different pressure limitations established in Section 7.2, it might be necessary to arrange piping so that separate pressure-regulating devices can be provided on the Class I and Class II hose connections.

For Class I standpipes, a pressure-regulating device should not be required where the static pressure of a hose connection exceeds 175 psi (12.1 bar), provided that downstream components, including fire hose and fire nozzles of the responding fire suppression company, are rated for the anticipated static and residual pressures, subject to approval by the AHJ. Due to the inherent safety issues of higher pressure, the AHJ should determine that the fire department is trained in the use of higher pressure hose and nozzles.

7.2.3.1 Where the residual pressure at a 1½ in. (40 mm) outlet on a hose connection available for trained personnel use exceeds 100 psi (6.9 bar), an approved pressure-regulating device shall be provided to limit the residual pressure at the flow required by Section 7.10 to 100 psi (6.9 bar).

> Paragraph 7.2.3.1 is intended to pertain to Class II standpipe systems for use by trained personnel. Pressures for a Class II standpipe system are generally limited to lower values due to the size of the hose, the nozzle type, and the lower flows needed for these systems.

7.2.3.2* Where the static pressure at a hose connection exceeds 175 psi (12.1 bar), an approved pressure-regulating device shall be provided to limit static and residual pressures at the outlet of the hose connection to 100 psi (6.9 bar) for 1½ in. (40 mm) hose connections available for trained personnel use and 175 psi (12.1 bar) for other hose connections.

A.7.2.3.2 Where the building fire department connections are inaccessible or inoperable, many fire departments lay a hoseline from the pumper into the building and connect to an accessible valve outlet using a double female swivel. To pressurize the standpipe, the hose valve is opened and the engine pumps into the system.

If the standpipe is equipped with pressure-reducing hose valves, the valve acts as a check valve, prohibiting pumping into the system when the valve is open.

A supplementary single-inlet fire department connection or hose valve with female threads at an accessible location on the standpipe allows pumping into that system.

> This requirement limits pressures on the outlet side of 2½ in. (65 mm) hose valves to 175 psi (12.1 bar) for Class I standpipe systems and 100 psi (6.9 bar) on the 1½ in. (40 mm) hose connection installed on Class III standpipe systems. This pressure limitation is based upon the type of equipment used in each standpipe system type and the training of the intended user.
>
> Although the pressure limitation is relatively high at 350 psi (24 bar), the designer or registered design professional (RDP) should be aware that pressures in the piping should never exceed those for which the pipe, fittings, or devices are rated.

7.2.3.3 The pressure on the inlet side of the pressure-regulating device shall not exceed the rated working pressure of the device.

7.2.4* When system pressure-regulating devices are used in lieu of providing separate pumps, multiple zones shall be permitted to be supplied by a single pump and pressure-regulating device(s) under the following conditions:

(1) Pressure-regulating device(s) shall be permitted to control pressure in the lower zone(s).
(2) A method to isolate the pressure-regulating device(s) shall be provided for maintenance and repair.

(3) Regulating devices shall be arranged so that the failure of any single device does not allow pressure in excess of 175 psi (12.1 bar) to more than two hose connections.
(4) An equally sized bypass around the pressure-regulating device(s), with a normally closed control valve, shall be installed.
(5) Pressure-regulating device(s) shall be installed not more than 7 ft 6 in. (2.31 m) above the floor.
(6) The pressure-regulating device shall be provided with inlet and outlet pressure gauges.
(7) The fire department connection(s) shall be connected to the system side of the outlet isolation valve.

> It is the intent that the fire department connection (FDC) always be installed downstream of pressure regulating devices installed in accordance with 7.2.4.

(8) The pressure-regulating device shall be provided with a pressure relief valve in accordance with the manufacturer's recommendations.
(9) Remote monitoring and supervision for detecting high pressure failure of the pressure-regulating device shall be provided in accordance with *NFPA 72, National Fire Alarm and Signaling Code*.

A.7.2.4 A small diameter pressure-reducing device can be required due to the minimum listed flow for large diameter pressure-reducing devices typically exceeding low flow conditions, to accommodate low flow conditions such as those created by the flow of a 1½ in. (40 mm) hose connection or a single sprinkler on a combined system. These should also be arranged such that the failure of a single device does not allow pressure in excess of 175 psi (12.1 bar) to more than two hose connections.

See Figure A.7.2.4 for methods to comply with 7.2.4.

> The requirement to install a separate fire pump for each zone of a standpipe system was added in the 1973 edition of NFPA 14 and remains as a design option today. The current edition permits the use of a pressure reducing valve for the lower zone of a standpipe system, allowing the installation of a single fire pump in many instances.
>
> NFPA 14 defines three types of pressure regulating devices in Chapter 3 and permits the installation of any of these devices, provided that the device selected is designed and installed in accordance with the requirements of Section 7.2. Previously, the standard provided detailed instructions for the installation of pressure regulating hose valves but did not provide any instructions for the installation of any other type of pressure regulating device.
>
> The design requirements for standpipe systems are complicated when considering zoning of standpipe systems in high-rise buildings. Unlike NFPA 13, *Standard for the Installation of*

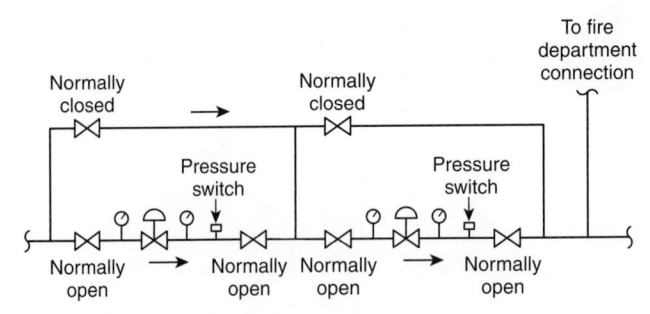

Note: FDC required downstream of pressure-regulating device but not required immediately adjacent thereto.

FIGURE A.7.2.4 *Pressure-Regulating Device Arrangement.*

Sprinkler Systems, NFPA 14 permits system pressures up to the pressure rating of the system components while limiting hose valve outlet pressures to a minimum to maximum range of 100 psi to 175 psi (6.9 bar to 12.1 bar). In many cases, this design results in a variety of pressure regulating device settings from floor to floor, which must be carefully monitored during construction to ensure that each device is set properly for the elevation at which it is installed in the system. In some situations, the use of a master pressure reducing valve to control pressures to the lower zone of a standpipe system may be convenient. Prior to the 2007 edition, the standard required a separate pump for each zone. In fact, a single pump could be used with a "master" pressure regulating device protecting the lower zone, but the standard was silent on this issue. With the 2007 edition, NFPA 14 permitted this alternate arrangement, which is a common practice in the field. Such an application is consistent with the listing of the pressure reducing valve.

Further compounding this application is the flow range needed for combined sprinkler and standpipe systems. Many pressure regulating devices on the market cannot provide the minimum and maximum flows needed in a combined system configuration. The estimated worst-case flow through a pressure regulating device in a combined sprinkler and standpipe system, with only partial sprinkler protection, can range from a minimum of 15 gpm (57 L/min) (for a single sprinkler) to a maximum of 1750 gpm (6623 L/min) [full standpipe system flow of 1250 gpm (4731 L/min) plus the sprinkler discharge of 500 gpm (1893 L/min), as required by NFPA 14, 7.10.1.3.2]. This wide range results in the need to install two devices in parallel: one for low-flow conditions and one for high-flow conditions (see Exhibit III.30).

Previously, the NFPA 14 technical committee rejected a proposal to permit a "master" pressure reducing valve as an option to the requirement for a single pump per zone. The Committee Statement in the Report on Comments, published by NFPA, indicated a concern on the part of the committee regarding device reliability and the potential for "entire segments of the standpipe system not being provided with proper water if the valve fails." The committee further stated that valves failing in an open position could present high pressures and a substantial risk to fire fighters. The installation requirements in 7.2.4 are intended to address both situations. In NFPA 14, pressure gauges and a pressure switch are required for constant

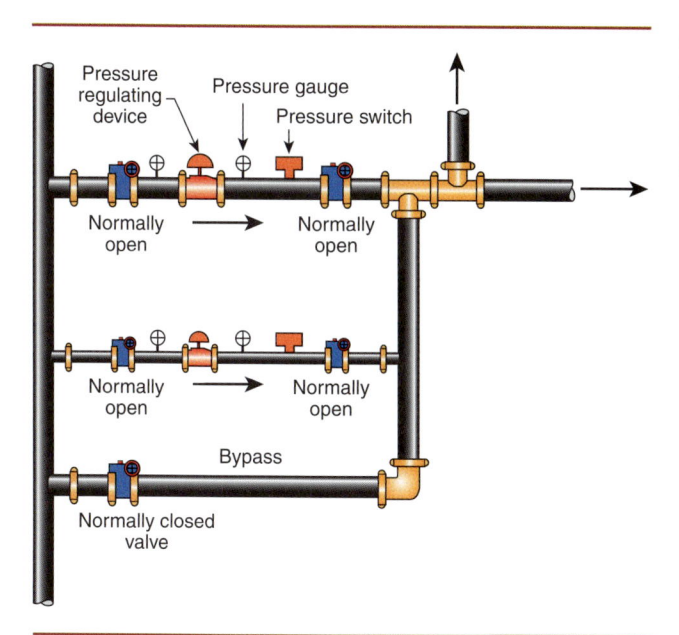

EXHIBIT III.30 *Parallel Systems. (Courtesy of Stephan Laforest, Summit Sprinkler Design Services, Inc.)*

> monitoring of the valve as shown in Figure A.7.2.4. A normally closed bypass valve is installed to permit bypass of the pressure reducing valve in the event of failure in the closed position and to allow for maintenance and repair.
>
> Additionally, to address the concern of failure in the open position, Figure A.7.2.4 provides an example of redundancy in which two valves can be installed in series. In this figure, since pressure reducing valves generally require a 10 psi (0.7 bar) pressure differential, the primary valve should be set for an outlet pressure of 165 psi (11.4 bar) [10 psi (0.7 bar) below the standard pressure rating for most system components]. The second, or redundant, valve can be set for an outlet pressure of 155 psi (10.7 bar), thereby preventing exposure of the end user to dangerous pressures at any point in the lower standpipe system zone.
>
> NFPA 25, *Standard for the Inspection, Testing, and Maintenance of Water-Based Fire Protection Systems*, includes weekly inspection requirements, with quarterly and annual flow test requirements to provide additional insurance of the operation of these devices.

7.8* Minimum and Maximum Pressure Limits

A.7.8 Where determining the pressure at the outlet of the remote hose connection, the pressure loss in the hose valve should be considered.

It is very important that fire departments choose an appropriate nozzle type for their standpipe fire-fighting operations. Constant pressure- (automatic-) type spray nozzles *(see NFPA 1964, Standard for Spray Nozzles)* should not be used for standpipe operations because many of these types require a minimum of 100 psi (6.9 bar) of pressure at the nozzle inlet to produce a reasonably effective fire stream. In standpipe operations, hose friction loss could prevent the delivery of 100 psi (6.9 bar) to the nozzle.

In high-rise standpipe systems with pressure-reducing hose valves, the fire department has little or no control over hose valve outlet pressure.

Many fire departments use combination (fog and straight stream) nozzles requiring 100 psi (6.9 bar) residual pressure at the nozzle inlet with 1½ in., 1¾ in., or 2 in. (40 mm, 44 mm, or 50 mm) hose in lengths of up to 150 ft (45.7 m). Some use 2½ in. (65 mm) hose with a smooth bore nozzle or a combination nozzle.

Some departments use 50 ft (15.2 m) of 2½ in. (65 mm) hose to a gated wye, supplying two 100 ft (30.5 m) lengths of 1½–2 in. (40–50 mm) hose with combination nozzles, requiring 120–149 psi (8.3–0.3 bar) at the valve outlet. *(See Table A.7.8.)*

Also see NFPA 1901, *Standard for Automotive Fire Apparatus*.

> Given the various nozzle and hose configurations available, the fire department's selection of equipment that is compatible with the flows and pressures specified in NFPA 14 is imperative. The 100 psi (6.9 bar) pressure requirement was selected from a wide variety of proposed pressures ranging from 65 psi to 150 psi (4.5 bar to 10.3 bar) to meet most of the nozzle and hose configurations listed in the NFPA standpipe survey. Note that friction loss through the hose valve should be included in the calculations, since the equivalent pipe length for a typical 2½ in. (65 mm) hose valve is 35 ft (10.7 m). When calculating a flow of 250 gpm (946 L/min) through 2½ in. (65 mm) pipe for a length of 35 ft (10.7 m), the pressure loss is as follows:

CALCULATION ▶

> Friction loss per foot of 2½ in. (65 mm) pipe while flowing at 250 gpm (946 L/min) =
> x psi/ft (x bar/m) × 35 ft (10.7 m) = y psi (y bar)

7.8.1 Minimum Design Pressure for Hydraulically Designed Systems. Hydraulically designed standpipe systems shall be designed to provide the waterflow rate required by

TABLE A.7.8 *Hose Stream Friction Losses Summary*

Calculation No.	Nozzle/Hose	Valve Outlet gpm	Valve Outlet L/min	Flow psi	Flow bar
1	2½ in. (65 mm) combination nozzle, with 150 ft (45.7 m) of 2½ in. (65 mm) hose	250	946	123	8.5
2	Two 1½ in. (40 mm) combination nozzles with 100 ft (30.5 m) of 1½ in. (40 mm) hose per nozzle, 2½ in. (65 mm) gated wye, and 50 ft (15.2 m) of 2½ in. (65 mm) hose	250	946	149	10.3
3	Same as calculation no. 2 with two 100 ft (30.5 m) lengths of 1½ in. (40 mm) hose	250	946	139	9.6
4	Same as calculation no. 3 with two 100 ft (30.5 m) lengths of 2 in. (50 mm) hose	250	946	120	8.3
5	1½ in. (40 mm) combination nozzle with 150 ft (45.7 m) of 2 in. (50 mm) hose	200	757	136	9.4
6	Same as calculation no. 5 with 1½ in. (40 mm) hose	200	757	168	11.6

Note: For a discussion of use by the fire department of fire department connections, see NFPA 13E, *Recommended Practice for Fire Department Operations in Properties Protected by Sprinkler and Standpipe Systems*.

Section 7.10 at a minimum residual pressure of 100 psi (6.9 bar) at the outlet of the hydraulically most remote 2½ in. (65 mm) hose connection and 65 psi (4.5 bar) at the outlet of the hydraulically most remote 1½ in. (40 mm) hose station.

It is the intent that automatic and manual standpipe systems be hydraulically calculated.

▶ **FAQ** Why is 100 psi (6.9 bar) required at the hydraulically most remote 2½ in. (65 mm) hose connection?

The 100 psi (6.9 bar) pressure requirement is based upon an in-depth study by the Technical Committee on Standpipes and was selected based on the type of equipment currently employed by most fire departments. Each fire department must carefully select the equipment used in the standpipe pack to ensure that it is compatible with the standpipe system. The 65 psi (4.5 bar) pressure requirement is intended for use in Class II standpipe systems, where the system is intended for use by industrial fire brigades only. The 65 psi (4.5 bar) pressure requirement was never an issue in the fires studied by the technical committee, and, therefore, no changes to this requirement are necessary.

7.8.1.1* Manual standpipe systems shall be designed to provide 100 psi (6.9 bar) at the topmost outlet with the calculations terminating at the fire department connection.

A.7.8.1.1 It is not the intent of this standard to provide an automatic water supply for manual standpipe systems. Manual standpipe systems are designed (sized) to provide 100 psi (6.9 bar) at the topmost outlet using a fire department pumper as the source of flow and pressure.

Stationary Fire Pumps Handbook 2013

7.9* Standpipe System Zones

CALCULATION ▶

Given the pressure limitation of 350 psi (24 bar) for systems with extra heavy pattern fittings, or 175 psi (12.1 bar) for systems utilizing standard weight fittings, the establishment of system zones becomes necessary to prevent the system design pressure from exceeding the aforementioned limitations. By sectioning systems into pressure zones and relay pumping, excessive pressures can be avoided. For example, assuming 12 ft (3.7 m) between floors in a high-rise building, the static head loss from floor to floor would be 5.196 psi (0.4 bar).

$$12 \text{ ft} \times 0.433 \text{ psi/ft} = 5.196 \text{ psi}$$

or, in SI units,

$$3.7 \text{ m} \times 0.108 \text{ bar/m} = 0.4 \text{ bar}$$

Given a design limitation of 175 psi (12.1 bar) and a starting pressure of 100 psi (6.9 bar) at the topmost outlet, only 75 psi (5.2 bar) remain for friction loss and static head loss:

$$175 \text{ psi} - 100 \text{ psi} = 75 \text{ psi}$$
$$12.1 \text{ bar} - 6.9 \text{ bar} = 5.2 \text{ bar}$$

Given a head loss of 5.196 psi (0.358 bar) per floor, 14 floors can be supplied by a single zone:

$$\frac{75}{5.196} \text{ psi} = 14 \text{ floors}$$

$$\frac{5.2}{0.358} \text{ bar} = 14 \text{ floors}$$

Accounting for friction loss, approximately 12 floors can be supplied by a single zone. Using the maximum pressure permitted by the standard,

$$350 \text{ psi} - 100 \text{ psi} = 250 \text{ psi}$$
$$24.1 \text{ bar} - 6.9 \text{ bar} = 17.2 \text{ bar}$$

Applying the head loss of 5.196 psi per floor,

$$\frac{250}{5.196} \text{ psi} = 48 \text{ floors}$$

$$\frac{17.2}{0.358} \text{ bar} = 48 \text{ floors}$$

Based on this information, each zone of a standpipe system would be 12 floors or 48 floors in height based on the design pressure limitations used. The zones exposed to high pressure would clearly require the use of pressure regulating devices to deliver pressure at safe levels for the end user.

A.7.9 Standpipe system zones are intended to limit system design pressures to not more than 350 psi (24 bar) or within the system component pressure ratings as required by Section 7.2. It is the intent of Section 7.9 to provide for this pressure limitation and to provide redundancy in the design of supply pipes and pumps to the upper zones of a standpipe system. When standpipe systems are subdivided to eliminate or avoid the use of pressure-reducing valves, such subdivisions should not constitute a standpipe system zone. *[See Figure A.7.1(d) for vertically staged pumps for two-zone systems.]*

7.9.1 Except as permitted by 7.2.4, each standpipe system zone shall be provided with a separate pump.

> The intent of 7.9.1 is to avoid the use of high-pressure pumps in combination with pressure regulating devices. Each standpipe system zone is required to have its own fire pump, except as permitted by 7.2.4.

7.9.1.1 The requirement in 7.9.1 shall not preclude the use of pumps arranged in series.

> Pumps arranged in series are needed to achieve the pressures necessary to deliver water to the upper floors of a high-rise building.

7.9.1.2 Pumps that are arranged in series shall be permitted to be, but are not required to be, located on the same level.

7.9.2 Each zone above the low zone shall have two or more separate and direct supply pipes sized to automatically and independently supply the flow and pressure requirements of Sections 7.8 and 7.10.

> The intent of 7.9.2 is to require redundancy in the supply pipe to the upper zone standpipe.

7.9.2.1 Standpipes from the lower zone shall be permitted to be used to serve as automatic and independent supplies to upper zones.

7.9.3* For systems with two or more zones in which any portion of the higher zones cannot be supplied by means of fire department pumpers through a fire department connection, an auxiliary means of supply in the form of high-level water storage with additional pumping equipment or other means acceptable to the AHJ shall be provided.

> It is vitally important that an auxiliary means of supply be available for buildings of heights in excess of the pumping capabilities of the fire department apparatus. An auxiliary means of supply would be the only existing supply for fire fighters during fire-fighting operations if the primary supply fails or becomes incapacitated.

A.7.9.3 An auxiliary means can also be in the form of pumping through the fire department connection in series with the low- or mid-zone fire pump, as approved by the AHJ.

7.10 Flow Rates

7.10.1 Class I and Class III Systems.

7.10.1.1* Flow Rate.

A.7.10.1.1 If a water supply system supplies more than one building or more than one fire area, the total supply can be calculated based on the single building or fire area requiring the greatest number of standpipes.

For a discussion of use by the fire department of fire department connections, see NFPA 13E, *Recommended Practice for Fire Department Operations in Properties Protected by Sprinkler and Standpipe Systems*.

> Where buildings are attached, complete fire separation must be provided if the systems are intended to be calculated separately. Otherwise, the two attached buildings must be considered as one and the standpipe systems calculated accordingly.

7.10.1.1.1 For Class I and Class III systems, the minimum flow rate for the hydraulically most remote standpipe shall be 500 gpm (1893 L/min), through two 2 ½ in. (65 mm) outlets at one location, and the calculation procedure shall be in accordance with 7.10.1.2.

> The hydraulically most demanding standpipe must be sized to handle this minimum flow rate. The calculation is based on flowing the two topmost outlets simultaneously. Each standpipe must be sized to accommodate a flow of 500 gpm (1893 L/min); however, only the hydraulically most demanding standpipe is required to produce that flow in the calculations. Additional standpipes must produce at least 250 gpm (946 L/min) as a supplemental measure. Each standpipe in the system is not anticipated to flow the 500 gpm (1893 L/min).

7.10.1.1.2* Where a horizontal standpipe on a Class I or Class III system supplies three or more hose connections on any floor, the minimum flow rate for the hydraulically most demanding horizontal standpipe shall be 750 gpm (2840 L/min), and the calculation procedure shall be in accordance with 7.10.1.2.2.

> In the case of horizontal standpipes, any pipe supplying more than three hose connections should be considered to be a standpipe by definition. Therefore, each of these pipes in a standpipe system must be sized to flow a total of 750 gpm (2840 L/min).

A.7.10.1.1.2 The intent of this section is to provide a different flow requirement for large area low-rise buildings and other structures protected by horizontal standpipes.

> Due to the problems associated with horizontal reach of hose lines in large-area low-rise buildings, a higher initial flow is required.

7.10.1.1.3 The minimum flow rate for additional standpipes shall be 250 gpm (946 L/min) per standpipe for buildings with floor areas that do not exceed 80,000 ft² (7432 m²) per floor. For buildings that exceed 80,000 ft² (7432 m²) per floor, the minimum flow rate for the additional standpipes shall be 500 gpm (1893 L/min) for the second standpipe and 250 gpm (946 L/min) for the third standpipe if the additional flow is required for an unsprinklered building.

> A higher flow is required for large-area buildings, because a larger quantity of water must be available to handle the larger fire area that is anticipated.
> It is intended that buildings that exceed 80,000 ft² (7432 m²) per floor be provided with minimum flow rates of 500 gpm (1893 L/min) for an additional second standpipe and 250 gpm (946 L/min) for a third standpipe if additional flow is required for an unsprinklered building.

7.10.1.1.4 Flow rates for combined systems shall be in accordance with 7.10.1.3.

7.10.1.1.5 The maximum flow rate shall be 1000 gpm (3785 L/min) for buildings that are sprinklered throughout, in accordance with NFPA 13, *Standard for the Installation of Sprinkler Systems*, and 1250 gpm (4731 L/min) for buildings that are not sprinklered throughout, in accordance with NFPA 13.

> Since a standpipe system is intended to provide a backup service to the sprinkler system, a lower flow rate is permitted when the building is provided with a sprinkler system that covers all floor areas. The higher flow is needed when the building is provided with a partial sprinkler system or for buildings without sprinkler systems.

7.10.1.2* **Hydraulic Calculation Requirements.**

> The characteristics of the attached water supply must be determined prior to calculating any system. For municipal or private water distribution systems, a hydrant flow test in accordance with NFPA 291, *Recommended Practice for Fire Flow Testing and Marking of Hydrants*, must be completed not more than 1 year prior to the commencement of design of the standpipe system. Where fire pumps are installed, see Part II of this handbook for determining fire pump output.
>
> Since manual standpipe systems are intended to be supplied by a fire department pumper through the FDC, the output from the fire department pumper should be ascertained. Lacking this information, the design specifications required by NFPA 1901, *Standard for Automotive Fire Apparatus*, can be used. According to NFPA 1901, a fire department pumper is required to include the following three performance points:
>
> 1. 100 percent rated capacity at 150 psi (10.3 bar)
> 2. 70 percent rated capacity at 200 psi (13.8 bar)
> 3. 50 percent rated capacity at 250 psi (17.2 bar)
>
> For example, a 1500 gpm (5677 L/min) pumper is required to include the following specifications:
>
> 1. 1500 gpm (5677 L/min) at 150 psi (10.3 bar)
> 2. 1050 gpm (3974 L/min) at 200 psi (13.8 bar)
> 3. 750 gpm (2840 L/min) at 250 psi (17.2 bar)
>
> The calculation procedures set forth in Chapter 23, Plans and Calculations, of NFPA 13 should be used to calculate the required flow and pressure for a standpipe system.
>
> The calculation must begin at the hydraulically most demanding point in the system. This point is usually furthest from the supply. NFPA 14 mandates the starting flow and pressure as 250 gpm at 100 psi (946 L/min at 6.9 bar). The accumulated system demand, accounting for friction loss through piping, fittings, and other devices, and the pressure adjustments due to elevation changes, must be less than the available water supply. If not, the calculation must be modified, usually by increasing the diameter of some piping.

A.7.10.1.2 See Section 14.4 of NFPA 13, *Standard for the Installation of Sprinkler Systems*.

When performing a hydraulic design, the hydraulic characteristics of each water supply need to be known. The procedure for determining the hydraulic characteristics of permanent water supplies, such as pumps, is fairly straightforward and is described in NFPA 20, *Standard for the Installation of Stationary Pumps for Fire Protection*. The procedure for determining the hydraulic characteristics of fire apparatus supplying a standpipe system are similar. Lacking better information about local fire apparatus, a conservative design would accommodate a 1000 gpm (3785 L/min) fire department pumper performing at the level of design specifications set forth in NFPA 1901, *Standard for Automotive Fire Apparatus*. NFPA 1901 specifies that fire department pumpers must be able to achieve three pressure/flow combinations. These are 100 percent of rated capacity at 150 psi (1034 kPa) net pump pressure, 70 percent of rated capacity at 200 psi (1379 kPa) net pump pressure, and 50 percent of rated capacity at 250 psi (1724 kPa) net pump pressure. Therefore, a 1000 gpm (3785 L/min) pumper can be expected to deliver no less than 1000 gpm (3785 L/min) at 150 psi (1034 kPa), 700 gpm (2650 L/min) at 200 psi (1379 kPa), and 500 gpm (1893 L/min) at 250 psi (1724 kPa). Residual supply pressure on the suction side of a pump from a municipal or other pressurized water supply can also be added.

To perform a hydraulic design, one should determine the minimum required pressure and flow at the hydraulically most remote hose connection and calculate this demand back through system piping to each water supply, accumulating losses for friction and elevation changes and adding flows for additional standpipes and sprinklers at each point where such standpipes or sprinklers connect to the hydraulic design path. When considering fire apparatus as a water supply, flows are calculated from system piping through the fire department connection and back through connecting hoses to the pump. If the pressure available at each supply source exceeds a standpipe system's pressure demand at the designated flow, the design is acceptable. Otherwise, the piping design or the water supply needs to be adjusted.

The intent of the standard is to require that each vertical standpipe serving two or more hose connections be capable of individually flowing 500 gpm (1893 L/min) and 250 gpm (946 L/min) at each of the two hydraulically most demanding connections at the required residual pressure. Given the requirement in 7.10.1.1.3 for the hydraulically most remote standpipe to supply this pressure and flow rate and given the minimum standpipe sizes in Section 7.6, the ability of standpipes that are not hydraulically most remote to satisfy this requirement is implicit and should not require additional hydraulic calculations.

> Note that a calculation of a manual standpipe system can be made based on the fire department pumper (see Section 23.4 of NFPA 13). Calculating the system demand back to the FDC is allowed for buildings where the building height is not beyond the pumping capacity of the fire department. In such cases, the standpipe system is calculated by providing 250 gpm (946 L/min) at each of the two topmost outlets on the hydraulically most demanding standpipe at 100 psi (6.9 bar) and adding 250 gpm (946 L/min) for each additional standpipe, up to the maximum flow required. This procedure is basically the same as that required for automatic standpipe systems, but, in this case, the calculations terminate at the FDC. Manual standpipe systems, therefore, do not require a fire pump.
>
> In cases where the sprinkler system portion of a combined sprinkler/standpipe system can dictate the need for a stationary fire pump, the stationary fire pump does not need to be sized to handle the standpipe system's demand. It cannot be overemphasized that this calculation method only applies to manual standpipe systems where the building height is not beyond the pumping capacity of the fire department.

7.10.1.2.1 Hydraulic calculations and pipe sizes for each standpipe shall be based on providing 250 gpm (946 L/min) at the two hydraulically most remote hose connections on the standpipe and at the topmost outlet of each of the other standpipes at the minimum residual pressure required by Section 7.8.

> This requirement was introduced into the 1993 edition of NFPA 14. At the time, the technical committee elected to retain the 500 gpm (1893 L/min) requirement for the hydraulically most demanding riser plus an additional 250 gpm (946 L/min) for other risers, unless the floor area dictated a higher flow rate. Most fire departments use a gated wye connection to supply the 250 gpm (946 L/min) flow with two 1¾ in. (45 mm) connections, each flowing 125 gpm (474 L/min). In order to flow the required 500 gpm (1893 L/min), four hose lines attached to two connections, with each connection located on a different floor of the building, would be necessary.
>
> Where a standpipe system has risers that terminate at different floor levels, it is the intent of NFPA 14 to allow separate hydraulic calculations for the standpipes that exist on each level. In each case, flow needs to be added only for standpipes that exist on the floor level addressed by the calculations.

7.10.1.2.1.1* Where a standpipe system has risers that terminate at different floor levels, separate hydraulic calculations shall be performed for the standpipes that exist on each level. In each case, flow shall be added only for standpipes that exist on the floor level of the calculations.

A.7.10.1.2.1.1 For example, consider the standpipe system shown in Figure A.7.10.1.2.1.1 with two risers that terminate at the 15th floor and two risers that terminate at the 10th floor of this fully sprinklered high-rise building. In this case, two separate hydraulic calculations

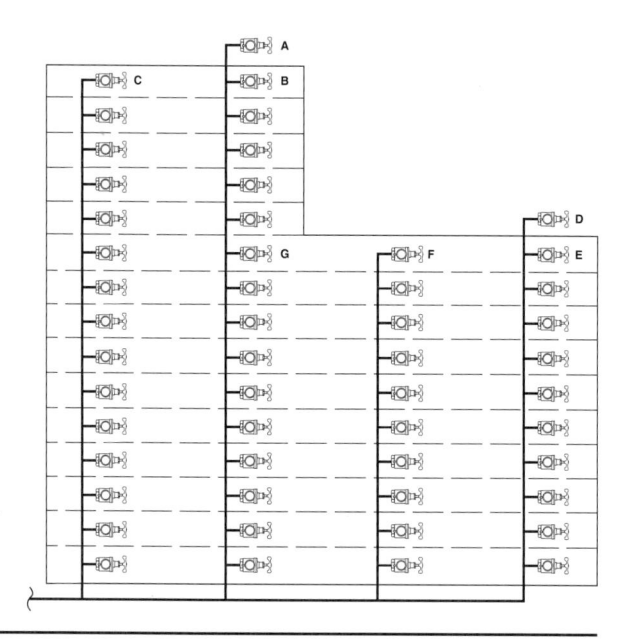

FIGURE A.7.10.1.2.1.1 *Standpipe System with Risers Terminating at Different Floor Levels.*

need to be performed. The first would verify that the system can deliver 100 psi (6.9 bar) to the top of the risers on the 15th floor with a total of 750 gpm (2840 L/min) flowing [250 gpm (946 L/min) each at points A, B, and C]. The second would need to prove that the system can deliver 100 psi (6.9 bar) to the 10th floor with a total of 1000 gpm (3785 L/min) flowing [250 gpm (946 L/min) each at points D, E, F, and G]. Note that since the building is sprinklered, there is no flow required from the fourth riser in this second calculation.

7.10.1.2.2 Where a horizontal standpipe on a Class I and Class III system supplies three or more hose connections on any floor, hydraulic calculations and pipe sizes for each standpipe shall be based on providing 250 gpm (946 L/min) at the three hydraulically most remote hose connections on the standpipe and at the topmost outlet of each of the other standpipes at the minimum residual pressure required by Section 7.8.

7.10.1.2.3* Common supply piping shall be calculated and sized to provide the required flow rate for all standpipes connected to such supply piping, with the total not to exceed the maximum flow demand in 7.10.1.1.5.

A.7.10.1.2.3 Flow is added at nodes in a standpipe system in 250 gpm (946 L/min) increments without requiring additional flow, which might occur from higher pressures at that node (balancing the system). The common supply piping should be hydraulically calculated based on the required flow rate [500, 750, 1000, or 1250 gpm (1893, 2840, 3785, or 4732 L/min)] for the standpipe system. The calculated pressure for the standpipe system does not have to be balanced at the point of connection to the common supply piping.

7.10.1.2.4 Flows from additional standpipes as required by 7.10.1.1 shall not be required to be balanced to the higher pressure at the point of connection.

7.10.1.3 Combined Systems.

7.10.1.3.1 For a building protected throughout by an approved automatic sprinkler system, the system demand established by Section 7.7 and 7.10.1 also shall be permitted to serve the sprinkler system.

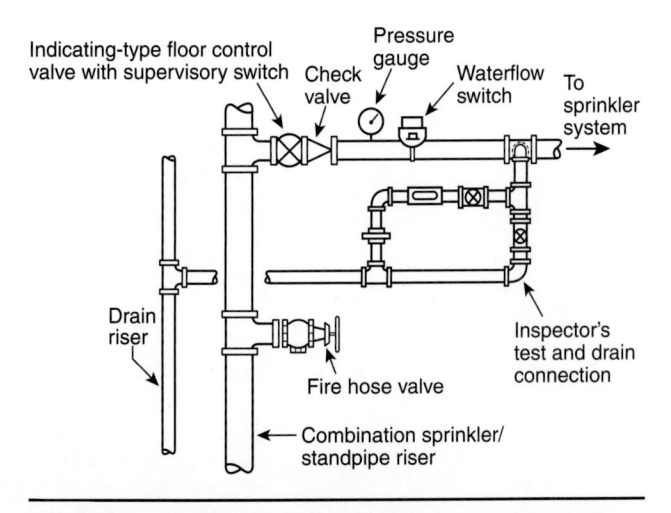

FIGURE A.6.3.5(a) *Acceptable Piping Arrangement for Combined Sprinkler/Standpipe System.*

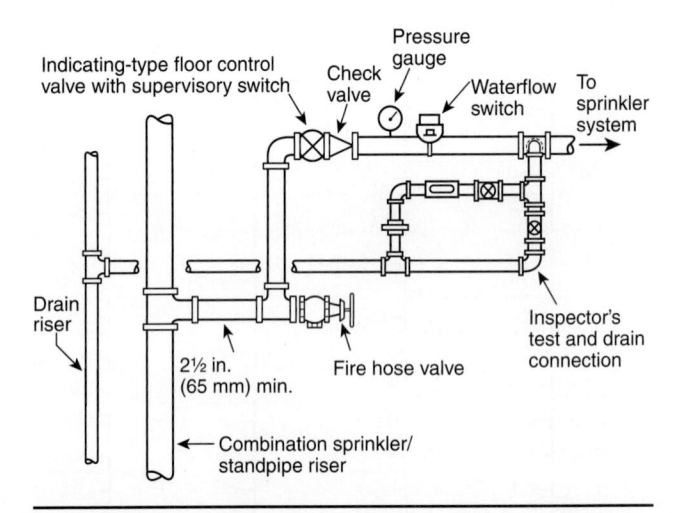

FIGURE A.6.3.5(b) *Combined Sprinkler/Standpipe System.*

> Figure A.6.3.5(a) and Figure A.6.3.5(b) from NFPA 14 illustrate the proper method for connecting a sprinkler system to a standpipe. The combination sprinkler/standpipe riser must be sized to accommodate the flow and pressure from either system. In cases where the building is not completely sprinklered, the combination sprinkler/standpipe riser must be sized to accommodate both system demands simultaneously.

7.10.1.3.1.1 Where the sprinkler system water supply requirement, including the hose stream allowance as determined in accordance with NFPA 13, *Standard for the Installation of Sprinkler Systems*, exceeds the system demand established by Section 7.7 and 7.10.1, the larger of the two values shall be provided.

> In most cases, the standpipe system demand will be much greater than the sprinkler demand. Therefore, the standpipe system demand should be calculated first to establish the appropriate size of the feed main piping and stationary fire pump.

7.10.1.3.1.2 A separate sprinkler demand shall not be required.

7.10.1.3.2 For a combined system in a building equipped with partial automatic sprinkler protection, the flow rate required by 7.10.1 shall be increased by an amount equal to the hydraulically calculated sprinkler demand or 150 gpm (568 L/min) for light hazard occupancies, or by 500 gpm (1893 L/min) for ordinary hazard occupancies, whichever is less.

> For situations where only partial sprinkler protection is provided, the standpipe system must be calculated as discharging simultaneously with the sprinkler system. Where complete sprinkler protection is provided, each system (sprinkler and standpipe) can be calculated as discharging separately.

WATER SUPPLY

9.1.5 Water supplies from the following sources shall be permitted:

(1) A public waterworks system where pressure and flow rate are adequate
(2) Automatic fire pumps connected to an approved water source in accordance with NFPA 20, *Standard for the Installation of Stationary Pumps for Fire Protection*
(3) Manually controlled fire pumps in combination with pressure tanks
(4) Pressure tanks installed in accordance with NFPA 22, *Standard for Water Tanks for Private Fire Protection*
(5) Manually controlled fire pumps operated by remote control devices at each hose station, supervised in accordance with *NFPA 72, National Fire Alarm and Signaling Code,* at each hose station
(6) Gravity tanks installed in accordance with NFPA 22, *Standard for Water Tanks for Private Fire Protection*

Summary

Part III, Section 5 of this handbook covers water supply requirements for standpipe systems and outlines the flow and pressure requirements for standpipes that provide fire fighters with sufficient tools to fight interior structure fires. The flow and pressure demands of standpipe systems often create the need for fire pumps, more so than in most other fire protection systems. For most projects, the standpipe system is calculated first to establish the size of feed main and supply main piping, and the standpipe calculations are frequently used to determine the size and capacity of the fire pump.

References Cited in Commentary

National Fire Protection Association, 1 Batterymarch Park, Quincy, MA 02169-7471.

NFPA 13, *Standard for the Installation of Sprinkler Systems*, 2013 edition.
NFPA 13E, *Recommended Practice for Fire Department Operations in Properties Protected by Sprinkler and Standpipe Systems*, 2010 edition.
NFPA 25, *Standard for the Inspection, Testing, and Maintenance of Water-Based Fire Protection Systems*, 2011 edition.
NFPA 291, *Recommended Practice for Fire Flow Testing and Marking of Hydrants*, 2013 edition.
NFPA 1901, *Standard for Automotive Fire Apparatus*, 2009 edition.

Extracts with Commentary from NFPA 13, *Standard for the Installation of Sprinkler Systems*, 2013 Edition

SECTION 6

> Part III, Section 6 outlines the hydraulic requirements for sprinkler systems. The systems discussed in this section often require a fire pump, because most water supplies alone are inadequate to meet the high demand placed on them by a sprinkler system. Sprinkler systems that are not combined with standpipe systems can create a significant demand on water and pressure; this demand varies, depending on the nature of the protected hazard.
>
> Section 11.1 of NFPA 13, *Standard for the Installation of Sprinkler Systems*, establishes the basic hydraulic criteria that apply to most sprinkler systems. A few exceptions include sprinkler systems for storage occupancies, which are covered in Chapters 12 through 21 of NFPA 13. This section of the handbook focuses on extra hazard hydraulic criteria, because the high flows and pressures needed to achieve the hydraulic demands of extra hazard areas typically require a fire pump. The room design method and light and ordinary hazard design approaches are not covered in this section, as these design approaches do not require a high flow or pressure demand and usually do not drive the need for a fire pump. Examples of hydraulic criteria that drive the need for a fire pump are selected from Chapter 11 of NFPA 13, because that chapter is the most frequently used for sprinkler systems. Other examples are taken from Section 14.4, because that section describes the application of early suppression fast-response (ESFR) sprinklers, which produce the highest system demand for storage applications and, therefore, create the greatest need for a fire pump.

Design Approaches

11.1 General

The requirements of Section 11.1 shall apply to all sprinkler systems unless modified by a specific section of Chapter 11 or Chapter 12.

11.1.1 A building or portion thereof shall be permitted to be protected in accordance with any applicable design approach at the discretion of the designer.

> In most cases, the designer selects the criteria that are the least demanding in an effort to produce a cost-effective design. Some occupancies — such as extra hazard occupancies — require a higher flow and pressure, and a cost-effective design is difficult to achieve. In any case, the designer has the responsibility to select the criteria that match the hazard. The authority having jurisdiction (AHJ) will confirm that the design criteria are appropriate.

11.1.2* Adjacent Hazards or Design Methods. For buildings with two or more adjacent hazards or design methods, the following shall apply:

(1) Where areas are not physically separated by a barrier or partition capable of delaying heat from a fire in one area from fusing sprinklers in the adjacent area, the required sprinkler protection for the more demanding design basis shall extend 15 ft (4.6 m) beyond its perimeter.

(2) The requirements of 11.1.2(1) shall not apply where the areas are separated by a barrier partition that is capable of preventing heat from a fire in one area from fusing sprinklers in the adjacent area.
(3) The requirements of 11.1.2(1) shall not apply to the extension of more demanding criteria from an upper ceiling level to beneath a lower ceiling level where the difference in height between the ceiling levels is at least 2 ft (0.6 m).

> Varying hazards and the resulting design areas often require multiple submissions of hydraulic calculations to prove that the most demanding area has been selected. The supply mains and feed mains should be sized for the most demanding area chosen.

A.11.1.2 The situation frequently arises where a small area of a higher hazard is surrounded by a lesser hazard. For example, consider a 600 ft² (55.7 m²) area consisting of 10 ft (3.05 m) high on-floor storage of cartoned unexpanded plastic commodities surrounded by a plastic extruding operation in a 15 ft (4.57 m) high building. In accordance with Chapter 12, the density required for the plastic storage must meet the requirements for extra hazard (Group 1) occupancies. The plastic extruding operation should be considered an ordinary hazard (Group 2) occupancy. In accordance with Chapter 11, the corresponding discharge densities should be 0.3 gpm/ft² (12.2 mm/min) over 2500 ft² (232 m²) for the storage and 0.2 gpm/ft² (8.1 mm/min) over 1500 ft² (139 m²) for the remainder of the area. *(Also see Chapter 11 for the required minimum areas of operation.)*

If the storage area is not separated from the surrounding area by a wall or partition *(see 11.1.2)*, the size of the operating area is determined by the higher hazard storage.

For example, the operating area is 2500 ft² (232 m²). The system must be able to provide the 0.3 gpm/ft² (12.2 mm/min) density over the storage area and 15 ft (4.57 m) beyond. If part of the remote area is outside the 600 ft² (55.7 m²) plus the 15 ft (4.57 m) overlap, only 0.2 gpm/ft² (8.1 mm/min) is needed for that portion.

If the storage is separated from the surrounding area by a floor-to-ceiling/roof partition that is capable of preventing heat from a fire on one side from fusing sprinklers on the other side, the size of the operating area is determined by the occupancy of the surrounding area. In this example, the design area is 1500 ft² (139 m²). A 0.3 gpm/ft² (12.2 mm/min) density is needed within the separated area with 0.2 gpm/ft² (8.1 mm/min) in the remainder of the remote area.

When the small higher hazard area is larger than the required minimum area dictated by the surrounding occupancy, even when separated by partitions capable of stopping heat, the size of the operating area is determined by the higher hazard storage.

11.1.3 For hydraulically calculated systems, the total system water supply requirements for each design basis shall be determined in accordance with the procedures of Section 23.4 unless modified by a section of Chapter 11 or Chapter 12.

11.1.4 Water Demand.

11.1.4.1* The water demand requirements shall be determined from the following:

(1) Occupancy hazard fire control approach and special design approaches of Chapter 11
(2) Storage design approaches of Chapter 12 through Chapter 20
(3) Special occupancy approaches of Chapter 22

A.11.1.4.1 See A.4.3.

11.1.4.2* The minimum water demand requirements for a sprinkler system shall be determined by adding the hose stream allowance to the water demand for sprinklers.

A.11.1.4.2 Appropriate area/density, other design criteria, and water supply requirements should be based on scientifically based engineering analyses that can include submitted fire testing, calculations, or results from appropriate computational models.

Recommended water supplies anticipate successful sprinkler operation. Because of the small but still significant number of uncontrolled fires in sprinklered properties, which have various causes, there should be an adequate water supply available for fire department use.

The hose stream demand required by this standard is intended to provide the fire department with the extra flow they need to conduct mop-up operations and final extinguishment of a fire at a sprinklered property. This is not the fire department manual fire flow, which is determined by other codes or standards. However, it is not the intent of this standard to require that the sprinkler demand be added to the manual fire flow demand required by other codes and standards. While the other codes and standards can factor in the presence of a sprinkler system in the determination of the manual fire flow requirement, the sprinkler system water demand and manual fire flow demand are intended to be separate stand-alone calculations. NFPA 1 emphasizes this fact by the statement in A.18.4.1 that "It is not the intent to add the minimum fire protection water supplies, such as for a sprinkler system, to the minimum fire flow for manual fire suppression purposes required by this section."

11.1.5 Water Supplies.

11.1.5.1 The minimum water supply shall be available for the minimum duration specified in Chapter 11.

> In addition to the sprinkler demand, hose demand must be included in the hydraulic calculations, because flow from building occupant hose and fire department hose may be employed simultaneously. Hose demand should be included in the calculations at the point of connection.

11.1.5.2* Tanks shall be sized to supply the equipment that they serve.

A.11.1.5.2 Where tanks serve sprinklers only, they can be sized to provide the duration required for the sprinkler system, ignoring any hose stream demands. Where tanks serve some combination of sprinklers, inside hose stations, outside hose stations, or domestic/process use, the tank needs to be capable of providing the duration for the equipment that is fed from the tank, but the demands of equipment not connected to the tank can be ignored. Where a tank is used for both domestic/process water and fire protection, the entire duration demand of the domestic/process water does not need to be included in the tank if provisions are made to segregate the tank so that adequate fire protection water is always present or if provisions are made to automatically cut off the simultaneous use in the event of fire.

> Normally, when sprinklers are supplied by water storage tanks, hose connections are not provided. In these cases, hose demand should not be included in the calculations to avoid unnecessarily oversizing the tank.

11.1.5.3* Pumps shall be sized to supply the equipment that they serve.

A.11.1.5.3 Where pumps serve sprinklers only, they can be sized to provide the flow required for the sprinkler system, ignoring any hose stream demands. Where pumps serve some combination of sprinklers, inside hose stations, or outside hose stations, the pump needs to be capable of providing the flow for the equipment that is fed from the pump, but the demands of equipment not connected to the pump can be ignored except for evaluating their impact on the available water supply to the pump.

> The intent of 11.1.5.3 is that the pump be capable of providing the flow for all of the equipment fed from the pump; however, the demands of equipment not connected to the pump can be ignored.

11.1.6 Hose Allowance.

11.1.6.1 Systems with Multiple Hazard Classifications. For systems with multiple hazard classifications, the hose stream allowance and water supply duration shall be in accordance with one of the following:

(1) The water supply requirements for the highest hazard classification within the system shall be used.
(2) The water supply requirements for each individual hazard classification shall be used in the calculations for the design area for that hazard.
(3)* For systems with multiple hazard classifications where the higher classification only lies within single rooms less than or equal to 400 ft^2 (37.2 m^2) in area with no such rooms adjacent, the water supply requirements for the principal occupancy shall be used for the remainder of the system.

A.11.1.6.1(3) When a light hazard occupancy, such as a school, contains separate ordinary hazard rooms no more than 400 ft^2 (37.2 m^2), the hose stream allowance and water supply duration would be that required for a light hazard occupancy.

11.1.6.2* Water allowance for outside hose shall be added to the sprinkler requirement at the connection to the city main or a private fire hydrant, whichever is closer to the system riser.

A.11.1.6.2 When the hose demand is provided by a separate water supply, the sprinkler calculation does not include the outside hose demand.

11.1.6.3 Where inside hose connections are planned or are required, the following shall apply:

(1) A total water allowance of 50 gpm (189 L/min) for a single hose connection installation shall be added to the sprinkler requirements.
(2) A total water allowance of 100 gpm (379 L/min) for a multiple hose connection installation shall be added to the sprinkler requirements.
(3) The water allowance shall be added in 50 gpm (189 L/min) increments beginning at the most remote hose connection, with each increment added at the pressure required by the sprinkler system design at that point.

> Paragraph 11.1.6.3 simply requires that the flow for a hose station be added to the calculations without consideration of pressure. This added demand creates additional friction loss in the system piping and so produces a higher flow demand at the water supply.

11.1.6.4* When hose valves for fire department use are attached to wet pipe sprinkler system risers in accordance with 8.17.5.2, the following shall apply:

(1) The sprinkler system demand shall not be required to be added to standpipe demand as determined from NFPA 14.
(2) Where the combined sprinkler system demand and hose stream allowance of Table 11.2.3.1.2 exceeds the requirements of NFPA 14, this higher demand shall be used.
(3) For partially sprinklered buildings, the sprinkler demand, not including hose stream allowance, as indicated in Figure 11.2.3.1.1 shall be added to the requirements given in NFPA 14.

CALCULATION ▶

> Only on rare occasions does the sprinkler demand need to be added to the standpipe system demand, because most buildings are provided with a complete sprinkler system throughout. As a result, calculating a sprinkler system and a standpipe system that discharge simultaneously is unusual. Sprinkler system demand approaches that of a standpipe system, beginning with Extra Hazard (Group I) criteria or higher. The following calculation is based on Table 11.2.3.1.2 from NFPA 13:
>
> $$0.3 \text{ gpm} \times 2500 \text{ ft}^2 = 750 \text{ gpm} + 500 \text{ gpm (hose)} = 1250 \text{ gpm}$$

or, in SI units,

$$12.2 \text{ L/min} \times 232 \text{ m}^2 = 2839 \text{ L/min} + 1893 \text{ L/min (hose)} = 4732 \text{ L/min}$$

As stated in the standpipe section discussed in part 3 section 5 of this standard, standpipe system demand is typically 1000 gpm (3785 L/min) for buildings that are protected throughout by an automatic sprinkler system and 1250 gpm (4731 L/min) for buildings with only partial protection. However, consideration of the pressure requirement for standpipe systems is included in this calculation. Even with the highest densities and spacing of sprinklers, the starting pressure for a sprinkler system when using the density/area curves is considerably below that of a standpipe system. Consider a sprinkler spacing of 100 ft² (12.1 m²) as permitted by Table 8.6.2.2.1(c) of NFPA 13 and a discharge density of 0.4 gpm/ft² (16.3 mm/min) from Figure 11.2.3.1.1 of NFPA 13:

$$100 \text{ ft}^2 \times 0.4 \text{ gpm/ft}^2 = 40 \text{ gpm}$$

$$9.3 \text{ m}^2 \times 16.3 \text{ L/min/m}^2 = 152 \text{ L/min}$$

To qualify as the most demanding sprinkler in this system, the sprinkler must discharge 40 gpm (152 L/min). To achieve this discharge with a ½ in. orifice (K-5.6) sprinkler, a starting pressure of 51 psi (3.5 bar) would be required as follows:

$$p = (Q/k)^2$$
$$p = (40 \text{ gpm}/5.6)^2$$
$$p = 51 \text{ psi (3.5 bar)}$$

where:

p = pressure (psi)

Q = flow (gpm)

k = K-factor of sprinkler

Depending on the amount of friction loss in the system piping, a standpipe system, in most cases, still presents the higher system demand.

ESFR SPRINKLERS

Extracts from Section 14.4 of NFPA 13 were selected as an example of sprinkler systems that may require fire pumps because of the high flow rates and pressures needed for the proper operation of ESFR sprinklers. A wide variety of criteria for storage applications exist that include standard spray sprinklers, control mode specific application (CMSA), and rack sprinklers. None of these criteria, however, approach those of the ESFR sprinkler in terms of flow and pressure demands. The calculation that follows is that of the hydraulic demand for a K-25.2 ESFR sprinkler, although other K-factor configurations from Table 14.4.1 can produce flow and pressure demands that drive the need for fire pumps.

A.11.1.6.4 For fully sprinklered buildings, if hose valves or stations are provided on a combination sprinkler riser and standpipe for fire department use in accordance with NFPA 14, the hydraulic calculation for the sprinkler system is not required to include the standpipe allowance.

11.1.7* High Volume Low Speed (HVLS) Fans. The installation of HVLS fans in buildings equipped with sprinklers, including ESFR sprinklers, shall comply with the following:

(1) The maximum fan diameter shall be 24 ft (7.3 m).
(2) The HVLS fan shall be centered approximately between four adjacent sprinklers.

(3) The vertical clearance from the HVLS fan to sprinkler deflector shall be a minimum of 3 ft (0.9 m).
(4) All HVLS fans shall be interlocked to shut down immediately upon receiving a waterflow signal from the alarm system in accordance with the requirements of *NFPA 72*.

A.11.1.7 A series of 10 full-scale fire tests and limited-scale testing were conducted to determine the impact of HVLS fan operation on the performance of sprinkler systems. The project, sponsored by the Property Insurance Research Group (PIRG) and other industry groups, was coordinated by the Fire Protection Research Foundation (FPRF). The complete test report, *High Volume/Low Speed Fan and Sprinkler Operation — Ph. 2 Final Report (2011)*, is available from the FPRF. Both control mode density area and early suppression fast response sprinklers were tested. Successful results were obtained when the HVLS fan was shut down upon the activation of the first sprinkler followed by a 90-second delay. Other methods of fan shutdown were also tested including shutdown by activation of air sampling–type detection and ionization-type smoke detectors. Earlier fan shutdown resulted in less commodity damage.

14.4 Early Suppression Fast-Response (ESFR) Sprinklers for Palletized or Solid-Piled Storage of Class I Through Class IV Commodities.

14.4.1 Protection of palletized and solid-piled storage of Class I through Class IV commodities shall be in accordance with Table 14.4.1.

14.4.2 ESFR sprinkler systems shall be designed such that the minimum operating pressure is not less than that indicated in Table 14.4.1 for commodity, storage height, and building height involved.

CALCULATION ▶

> Assuming a maximum ceiling/roof height of 45 ft (13.7 m), a maximum storage height of 40 ft (12.2 m), and the use of a K-25.2 ESFR sprinkler, a minimum operating pressure of 40 psi (2.8 bar) is required (Table 14.4.1). Therefore,
>
> $$Q = k \times \sqrt{p}$$
> $$Q = 25.2 \times \sqrt{40}$$
> $$Q = 159.37 \text{ gpm } (603.2 \text{ L/min})$$

14.4.3 The design area shall consist of the most hydraulically demanding area of 12 sprinklers, consisting of four sprinklers on each of three branch lines.

> This subsection specifies the exact number of sprinklers that must be considered to be discharging in the calculation. The following calculation will provide the minimum flow rate including hose demand from Table 14.4.1. Therefore,
>
> $$159.2 \text{ gpm} \times 12 = 1910 \text{ gpm } (7230.8 \text{ L/min}) + 250 \text{ gpm (hose)}$$
> $$= 2160 \text{ gpm } (8175 \text{ L/min})$$
>
> This flow is a minimum rate; it does not include overdischarge due to friction loss in the branch lines or between branch lines. Usually, a sprinkler system discharges between 110 percent to 120 percent of the minimum specified discharge, due to friction and balancing. The calculated flow would be needed at a pressure in excess of 59 psi (4.1 bar) derived from the pressure loss from elevation and the end head pressure as stated in Table 14.4.1, see as follows:
>
> $$44 \text{ ft } (13.4 \text{ m}) \times 0.433 \text{ psi/ft } (0.097 \text{ bar/m})$$
> $$= 19 \text{ psi } (1.3 \text{ bar}) + 40 \text{ psi } (2.8 \text{ bar}) \text{ (starting pressure)}$$
> $$= 59 \text{ psi } (4.1 \text{ bar})$$

TABLE 14.4.1 ESFR Protection of Palletized and Solid-Piled Storage of Class I Through Class IV Commodities

Commodity	Maximum Storage Height ft	Maximum Storage Height m	Maximum Ceiling/Roof Height ft	Maximum Ceiling/Roof Height m	Nominal K-Factor	Orientation	Minimum Operating Pressure psi	Minimum Operating Pressure bar
Class I, II, III, or IV, encapsulated and nonencapsulated (no open-top containers)	20	6.1	25	7.6	14.0 (200)	Upright/pendent	50	3.4
					16.8 (240)	Upright/pendent	35	2.4
					22.4 (320)	Pendent	25	1.7
					25.2 (360)	Pendent	15	1.0
	25	7.6	30	9.1	14.0 (200)	Upright/pendent	50	3.4
					16.8 (240)	Upright/pendent	35	2.4
					22.4 (320)	Pendent	25	1.7
					25.2 (360)	Pendent	15	1.0
			32	9.8	14.0 (200)	Upright/pendent	60	4.1
					16.8 (240)	Pendent	42	2.9
	30	9.1	35	10.7	14.0 (200)	Upright/pendent	75	5.2
					16.8 (240)	Upright/pendent	52	3.6
					22.4 (320)	Pendent	35	2.4
					25.2 (360)	Pendent	20	1.4
	35	10.7	40	12.2	16.8 (240)	Upright/pendent	52	3.6
					22.4 (320)	Pendent	40	2.8
					25.2 (360)	Pendent	25	1.7
	35	10.7	45	13.7	22.4 (320)	Pendent	40	2.8
					25.2 (360)	Pendent	40	2.8
	40	12.2	45	13.7	22.4 (320)	Pendent	40	2.8
					25.2 (360)	Pendent	40	2.8

The 44 ft (13.4 m) elevation is estimated to be the elevation of the sprinkler piping [approximately 1 ft (0.3 m) below the roof height], and the 40 psi (2.8 bar) pressure is the minimum operating pressure for the K-25.2 sprinkler (from Table 14.4.1). This estimated calculation does not include friction loss through the piping system, which can be substantial based on the overall size of the system. Therefore, initially, for this system, a total system demand of 2160 gpm (8175 L/min) at a minimum pressure of 59 psi (4.1 bar) is needed. Most municipal water supply systems and private water supply systems cannot meet this demand without the assistance of a fire pump.

Summary

Part III, Section 6 of this handbook covers water supply requirements for sprinkler systems meeting extra hazard hydraulic criteria. In systems protecting extra hazard occupancies, or in ESFR systems protecting storage of Class I through Class IV commodities, high flows and sometimes high pressures are needed. In most cases, a fire pump is necessary to meet the hydraulic demands of these systems.

PART FOUR

Supplements

In addition to the commentary presented in Parts I through III, the *Stationary Fire Pumps Handbook* includes four supplements, which explore the background of selected topics related to fire pumps in more detail than the commentary in Parts I through III. These supplements are not part of the standards; they are included as additional information for handbook users.

The four supplements in Part IV are as follows:

1. Interrelationship of NFPA Standards Pertaining to a Fire Pump Installation
   (This Supplement includes a poster, packaged with the book, which illustrates the installation referenced.)

2. Commissioning and Inspection, Testing, and Maintenance Forms for Fire Pumps

3. Article 695, Fire Pumps, from *NFPA 70®*, *National Electrical Code®*, 2011 Edition

4. Technical/Substantive Changes from the 2010 Edition to the 2013 Edition of NFPA 20

SUPPLEMENT 1

Interrelationship of NFPA Standards Pertaining to a Fire Pump Installation

INTRODUCTION

Supplement 1 works with the enclosed poster, which depicts a hypothetical fire pump installation and the equipment and components involved, to illustrate that the installation of a fire pump requires the application of standards and recommendations in addition to those of NFPA 20, *Standard for the Installation of Stationary Pumps for Fire Protection*. The fire pump installation shown in the poster relies on the interrelationship of several NFPA publications that are associated with such an installation. Along with NFPA 20, these publications include NFPA 13, *Standard for the Installation of Sprinkler Systems*; NFPA 22, *Standard for Water Tanks for Private Fire Protection*; NFPA 24, *Standard for the Installation of Private Fire Service Mains and Their Appurtenances*; NFPA 25, *Standard for the Inspection, Testing, and Maintenance of Water-Based Fire Protection Systems*; and NFPA 291, *Recommended Practice for Fire Flow Testing and Marking of Hydrants*. The key equipment and component provisions of these NFPA publications are identified for key equipment and components identified on the poster. To obtain the most benefit from this supplement and poster, the reader should possess a fundamental understanding of fire protection system hydraulics.

The hypothetical fire pump shown in the poster serves a sprinkler system for a rural storage facility (not shown) used to warehouse rubber tires. The tires are stored in a laced configuration on open portable steel racks. The portable racks can be stacked to a maximum height of 25 ft (7.6 m), and the underside of the building's ceiling deck reaches 30 ft (9.1 m). Municipal water is not available for the sprinkler system, so a suction tank is provided. In the example shown in the poster, the tank supplies water for the sprinkler system only and does not supply any associated yard hydrants.

The sprinkler system design consists of K16.8 early suppression fast-response (ESFR) pendent sprinklers. In accordance with Table 18.4(d) of NFPA 13, the design area is the most hydraulically demanding area of 20 sprinklers — 5 sprinklers on each of four branch lines. Each of the 20 sprinklers in the design area is to operate at a minimum discharge pressure of 52 psi (3.6 bar) — a minimum flow rate of roughly 121.2 gpm (463 L/min) from the single hydraulically most demanding sprinkler calculated as follows:

$$Q = K\sqrt{P}$$
$$Q = 16.8\sqrt{52} \text{ psi (3.6 bar)} = 121.2 \text{ gpm (459 L/min)}$$

where:

Q = flow rate
K = K-factor
P = pressure

Ignoring the effects of friction and elevation losses, the minimum flow rate for the 20-sprinkler design area is roughly 2424 gpm (9175 L/min) [20 sprinklers × 121.2 gpm (459 L/min)]. The minimum pressure requirement for the system at the base of the riser should be not less than 65 psi (4.5 bar). This pressure requirement includes the 52 psi (3.6 bar) required at each sprinkler and the 30 ft (9.1 m) elevation difference corresponding to the height of the ceiling, which results in an elevation loss of 13 psi (0.9 bar) [30 ft × 0.433 psi/ft (9.1 m × 0.097 bar/m)]. When reviewing the initial design, the sprinkler system demand at the base of the riser should be not less than 2424 gpm (9175 L/min) at 65 psi (4.5 bar).

In an actual design project, a hydraulic analysis (with a commercially available software program) would be conducted; this analysis might reveal friction losses and elevation losses, which could result in a substantially greater system demand at the base of the riser. This increase in system demand is due to the fact that actual system demand is not only dependent on the initial sprinkler system discharge criteria but also on the spacing of the sprinklers; pipe material used; pipe sizes; arrangement and type of fittings, valves, and other devices on the system; or number of sprinklers. If using an ESFR configuration requiring 12 sprinklers, more than 12 sprinklers might be needed in the design area due to building elements that obstruct the sprinkler's discharge pattern. See Note 2 to Table 18.4(d) of NFPA 13.

In the case of the installation shown in the poster, assume that the hydraulic calculations indicate a sprinkler system demand of 2600 gpm (9841 L/min) at 95 psi (6.6 bar) at the base of the riser. The sprinkler system discharge criteria in Table 18.4(d) of NFPA 13 also requires a 3-hour water supply duration and requires that 500 gpm (1893 L/min) be available for hose streams.

Water Storage Tank

In the installation depicted in the poster, a municipal water supply is not available for fire protection purposes, so the water supply for the sprinkler system needs to be provided on-site via a fixed water supply. For purposes of this example, it was decided to use a suction tank with a fire pump that provides the necessary flow and pressure for the water supply. NFPA 22 addresses the design, construction, and installation of the suction tank.

The quantity of water needed for the sprinkler system and the size of the tank are to be determined based on sprinkler system demand and water supply duration. In accordance with 4.1.1 of NFPA 22, the size and elevation of the tank must be determined by conditions at each individual property after consideration of all factors involved, and A.4.1.1 of NFPA 22 makes specific reference to NFPA 13.

The overall sprinkler system demand of 2600 gpm (9841 L/min) at the base of the riser is used for tank sizing. While the base of the riser might be located at some distance from the pump and tank location, no water is expected to flow from the piping system between the tank and the activated sprinklers during a fire (e.g., there are no provisions for outside hose streams). The only waterflow expected is from the 20 operated sprinklers [2600 gpm (9841 L/min)]. For this example, no inside hose stations are provided, because none are required.

Subsection 12.8.1 of NFPA 13 indicates that tanks can be sized to supply only the equipment that they serve. In the example shown in this supplement, the only equipment served by the tank is the sprinkler system. The text of A.12.8.1 of NFPA 13 indicates that, where tanks serve sprinklers only, they can be sized to provide the flow duration required for the sprinkler system — in this case, 3 hours — ignoring any hose stream demands. (Note that this hypothetical example includes no hose stream demand). As such, the 500 gpm (1900 L/min) requirement for hose streams does not need to be included in the overall water quantity to be provided by the suction tank.

If the tank in the example was used for both domestic/process water and fire protection, additional considerations would be needed per A.12.8.1 of NFPA 13. Where tanks serve some combination of fire protection system and domestic/process use, the tank needs to be capable of providing the duration of flow for all equipment supplied by the tank for both fire protection and non-fire protection. However, the entire duration demand of the domestic/process water can be excluded when sizing the tank under either of the following conditions:

(1) Arrangements are made to appropriately segregate the tank so that the necessary amount of fire protection water is always present.
(2) It is possible for simultaneous use of the domestic/process water to be automatically shut off in case of fire, so that the total water supply is available for fire protection purposes.

While provisions of NFPA 13 address tanks supplying water for both fire protection and non-fire protection use, the scope of NFPA 22 limits the document to tanks used for private fire protection purposes only. NFPA 22 does not include provisions for chlorination, disinfection, or other water treatments necessary for conditioning water for domestic or process use. See Section 1.1 of NFPA 22.

In the case of the hypothetical pump shown in the poster, since 2600 gpm (9841 L/min) are required for the sprinkler system for a duration of 3 hours, the total quantity of water to be stored on-site for fire protection purposes is 468,000 gallons (1772 m³) [2600 gpm (9841 L/min) × 180 min]. Therefore, the suction tank needs to be large enough to hold at least 468,000 gal (1772 m³) of water. Tanks are manufactured in standard sizes, so a 500,000 gal (1892.50 m³) tank would be used. Depending on the specific manufacturer, such a tank would be about 46½ ft (12 m) in diameter and 32 ft (9.8 m) in height. Subsection 4.1.2 of NFPA 22 addresses the use of standard sizes of tanks and their heights.

When sizing the suction tank, it should be recognized that not all the water in the tank would be available for fire protection purposes. As noted in 4.1.3 of NFPA 22, the net capacity of the tank is the amount of water between the inlet of the overflow device and the level of the vortex plate. The vortex plate sits near the bottom of the tank over the outlet that feeds the piping to the fire pump. The purpose of the vortex plate is to minimize the likelihood of vortex flow of the water discharging from the tank. NFPA 20 also addresses vortex plates in 4.14.10. It is best for water to be in a laminar condition as it enters the fire pump. The vortex plate keeps the water laminar.

Subsections 4.6.2 and 4.6.4 of NFPA 20 also address the water supply for fire protection systems. Specifically, any source of water that is adequate in quality, quantity, and pressure is permitted as the supply for the fire pump. The adequacy of the water supply should be determined and evaluated prior to the specification and installation of the fire pump. As previously noted, the suction tank must be appropriately sized. While the suction tank is not expected to provide an adequate source of pressure for the sprinkler system, its elevation in relationship to the fire pump must be such that the gauge pressure at the pump suction flange does not drop to −3.0 psi (−0.2 bar). See 4.14.3.2 of NFPA 20.

NFPA 20 requires that a stored supply, such as a suction tank, must include a reliable method for replenishing the supply. Both the tank and the refill arrangement must be sufficient to meet the system demand for the design duration of flow. A permanent automatic refill connection is preferable and is to be designed and installed with sufficient flow capacity to refill the stored supply within 8 hours in accordance with the provisions of Section 14.4 of NFPA 22. In addition, the suction tank is to be kept filled, and the water level should never drop more than 4 in. (102 mm) below the designated fire service level.

Although not specifically covered in this supplement, the overall design, construction, and installation of the suction tank also need to address the tank foundation; pipe connections, fittings, and support; valve enclosures; tank filling arrangement; and protection against freezing, depending on the tank's geographic location, as well as provisions to facilitate inspection, testing, and maintenance activities. See NFPA 22 for further information.

Pump Size

The size of the pump needed affects the selection and installation of other necessary components such as suction and discharge piping, the relief valve, and the fire pump test header.

When determining the pressure requirements for the pump, the sprinkler system demand of 2600 gpm (9841 L/min) at 95 psi (6.6 bar) at the base of the riser serves as a starting point. However, since the fire pump will not be located at the base of the riser, and since additional fittings and valves are expected between the riser and the pump, pressure and elevation losses between the base of the riser and the fire pump discharge flange need to be considered though hydraulic analysis. Specific pump curves developed by pump manufacturers need to be considered, and a pump that produces the pressure required at the pump discharge flange at 2600 gpm (9841 L/min) needs to be selected. This pressure will be in excess of the 95 psi (6.6 bar) required, as previously noted.

When determining capacity, no flow in addition to the 2600 gpm (9841 L/min) is expected. Paragraph A.4.8 of NFPA 20 indicates that fire pumps should be selected so that maximum flow occurs within the range of 140 percent to 90 percent of the rated pump capacity. With a flow demand of 2600 gpm (9841 L/min), an 1857 gpm (7029 L/min) rated pump could be considered [2600 gpm (1729 L/min)/1.4]. Since pumps are manufactured in standard sizes as indicated in Table 4.8.2 of NFPA 20, a 2000 gpm (7570 L/min) pump could be used. At 140 percent capacity, a 2000 gpm (7570 L/min) pump could produce a flow of 2800 gpm (10,598 L/min). Note that this calculation pertains to flow only and not to pressure.

As indicated in Section 6.2 of NFPA 20, pump manufacturers can produce pumps with a wide range of flow and pressure characteristics. NFPA 20 places limits on this operating range. At the overload point, or 150 percent of rated capacity, the pump must produce a pressure of not less than 65 percent of its rated capacity. Therefore, if the 2000 gpm (7570 L/min) pump is rated at 150 psi (10.3 bar), when operating at its overload point of 3000 gpm (11,355 L/min) [2000 gpm (7570 L/min) × 1.5], the pump must produce a pressure of at least 97.5 psi (6.7 bar) [150 psi (10.3 bar) × 0.65]. Note that this is the lower pressure limit. Manufacturers could produce a 2000 gpm/150 psi (7570 L/min/10.3 bar) rated pump that develops a pressure of 110 psi (7.6 bar) when operating at its overload point.

The upper pressure limit pertains to the churn, or shutoff head, condition. Churn is the condition under which the pump is operating but no water is flowing through the pump (zero flow); that is, the pump is activated, but a valve downstream of the discharge flange is closed, or no sprinklers have fused. Under churn condition, the pump cannot develop a pressure in excess of 140 percent of its rated pressure. Therefore, the 2000 gpm/150 psi (7570 L/min/10.3 bar) rated pump cannot develop a shutoff pressure of more than 210 psi (14.5 bar) [150 psi (10.3 bar) × 1.4]. A discharge pressure of 210 psi (14.5 bar) exceeds the pressure rating of standard system components, such as pipe and sprinklers, which are typically rated for a maximum pressure of 175 psi (12.1 bar). The use of higher-rated system components, such as pressure relief valves and overspeed shutdown devices, should be considered in such cases. Pressure relief devices are further discussed in the text that follows.

Suction Pipe – Piping Between the Suction Tank and Fire Pump

Provisions of NFPA 20, NFPA 22, and NFPA 24 address the piping connecting the suction tank and the fire pump. This pipe is referred to as suction piping, because it feeds water to the suction flange of the fire pump. Subsection 4.14.1 of NFPA 20 specifically notes that all suction components are pipe, valves, and fittings from the pump suction flange to the connection to the storage tank. The selection and installation of such suction pipe material should be in compliance with NFPA 24. Chapter 10 of NFPA 24 identifies the pipe and fitting materials that can be used. These materials include specific types of iron, steel, concrete, plastic, and copper. Chapter 10 of NFPA 24 also specifies how pipe and fittings are to be joined together, depth of cover for buried pipe, protection of pipe against freezing and other conditions, joint restraint, and acceptance testing, including flushing and hydrostatic tests.

Generally, the suction pipe, and the devices placed within the suction pipe, need to be arranged in accordance with 4.14.3 of NFPA 20 to minimize the likelihood of turbulent and imbalanced flow of water as it enters the pump suction flange. The smaller the pipe used, the faster the flow of water, and the more turbulent the flow. Increasing the pipe size slows the flow velocity and reduces the occurrence of turbulence. Turbulent conditions decrease overall pump performance, can result in a sudden system failure, and cause premature wear of system components. For instance, if the flow at the suction flange is overly turbulent, greater waterflow could occur on one side of the pump impeller, causing an imbalance in impeller rotation affecting flow to the sprinklers and shortening system life. Section 4.14 of NFPA 20 addresses the installation of the suction pipe and the devices that can be installed in the suction piping.

The size of the suction pipe is influenced mostly by the fire protection system demand (in the hypothetical example here, the demand is from an ESFR sprinkler system) and the size of the fire pump selected. NFPA 24 provides some guidance on pipe sizes; generally, the pipe should be at least 6 in. (152 mm) (nominal) in diameter in accordance with 5.2.2 of NFPA 24. However, in some cases, pipe sizes should be verified through hydraulic calculations that demonstrate that the pipe can supply the total demand at the appropriate pressure. NFPA 22 provides more specific guidance, especially with regard to the suction pipe connected to the suction tank. In accordance with 14.2.2.2 of NFPA 22, the size of the suction pipe in the example system needs to be at least 10 in. (250 mm) (nominal) in diameter, because the suction tank exceeds 100,000 gal (378.50 m³) in capacity.

The size of the suction pipe is required to be such that, with the pump operating at its maximum flow (i.e., 150 percent of its rated capacity or the maximum flow available from the water supply), the gauge pressure at the pump suction flange does not drop below −3 psi (−0.2 bar). Additionally, the suction pipe is to be sized such that, with the pump operating at 150 percent of its rated capacity (the overload point), the velocity in that portion of the suction pipe located within 10 pipe diameters upstream of the pump suction flange does not exceed 15 ft/sec (4.57 m/sec).

Table 4.26(a) and Table 4.26(b) of NFPA 20 specify pipe sizes based on the rating of the fire pump. In the example shown here, since a 2000 gpm (7570 L/min) pump would be sufficient to satisfy the system demand of 2600 gpm (9841 L/min), the suction pipe needs to be at least 10 in. (250 mm) in diameter, although NFPA allows the suction flange to be smaller than 10 in. (250 mm), provided that an eccentric tapered reducer is used. See the footnote a to Table 4.26(a) and Table 4.26(b) and 4.14.6.4 of NFPA 20.

NFPA 20 addresses only the size of the suction pipe within 10 pipe diameters of the pump suction flange, while NFPA 22 addresses the size of the pipe connected to the tank. Therefore, if we had a pump that permitted the use of a suction pipe smaller than 10 in. per Table 4.26(a) and Table 4.26(b) of NFPA 20, then the suction pipe would need to be 10 in. (250 mm) in diameter up to a point 80 in. (2000 mm), or approximately 7 ft (2.1 m), from the pump suction flange at which point the pipe size can be reduced to a size permitted in the tables. However, the larger diameter pipe can be extended up to the suction flange, provided it exceeds the minimum criteria for the pump being used.

Discharge Pipe – Piping Between the Fire Pump and the Base of the Sprinkler Riser

Section 4.15 of NFPA 20 defines discharge pipe and equipment as the pipe, valves, and fittings that extend from the pump discharge flange to the system side of the discharge valve. Practically, any pipe, valve, or fitting downstream of the discharge control valve is no longer considered to be part of the discharge piping but, instead, is considered part of the supply piping for the fire protection system or riser assembly in question. In the case of our example, this supply piping is for the ESFR sprinkler system riser. The requirements of NFPA 13 would apply from the point of the pump discharge control valve. Furthermore, any underground piping between the fire pump discharge control valve and the sprinkler system riser would need to be in accordance with NFPA 24 or Chapter 10 of NFPA 13.

Subsection 4.15.3 of NFPA 20 requires all of the aboveground piping to be steel pipe, while 4.15.5 addresses the size of the discharge pipe and associated fittings. For our example, where a 2000 gpm (7570 L/min) pump has been selected, the discharge pipe is required to be a minimum diameter of 10 in. (250 mm). In certain cases, the discharge pipe is permitted to be smaller in diameter than the suction pipe, because the velocity of the pump discharge is not of as great a concern on the discharge side of the pump. The size of the discharge pipe has an effect on friction loss, but that effect can be accounted for through hydraulic analysis.

A control valve is required to be installed on the discharge piping, so that the pump can be isolated for maintenance. Additional valves are discouraged to minimize the possibility that a valve is accidentally shut and not reopened. The control valve is permitted to be any type of valve listed for fire protection use, including a butterfly valve, because, as previously noted, turbulence is not as important an issue on the discharge side of the pump.

A check valve is required to be installed on the discharge piping. It is to be installed between the fire pump and the discharge control valve. The discharge check valve traps the higher pressure in the fire protection system after the fire pump has been shut down. It also prevents any other source of waterflow into the system, such as flow through a fire department connection, from flowing back into the pump discharge.

Section 4.15 of NFPA 20 requires that the pressure rating of the discharge components — including all piping, fittings, and valves — be adequate for the maximum total discharge pressure with the pump operating at churn conditions at rated speed. As is the case when sizing the pump, where discharge pressures exceed the rating of the components, certain precautions need to be taken.

Pump Bypass Piping

A bypass is an arrangement of piping around the fire pump that can be used to supply water to the fire protection system if the pump is taken out of service for any reason. Bypass piping is required to be the same minimum size as that required for the discharge pipe. Bypass piping should be arranged so that, if the control valves on the suction and discharge sides of the pump are closed, water can still flow through the bypass.

A check valve needs to be installed in the bypass piping so that the flow from the pump discharge does not recirculate back to the pump suction, and control valves need to be installed on either side of the check valve so that the check valve can be isolated for maintenance. The control valves are to be kept open so that the bypass is available if the pump does not start or if an obstruction becomes lodged in the pump impeller.

A bypass line is required on a fire pump assembly where the water supply is considered to have material value to the fire protection system without the pump. (See 4.14.4 of NFPA 20.) Bypass lines are typically required where the water supply is provided by a pressurized fire service main (as in municipal waterworks). Where the water supply is from a private fixed supply such as the suction tank in the installation shown in the poster, a minimum flow pressure is available and is not usually considered to be of material value. In the case of the 32 ft (9.8 m) high tank in the example, a residual pressure of approximately 14 psi (1.0 bar) would be available when the tank is full. Therefore, a bypass line is not required to be installed.

Pressure Maintenance Pump

Fire protection systems, such as the ESFR sprinkler system supplied by the pump installation shown in our poster, are designed so that they are pressurized upon their installation. A check valve serves to maintain system pressures. During a fire, the activation of a sprinkler causes a drop in system pressure, which is sensed by the pressure switch in a fire pump controller. In turn, activation of the fire pump is initiated.

However, pressure losses can occur downstream of the fire pump check valve under non-fire conditions due to water seepage across check valves or leaky fittings, or due to changes in system temperature. With changes in ambient temperature, air pockets that are usually trapped in the system piping fluctuate in size, which changes the relative pressure in the system piping. A large decrease in ambient temperature in the warehouse housing the installation in the example, which can occur in an unconditioned space over a 24-hour period, can cause a notable pressure drop, which is sensed by the fire pump pressure switch.

A fire pump should operate only during a fire or when it is being tested. A fire pump should not be used to maintain system pressure under non-fire conditions. The activation of a fire pump provides a fire alarm signal, because it indicates the operation of a sprinkler, and such fire pump activation under non-fire conditions would constitute a false alarm. Pressure maintenance pumps, also referred to as jockey pumps, are used to maintain pressures within the fire protection system under non-fire conditions, which helps prevent false alarms.

As in the case of a fire pump, the jockey pump installation includes a controller with a pressure switch. The jockey pump pressure switch is normally set at a higher pressure than the fire pump. The higher pressure setting is used so that the jockey pump starts before the fire pump, and the controller for each of the pumps must have an independent pressure sensing line that connects the sprinkler system with the pressure switches in each controller.

Jockey pumps are high-pressure/low-flow pumps that typically cannot sustain system pressures after the activation of a single sprinkler. When a sprinkler operates, the jockey pump operates but cannot maintain adequate system pressure, because an operating sprinkler demands a higher volume of flow than that of a leaky fitting. The pressure within the system continues to fall until the fire pump starts and produces the required flow and pressure for the operating system. Jockey pumps do not require a listing as fire protection equipment. Any pump that can produce the necessary pressure is acceptable. In general, Jockey pumps are sized so that their flow is lower than that expected from the smallest orifice sprinkler on the system, allowing for system pressure to fall and the fire pump to properly activate.

Although jockey pumps and their controllers do not require a listing, there are a number of requirements addressing their installation. Jockey pumps are not required as part of fire pump installation. However, some means of maintaining system pressure under non-fire conditions without relying on the fire pump as a pressure maintenance pump is needed. See Section 4.25 of NFPA 20 for more information.

Test Header and Flow Meter

Every fire pump installation needs to be provided with a means for testing the associated fire pump to ensure proper operation of the pump during its lifetime. Arrangements must be made to evaluate the pump when it operates at its rated conditions as well as at its overload condition (150 percent of its rated capacity). The means of testing must be arranged to allow for the flow and discharge of large quantities of water. NFPA 20 includes provisions for sizing the pipe used for testing. Tests are conducted during the initial acceptance or commissioning of the fire pump installation and on an annual basis in accordance with Chapter 8 of NFPA 25.

Section 4.20 of NFPA 20 allows for three different types of testing arrangements. The arrangements include the use of a discharge outlet, such as a test header, where water is discharged to atmosphere through connected hoses and nozzles, with appropriate pressure and flow readings taken. The other two methods involve a metering device used to measure the flow produced by the fire pump. The metering device is installed on a pipe loop that is arranged so the pump discharge is circulated back to the water supply tank or is arranged so that the pump discharge is circulated directly back to the suction line feeding the fire pump. This latter arrangement, often referred to as closed-loop metering, is shown in the example.

For closed-loop metering arrangements, NFPA 20 now requires that an alternate means of measuring flow, such as a test header, be provided. This arrangement is depicted in our example. Alternate means of measuring flow must be installed downstream of, and in series with, the flowmeter. Chapter 8 of NFPA 25 includes the provision that fire pump metering devices be recalibrated every 3 years. Locating the alternate means of measuring flow (test header) in the manner required by NFPA 20 facilitates this calibration activity and better ensures an accurate assessment of fire pump performance.

A test header can be installed without the use of a metering device and loop. Located on the discharge side of the pump, the test header must be installed on an exterior wall of the pump room or pump house, or it must be installed in another location outside the pump room to allow for adequate water discharge during testing. Hoses are connected to the test header during testing to allow for proper discharge and measurement of the waterflow. Flow from the test header is usually measured by using a pitot gauge or other flow measuring device placed in the flow stream. The pitot gauge registers a velocity pressure from the flow discharge, which can then be converted to flow rate using a conversion formula or table. See NFPA 291 for further information on flow testing procedures.

The connection for the test header should be located between the discharge check valve and the discharge control valve for the pump assembly, as indicated in the example on the poster, to allow the pump to be tested even when the control valve is closed, isolating the pump from the rest of the system.

The size of the pipe leading to the test header, and the number of hose connections, depends on the size of the pump. This issue is specifically addressed in Table 4.26(a) and Table 4.26(b) and 4.15.5 of NFPA 20. In the case of the 2000 gpm (7570 L/min) pump shown in the poster, a pipe at least 8 in. (200 mm) in diameter is required. The test header itself is to consist of six 2½ in. (65 mm) (nominal) hose valves and outlets. Where the length of pipe leading to the hose valve test header is more than 15 ft (4.5 m), the next pipe size, as indicated in Table 4.26(a) and Table 4.26(b), is to be used. In the case of the installation shown in the poster, if the pipe length to the test header did exceed 15 ft (4.5 m), an increase in the number and size of hose valves would be required, the next pump size is a 2500 gpm (9462 L/min) pump requiring an increase from 6 to 8 hose valves and a pipe size increase from 8 in. to 10 in. If we had for instance, a 1250 gpm (4731 L/min) pump and the length of the test header exceeded 15 ft (4.5 m), an increase in size would not be necessary because the same number and sizes of hose valves are required for a 1500 gpm (5678 L/min) pump. The pipe can also be sized through the use of hydraulic calculations based on a total flow of 150 percent of the rated pump capacity. The hydraulic calculations should include the friction loss for the total length of pipe plus any equivalent lengths of fittings, control valves, and hose valves. They also should include elevation losses between the pump discharge flange and the hose valve outlets. The hydraulic calculations then need to be verified by a flow test. See 4.20.3.4 of NFPA 20 for more details.

Pump Driver

A means of driving the fire pump is necessary and is typically provided through an electric motor or a diesel engine. Both options offer advantages and limitations. Site conditions and overall objectives for the facility need to be fully considered when making the final determination. For the example here, a diesel engine was selected.

Diesel engines serve as a good option where a dependable, self-contained power source is desired or where the electric power supply is unreliable. However, diesel engine installations require a larger, properly conditioned space and more associated equipment; produce more noise and pollutants; and necessitate greater maintenance efforts than electric motor drive installations. Chapter 11 of NFPA 20 addresses the selection and installation of diesel engine drives for fire pumps.

Before selecting a driver, proper consideration needs to be given to the location of the installation, the reserve on-site supply of diesel fuel, and the means for starting the engine. Because the engine produces a significant amount of noise when it operates, and because NFPA 25 requires testing and maintenance activities on a weekly basis, the fire pump and its diesel engine driver should be installed in a somewhat isolated location. If not, it should at least be located in a space that is sufficiently insulated against noise. The sound level produced by an operating diesel-driven fire pump can be bothersome to building occupants during the weekly testing activities.

Battery Units

A reliable means of starting the diesel engine must be provided. While NFPA 20 permits hydraulic starting and air starting, most diesel engine–driven fire pumps employ electric starter motors powered by battery units. Once the engine is running under its own power, the starter motor should cease its cranking cycle.

The term *overcranking* is used to describe the condition where the starter motor continues to crank the engine even after the engine has started. The term has also been used in the fire pump industry to refer to the situation where a failure to start condition occurs because the starter motor exceeds its time limit for cranking the engine without the engine starting.

Storage battery units are the only means allowed for powering an electric starter motor(s) for diesel engine–driven fire pumps. NFPA 20 requires each engine to be provided with two battery units, with each battery unit capable of providing sufficient power to the starter motor to crank (start) the engine. Two battery units are required as a means of redundancy and to improve overall reliability of the fire pump installation.

To ensure that the battery units retain a sufficient level of power, Section 12.5 of NFPA 20 requires two means of recharging the battery units. One method should be the generator or alternator furnished with the motor, and the other should be an automatically controlled charger taking power from an alternating current (ac) power source. If an ac power source is not available or is not considered to be reliable, another charging method in addition to the generator or alternator should be furnished with the engine.

Engine Cooling

When the diesel engine operates, it burns fuel and generates heat. A cooling system needs to be provided to regulate the engine operating temperatures. Two types of cooling systems are typically used with diesel engine drives. The first type uses water from the discharge side of the pump, which is delivered to the engine through a coolant loop. When passing through the heat exchange elements, the coolant water is then discharged to drain. The second type of cooling system consists of a radiator recirculation system similar to that used in an automobile or a truck engine. For the radiator system, coolant fluid is recirculated and not discharged to drain during pump operation. Subsection 11.2.8 of NFPA 20 addresses engine cooling systems.

Pump Room Ventilation

Proper ventilation of the pump room is an extremely important consideration, because it facilitates the availability of air for combustion of the diesel fuel within the engine and helps control the engine's operating temperatures. An adequate supply of combustion air minimizes the likelihood of an air pressure drop within the engine. Such a drop in air pressure reduces engine power output capacity, which limits the fire pump's ability to produce the necessary flow and pressure. In addition, if the pump room enclosure does not receive adequate ventilation, temperatures in the room will rise to unacceptable levels, eventually causing the engine to lose power and possibly overheat. Monitoring of pump room ventilation and overall room temperatures is especially important for engines using a radiator recirculation cooling system. As part of the heat exchange process, air passing through the radiator is discharged at elevated temperatures, which increases the overall temperature of the fire pump room to unacceptable levels unless the room is properly ventilated. See 11.3.2 of NFPA 20 for specific provisions addressing ventilation of the fire pump room.

Fuel Tank

Diesel engines require a dedicated supply of fuel on-site. The amount of fuel needed and, therefore, the size of the fuel tank are dependent on a number of factors, including the following:

(1) Power rating of the diesel engine selected
(2) Length of time for which the fire protection system is required to operate, both under fire conditions and during routine testing in accordance with NFPA 25
(3) Pump room temperatures
(4) Expected fuel delivery schedule

In the example shown in the poster, the fuel tank must be sized so that sufficient fuel remains in the tank before the next scheduled fuel delivery and after all periodic tests conducted since the last delivery. If a fire does occur before the next scheduled fuel delivery, sufficient tank fuel must be available to properly operate the fire pump for the ESFR sprinkler system for 3 hours as required by NFPA 13.

In addition to the amount of fuel needed for operation of the diesel engine, NFPA 20 requires the fuel tank capacity to be increased by an additional 10 percent. This includes 5 percent to accommodate fuel expansion due to anticipated temperature fluctuations in the pump room and 5 percent for sump volume, which is the amount of fuel in the tank that is not usable by the engine. Furthermore, NFPA 20 requires the fuel storage tank to be designed and installed so it can be kept as full as practical at all times but never below 66 percent (2/3) of tank capacity. If the tank is not a double-wall tank, it needs to be enclosed by a means such as a dike or curb to hold the entire capacity of the tank in the event of a fuel leak. See Section 11.4 and 11.6.4 of NFPA 20 for specific provisions on fuel supply and arrangement.

Engine Exhaust

Each diesel engine is to be equipped with an independent exhaust system. The exhaust must discharge outside the pump room to a safe location where it will not endanger people or structures. Because the exhaust includes hot gases and has the potential to spark, the exhaust outlet cannot be located near combustible materials or where concentrations of ignitable gases or dusts could be present. In certain situations, a spark-arresting muffler should be considered. In addition, exhaust outlets should not terminate under loading platforms or other structures where the engine exhaust can accumulate. Engine exhaust outlets should not be installed near the ventilation air intakes where engine exhaust could be drawn back into the fire pump room. As previously noted, the fire pump room requires sufficient ventilation for proper diesel engine operation. A means of preventing the accumulation of water in the engine exhaust outlet also needs to be provided. See Section 11.5 of NFPA 20 for specific provisions addressing diesel engine exhaust.

Pressure Relief Device

Because they are useful in a broad range of applications, diesel engines are designed and built to operate over a range of speeds. For the purposes of driving a fire pump, a diesel engine should run at or near its rated speed so that the fire pump produces the desired flows and pressures. However, situations can occur in which the diesel engine operates faster than its rated speed, creating an overspeed condition that produces excessive system pressures. Such pressures can cause a catastrophic system failure or shorten the life of system components. According to pump affinity laws, a small increase in speed can cause an exponentially larger increase in pressure. A pressure relief valve is needed on the discharge side of the fire pump to prevent overpressurization of the system.

The pressure relief valve operates when the pressure in the system reaches an unacceptably high level, such as can occur during an engine overspeed condition. Operation of the pressure relief valve results in discharge of water from the fire protection system (in the hypothetical pump here, the ESFR sprinkler system), causing the pressure in the system to drop. One type of pressure relief valve employs an adjustable spring-loaded mechanism. When the pressure in the system reaches a predetermined level, the system pressure overcomes the force of the spring and forces the valve open. Another type of pressure relief valve uses a pilot-operated diaphragm, which forces the valve open when the pressure in the system reaches a predetermined level. With the use of either of these valves, a substantial discharge flow, which needs to be accounted for, is expected.

As noted in NFPA 20, Section 4.18, pressure relief valves should be used only where an overspeed condition in a diesel engine results in system pressure in excess of the pressure rating of the system components [typically 175 psi (12.1 bar)]. Paragraph 4.18.1.2 of NFPA 20 requires a pressure relief valve where a total of 121 percent of the net rated shutoff (churn) pressure plus the maximum static suction pressure (adjusted for elevation) exceeds the pressure for which the system components are rated. See Section 4.18 of NFPA 20 for provisions regarding relief valves.

NFPA 20 requires the installation of an engine governor to regulate engine speed and thus avoid an overspeed condition. The governor is required to be capable of limiting the maximum engine speed to 110 percent of its rated speed, resulting in a maximum system pressure of 121 percent of the fire pump churn pressure. Because failure of the governor would result in an even more critical overspeed condition, an overspeed shutdown device is also required. The device must be equipped to sense the speed of the engine and shut down the engine when it operates at a speed greater than 20 percent over its rated speed. When the overspeed shutdown device operates, it sends a signal to the fire pump controller that prevents automatic restarting of the engine until the situation is investigated and a knowledgeable person manually resets the switch in the controller. See 11.2.4 of NFPA 20 for details regarding engine speed controls.

Engine speed and system pressurization can also be regulated though the use of a controller equipped with a variable speed pressure limiting control (VSPLC), which limits the total discharge pressure produced by the fire pump by reducing the pump driver speed. See 11.2.4.3 of NFPA 20 for requirements on the use of such devices with diesel engine drives. Note that similar devices are available for use with electric motor drives.

Controllers

The fire pump controller ensures that the diesel engine starts and operates as intended. It receives all signals associated with the fire pump, such as pressure loss or another demand signal needed to start the fire pump engine. The fire pump controller also monitors the pump and diesel engine's operation and certain fire protection system conditions, and it annunciates information on these conditions. For example, the controller monitors and annunciates a situation in which the engine overspeed shutdown device has operated or the engine has failed to start. The fire pump controller is also wired to relay alarm messages to a remote alarm panel and/or a central station monitoring facility. Controllers for diesel engine–driven fire pumps must be specifically listed for the purposes outlined in this paragraph. See Chapter 12 of NFPA 20 for specific details and functions for engine-drive controllers.

SUMMARY

This supplement highlights some of the key equipment associated with a fire pump installation for the specific hypothetical example shown in the associated poster. It references NFPA documents that need to be considered by fire protection system designers, installers, and authorities having jurisdiction. The application and enforcement of documents in addition to NFPA 20 are necessary for proper fire pump installation. For other types of installations requiring fire pumps — such as high-rise buildings requiring standpipe systems or transformers protected with water spray systems — other NFPA documents, such as NFPA 14, *Standard for the Installation of Standpipe and Hose Systems,* and NFPA 15, *Standard for Water Spray Fixed Systems for Fire Protection,* also need to be considered. The user is encouraged to develop a working knowledge of all applicable NFPA documents that affect a fire pump design and installation. All fire pump components discussed in this supplement can be viewed on the related poster, which serves as a visual aid in understanding the applications and requirements of the interrelated referenced NFPA standards. This supplement can help the user understand the steps required in designing an acceptable fire pump configuration in accordance with NFPA standards. Several of the referenced NFPA standards — either in part, or in their entirety — can be found in Part III of this handbook.

SUPPLEMENT 2

Commissioning and Inspection, Testing, and Maintenance Forms for Fire Pumps

PLAN REVIEW CHECKLIST

Date Documents Submitted: _____

Log No.: _____

File No.: _____

Property Information

Building Name: _____

Building Address: _____

Owner's Name: _____

Owner's Address: _____

Owner's Phone: _____ Fax: _____ E-mail: _____

System Designer/Contractor

Company Name: _____

Company Address: _____

Contact Person (Designer): _____

Designer Qualifications: _____

Phone: _____ Fax: _____ E-mail: _____

General

Building type:

❑ New ❑ Existing ❑ Renovation

Pump make: _____ Drive: ❑ Electric ❑ Diesel

Model No.: _____ Pump rating: _____ gpm @ _____ psi

Rated speed: _____ rpm

❑ Yes ❑ No Area/building protected

What is fire pump feeding?

❑ Automatic sprinkler system ❑ Standpipe system

❑ Fire hydrants ❑ Other _____

Building Use and Occupancy Classification

Applicable building code: _____ Edition: _____

❑ NFPA 20, *Standard for the Illustration of Stationary Pumps for Fire Protection* Edition: _____

❑ Other _____

Copyright © 2013 National Fire Protection Association.
This form may be copied for individual use other than for resale. It may not be copied for commercial sale or distribution.

(p. 1 of 4)

FORM S2.1 *Sample Plan Review Checklist.*

Building Use and Occupancy Classification *(continued)*

❑ Fire pump system required by building or fire code

❑ Fire pump system required by local ordinances

❑ Fire pump system required for equivalency, alternative level of protection, etc.

❑ Fire pump system not required, property owner voluntary safety improvements

❑ Other _____

Fire Pump Plan Review

Installation Features

❑ Yes ❑ No		Water supply to the fire pump adequate to meet fire pump requirements
❑ Yes ❑ No		Water supply to the fire pump added to the fire pump rating meets or exceeds the demands placed on it
❑ Yes ❑ No		Centrifugal fire pump listed for, and used exclusively for, fire protection service
❑ Yes ❑ No		Horizontal pump/driver on common base plate and connected by a listed flexible coupling
❑ Yes ❑ No		Indoor fire pump units separated from all other areas of the building by 2-hour fire-rated construction; 1-hour fire-rated construction if fire pump buildings is protected with an automatic sprinkler system
❑ Yes ❑ No		Fire pump units located outdoors and fire pump installations in buildings other than that building being protected by the fire pump located at least 50 feet away from the protected building
❑ Yes ❑ No		Pump room/house can be inundated by water
❑ Yes ❑ No		Suction piping the proper size: (5 inch for 500 gpm) (6 inch for 750 gpm) (8 inch for 1000 or 1500 gpm) (10 inch for 2000 or 2500 gpm)
❑ Yes ❑ No		Suction pipe galvanized or painted on the inside
❑ Yes ❑ No		An OS&Y valve provided in the suction piping *(Butterfly valves not permitted in suction piping)*
❑ Yes ❑ No		Backflow prevention or other device in the suction piping
❑ Yes ❑ No		Elbows perpendicular to impeller of horizontal pump within 10 pipe diameters of the intake flange
❑ Yes ❑ No		Reducer at pump intake, if provided, eccentric and installed with flat side up
❑ Yes ❑ No		If the suction supply is of sufficient pressure to be of material value without the pump, a bypass at least the required size of the discharge pipe provided
❑ Yes ❑ No		A listed indicating-type valve on each side of the check valve in the bypass and they are normally open
❑ Yes ❑ No		A 3½ inch compound gauge, having a rating of at least 100 psi and a range of at least twice the maximum suction pressure, provided on the suction piping
❑ Yes ❑ No		A 3½ inch pressure gauge, with a rating of at least 200 psi and a range of at least twice the working pressure of the pump, provided near the discharge casting
❑ Yes ❑ No		A ¾ inch circulating relief valve (1 inch if pump is rated over 2500 gpm) provided and piped to a drain *(Not needed for engine driven pumps cooled by water from pump discharge)*

Copyright © 2013 National Fire Protection Association.
This form may be copied for individual use other than for resale. It may not be copied for commercial sale or distribution.

FORM S2.1 *Continued.*

Installation Features (continued)

☐ Yes ☐ No Listed, float-operated, automatic, air release valve (no less than ½ inch in size) provided

☐ Yes ☐ No Discharge piping the proper size: (5 inch for 500 gpm) (6 inch for 750 or 1000 gpm) (8 inch for 1250 or 1500 gpm) (10 inch for 2000 or 2500 gpm)

☐ Yes ☐ No A listed indicating valve installed on the fire protection system side of the pump

☐ Yes ☐ No A check valve provided between discharge valve and the pump

☐ Yes ☐ No Pump driver, regardless of diesel or electric, listed for fire pump service

☐ Yes ☐ No Pump controller, regardless of diesel or electric, listed

☐ Yes ☐ No If pump is electric motor driven, wiring, wiring elements, and components arranged in approved manner

☐ Yes ☐ No If pump is diesel driven or if churn pressure can exceed rating of system components, a properly sized relief valve provided (5 inch for 500 gpm) (6 inch for 750 gpm) (8 inch for 1000 and 1500 gpm) (10 inch for 2000 gpm)

☐ Yes ☐ No Relief valve piping installed with no valves

☐ Yes ☐ No For a diesel engine driver, storage battery units provided with battery chargers specifically listed for fire pump service, arranged to automatically charge at the maximum rate whenever required by the state of charge of the battery, and arranged to indicate loss of current

☐ Yes ☐ No For a diesel engine driver cooled by a heat exchanger, the cooling water supply from the discharge of the pump taken is prior to the discharge check valve

☐ Yes ☐ No The heat exchanger piping for a diesel engine driver is equipped with an indicating manual shutoff valve, an approved flushing-type strainer, a pressure regulator, an automatic valve listed for fire protection service, and a second indicating manual shutoff valve

☐ Yes ☐ No The heat exchanger piping for a diesel engine driver is equipped with a pressure gauge installed in the cooling water supply system on the engine side of the last manual valve

☐ Yes ☐ No The heat exchanger piping for a diesel engine driver is provided with a bypass line

☐ Yes ☐ No An outlet is provided for the wastewater line from the heat exchanger with a discharge line not less than one size larger than the inlet line, discharging into a visible open waste cone, and having no valves

☐ Yes ☐ No A diesel fuel supply tank provided with a capacity of 1 gallon per engine horsepower plus 10%

☐ Yes ☐ No A diesel fuel supply tank located aboveground

☐ Yes ☐ No A test header or flowmeter tapped between the discharge check valve and the discharge valve is provided for annual fire pump flow testing *(If a flowmeter is used, it is to be arranged so as to test both pump performance and suction supply)*

☐ Yes ☐ No Proper number of listed 2½ inch hose valves provided on test header (2 for 500 gpm) (3 for 750 gpm) (4 for 1000 gpm) (6 for up to 2500 gpm)

☐ Yes ☐ No Test header piping is the proper size (4 inch for 500 gpm) (6 inch for 750 & 1000 gpm) (8 inch up to 2500 gpm) (10 inch for 2500 gpm)

☐ Yes ☐ No For test header piping over 15 feet in length, next larger piping size is used

Copyright © 2013 National Fire Protection Association.
This form may be copied for individual use other than for resale. It may not be copied for commercial sale or distribution.

FORM S2.1 Continued.

Installation Features (continued)

☐ Yes ☐ No Drain valve located at a low point of the test header pipe between the normally closed test header valve and the test header

☐ Yes ☐ No If a flowmeter provided, is the meter system piping is the proper size (5 inch for 500 and 750 GPM) (6 inch for 1000 and 1250 GPM) (8 inch for up to 2500 GPM) *(If the meter system piping exceeds 100 feet equivalent of pipe is next larger size pipe used)*

☐ Yes ☐ No Jockey pump takes suction upstream of the main pump suction control valve and discharges downstream of the installation, typically the main pump discharge valve

☐ Yes ☐ No Jockey pump provided with sensing line totally independent from that of main pump sensing line

☐ Yes ☐ No Sensing lines both tap the discharge pipes between the check valve and the discharge control valve of the pumps they respectively serve

☐ Yes ☐ No Both sensing lines ½ inch nominal and brass, copper, or series 300 stainless steel piping, tube, and fittings

☐ Yes ☐ No Two check valves are installed in each pressure sensing line at least 5 feet apart with $^3/_{32}$ inch holes drilled in the clappers

☐ Yes ☐ No There are no shutoff valves in the sensing lines

☐ Yes ☐ No Valves supervised open *(Test header and flow meter valves should be supervised shut)*

Location of test header: _____

Approval

Reviewer: _____ Date: _____

Approved ☐ Yes ☐ No

If no, reason(s): _____

_____

_____

_____

Notes:

_____

_____

_____

_____

Copyright © 2013 National Fire Protection Association.
This form may be copied for individual use other than for resale. It may not be copied for commercial sale or distribution.

(p. 4 of 4)

FORM S2.1 *Continued.*

ACCEPTANCE TEST CHECKLIST

Date Documents Submitted: _____

Log No.: _____

File No.: _____

Plan Examiner: _____

Date of Approval: _____

Permit No.: _____

Property Information

Building Name: _____

Building Address: _____

Owner's Name: _____

Owner's Address: _____

Owner's Phone: _____ Fax: _____ E-mail: _____

System Designer/Contractor

Company Name: _____

Company Address: _____

Contact Person (Designer): _____

Designer Qualifications: _____

Phone: _____ Fax: _____ E-mail: _____

General

Type of building:

❏ New   ❏ Existing   ❏ Renovation

Pump make: _____   Drive: ❏ Electric   ❏ Diesel

Model No.: _____   Pump rating: _____ gpm @ _____ psi

Rated speed: _____ rpm

What is fire pump feeding?

❏ Automatic sprinkler system   ❏ Standpipe system

❏ Fire hydrants   ❏ Other _____

Copyright © 2013 National Fire Protection Association.
This form may be copied for individual use other than for resale. It may not be copied for commercial sale or distribution.

(p. 1 of 5)

FORM S2.2 *Sample Acceptance Test Checklist.*

General *(continued)*

Present at test:

	Authorized Representative	**Manufacturer**
Pump		
Engine (if diesel)		
Controller		
Transfer switch		

Date the suction piping was flushed prior to hydrostatic test: _____

Flow rate: _____ gpm

Pressure at which piping hydrostatic tested: _____ psi

Fire Pump Acceptance Test

Installation

❑ Yes ❑ No Certificate for flushing and hydrostatic testing furnished

❑ Yes ❑ No Centrifugal fire pump listed for fire protection service

❑ Yes ❑ No Horizontal pump/driver on common base plate and connected by a listed flexible coupling

❑ Yes ❑ No Guards provided for flexible couplings and flexible connecting shafts

❑ Yes ❑ No Base plate securely attached to a solid reinforced concrete foundation

❑ Yes ❑ No Indoor fire pump units separated from all other areas of the building by 2-hour fire-rated construction; 1-hour fire-rated construction in buildings protected with an automatic sprinkler system

❑ Yes ❑ No If fire pump unit is located outdoors or if fire pump installation is in a building other than that building being protected by the fire pump, it is located at least 50 feet away from the protected building

❑ Yes ❑ No A suitable means for maintaining 40°F provided; 70°F if driver is diesel engine. *(Portable units, plug-in units, and hardwired electric units without secured circuit breakers are not reliable)*

❑ Yes ❑ No Pump room/house provided with normal lighting and emergency lighting

❑ Yes ❑ No Pump room/house adequately ventilated and floor is pitched toward drain

❑ Yes ❑ No Suction piping is the proper size. (5 inch for 500 gpm) (6 inch for 750 gpm) (8 inch for 1000 or 1500 gpm) (10 inch for 2000 or 2500 gpm)

❑ Yes ❑ No OS&Y valve provided in the suction piping *(Butterfly valves not permitted in suction piping)*

❑ Yes ❑ No No backflow prevention or other devices are in the suction piping

❑ Yes ❑ No No elbows perpendicular to impeller of horizontal pump are within 10 pipe diameters of the intake flange

Copyright © 2013 National Fire Protection Association.
This form may be copied for individual use other than for resale. It may not be copied for commercial sale or distribution.

FORM S2.2 *Continued.*

Installation (continued)

❑ Yes ❑ No		Reducer at pump intake is eccentric and installed with flat side up
❑ Yes ❑ No		A bypass, at least the required size of the discharge pipe, is provided if the suction supply is of sufficient pressure to be of material value without the pump
❑ Yes ❑ No		Listed indicating type valves are on each side of the check valve in the bypass and are normally open
❑ Yes ❑ No		A 3½ inch compound gauge, having a rating of at least 100 psi and a range of at least twice the maximum suction pressure, is provided on the suction piping
❑ Yes ❑ No		A 3½ inch pressure gauge, with a rating of at least 200 psi and a range of at least twice the working pressure of the pump, is provided near the discharge casting
❑ Yes ❑ No		A ¾ inch circulating relief valve (1 inch if pump is rated over 2500 gpm) is provided and piped to a drain. *(Not needed for engine driven pumps cooled by water from pump discharge)*
❑ Yes ❑ No		A listed, float-operated, automatic, air release valve (no less than ½ inch in size) is provided
❑ Yes ❑ No		Discharge piping is of the proper size. (5 inch for 500 gpm) (6 inch for 750 or 1000 gpm) (8 inch for 1250 or 1500 gpm) (10 inch for 2000 or 2500 gpm)
❑ Yes ❑ No		A listed indicating valve is installed on the fire protection system side of the pump
❑ Yes ❑ No		A check valve is provided between the discharge valve and the pump
❑ Yes ❑ No		The pump driver, regardless of diesel or electric, is listed for fire pump service
❑ Yes ❑ No		A properly sized relief valve has been provided if pump is diesel driven or if churn pressure can exceed rating of system components. (5 inch for 500 gpm) (6 inch for 750 gpm) (8 inch for 1000 and 1500 gpm) (10 inch for 2000 gpm)
❑ Yes ❑ No		No valves are installed in the relief valve piping
❑ Yes ❑ No		For diesel engine drivers, there are two storage battery units provided and rack-supported above the floor, secured against displacement, and located where they are readily accessible for servicing and not subject to excessive temperature, vibration, mechanical injury, or flooding
❑ Yes ❑ No		For diesel engine driver, storage battery units are provided with battery chargers specifically listed for fire pump service, arranged to automatically charge at the maximum rate whenever required by the state of charge of the battery, and arranged to indicate loss of current
❑ Yes ❑ No		For diesel engine driver cooled by a heat exchanger, the cooling water supply is from the discharge of the pump and taken prior to the discharge check valve
❑ Yes ❑ No		The heat exchanger piping for a diesel engine driver is equipped with an indicating manual shutoff valve, an approved flushing-type strainer, a pressure regulator, an automatic valve listed for fire protection service, and a second indicating manual shutoff valve
❑ Yes ❑ No		Heat exchanger piping of a diesel engine driver is equipped with a pressure gauge installed in the cooling water supply system on the engine side of the last manual valve
❑ Yes ❑ No		Heat exchanger piping of a diesel engine driver is provided with a bypass line
❑ Yes ❑ No		The outlet provided for the wastewater line from the heat exchanger has a discharge line not less than one size larger than the inlet line, discharges into a visible open waste cone, and has no valves

Copyright © 2013 National Fire Protection Association.
This form may be copied for individual use other than for resale. It may not be copied for commercial sale or distribution.

FORM S2.2 Continued.

Installation (continued)

☐ Yes ☐ No		Diesel fuel supply tank has a capacity of 1 gallon per engine horsepower plus 10%
☐ Yes ☐ No		Diesel fuel supply tank is located aboveground
☐ Yes ☐ No		Exposed fuel lines are provided with guard or protecting pipe
☐ Yes ☐ No		The test header or flowmeter is tapped between the discharge check valve and the discharge valve provided for annual fire pump flow testing (If a flowmeter is used, verify that it is arranged so as to test both pump performance and suction supply)
☐ Yes ☐ No		Proper number of listed 2½ inch hose valves is provided on test header (2 for 500 gpm) (3 for 750 gpm) (4 for 1000 gpm) (6 for up to 2500 gpm)
☐ Yes ☐ No		Test header piping is of the proper size (4 inch for 500 gpm) (6 inch for 750 and 1000 gpm) (8 inch for up to 2500 gpm) (10 inch for 2500 gpm)
☐ Yes ☐ No		If test header piping is over 15 feet in length, the next larger piping size is used
☐ Yes ☐ No		A drain valve is located at a low point of the test header pipe between the normally closed test header valve and the test header
☐ Yes ☐ No		If a flowmeter is provided, meter system piping is of the proper size. (5 inch for 500 and 750 gpm) (6 inch for 1000 and 1250 gpm) (8 inch for up to 2500 gpm)
☐ Yes ☐ No		If the meter system piping exceeds 100 feet equivalent of pipe, the next larger size pipe is used
☐ Yes ☐ No		Jockey pump is provided with sensing line totally independent from that of main pump sensing line
☐ Yes ☐ No		The sensing lines both tap the discharge pipes between the check valve and the discharge control valve of the pumps they respectively serve
☐ Yes ☐ No		Both sensing lines are ½ inch and brass, copper, or series 300 stainless steel piping, tube, and fittings
☐ Yes ☐ No		Two check valves are installed in each pressure sensing line at least 5 feet apart
☐ Yes ☐ No		No shut off valves in the sensing lines

For diesel driven pumps, verify that the following alarms are provided on the controller and operative:

- ☐ Low oil pressure
- ☐ High engine temperature
- ☐ Failure to start
- ☐ Shutdown on overspeed
- ☐ Battery failure/battery missing
- ☐ Battery charger failure
- ☐ Low (less than X\c) fuel level

For diesel driven pumps, verify that the following alarms are provided and transmit to a constantly attended location:

- ☐ Pump running
- ☐ Controller main switch in a position other than "AUTOMATIC"
- ☐ Trouble on controller or engine

Copyright © 2013 National Fire Protection Association.
This form may be copied for individual use other than for resale. It may not be copied for commercial sale or distribution.

FORM S2.2 *Continued.*

Installation (continued)

For electric driven pumps, verify that are the following alarms are operative:

- ❏ Loss of power
- ❏ Phase reversal
- ❏ Pump running
- ❏ Other _____

❏ Yes ❏ No   Verify that the cut-in and cut-out of the jockey pump is properly set.

❏ Yes ❏ No   Verify that the cut-in of the main pump is properly set.

❏ Yes ❏ No   Verify that all valves are supervised open. *(Test header and flowmeter valves should be supervised shut)*

❏ Yes ❏ No   Verify that the pump performance met or exceeds the demands of the systems supplied by pump. *(See results below.)*

Test Results

Test	Discharge Pressure	Intake pressure	Net pressure	Speed	Nozzle Size and Pitot Pressures	gpm
1						
2						
3						
4						
5						

Inspector: _____   Date: _____

Approval

Inspector: _____   Date: _____

Approved   ❏ Yes   ❏ No

If no, reason(s): _____
_____
_____

Notes:
_____
_____
_____

Copyright © 2013 National Fire Protection Association.
This form may be copied for individual use other than for resale. It may not be copied for commercial sale or distribution.

FORM S2.2 *Continued.*

ROUTINE INSPECTION CHECKLIST

Date of Inspection: _____

Inspector: _____

Date of Last Inspection: _____

Property Information

Building Name: _____

Building Address: _____

Contact Person (Owner/Tenant): _____

Address: _____

Phone: _____ Fax: _____ E-mail: _____

General

Type of building:

❏ New   ❏ Existing   ❏ Renovation

Pump make: _____   Drive: ❏ Electric   ❏ Diesel

Model No.: _____   Pump rating: _____ gpm @ _____ psi

Rated speed: _____ rpm

What is fire pump feeding?

❏ Automatic sprinkler system   ❏ Standpipe system

❏ Fire hydrants   ❏ Other _____

❏ Yes   ❏ No   Area protected

Fire Pump Inspection

Installation

❏ Yes ❏ No Water supply to the fire pump adequate to meet fire pump requirements

❏ Yes ❏ No Water supply to the fire pump added to the fire pump rating meets or exceeds the demands placed on it

❏ Yes ❏ No Centrifugal fire pump listed for, and used exclusively for, fire protection service

❏ Yes ❏ No Change in installation since last inspection

❏ Yes ❏ No Guards provided for the flexible couplings and flexible connecting shafts in good order

❏ Yes ❏ No Required rated building construction housing the fire pump intact

❏ Yes ❏ No Suitable means for maintaining 40°F being provided; 70°F if driver is diesel engine *(Portable units, plug-in units, and hardwired electric units without secured circuit breakers are not reliable)*

Copyright © 2013 National Fire Protection Association.
This form may be copied for individual use other than for resale. It may not be copied for commercial sale or distribution.   (p. 1 of 2)

FORM S2.3 *Sample Routine Inspection Checklist.*

Installation (continued)

- ❏ Yes ❏ No Both normal lighting and emergency lighting maintained for pump room/house
- ❏ Yes ❏ No Pump room/house adequately ventilated
- ❏ Yes ❏ No All valves in the fire pump piping (except the test header valve) normally open
- ❏ Yes ❏ No Suction piping compound and the discharge pressure gauges appear operative
- ❏ Yes ❏ No Circulating relief valve functions properly
- ❏ Yes ❏ No For diesel engine driver, storage battery units maintained
- ❏ Yes ❏ No For diesel engine driver, battery charger units maintained
- ❏ Yes ❏ No For diesel driver cooled by heat exchanger, cooling water able to discharge through the waste cone, manual shutoff valves in the bypass line normally closed, and flushing-type strainer being maintained
- ❏ Yes ❏ No For diesel driven pumps, fuel level is appropriate
- ❏ Yes ❏ No All alarms functional
- ❏ Yes ❏ No Approved vendor serviced fire pump in the past 12 months
- ❏ Yes ❏ No Annual fire pump test conducted

  Date of last certification _____

- ❏ Yes ❏ No Copy of annual inspection by approved vendor provided
- ❏ Yes ❏ No Pump performance meets or exceeds the demands of the systems supplied by pump

Approval

Inspector: _____ Date: _____

System inspection considered satisfactory   ❏ Yes   ❏ No

If no, reason(s): _____

_____

_____

_____

Notes:

_____

_____

_____

_____

Copyright © 2013 National Fire Protection Association.
This form may be copied for individual use other than for resale. It may not be copied for commercial sale or distribution.

FORM S2.3 *Continued.*

Fire Pumps

Weekly Inspection

This form covers a 3-month period.

Year: _____ System: _____
Location: _____

Y = Satisfactory   N = Unsatisfactory (explain on reverse)   N/A = Not applicable

Date												
Inspector												
Heat in pump room is 40°F or higher.												
Intake air louvers in pump room appear operational.												
Pump suction, discharge, and bypass valves are open.												
No piping or hoses leak.												
Fire pump leaking one drop of water per second at seals.												
Suction line pressure is normal.												
System line pressure is normal.												
Wet pit suction screens are unobstructed and in place												
Water flow test valves are in the closed position												
Suction reservoir is full.												
Controller pilot light (power on) is illuminated.												
Transfer switch normal power light is illuminated.												
Isolating switch for standby power is closed.												
Reverse-phase alarm light is not illuminated.												
Normal-phase rotation light is illuminated.												

Copyright © 2013 National Fire Protection Association.
This form may be copied for individual use other than for resale. It may not be copied for commercial sale or distribution.

FORM S2.4 *Weekly Inspection Checklist.*

Stationary Fire Pumps Handbook   2013

Fire Pumps
Weekly Inspection—Continued

Y = Satisfactory   N = Unsatisfactory (explain on reverse)   N/A = Not applicable

Item											
Oil level in vertical motor sight glass is normal.											
Diesel fuel tank is at least ²⁄₃ full.											
Controller selector switch is in "auto" position.											
Voltage readings for batteries (2) are normal.											
Charging current readings are normal for batteries.											
Pilot lights for batteries are on or battery failure pilot lights are "off."											
All alarm pilot lights are "off."											
Record engine running time from meter: _____											
Oil level is normal in right-angle gear-drive pumps.											
Power to jockey pump is provided.											
Crankcase oil level is normal.											
Cooling water level is normal.											
Electrolyte level in batteries is normal.											
Battery terminals are free of corrosion.											
Water-jacket heater is operational.											
For steam-driven pumps, steam pressure is normal.											
Examine exhaust system for leaks.											
Check lube oil heater for operation (diesel pumps).											
Drain condensate trap of cooling system.											
Check for water in diesel fuel tank.											

Copyright © 2013 National Fire Protection Association.
This form may be copied for individual use other than for resale. It may not be copied for commercial sale or distribution.

FORM S2.4 *Continued.*

Fire Pumps
Monthly Inspection

This form covers a 1-year period.

Year: _____ System: _____
Location: _____

Y = Satisfactory   N = Unsatisfactory (explain on reverse)   N/A = Not applicable

Date												
Inspector												
Remove battery corrosion; clean and dry battery case.												
Check battery charger and charger rate.												
Equalize charge in battery system.												
Exercise isolating switch and circuit breaker.												
Inspect, clean, and test circuit breakers.												

Notes
_____
_____
_____
_____
_____
_____
_____
_____
_____
_____
_____
_____
_____
_____
_____

Copyright © 2013 National Fire Protection Association.
This form may be copied for individual use other than for resale. It may not be copied for commercial sale or distribution.

FORM S2.5 *Monthly Inspection Checklist.*

Fire Pumps
Quarterly Inspection

This form covers a 1-year period.

Year: _____ System: _____
Location: _____

Y = Satisfactory   N = Unsatisfactory (explain on reverse)   N/A = Not applicable

Date				
Inspector				
Check crankcase breather on diesel pump for proper operation.				
Clean water strainer in cooling system for diesel fire pump.				
Check exhaust system insulation for integrity.				
Check exhaust system clearance to combustibles to prevent fire hazard.				
Check battery terminals to ensure they are clean and tight.				
Check electrical wiring for chafing where subject to movement.				
Check operation of safety devices and alarms (semi-annually).				

Notes
_____
_____
_____
_____
_____
_____
_____
_____
_____
_____
_____
_____

Copyright © 2013 National Fire Protection Association.
This form may be copied for individual use other than for resale. It may not be copied for commercial sale or distribution.

FORM S2.6 *Quarterly Inspection Checklist.*

Fire Pumps
Annual Inspection

Date: _____ Inspector: _____ System: _____
Location: _____

Y = Satisfactory   N = Unsatisfactory (explain on reverse)   N/A = Not applicable

All Pumps—Hydraulic System	
Suction pressure gauge: _____ psi (bar).	
Discharge pressure gauge: _____ psi (bar).	
Pump starting pressure from pressure switch in controller: _____ psi (bar).	
Pump run time from controller: _____ minutes.	
Suction line control valves are sealed open.	
Discharge line control valves are sealed open.	
Bypass line control valves are sealed open.	
All control valves are accessible.	
Suction reservoir is full.	
Pump shaft seals dripping water (1 drop per second).	
System is free of vibration or unusual noise when running.	
Packing boxes, bearings, and pump casing are free of overheating.	
Check pump shaft end play.	
Check pump coupling alignment.	
Electric Fire Pumps Only	
Isolating switch is monitoring abnormal power.	
Normal-phase rotation pilot light is "on."	
Reverse-phase pilot light is "off."	
Trip circuit breaker.	
Check voltmeter and ammeter for accuracy (5%).	
Check for corrosion on any circuit boards.	
Check for cracked electrical insulation.	
Check for leaks in plumbing parts.	
Check for water in electrical parts.	
Oil level in vertical motor sight glass is in normal range.	
Steam Fire Pumps Only	
Steam pressure gauge reading normal: _____ psi (bar).	
Record time to reach running speed: _____ min, _____ sec.	
Diesel Fire Pumps Only	
Diesel tank is $2/3$ full.	
Batteries are fully charged.	

Copyright © 2013 National Fire Protection Association.
This form may be copied for individual use other than for resale. It may not be copied for commercial sale or distribution.

FORM S2.7 *Annual Inspection Checklist.*

Fire Pumps
Annual Inspection—*Continued*

Y = Satisfactory   N = Unsatisfactory (explain on reverse)   N/A = Not applicable

Battery charger is operating properly.	
Battery terminals are clean.	
Battery state of charge is checked.	
Battery pilot lights are "on."	
Battery-failure pilot lights are "off."	
Engine-running-time meter is recording pump operation properly.	
Oil level in right-angle gear drive is normal.	
Diesel engine oil level is full.	
Check piping.	
Check tank vents and overflow piping for obstructions.	
Check antifreeze protection level.	
Check deisel exhaust system hangers and supports.	
Tighten control and power wiring connections.	
Diesel engine water level is full.	
Water-jacket heater appears to be working properly.	
Water-jacket piping is drip tight.	
Diesel engine water hose is in good condition.	
Coolant antifreeze protection is adequate.	
Cooling line strainer is clean.	
Solenoid valve is operating correctly.	
Bearings and valves are lubricated.	
System Components	
Casing relief valve is free of damage.	
Pressure-relief valve is free of damage.	
All valves, fittings, and pipe are leak tight.	
Condensate drain trap is clean.	
Fire pump controller power is "on."	
Transfer-switch normal pilot light is "on."	
Jockey pump is operational.	
Jockey pump controller power is "on."	
Jockey pump controller is set on "auto."	
Fire pump shaft coupling appears properly aligned.	
Packing glands appear properly adjusted.	
Test header control valve is closed.	

Copyright © 2013 National Fire Protection Association.
This form may be copied for individual use other than for resale. It may not be copied for commercial sale or distribution.

FORM S2.7 *Continued.*

Fire Pumps
Annual Inspection—*Continued*

Y = Satisfactory   N = Unsatisfactory (explain on reverse)   N/A = Not applicable

Test header is in good condition.	
Test header valves and caps are in good condition.	
Test header valve handles are in good condition.	
Test header valve swivel rotation is nonbinding.	
Bypass control valves are open.	
Control valves are sealed/not tampered.	
Control valves are locked/tampered.	
Control valves are properly tagged and identified.	
Flow meter control valves are closed.	
Relief valve and cone are operational.	
Relief-valve pressure appears properly adjusted.	

Notes

Copyright © 2013 National Fire Protection Association.
This form may be copied for individual use other than for resale. It may not be copied for commercial sale or distribution.

FORM S2.7 *Continued.*

Fire Pumps (Diesel Only)

Quarterly and Semi-Annual Maintenance

This form covers a 1-year period.

Year: _____ System: _____
Location: _____

Y = Satisfactory   N = Unsatisfactory (explain on reverse)   N/A = Not applicable

Quarterly

Date				
Inspector				
Clean strainer.				
Clean filter.				
Clean dirt leg.				
Clean crankcase breather.				
Clean water strainer of cooling system.				
Clean and tighten battery terminals.				
Examine wire insulation for breaks, cracks, or chafing.				

Semi-Annually (diesel pumps only)

Date				
Inspector				
Test antifreeze level.				
Inspect flexible exhaust section of diesel exhaust piping.				
Clean boxes, panels, and cabinets.				
Test all safeties and alarms for proper operation.				

Notes
_____
_____
_____
_____
_____

Copyright © 2013 National Fire Protection Association.
This form may be copied for individual use other than for resale. It may not be copied for commercial sale or distribution.

FORM S2.8 *Quarterly and Semi-Annual Maintenance Checklist.*

Fire Pumps
Annual Maintenance

Date: _____ Inspector: _____ System: _____
Location: _____

Y = Satisfactory   N = Unsatisfactory (explain on reverse)   N/A = Not applicable

Lubricate pump bearings.

Lubricate coupling.

Lubricate right-angle gear drive.

Grease motor bearings.

Replace flexible hoses and connector.

Replace oil at 50 hours or annually.

Replace oil filter at 50 hours or annually.

Calibrate pressure switch settings.

Check accuracy of pressure sensors.

Clean pump room louvers.

Remove water and foreign material from diesel fuel tank.

Rod out the heat exchanger or cooling system.

Fire pump controller in service.

Jockey pump controller in service.

Fire alarm panel "normal."

Notes

_____
_____
_____
_____

Copyright © 2013 National Fire Protection Association.
This form may be copied for individual use other than for resale. It may not be copied for commercial sale or distribution.

FORM S2.9 *Annual Maintenance Checklist.*

Fire Pumps
Weekly Operating Tests

This form covers a 1-month period.

- Year: _____ System: _____
- Location: _____

Y = Satisfactory   N = Unsatisfactory (explain on reverse)   N/A = Not applicable

Date					
Inspector					
Operate electric fire pumps for 10 minutes monthly.					
Operate diesel fire pump for 30 minutes weekly.					
Check packing gland tightness (slight leak at no flow).					
Record suction pressure from gauge in psi (bar).					
Record discharge pressure from gauge in psi (bar).					
Adjust gland nuts if necessary.					
Check for unusual noise or vibration.					
Check packing boxes, bearings, or pump casing for overheating.					
Record pump starting pressure.					
Observe time for motor to accelerate to full speed (diesel and steam pumps).					
For reduced-voltage or reduced-current starting, record time controller is on first step.					
Record time pump runs after starting for pumps having automatic stop feature.					
Record time for diesel engine to crank.					
Record time for diesel engine to reach running speed.					
Check oil pressure gauge, speed indicator, water and oil temperatures while engine is running.					
Check heat exchanger for cooling water flow.					
Record steam pressure for steam-operated pumps.					
Check water tank float switch.					
Check solenoids for proper operation.					
Operate speed governor (internal combustion engine only).					
Check controller alarms.					
Record any notes that the inspector believes to be significant. Place a number in the block and number the corresponding note on the reverse of this form.					

Copyright © 2013 National Fire Protection Association.
This form may be copied for individual use other than for resale. It may not be copied for commercial sale or distribution.

FORM S2.10 *Weekly Operating Tests Checklist.*

Fire Pumps
Monthly and Semi-Annual Tests

This form covers a 1-year period.

Year: _____ System: _____
Location: _____

Y = Satisfactory   N = Unsatisfactory (explain on reverse)   N/A = Not applicable

Monthly

Date										
Inspector										
Exercise isolating switch and circuit breaker.										
Test antifreeze to determine protection level.										
Test batteries for specific gravity or state of charge.										
Test circuit breakers and fuses for proper operation.										

Semi-Annually

Date										
Inspector										
Operate manual starting means.										
Operate safety devices and alarms.										

Notes
_____
_____
_____
_____
_____

Copyright © 2013 National Fire Protection Association.
This form may be copied for individual use other than for resale. It may not be copied for commercial sale or distribution.

FORM S2.11 *Monthly and Semi-Annual Tests Checklist.*

Fire Pumps

Annual Performance Tests

Date: _____ Inspector: _____ System: _____
Location: _____

Y = Satisfactory   N = Unsatisfactory (explain on reverse)   N/A = Not applicable

Pump manufacturer and model: _____
Type: ☐ Centrifugal ☐ Turbine
Controller manufacturer and model: _____
Rated capacity: _____ gpm (L/min)
Water supply source: _____
Rated pressure: _____ psi (bar)   Rated speed: _____ rpm
Power: ☐ Electric ☐ Diesel ☐ Steam

Automatic starts performed 10 times.	Remote stop: _____ psi (bar)
Automatic start functions properly.	Timer indicates total run time: _____ min.
Automatic stop functions properly.	Timer reset and graph paper changed?
Automatic start: _____ psi (bar)	Test data and flow charts completed. (Attach all water flow charts, electrical power charts, performance curves, etc.)
Automatic stop: _____ psi (bar)	
Manual starts performed 10 times.	Fire pump electrical power readings recorded at each flow condition?
Manual start functions properly.	Fire pump motor speed: _____ rpm
Manual stop functions properly.	Fire pump discharge flow: _____ gpm (L/min)
Manual start: _____ psi (bar)	Jockey pump operational.
Manual stop: _____ psi (bar)	Jockey pump appears properly aligned.
Remote start functions properly.	Jockey pump valves open.
Remote stop functions properly.	Jockey pump "turn-on": _____ psi (bar)
Remote start: _____ psi (bar)	Jockey pump "turn-off": _____ psi (bar)

Notes
_____
_____
_____
_____

Copyright © 2013 National Fire Protection Association.
This form may be copied for individual use other than for resale. It may not be copied for commercial sale or distribution.

FORM S2.12 *Annual Performance Tests Checklist.*

Fire Pumps
Annual Test Summary Page

Date: _____ **Inspector:** _____ **System:** _____
Location: _____

	Test 1	Test 2	Test 3
Approximate percent of rated pump discharge (gpm)/(L/min)	0	100%	150%
Nozzle size in inches (mm)	No flow		
Pitot pressure in psi (bar)	None		
Flow in gpm (L/min)	None		
Pump suction in psi (bar)			
Pump discharge in psi (bar)			
Net pump head (discharge pressure minus suction pressure)			
Pump speed (rpm)			
Operate electric circuit breaker.			
Test emergency power supply.			
Check for excessive back pressure in exhaust system.			

Notes
_____
_____
_____
_____
_____
_____
_____
_____
_____
_____

Copyright © 2013 National Fire Protection Association.
This form may be copied for individual use other than for resale. It may not be copied for commercial sale or distribution.

FORM S2.13 *Annual Test Summary Page.*

628 Part IV • Supplement 2 • Commissioning and Inspection, Testing, and Maintenance Forms for Fire Pumps

Fire Pumps
Flow and Pressure Record

Date: _____ Inspector: _____ System: _____ Location: _____

N^1.85

Flow—gal. per min (gpm)

Pressure—lb. per sq. in. (psi)

Copyright © 2013 National Fire Protection Association.
This form may be copied for individual use other than for resale. It may not be copied for commercial sale or distribution.

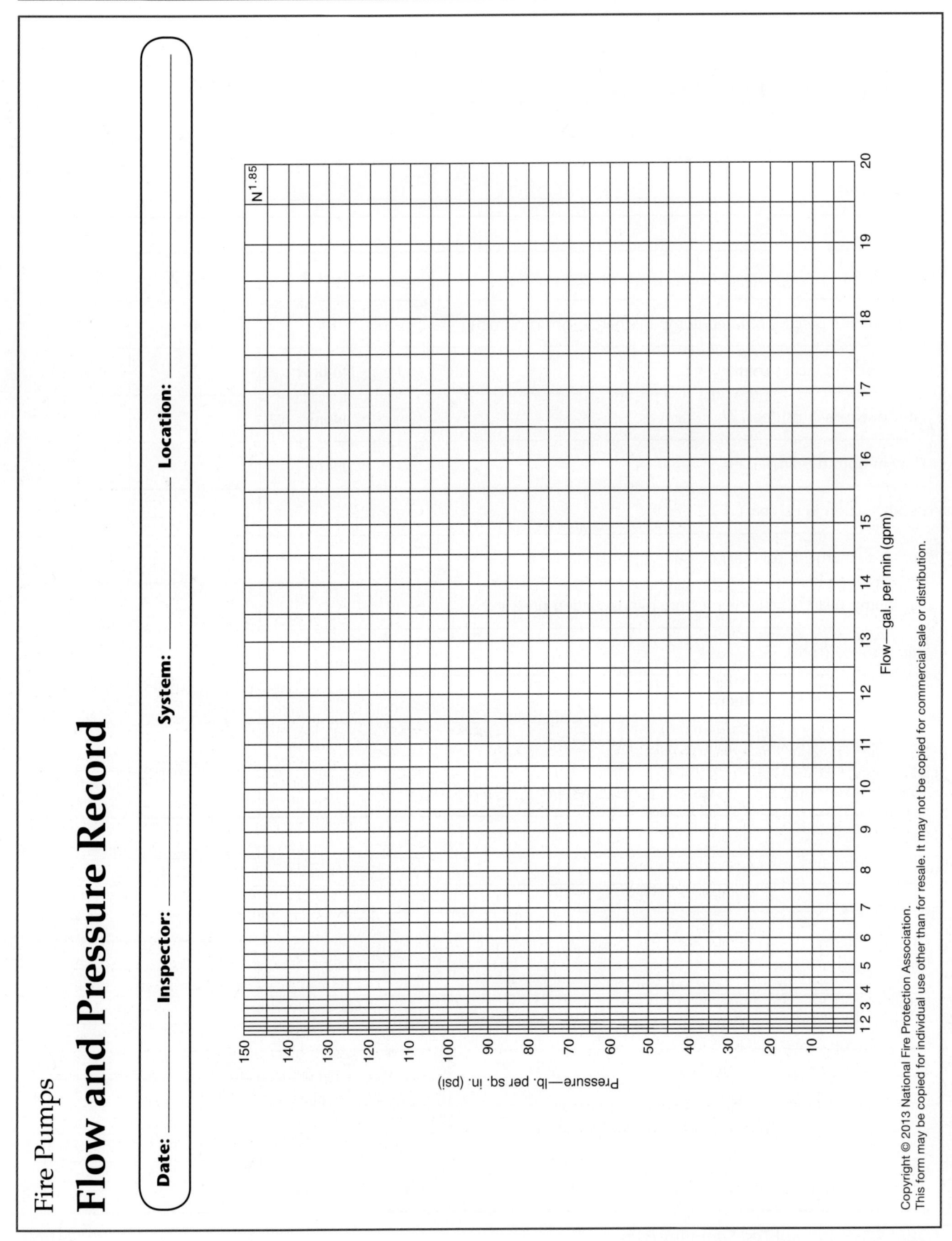

FORM S2.14 *Flow and Pressure Record.*

2013 *Stationary Fire Pumps Handbook*

SUPPLEMENT 3

NEC® Article 695, Fire Pumps

Editor's Note: Supplement 3 extracts Article 695 from the 2011 edition of NFPA 70®, National Electrical Code® (NEC®) Handbook, which contains the installation requirements for electric power sources and interconnecting circuitry for electric motor–driven fire pumps and switching and control equipment for electric fire pump motors. Article 695 does not cover performance, maintenance, or acceptance testing of the fire pump system, nor does it cover the internal wiring of the system components or wiring for diesel fire pumps or pressure maintenance pumps. This extracted text from the 2011 NEC Handbook references the 2010 edition of NFPA 20. No attempt has been made to update NFPA 20 references because they are current with the extracted text.

ARTICLE 695 FIRE PUMPS

695.1 Scope.

Informational Note: Text that is followed by a reference in brackets has been extracted from NFPA 20-2010, *Standard for the Installation of Stationary Pumps for Fire Protection*. Only editorial changes were made to the extracted text to make it consistent with this *Code*.

(A) Covered. This article covers the installation of the following:

(1) Electric power sources and interconnecting circuits
(2) Switching and control equipment dedicated to fire pump drivers

(B) Not Covered. This article does not cover the following:

(1) The performance, maintenance, and acceptance testing of the fire pump system, and the internal wiring of the components of the system
(2) Pressure maintenance (jockey or makeup) pumps

Informational Note: See NFPA 20-2010, *Standard for the Installation of Stationary Pumps for Fire Protection*, for further information.

> Since it was first included in the 1996 *NEC*, Article 695 has been the subject of a number of important revisions, particularly in the requirements covering reliable power supplies for electric fire pump motors. Those revisions correlate the requirements in the *NEC* with those in NFPA 20, *Standard for the Installation of Stationary Pumps for Fire Protection*.
>
> One of the most significant changes that has occurred in the evolution of Article 695 is the requirement of 695.3(C) covering reliable power for electric fire pumps installed in multibuilding campus-style arrangements such as those found in many institutional and industrial settings. This revision allows for fire pumps to be supplied by feeder circuits that are part of a high or medium voltage premises wiring system. This distribution arrangement is quite common in industrial and institutional campus settings. The conductors supplied by the higher voltage level distribution systems are not service conductors, because the service point and service-disconnecting means is generally located at a campus distribution switchyard or distribution building. Also, all of the distribution conductors on the load side of the service equipment, even though they resemble electric utility–type distribution, are considered to be feeders or in some cases where the circuit supplies a single piece of utilization equipment, branch circuits.

Noteworthy is the distinct division of responsibility for fire pump requirements between the *NEC* and NFPA 20. Performance issues, including the determination of power-supply reliability, are under the jurisdiction of the NFPA Technical Committee on Fire Pumps, while electrical installation requirements are within the purview of the National Electrical Code Committee. The scope of Article 695, stated in 695.1, speaks to this division of responsibility.

Article 695 does not apply to pumps used to supply sprinkler systems in one- and two-family dwellings. NFPA 13D, *Standard for the Installation of Sprinkler Systems in One- and Two-Family Dwellings and Manufactured Homes*, does not require the use of a listed pump; thus, neither NFPA 20 nor Article 695 is applicable. Although the installation requirements for pressure maintenance (jockey) pumps are not covered by Article 695, there are provisions in Article 695 permitting these pumps to be supplied by a fire pump service or feeder.

Other than the specific reference in 695.4(B)(3)(b), the requirements of Article 695 are independent of those in Article 700 unless otherwise mandated by the authority having jurisdiction. Installation of the supply wiring to an electric motor–driven fire pump such as the one shown in Exhibit 695.1 is covered by the requirements of Article 695.

EXHIBIT 695.1 *An Electric Motor Specifically Listed for Fire Pump Service. (Courtesy of Liberty Mutual Risk Engineering)*

695.2 Definitions.

Fault-Tolerant External Control Circuits. Those control circuits either entering or leaving the fire pump controller enclosure, which if broken, disconnected, or shorted will not prevent the controller from starting the fire pump from all other internal or external means and may cause the controller to start the pump under these conditions.

On-Site Power Production Facility. The normal supply of electric power for the site that is expected to be constantly producing power.

On-Site Standby Generator. A facility producing electric power on site as the alternate supply of electric power. It differs from an on-site power production facility, in that it is not constantly producing power.

695.3 Power Source(s) for Electric Motor-Driven Fire Pumps.

Electric motor-driven fire pumps shall have a reliable source of power.

(A) Individual Sources. Where reliable, and where capable of carrying indefinitely the sum of the locked-rotor current of the fire pump motor(s) and the pressure maintenance pump motor(s) and the full-load current of the associated fire pump accessory equipment when connected to this power supply, the power source for an electric motor driven fire pump shall be one or more of the following.

The power source for an electric motor–driven fire pump must be reliable and have adequate capacity to carry the locked-rotor currents of the fire pump motor and accessory equipment. These two main requirements ensure that the fire pump will operate in the event of a fire without being accidentally disconnected, and that the fire pump will continue to operate until the fire is extinguished, the fire pump is purposely shut down, or the pump itself is destroyed.

The determination of whether the serving electric utility is a reliable source of power is an issue for the authority having jurisdiction. Section A.9.3.2 in Annex A of NFPA 20-2010, *Standard for the Installation of Stationary Pumps for Fire Protection*, elaborates on several key characteristics of a reliable power supply.

(1) Electric Utility Service Connection. A fire pump shall be permitted to be supplied by a separate service, or from a connection located ahead of and not within the same cabinet, enclosure, or vertical switchboard section as the service disconnecting means. The connection shall be located and arranged so as to minimize the possibility of damage by fire from within the premises and from exposing hazards. A tap ahead of the service disconnecting means shall comply with 230.82(5). The service equipment shall comply with the labeling requirements in 230.2 and the location requirements in 230.72(B). [20:9.2.2(1)]

The top example in Exhibit 695.2 shows a single service where a dedicated set of service conductors to supply the fire pump is tapped to the incoming service conductors. The tap cannot be made in the section of the equipment that contains the service disconnecting means. This tap is permitted under the conditions specified in 230.40, Exception No. 5 and 230.82(5). The bottom example shows a dedicated service to supply the fire pump. A separate service for the fire pump is permitted by 230.2(A)(1).

(2) On-Site Power Production Facility. A fire pump shall be permitted to be supplied by an on-site power production facility. The source facility shall be located and protected to minimize the possibility of damage by fire. [20:9.2.2(3)]

On-site power production facilities are defined in 695.2 and are permitted to be used as the sole power source for an electrically driven fire pump motor. An on-site power production facility differs in normal application from an on-site standby generator, in that it is the normal source of electrical supply for a building, structure, or campus and is not a utility-owned generating facility. For some small installations, the normal use of the generator is the distinguishing feature between whether the equipment is defined as an on-site power production facility or an on-site standby generator. Exhibit 695.3 illustrates generating

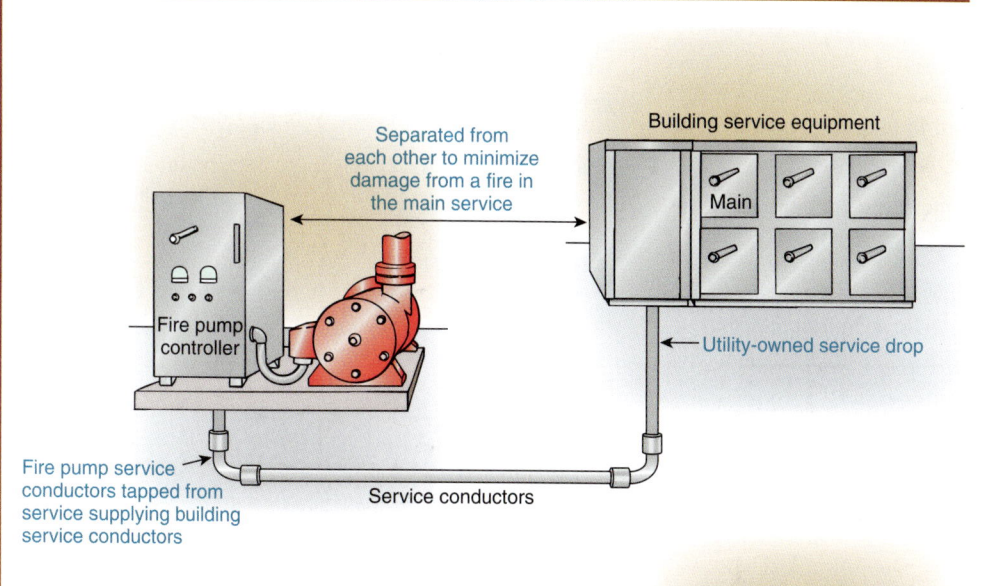

EXHIBIT 695.2 Two Permitted Configurations for Connecting to an Electric Utility–Supplied Service.

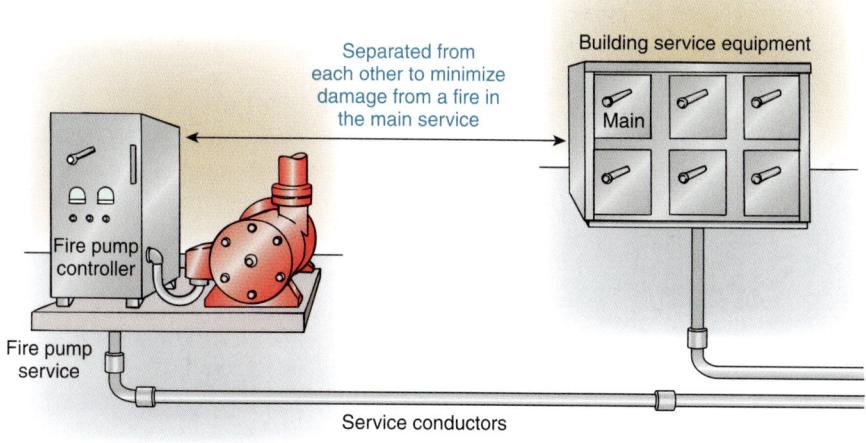

equipment that is the normal source of power for the premises wiring system and meets the definition of on-site power production facility.

(3) Dedicated Feeder. A dedicated feeder shall be permitted where it is derived from a service connection as described in 695.3(A)(1). [20:9.2.2(3)]

One or more of the following types of power sources are permitted to individually supply an electrically driven fire pump. If only one source is used, that source must be capable of supplying locked-rotor current to the fire pump and accessory equipment. Power may be supplied by the following:

1. A separate utility service or connection ahead of the main disconnecting means
2. An on-site power production system
3. A feeder dedicated to the fire pump that is supplied by a service meeting the requirements of 695.3(A)(1).

(B) Multiple Sources. If reliable power cannot be obtained from a source described in 695.3(A), power shall be supplied by one of the following: [20:9.3.2]

(1) Individual Sources. An approved combination of two or more of the sources from 695.3(A).

Where the alternate source of power is an on-site standby generator, the alternate source disconnecting means and the alternative source overcurrent protective device(s) for the electric-drive fire pump are not required to be sized for locked-rotor current of the fire pump motor(s). The generator is required to be sized to allow the fire pump to start and run per 9.6.1.1 in NFPA 20-2010, *Standard for the Installation of Stationary Pumps for Fire Protection*. The associated circuit components are required to be sized according to Article 430 and to "allow instantaneous pickup of the full [fire] pump room load." This requirement is also found in 9.6.5 in NFPA 20-2010.

(2) Individual Source and On-site Standby Generator. An approved combination of one or more of the sources in 695.3(A) and an on-site standby generator complying with 695.3(D). [20:9.3.4]

Exception to (B)(1) and (B)(2): An alternate source of power shall not be required where a back-up engine-driven or back-up steam turbine-driven fire pump is installed. [20:9.3.3]

If none of the power supply sources specified in 695.3(A)(1) through (3) can individually provide reliable power with adequate capacity, 695.3(B) permits an approved combination (two or more) of these sources or a combination of one or more of these sources with an on-site standby generator. A back-up engine– or steam turbine–driven fire pump is permitted in lieu of installing an on-site standby generator; therefore, the electric fire pump is required to be supplied by only a single power source. This allowance provides some design options on how to augment an electric fire pump that is supplied by a source that has not been deemed to be reliable.

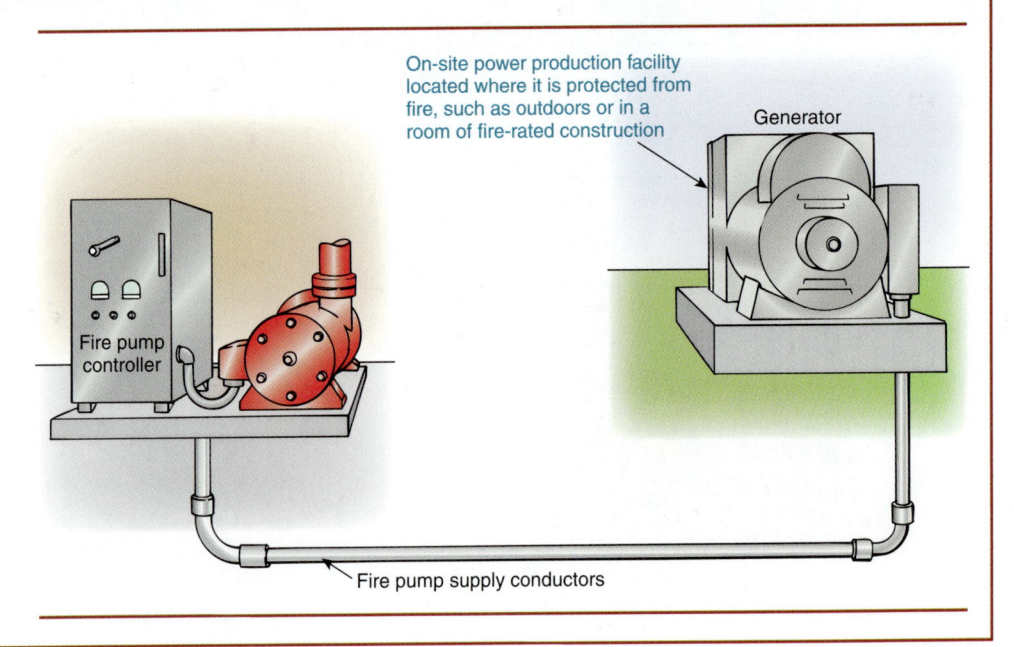

EXHIBIT 695.3 On-Site Power Production Facility as a Power Source for a Fire Pump Installation.

(C) Multibuilding Campus-Style Complexes. If the sources in 695.3(A) are not practicable and the installation is part of a multibuilding campus-style complex, feeder sources shall be permitted if approved by the authority having jurisdiction and installed in accordance with either (C)(1) and (C)(3) or (C)(2) and (C)(3).

(1) Feeder Sources. Two or more feeders shall be permitted as more than one power source if such feeders are connected to, or derived from, separate utility services. The connection(s), overcurrent protective device(s), and disconnecting means for such feeders shall meet the requirements of 695.4(B).

(2) Feeder and Alternate Source. A feeder shall be permitted as a normal source of power if an alternate source of power independent from the feeder is provided. The connection(s), overcurrent protective device(s), and disconnecting means for such feeders shall meet the requirements of 695.4(B).

(3) Selective Coordination. The overcurrent protective device(s) in each disconnecting means shall be selectively coordinated with any other supply-side overcurrent protective device(s).

The requirements of 695.3(C) permit the use of feeder (see definition in Article 100) sources for multibuilding campus style applications. This method of power supply requires the approval of the authority having jurisdiction. A fire pump supplied by a radial loop type of distribution system (commonly used for medium and high voltage distribution) where the two feeders originate from a single substation has to be augmented by an on-site standby generator in order for the power supply to be deemed reliable. In the system shown in Exhibit 695.4, the two feeders originate from different utility services, a distribution arrangement that allows the two feeders, without an on-site standby generator, to be the single source for the electric fire pump as permitted by 695.3(C)(1).

Where there is more than one overcurrent device in series with the fire pump supply, 695.3(C)(3) requires each device to be selectively coordinated with supply-side overcurrent protective devices.

(D) On-Site Standby Generator as Alternate Source. An on-site standby generator(s) used as an alternate source of power shall comply with (D)(1) through (D)(3). [20:9.6.2.1]

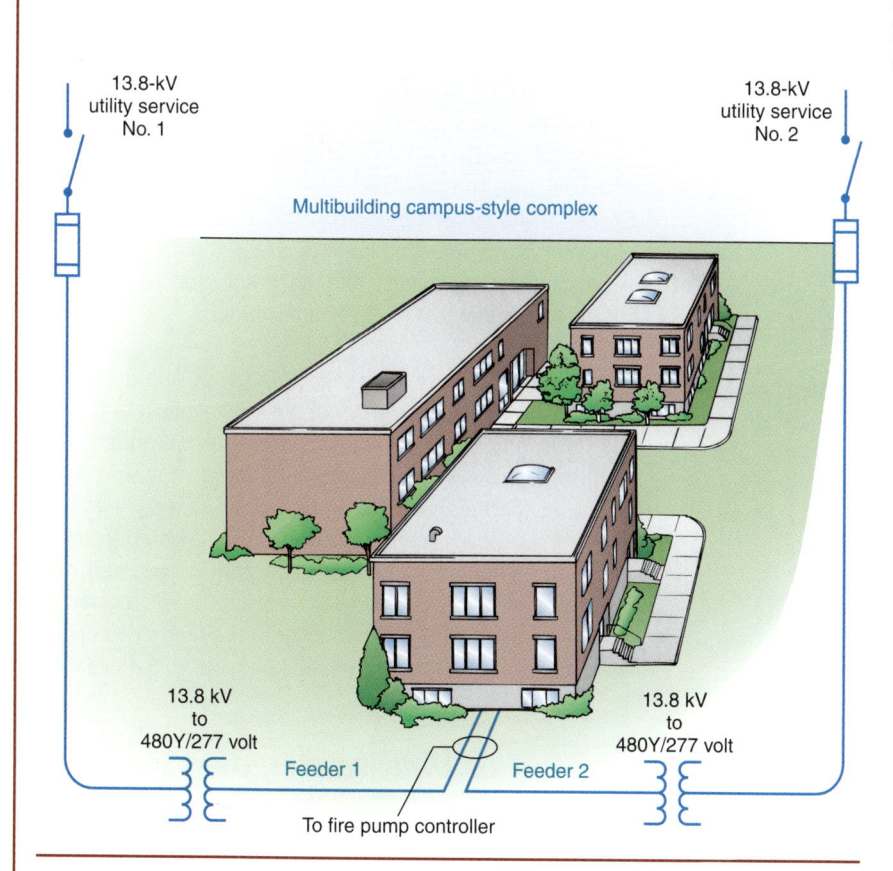

EXHIBIT 695.4 *Multiple Feeder Sources for Campus-Style Application, as Described in 695.3(C)(1).*

(1) Capacity. The generator shall have sufficient capacity to allow normal starting and running of the motor(s) driving the fire pump(s) while supplying all other simultaneously operated load(s). [**20**:9.6.1.1]

Automatic shedding of one or more optional standby loads in order to comply with this capacity requirement shall be permitted.

> On-site standby generators are required only to be capable of carrying the starting and running current of the fire pump motor. Only the sources specified in 695.3(A)(1) through (3) are required to be capable of indefinitely carrying the locked rotor current of the fire pump motor.

(2) Connection. A tap ahead of the generator disconnecting means shall not be required. [**20**:9.6.1.2]

(3) Adjacent Disconnects. The requirements of 430.113 shall not apply.

(E) Arrangement. All power supplies shall be located and arranged to protect against damage by fire from within the premises and exposing hazards. [**20**:9.1.4]

Multiple power sources shall be arranged so that a fire at one source does not cause an interruption at the other source.

> Determining compliance with this performance requirement has to take into account individual building or structure characteristics including, but not limited to, the type of construction, type of content, proximity of building to other hazard exposures, and the location of the primary and alternate power sources for the fire pump.

(F) Phase Converters. Phase converters shall not be permitted to be used for fire pump service. [**20**:9.1.7]

> Concern over voltage imbalance between phases under unloaded or lightly loaded conditions is the reason phase converters are not permitted for fire pump service. This condition could adversely impact electronics integral to the controller. Because the controller has to be constantly powered, a phase converter used in a fire pump circuit would be in continuous operation.

695.4 Continuity of Power.

Circuits that supply electric motor-driven fire pumps shall be supervised from inadvertent disconnection as covered in 695.4(A) or (B).

(A) Direct Connection. The supply conductors shall directly connect the power source to either a listed fire pump controller or listed combination fire pump controller and power transfer switch.

(B) Connection Through Disconnecting Means and Overcurrent Device.

> Section 695.4(B) permits, but does not require, the installation of a disconnecting means and associated overcurrent protection between a power source and the fire pump control devices described in 695.4(B)(1). Other requirements in the *Code* such as 230.70 and 225.31 may necessitate the installation of the disconnecting means and overcurrent protection covered in the requirements of 695.4(B). Ideal, but not always possible, would be the direct connection of the source to the fire pump control equipment in accordance with 695.4(A).

(1) Number of Disconnecting Means.

(a) *General.* A single disconnecting means and associated overcurrent protective device(s) shall be permitted to be installed between the fire pump power source(s) and one of the following: [**20**:9.1.2]
  (1) A listed fire pump controller
  (2) A listed fire pump power transfer switch
  (3) A listed combination fire pump controller and power transfer switch

(b) *Feeder Sources.* For systems installed under the provisions of 695.3(C) only, additional disconnecting means and the associated overcurrent protective device(s) shall be permitted as required to comply with other provisions of this *Code*.

(c) *On-Site Standby Generator.* Where an on-site standby generator is used to supply a fire pump, an additional disconnecting means and an associated overcurrent protective device(s) shall be permitted.

> This provision allows an on-site standby generator to be equipped with an integral disconnecting means and overcurrent protection that is in addition to a disconnecting means and overcurrent protection installed elsewhere in the alternate supply circuit to the fire pump. Alternatively, this disconnecting means could be located in distribution equipment that is supplied from a generator or multiple generators connected in parallel. The disconnecting means permitted by this section is required to comply with the requirements of 695.4(B)(2)(a) and 695.4(B)(3)(b) through (e).

(2) Overcurrent Device Selection. Overcurrent devices shall comply with (a) or (b).

(a) *Individual Sources.* The overcurrent protective device(s) shall be rated to carry indefinitely the sum of the locked-rotor current of the fire pump motor(s) and the pressure maintenance pump motor(s) and the full-load current of the associated fire pump accessory equipment when connected to this power supply. Where the locked-rotor current value does not correspond to a standard overcurrent device size, the next standard overcurrent device size shall be used in accordance with 240.6. The requirement to carry the locked-rotor currents indefinitely shall not apply to conductors or devices other than overcurrent devices in the fire pump motor circuit(s). [**20**:9.2.3.4]

> The requirement for sizing the overcurrent protection in a supervised fire pump disconnecting means so it is able to carry locked-rotor current indefinitely is a key factor in the reliable power source equation. Due to the critical life safety and property protection function of a fire pump, opening of the circuit by an overcurrent device installed between the point of connection to the power source and the fire pump controller cannot be tolerated and the circuit has to perform as if there were a direct connection to the power source. Sizing for locked-rotor current applies only to overcurrent protective devices and does not extend to conductors or other devices in the fire pump motor circuit. Similar provisions to this effect are contained in 695.5(B) and 695.5(C)(2).

(b) *On-Site Standby Generators.* Overcurrent protective devices between an on-site standby generator and a fire pump controller shall be selected and sized to allow for instantaneous pickup of the full pump room load, but shall not be larger than the value selected to comply with 430.62 to provide short-circuit protection only. [**20**:9.6.1.1]

> This requirement correlates with 695.3(D)(1) covering the required capacity of an on-site standby generator. Overcurrent protective devices supplied by an on-site standby generator are not required to be sized to carry the locked rotor current of the fire pump(s) indefinitely. The on-site standby generator is not limited to supplying only the fire pump, but where other pump room loads are supplied, it must have sufficient capacity to instantaneously carry the entire load supplied. The overcurrent protective devices are not required to provide overload protection and are required to be sized per 430.62, which covers devices supplying multiple motors or motors and other loads.

(3) Disconnecting Means. All disconnecting devices that are unique to the fire pump loads shall comply with items (a) through (e).

(a) *Features and Location — Normal Power Source.* The disconnecting means for the normal power source shall comply with all of the following: [**20**:9.2.3.1]
 (1) Be identified as suitable for use as service equipment
 (2) Be lockable in the closed position
 (3) Not be located within equipment that feeds loads other than the fire pump
 (4) Be located sufficiently remote from other building or other fire pump source disconnecting means such that inadvertent operation at the same time would be unlikely

> A supervised disconnecting means supplied by one of the individual sources specified in 695.3(A) cannot be installed in distribution equipment that supplies other than fire pump loads. This requirement provides clear direction on how compliance with 695.4(B)(3)(a)(4) can be achieved. "Sufficiently remote from other building or other fire pump source disconnecting means" cannot be interpreted as permitting a supervised fire pump disconnecting means to be located in a switchboard or panelboard that supplies other than fire pump loads.

(b) *Features and Location — On-Site Standby Generator.* The disconnecting means for an on-site standby generator(s) used as the alternate power source shall be installed in accordance with 700.10(B)(5) for emergency circuits and shall be lockable in the closed position.

> A disconnecting means supplied by an on-site standby generator is permitted to be installed in equipment that supplies other loads. However, the requirements of 700.10(B)(5) must be followed. The effect of this requirement is that a fire pump feeder cannot be supplied from equipment in which the fire pump conductors are installed in the same enclosure or vertical switchboard section with conductors supplying loads that are designated or classed as legally required standby (Article 701) or optional standby (Article 702) loads. To help minimize inadvertent opening of the fire pump circuit, the disconnecting means is required to be capable of being locked in the closed (on) position.

(c) *Disconnect Marking.* The disconnecting means shall be marked "Fire Pump Disconnecting Means." The letters shall be at least 25 mm (1 in.) in height, and they shall be visible without opening enclosure doors or covers. [**20**:9.2.3.1(5)]

(d) *Controller Marking.* A placard shall be placed adjacent to the fire pump controller, stating the location of this disconnecting means and the location of the key (if the disconnecting means is locked). [**20**:9.2.3.2]

(e) *Supervision.* The disconnecting means shall be supervised in the closed position by one of the following methods:
   (1) Central station, proprietary, or remote station signal device
   (2) Local signaling service that causes the sounding of an audible signal at a constantly attended point
   (3) Locking the disconnecting means in the closed position
   (4) Sealing of disconnecting means and approved weekly recorded inspections when the disconnecting means are located within fenced enclosures or in buildings under the control of the owner [**20**:9.2.3.3]

Minimizing the locations where the power supply to the fire pump can be automatically or manually interrupted is the objective of 695.4. Ideally, the supply conductors from the power source are run directly to the listed fire pump control and/or transfer equipment, without the need for an additional service disconnecting means and overcurrent protection. However, this arrangement is not always possible; therefore, a single supervised disconnecting means is permitted, provided all the conditions in 695.4(B) are met. Where installed in the fire pump circuit, the supervised disconnecting means with overcurrent protection must meet the following conditions:

1. Overcurrent device is sized to carry the locked-rotor currents of all the fire pump motors.
2. Fire pump disconnecting means is located away from the other service disconnecting means. This disconnecting means is rated as service equipment and is lockable in the "on" position.
3. Equipment containing the disconnecting means supplies only the fire pump and associated loads.
4. Disconnect and controller are marked and placarded as described.
5. Circuit is supervised in the closed position, in accordance with 695.4(B)(3)(e).

The size of the overcurrent protective device is the sum of the locked-rotor currents of all the permitted motors plus the sum of any other fire pump auxiliary loads.

Calculation Example

A fusible service disconnect switch supplies power to a 100-hp, 460-volt, 3-phase fire pump and to a 1½-hp, 460-volt, 3-phase jockey pump. The fire pump feeder circuit will be installed in a raceway between the disconnecting means and fire pump controller. The raceway is considered outside the building per 230.6. Using the requirements of Article 695, determine the sizes of the disconnecting means and overcurrent protection device. Also determine the minimum ampacity of the feeder conductors.

Solution

STEP 1. Determine the minimum ratings of the disconnecting means and the overcurrent protective device.

According to the motor nameplates, the locked-rotor current (LRC) is 725 amperes for the 100-hp motor and 20 amperes for the 1½-hp motor. If the locked-rotor amperes are not on the nameplates, the locked-rotor currents found in Table 430.251(B) must be used. Calculate the size by summing the locked-rotor currents of both motors and then going to the next larger standard-size overcurrent device, as follows:

$$100\text{-hp, 3-phase LRC} = 725 \text{ A}$$
$$1\tfrac{1}{2}\text{-hp, 3-phase LRC} = 20 \text{ A}$$
$$\text{Total LRC} = 745 \text{ A}$$

The next larger standard-size disconnect switch and overcurrent device is 800 amperes. An adjustable-trip circuit breaker of 750 amperes is also permitted, because it, too, will carry the locked-rotor current indefinitely.

STEP 2. Determine the minimum ampacity for the fire pump feeder conductor.

Even though the disconnect switch, fuse, and circuit breakers are sized according to locked-rotor currents, the feeder conductors to the fire pump and associated equipment are required to have an ampacity not less than 125 percent of the full-load current (FLC) rating of the fire pump motor(s) and pressure maintenance pump motor(s), plus 100 percent of associated accessory equipment. Using the same motors as in Step 1, calculate the size of the feeder to the fire pump controller as follows, using 430.6(A)(1) and Table 430.250 for the full-load currents of the motors:

$$100\text{-hp, 3-phase FLC}$$
$$124 \text{ A} \times 1.25 = 155.00 \text{ A}$$
$$1\tfrac{1}{2}\text{-hp, 3-phase FLC}$$
$$3 \text{ A} \times 1.25 = 3.75 \text{ A}$$
$$\text{Total FLC} = 158.75 \text{ A or } 159 \text{ A}$$

Thus, the minimum ampacity for the feeder conductors is 159 amperes. Using the 75°C column from Table 310.15(B)16, a 2/0 copper conductor is the minimum size required, per 110.14(C)(1)(b).

Supervision of the disconnecting means by a local (protected premises) fire alarm system, central station, proprietary supervising station, or remote supervising station requires a connection to the premises fire alarm system. The connection is generally made through the use of dry contacts in the fire pump controller that are connected to a fire alarm system initiating device circuit. This circuit is programmed to generate a supervisory signal at the fire alarm control unit on loss of voltage to the fire pump controller. A supervisory signal indicates that the suppression system is "off-normal." Connection of this device must initiate a supervisory signal and may also initiate a trouble signal. For more information on this interface with the fire alarm system, see *NFPA 72®, National Fire Alarm and Signaling Code*.

695.5 Transformers.

Where the service or system voltage is different from the utilization voltage of the fire pump motor, transformer(s) protected by disconnecting means and overcurrent protective devices shall be permitted to be installed between the system supply and the fire pump controller in accordance with 695.5(A) and (B), or with (C). Only transformers covered in 695.5(C) shall be permitted to supply loads not directly associated with the fire pump system.

(A) Size. Where a transformer supplies an electric motor driven fire pump, it shall be rated at a minimum of 125 percent of the sum of the fire pump motor(s) and pressure maintenance pump(s) motor loads, and 100 percent of the associated fire pump accessory equipment supplied by the transformer.

(B) Overcurrent Protection. The primary overcurrent protective device(s) shall be selected or set to carry indefinitely the sum of the locked-rotor current of the fire pump motor(s) and the pressure maintenance pump motor(s) and the full-load current of the associated fire pump accessory equipment when connected to this power supply. Secondary overcurrent protection shall not be permitted. The requirement to carry the locked-rotor currents indefinitely shall not apply to conductors or devices other than overcurrent devices in the fire pump motor circuit(s).

Dedicated transformer and overcurrent protection sizing can be broken down into three basic requirements. Generally stated, they are as follows:

1. The transformer must be sized to at least 125 percent of the sum of the loads.
2. The transformer primary overcurrent device must be at least a specified minimum size.
3. The transformer secondary must not contain any overcurrent devices whatsoever.

See Exhibit 695.5 for a simple one-line diagram on applying the dedicated fire pump transformer overcurrent protection requirements.

Calculation Example

A 4160/480-volt, 3-phase, dedicated transformer supplies power to a 100-hp, 460-volt, 3-phase, code letter G fire pump and to a 1½-hp, 460-volt, 3-phase, code letter H jockey pump. Using the requirements of Article 695, determine the sizes of the dedicated transformer and its primary overcurrent protection.

Solution

STEP 1. Determine the minimum standard-size transformer. According to 695.5(A), to determine the minimum current value for use in the 3-phase power calculation, add the fullload currents of the fire pump motor(s) and jockey pump motor(s), then increase the total to 125 percent. The result of this calculation provides the minimum kVA rating of a transformer dedicated to a fire pump installation. Any transformer with a kVA rating equal to or greater than the value determined by this calculation is acceptable. First, add the full-load currents of the two motors, using the full-load current (FLC) values from Table 430.250, as follows:

$$100\text{-hp, 3-phase FLC} = 124 \text{ A}$$
$$1\tfrac{1}{2}\text{-hp, 3-phase FLC} = \underline{3 \text{ A}}$$
$$\text{Total FLC} = 127 \text{ A}$$

Now, increase the sum of the fire pump motor and the jockey pump motor to 125 percent, as follows:

$$127 \text{ A} \times 1.25 = 158.75 \text{ A}$$

Then, size the transformer as follows:

$$\text{Transformer kVA} = \frac{\text{volts} \times \text{amperes} \times \sqrt{3}}{1000}$$
$$= \frac{480 \times 158.75 \times \sqrt{3}}{1000}$$
$$= 131.98 \text{ kVA}$$

Thus, the minimum-size transformer permitted by 695.5(A) is 131.98 kVA. The next larger standard-size transformer available is 150 kVA, but any larger size is permitted.

STEP 2. Calculate the minimum-size primary overcurrent protection device permitted for this transformer. According to 695.5(B), the minimum primary overcurrent protection device must allow the transformer secondary to supply the locked-rotor current (LRC) to the fire pump and, in this case, the jockey pump. The locked-rotor current of each motor must be individually calculated if it is not available on the motor nameplate. In this example, however, we are assuming that only the kVA code letters are available. According to 430.7(B) and using the maximum values for the individual code letters per Table 430.7(B), calculate the maximum locked-rotor currents, as follows.

For the 100-hp motor, code letter G:

$$\text{LRC} = \text{motor hp} \times \text{max. code letter value}$$
$$\times \frac{1000}{\text{motor voltage} \times \text{3-phase factor}}$$
$$= 100 \text{ hp} \times \frac{6.29 \text{ kVA}}{\text{hp}} \times \frac{1000}{460 \times \sqrt{3}} = 789.49 \text{ A}$$

For the 1½-hp motor, code letter H (using the same formula):

$$\text{LRC} = 1\tfrac{1}{2} \text{ hp} \times 7.09 \frac{\text{kVA}}{\text{hp}} \times \frac{1000}{460 \times \sqrt{3}} = 13.35 \text{ A}$$

For the total LRC:

100-hp LRC = 789.49 A
1½-hp LRC = 13.35 A
Total LRC = 802.84 A or 803 A

Now, calculate the equivalent LRC on the primary side of the transformer, based on the calculated LRC of the secondary of the transformer, as follows:

$$\text{LRC}_{\text{primary}} = \frac{\text{secondary voltage}}{\text{primary voltage}} \times \text{LRC}_{\text{secondary}}$$
$$= \frac{480 \text{ V}}{4160 \text{ V}} \times 803 \text{ A}$$
$$= 92.65 \text{ A or } 93 \text{ A}$$

This value of 93 amperes represents the secondary LRC reflected to the primary side of the transformer. Because this value is the absolute smallest overcurrent protective device permitted, the next larger standard size, according to 240.6, is 100 amperes. Thus, the minimum standard-size overcurrent protective device is 100 amperes.

Conclusion: The calculation for a 4160/480-volt, 3-phase transformer supplying a 100-hp fire pump and a 1½-hp jockey pump, both at 460 volts, 3 phase, can be summarized as follows:

1. The smallest standard-size transformer that is permitted is 150 kVA.
2. The smallest standard-size overcurrent protective device permitted on the primary of the transformer is 100 amperes.
3. A secondary overcurrent protective device is not permitted.

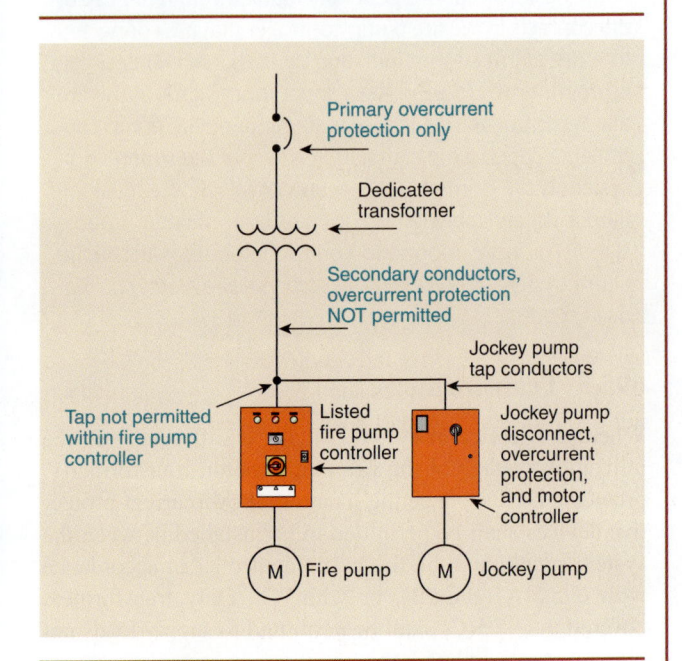

EXHIBIT 695.5 *Overcurrent protection for a transformer supplying a fire pump and associated equipment. The device is permitted only in the primary circuit and must be capable of carrying locked-rotor currents of fire pump motor and jockey pump motor indefinitely.*

(C) Feeder Source. Where a feeder source is provided in accordance with 695.3(C), transformers supplying the fire pump system shall be permitted to supply other loads. All other loads shall be calculated in accordance with Article 220, including demand factors as applicable.

(1) Size. Transformers shall be rated at a minimum of 125 percent of the sum of the fire pump motor(s) and pressure maintenance pump(s) motor loads, and 100 percent of the remaining load supplied by the transformer.

(2) Overcurrent Protection. The transformer size, the feeder size, and the overcurrent protective device(s) shall be coordinated such that overcurrent protection is provided for the transformer in accordance with 450.3 and for the feeder in accordance with 215.3, and such that the overcurrent protective device(s) is selected or set to carry indefinitely the sum of the locked-rotor current of the fire pump motor(s), the pressure maintenance pump motor(s), the full-load current of the associated fire pump accessory equipment, and 100 percent of the remaining loads supplied by the transformer. The requirement to carry the locked-rotor currents indefinitely shall not apply to conductors or devices other than overcurrent devices in the fire pump motor circuit(s).

695.6 Power Wiring.

Power circuits and wiring methods shall comply with the requirements in 695.6(A) through (J), and as permitted in 230.90(A), Exception No. 4; 230.94, Exception No. 4; 240.13; 230.208; 240.4(A); and 430.31.

(A) Supply Conductors.

(1) Services and On-Site Power Production Facilities. Service conductors and conductors supplied by on-site power production facilities shall be physically routed outside a building(s) and shall be installed as service-entrance conductors in accordance with 230.6, 230.9, and Parts III and IV of Article 230. Where supply conductors cannot be physically routed outside of buildings, the conductors shall be permitted to be routed through the building(s) where installed in accordance with 230.6(1) or (2).

(2) Feeders. Fire pump supply conductors on the load side of the final disconnecting means and overcurrent device(s) permitted by 695.4(B), or conductors that connect directly to an on-site standby generator, shall comply with all of the following:

(a) *Independent Routing.* The conductors shall be kept entirely independent of all other wiring.
(b) *Associated Fire Pump Loads.* The conductors shall supply only loads that are directly associated with the fire pump system.
(c) *Protection from Potential Damage.* The conductors shall be protected from potential damage by fire, structural failure, or operational accident.
(d) *Inside of a Building.* Where routed through a building, the conductors shall be installed using one of the following methods:
  (1) Be encased in a minimum 50 mm (2 in.) of concrete
  (2) Be protected by a fire-rated assembly listed to achieve a minimum fire rating of 2 hours and dedicated to the fire pump circuit(s)
  (3) Be a listed electrical circuit protective system with a minimum 2-hour fire rating

Informational Note: UL guide information for electrical circuit protective systems (FHIT) contains information on proper installation requirements to maintain the fire rating.

Exception to (A)(2)(d): The supply conductors located in the electrical equipment room where they originate and in the fire pump room shall not be required to have the minimum 2-hour fire separation or fire resistance rating, unless otherwise required by 700.10(D) of this Code.

The feeder conductors covered by 695.6(A)(2) are those that are installed from the load side of the supervised disconnecting means permitted by 695.4(B)(1)(a), (b), or (c) to the fire pump controller or combination controller/transfer switch. In addition, conductors directly connected to the output of an on-site standby generator are also subject to the installation requirements of 695.6(A)(2). Unlike service conductors, feeder conductors are installed on the load side of a short-circuit protective device and are not required to be installed on the outside of a building or structure. However, if the feeder conductors are run through a building from the disconnecting means to the fire pump equipment room, they are required to be protected against the effects of a fire to ensure that power to the fire pump is not compromised.

Encasement in 2 in. of concrete, as recognized in 230.6 for service conductors, is one method permitted to isolate and protect feeder conductors. In addition, enclosure of a wiring method within 2-hour fire-resistive building construction and the use of a wiring method listed as an electrical circuit protective system with a minimum 2-hour fire rating are methods permitted for these circuit conductors that do have short-circuit protection. These latter two wiring methods are recognized for protecting conductors against the effects of fire, but they are not recognized as means by which conductors can be considered as being outside a building or other structure. Therefore, if the installation involves service conductors that have to be run through a building, the only available options, short of installing a supervised disconnecting means, are to install the conductors beneath not less than 2 in. of concrete or to encase the conductors in 2 in. of concrete.

The difference between a 2-hour fire rating of an electrical circuit, such as a conduit with wires, and a 2-hour fire-resistance rating of a structural member, such as a wall, is that at the end of a 2-hour fire test on an electrical conduit with wires, the circuit must function electrically (no short circuits, grounds, or opens are permitted). The circuit and its insulation must be intact and electrically functioning. A wall subjected to a 2-hour fire-resistance test must only prevent a fire from passing through or past the wall, without regard to damage to the wall. All fire ratings and fire-resistance ratings are based on the assumption that the structural supports for the assembly are not impaired by the effects of the fire.

The UL 2004 *Fire Resistance Directory*, Volume 2, describes three categories of products that can be used in the fire protection of electrical circuits for fire pumps: various electrical circuit protective systems (FHIT), electrical circuit protective materials (FHIY), and fire-resistive cables (FHJR). (The four-letter codes in parentheses are the UL product category guide designations.) For information on electrical circuit protective systems, see UL Subject 1724, *Fire Tests for Electrical Circuit Protective Systems*.

(B) Conductor Size.

(1) Fire Pump Motors and Other Equipment. Conductors supplying a fire pump motor(s), pressure maintenance pumps, and associated fire pump accessory equipment shall have a rating not less than 125 percent of the sum of the fire pump motor(s) and pressure maintenance motor(s) full-load current(s), and 100 percent of the associated fire pump accessory equipment.

(2) Fire Pump Motors Only. Conductors supplying only a fire pump motor shall have a minimum ampacity in accordance with 430.22 and shall comply with the voltage drop requirements in 695.7.

Listed fire pump controller and pump combinations are available in a wye-start, delta-run configuration. From the controller to the motor are six circuit conductors, and when the motor is in the run mode, the conductors that supply each winding are connected in parallel. The minimum conductor ampacity for each of the six leads between the controller and the motor is calculated as shown in the following example.

Calculation Example
Given a fire pump with a 50-hp, 3-phase motor, 460-volt motor, determine the minimum size for the conductors between the controller and the pump. The pump motor and controller are configured for a wye-start, delta-run operation.

- Fire pump motor nameplate information: 50 hp, 3 phase, 460 volt
- Table 430.250 full-load current for 50-hp motor: 65 amperes
- Section 430.22(A) specifies motor-circuit conductor minimum ampacity based on 125 percent of motor full-load current
- Section 430.22(C) specifies full-load current for conductors on the load side of controller based on 58 percent of motor full-load current

Solution
STEP 1. Determine minimum conductor ampacity.

$$65 \text{ amperes} \times 0.58 \times 1.25 = 47 \text{ amperes}$$

STEP 2. Determine minimum conductor size.

Assuming 75°C terminations in the controller and conductor insulation type THWN, ampacity selection from Table 310.15(B)16 equals 8 AWG copper circuit conductors (50 amperes)

- There will be six 8 AWG copper conductors between the controller and the motor.
- The combined ampacity of the two circuit conductors connected in parallel to each winding in the run mode is 100 amperes (combined ampacity of two 8 AWG conductors).
- Comparison of controller load-side circuit conductors with minimum ampacity for circuit conductors from the power source to the line side of fire pump controller:

$$65 \text{ amperes} \times 1.25 = 81 \text{ amperes}$$

4 AWG THWN copper circuit conductors

- The minimum size for the conductors as determined here may have to be increased to comply with the mandatory voltage drop performance requirements specified in 695.7.

(C) Overload Protection. Power circuits shall not have automatic protection against overloads. Except for protection of transformer primaries provided in 695.5(C)(2), branch-circuit and feeder conductors shall be protected against short circuit only. Where a tap is made to supply a fire pump, the wiring shall be treated as service conductors in accordance with 230.6. The applicable distance and size restrictions in 240.21 shall not apply.

Exception No. 1: Conductors between storage batteries and the engine shall not require overcurrent protection or disconnecting means.

Exception No. 2: For an on-site standby generator(s) rated to produce continuous current in excess of 225 percent of the full-load amperes of the fire pump motor, the conductors between the on-site generator(s) and the combination fire pump transfer switch controller or separately mounted transfer switch shall be installed in accordance with A(695.6)(2). The protection provided shall be in accordance with the short-circuit current rating of the combination fire pump transfer switch controller or separately mounted transfer switch.

(D) Pump Wiring. All wiring from the controllers to the pump motors shall be in rigid metal conduit, intermediate metal conduit, electrical metallic tubing, liquidtight flexible metal conduit, or liquidtight flexible nonmetallic conduit Type LFNC-B, listed Type MC cable with an impervious covering, or Type MI cable.

> Section 695.6(D) does not apply to light switches, convenience receptacles, telephone outlets, fire detectors, and similar equipment located in the fire pump room.

(E) Loads Supplied by Controllers and Transfer Switches. A fire pump controller and fire pump power transfer switch, if provided, shall not serve any load other than the fire pump for which it is intended.

(F) Mechanical Protection. All wiring from engine controllers and batteries shall be protected against physical damage and shall be installed in accordance with the controller and engine manufacturer's instructions.

(G) Ground-Fault Protection of Equipment. Ground-fault protection of equipment shall not be permitted for fire pumps.

> Ground-fault protection of equipment as specified in several sections of the *NEC* is not permitted to be used to protect components of a fire pump installation. (The function of ground-fault protection of equipment protection should not be confused with the function of GFCI protection for personnel.) Although the protection afforded to equipment by this type of ground-fault protection is a mandatory safety requirement in certain circumstances, the need for an uninterrupted source of power takes precedence for fire pump installations. See 240.13(3).

(H) Listed Electrical Circuit Protective System to Controller Wiring. Electrical circuit protective system installation shall comply with any restrictions provided in the listing of the electrical circuit protective system used and the following also shall apply:

(1) A junction box shall be installed ahead of the fire pump controller a minimum of 300 mm (12 in.) beyond the fire-rated wall or floor bounding the fire zone.

> The required junction box allows for a transition between solid conductors that are used in some electrical circuit protective systems (Type MI cable, for example) and stranded conductors that are required at the supply terminals of the controller by the controller manufacturer and its listing without having to make the splice in the controller enclosure. In addition, where an electrical circuit protective system employs single conductor cables, such as Type MI cable, the necessity to modify enclosures to prevent inductive heating can result in a compromise of the controller enclosure's resistance to water infiltration.

(2) Where required by the manufacturer of a listed electrical circuit protective system or by the listing, or as required elsewhere in this *Code*, the raceway between a junction box and the fire pump controller shall be sealed at the junction box end as required and in accordance with the instructions of the manufacturer. [**20:**9.8.2]

> Sealing of the raceway between the junction box and the enclosure may be required by manufacturers of some electrical circuit protective systems. Sealing prevents any conductive material or gases that emanate in the electrical circuit protective system from entering the controller enclosure and compromising the controller operation.

(3) Standard wiring between the junction box and the controller shall be permitted. [**20:**9.8.3]

(I) Junction Boxes. Where fire pump wiring to or from a fire pump controller is routed through a junction box, the following requirements shall be met:

(1) The junction box shall be securely mounted. [**20:**9.7(1)]
(2) Mounting and installing of a junction box shall not violate the enclosure type rating of the fire pump controller(s). [**20:**9.7(2)]
(3) Mounting and installing of a junction box shall not violate the integrity of the fire pump controller(s) and shall not affect the short-circuit rating of the controller(s). [**20:**9.7(3)]
(4) As a minimum, a Type 2, drip-proof enclosure (junction box) shall be used where installed in the fire pump room. The enclosure shall be listed to match the fire pump controller enclosure type rating. [**20:**9.7(4)]
(5) Terminals, junction blocks, wire connectors, and splices, where used, shall be listed. [**20:**9.7(5)]
(6) A fire pump controller or fire pump power transfer switch, where provided, shall not be used as a junction box to supply other equipment, including a pressure maintenance (jockey) pump(s).

(J) Raceway Terminations. Where raceways are terminated at a fire pump controller, the following requirements shall be met: [**20:**9.9]

(1) Listed conduit hubs shall be used. [**20:**9.9.1]
(2) The type rating of the conduit hub(s) shall be at least equal to that of the fire pump controller. [**20:**9.9.2]
(3) The installation instructions of the manufacturer of the fire pump controller shall be followed. [**20:**9.9.3]

(4) Alterations to the fire pump controller, other than conduit entry as allowed elsewhere in this *Code*, shall be approved by the authority having jurisdiction. [**20**:9.9.4]

> The requirements of this section are to ensure that the controller enclosure's environmental rating is not compromised by raceway entries into the enclosure. Use of conduit hubs having the same environmental rating as the controller enclosure will minimize the entry of water or other liquids into the enclosure.

695.7 Voltage Drop.

(A) Starting. The voltage at the fire pump controller line terminals shall not drop more than 15 percent below normal (controller-rated voltage) under motor starting conditions.

Exception: This limitation shall not apply for emergency run mechanical starting. [**20**:9.4.2]

(B) Running. The voltage at the motor terminals shall not drop more than 5 percent below the voltage rating of the motor when the motor is operating at 115 percent of the full-load current rating of the motor.

695.10 Listed Equipment.

Diesel engine fire pump controllers, electric fire pump controllers, electric motors, fire pump power transfer switches, foam pump controllers, and limited service controllers shall be listed for fire pump service. [**20**:9.5.1.1, 10.1.2.1, 12.1.3.1]

> This requirement parallels those provisions in NFPA 20, *Standard for the Installation of Stationary Pumps for Fire Protection*, that require components used in fire pump systems to be listed. Prior to being shipped to the installation site, listed fire pump controllers are matched with the listed electric motor(s) they will control to ensure compatibility of the individually listed components. The fire pump controller and transfer switch shown in Exhibit 695.6 is an example of the listed equipment covered in 695.10.

695.12 Equipment Location.

(A) Controllers and Transfer Switches. Electric motor-driven fire pump controllers and power transfer switches shall be located as close as practicable to, and within sight of, the motors that they control.

(B) Engine-Drive Controllers. Engine-drive fire pump controllers shall be located as close as is practical to, and within sight of, the engines that they control.

EXHIBIT 695.6 *Listed Fire Pump Controller and Power Transfer Switch.*

(C) Storage Batteries. Storage batteries for fire pump engine drives shall be supported above the floor, secured against displacement, and located where they are not subject to physical damage, flooding with water, excessive temperature, or excessive vibration.

(D) Energized Equipment. All energized equipment parts shall be located at least 300 mm (12 in.) above the floor level.

(E) Protection Against Pump Water. Fire pump controller and power transfer switches shall be located or protected so that they are not damaged by water escaping from pumps or pump connections.

(F) Mounting. All fire pump control equipment shall be mounted in a substantial manner on noncombustible supporting structures.

> The *Code* does not mandate a dedicated room for the fire pump. However, NFPA 20 does specify a suitable space for this equipment that is free from hazards that may impair the operation of the fire pump. Compliance with the 695.12(A) requirement that fire pump controllers and transfer switches be located "as close as practicable" to the fire pump motor that they control may require that additional space be available to achieve the minimum maintenance working space set forth in 110.26.
>
> Generally, fire pump controllers are housed in substantial enclosures suitable to protect the contents against limited amounts of falling water and dirt. In addition, all energized parts in the enclosure must be mounted at least 12 in. above the floor. Typically, the floor space for this area is equipped with a floor drain.

> The requirement of 695.12(F) does not permit fire pump control equipment to be mounted on combustible backboards (such as plywood).

695.14 Control Wiring.

(A) Control Circuit Failures. External control circuits that extend outside the fire pump room shall be arranged so that failure of any external circuit (open or short circuit) shall not prevent the operation of a pump(s) from all other internal or external means. Breakage, disconnecting, shorting of the wires, or loss of power to these circuits could cause continuous running of the fire pump but shall not prevent the controller(s) from starting the fire pump(s) due to causes other than these external control circuits. All control conductors within the fire pump room that are not fault tolerant shall be protected against physical damage. [**20:**10.5.2.6, 12.5.2.5]

(B) Sensor Functioning. No undervoltage, phase-loss, frequency-sensitive, or other sensor(s) shall be installed that automatically or manually prohibits actuation of the motor contactor. [**20:**10.4.5.6]

Exception: A phase loss sensor(s) shall be permitted only as a part of a listed fire pump controller.

(C) Remote Device(s). No remote device(s) shall be installed that will prevent automatic operation of the transfer switch. [**20:**10.8.1.3]

(D) Engine-Drive Control Wiring. All wiring between the controller and the diesel engine shall be stranded and sized to continuously carry the charging or control currents as required by the controller manufacturer. Such wiring shall be protected against physical damage. Controller manufacturer's specifications for distance and wire size shall be followed. [**20:**12.3.5.1]

(E) Electric Fire Pump Control Wiring Methods. All electric motor–driven fire pump control wiring shall be in rigid metal conduit, intermediate metal conduit, liquidtight flexible metal conduit, liquidtight flexible nonmetallic conduit Type B (LFNC-B), listed Type MC cable with an impervious covering, or Type MI cable.

> The wiring methods described in 695.14(E) apply only to the control wiring for electric motor–driven fire pumps. They do not apply to the control wiring for engine-driven fire pumps, because there are no similar requirements in NFPA 20, *Standard for the Installation of Stationary Pumps for Fire Protection*.

(F) Generator Control Wiring Methods. Control conductors installed between the fire pump power transfer switch and the standby generator supplying the fire pump during normal power loss shall be kept entirely independent of all other wiring. They shall be protected to resist potential damage by fire or structural failure. They shall be permitted to be routed through a building(s) using one of the following methods:

(1) Be encased in a minimum of 50 mm (2 in.) of concrete.
(2) Be protected by a fire-rated assembly listed to achieve a minimum fire rating of 2 hours and dedicated to the fire pump circuits.
(3) Be a listed electrical circuit protective system with a minimum 2-hour fire rating. The installation shall comply with any restrictions provided in the listing of the electrical circuit protective system used.

Informational Note: UL guide information for electrical circuit protective systems (FHIT) contains information on proper installation requirements to maintain the fire rating.

> To ensure operation of an on-site standby generator used to supply an electric fire pump motor, 695.14(F) requires that the conductors be installed using one of the wiring methods that are required for the fire pump power circuit in 695.6(A) (2). Having the power wiring protected against the effects of fire is only one piece of the power reliability consideration. In order for the generator to provide power, it has to receive the necessary signal to start, hence the requirement to provide protection against fire and other physical damage for the control circuit wiring between the fire pump transfer switch/controller and the on-site standby generator.

SUPPLEMENT 4

Technical/Substantive Changes from the 2010 to 2013 Edition of NFPA 20

2013 Edition Text	Reason for Change
Chapter 3 Definitions	
3.3.25 In Sight From (Within Sight From, Within Sight). Where this Code specifies that one equipment shall be "in sight from," "within sight from," or "within sight of," and so forth, another equipment, the specified equipment is to be visible and not more than 15 m (50 ft) distant from the other. [**70:** Art. 100]	The standard refers to "in sight from" within the electrical sections, so it is important to add this definition to maintain consistency with *NFPA 70*.
3.3.38.18* *Water Mist Positive Displacement Pumping Unit.* Multiple positive displacement pumps designed to operate in parallel that discharges into a single common water mist distribution system. **A.3.3.38.18 Water Mist Positive Displacement Pumping Unit.** It is not the intent of this standard to apply this term to individual pumps used to supply water mist systems. This term is intended to apply to water mist systems designed with multiple pumps where a pump operates individually or multiple pumps operate in parallel based on the demand of the system downstream and the number of nozzles that discharge. These pumps work together as a single unit to provide the necessary flow and pressure of the water mist system.	This new technology was brought before the NFPA 20 technical committee. A task group was structured and spent many hours determining the definition and criteria for this new unit as it pertains to water mist systems.
3.3.41 Record Drawing. A design, working drawing, or as-built drawing that is submitted as the final record of documentation for the project.	The term *record drawing* has been added to the standard, as it is important that records be maintained and kept for the life of the fire pump.
4.5.1 Certified shop test curves showing head capacity and brake horsepower of the pump shall be furnished by the manufacturer to the purchaser. **4.5.1.1** For water mist positive displacement pumping units, certified shop test data, including flow, pressure, and horsepower, shall be provided for each independent pump. **4.5.1.2** For water mist positive displacement pumping units, certified shop test data, including flow, pressure, and horsepower, shall also be provided for the fire pump unit with variable speed features deactivated. **4.5.1.2.1** The certified fire pump unit shop test data shall be developed by activating the individual fire pumps in the same operating sequence that the controller will utilize.	Subsection 4.5.1 has been modified to include specific requirements related to water mist positive displacement pumping units. These requirements were necessary, because the term *water mist positive displacement pumping unit* is new to the 2013 edition.

2013 Edition Text	Reason for Change
4.5.1.3 For water mist positive displacement pumping units with variable speed features, certified shop test data, including flow, pressure, and horsepower, shall also be provided for the fire pump unit with variable speed features activated. **4.5.1.3.1** The certified fire pump unit shop test data shall be developed by activating the individual fire pumps in the same operating sequence that the controller will utilize.	
4.6.1.2 Where a water flow test is used to determine the adequacy of the attached water supply, the test shall have been completed not more than 12 months prior to the submission of working plans, unless otherwise permitted by the authority having jurisdiction.	This requirement has been updated to correlate with the flow test requirements of NFPA 13 and NFPA 14.
4.6.5.2 Where the water supply cannot provide a flow of 150 percent of the rated flow of the pump but the water supply can provide the greater of 100 percent of the rated flow or the flow demand of the fire protection system(s), the head available from the water supply shall be permitted to be calculated on the basis of the maximum flow available as allowed by 4.6.2.3.1.	This section was revised to make it consistent with requirements of section 4.6.2.3.1, and to clarify that at a minimum the water supply must be capable of providing a flow of 100 percent of the pump rating or the fire protection demand, whichever is greater.
4.7.5 Each driver or water mist positive displacement pumping unit shall have its own dedicated controller.	The allowance for a single controller to operate multiple pumps within the unit was added, because water mist positive displacement pumping units differ from the standard pump configuration.
4.9.2 The name plate shall be made of and attached using corrosion resistant material.	This requirement was added to the 2013 edition because during many inspections, fire pump nameplates were found to be corroding to the point of being unreadable.
4.12.1.1.2* Indoor fire pump rooms in non-high-rise buildings or in separate fire pump buildings shall be physically separated or protected by fire-rated construction in accordance with Table 4.12.1.1.2. **A.4.12.1.1.2** The purpose for the "Not Sprinklered" column in Table 4.12.1.1.2 is to provide guidance for unsprinklered buildings. This does not permit sprinklers to be omitted from pump rooms in fully sprinklered buildings.	New annex language was added to this section to clarify the purpose of the "Not Sprinklered" column in Table 4.12.1.1.2, which is often misinterpreted to mean that sprinklers can be omitted from fully sprinklered buildings.
4.12.2.1 The location of and access to the fire pump room(s) shall be pre-planned with the fire department. **4.12.2.1.1** Fire pump rooms not directly accessible from the outside shall be accessible through an enclosed passageway from an enclosed stairway or exterior exit. **4.12.2.1.2** The enclosed passageway shall have a fire-resistance rating not less than the fire-resistance rating of the fire pump room.	This section was updated to verify that both the location and the access need to be pre-planned with the fire department. The 2-hour fire resistance-rating requirement for the passageway was removed from 4.12.2.1.1, and the new requirement of 4.12.2.1.2 was added to clarify that if a 1-hour fire pump room were required, only a 1-hour passageway would be needed.
4.14.5.2 No control valve other than a listed OS&Y valve and the devices as permitted in 4.27.3 shall be installed in the suction pipe within 50 ft (15.3 m) of the pump suction flange.	The reference to 4.27.3 was added to provide the user guidance on what other requirements need to be reviewed for devices in the suction piping.
4.14.6.3.3 Elbows and tees with a centerline plane perpendicular to the horizontal split-case pump shaft shall be permitted at any location in the pump suction intake.	The term *tees* was added to this requirement, clarifying that it applies to both elbows and tees.

2013 Edition Text	Reason for Change
4.14.9.2... (1) Check valves and backflow prevention devices and assemblies shall be permitted where required by other NFPA standards or the authority having jurisdiction and installed in accordance with Section 4.27.	A reference to Section 4.27 was added, since 4.27 provides criteria that must be met if a check valve or backflow is installed in the fire pump suction pipe.
4.15.7* A listed indicating gate or butterfly valve shall be installed on the fire protection system side of the pump discharge check valve. **A.4.15.7** See A.4.15.6 for circumstances where a backflow preventer can be substituted for the discharge control valve.	Annex language was added to provide guidance regarding this requirement when a backflow is installed. The control valve on the backflow meets this requirement, and an additional valve is not required.
4.15.9.4 The friction loss through a low suction throttling valve in the fully open position shall be taken into account in the design of the fire protection system. **4.15.9.5** System design shall be such that the low suction throttling valve is in the fully open position at system design point.	This text was moved from the annex to the body of the standard, making it a requirement to account for the friction loss through a low suction throttle valve when performing the design of a system.
4.17* Protection of Piping Against Damage Due to Movement. A clearance shall be provided around pipes that pass through walls, ceilings, or floors of the fire pump room enclosure. **4.17.1** Unless the requirements of 4.17.2 through 4.17.4 are met, where pipe passes through walls, ceilings, or floors of the fire pump room enclosure, the holes shall be sized such that the diameter of the hole is nominally 2 in. (50 mm) larger than the pipe. **4.17.2** Where clearance is provided by a pipe sleeve, a nominal diameter 2 in. (50 mm) larger than the nominal diameter of the pipe shall be acceptable. **4.17.3** No clearance is required if flexible couplings are located within 1 ft (305 mm) of each side of the wall, ceiling, or floor. **4.17.4** Where protection of piping against damage caused by earthquakes is required, the provisions of Section 4.28 shall apply. **4.17.5** Where required, the clearance shall be filled with flexible material that is compatible with the piping materials and maintains any required fire resistance rating of the enclosure.	Language was added to clarify that only those pipes passing though the fire pump room enclosure require a clearance, as also revised for consistency with NFPA 13, and to allow for the use of a flexible coupling on each side of the penetrated assembly as a means of protection. This section indicates that additional clearance criteria is required where the installation requires protection against seismic forces.
4.18.7.2 Where pump discharge water is piped back to pump suction and the pump is driven by a diesel engine with heat exchanger cooling, a high cooling water temperature signal at 104°F (40°C) from the engine inlet of the heat exchanger water supply shall be sent to the fire pump controller, and the controller shall stop the engine provided that there are no active emergency requirements for the pump to run. **4.18.7.2.1** The requirements of 4.18.7.2 shall not apply when pump discharge water is being piped back to a water storage reservoir.	This requirement was added due to the growing problem of pump discharge being piped back to suction, which overheats the engine intake air temperature and causes the engine to go above EPA engine emission requirements. It is important that the cooling water taken from pump suction be kept below 104°F (40°C).
4.20.1.4* Where a test header is installed, it shall be installed on an exterior wall or in another location outside the pump room that allows for water discharge during testing. **A.4.20.1.4** The hose valves of the fire pump test header should be located on the building exterior. This is because the test discharge needs to be directed to a safe outdoor location, and to protect the fire pumps, controllers, and so forth, from accidental water spray. In instances where damage from theft or vandalism is a concern, the test header hose valves can be located within the building but outside of the fire pump room if, in the judgment of the AHJ, the test flow can be safely directed outside the building without undue risk of water spray to the fire pump equipment.	Annex language was added to provide guidance for installations where freezing or theft could be a prevalent issue.

2013 Edition Text	Reason for Change
4.20.2.10* Where a metering device is installed in a loop arrangement for fire pump flow testing, an alternate means of measuring flow shall be provided. **A.4.20.2.10** The testing arrangement should be designed to minimize the length of fire hose needed to discharge water safely [approximately 100 ft (30 m)]. Where a flow test meter is installed, an alternate means of testing, such as hydrants, hose valves, test header(s), and so forth is needed as an alternate means of testing the performance of the fire pump, and to verify the accuracy of the metering device. **4.20.2.10.1** The alternate means of measuring flow shall be located downstream of and in series with the flow meter. **4.20.2.10.2** The alternate means of measuring flow shall function for the range of flows necessary to conduct a full flow test. **4.20.2.10.3** An appropriately sized test header shall be an acceptable alternate means of measuring flow.	This requirement was added to provide a provision for verifying the accuracy of a flow meter. This provision was missing from the 2010 edition.
4.20.3.3.1 A listed indicating butterfly or gate valve shall be located in the pipeline to the hose valve header. **4.20.3.3.2** A drain valve or automatic ball drip shall be located in the pipeline at a low point between the valve and the header. *[See Figure A.6.3.1(a) and Figure A.7.2.2.1.]*	Language was removed from this requirement; "where the hose valve is subject to freezing" was deleted, since a control valve is always required to be installed in the test header piping, not just when the header is subject to freezing. This new language also prevents the test header piping from being under pressure all the way to the test header.
4.25.6 Except as permitted in Chapter 8, the primary or standby fire pump shall not be used as a pressure maintenance pump.	This paragraph was revised to reflect the addition of "water mist positive displacement pumping units" to the 2013 edition.
4.28.1 General. **4.28.1.1** Where local codes require fire protection systems to be protected from damage subject to earthquakes, 4.28.2 and 4.28.3 shall apply. **4.28.1.2*** Horizontal seismic loads shall be based on NFPA 13, *Standard for the Installation of Sprinkler Systems,* SEI/ASCE 7, *Minimum Design Loads for Buildings and Other Structures,* local, state, or international codes, or other sources acceptable to the authority having jurisdiction. **A.4.28.1.2** NFPA 13, *Standard for the Installation of Sprinkler Systems,* contains specific requirements for seismic design of fire protection systems. It is a simplified approach that was developed to coincide with SEI/ASCE 7, *Minimum Design Loads for Buildings and Other Structures,* and current building codes. **4.28.2 Pump Driver and Controller.** The fire pump, driver, diesel fuel tank (where installed), and fire pump controller shall be attached to their foundations with materials capable of resisting lateral movement from horizontal loads. • **4.28.2.1** Pumps with high centers of gravity, such as vertical inline pumps, shall be mounted at their base and braced above their center of gravity.	The seismic requirements in this section have been revised to provide a more comprehensive set of requirements, and to correlate with the requirements of NFPA 13.
Chapter 5 Fire Pumps for High-Rise Buildings	
5.1.1.2... • **5.2* Equipment Access.** Location and access to the fire pump room shall be preplanned with the fire department.	Chapter 5 has been rearranged into a more logical sequence.

2013 Edition Text	Reason for Change
A.5.2 The location of a pump room in a high-rise building requires a great deal of consideration. Personnel are required to be sent to the pump room to monitor the operation of the pump during fire-fighting activities in the building. The best way to protect these people who are being sent to the pump room is to have the pump room directly accessible from the outside, but that is not always possible in high-rise buildings. In many cases, pump rooms in high-rise buildings will need to be located many floors above grade or at a location below grade, or both. For the circumstance where the pump room is not at grade level, this standard requires protected passageways of a fire resistance rating that meets the minimum requirements for exit stairwells at the level of the pump room from the exit stairwell to the pump room. Many codes do not allow the pump room to open directly onto an exit stairwell, but the distance between an exit stairwell and the pump room on upper or lower floors needs to be as short as possible with as few openings to other building areas as possible to afford as much protection as possible to the people going to the pump room and staying in the pump room during a fire in the building. In addition to the need for access, pump rooms need to be located such that discharge from pump equipment including packing discharge and relief valves is adequately handled. **5.3 Water Supply Tanks.** **5.3.1** Where provided, water tanks shall be installed in accordance with NFPA 22, *Standard for Water Tanks for Private Fire Protection*.	
5.6* Very Tall Buildings. **A.5.6** When trying to determine the pumping capability of the fire department, the concern is for the height of the building. There are some buildings that are so tall that it is impossible for the fire department apparatus to pump into the fire department connection at the street and overcome the elevation loss and friction loss in order to achieve 100 psi at the hose outlets up in the building. In these cases, the fire protection system in the building needs to have additional protection, including sufficient water supplies within the building to fight a fire and sufficient redundancy to be safe. Since fire departments all purchase different apparatus with different pumping capabilities, this standard has chosen to address this concern with performance-based criteria rather than a specified building height. Most urban fire departments have the capability of getting sufficient water at sufficient pressure up to the top of 200 ft (61 m) tall buildings. Some have the ability to get water as high as 350 ft (107 m). The design professional will need to check with the local fire department to determine its capability. **5.6.1 Water Supply Tanks for Very Tall Buildings.** **5.6.1.1** Where the primary supply source is a tank, two or more water tanks shall be provided. **5.6.1.1.1** A water tank shall be permitted to be divided into compartments such that the compartments function as individual tanks. **5.6.1.1.2** The total volume of all tanks or compartments shall be sufficient for the full fire protection demand. **5.6.1.1.3** Each individual tank or compartment shall be sized so that at least 50 percent of the fire protection demand is stored with any one compartment or tank out of service. **5.6.1.2** An automatic refill valve shall be provided for each tank or tank compartment. **5.6.1.3** A manual refill valve shall be provided for each tank or tank compartment.	This new section on very tall buildings has been added to Chapter 5, providing requirements for where the height of buildings is above the pumping capabilities of the fire department apparatus.

2013 Edition Text	Reason for Change
5.6.1.4 Each refill valve shall be sized and arranged to independently supply the system fire protection demand. **5.6.1.5** The automatic and manual fill valve combination for each tank or tank compartment shall have its own connection to one of the following: (1) A standpipe riser that is supplied with a backup fire pump (2) A reliable domestic riser sized to meet the requirements of 5.6.1.4 **5.6.1.5.1*** Each connection shall be made to a different riser. **A.5.6.1.5.1** The different connections should be arranged so that the tank refill rate required in 5.6.1.4 can be maintained even with the failure of any single valve, pipe, or pump. **5.6.2 Fire Pump Backup.** Fire pumps serving zones that are partially or wholly beyond the pumping capability of the fire department apparatus shall be provided with one of the following: (1) A fully independent and automatic backup fire pump unit(s) arranged so that all zones can be maintained in full service with any one pump out of service. (2) An auxiliary means that is capable of providing the full fire protection demand and that is acceptable to the authority having jurisdiction.	
Chapter 7 Vertical Shaft Turbine–Type Pumps	
7.5.1.6.4 For vertical turbine pumps using angle gear drives driven by a diesel engine, a torsional vibration damping type coupling shall be used and mounted on the engine side of the driver shaft. **7.5.1.6.4.1** The torsional vibration damping type coupling shall be permitted to be omitted when a mass elastic system torsional analysis is provided and accepted by the authority having jurisdiction.	These new paragraphs were added to address the harmonic vibrations created when a diesel engine drive is used as the driver for a vertical turbine fire pump.
Chapter 8 Positive Displacement Pumps	
8.4 Water Mist Positive Displacement Pumping Units. **8.4.1** Water mist positive displacement pumping units shall be dedicated to and listed as a unit for fire protection service. **8.4.2** Except as provided in 8.4.3 through 8.4.8, all the requirements of this standard shall apply.	This section is new to the 2013 edition. A task group was formed by the Technical Committee on Fire Pumps, and the task group spent a substantial amount of time researching this new technology and providing guidance to the TC, which assisted in the development of the new requirements.
8.4.3 Water mist positive displacement pumping units shall include pumps, driver(s), and controller as a complete operating unit. **8.4.4** The pump controller shall manage the performance of all pumps and drivers to provide continuous and smooth operation without intermittent pump cycling or discharge pressure varying by more than 10 percent during pump sequencing after rated pressure has been achieved. **8.4.5** Redundancy shall be built into the units such that failure of a line pressure sensor or primary control board will not prevent the system from functioning as intended. **8.4.6** Where provided with a variable speed control, failure of the variable speed control feature shall cause the controller to bypass and isolate the variable speed control system. **8.4.7** The unit controller shall be arranged so that each pump can be manually operated individually without opening the enclosure door. **8.4.8** The requirement in 10.3.4.3 shall apply to each individual motor and the entire unit.	In order to be classified as a *water mist positive displacement pumping unit*, the pumps, driver(s), and controller must all be listed together as a unit; components cannot be mixed and matched. These units have some unique features beyond that of a normal pump, driver, and controller arrangement (such as the allowance for one controller to operate multiple pumps).

2013 Edition Text	Reason for Change
8.5.7 Pressure Maintenance. **8.5.7.1** Except as permitted in 8.5.7.2, the primary or standby fire pump shall not be used as a pressure maintenance pump. **8.5.7.2** Water mist positive displacement pumping units that are designed and listed to alternate pressure maintenance duty between two or more pumps with variable speed pressure limiting control, and that provide a supervisory signal wherever pressure maintenance is required more than two times in one hour, shall be permitted to maintain system pressure. **8.5.7.3** When in the pressure maintenance mode, water mist positive displacement pumping units used for pressure maintenance shall not provide more than half of the nozzle flow of the smallest system nozzle when the standby pressure is applied at the smallest nozzle. **8.5.7.4** A single sensing line shall be permitted to be used for a water mist positive displacement pumping unit controller where the unit also serves for pressure maintenance on a water mist system.	Unlike a conventional fire pump design, which requires a means of maintaining system pressure without the use of the fire pump, a water mist positive displacement pumping unit has the ability to use the fire pumps as the pressure maintenance for a system.
Chapter 9 Electric Drive for Pumps	
9.1.8* Interruption. **A.9.1.8** Ground fault alarm provisions are not prohibited. **9.1.8.1** No ground fault interruption means shall be installed in any fire pump control or power circuit. **9.1.8.2** No arc fault interruption means shall be installed in any fire pump control or power circuit. **9.2* Normal Power.** **A.9.2** See Figure A.9.2 for typical power supply arrangements from source to the fire pump motor.	The requirements on interruption have been added to clarify that these types of protection schemes are not permitted in the fire pump design and installation. Figure A.9.2 has been revised to include a third option for the arrangement of the power supply from the source to the pump motor.
9.4.4 The voltage at the contactor(s) load terminals to which the motor is connected shall not drop more than 5 percent below the voltage rating of the motor when the motor is operating at 115 percent of the full-load current rating of the motor. **9.4.4.1** Wiring from the controller(s) to the pump motor shall be in rigid metal conduit, intermediate metal conduit, electrical metallic tubing, liquidtight flexible metal conduit, or liquid tight flexible nonmetallic conduit Type LFNC-B, listed Type MC cable with an impervious covering, or Type MI cable. **9.4.4.2** Electrical connections at motor terminal boxes shall be made with a listed means of connection. **9.4.4.3** Twist-on insulation-piercing type and soldered wire connectors shall not be permitted to be used for this purpose.	Additional requirements have been added to 9.4.4 in order to correlate with NFPA 79. In addition, the new requirements add an increased level of safety when performing the voltage test.
9.5.1.4* Motors Used with Variable Speed Controllers. **A.9.5.1.4** Variable fire pump motors must be of the inverter duty type for the installation to be reliable. Inverter duty motors have higher insulation voltage rating, suitable temperature rise rating, and protection from bearing damage. **9.5.1.4.1** Motors shall meet the applicable requirements of NEMA MG-1, *Motors and Generators,* Part 30 or 31. **9.5.1.4.2** Motors shall be listed, suitable, and marked for inverter duty. **9.5.1.4.3** Listing shall not be required if 9.5.1.2 applies.	This paragraph has been revised to clarify that motors meeting the requirements of 9.5.1.2 do not require a listing.

Stationary Fire Pumps Handbook 2013

2013 Edition Text	Reason for Change
9.7 Junction Boxes.... (6) A fire pump controller or fire pump power transfer switch, where provided, shall not be used as a junction box to supply other equipment, including a pressure maintenance (jockey) pump(s). *(See 10.3.4.5.1 and 10.3.4.6.)*	This section was revised to correlate with the requirements of *NFPA 70, National Electric Code*.
9.8.1* Where single conductors (individual conductors) are used, they shall be terminated in a separate junction box. **9.8.1.1** The junction box shall be installed ahead of the fire pump controller, a minimum of 12 in. (305 mm) beyond the fire-rated wall or floor bounding the fire zone. **9.8.1.2** Single conductors (individual conductors) shall not enter the fire pump enclosure separately. **9.8.2*** Where required by the manufacturer of a listed electrical circuit protective system, by *NFPA 70, National Electrical Code,* or by the listing, the raceway between a junction box and the fire pump controller shall be sealed at the junction box end as required and in accordance with the instructions of the manufacturer. *(See NFPA 70, National Electrical Code, Article 695.)*	This section was revised to correlate with the requirements of *NFPA 70, National Electric Code*.
9.9.5 Where the raceway (conduit) between the controller and motor is not capable of conducting ground fault current sufficient to trip the circuit breaker when a ground fault occurs, a separate equipment grounding conductor shall be installed between the controller and motor.	This paragraph has been added to the 2013 edition to address concerns about the use of conduit that is not capable of conducting ground fault current sufficient to trip the circuit breaker.
Chapter 10 Electric-Drive Controllers and Accessories	
10.3.7.4 The installation instructions of the manufacturer of the fire pump controller shall be followed.	This paragraph was added to clarify that controllers must be installed in accordance with manufacturer's instructions. It has been noted over the years that sometimes these instructions were not followed, and incorrect installations occurred.
10.4.1.4 The requirements of 10.4.1.1 and 10.4.1.2 shall not apply where the controller can withstand without damage a 10 kV impulse in accordance with ANSI/IEEE C62.41, *IEEE Recommended Practice for Surge Voltages in Low-Voltage AC Power Circuits,* or where the controller is listed to withstand surges and impulses in accordance with ANSI/UL 1449, *Standard for Surge Protective Devices*.	A reference to ANSI/UL 1449 has been added clarifying that controllers now meet that testing criteria. This has led to a more robustly built controller that is better able to handle surge currents.
10.4.3.3.1.1* The circuit breaker shall not trip when starting a motor from rest in the across-the-line (direct-on-line) mode, whether or not the controller is of the reduced inrush starting type. **A.10.4.3.3.1.1** The isolating switch is not allowed to trip either. See also 10.4.2.1.3. **10.4.3.3.1.2*** The circuit breaker shall not trip when power is interrupted from a running pump, or if the pump is restarted in less than 3 seconds after being shut down. If a control circuit preventing a re-start within 3 seconds is provided, this requirement shall not apply. **A.10.4.3.3.1.2** See also A.10.4.3.3.1.1.	This section was revised to address those situations when the manual operator is engaged and power is lost, and power is transferred from or to an emergency power source within or ahead of the fire pump controller. This section was revised to address those situations when magnetic flux decays sufficiently, and the back EMF combined with a phase angle difference with the inline voltage causes current transients and results in the tripping of the breaker.
10.5.2.1.7.2* A pressure recording device shall record the pressure in each fire pump controller pressure-sensing line at the input to the controller.	The listing requirements were clarified as pertaining to pressure recording devices.

2013 Edition Text	Reason for Change
A.10.5.2.1.7.2 The pressure recorder should be able to record a pressure at least 150 percent of the pump discharge pressure under no-flow conditions. In a high-rise building, this requirement can exceed 400 psi (27.6 bar). This pressure recorder should be readable without opening the fire pump controller enclosure. This requirement does not mandate a separate recording device for each controller. A single multichannel recording device can serve multiple sensors. If the pressure recording device is integrated into the pressure controller, the pressure sensing element should be used to record system pressure. **10.5.2.1.7.3** The pressure recorder shall be listed as part of the controller or shall be a separately listed unit installed to sense the pressure at the input of the controller.	
10.5.3.2.1.3 The mechanical latching shall be designed to be automatic or manual.	This requirement was revised to clarify that the mechanical latch may be automatic or manual. The previous edition stated that this latch could not be automatic.
10.5.3.2.4 The operating handle shall be marked or labeled as to function and operation.	This new requirement was added for operating handles on controllers to be marked or labeled as to their function and operation.
10.7.2 Requirements. The provisions of Sections 10.1 through 10.5 shall apply, unless specifically addressed in 10.8.2.1 through 10.8.2.3. • **10.7.2.1** In lieu of 10.1.2.5.1, each controller shall be marked "Limited Service Controller" and shall show plainly the name of the manufacturer, the identifying designation, the maximum operating pressure, the enclosure type designation, and the complete electrical rating.	Old paragraph 10.7.2.1 was removed, because limited service controllers use the same circuit breaker used in a full service controller. According to this revised requirement, markings are now required on limited service controllers, indicating maximum operating pressure and enclosure type designation. This correlates with the requirements for a full service controller.
10.8.2.2 Arrangement II.... (2) The transfer switch overcurrent protection for both the normal and alternate sources shall comply with 9.2.3.4 or 9.2.3.4.1.	This paragraph has been clarified to indicate that the overcurrent protection requirements that apply to normal power also apply to alternate power sources.
10.8.3.7.3 Where the fire pump controller is marked to indicate that the alternate source is provided by a second utility power source, the requirements of 10.8.3.7.1 and 10.8.3.7.2 shall not apply, and under voltage-sensing devices shall monitor all ungrounded conductors in lieu of a frequency-sensing device.	The addition of a marking on the controller is necessary in order to determine if the requirements of 10.8.3.7.3 apply.
10.8.3.10.1 The use of an "in-phase monitor" or an intentional delay via an open neutral position of the transfer switch to comply with the requirements of 10.8.3.10 shall be prohibited. **10.8.3.11* Overcurrent Protection.** The power transfer switch shall not have short circuit or overcurrent protection as part of the switching mechanism of the transfer switch. ments contained within the switching mechanism of the transfer switch. This is to prevent a switching mechanism from inhibiting transfer of power.	This new requirement was added to avoid a malfunction that may result in the fire pump not receiving power.
10.10.3.4 Automatic Shutdown. When the variable speed pressure limiting control is bypassed, automatic shutdown of the controller shall be as permitted by 10.5.4.2. **10.10.3.5** When the manual selection means required in 10.10.7.3 is used to initiate a switchover from variable speed to bypass mode, if the pump is running in the variable speed mode and none of the conditions in 10.10.3 that require the controller to initiate the bypass operation exist, the controller shall be arranged to provide a restart delay to allow the motor to be deenergized before it is re-energized in the bypass mode.	This new text was added to recognize industry practice allowing automatic shutdown of a VSPLC when it is bypassed. Also, 10.10.3.5 was added to reduce the risk of motor damage during switchover.

2013 Edition Text	Reason for Change
10.10.7.2 Except as provided in 10.10.7.2.1, the variable speed pressure sensing element connected in accordance with 10.5.2.1.7.6 shall only be used to control the variable speed drive. **10.10.7.2.1** Where redundant pressure sensing elements are provided as part of a water mist positive displacement pumping unit, they shall be permitted for other system functions.	These requirements were revised and added to address the unique nature of water mist positive displacement pumping units.
10.10.7.4 Except as provided in 10.10.7.4.2, common pressure control shall not be used for multiple pump installations. **10.10.7.4.1** Each controller pressure sensing control circuit shall operate independently. **10.10.7.4.2** A common pressure control shall be permitted to be used for a water mist positive displacement pumping unit controller.	These requirements were revised and added to address the unique nature of water mist positive displacement pumping units.
Chapter 11 Diesel Engine Drive	
11.2.4.2.7.1* In the standby mode, the engine batteries or battery chargers shall be used to power the ECM. **A.11.2.4.2.7.1** When engines are in standby and the battery chargers have the batteries in float, it is actually the chargers that are providing the current to support the engine, controller, and pump room as defined in 11.2.7.2.3.2. **11.2.4.2.7.2** Engines shall not require more than 0.5 ampere from the battery or battery charger while the engine is not running.	Annex language was added clarifying that the charger provides the current to support the engine, controller, and pump room when in standby mode.
11.2.4.3.4 Pressure Sensing Line. **11.2.4.3.4.1** A pressure sensing line shall be provided to the engine with a ½ in. (12.7 mm) nominal size inside diameter line. **11.2.4.3.4.2** For pressure limiting control, a sensing line shall be installed from a connection between the pump discharge flange and the discharge check valve to the engine. **11.2.4.3.4.3** For suction limiting control, a sensing line shall be installed from a connection at the pump inlet flange to the engine.	This new language was added to provide guidance for the installation of a sensing line for suction pressure control systems.
11.2.4.4.5 Means shall be provided for signaling critically low oil pressure in the engine lubrication system to the controller. **11.2.4.4.5.1** Means shall be provided on the engine for testing the operation of the oil pressure signal to the controller resulting in visible and common audible alarm on the controller as required in 12.4.1.3. **11.2.4.4.5.2** Instructions for performing the test in 11.2.4.4.5.1 shall be included in the engine manual. **11.2.4.4.6** Means shall be provided for signaling high engine temperature to the controller. **11.2.4.4.6.1** Means shall be provided on the engine for testing the operation of the high engine temperature signal to the controller, resulting in visible and common audible alarm on the controller as required in 12.4.1.3. **11.2.4.4.6.2** Instructions for performing the test in 11.2.4.4.6.1 shall be included in the engine manual. **11.2.4.4.7** Means shall be provided for signaling low engine temperature to the controller.	These requirements were added to the standard to assure proper monitoring and testing of engine fluid levels and temperatures under operating conditions.

2013 Edition Text	Reason for Change
11.2.4.4.7.1 Means shall be provided on the engine for testing the operation of the low engine temperature signal to the controller, resulting in visible and common audible alarm on the controller as required in 12.4.1.3. **11.2.4.4.7.2** Instructions for performing the test in 11.2.4.4.7.1 shall be included in the engine manual.	
11.2.6.2.1.1 Interconnection wire size shall be based on length as recommended for each terminal by the controller manufacturer.	This new language was added to provide guidance for wire sizing.
11.2.7.3.5... (2) Electrically operated means shall automatically recharge and maintain the stored hydraulic pressure within the predetermined pressure limits. … (4) Engine driven means shall be provided to recharge the hydraulic system when the engine is running.	This new language was added to provide guidance for engine recharge capability of the hydraulic system.
11.4.1.2.4 Single wall fuel tanks shall be enclosed with a wall, curb, or dike sufficient to hold the entire capacity of the tank.	This paragraph was revised to clarify that only single wall tanks are required to be enclosed by a wall or dike.
11.4.4.5 A manual shut off valve shall be provided within the tank fuel supply line. **11.4.4.5.1** The valve shall be locked in the open position. **11.4.4.5.2** No other valve than a manual locked open valve shall be put in the fuel line from the fuel tank to the engine.	These paragraphs were revised to specify the number and type of valves permitted in the fuel piping.
11.4.6* Static Electricity. **A.11.4.6** The prevention of electrostatic ignition in equipment is a complex subject. Refer to NFPA 77, *Recommended Practice on Static Electricity*, for more guidance. **11.4.6.1** The tank, pump, and piping shall be designed and operated to prevent electrostatic ignitions. **11.4.6.2** The tank, pump, and piping shall be bonded and grounded. **11.4.6.3** The bond and ground shall be physically applied or shall be inherently present by the nature of the installation. **11.4.6.4** Any electrically isolated section of metallic piping or equipment shall be bonded and grounded to prevent hazardous accumulation of static electricity.	This subsection was added to assure that the pump, tank, and piping will be grounded for all installations.
Chapter 12 Engine Drive Controllers	
12.3.6.4 The installation instructions of the manufacturer of the fire pump controller shall be followed.	This new paragraph was added to clarify that controllers must be installed in accordance with the manufacturer's instructions. Over the years, sometimes these instructions were not followed, and incorrect installations occurred.
12.4.1.4... (1)* Battery failure or missing battery. Each controller shall be provided with a separate visible indicator for each battery. The battery failure signal shall initiate at no lower than two-thirds of battery nominal voltage rating (8.0 V dc on a 12 V dc system). Sensing shall be delayed to prevent nuisance signals. … (10) Supervisory signal for interstitial space liquid intrusion. (11) High cooling water temperature.	This text was revised to quantify the intention of the battery failure alarm.

2013 Edition Text	Reason for Change
12.4.1.5 A separate signal silencing switch or valve, other than the controller main switch, shall be provided for the conditions reflected in 12.4.1.3 and 12.4.1.4. **12.4.1.5.1** The switch or valve shall allow the audible device to be silenced for up to 4 hours and then re-sound repeatedly for the conditions in 12.4.1.3. **12.4.1.5.2** The switch or valve shall allow the audible device to be silenced for up to 24 hours and then re-sound repeatedly for the conditions in 12.4.1.4. **12.4.1.5.3** The audible device shall re-sound until the condition is corrected or the main switch is placed in the off position. **12.4.1.6*** The controller shall automatically return to the nonsilenced state when the alarm(s) have cleared (returned to normal). **12.4.1.6.1** This switch shall be clearly marked as to its function.	These new requirements provide the ability to silence audible alarms for certain alarm conditions, while maintaining alarm capabilities in the event of a fire.
12.6 Battery Chargers.... (2) Additional chargers also listed for fire pump service shall be permitted to be installed external to the diesel fire pump controller for added capacity or redundancy.	If larger batteries are required based on the requirements of 11.2.7.2.1.5, this new requirement provides a method of meeting the requirement.
12.7.2.2.3 Starting of the engine shall be initiated by the opening of the control circuit loop containing this fire protection equipment.	This paragraph was added to align Chapter 12 with Chapter 10.
Chapter 14 Acceptance Testing, Performance, and Maintenance	
14.1.1* Flushing. **A.14.1.1** The suction piping to a fire pump needs to be adequately flushed to make sure stones, silt, and other debris will not enter the pump or the fire protection system. The flow rates in Table 14.1.1.1 are the minimum recommended, which will produce a velocity of approximately 15 ft/sec (4.6 m/sec). If the flow rate cannot be achieved with the existing water supply, a supplemental source such as a fire department pumper could be necessary. The procedure is to be performed, witnessed, and signed off before connection to the suction piping is completed. **14.1.1.1** Suction piping shall be flushed at a flow rate not less than indicated in Table 14.1.1.1 or at the hydraulically calculated water demand rate of the system, whichever is greater.	This new annex language was added to help the user understand the purpose and requirements of the flushing criteria of Table 14.1.1.1.
14.1.1.3 Where the maximum flow available from the water supply cannot provide the flow rate provided in Table 14.1.1.1, the flushing flow rate shall be the greater of 100 percent of rated flow of the connected fire pump or the maximum flow demand of the fire protection system(s). **14.1.1.3.1** This reduced flushing flow capacity shall constitute an acceptable test, provided that the flow rate exceeds the fire protection system design and flow rate.	This new language was added to provide an alternative method of flushing, should the flow rates of Table 14.1.1.1 not be obtainable.
Table 14.1.1.1 Minimum Flow Rates for Flushing Suction Piping	Two separate tables have been combined here to simplify the flushing requirements.
14.2.4.1.1 For water mist positive displacement pumping units, a copy of the manufacturer's certified shop test data for both variable speed and non-variable speed operation shall be available for comparison of the results of the field acceptance test. **14.2.4.2** At all flow conditions, including those required to be tested in 14.2.6.2, the fire pump as installed shall equal the performance as indicated on the manufacturer's certified shop test characteristic curve within the accuracy limits of the test equipment.	These paragraphs were added, introducing new testing criteria specific to water mist positive displacement pumping units based on the work of the NFPA 20 TC Task Group on Water Mist Positive Displacement Pumping Units.

2013 Edition Text	Reason for Change
14.2.4.2.1 For water mist positive displacement pumping units with variable speed features, the pump unit as installed shall equal the performance as indicated on the fire pump unit manufacturer's certified shop test data, with variable speed features deactivated within the accuracy limits of the test equipment. **14.2.4.2.2** For water mist positive displacement pumping units, the pump unit as installed shall equal the performance as indicated on the fire pump unit manufacturer's certified shop test data, with variable speed features activated within the accuracy limits of the test equipment.	These paragraphs were added, introducing new testing criteria specific to water mist positive displacement pumping units based on the work of the NFPA 20 TC Task Group on Water Mist Positive Displacement Pumping Units.
14.2.6.1.3 Requirements for personal protective equipment and procedures in accordance with *NFPA 70E, Standard for Electrical Safety in the Workplace*, shall be followed when working near energized electrical or rotating equipment.	This text was added to reinforce that proper PPE should be worn when working with energized electrical equipment, due to the life threatening dangers of electricity, such as arc flash.
14.2.6.2.1 For water mist positive displacement pumping units, each pump shall be operated manually a minimum of six times during the acceptance test. **14.2.6.2.2** For water mist positive displacement pumping units, each of the required automatic operations shall operate all pumps, except as provided in 14.2.6.2.2.1 and 14.2.6.2.2.2. **14.2.6.2.2.1** Where redundant pumps are provided, each of the automatic operations shall operate the number of pumps required to meet system demand. **14.2.6.2.2.2** Where redundant pumps are provided, each pump shall operate for a minimum of three automatic operations.	These paragraphs were added, introducing new testing criteria specific to water mist positive displacement pumping units based on the work of the NFPA 20 TC Task Group on Water Mist Positive Displacement Pumping Units.
14.2.6.3.1 Pumps with variable speed pressure limiting control shall be tested at no-flow, 25 percent, 50 percent, 75 percent, 100 percent, 125 percent, and 150 percent of rated load in the variable speed mode. **14.2.6.3.1.1** They shall also be tested at minimum, rated, and peak loads, with the fire pump operating at rated speed. **14.2.6.3.2** The fire protection system shall be isolated and the pressure relief valve closed for the rated speed tests required in 14.2.6.3.1. **14.2.6.3.3** The fire protection system shall be open and the relief valve set for the variable speed tests required in 14.2.6.3.1.	Additional test points have been added for pumps with variable speed pressure limiting controls, to verify there are not stability issues throughout the flow range of the system. Also, 14.2.6.3.3 was added to ensure that the system and the relief valve are left in the open position at the conclusion of the testing, and to confirm there is no interaction with the fire protection system throughout the entire range of flow testing.
14.2.6.4.6 The voltage at the motor contacter ouput lugs shall not vary more than 5 percent below or 10 percent above rated (nameplate) voltage during the test. *(See Section 9.4.)*	This revised text provides a safer method of testing the voltage output by avoiding the practice of exposing bare wires.
14.3* Record Drawings, Test Reports, Manuals, Special Tools, and Spare Parts. **A.14.3** It is the intent to retain the record drawing, equipment manual, and completed test report for the life of the fire pump system. **14.3.1** One set of record drawings shall be provided to the building owner. **14.3.2** One copy of the completed test report shall be provided to the building owner. **14.3.3*** One set of instruction manuals for all major components of the fire pump system shall be supplied by the manufacturer of each major component. **A.14.3.3** Consideration should be given to stocking spare parts for critical items not readily available.	This adds a requirement to provide as-built plans and acceptance test documentation as part of the closeout package.

2013 Edition Text	Reason for Change
14.5.1.3.1 For water mist positive displacement pumping units, the retest shall include the pump unit as a whole.	This new criteria was added, based on the addition of water mist positive displacement pumping units to the standard.
14.5.1.4.1 The field retest results shall be compared to the original pump performance as indicated by the fire pump manufacturer's original factory-certified test curve, whenever it is available.	This text was added to clarify that the curve that is to be used is the one provided by the manufacturer.
14.5.2.4 Component Replacement. The requirements of Table 8.6.1 of NFPA 25, *Standard for the Inspection, Testing, and Maintenance of Water-Based Fire Protection Systems*, shall be followed for component replacement testing.	The component replacement table has been removed, because it was extracted from NFPA 25. Since NFPA 25 and NFPA 20 are in different revision cycles, it is difficult to maintain correlation of the table between the two standards; therefore, NFPA 20 now directs the user to NFPA 25.
14.5.2.6 Whenever replacement, change, or modification to a critical path component is performed on a fire pump, driver, or controller, as described in 14.5.2.5, a new acceptance test shall be conducted by the pump manufacturer, factory-authorized representative, or qualified persons acceptable to the authority having jurisdiction.	This paragraph has been modified to emphasize the importance of performing a complete acceptance test if a critical path component has been replaced.

NFPA 20 Index

A

Additive
   Definition 3.3.1
   Piping 4.13.3
Additive pumps, 8.2, 8.10.2, A.8.2.2 to A.8.2.6; *see also* Foam concentrate pumps
   Definition, 3.3.38.1
   Motors, controllers for, 10.9
Air leaks/pockets, B.1.1, B.1.3, B.1.7
Air release fittings, automatic, 6.3.1, 6.3.3, 7.3.5.1(1), 7.3.5.2, A.6.3.1
Air starting, 11.2.7.4, 12.4.1.4(3), 12.8, A.11.2.7.4.4
Alarms, fire pump, *see* Fire pump alarms
Application of standard, 1.3
Approved/approval
   Definition, 3.2.1, A.3.2.1
   Requirements, 4.2, A.4.2
Aquifer (definition), 3.3.2
Aquifer performance analysis, A.7.2.1.2
   Definition, 3.3.3
Authority having jurisdiction
   Definition, 3.2.2, A.3.2.2
   Fuel system plan review, 11.4.1.1
Automatic air release fittings, *see* Air release fittings, automatic
Automatic transfer switches, 10.8.1.3
   Definition, 3.3.50.2.1

B

Backflow preventers, 4.14.9.2(1), 4.15.6, 4.16.1, 4.27, 4.31.1, 11.2.8.5.3.7, A.4.14.9, A.4.15.6, A.11.2.8.5.3.7(A), A.11.2.8.5.3.7(B)
Backup fire pump, high-rise buildings, 5.6.2
Batteries, storage, 11.2.7.2, 12.7.4, 14.2.6.4.7.1, 14.2.7.7, A.11.2.7.2.1.5 to A.11.7.2.4
   Failure indicators, 12.4.1.4(1), 12.4.1.4(2), A.12.4.1.4(1)
   Isolation, 11.2.7.2.2, A.11.2.7.2.2
   Load, 11.2.7.2.3
   Location, 11.2.7.2.4, A.11.2.7.2.4
   Maintenance, 11.6.3, B.1.32
   Recharging, 12.5, A.12.5
   Voltmeter, 12.4.5
Battery cables, 11.2.6.3
Battery chargers, 12.4.1.4(2), 12.5.3, 12.6
Bowl assembly, vertical shaft turbine pumps, 7.3.3
Branch circuit (definition), 3.3.7.1
Break tanks, 4.31
   Definition, 3.3.6
   Refill mechanism, 4.31.3
   Refill rate test, 14.2.6.2.7
   Size, 4.31.2
Butterfly valves, 4.15.7, 4.15.8, 4.27.3.1, A.4.15.7
Bypass line, 4.14.4, 11.2.8.6, A.4.14.4, A.11.2.8.6, A.13.2.1.1, B.1.25
Bypass operation, variable speed pressure limiting control, 10.10.3, A.10.10.3
Bypass valves, 4.16.1

C

Can pump (definition), 3.3.38.2
Capacity, pump, 4.8, 6.2.1, A.4.8, A.14.2.6.4
Centrifugal pumps
   Capacity, 4.8, 6.2.1, A.4.8, A.14.2.6.4
   Component replacement, 14.5.2, A.14.5.2.3
   Connection to driver and alignment, 6.5, A.6.5, B.1.23
   Definition, 3.3.38.3
   End suction, 6.1.1.2
   Factory and field performance, 6.2, A.6.2
   In-line, 6.1.1.2
   Maximum pressure for, 4.7.7, A.4.7.7
   Pressure maintenance pumps, 4.25.4.1, A.4.25.4
   Relief valves, 4.18, A.4.18.1 to A.4.18.8
   Sole supply, 12.7.2.6
   Types, 6.1.1, A.6.1.1
Check valves, 4.14.9.2(1), 4.15.6, 4.25.5.4, 4.27, 4.30.2, 4.30.4.1, 11.2.8.5.3.2, 11.2.8.5.3.7, A.4.15.6, A.11.2.8.5.3.7(A), A.11.2.8.5.3.7(B)
Circuit breakers, *see* Disconnecting means
Circuits, *see also* Fault tolerant external control circuits
   Branch circuit (definition), 3.3.7.1
   Internal conductors, B.3.2.3
   Protection of, *see* Protective devices
Circulation relief valves, *see* Relief valves
Columns, vertical shaft turbine pumps, 7.3.2, A.7.3.2.1
Contractor's material and test certificate, 14.1.3, A.14.1.3
Controllers, fire pump, *see* Fire pump controllers
Control valves, *see* Gate valves
Coolant, engine, 11.2.8.3, 11.2.8.4
Corrosion-resistant material (definition), 3.3.9
Couplings, flexible, *see* Flexible couplings
Cutting, torch, 4.13.6, A.4.13.6

D

Definitions, Chap. 3
Designer, system, 4.3.2, A.4.3.2(2)
Detectors, liquid level, *see* Liquid level
Diesel engines, 4.7.2, 4.7.6, Chap. 11, A.4.7.6; *see also* Engine drive controllers
   Backup, 9.3.3
   Connection to pump, 11.2.3
   Cooling, 11.2.8, A.11.2.8.5 to A.11.2.8.6, B.1.25
   Critical path components, 14.5.2.5(3)
   Definition, 3.3.13.1, A.3.3.13.1
   Emergency starting and stopping, 11.6.6
   Exhaust, 11.5, A.11.5.2
   Fire pump buildings or rooms with, 4.12.1.3
   Fuel supply and arrangement, 11.4, A.11.4.2 to A.11.4.6
   Instrumentation and control, 11.2.4, A.11.2.4.2 to A.11.2.4.2.7.1
   Listing, 11.2.1
   Lubrication, 11.2.9
   Operation and maintenance, 11.6, A.11.6, A.14.2.6
   Pump room, 11.3, A.11.3

659

Ratings, 11.2.2, A.11.2.2.2 to A.11.2.2.5
Starting methods, 11.2.7, 11.6.6, A.11.2.7.2.1.5 to A.11.2.7.4.4, A.14.2.6
Tests, 14.2.6 to 14.2.8, A.14.2.6, A.14.2.7.1
Type, 11.1.3, A.11.1.3
Vertical lineshaft turbine pumps, 7.5.1.3, 7.5.1.4, 7.5.2
Discharge, exhaust, 11.5.3
Discharge cones, 4.18.5.1, 4.18.5.4, 6.3.2(4), 7.3.5.1(4), A.4.18.5
Discharge pipe and fittings, 4.2.3, 4.15, A.4.15.3 to A.4.15.10; *see also* Drainage
From dump valves, 8.1.6.5
Pressure maintenance pumps, 4.25.5.1
From relief valves, 4.18.5 to 4.18.9, A.4.18.7, A.4.18.8
Valves, 4.15.6 to 4.15.9, 4.16.1, 4.25.4.1, 4.30.2, A.4.15.6, A.4.15.7
Discharge pressure gauges, 4.10.1, 8.5.1
Disconnecting means
Definition, 3.3.11
Electric-drive controllers, 10.4.3, 10.6.3, 10.6.8, 10.8.2.1.3, A.10.4.3.1 to A.10.4.3.3.2, B.2, B.3.1, B.3.2.2
Variable speed drives, 10.10.5, A.10.10.5
Electric drivers, 9.2.3, 9.3.6, A.9.2.3.1(3), A.9.2.3.1(4), B.2
Domestic water supply, 4.12.1.1.5
Drainage
Pump room/house, 4.12.7, 11.3.1, A.4.12.7
Relief valve, 4.27.2
Water mist system pumps, 8.5.4.1
Drain piping, 4.13.4
Drawdown (definition), 3.3.12
Dripproof motors, 9.5.1.9.1, 9.5.2.4
Definition, 3.3.33.2
Guarded (definition), 3.3.33.1
Drivers, 4.2.3, 4.2.4, 4.4.1, 4.7, A.4.4.1, A.4.7.1 to A.4.7.7.3.2, B.1.23, B.1.26; *see also* Diesel engines; Electric motors (drivers); Steam turbines
Critical path components, 14.5.2, A.14.5.2.3
Earthquake protection, 4.28.2
Positive displacement pumps, 8.4.3, 8.6, 8.9, A.8.6.1
Pump connection and alignment, 6.5, A.6.5, B.1.23
Speed, B.1.28, B.1.30
Vertical lineshaft turbine pumps, 7.5, 7.6.1.6

Dual-drive pump units, A.4.7.3
Dump valves, 8.1.6
Control, 8.1.6.3
Definition, 3.3.57.1
Dust-ignition-proof motor (definition), 3.3.33.3

E

Earthquake protection, 4.12.1, 4.13.5, 4.28, A.4.12.1, A.4.28.1.2
Eccentric tapered reducer or increaser, 4.14.6.4, 6.3.2(1)
Electric-drive controllers, 9.1.2, Chap. 10; *see also* Electric motors (drivers)
Additive pump motors, 10.9
Automatic, 10.5.1, 10.5.2, 10.9.2, A.10.5.1, A.10.5.2.1
Components, 10.4, A.10.4.1 to A.10.4.7
Connections and wiring, 10.3.4
Continuous-duty basis, 10.3.4.4
Control circuits, protection of, 10.3.5
Construction, 10.3, A.10.3.3.1 to A.10.3.7.3
Design, 10.1.3, A.10.1.3
Electrical diagrams and instructions, 10.3.7, A.10.3.7.3
Electrical power supply connection, 9.6.4
Emergency-run control, 10.5.3.2, 10.6.10, A.10.5.3.2
External operations, 10.3.6, 10.5.2.6, A.10.3.6
Indicating devices
Available power, 10.6.6
Power source, 10.8.3.8
Remote indication, 10.4.8, 10.8.3.14, 10.10.9, A.10.4.5.7
Variable speed drive operations, 10.10.8, 10.10.9
Listing, 10.1.2.1, 10.1.2.4
Location, 10.2, A.10.2.1
Low-voltage control circuit, 10.6.5
Nonautomatic, 10.5.1, 10.5.2.4, 10.5.3, A.10.5.1, A.10.5.3.2
Power transfer for alternate power supply, 10.8, A.10.8
Rated in excess of 600 Volts, 10.6
Service arrangements, 10.1.2.6
Starting and control, 10.5, A.10.5.1 to A.10.5.3.2
Motor starting circuitry, 10.4.5, A.10.4.5.7
Operating coils, 10.4.5.6
Single-phase sensors in controller, 10.4.5.7, A.10.4.5.7
Soft start units, 10.4.5.5
Starting reactors and autotransformers, 10.4.5.4

Starting resistors, 10.4.5.3
Timed acceleration, 10.4.5.2
State of readiness, 10.1.2.7
Stopping methods, 10.5.4, 10.9.3
Variable speed pressure limiting control, 10.5.2.1.7.6, 10.10, A.10.10
Voltage drop, 9.1.6, 9.4, A.9.4
Electric motors (drivers), 4.7.2, 4.7.6, 6.5.1.1, Chap. 9, A.4.7.6; *see also* Electric-drive controllers
Backup, 9.3.3
Critical path components, 14.5.2.5(3)
Current limits, 9.5.2
Definition, 3.3.33.4
Operation, A.14.2.6
Phase reversal test, 14.2.6.6, A.14.2.6.6
Power sources and supply, *see* Power supply
Problems of, B.1.26, B.1.28 to B.1.30
Speed, B.1.28, B.1.30
Starting methods, A.14.2.6
Tests, 14.2.6 to 14.2.8, A.14.2.6, A.14.2.7.1
Vertical lineshaft turbine pumps, 7.5.1.3, 7.5.1.5, 7.5.2
Voltage drop, 9.1.6, 9.4, A.9.4
Electric starting, diesel engine, 11.2.7.2, A.11.2.7.2.1.5 to A.11.7.2.4
Electric supply, *see* Power supply
Electronic fuel management control, 11.2.4.2, 12.4.1.4(5), 12.4.1.4(6), 14.2.12, 14.5.2.3, A.11.2.4.2, A.14.2.12, A.14.5.2.3
Electrostatic ignitions, preventing, 11.4.6, A.11.4.6; *see also* Grounding
Emergency boiler feed, steam turbine, A.13.3
Emergency control for engine drive controllers, 12.7.6
Emergency governors, 13.2.2.4, 13.2.2.5, 14.2.9
Emergency lighting, 4.12.5
Emergency-run mechanical control, 10.5.3.2, 10.6.10, A.10.5.3.2
Enclosures, pump, *see* Pump rooms/houses
Enclosures for controllers, 9.7(2), 10.3.3, 12.3.3, A.9.7(2), A.10.3.3.1, A.12.3.3.1, B.3.2.1
End suction pumps, 6.1.1.2, Fig. A.6.1.1(a)
Definition, 3.3.38.4
Engine drive controllers, Chap. 12
Air starting, 11.2.7.4, 12.8, A.11.2.7.4.4
Automatic, 12.7.1, 12.7.2
In-factory wiring, 11.2.6.1, 11.2.7.4.2, A.11.2.6.1.1
In-field wiring, 11.2.6.2, A.11.2.6.2

Battery charging and recharging, 12.5, 12.6, A.12.5
Components, 12.4, A.12.4.1.2 to A.12.4.4
Connections and wiring, 11.2.6, 11.2.7.4.2, 12.3.5, A.11.2.6.1.1, A.11.2.6.2
Construction, 12.3, A.12.3.1.1 to A.12.3.8
Contacts, 12.4.3
Electrical diagrams and instructions, 12.3.6, 12.3.8, A.12.3.8
External operations, 12.3.6.3, 12.7.2.5
Indicators, 12.4.1, A.12.4.1.2 to A.12.4.1.6
 Remote, 12.4.2, 12.4.3, 14.2.7.6, A.12.4.2.3(3)
Location, 12.2, A.12.2.1
Locked cabinet for switches, 12.3.4
Nonautomatic, 12.7.1, 12.7.2.3, 12.7.3
Starting and control, 12.7, 12.8.2, A.12.7
Stopping methods, 12.7.5, 12.8.3, A.12.7.5.2
 Automatic shutdown, 12.7.5.2, A.12.7.5.2
 Manual shutdown, 12.7.5.1, 12.8.3
Tests, 12.7.3.2, 14.2.7.7
Engines
Diesel, *see* Diesel engines
Internal combustion, *see* Internal combustion engines
Equivalency to standard, 1.5
Exhaust system
Diesel engine, 11.5, A.11.5.2
Steam turbines, A.13.3
Explosionproof motor (definition), 3.3.33.5

F

Fan-cooled motor, totally enclosed, 9.5.2.4
Definition, 3.3.33.8
Fault tolerant external control circuits, 10.5.2.6, 12.7.2.5
Definition, 3.3.7.2
Feeder, 9.2.2, 9.3.6
Definition, 3.3.15
Inadequate capacity, B.1.28
Field acceptance tests, 4.4.2, 4.32, 14.2, 14.3.5, 14.5, 14.5.2.7, A.14.2.1 to A.14.2.12, A.14.5.2.3
Component replacement, 14.5.1
Fill/refill valves, water tank, 5.6.1.2 to 5.6.1.5
Fire protection equipment control, 10.5.2.3, 12.7.2.2
Fire pump alarms
Definition, 3.3.16
Electric-drive controllers, 10.1.1.2, 10.4.7, A.10.4.7

Engine drive controllers, 12.1.1, 12.4.1.3 to 12.4.1.5, 12.4.2, 12.4.3, A.12.4.1.4(1), A.12.4.2.3(3)
Tests, 14.2.10
Fire pump controllers, 4.2.3, 4.2.4, 4.4.1, 8.7, A.4.4.1, A.8.7; *see also* Electric-drive controllers; Engine drive controllers
Acceptance test, 14.2.6, A.14.2.6
Additive pump motors, 10.9
Critical path components, 14.5.2, A.14.5.2.3
Dedicated controller for each driver, 4.7.5
Definition, 3.3.17
Earthquake protection, 4.28.2
Maintenance, after fault condition, B.3
Positive displacement pump, 8.1.6.3, 8.4.3, 8.4.4
Pressure actuated controller pressure sensing lines, 4.30, A.4.30
Pressure maintenance pump, 4.25.7
Protection of, 4.12.1, A.4.12.1
Vertical lineshaft turbine pumps, 7.5.2, 10.10.12
Fire pumps, *see also* Pumps
Backup, 9.3.3
Critical path components, 14.5.2, A.14.5.2.3
Definition, 3.3.38.5
In high-rise buildings, *see* High-rise buildings
Operations, *see* Operations
Packaged fire pump assemblies, 4.29, A.4.29.9
Redundant, 14.2.6.2.2.1, 14.2.6.2.2.2
Standby, 4.25.6
Summary of data, Table 4.26(a), Table 4.26(b)
Fire pump units
Definition, 3.3.18
Dual-drive, A.4.7.3
Field acceptance tests, 4.32
Location and protection
 Indoor units, 4.12.1.1, A.4.12.1.1
 Outdoor units, 4.12.1.2
Performance, 4.4, 14.2.6.5, A.4.4.1
Series, 4.19, A.4.19.2.1, A.4.19.2.7.1
 Definition, 3.3.42, A.3.3.42
Fittings, 4.2.3, 4.13, 6.3, A.4.13.1 to A.4.13.6, A.6.3.1
Discharge, 4.15, A.4.15.3 to A.4.15.10
Earthquake protection, 4.28.3
Joining method, 4.13.2
Maintenance, B.1.13
Positive displacement pumps, 8.5, A.8.5.2 to A.8.5.5
Suction, 4.14, A.4.14.1 to A.4.14.10
Vertical lineshaft turbine pumps, 7.3.5, A.7.3.5.3

Flexible connecting shafts, 6.5.1.1, 11.2.3.1.1, 11.2.3.2.1, A.6.5
Definition, 3.3.19
Guards for, 4.12.8
Vertical lineshaft turbine pumps, 7.5.1.7.1, 7.5.1.8
Flexible couplings, 6.5.1.1, Fig. A.6.1.1(e), A.6.5
Definition, 3.3.20
Diesel engine pump connection, 11.2.3.1
Earthquake protection, 4.28.3.2
Guards, 4.12.8
Positive displacement pumps, 8.9.1, 8.9.2
Flooded suction (definition), 3.3.21
Flow-measuring devices, 5.4, 6.3.2(3), 8.10.2, 8.10.3, 14.2.6.4.3.2, 14.2.6.4.3.3
Flow tests, *see* Tests
Foam concentrate piping, 4.13.3
Foam concentrate pumps, 8.2, 8.5.3, 14.2.11, A.8.2.2 to A.8.2.6, A.8.5.3, A14.2.11; *see also* Additive pumps
Foundations, 6.4, A.6.4.1, A.6.4.4, B.1.24
Positive displacement pumps, 8.8
Vertical lineshaft turbine pumps, 7.4.3
Frequency-sensing devices, 10.8.3.7
Fuel supply, 11.4, A.11.4.2 to A.11.4.6; *see also* Electronic fuel management control
Earthquake protection, 4.28.2
Fuel types, 11.4.5, A.11.4.5
Location, 11.4.3, A.11.4.3
Maintenance, 11.6.4, A.11.6.4.3
Obstructed system, B.1.32
On-site standby generators, 9.6.2.3, A.9.6.2

G

Gate valves, 4.14.5, 4.15.7, A.4.14.5, A.4.15.7, A.13.3
Gear drives, 7.5.1.7, 11.2.3.2.1, 14.2.6.4.9
Gear pump (definition), 3.3.38.7
Generator, *see* On-site standby generator
Governors, B.1.30
Diesel engine, 11.2.4.1, 14.2.6.4.7.3, 14.2.6.4.7.4
Emergency, 13.2.2.4, 13.2.2.5, 14.2.9
Speed, steam turbine, 13.2.2
Ground-face unions, 4.30.4.2
Grounding
Diesel engines, 11.4.6.2 to 11.4.6.4
Electric drive controller enclosures, 10.3.3.3
Engine drive controllers, 12.3.3.2
Groundwater (definition), 3.3.22

Guarded motors
  Definition, 3.3.33.6
  Dripproof (definition), 3.3.33.1

H

Hangers, 4.13.5
Head, 4.5.1, 6.2.2; *see also* Net positive suction head (NPSH) (h_sv); Total head (H)
  Available from water supply, 4.6.5
  Definition, 3.3.23, A.3.3.23
  Net head lower than rated, B.1.12
  Static, A.6.1.2
  Total discharge head (h_d) (definition), 3.3.23.2
  Total rated head (definition), 3.3.23.4
  Total suction head (definition), 3.3.23.5
  Velocity head (h_v) (definition), 3.3.23.6
  Vertical turbine pump head component, 7.3.1, A.7.3.1
Heat exchangers, 11.2.8.1(1), 11.2.8.5, 11.2.8.6, A.11.2.8.5, A.11.2.8.6, A.11.3.2.3, A.11.3.2.4, B.1.25
Heat source, pump room/house, 4.12.3
High-rise buildings
  Definition, 3.3.24
  Fire pumps for, 4.12.1.1.1, Chap. 5, 9.3.1
Horizontal pumps, 4.1
  Definition, 3.3.38.8
  Diesel engine drive connection, 11.2.3.1
  Installation, Fig. A.6.3.1(a)
  Shaft rotation, A.4.23
  Split-case, 6.1.1.3, Fig. A.6.1.1(f), Fig. A.6.3.1(a)
    Definition, 3.3.38.9
    Suction pipe and fittings, 4.14.6.3.1 to 4.14.6.3.3
  Total head (H), A.14.2.6.4
    Definition, 3.3.23.3.1, A.3.3.23.3.1
Hose valves, 4.20.3, 6.3.2(2), 7.3.5.1(5), A.4.20.3.1 to A.4.20.3.4(2), A.14.2.6.2.5
Hydraulic starting, 11.2.7.3, 12.4.1.4(3)

I

Impellers
  Impeller between bearings design, 6.1.1.3, 6.4.1, A.6.1.1, A.6.4.1
  Overhung impeller design, 6.1.1.2, 6.4.1, 6.4.2, A.6.1.1, A.6.4.1
  Problems, B.1.8, B.1.10 to B.1.12, B.1.15, B.1.16, B.1.29
  Vertical lineshaft turbine pumps, 7.3.3.2, 7.5.1.1, 7.5.1.2, 7.5.1.6.2, 7.5.1.7.2, 7.6.1.3, A.7.6.1.1

Indicating valves, 4.15.7, 4.25.5.5, A.4.15.7
In-line pumps, 6.1.1.2, Figs. A.6.1.1(c) to (e)
  Definition, 3.3.38.10
In-rush currents, 10.8.3.10
In sight from (within sight from, within sight)
  Definition, 3.3.25
Inspection, periodic, 14.3.5, 14.4
Installer, system, 4.3.3, A.4.3.3(2)
Instrument panel, 11.2.5.1
Internal combustion engines, B.1.26; *see also* Diesel engines
  Definition, 3.3.13.1.1
  Pump room or house, heat for, 4.12.3.2
  Spark-ignited, 11.1.3.2
  Speed, B.1.28
Isolating switches, 10.4.2, 10.8.2.1.2, 10.8.2.2, 10.8.3.12.1, A.10.4.2.1.2, A.10.4.2.3, B.3.1, B.3.2.2
  Definition, 3.3.50.1
Isolation valves, 4.16.1, 4.25.5.3, 4.25.5.4, 4.25.5.7

J

Jockey pumps, *see* Pressure maintenance (jockey or make-up) pumps
Junction boxes, 9.7, 10.3.4.5.1, A.9.7(2), A.9.7(3)

L

Lighting
  Emergency, 4.12.5
  Normal artificial, 4.12.4
Limited service controllers, 10.7
Liquid (definition), 3.3.28
Liquid level
  Detectors, 7.3.5.1(2), 7.3.5.3, 14.2.6.2.8, A.7.3.5.3
  Pumping, 4.6.3, B.1.19
    Definition, 3.3.29.1
  Static, 6.1.2, A.6.1.2
    Definition, 3.3.29.2
  Well or wet pit, 4.6.3
Liquid supplies, 4.2.3, 4.2.4, 4.6, A.4.6.1 to A.4.6.4
  Domestic water supply, 4.12.1.1.5, 5.3.2, 5.6.1.5(2)
  Head, 4.6.5, A.6.1.2
  Heat exchanger, 11.2.8.5, 11.2.8.6, A.11.2.8.5, A.11.2.8.6
  Potable water, 8.5.6, 11.2.8.5.3.7
  Pumps, priming of, B.1.19
  Reliability, 4.6.1, A.4.6.1
  Sources, 4.6.2, 7.2.1, A.4.6.2, A.7.1.1, A.7.2.1.1, A.7.2.1.2
    Discharge to, 4.18.7, A.4.18.7

Stored supply, 4.6.4, 5.3, 5.4, 5.6.1, A.4.6.4, A.5.6.1.5.1
Vertical lineshaft turbine pumps, 7.1.1, 7.2, A.7.1.1, A.7.2.1.1 to A.7.2.7, B.1.19
Listed
  Break tank refill mechanism, 4.31.3
  Component replacement parts, 14.5.2.4.1, 14.5.2.4.2
  Controllers and transfer switches, 4.25.7, 10.1.2.1, 10.1.2.4, 10.10.1.2, 12.1.3.1
  Definition, 3.2.3, A.3.2.3
  Dump valves, 8.1.6.4
  Electrical circuit protective system to controller wiring, 9.8, A.9.8.1, A.9.8.2
  Engines, 11.2.1
  Pumps, 4.7.1, 6.5.1.2, A.4.7.1, A.6.2
Loads start test, 14.2.6.5
Locked rotor overcurrent protection, 10.4.4, 10.6.9, A.10.4.4(3)
Lockout, additive pump motors, 10.9.4
Loss of phase, 10.4.7.2.2
  Definition, 3.3.30
Low suction throttling valves, 4.14.9.2(2), 4.15.9
  Definition, 3.3.57.2
Lubrication, pump, B.1.21, B.1.27

M

Maintenance
  Batteries, storage, 11.6.3, B.1.32
  Controllers, B.3
  Diesel engines, 11.6, A.11.6
  Fittings, B.1.13
  Fuel supply, 11.6.4, A.11.6.4.3
  Pumps, 7.6.2, 14.3.4(2), 14.3.5, 14.4
  Water seals, B.1.6
Main throttle valve, steam turbine, 13.2.1.3
Make-up pumps, *see* Pressure maintenance (jockey or make-up) pumps
Manuals, instruction, 14.3.3, 14.3.4, A.14.3
Manual transfer switches, 10.8.1.2, 10.8.3.5
  Definition, 3.3.50.2.2
Marking
  Additive pump motor controllers, 10.9.5
  Disconnecting means, 9.2.3.1(5), 9.2.3.2
  Electric-drive controllers, 10.1.2.2, 10.1.2.5, 10.3.8, 10.9.5, 10.10.2, 10.10.11, A.10.1.2.2
  Electric drivers, 9.5.3
  Engine drive controllers, 12.1.3.3, 12.3.7
  Transfer switches, 10.1.2.2, A.10.1.2.2

Mass elastic system, 7.5.1.6
Maximum pump brake horsepower, 4.5.1
  Definition, 3.3.32
Measurement, units of, 1.6
Meters, 4.20.2, 8.10.2, 8.10.3, A.4.20.2.1, A.4.20.2.10
Motor contactors, 10.4.5.1, B.3.2.4
Motors
  Dripproof, *see* Dripproof motors
  Dust-ignition-proof (definition), 3.3.33.3
  Electric, *see* Electric motors (drivers)
  Explosionproof (definition), 3.3.33.5
  Guarded (definition), 3.3.33.6
  Open, 9.5.2.4
    Definition, 3.3.33.7
  Speed, *see* Speed
  Totally enclosed, *see* Totally enclosed motors
Multistage pumps, 4.1, 6.1.1.2, 6.1.1.3

N

Nameplates, on pumps, 4.9
Net positive suction head (NPSH) ($h_s v$), 7.2.2.2.2, 8.2.2, 8.3.2, A.8.2.2
  Definition, 3.3.23.1
Nonreverse ratchets, 7.5.1.4, 7.5.1.7.3, 9.5.1.9.2, 11.2.3.2.2
Nonventilated motor, totally enclosed, 9.5.2.4
  Definition, 3.3.33.10

O

Oil pressure gauge, 11.2.5.3
On-site power production facility, 9.2.2(2), 9.2.3, A.9.2.3.1(3), A.9.2.3.1(4); *see also* Power supply
  Definition, 3.3.35
On-site standby generator, 5.5, 9.3.4, 9.6, 10.8.3.6.2, 10.8.3.12, A.9.6.2, A.9.6.5; *see also* Power supply, Alternate power sources
  Definition, 3.3.36, A.3.3.36
Open motors, 9.5.2.4
  Definition, 3.3.33.7
Operations
  Controllers, external operations, 10.3.6, 10.5.2.6, 12.3.6.3, 12.7.2.5, A.10.3.6
  Diesel engines, 11.6, A.11.6
  Pumps, 4.3, 7.6.1, A.4.3.2(2) to A.4.3.4.1(2), A.7.6.1.1, A.7.6.1.4, A.14.2.6
  Vertical lineshaft turbine pumps, 7.6.1, A.7.6.1.1, A.7.6.1.4
Outdoor setting
  Fire pump units, 4.12.1.2, A.4.12.1
  Vertical lineshaft turbine pumps, 7.4.2

Outside screw and yoke gate valves, *see* Gate valves
Overcurrent protection, 9.2.2(4), 9.2.3, 9.3.4.1, 9.3.6, 10.3.5.1, 10.8.2.2(3), 10.8.3.11, 14.2.7.8, A.9.2.3.1(3), A.9.2.3.1(4), A.10.8.3.11; *see also* Disconnecting means
  Isolating switch, 10.4.2.1.3
  Locked rotor, 10.4.4, 10.6.9, A.10.4.4(3)
Overspeed shutdown devices, 11.2.4.4

P

Packaged fire pump assemblies, 4.29, A.4.29.9
  Definition, 3.3.38.11
Phase converters, 9.1.7, A.9.1.7
Phase reversal, 10.4.6.2, 10.4.7.2.3, 14.2.6.6, A.14.2.6.6
Pipe, 4.13, A.4.13.1 to A.4.13.6; *see also* Discharge pipe and fittings; Suction pipe and fittings
  Break tank refill, 4.31.3.2
  Earthquake protection, 4.28.3
  Exhaust, 11.5.2, A.11.5.2
  Flushing, 14.1, A.14.1.1, A.14.1.3
  Fuel, 11.4.4, A.11.4.4
  Joining method, 4.13.2
  Problems, causes of, B.1.1 to B.1.3, B.1.6
  Protection
    Against damage due to movement, 4.17, A.4.17
    For fuel line, 11.4.4.6, A.11.4.4.6
    Seismic bracing, 4.13.5
  Size
    Minimum pipe size data, Table 4.26(a), Table 4.26(b)
    Suction pipe, 4.14.3, A.4.14.3.2
  Steel, 4.13.1, 4.13.2.1, 4.25.5.1, 4.30.3, A.4.13.1, A.4.30.3
Pipeline strainers, 6.3.2(5)
Piston plunger pump (definition), 3.3.38.12
Positive displacement pumps, 4.1, Chap. 8
  Application, 8.1.3
  Component replacement, 14.5.1
  Connection to driver and alignment, 8.9
  Controllers, 8.7, A.8.7
  Definition, 3.3.38.13
  Drivers, 8.6, 8.9, A.8.6.1
  Fittings, 8.5, A.8.5.2 to A.8.5.5
  Flow tests, 8.10, 14.2.6.4.3
  Foam concentrate and additive pumps, 8.2, 8.5.3, 8.10.2, A.8.2.2 to A.8.2.6, A.8.5.3
  Foundation and setting, 8.8
  Materials, 8.1.5, A.8.1.5

  Seals, 8.1.4, 8.2.3
  Suitability, 8.1.2, A.8.1.2
  Types, 8.1.1
  Water mist system pumps, 8.3, A.8.3.1
Power supply, 4.2.3, 4.2.4
  Alternate power sources, 9.2.2(4), 9.3, 9.6.4, 10.4.7.2.4, 10.8, 14.2.3, 14.2.8, A.9.3.2, A.10.8; *see also* On-site standby generator
    Momentary test switch, 10.8.3.13
    Overcurrent protection, 10.8.3.11, A.10.8.3.11
    Transfer of power, 9.6.4, 10.8, A.10.8
  Electric drive for pumps, 9.1.4 to 9.1.6, 9.2, A.9.1.4, A.9.2, B.1.28, B.1.30
    Alternate power sources, *see* subhead: Alternate power sources
    High-rise buildings, auxiliary power for, 5.5
    Normal power, 9.2, A.9.2
    On-site power production facility, A.9.2.3.1(3), A.9.2.3.1(4)
      Definition, 3.3.35
    Service-supplied, 9.2.2
    Variable speed drives, 10.10.6, A.10.10.6.3
  Steam supply, *see* Steam supply
  Tests, 10.8.3.13, 14.2.7.9, 14.2.8, A.14.2.6.4
Pressure control valves, A.13.2.1.1
  Definition, 3.3.57.3
Pressure gauges, 4.10, 5.4, 6.3.1, A.4.10.2, A.6.3.1, B.1.14
  Oil, 11.2.5.3
  Positive displacement pumps, 8.5.1
  Steam, 13.2.3
  Vertical lineshaft turbine pumps, 7.3.5.1(3)
Pressure maintenance (jockey or make-up) pumps, 4.25, 10.3.4.6, 14.2.3, A.4.14.4, A.4.14.9, A.4.25, A.14.2.6(4)
  Definition, 3.3.38.14
  Overcurrent protection, 9.2.3.4
  Pressure sensing lines, 4.30.1, 4.30.2
  Valves, 4.25.5.2 to 4.25.5.5, 4.25.5.7, A.4.25.5.5
Pressure recorders, 12.4.4, A.12.4.4
Pressure-reducing valves, A.13.3
  Definition, 3.3.57.4
Pressure regulating devices, 4.15.10, A.4.15.10, A.13.2.1.1
  Definition, 3.3.37
Pressure relief valves, *see* Relief valves
Pressure sensing lines, pressure actuated controllers, 4.30, A.4.30
Protection
  Equipment, 4.12, A.4.12

Personnel, 10.6.7, B.2
Piping
　Fuel line, 11.4.4.6, A.11.4.4.6
　Movement, damage due, 4.17, A.4.17
　Seismic bracing, 4.13.5
Protective devices, *see also* Overcurrent protection
　Control circuits, 10.3.5
　Controller, 10.4.2 to 10.4.4, 10.6.9, A.10.4.2.1.2, A.10.4.2.3
　Listed electrical circuit protective system to controller wiring, 9.8, A.9.8.1, A.9.8.2
　On-site power source circuits, 9.6.5, A.9.6.5
　Overspeed shutdown device, 11.2.4.4
　Variable speed drive circuit protection, 10.10.5, A.10.10.5
Pump brake horsepower, maximum (definition), 3.3.32
Pumping liquid level, 4.6.3, A.7.1.1
　Definition, 3.3.29.1
Pump manufacturers, 10.1.2.6, 12.1.4, 14.2.1, A.14.2.1
Pump rooms/houses, 4.12, A.4.12
　Access to, 4.12.2, 5.2, A.4.12, A.5.2
　Controller, transfer of power to, 9.6.4
　Diesel engine drive, 11.3, A.11.3
　Drainage, 4.12.7, 11.3.1, A.4.12.7
　Lighting, 4.12.4, 4.12.5
　Temperature of, 4.12.3, 11.6.5, A.11.6.5, B.1.17
　Torch-cutting or welding in, 4.13.6, A.4.13.6
　Ventilation, 4.12.6, 11.3.2, A.11.3.2
　Vertical lineshaft turbine pumps, 7.4.1
Pumps
　Additive, *see* Additive pumps
　Bypass, with, 4.14.4, 11.2.8.6, A.4.14.4, A.11.2.8.6
　Can (definition), 3.3.38.2
　Centrifugal, *see* Centrifugal pumps
　End suction, *see* End suction pumps
　Fire, *see* Fire pumps
　Foam concentrate, *see* Foam concentrate pumps
　Gear (definition), 3.3.38.7
　Horizontal, *see* Horizontal pumps
　In-line, *see* In-line pumps
　Listed, 4.7.1, A.4.7.1
　Lubrication, B.1.21, B.1.27
　Multiple, 4.14.3.1, 4.14.7, 12.1.3.3.2, 14.2.3, 14.2.6.2.5.1
　　Sequence starting of, 9.6.3, 10.5.2.5, 12.7.2.4
　Multistage, 4.1, 6.1.1.2, 6.1.1.3
　Piston plunger (definition), 3.3.38.12
　Positive displacement, *see* Positive displacement pumps
　Pressure maintenance, *see* Pressure maintenance (jockey or make-up) pumps
　Pressure maintenance (jockey or make-up), *see* Pressure maintenance (jockey or make-up) pumps
　Priming, B.1.19
　Problems, causes of, Annex B
　Rotary lobe (definition), 3.3.38.15
　Rotary vane (definition), 3.3.38.16
　Single-stage, 4.1, 6.1.1.2, 6.1.1.3, A.6.1.1
　Speed, 4.7.6, A.4.7.6, B.1.30
　Vertical lineshaft turbine, *see* Vertical lineshaft turbine pumps
Pump shaft rotation, 4.23, A.4.23, B.1.18
Purpose of standard, 1.2

Q
Qualified person (definition), 3.3.40

R
Radiators, 11.2.8.8, 11.3.2.1(4), 11.3.2.4.3, A.11.3.2, A.11.3.2.4.3.4
Record drawings, 14.3.1, A.14.3
　Definition, 3.3.41
References, Chap. 2, Annex C
Relief valves, 4.7.7.2, 4.18, 4.25.4.1, 6.3.1, 6.3.2(4), 14.2.6.3.2, 14.2.6.3.3, A.4.7.7.2, A.4.18.1 to A.4.18.8, A.6.3.1, A.14.2.6.1
　Circulation relief valves, 6.3.1, A.6.3.1, A.14.2.6.1
　　Automatic, 4.11.1
　　Definition, 3.3.57.5.1
　Definition, 3.3.57.5
　Drainage, backflow prevention device, 4.27.2
　Positive displacement pumps, 8.5.2 to 8.5.4, A.8.5.2 to A.8.5.4
　Vertical lineshaft turbine pumps, 7.3.5.1(4)
Remote station, manual control at, 10.5.2.4, 12.7.2.3
Retroactivity of standard, 1.4
Rotary lobe pump (definition), 3.3.38.15
Rotary vane pump (definition), 3.3.38.16

S
Scope of standard, 1.1, A.1.1
Screens
　Suction pipe, 4.14.8, 7.3.4.3, A.4.14.8
　Well, 7.2.4.5 to 7.2.4.11

Seals
　Positive displacement pumps, 8.1.4, 8.2.3
　Rings improperly located in stuffing box, B.1.20
　Water seals, maintenance of, B.1.6
Seismic bracing, 4.13.5
Sequence starting of pumps, 9.6.3, 10.5.2.5, 12.7.2.4
Series fire pump units, 4.19, A.4.19.2.1, A.4.19.2.7.1
　Definition, 3.3.42
Service (power source), 9.2.2
　Definition, 3.3.43, A.3.3.43
Service equipment, 9.2.3.1(1)
　Definition, 3.3.44, A.3.3.44
Service factor (definition), 3.3.45
Service personnel, 4.3.4, A.4.3.4
Set pressure, 4.7.7.3.2, A.4.7.7.3.2
　Definition, 3.3.46
　Variable speed pressure limiting control, 10.10.3.1, 10.10.8.3, A.10.10.3.1
Shaft rotation, *see* Pump shaft rotation
Shafts, flexible connecting, *see* Flexible connecting shafts
Shall (definition), 3.2.4
Shop tests, *see* Tests
Should (definition), 3.2.5
Shutoff valves, 4.18.9, 4.30.5, 11.2.8.5.3.2, 11.2.8.5.3.3, 11.2.8.6.2, 11.2.8.6.3, 11.4.4.4, 11.4.4.5
Signal devices
　Electric-drive controllers, 10.1.1.2, 10.4.6, 10.4.7, A.10.4.6, A.10.4.7
　Engine drive controllers, 12.1.2, 12.4.2, 12.4.3, A.12.4.2.3(3)
　Phase reversal indicator, 10.4.6.2, 10.4.7.2.3
Signals, 4.24, A.4.24
　Break tank low liquid level, 4.31.3.1.5
　Definition, 3.3.47, A.3.3.47
　Electric-drive controllers, 10.4.5.7.3, 10.4.7.1, 10.8.3.12.2
　Engine drive controllers, 12.4.1.3 to 12.4.1.5, 14.2.7.6, A.12.4.1.4(1)
　Engine running and crank termination, 11.2.7.4.3
　Tests, 14.2.10
　Visible, 10.4.7.1; *see also* Visible indicators
Single-stage pumps, 4.1, 6.1.1.2, 6.1.1.3, A.6.1.1
SI units, 1.6
Spare parts, 14.3.4(6), 14.5.2.4, A.14.3.3, A.14.3.4(6)
Speed, *see also* Variable speed pressure limiting control
　Engine
　　Definition, 3.3.48.1

Internal combustion engine, B.1.28
Overspeed shutdown device, 11.2.4.4
Motor, B.1.28, B.1.30
Definition, 3.3.48.2
Rated, 4.7.6, A.4.7.6, A.14.2.6.4
Definition, 3.3.48.3
Steam turbine, B.1.28
Speed governor, steam turbine, 13.2.2
Split-case pumps, *see* Horizontal pumps
Sprinkler systems, 4.12.1.3
Standard (definition), 3.2.6
Static liquid level, 6.1.2, A.6.1.2
Definition, 3.3.29.2
Steam supply, 4.21, 13.1.3.1, 13.3, A.13.3
Steam turbines, 4.7.2, 4.7.6, Chap. 13, A.4.7.6, B.1.26
Acceptability, 13.1.1
Backup, 9.3.3
Bearings, 13.2.6
Capacity, 13.1.2
Casing and other parts, 13.2.1, A.13.2.1.1
Critical path components, 14.5.2.5(3)
Gauge and gauge connections, 13.2.3
Installation, 13.3, A.13.3
Obstructed pipe, B.1.32
Rotor, 13.2.4
Shaft, 13.2.5
Speed, B.1.28
Speed governor, 13.2.2
Starting methods, A.14.2.6
Steam consumption, 13.1.3, A.13.1.3
Tests, 14.2.6, 14.2.9, A.14.2.6
Vertical lineshaft turbine pumps, 7.5.1.3, 7.5.1.4, 7.5.2
Steel pipe, *see* Pipe
Storage batteries, *see* Batteries, storage
Strainers, *see also* Suction strainers
Engine cooling system, B.1.25
Pipeline, 6.3.2(5)
Suction, vertical shaft turbine pumps, 7.3.4
Stuffing boxes, B.1.5, B.1.7, B.1.20
Suction, *see also* Net positive suction head
Backflow preventers, evaluation of, 4.27.4.1, 4.27.4.2
Static suction lift, 6.1.2, A.6.1.2
Total suction head (definition), 3.3.23.5
Total suction lift (h_l) (definition), 3.3.56
Suction pipe and fittings, 4.2.3, 4.14, A.4.14.1 to A.4.14.10
Devices in, 4.14.9, 4.27.3, 4.27.4.3, A.4.14.9
Elbows and tees, 4.14.6.3
Freeze protection, 4.14.6.2
Pressure maintenance pumps, 4.25.5.1
Problems, causes of, B.1.1 to B.1.3
Strain relief, 4.14.6.5
Valves, 4.16.1

Suction pressure gauges, 4.10.2, 4.27.4.2, 8.5.1, A.4.10.2
Suction reservoir, discharge to, 4.18.8, A.4.18.8
Suction screening, 4.14.8, A.4.14.8
Suction strainers
Positive displacement pumps, 8.5.5, A.8.5.5
Vertical shaft turbine pumps, 7.3.4
Sump, vertical shaft turbine-type pumps, 7.4.3.7
Switches, *see also* Isolating switches; Transfer switches
Electronic control module (ECM) selector switch, 11.2.4.2.3
Locked cabinets for, 12.3.4
Pressure switch, pressure sensing line, 4.30.6
System designer, 4.3.2, A.4.3.2(2)
System installer, 4.3.3, A.4.3.3(2)

T

Tachometer, 11.2.5.2
Tanks
Break, *see* Break tanks
Fuel supply, 4.28.2, 11.4.1.2, 11.4.2, A.11.4.2
Water supply, high-rise buildings, 5.3, 5.4, 5.6.1, A.5.6.1.5.1
Temperature gauge, 11.2.5.4
Tests, *see also* Field acceptance tests; Water flow test devices
Aquifer performance analysis, A.7.2.1.2
Definition, 3.3.3
Component replacement, 14.5, A.14.5.2.3
Controllers, 10.1.2.3, 10.1.2.6
Acceptance, 14.2.6, A.14.2.6
Engine drive controllers, manual testing of, 12.7.3.2
Pressure switch-actuated automatic controller, 10.5.2.1.6
Rated in excess of 600 Volts, 10.6
Duration, 14.2.11, A14.2.11
Flow, 5.4, 8.10, 14.2.6.2, 14.2.6.4.3, A.14.2.6.2.5
High-rise buildings, fire pumps for, 5.4
Hydrostatic, 4.22.2, 14.1, A.14.1.1, A.14.1.3
Loads start test, 14.2.6.5
Metering devices or fixed nozzles for, 4.20.2.1, 8.10.2, 8.10.3, A.4.20.2.1
Momentary test switch, alternate power source, 10.8.3.13
Periodic, 14.4
Reports, 14.3.2, A.14.3
Shop (preshipment), 4.5, 4.22, 10.1.2.3, 14.2.4.2

Suction pipe, 4.14.2
Vertical lineshaft turbine pump wells, 7.2.7, A.7.2.7
Timer, weekly program, 12.7.2.7
Tools, special, 14.3.5
Total discharge head (h_d) (definition), 3.3.23.2
Total head (H)
Horizontal pumps, A.14.2.6.4
Definition, 3.3.23.3.1, A.3.3.23.3.1
Vertical turbine pumps, 7.1.2, A.14.2.6.4
Definition, 3.3.23.3.2, A.3.3.23.3.2
Totally enclosed motors
Definition, 3.3.33.9
Fan-cooled, 9.5.2.4
Definition, 3.3.33.8
Nonventilated, 9.5.2.4
Definition, 3.3.33.10
Total rated head, 7.1.2
Definition, 3.3.23.4
Total suction head, 4.22.2.2
Definition, 3.3.23.5
Total suction lift (h_l) (definition), 3.3.56
Trade sizes, 1.6.5
Transfer switches, 9.1.2, 10.1.1.1, 10.4.8, 10.8, A.10.8
Automatic, *see* Automatic transfer switches
Listing, 10.1.2.1, 10.1.2.4
Manual, *see* Manual transfer switches
Marking, 10.1.2.2, A.10.1.2.2
Nonpressure-actuated, 10.5.1.2, 10.5.2.2
Pressure-actuated, 10.5.1.2, 10.5.2.1.1, 10.6.4, 12.7.2.1.1, 14.2.7.6
Service arrangements, 10.1.2.6
Transformers, 9.2.2(5)
Troubleshooting, Annex B
Turbine pumps, vertical lineshaft, *see* Vertical lineshaft turbine pumps
Turbines, steam, *see* Steam turbines

U

Undervoltage-sensing devices, 10.8.3.6
Units, pump, *see* Fire pump units
Units of measurement, 1.6
Unloader valves, 8.3.4, 8.3.5
Definition, 3.3.57.6

V

Valves, *see also* Butterfly valves; Check valves; Dump valves; Gate valves; Hose valves; Relief valves; Shutoff valves
Bypass, 4.16.1
Discharge pipe, 4.15.6 to 4.15.9, A.4.15.6, A.4.15.7

Emergency governor, 14.2.9
Fill/refill, water tank, 5.6.1.2 to 5.6.1.5
Fuel solenoid, 11.4.4.7, 12.7.3.2
Isolation, 4.16.1
Low suction throttling, 4.14.9.2(2), 4.15.9
   Definition, 3.3.57.2
Main throttle, steam turbine, 13.2.1.3
Pressure control, A.13.2.1.1
   Definition, 3.3.57.3
Pressure-reducing, A.13.3
   Definition, 3.3.57.4
Supervision of, 4.16, A.4.16
Unloader, 8.3.4, 8.3.5
   Definition, 3.3.57.6

Variable speed controllers, *see also* Variable speed pressure limiting control; Variable speed suction limiting control
   Motors used with, 9.5.1.4, A.9.5.1.4
   Positive displacement pumps, 8.4.6, 8.5.7.2

Variable speed pressure limiting control, 4.7.7.3, 4.18.1.3, A.4.7.7.3.2
   Definition, 3.3.58
   Diesel engines, 11.2.4.3
   Electric drive controllers, 10.5.2.1.7.6, 10.10, A.10.10
   Positive displacement pumps, 8.5.7.2
   Tests, 14.2.6.3, 14.2.6.4.7.4

Variable speed suction limiting control
   Definition, 3.3.59
   Diesel engines, 11.2.4.3
   Electric drive controllers, 10.10, A.10.10

Variable speed vertical pumps, 7.5.3, 10.10.12

Velocity head (h_v) (definition), 3.3.23.6, A.3.3.23.6

Ventilation of pump room/house, 4.12.6, 11.3.2, A.11.3.2

Vertical hollow shaft motors, 7.5.1.5, 7.5.1.7.2

Vertical in-line pumps, 4.28.2.1

Vertical lineshaft turbine pumps, 4.1, Chap. 7; *see also* Wells, vertical shaft turbine pumps
   Bowl assembly, 7.3.3
   Characteristics, 7.1.2
   Column, 7.3.2, A.7.3.2.1
   Consolidated formations, 7.2.5, A.7.2.5
   Controllers, 7.5.2, 10.10.12
   Definition, 3.3.38.17
   Drivers, 7.5, 7.6.1.6, 9.5.1.9, 11.2.3.2
   Fittings, 7.3.5, A.7.3.5.3
   Head, 7.3.1, A.7.3.1
      Total head (H), 7.1.2, A.14.2.5.4
         Definition, 3.3.23.3.2, A.3.3.23.3.2
   Installation, 7.4, A.7.1, A.7.4
   Maintenance, 7.6.2
   Oil-lubricated type, 7.3.2.4 to 7.3.2.6, A.7.1.1
   Operation, 7.6.1, A.7.6.1.1, A.7.6.1.4
   Pump house, A.4.12
   Shaft rotation, A.4.23
   Shop test, 4.22.2.5
   Submergence, 7.2.2, A.7.2.2.1, A.7.2.2.2
   Suction strainer, 7.3.4
   Suitability, A.7.1.1
   Unconsolidated formations, 7.2.4
   Water supply, 7.1.1, 7.2, A.7.1.1, A.7.2.1.1 to A.7.2.7, B.1.19

Vibration, pump, 7.6.1.5

Visible indicators
   Electric-drive controllers, 10.4.5.7.3, 10.4.6.1, 10.6.6.2 to 10.6.6.4, 10.8.3.8, 10.8.3.12.2, A.10.4.5.7
   Engine drive controllers, 12.4.1.1 to 12.4.1.4, A.12.4.1.2, A.12.4.1.4(1)

Voltage
   Low, B.1.28
   Rated motor voltage different from line voltage, B.1.31
   Tests, 14.2.6.4.4 to 14.2.6.4.6

Voltage drop, 9.1.6, 9.4, A.9.4

Voltage-sensing devices, 10.8.3.7

Voltage surge arresters, 10.4.1, A.10.4.1

Voltmeter, 12.4.5, A.14.2.6.4

Vortex plate, 4.14.10, A.4.14.10

W

Waste outlet, heat exchanger, 11.2.8.7

Water flow test devices, 4.20, A.4.20.1.1 to A.4.20.3.4(2)

Water level, *see* Liquid level

Water mist positive displacement pumping units, 8.4
   Definition, 3.3.38.18, A.3.3.38.18
   Pressure maintenance, 8.5.7.2 to 8.5.7.4
   Tests, 14.2.4.1.1, 14.2.4.2.1, 14.2.4.2.2, 14.2.6.2.1, 14.2.6.2.2, 14.5.1.3.1, A.14.2.6.1

Water mist positive displacement pumps, 8.3, 8.5.4, A.8.3.1, A.8.5.4

Water pressure control, 10.5.2.1, 10.6.4, 12.7.2.1, A.10.5.2.1

Water supplies, *see* Liquid supplies

Wearing rings, B.1.9, B.1.19

Weekly program timer, 12.7.2.7

Welding, 4.13.6, A.4.13.6

Wells, vertical shaft turbine pumps
   Construction, 7.2.3
   Developing, 7.2.6
   Installations, 7.2.2.1, A.7.1.1, A.7.2.2.1
   Problems, causes of, B.1.4
   Screens, 7.2.4.5 to 7.2.4.11
   Test and inspection, 7.2.7, A.7.2.7
   Tubular wells, 7.2.4.16
   In unconsolidated formations, 7.2.4
   Water level, 4.6.3

Wet pits
   Definition, 3.3.61
   Installation of vertical shaft turbine pumps, 7.2.2.2, 7.4.3.7, A.7.2.1.1, A.7.2.2.2
   Suction strainer requirement, 7.3.4.3
   Water level, 4.6.3

Wiring, *see also* Disconnecting means
   Electric-drive controllers, 10.3.4
   Engine drive controllers, 11.2.6, 12.3.5, A.11.2.6.1.1, A.11.2.6.2
   Field acceptance tests, 14.2.3
   Packaged fire pump assemblies, 4.29.2
   Problems of, B.1.32

IMPORTANT NOTICES AND DISCLAIMERS CONCERNING NFPA® DOCUMENTS

NOTICE AND DISCLAIMER OF LIABILITY CONCERNING THE USE OF NFPA DOCUMENTS

NFPA® codes, standards, recommended practices, and guides ("NFPA Documents"), including the NFPA Documents contained herein, are developed through a consensus standards development process approved by the American National Standards Institute. This process brings together volunteers representing varied viewpoints and interests to achieve consensus on fire and other safety issues. While the NFPA administers the process and establishes rules to promote fairness in the development of consensus, it does not independently test, evaluate, or verify the accuracy of any information or the soundness of any judgments contained in NFPA Documents.

The NFPA disclaims liability for any personal injury, property or other damages of any nature whatsoever, whether special, indirect, consequential or compensatory, directly or indirectly resulting from the publication, use of, or reliance on NFPA Documents. The NFPA also makes no guaranty or warranty as to the accuracy or completeness of any information published herein.

In issuing and making NFPA Documents available, the NFPA is not undertaking to render professional or other services for or on behalf of any person or entity. Nor is the NFPA undertaking to perform any duty owed by any person or entity to someone else. Anyone using this document should rely on his or her own independent judgment or, as appropriate, seek the advice of a competent professional in determining the exercise of reasonable care in any given circumstances.

The NFPA has no power, nor does it undertake, to police or enforce compliance with the contents of NFPA Documents. Nor does the NFPA list, certify, test, or inspect products, designs, or installations for compliance with this document. Any certification or other statement of compliance with the requirements of this document shall not be attributable to the NFPA and is solely the responsibility of the certifier or maker of the statement.

ADDITIONAL NOTICES AND DISCLAIMERS

Updating of NFPA Documents

Users of NFPA codes, standards, recommended practices, and guides ("NFPA Documents") should be aware that these documents may be superseded at any time by the issuance of new editions or may be amended from time to time through the issuance of Tentative Interim Amendments. An official NFPA Document at any point in time consists of the current edition of the document together with any Tentative Interim Amendments and any Errata then in effect. In order to determine whether a given document is the current edition and whether it has been amended through the issuance of Tentative Interim Amendments or corrected through the issuance of Errata, consult appropriate NFPA publications such as the National Fire Codes® Subscription Service, visit the NFPA website at www.nfpa.org, or contact the NFPA at the address listed below.

Interpretations of NFPA Documents

A statement, written or oral, that is not processed in accordance with Section 6 of the Regulations Governing Committee Projects shall not be considered the official position of NFPA or any of its Committees and shall not be considered to be, nor be relied upon as, a Formal Interpretation.

Patents

The NFPA does not take any position with respect to the validity of any patent rights referenced in, related to, or asserted in connection with an NFPA Document. The users of NFPA Documents bear the sole responsibility for determining the validity of any such patent rights, as well as the risk of infringement of such rights, and the NFPA disclaims liability for the infringement of any patent resulting from the use of or reliance on NFPA Documents.

NFPA adheres to the policy of the American National Standards Institute (ANSI) regarding the inclusion of patents in American National Standards ("the ANSI Patent Policy"), and hereby gives the following notice pursuant to that policy:

> **NOTICE:** The user's attention is called to the possibility that compliance with an NFPA Document may require use of an invention covered by patent rights. NFPA takes no position as to the validity of any such patent rights or as to whether such patent rights constitute or include essential patent claims under the ANSI Patent Policy. If, in connection with the ANSI Patent Policy, a patent holder has filed a statement of willingness to grant licenses under these rights on reasonable and nondiscriminatory terms and conditions to applicants desiring to obtain such a license, copies of such filed statements can be obtained, on request, from NFPA. For further information, contact the NFPA at the address listed below.

Law and Regulations

Users of NFPA Documents should consult applicable federal, state, and local laws and regulations. NFPA does not, by the publication of its codes, standards, recommended practices, and guides, intend to urge action that is not in compliance with applicable laws, and these documents may not be construed as doing so.

Copyrights

NFPA Documents are copyrighted by the NFPA. They are made available for a wide variety of both public and private uses. These include both use, by reference, in laws and regulations, and use in private self-regulation, standardization, and the promotion of safe practices and methods. By making these documents available for use and adoption by public authorities and private users, the NFPA does not waive any rights in copyright to these documents.

Use of NFPA Documents for regulatory purposes should be accomplished through adoption by reference. The term "adoption by reference" means the citing of title, edition, and publishing information only. Any deletions, additions, and changes desired by the adopting authority should be noted separately in the adopting instrument. In order to assist NFPA in following the uses made of its documents, adopting authorities are requested to notify the NFPA (Attention: Secretary, Standards Council) in writing of such use. For technical assistance and questions concerning adoption of NFPA Documents, contact NFPA at the address below.

For Further Information

All questions or other communications relating to NFPA Documents and all requests for information on NFPA procedures governing its codes and standards development process, including information on the procedures for requesting Formal Interpretations, for proposing Tentative Interim Amendments, and for proposing revisions to NFPA documents during regular revision cycles, should be sent to NFPA headquarters, addressed to the attention of the Secretary, Standards Council, NFPA, 1 Batterymarch Park, P.O. Box 9101, Quincy, MA 02269-9101; email: stds_admin@nfpa.org

For more information about NFPA, visit the NFPA website at www.nfpa.org.

A Guide to Using the *Stationary Fire Pumps Handbook*

This fourth edition of *Stationary Fire Pumps Handbook* contains the complete mandatory text of the 2013 edition of NFPA 20, *Standard for the Installation of Stationary Pumps for Fire Protection,* and the associated nonmandatory annex material. Commentary is provided in this handbook to explain the reasoning behind the Standard's requirements.

▶ New or revised Standard text, figures, and tables are indicated by a vertical rule.

▶ A bullet indicates where a portion of the Standard has been removed.

▶ AHJ FAQs offer snapshots of typical situations faced by Authorities Having Jurisdiction.

▶ Commentary includes answers to frequently asked questions.